ELECTRODIAGNOSIS IN DISEASES OF NERVE AND MUSCLE:
PRINCIPLES AND PRACTICE

EDITION 2

ELECTRODIAGNOSIS IN DISEASES OF NERVE AND MUSCLE:
PRINCIPLES AND PRACTICE

EDITION 2

JUN KIMURA, M.D.

Professor and Chairman
Department of Neurology
University of Kyoto School of Medicine
Kyoto, Japan

Professor
Division of Clinical Electrophysiology
Department of Neurology
University of Iowa College of Medicine
Iowa City, Iowa

F.A. DAVIS COMPANY · Philadelphia

Printed in the United States of America

Last digit indicates print number: 10 9 8 7 6 5 4 3 2 1

NOTE: As new scientific information becomes available through basic and clinical research, recommended procedures undergo changes. The author and publisher have done everything possible to make this book accurate, up-to-date, and in accord with accepted standards at the time of publication. Nonetheless, the reader is advised always to check changes and new information regarding the current practice and contraindications before conducting any tests. Caution is especially urged with new or infrequently used equipment.

Library of Congress Cataloging-in-Publication Data

Kimura, Jun.
    Electrodiagnosis in diseases of nerve and muscle.

    Includes bibliographies and index.
    1. Neuromuscular diseases—Diagnosis. 2. Electromyography. 3. Electrodiagnosis. I. Title. [DNLM: 1. Electrodiagnosis. 2. Neuromuscular diseases—diagnosis. WE 550 K49e]
RC925.7.K55 1989   616.7′407547   88-31031
ISBN 0−8036−5342−5

**AGAIN, TO JUNKO
AND
OUR PARENTS**

PREFACE AND ACKNOWLEDGMENTS FOR THE SECOND EDITION

The preparation for the second edition began in 1983 with the original volume still in press, literally before the ink had dried. Kind encouragements and constructive criticisms received from different corners of the world added further incentive for early revision. Most suggestions proved helpful in improving the contents and style. A few requests, however, posed problems because they represented mutually exclusive views: for example, inclusion or exclusion of expanded coverage of evoked potential studies. Here, I had to accept the old maxim that, however much one wishes, one cannot please everybody all the time (or even most people much of the time!). Thus, I followed my own bias as to the relative importance of a topic for the principles and practice of electrodiagnosis.

This revision, though initially conceived as routine and minor, eventually required major changes, in part reflecting the rapid medical and technologic advances in the field during the past five years. The sections rewritten in their entirety include Facts, Fallacies, and Fancies of Nerve Stimulation Techniques (Chapter 7), Single-Fiber and Macro Electromyography (Chapter 15), Somatosensory and Motor Evoked Potentials (Chapter 19), Polyneuropathies (Chapter 22), Myasthenia Gravis and Other Disorders of Neuromuscular Transmission (Chapter 24), Myopathies (Chapter 25), and Fundamentals of Electronics and Electrical Safety (Appendices 2 and 3). Most other sections also underwent substantial changes to update, clarify, and tighten the contents. The book now cites more than 1200 additional references selected from some 2500 recent publications that I personally reviewed, with the hope that the inclusive bibliography helps encourage further research in the area of electrodiagnostic medicine.

The hustle and bustle inherent to the preparation of voluminous manuscripts, by necessity, involve directly or indirectly those who share the work environment with the author. I could not have completed the job without assistance from my colleagues, who endured the fate of "galley" slaves over an extended period of time. Drs. Thoru Yamada and Stokes Dickins ran the busy services of the Division despite my preoccupation with writing. D. David Walker, M.S.E.E., rewrote the appendix on electronics previously coauthored by Pete Seaba, M.S.E.E., who left the ranks for private enterprise. Sheila Mennen, our Chief Technologist, Deborah Gevock, and Cheri Doggett played major roles in maintaining the daily clinical operation and organizing technical as well as secretarial needs. A number of clinical fellows and residents participated in teaching sessions, shedding new insights into the type of coverage essential in an electrodiagnostic text. A total of 35 research fellows from Japan and elsewhere spent one to two years with us during this interim contributing original data in clinical electrophysiology, much of which found its way into the revised text.

Dr. Maurice Van Allen, who had provided a kind foreword for the first edition, continued to support my literary endeavor until his untimely death in 1986. I lost a teacher and friend, and a new foreword, which he had promised. He had jokingly, but perhaps with good reason, attributed the success of the first edition to his opening remarks, which are retained in his honor. Dr. A. L. Sahs, who initiated me into neurology, rendering help when I needed it most, also passed away later in the same year. It was my good fortune that the Department prospered under the direction of Dr. Antonio Damasio, who, together with Dr. Robert L. Rodnitzky, provided the kind of environment enticing to academic pursuit. I owe my thanks to Mr. Robert H. Craven, Sr., Mr. Robert H. Craven, Jr., Dr. Sylvia Fields, Ms. Linda Weinerman, Ms. Jessie Raymond, and Mr. Herbert Powell of F. A. Davis for their patience and encouragement. I am indebted to the American Association of Electromyography and Electrodiagnosis and its Nomenclature Committee, who granted permission to reprint the AAEE Glossary of Terms in Clinical Electromyography (1987) as Appendix 4.

The work turned into a family project of sorts over the past several years. Our three sons, five years older and perhaps wiser, if not quieter, could now assist in substance by typing the book, cover to cover, into a word processor to facilitate rewriting. I acknowledge the yeomans' service by honoring their request again to dedicate the book to their mother, who, I know, has funded the teenagers' operation from time to time to boost their spirit of devotion. We lost her father and mine during the preparation of the first edition and my mother in this interim. I salute them for their constant support of our venture abroad, with the credit given to whom it is most justifiably due.

J.K.

FOREWORD FOR THE FIRST EDITION

I found particular pleasure in preparing this foreword to the work of a colleague whose professional development and scientific accomplishments I have followed very closely indeed for some twenty years.

Dr. Kimura, very early after his training in neurology, expressed an interest in clinical electrophysiology. His energy and talents led to full-time assignment and responsibility for the development and application of electrodiagnostic techniques in our laboratory of electromyography and then to direction of the Division of Clinical Electrophysiology.

From his early assignment, Dr. Kimura has exploited the possibilities for the applications of clinical electrophysiologic techniques to their apparent limits, which, however, seem to continually advance to the benefit of us all. This volume is based on very extensive personal experience with application of all of the now recognized procedures.

The beginner will be able to follow this discipline from its historical roots to the latest techniques with the advantage of an explanatory background of the clinical, physiologic, anatomic, and pathologic foundations of the methods and their interpretation. The instrumentation, so essential to any success in application of techniques, is further described and explained. The more experienced diagnostician will both appreciate and profit from this pragmatic, well-organized, and authoritative source with its important bibliographic references; the beginner will find it a bible.

There are few areas in electrodiagnosis that Dr. Kimura does not address from his own extensive experience, backed by clinical and pathologic confirmation. The sections on the blink reflex and the F wave reflect his own pioneering work. He has closely followed the application of new techniques to the study of disease of the central nervous system by evoked cerebral potentials from the beginning.

These sections reflect a substantial personal experience in establishment of standards and in interpretation of changes in disease.

So important are the findings of electrodiagnostic methods that the clinical neurologist must himself be an expert in their interpretation. Preferably he should perform tests on his own patients or closely supervise such tests. Only in this way can he best derive the data that he needs or direct the examination in progress to secure important information as unexpected findings appear. To acquire the knowledge to guide him either in supervised training or in self-teaching, he needs first an excellent and comprehensive guide such as this text by Dr. Kimura.

Dr. Kimura is justifiably regarded as a leader in clinical electrophysiology both nationally and internationally. Those of us who profit from daily contact with him should be pardoned for our pride in this substantial and authoritative work.

Maurice W. Van Allen, M.D.

PREFACE FOR THE FIRST EDITION

This book grew out of my personal experience in working with fellows and residents in our electromyography laboratory. It is intended for clinicians who perform electrodiagnostic procedures as an extension of their clinical examination. As such, it emphasizes the electrical findings in the context of the clinical disorder. Although the choice of material has been oriented toward neurology, I have attempted to present facts useful to practicing electromyographers regardless of their clinical disciplines. I hope that the book will also prove to be of value to neurologists and physiatrists who are interested in neuromuscular disorders and to others who regularly request electrodiagnostic tests as an integral part of their clinical practice.

The book is divided into seven parts and three appendices. Part 1 provides an overview of basic anatomy and physiology of the neuromuscular system. Nerve conduction studies, tests of neuromuscular transmission, and conventional and single-fiber electromyography are described in the next three parts. Part 5 covers supplemental methods designed to test less accessible regions of the nervous system. The last two parts are devoted to clinical discussion. The appendices consist of the historical review, electronics and instrumentation, and a glossary of terms.

The selection of technique is necessarily influenced by the special interest of the author. Thus, in Part 5, the blink reflex, F wave, H reflex, and somatosensory evoked potential have been given more emphasis than is customary in other texts. I hope that I am not overestimating their practical importance and that these newer techniques will soon find their way into routine clinical practice. This is, of course, not to de-emphasize the conventional methods, which I hope are adequately covered in this text. The ample space allocated for clinical discussion in Parts 6 and 7 reflects my personal conviction that clinical acumen is a prerequisite for meaningful electrophysiologic evaluations. Numerous references are provided to document the statements made in the text. I hope that use of these references will promote interest and research in the field of electrodiagnosis.

J.K.

ACKNOWLEDGMENTS FOR THE FIRST EDITION

I came from the Island of the Rising Sun, where English is not the native language. It was thus with trepidation that I undertook the task of writing an English text. Although its completion gives me personal pride and satisfaction, I hasten to acknowledge that the goal could not have been achieved without help from others.

Dr. M. W. Van Allen has provided me with more than a kind foreword. I wish to thank him for his initial encouragement and continued support and advice. He was one of the first to do electromyography in Iowa. During my early years of training I had the pleasure of using his battery-operated amplifier and a homemade loudspeaker (which worked only in his presence). I am indebted to Dr. A. L. Sahs, who initiated me into the field of clinical neurology, and Dr. J. R. Knott, who taught me clinical neurophysiology. I am grateful to Drs. T. Yamada and E. Shivapour for attending the busy service of the Division of Clinical Electrophysiology while I devoted myself to writing. Dr. Yamada also gave me most valuable assistance in preparing the section on central somatosensory evoked potentials, which includes many of his original contributions. Drs. R. L. Rodnitzky, E. P. Bosch, J. T. Wilkinson, A. M. Brugger, F. O. Walker, and H. C. Chui read the manuscript and gave most helpful advice. Peter J. Seaba, M.S.E.E., and D. David Walker, M.S.E.E., our electrical engineers, contributed Appendix 2 and reviewed the text.

My special thanks go to the technicians and secretaries of the Division of Clinical Electrophysiology. Sheila R. Mennen, the senior technician of our electromyography laboratory, typed (and retyped time and time again) all the manuscript with devotion and dedication. Deborah A. Gevock, Cheri L. Doggett, Joanne M. Colter, Lauri Longnecker, Jane Austin, Sharon S. Rath, Lori A. Garwood, and Allen L. Frauenholtz have all given me valuable technical or secretarial assistance. Linda C. Godfrey and her staff in the Medical Graphics Department have been most helpful in preparing illustrations.

I owe my gratitude to Mr. Robert H. Craven, Sr., Mr. Robert H.

Craven, Jr., Dr. Sylvia K. Fields, Miss Agnes A. Hunt, Ms. Sally Burke, Miss Lenoire Brown, Mrs. Christine H. Young, and two anonymous reviewers of the F. A. Davis Company for their interest and invaluable guidance. A number of previously published figures and tables have been reproduced with permission from the publishers and authors. I wish to express my sincere appreciation for their courtesy. The sources are acknowledged in the legends. The Glossary of Terms Commonly Used in Electromyography of the American Association of Electromyography and Electrodiagnosis is reprinted in its entirety as Appendix 3, with kind permission from the Association and the members of the Nomenclature Committee.

My sons asked if the book might be dedicated to them for having kept mostly, though not always, quiet during my long hours of writing at home. However, the honor went to their mother instead, a decision enthusiastically approved by the children, in appreciation for her effort to keep peace at home. In concluding the acknowledgment, my heart goes to the members of my family in Nagoya and those of my wife's in Takayama, who have given us kind and warm support throughout our prolonged stay abroad. The credit is certainly theirs for my venture finally coming to fruition.

J.K.

CONTENTS

Chapter 2
ELECTRICAL PROPERTIES OF NERVE
AND MUSCLE **25**

Chapter 3
ELECTRODES AND RECORDING APPARATUS....... **37**

Part III
ASSESSMENT OF NEUROMUSCULAR TRANSMISSION **167**

Part IV

Part V
TESTS FOR LESS ACCESSIBLE
REGIONS OF THE NERVOUS

Chapter 16

Part I

BASICS OF ELECTRODIAGNOSIS

Chapter 1

ANATOMIC BASIS
FOR LOCALIZATION

1 INTRODUCTION

Electrodiagnosis, as an extension of the neurologic examination, employs the same anatomic principles of localization, searching for evidence of motor and sensory compromise. Neurophysiologic studies supplement the clinical examination by providing additional precision, detail, and objectivity. They delineate a variety of pathologic changes that may otherwise escape detection and help examine atrophic, deeply situated, or paretic muscles, which tend to defy clinical evaluation. Specialized techniques provide means of isolating the neuromuscular junction and nerve segments for separate testing. Electrical studies allow quantitative assessment of reflex amplitude and latencies as well as complex motor phenomena.

Individual tests can be used in evaluating different groups of overlapping neural circuits. Meaningful analysis of electrophysiologic findings, therefore, demands adequate knowledge of neuroanatomy. In addition, superficial anatomy of skeletal muscles and peripheral nerves is a prerequisite for accurate placement of recording and stimulating electrodes. The first section contains a review of peripheral neuroanatomy important for the performance and interpretation of electrodiagnostic studies. A concise summary of clinically useful information serves as a framework for the rest of the text.

Despite the recognized importance of understanding muscle and nerve anatomy, written descriptions render the subject complicated and rather dry. The use of schematic illustrations in this chapter simplifies the discussion to compensate for this inherent problem. Existing texts provide more detailed accounts with regard to the superficial anatomy of skeletal muscles[3,6,7] or the general peripheral neuromuscular anatomy.[1,2,8–10,12–14]

2 CRANIAL NERVES

Nine cranial nerves innervate voluntary muscles, as summarized in Table 1–1. The oculomotor, trochlear, and abducent nerves control the movement of the eyes. The trigeminal and facial nerves innervate the muscles of mastication and those of facial expression, respectively. The laryngeal muscles receive innervation from the glossopharyngeal and vagal nerves and the cranial root of the accessory nerve. The hypoglossal nerve supplies the tongue. The spinal root of the accessory nerve innervates the sternocleidomastoid and upper portion of the trapezius. Of these, the most commonly tested in an electromyographic laboratory include the facial, trigeminal, and accessory nerves.

Facial Nerve

The course of the facial nerve, from the nucleus to the distal trunk, consists of four arbitrarily subdivided segments (Fig. 1–1). The central component, referred to as the intrapontine portion, initially courses posteriorly to loop around the sixth nerve nucleus. Its elongated course makes it vulnerable to various pontine lesions, which cause a peripheral, rather than central, type of facial palsy. The facial nerve complex exits the brainstem ventrolaterally at the caudal pons. Acoustic neuromas or other cerebellopontine angle masses may compress the nerve in this area. After traversing the subarachnoid space, the facial nerve enters the internal auditory meatus. Here it begins the longest and most complex intraosseous course of any nerve in the body. Within this segment lies the presumed site of the lesion in Bell's palsy. Upon exiting from the skull through the stylomastoid foramen, the facial nerve penetrates the superficial and deep lobes of the parotid gland. It then branches with some variation into five distal segments (Fig. 1–2).

Trigeminal Nerve

The trigeminal nerve subserves all superficial sensation to the face and buccal and nasal mucosa. It also supplies the muscles of mastication, which consist of the masseter, temporalis, and pterygoids.

**Table 1–1 MUSCLES INNERVATED BY THE CRANIAL
NERVES AND CERVICAL PLEXUS**

Nerve	Mesencephalon	Pons	Medulla	C-2	C-3	C-4
Oculomotor	_ Levator palpebrae _ _ Superior rectus _____ _ Medial rectus _____ _ Inferior rectus _____ _ Inferior oblique ____					
Trochlear	_ Superior oblique ____					
Trigeminal	_ Masseter _____ _ Temporalis _ _ Pterygoid _____					
Abducens		_ Lateral rectus _____				
Facial		_ Frontalis _____ _ Orbicularis oculi _____ _ Orbicularis oris _____ _ Platysma _____ _ Digastric & stylohyoid muscles __				
Glosso-pharyngeal			_ Laryngeal muscles _			
Vagus			_ Laryngeal muscles _			
Accessory (cranial root)			_ Laryngeal muscles _			
Hypoglossal			_ Tongue _____			
Accessory (spinal root)				_ Sternocleidomastoid _ _____Trapezius Upper _____ Middle _____		
Cervical plexus					_____Trapezius Lower _____	
Phrenic					_____ Diaphragm _____	

The ophthalmic and maxillary divisions of the trigeminal nerve supply sensation to the upper and middle parts of the face, respectively, whereas the mandibular division carries the sensory fibers to the lower portion of the face as well as the motor fibers (see Fig. 1–1). The first-order neurons, concerned primarily with tactile sensation, have their cell bodies in the gasserian ganglion. Their proximal branches enter the lateral portion of the pons and ascend to reach the main sensory nucleus. Those fibers subserving pain and temperature sensation also have cell bodies in the gasserian ganglion. However, upon entering the pons, their fibers descend to reach the spinal nucleus of the trigeminal nerve.

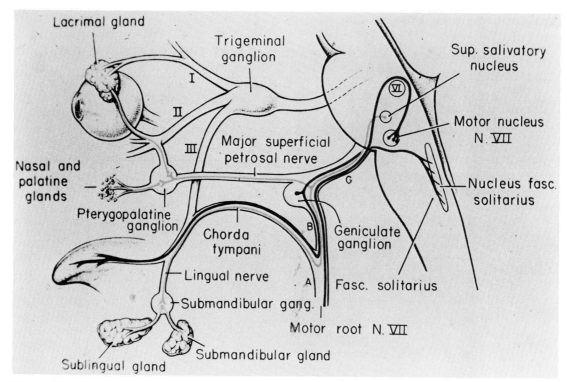

Figure 1–1. Functional components of the facial nerve and the three major divisions of the trigeminal nerve. The facial nerve (N. VII), consists of the portion at the stylomastoid foramen (*A*), middle segment distal to the geniculate ganglion (*B*), and a more proximal segment that includes extra and intrapontine pathways. (From Carpenter,[2] with permission.)

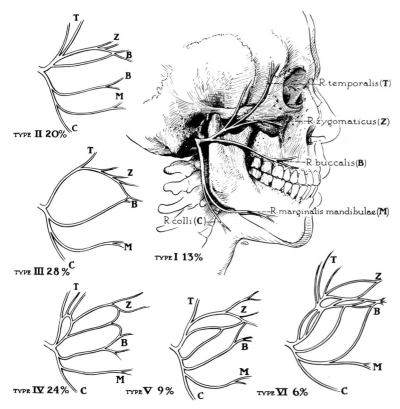

R·temporalis(**T**)

R·zygomaticus(**Z**)

R·buccalis(**B**)

R·marginalis mandibulae(**M**)

R·colli(**C**)

TYPE I 13%

TYPE II 20%

TYPE III 28%

TYPE IV 24%

TYPE V 9%

TYPE VI 6%

Figure 1–2. Major types of branching and intercommunication with percentage occurrence of each pattern in 350 recordings. (From Anson,[1] with permission.)

6

The first-order afferent fibers, subserving proprioception from the muscles of mastication, have their cell bodies in the mesencephalic nucleus. They make monosynaptic connection with the motor nucleus of the trigeminal nerve located in the mid pons, medial to the main sensory nucleus. This pathway provides the anatomic substrate for the masseter reflex. The first component of the blink reflex probably follows a disynaptic connection from the main sensory nucleus to the ipsilateral facial nucleus. The pathway for the second component, relayed through polysynaptic connections, includes the ipsilateral spinal nucleus and the facial nuclei on both sides (see Fig. 16–1).

Accessory Nerve

The cranial accessory nerve has the cell bodies in the nucleus ambiguus. The fibers join the vagus nerve and together distribute to the striated muscles of the pharynx and larynx. Thus, despite the traditional name, the cranial portion of the accessory nerve functionally constitutes a part of the vagus nerve. The spinal accessory nerve has its cells of origin in the spinal nucleus located in the first five or six cervical segments of the spinal cord (Figs. 1–3 and 1–4). The fibers ascend in the spinal canal to enter the cranial cavity through the foramen magnum and then leave by the jugular foramen to end in the trapezius and the sternocleidomastoid muscles. These two muscles receive additional nerve supply directly from C-2 through C-4 roots. The spinal accessory nerve provides the sole motor function, whereas the cervical roots subserve purely proprioceptive sensation (Fig. 1–5). The accessory nucleus consists of several separate portions. Thus, a lesion in the spinal cord may affect it only in part, causing partial paraly-

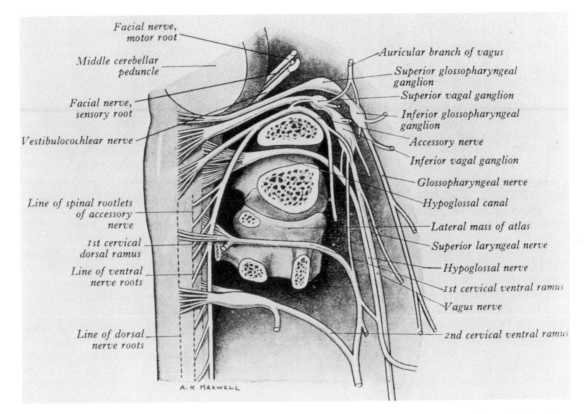

Figure 1–3. Communication between the last four cranial nerves on the right side viewed from the dorsolateral aspect. Note the division of the accessory nerve into the cranial accessory nerve, which joins the vagal nerve, and the spinal accessory nerve, which supplies the trapezius and sternocleidomastoid muscles. (From Williams and Warwick,[14] with permission.)

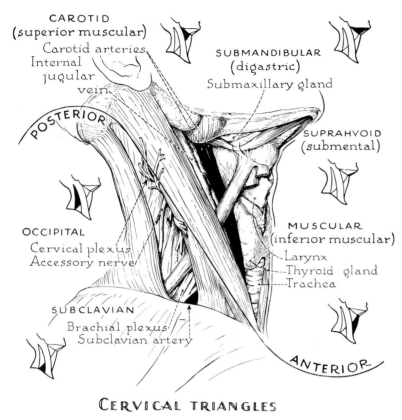

CAROTID
(superior muscular)
Carotid arteries
Internal
jugular
vein

POSTERIOR

SUBMANDIBULAR
(digastric)
Submaxillary gland

SUPRAHYOID
(submental)

OCCIPITAL
Cervical plexus
Accessory nerve

MUSCULAR
(inferior muscular)
Larynx
Thyroid gland
Trachea

SUBCLAVIAN
Brachial plexus
Subclavian artery

ANTERIOR

CERVICAL TRIANGLES

Figure 1–4. The sternocleidomastoid divides the field bounded by the trapezius, mandible, midline of the neck, and clavicle into anterior and posterior triangles. The obliquely coursing omohyoid further subdivides the posterior triangle into occipital and subclavian triangles. The contents of the occipital and subclavian triangles include the cervical plexus, spinal accessory nerve, and brachial plexus. The spinal accessory nerve becomes relatively superficial in the middle portion of the sternocleidomastoid along its posterior margin, where it is accessible to percutaneous stimulation. (From Anson,[1] with permission.)

sis of the muscle groups innervated by this nerve. This central dissociation could mimic a peripheral lesion affecting individual branches.

3   ANTERIOR AND POSTERIOR RAMI

The anterior and posterior roots, each composed of several rootlets, emerge from the spinal cord carrying motor and sensory fibers, respectively (Fig. 1–6). They join to form the spinal nerve that exits from the respective spinal canal through the intervertebral foramina. A small ganglion, containing the cell bodies of sensory fibers, lies on each posterior root in the intervertebral foramina just proximal to its union with the anterior root but distal to the cessation of the dural sleeve. There are 31 spinal nerves on each side: 8 cervical, 12 thoracic, 5 lumbar, 5 sacral, and 1 coccygeal nerve. After passing through the foramina, the spinal nerve

branches into two divisions, the anterior and posterior primary rami.

The posterior rami supply the posterior part of the skin and the paraspinal muscles, which include the rectus capitis posterior, obliquus capitis superior and inferior, semispinalis capitis, splenius capitis, longus capitis, and sacrospinalis. These muscles extend the head, neck, trunk, and pelvis, respectively. The anterior rami supply the skin of the anterolateral portion of the trunk and the extremities. They also form the brachial and lumbosacral plexuses, which, in turn, give rise to peripheral nerves in the upper and lower extremities. The anterior rami of the thoracic spinal nerves become 12 pairs of intercostal nerves supplying the intercostal and abdominal muscles. At least two adjoining intercostal nerves supply each segmental level in both the thoracic and abdominal regions.

The diagnosis of a root lesion depends on abnormalities confined to a single spinal nerve without affecting adjacent higher or lower levels. The posterior rami

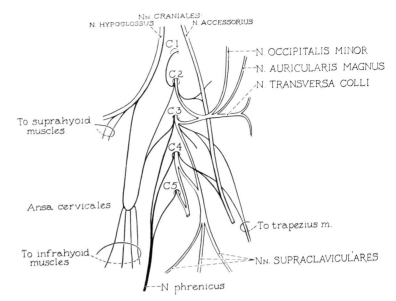

Figure 1–5. Anterior rami of the cervical spinal nerves, forming the cervical plexus. Note the phrenic nerve supplying the diaphragm, and the branches from the C-2, C-3, and C-4 roots and the accessory nerve, both innervating the trapezius. (From Anson,[1] with permission.)

that supply the paraspinal muscles branch off the spinal nerve just distal to the intervertebral foramen. Hence, denervation found at this level differentiates radiculopathy from more distal lesions of the plexus or peripheral nerve. Radiculopathy, however does not necessarily affect the paraspinal muscles in early stages of the disease, because the compressing lesions may only irritate the root without causing structural damage. Furthermore, spontaneous discharges appear in the denervated muscles 2 to 3 weeks after nerve injury. As in the innervation of the intercostal muscles by the anterior rami, the paraspinal muscles receive supplies from multiple posterior rami with substantial overlap. Therefore, the distribution of abnormalities in the extremities, rather than the paraspinal muscles, determines the level of a radicular lesion.

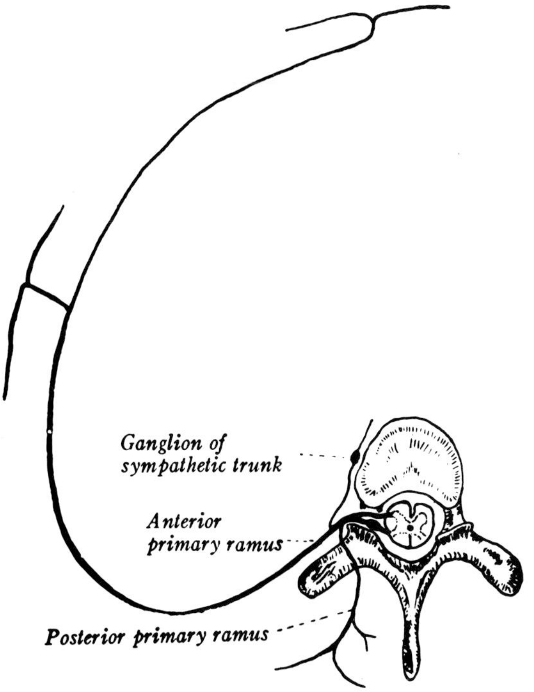

4 CERVICAL AND BRACHIAL PLEXUSES

The anterior rami of the upper four cervical nerves, C-1 through C-4, form the cervical plexus (see Fig. 1–5). It innervates the lateral and anterior flexors of the head, which consist of the rectus capitis lateralis, anterior longus capitis, and anterior longus colli. The brachial plexus, formed by the anterior rami of C-5 through T-1 spinal nerves, supply the muscles of the upper limb. Occasional variations of innervation include the prefixed brachial plexus, with main contributions from C-4 through C-8, and the

Figure 1–6. Ventral and dorsal roots forming the spinal nerve, which divides into the anterior and posterior rami. The sensory ganglion of the dorsal root lies within the respective intervertebral foramen. (From Ranson and Clark,[11] with permission.)

Ganglion of sympathetic trunk

Anterior primary ramus

Posterior primary ramus

Table 1–2 MUSCLES OF THE SHOULDER GIRDLE AND UPPER EXTREMITY

Nerve	C-4	C-5	C-6	C-7	C-8	T-1
Dorsal scapular	——— Levator scapulae ———					
		— Rhomboideus major & minor —				
Supra-scapular		— Supraspinatus —				
		— Infraspinatus —				
Axillary		— Teres minor —				
		— Deltoid —— Anterior —— Middle —— Posterior —				
Subscapular		— Teres major —				
Musculo-cutaneous		— Brachialis —				
		— Biceps brachii —				
			— Coraco-brachialis —			
Long thoracic		— Serratus - - - - - - - anterior —				
Anterior thoracic		— Pectoralis major (clavicular part) —				
				— Pectoralis major (sternocostal part) —		
				Pectoralis - - - - - minor		
Thoracodorsal			— Latissimus dorsi —			
Radial nerve		— Brachioradialis —				
			— Extensor carpi radialis longus & brevis —			
			— Triceps—long, lateral & medial heads —			
				— Anconeus —		
Posterior interosseous nerve			— Supinator —			
				— Extensor carpi ulnaris —		
				— Extensor digitorum —		
				— Extensor digiti minimi —		
				— Abductor pollicis longus —		
				— Extensor pollicis longus —		
				— Extensor pollicis brevis —		
				— Extensor indicis —		

Median nerve
- Pronator teres
- Flexor carpi radialis
- Palmaris longus
- Flexor digitorum sublimis
- Abductor pollicis brevis
- Flexor pollicis brevis (superficial head)
- Lumbricals I & II
- Opponens pollicis

Anterior interosseous nerve
- Flexor digitorum profundus I & II
- Flexor pollicis longus
- Pronator quadratus

Ulnar nerve
- Flexor digitorum profundus III & IV
- Flexor carpi ulnaris
- Adductor pollicis
- Flexor pollicis brevis (deep head)
- Abductor digiti minimi
- Opponens digiti minimi
- Flexor digiti minimi
- Volar interossei
- Dorsal interossei
- Lumbricals III & IV

postfixed brachial plexus, derived primarily from C-6 through T-2. Tables 1–1 and 1–2 present a summary of the anatomic relationship between the nerves derived from cervical and brachial plexuses and the muscles of the shoulder, arm, and hand.

Topographic divisions of the brachial plexus include root, trunk, cord, and peripheral nerve (Fig. 1–7). The roots combine to form three trunks. The union of C-5 and C-6 roots forms the upper trunk, and that of C-8 and T-1 roots, the lower trunks; whereas the C-7 root alone continues as the middle trunk. Each of the three trunks divides anterior and poste-

rior divisions. The posterior cord, formed by the union of all three posterior divisions, gives off the axillary nerve at the axilla and continues as the radial nerve. The anterior divisions of the upper and middle trunks form the lateral cord, which gives rise to the musculocutaneous nerve and the outer branch of the median nerve. The anterior division of the lower trunk, forming the medial cord, gives off the ulnar nerve, the inner branch of the median nerve, and cutaneous nerves.

The trunks pass through the supraclavicular fossa under the cervical and scalene muscles, forming the cords just above the clavicle at the level of the first

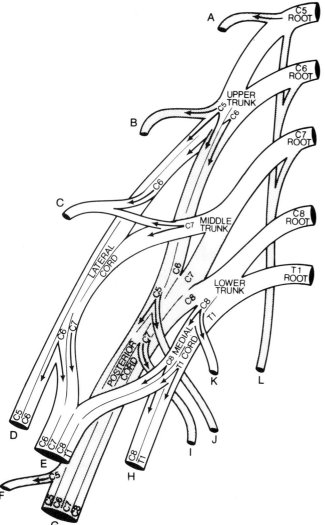

Figure 1–7. Anatomy of the brachial plexus with eventual destination of all root components. The posterior divisions of the trunks and their derivatives are shaded. The brachial plexus gives rise to dorsal scapular (*A*), suprascapular (*B*), lateral pectral (*C*), musculocutaneous (*D*), median (*E*), axillary (*F*), radial (*G*), ulnar (*H*), thoracodorsal (*I*), subscapular (*J*), medial pectral (*K*) and long thoracic (*L*) nerves. (Modified from Patten.[10])

rib. Accompanied by the subclavian artery, the cords traverse the space known as the thoracic outlet between the first rib and the clavicle. Consequently, injuries above or below the clavicle affect the trunks or cords, respectively. A more distal lesion involves the peripheral nerves that emerge from the cords between the clavicle and the axilla.

Phrenic Nerve

The phrenic nerve, one of the most important branches of the cervical plexus, arises from C-3 and C-4 roots and innervates the ipsilateral hemidiaphragm (see Table 1–1).

Dorsal Scapular Nerve

The dorsal scapular nerve, derived from C-4 and C-5 roots through the upper trunk of the brachial plexus, supplies the rhomboid major and minor and a portion of the levator scapulae, which keeps the scapula attached to the posterior chest wall during arm motion.

Suprascapular Nerve

The suprascapular nerve arises from C-5 and C-6 roots through the upper trunk of the brachial plexus. It reaches the upper border of the scapula behind the brachial plexus to enter the suprascapular notch, a possible site of entrapment. The nerve supplies the supraspinatus and infraspinatus (see Fig. 1–7).

Musculocutaneous Nerve

The musculocutaneous nerve arises from the lateral cord of the brachial plexus near the lower border of the pectoralis minor (Fig. 1–8). Its axons, chiefly originating from C-5 and C-6 roots, reach the biceps, brachialis, and coracobrachialis, with some variations of innervation for the last two muscles. Its terminal sensory branch, called the lateral cutaneous nerve, supplies the skin over the lateral aspect of the forearm.

Axillary Nerve

The axillary nerve arises from the posterior cord of the brachial plexus, originating from C-5 and C-6 roots. It supplies the skin of the lateral aspect of the arm and the deltoid and teres minor muscles (Fig. 1–9).

5 PRINCIPAL NERVES OF THE UPPER EXTREMITY

Radial Nerve

The radial nerve, as a continuation of the posterior cord, derives its axons from the C-5 through T-1, or all the spinal roots contributing to the brachial plexus (see Fig. 1–7). The nerve gives off its supply to the three heads of the triceps and the anconeus, which originates from the lateral epicondyle of the humerus as an extension of the medial head. The radial nerve then enters the spiral groove, winding around the humerus posteriorly from the medial to the lateral side (see Fig. 1–9). As the nerve emerges from the spiral groove, it supplies the brachioradialis and, slightly more distally, the extensor carpi radialis longus. Located lateral to the biceps at the level of the lateral epicondyle, it enters the forearm between the brachialis and brachioradialis. At this point, it divides into a muscle branch, the posterior interosseous nerve, and a sensory branch, which surfaces in the distal third of the forearm. The muscle branch innervates the supinator, the abductor pollicis longus, and all the extensor muscles in the forearm: extensor carpi radialis brevis, extensor carpi ulnaris, extensor digitorum, extensor digiti minimi, extensor pollicis longus and brevis, and extensor indicis. The sensory fibers, originating from the C-6 and C-7 roots, pass through the upper and middle trunks and the posterior cord and supply the skin over the lateral aspect of the dorsum of the hand.

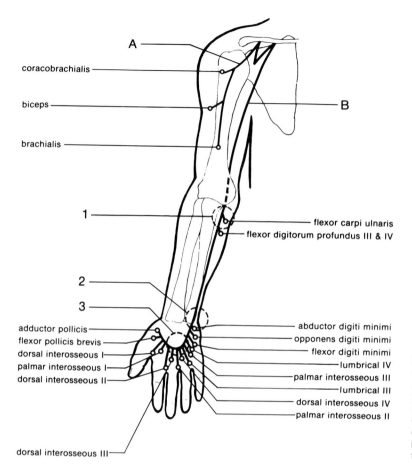

coracobrachialis

biceps

brachialis

1 —— flexor carpi ulnaris
—— flexor digitorum profundus III & IV

2

3

adductor pollicis
flexor pollicis brevis
dorsal interosseous I
palmar interosseous I
dorsal interosseous II

abductor digiti minimi
opponens digiti minimi
flexor digiti minimi
lumbrical IV
palmar interosseous III
lumbrical III
dorsal interosseous IV
palmar interosseous II

dorsal interosseous III

Figure 1–8. Musculocutaneous nerve (*A*) and ulnar nerve (*B*) and the muscles they supply. The common sites of lesion include ulnar groove and cubital tunnel (*1*), Guyon's canal (*2*), and midpalm (*3*). (Modified from The Guarantors of Brain.[7])

Median Nerve

The median nerve arises from the lateral and median cords of the brachial plexus as a mixed nerve derived from the C-6 and T-1 roots (see Fig. 1–7). It supplies most forearm flexors and the muscles of the thenar eminence. It also subserves sensation to the skin over the lateral aspect of the palm and the dorsal surfaces of the terminal phalanges, along with the volar surfaces of the thumb, the index and middle fingers, and half of the ring finger. The sensory fibers of the middle finger enter the C-7 root through the lateral cord and middle trunk, whereas the skin of the thumb and the index finger receives fibers from the C-6 or C-7 root through the lateral cord and upper or middle trunk. The median nerve inner-

vates no muscles in the upper arm (Fig. 1–10). It enters the forearm between the two heads of the pronator teres, which it supplies along with the flexor carpi radialis, palmaris longus, and flexor digitorum superficialis. It then gives rise to a pure muscle branch called the anterior interosseous nerve, which innervates the flexor pollicis longus, pronator quadratus, and flexor digitorum profundus I and II. The main nerve descends the forearm and passes through the carpal tunnel between the wrist and palm. It supplies lumbricals I and II after giving off the recurrent thenar nerve at the distal edge of the carpal ligaments. This muscle branch to the thenar eminence innervates the abductor pollicis brevis, the lateral half of the flexor pollicis brevis, and the opponens pollicis.

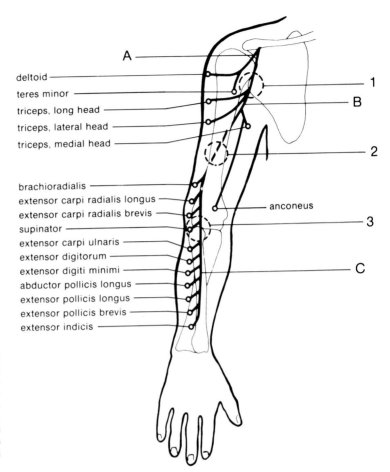

deltoid
teres minor
triceps, long head
triceps, lateral head
triceps, medial head

brachioradialis
extensor carpi radialis longus
extensor carpi radialis brevis
supinator
extensor carpi ulnaris
extensor digitorum
extensor digiti minimi
abductor pollicis longus
extensor pollicis longus
extensor pollicis brevis
extensor indicis

A

anconeus

1
B
2
3
C

Figure 1–9. Axillary nerve (*A*) and radial nerve (*B*) with its main terminal branch, posterior interosseous nerve (*C*), and the muscles they supply. The nerve injury may occur at the axilla (*1*), spiral groove (*2*), or elbow (*3*), as in the posterior interosseous nerve syndrome. (Modified from The Guarantors of Brain.[7])

Ulnar Nerve

The ulnar nerve, as a continuation of the medial cord of the brachial plexus, derives its fibers from the C-8 and T-1 roots (see Fig. 1–7). It lies in close proximity to the median nerve and brachial artery at the axilla. In this position the ulnar nerve passes between the biceps and triceps, then deviates posteriorly at the midportion of the upper arm and becomes superficial behind the medial epicondyle (see Fig. 1–8). After entering the forearm, it supplies the flexor carpi ulnaris and flexor digitorum profundus III and IV. It passes along the medial aspect of the wrist to enter the hand, where it gives off two branches. The superficial sensory branch supplies the skin over the medial aspect of the hand from the wrist distally, including the hypothenar emi-

nence, the fifth digit, and half of the fourth digit. The deep muscle branch first supplies hypothenar muscles: the abductor, opponens, and flexor digiti minimi. It then deviates laterally around the hamate bone to reach the lateral aspect of the hand, where it reaches the adductor pollicis and medial half of the flexor pollicis brevis. Along its course from hypothenar to thenar eminence the deep branch also innervates the interossei and lumbricals III and IV.

General Rules and Anomalies

Table 1–2 summarizes the pattern of nerve supply in the upper extremities. One cannot memorize the exact innervation for all the individual muscles, but learning certain rules helps broadly cate-

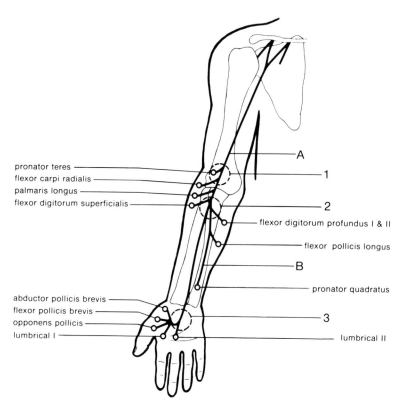

pronator teres
flexor carpi radialis
palmaris longus
flexor digitorum superficialis

A
1
2
flexor digitorum profundus I & II
flexor pollicis longus
B
pronator quadratus

abductor pollicis brevis
flexor pollicis brevis
opponens pollicis
lumbrical I

3
lumbrical II

Figure 1–10. The median nerve (*A*) with its branch, the anterior interosseous nerve (*B*), and the muscles they supply. The nerve may undergo compression at the elbow between the two heads of pronator teres (*1*), or slightly distally (*2*) as in the anterior interosseous syndrome, or at the palm (*3*) as in the carpal tunnel syndrome. (Modified from The Guarantors of Brain.[7])

gorize muscles. The radial nerve innervates the brachioradialis, triceps, and with its main terminal branch, posterior interosseous nerve, all the extensors in the forearm, but none of the intrinsic hand muscles. The median nerve supplies most flexors in the forearm, in addition to the intrinsic hand muscles of the thenar eminence, and lumbricals I and II. The anterior interosseous nerve branches off the median nerve trunk in the forearm to innervate the flexor digitorum profundus I and II, flexor pollicis longus, and pronator quadratus. The ulnar nerve supplies only intrinsic hand muscles, with the exception of the flexor carpi ulnaris and the flexor digitorum profundus III and IV.

The most common anomaly of innervation in the upper limb results from the presence of a communicating branch from the median to the ulnar nerve in the forearm. The fibers involved in this crossover, called the Martin-Gruber anastomosis, usually supply ordinarily ulnar-innervated intrinsic hand muscles.

Thus, the anomalous fibers form a portion of the ulnar nerve that, instead of branching off from the medial cord of the brachial plexus, takes an aberrant route distally along with the median nerve and then reunites with the ulnar nerve proper in the distal forearm. Other anomalies reported in the literature include communication from the ulnar to median nerve in the forearm, and all median or all ulnar hands, in which one or the other nerve supplies all the intrinsic hand muscles. These extremely rare patterns stand in contrast to the high incidence of the median-to-ulnar anastomosis. Failure to recognize an anomaly is a common source of error in clinical electrophysiology (see Chapter 7.4).

6 LUMBAR PLEXUS AND ITS PRINCIPAL NERVES

The spinal cord ends at the level of the L1-2 intervertebral space as the pre-conus, which consists of the L-5 and S-1 cord segments and conus medullaris, which contains the S-2 through S-5 levels. The cauda equina, formed by the lumbar and sacral roots, assumes a downward direction from the conus toward their respective exit foramina. The fibrous filum terminale interna extends from the lowermost end of the spinal cord to the bottom of the dural sac at the level of the S-2 vertebra. Table 1–3 summarizes the nerves derived from the lumbar plexus and the muscles they innervate.

The anterior rami of the first three lumbar spinal nerves from the L-1, L-2, L-3, and part of the L-4 roots unite to form the lumbar plexus within the psoas major muscle (Figs. 1–11 to 1–13). The iliohypogastric and ilioinguinal nerves arise from the L-1 root and supply the skin of the hypogastric region and medial thigh, respectively. The genitofemoral nerve, derived from the L-1 and L-2 roots, supplies the cremasteric muscle and the skin of the scrotum. The lateral femoral cutaneous nerve originates from the L-2 and L-3 roots. It leaves the psoas muscle laterally to supply the lateral and anterior thigh. The anterior divisions of the L-2 through L-4 anterior rami join to form the obturator nerve, which exits the psoas muscle medially to innervate the adductor muscles of the thigh. The posterior divisions of the same rami give rise to the femoral nerve, which leaves the psoas muscle laterally. It then descends under the iliacus fascia to reach the femoral triangle beneath the inguinal ligament. Though primarily a muscle nerve, it also gives off sensory branches—the intermediate and medial cutaneous nerves and the saphenous nerve.

Iliohypogastric Nerve

The iliohypogastric nerve arises from the L-1 root and supplies the skin of the upper buttock and hypogastric region.

Ilioinguinal Nerve

The ilioinguinal nerve, arising from the L-1 and L-2 roots, supplies the skin over the upper and medial part of the thigh, the root of the penis, and the upper part of the scrotum or labium major. It also innervates the transversalis and internal oblique muscles. The nerve follows the basic pattern of an intercostal nerve, winding around the inner side of the trunk to the medial anterior iliac spine.

Genitofemoral Nerve

The genitofemoral nerve, arising from the L-1 and L-2 roots, branches into lumboinguinal and external spermatic nerves. The lumboinguinal nerve supplies the skin over the femoral triangle. The external spermatic nerve innervates the cremasteric muscle and the skin of the inner aspect of the upper thigh, scrotum, or labium.

Lateral Femoral Cutaneous Nerve

The lateral femoral cutaneous nerve, the first sensory branch of the lumbar plexus, receives fibers from the L-2 and L-3 roots. It emerges from the lateral border of the psoas major muscle and runs forward, coursing along the brim of the pelvis to the lateral end of the inguinal ligament. The nerve reaches the upper thigh after passing through a tunnel formed by the lateral attachment of the inguinal ligament and the anterior superior iliac spine. About 12 cm below its exit from the tunnel, the nerve gives off an anterior branch, which supplies the skin over the lateral and anterior surface of the thigh, and a posterior branch, which innervates the lateral and posterior portion of the thigh.

Femoral Nerve

The femoral nerve, formed near the vertebral canal, arises from the anterior rami of the L-2 through L-4 roots (Fig. 1–14). The nerve reaches the front of the leg, passing along the lateral edge of the psoas muscle, which it supplies together with the iliacus. It then exits the pelvis

Table 1–3 MUSCLES OF THE PELVIC GIRDLE AND LOWER EXTREMITY

Nerve	Muscle	L-2	L-3	L-4	L-5	S-1	S-2
Femoral nerve	Iliopsoas	—	—	—			
	Pectineus	—	—	—			
	Sartorius	—	—				
	Vastus intermedius	—	—	—			
	Rectus femoris		—	—			
	Vastus lateralis		—	—			
	Vastus medialis		—	—			
Obturator nerve	Gracilis	—	—	—			
	Adductor longus & brevis	—	—	—			
	Adductor magnus		—	—			
	Obturator externus	—	—	—	—		
Sacral plexus	Obturator internus			—	—	—	—
	Inferior gemelli			—	—	—	
	Piriformis					—	—
Superior gluteal nerve	Gluteus medius			—	—	—	
	Gluteus minimus			—	—	—	
	Tensor fasciae latae			—	—	—	
Inferior gluteal nerve	Gluteus maximus				—	—	—
Sciatic nerve trunk — Peroneal division	Biceps femoris short head				—	—	—
Sciatic nerve trunk — Tibial division	Quadratus femoris			—	—	—	
	Semitendinosus			—	—	—	—
	Semimembranosus			—	—	—	—

Common peroneal nerve

Deep peroneal nerve

Tibialis anterior
Extensor digitorum longus
Extensor digitorum brevis
Peroneus tertius
Extensor hallucis longus

Superficial peroneal nerve

Peroneus longus
Peroneus brevis

Biceps femoris long head

Tibial nerve

Tibialis posterior
Popliteus
Flexor digitorum longus
Flexor hallucis longus
Gastrocnemius — Medial head — Lateral head
Soleus

Medial plantar nerve

Flexor digitorum brevis
Flexor hallucis brevis
Abductor hallucis
Lumbrical I

Lateral plantar nerve

Abductor digiti minimi
Abductor hallucis
Flexor digiti minimi
Interossei
Quadratus plantae
Lumbricals II, III, IV

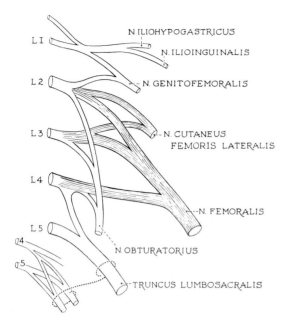

Figure 1–11. Anterior rami of the lumbar spinal nerve forming the lumbar plexus, with the major nerves derived from this plexus. The shaded portion indicates the dorsal divisions. (From Anson,[1] with permission.)

under the inguinal ligament just lateral to the femoral artery and vein. Its sensory branches supply the skin of the anterior thigh and medial aspect of the calf. The muscle branch innervates the pectineus, sartorius, and quadriceps femoris, which consists of the rectus femoris, vastus lateralis, vastus intermedius, and vastus medialis. Of the muscles innervated by this nerve, the iliopsoas flexes the hip at the thigh, the quadriceps femoris extends the leg at the knee, the sartorius flexes the leg and the thigh, and the pectineus flexes the thigh.

Saphenous Nerve

The saphenous nerve, the largest and longest sensory branch of the femoral nerve, supplies the skin over the medial aspect of the thigh, leg, and foot. It accompanies the femoral artery in the femoral triangle, then descends medially under the sartorius muscle. The nerve

gives off the infrapatellar branch at the lower thigh, which supplies the medial aspect of the knee. The main terminal branch descends along the medial aspect of the leg, accompanied by the long saphenous vein. It passes just anterior to the medial malleolus supplying the medial side of the foot.

Obturator Nerve

The obturator nerve arises from the anterior divisions of the L-2 through L-4 roots (see Fig. 1–14). Formed within the psoas muscle, it enters the pelvis immediately anterior to the sacroiliac joint. As it passes through the obturator canal, the obturator nerve gives off an anterior branch, which supplies the adductor longus and brevis and the gracilis, and a posterior branch, which innervates the obturator externus and half of the adductor magnus muscle. The sensory fibers supply the skin of the upper thigh over the medial aspect and send anastomoses to the saphenous nerve.

7 SACRAL PLEXUS AND ITS PRINCIPAL NERVES

The sacral plexus arises from the L-5, S-1, and S-2 roots in front of the sacroiliac joint (see Figs. 1–12 and 1–13). Designation as the lumbosacral plexus implies an interconnection between the sacral and lumbar plexuses. Common anomalous derivations include a prefixed pattern with a major contribution of the L-4 root to the sacral plexus and a postfixed form with the L-5 root supplying mainly the lumbar plexus. The sacral plexus gives rise to the superior gluteal nerve, derived from the L-4, L-5, and S-1 roots; and the inferior gluteal nerve, from L-5, S-1, and S-2 roots. The sciatic nerve, the largest nerve in the body, arises from the L-4 through S-2 roots. After giving off branches to the hamstring muscles, it divides into the tibial and the common peroneal nerves. Table 1–3 summarizes the

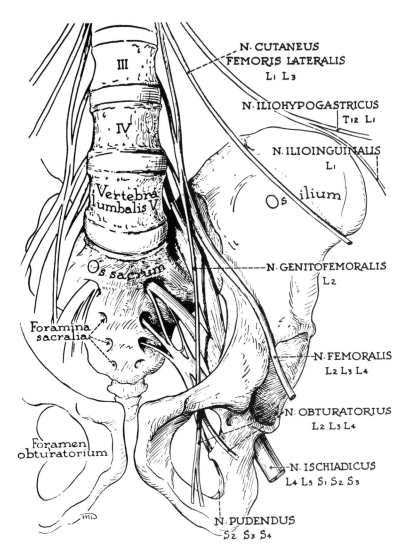

III

N. CUTANEUS
FEMORIS LATERALIS
L₁ L₃

N. ILIOHYPOGASTRICUS
T₁₂ L₁

IV

N. ILIOINGUINALIS
L₁

Vertebra
lumbalis V

Os ilium

Os sacrum

N. GENITOFEMORALIS
L₂

Foramina
sacralia

N. FEMORALIS
L₂ L₃ L₄

N. OBTURATORIUS
L₂ L₃ L₄

Foramen
obturatorium

N. ISCHIADICUS
L₄ L₅ S₁ S₂ S₃

N. PUDENDUS
S₂ S₃ S₄

Figure 1–12. Lumbosacral plexus and the courses of the femoral, obturator, and sciatic nerves. (From Anson,[1] with permission.)

nerves derived from the sacral plexus and the muscles they innervate.

Superior and Inferior Gluteal Nerves

The superior gluteal nerve, derived from the L-4 through S-1 roots, innervates the gluteus medius and minimus and tensor fascia lata, which together abduct and rotate the thigh internally. The inferior gluteal nerve, arising from the L-5 through S-2 roots, innervates the gluteus maximus, which extends, abducts, and rotates the thigh externally.

Sciatic Nerve

The union of all of the L-4 to S-2 roots gives rise to the sciatic nerve, which leaves the pelvis through the greater sciatic foramen (Fig. 1–15). The nerve consists of a peroneal portion derived from the posterior division of the anterior rami and a tibial portion composed of the anterior divisions. These two components eventually form separate peroneal and tibial nerves in the lower third of the thigh. In the posterior aspect of the thigh, the tibial component of the sciatic trunk gives off a series of short branches to innervate the bulk of the hamstring mus-

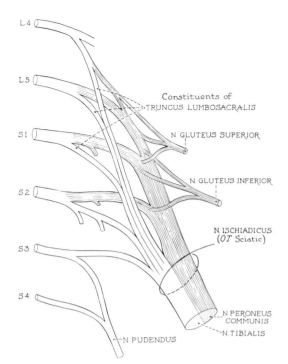

Figure 1–13. Anterior rami of the lumbosacral spinal nerve forming the sacral plexus with the major nerves derived from this plexus. The *shaded portion* indicates the dorsal divisions. (From Anson,[1] with permission.)

cles, which consist of the long head of the biceps femoris, semitendinosus, and semimembranosus. The peroneal component supplies the short head of the biceps femoris. The adductor magnus, primarily supplied by the obturator nerve, also receives partial innervation from the sciatic trunk.

Tibial Nerve

The tibial nerve, sometimes called posterior tibial nerve, arises as the extension of the medial popliteal nerve that bifurcates from the sciatic nerve in the popliteal fossa (see Fig. 1–15). After giving off branches to the medial and lateral heads of the gastrocnemius and soleus, it supplies the tibialis posterior, flexor digitorum longus, and flexor hallucis longus in the leg. The nerve enters the foot, passing through the space between the medial malleolus and the flexor retinaculum. Here it splits into medial and lateral plantar nerves after giving off a small

calcaneal nerve. This bifurcation occurs within 1 cm of the malleolar-calcaneal axis in 90 percent of feet.[4]

The medial plantar artery, which accompanies the medial plantar nerve, serves as the landmark to locate the nerve just below the medial malleolus. The muscle branches innervate the abductor hallucis, flexor digitorum brevis, and flexor hallucis brevis. The sensory fibers of the medial plantar nerve supply the medial anterior two thirds of the sole and the plantar skin of the first three toes and part of the fourth toe. The lateral plantar nerve winds around the heel to the lateral side of the sole to innervate the abductor digiti minimi, flexor digiti minimi, abductor hallucis, and interossei. It supplies the skin over the lateral aspect of the sole, lateral half of the fourth toe, and the fifth toe.

Common Peroneal Nerve

The common peroneal nerve arises as an extension of the lateral popliteal nerve, which branches off laterally from the sciatic trunk in the popliteal fossa (see Fig. 1–14). It consists of fibers derived from the L-4, L-5, and S-1 roots. Immediately after its origin, the nerve becomes superficial as it winds laterally around the head of the fibula. After entering the leg at this position, it gives off a small recurrent nerve that supplies sensation to the patella. It then bifurcates into the superficial and deep peroneal nerves.

The superficial peroneal nerve, also known as the musculocutaneous nerve, supplies the peroneus longus and the peroneus brevis, which allow plantar flexion and eversion of the foot. The nerve descends between the peroneal muscles, which it innervates, then divides into medial and intermediate dorsal cutaneous nerves. These sensory branches pass anterior to the extensor retinaculum to supply the anterolateral aspect of the lower half of the leg and dorsum of the foot and toes.

The deep peroneal nerve, sometimes called the anterior tibial nerve, innervates the muscles that allow dorsiflexion

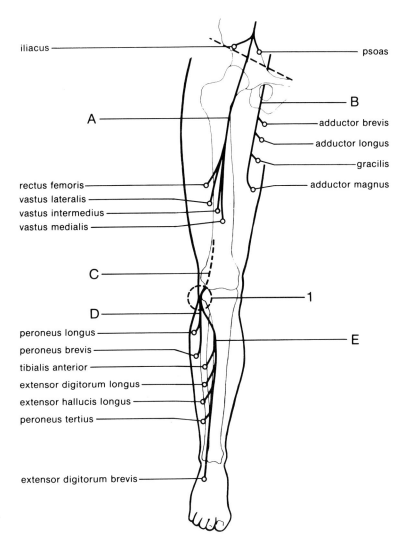

Figure 1–14. Femoral nerve (*A*), obturator nerve (*B*), and common peroneal nerve (*C*) branching into superficial (*D*) and deep peroneal nerve (*E*) and the muscles they supply. The compression of the peroneal nerve commonly occurs at the fibular head (*1*). (Modified from The Guarantors of Brain.[7])

and eversion of the foot. These muscles include the tibialis anterior, extensor digitorum longus, extensor hallucis longus, peroneus tertius, and extensor digitorum brevis. An anomalous communicating branch called the accessory deep peroneal nerve may arise from the superficial peroneal nerve at the knee to innervate the lateral portion of the extensor digitorum brevis (see Chapter 7.4). The deep peroneal nerve also supplies the skin over a small, wedge-shaped area between the first and second toes.

Sural Nerve

The sural nerve originates from the union of the medial sural cutaneous branch of the tibial nerve and the sural communicating branch of the common peroneal nerve. It arises below the popliteal space, descends between the two bellies of the gastrocnemius, winds behind the lateral malleolus, and reaches the dorsum of the fifth toe. It supplies the skin over the posterolateral aspect of the distal leg and lateral aspect of the foot. As one of the few readily accessible sensory nerves in the lower extremities, the sural nerve offers an ideal site for biopsy, especially because its removal induces only minimal sensory changes. A fascicular biopsy of the sural nerve allows in vitro recording of nerve action potentials (see Chapter 4.4). Therefore, in vivo studies of the sural nerve before such a procedure

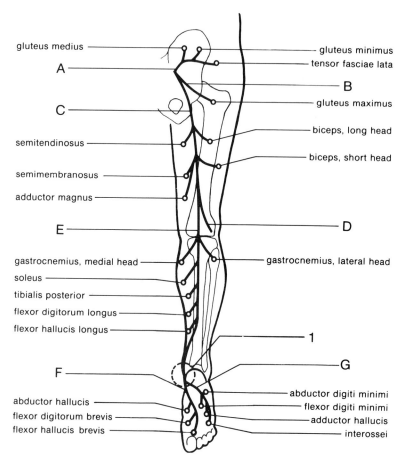

gluteus medius

A

C

semitendinosus

semimembranosus

adductor magnus

E

gastrocnemius, medial head
soleus
tibialis posterior
flexor digitorum longus
flexor hallucis longus

F

abductor hallucis
flexor digitorum brevis
flexor hallucis brevis

gluteus minimus
tensor fasciae lata

B

gluteus maximus

biceps, long head

biceps, short head

D

gastrocnemius, lateral head

1

G

abductor digiti minimi
flexor digiti minimi
adductor hallucis
interossei

Figure 1–15. Superior gluteal nerve (*A*), inferior gluteal nerve (*B*), and sciatic nerve trunk (*C*) and the muscles they supply. The sciatic nerve bifurcates to form the common peroneal nerve (*D*) and the tibial nerve (*E*). The tibial nerve in turn gives rise to the medial (*F*) and lateral plantar nerve (*G*). The compression of the tibial nerve may occur at the medial malleolus in the tarsal tunnel (*1*). (Modified from The Guarantors of Brain.[7])

provide an interesting opportunity to correlate the data directly with in vitro conduction characteristics and the histologic findings of the biopsy specimen.[5]

REFERENCES

1. Anson, BJ: An Atlas of Human Anatomy, ed 2. WB Saunders, Philadelphia, 1963.
2. Carpenter, MB: Human Neuroanatomy, ed 7. Williams & Wilkins, Baltimore, 1976.
3. Delagi, EF, Perotto, A, Iazzetti, J, and Morrison, D: Anatomic Guide for the Electromyographer: The Limbs. Charles C Thomas, Springfield, Ill, 1975.
4. Dellon, AL, and Mackinnon, SE: Tibial nerve branching in the tarsal tunnel. Arch Neurol 41:645–646, 1984.
5. Dyck, PJ, Lambert, EH, and Nichols, PC: Quantitative measurement of sensation related to compound action potential and number and size of myelinated fibers of sural nerve in health, Friedrich's ataxia, hereditary sensory neuropathy and tabes dorsalis. In Remond, A (ed): Handbook of Electroencephalography and Clinical Neurophysiology, Vol 9. Elsevier, Amsterdam, 1971, pp 83–118.
6. Goodgold, J: Anatomical Correlates of Clinical Electromyography. Williams & Wilkins, Baltimore, 1974.
7. The Guarantors of Brain: Aids to the Examination of the Peripheral Nervous System. Bailliere Tindall, East Sussex, England, 1986.
8. Haymaker, W, and Woodhall, B: Peripheral Nerve Injuries: Principles of Diagnosis, ed 2. WB Saunders, Philadelphia, 1953.
9. Hollinshead, WH: Functional Anatomy of the Limbs and Back, ed 3. WB Saunders, Philadelphia, 1969.
10. Patten, J: Neurological Differential Diagnosis. Harold Starke Ltd, London, 1977.
11. Ranson, SW, and Clark, SL: The Anatomy of the Nervous System: Its Development and Function, ed 10. WB Saunders, Philadelphia, 1959.
12. Romanes, GJ: Cunningham's Textbook of Anatomy, ed 11. Oxford University Press, London, 1972.
13. Sunderland, S: Nerves and Nerve Injuries, ed 2. Churchill Livingstone, New York, 1978.
14. Williams, PL, and Warwick, R: Gray's Anatomy, ed 36 (British). Churchill Livingstone, Edinburgh, 1980.

Chapter 2

ELECTRICAL PROPERTIES OF NERVE AND MUSCLE

1 INTRODUCTION

The nervous system conveys information by means of action potentials, which originate in the cell body or axon terminal and propagate along the nerve fibers. An electrophysiologic study of sensory conduction depends on recording such neural impulses after stimulation of the nerve through surface or needle electrodes. To assess the motor conduction, one ordinarily records a muscle action potential elicited by stimulation of the mixed nerve. Electromyography permits analysis of electrical properties in the skeletal muscle at rest and during voluntary contraction. It is apparent, therefore,

that the proper interpretation of electrodiagnostic data in the clinical domain requires an understanding of the electrical properties of nerve and muscle.

Despite different anatomic substrates subserving electrical impulses, the same basic membrane physiology applies to both nerve and muscle. Excitability of the tissues reflects the magnitude of the transmembrane potential in a steady state. When stimulated electrically or by other means, the cell membrane undergoes an intensity-dependent depolarization. If the change reaches the critical level, called threshold, it results in generation of an action potential, which then propagates across the membrane. In contrast to intracellular recording in animal experiments, clinical electrodiagnostic procedures analyze the extracellular potentials by surface or needle electrodes. Here the interstitial tissues act as a volume conductor, which may substantially distort the waveform of the electrical potentials, depending on the position of the recording electrode relative to the generator source.

2 TRANSMEMBRANE POTENTIAL

Understanding the membrane physiology at the cellular level forms the basis for an electrophysiologic examination in the clinical domain. This section deals with the ionic concentration of cell plasma and its role in maintaining transmembrane potentials. The next sections summarize the basic physiology of the propagating action potential recorded through volume conductors. The following discussion, intended merely as a background for subsequent discussion, covers only the fundamental principles relevant to clinical electrophysiology. Interested readers can find a more detailed account of basic cell physiology in established texts.[4,16–18,22,23,31,35,38,42,43]

Ionic Concentration of Cells

The muscle membrane constitutes the boundary between intracellular fluid in cell cytoplasm and extracellular intersti-

tial fluids. Both contain approximately equal numbers of ions dissolved in water but differ in two major aspects. First, an electrical potential exists across the cell membrane, with a relative negativity inside the cell as compared with outside the cell. This steady transmembrane potential measures approximately −90 mV in human skeletal muscle cells,[37] but it varies from one tissue to another, ranging from −20 to −100 mV. Second, intracellular fluid has a much higher concentration of potassium (K^+) and a lower concentration of sodium (Na^+) and chloride (Cl^-) relative to extracellular fluid (Table 2–1).

Nernst Equation

In the steady state, the influx of an ion precisely counters the efflux, maintaining an equilibrium. Thus, various factors that determine the direction and the rate of the ionic flow must exert balanced force together. If one measures the ionic concentration, one can calculate the equilibrium potential, that is, the transmembrane potential theoretically required to establish such a balance (Fig. 2–1).

In the case of potassium, for example, the ionic difference tends to push potassium from inside to outside the cell, reflecting a higher concentration inside. This force per mole of potassium, or the chemical work (W_c), increases in proportion to the logarithm of the ratio between the internal and external concentration

Table 2–1 COMPOSITIONS OF EXTRACELLULAR AND INTRACELLULAR FLUIDS OF MAMMALIAN MUSCLE

	Extracellular (mmol/l)	Intracellular (mmol/l)	Equilibrium Potential (mV)
Cations			
Na^+	145	12	66
K^+	4	155	−97
Others	5	—	—
Anions			
Cl^-	120	4	−90
HCO_3^-	27	8	−32
Others	7	155	—
Potential	0 mV	−90 mV	

From Patton,[35] with permission.

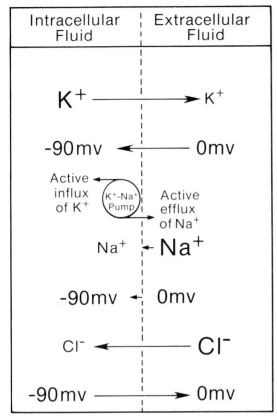

Intracellular Fluid	Extracellular Fluid

Figure 2–1. Simplified scheme of active and passive fluxes of potassium (K$^+$), sodium (Na$^+$) and chloride (Cl$^-$) in the steady state with driving force on each ion shown by vectors. For potassium, the efflux along the concentration gradient equals the influx caused by the electrical force plus the active influx by the sodium-potassium pump. For sodium, the electrical and chemical gradient produces only a small influx because of membrane resistance. The sum of the two equals the active efflux by the sodium-potassium pump. For chloride, the concentration gradient almost exactly counters the electrical force. The ratio of sodium and potassium exchange averages 3:2, although this diagram illustrates a neutral pump with a ratio of 1:1.

of the cation, (K$^+$)$_i$ and (K$^+$)$_o$, according to the equation

$$W_c = RT \, \log_e (K^+)_i/(K^+)_o,$$

where R represents the universal gas constant and T, the absolute temperature.

The energy required to counter this force must come from the negative equilibrium potential (E$_K$) pulling the positively charged potassium from outside to

inside the cell. This force per mole of potassium, or the electrical work (W$_e$), increases in proportion to the transmembrane voltage E$_K$, according to the equation

$$W_e = Z_K \, F \, E_K,$$

where F represents the number of coulombs per mole of charge and Z$_K$, the valence of the ion.

In the steady state, the sum of these two energies, W$_c$ and W$_e$, must equal zero, because they represent forces with opposite vectors. Therefore,

$$Z_k \, F \, E_K + RT \, \log_e (K^+)_i/(K^+)_o = 0.$$

Thus, the Nernst equation provides the theoretic potassium equilibrium potential E$_K$ as follows

$$E_K = - (RT/Z_K F) \, \log_e (K^+)_i/(K^+)_o.$$

The same equation applies to calculate the sodium and chloride equilibrium potentials, E$_{Na}$ and E$_{Cl}$, as follows:

$$E_{Na} = - (RT/Z_{Na} \, F) \, \log_e (Na^+)_i/(Na^+)_o$$

and

$$E_{Cl} = - (RT/Z_{Cl} F) \, \log_e (Cl^-)_o/(Cl^-)_i$$

Table 2–1 shows the values of E$_K$ (−97 mV), E$_{Na}$ (+66 mV), and E$_{Cl}$ (−90 mV) determined on the basis of the ionic concentrations. These compare with the actual transmembrane potential (−90 mV) in the example under consideration. Thus, ionic concentration and transmembrane potential alone can maintain chloride ions in perfect balance. To keep potassium and sodium in equilibrium at transmembrane potentials of −90 mV, therefore, other factors must exert a substantial influence on ionic movements. These include selective permeability of the cell membrane to certain ions and the energy-dependent sodium-potassium pump.

Sodium-Potassium Pump

In the case of potassium, an additional factor, the active transport of potassium by an energy-dependent pump, explains the small discrepancy between E$_K$ (−97 mV) and E$_m$ (−90 mV). Here, the forces pulling potassium from outside to inside the cell consist of potential difference (−90 mV) and active potassium transport (approximately equivalent to −7 mV). Together they counter almost exactly the concentration gradient pushing potassium from inside to outside the cell. In the case of sodium, both the concentration gradient and potential difference (−90 mV) pull the ion from outside to inside the cell. Nonetheless, this cation remains in equilibrium because of its impermeability through a mechanical barrier imposed by the structure of the cell membrane. Active transport of sodium from inside to outside counters the small amount of sodium that does leak inward.

This energy-dependent process, known as the sodium-potassium pump, transports sodium outward in exchange for inward movement of potassium. Figure 2–1 depicts a neutral pump that exchanges one sodium ion for every potassium ion actively transported inward. The ratio of sodium and potassium exchange, however, averages 3:2 in most tissues.[34] Such an imbalanced arrangement, called an electrogenic sodium-potassium pump, directly contributes to the membrane potential, but only minimally, compared with changes in membrane permeability.

Goldman-Hodgkin-Katz Equation

The Nernst equation closely predicts the membrane potential for highly diffusible chloride and potassium ions. It does not fit well with much less permeable sodium ions because it ignores the relative membrane permeability. The addition of this factor leads to the more comprehensive Goldman-Hodgkin-Katz formula, derived from the concentration gradients and membrane permeabilities of all ions.

$$EM = (RT/F) \log_e$$
$$\frac{P_{Na}(Na^+_o) + P_K(K^+_o) ----- + P_{Cl}(Cl^-_i)}{P_{Na}(Na^+_i) + P_K(K^+_i) ----- + P_{Cl}(Cl^-_o)}$$

where P_{Na}, P_K, and P_{Cl} represent permeabilities of the respective ions.

According to this equation, the concentration gradient of the most permeable

ions dictates the transmembrane potentials. In the resting membrane with very high P_K relative to negligible P_{Na}, the Goldman-Hodgkin-Katz equation would approximate the Nernst equation with the potassium concentration gradients. The transmembrane potentials calculated with either equation range from −80 to −90 mV. Conversely, the Goldman-Hodgkin-Katz potential would nearly equal the Nernst potential for sodium, with negligible P_K relative to high P_{Na}. In this situation, the calculated membrane potentials range from +50 to +70 mV. This reversal of polarity, in fact, characterizes the generation of an action potential as outlined below.

3 GENERATION OF ACTION POTENTIAL

Generation of the action potential consists of two phases: subthreshold and threshold events. Essential characteristics of subthreshold activation include a graded response that produces local changes in transmembrane potential. Subthreshold activation gives rise to a self-limiting local potential change that diminishes with distance. If, on the other hand, the membrane potential reaches the critical level with about 15 to 25 mV of depolarization, from −90 mV to −65 to −75 mV in the case of human muscle cell,[37] the action potential develops in an all-or-none fashion; that is, the same maximal response occurs through a complex energy-dependent process regardless of the kind or magnitude of the stimulus (Fig. 2–2).

All-or-None Response

In the living cell, molecular structure regulates the conductance of sodium and potassium ions across the membrane. One set of channels controls the movement of sodium ions and another set, potassium ions. Depending on the transmembrane potential, they either open or close pathways for that specific ion through the membrane. In the resting

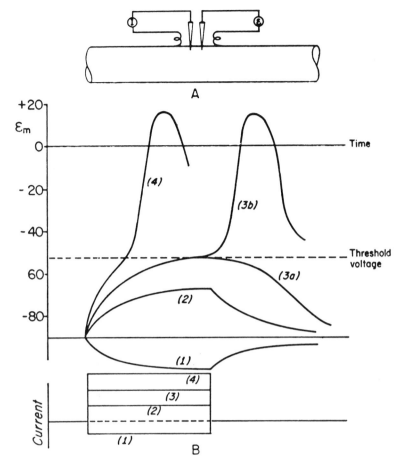

Figure 2-2. Schematic diagram of graded responses after subthreshold stimuli and generation of action potentials after suprathreshold stimuli. The experimental arrangement shows intracellular stimulation (*I*) and recording electrodes (*E*) in **A** and polarity, strength, and duration of a constant current in **B**: Hyperpolarizing (*1*) and subthreshold depolarizing current (*2*) induce a nonpropagating local response. Current of just threshold strength will produce either local change (*3a*) or an action potential (*3b*). Suprathreshold stimulation (*4*) also generates an action potential, but with a more rapid time course of depolarization. (From Woodbury,[42] with permission.)

stage, potassium ions move freely, whereas sodium ions remain static. Depolarization to the critical level opens the sodium channels, giving rise to a 500-fold increase in sodium permeability.

This intrinsic property of the nerve and muscle underlies the all-or-none response: regardless of the nature of the stimulus, the same action potential occurs as long as depolarization reaches the critical level. The increased conductance of permeability allows sodium ions to enter the cell seeking a new steady state. Sodium entry further depolarizes the cell, which in turn accelerates the inward movement of this ion. Because of this regenerative sequence, changing the membrane toward the sodium equilibrium potential, an action potential develops explosively to its full size. Dramatic change in sodium permeability

during the course of the action potential results in a reversal of membrane potential from −80 or −90 mV to +20 or +30 mV. In other words, a switch from the potassium to the sodium equilibrium constitutes the generation of an action potential.

In the depolarized membrane, the permeability to potassium ions also increases as a result of a molecular change, but only after a delay of about 1 msec. At about the same time, the increased permeability to sodium falls again to near the resting value. This inactivation of sodium conductance, together with increased potassium permeability, results in rapid recovery of the cell membrane from depolarization. After the potential falls precipitously toward the resting level, a transient increase in potassium conductance hyperpolarizes the membrane, which then returns slowly to the resting value, completing the cycle of repolarization. The amount of sodium influx and potassium efflux during the course of an action potential alters the concentration gradients of these two ions very little. Although repolarization primarily results from a delayed increase in potassium conductance in squid giant axons,[19] this may not apply to mammalian peripheral or central myelinated axons.

Voltage clamp experiments indicate that sodium channels abound at the nodes of Ranvier, where potassium conductance may be minimal or absent in the intact mammalian peripheral myelinated axons[7] or mammalian dorsal column axons.[29] In contrast, potassium channels are distributed all along the internodes, although paranodal regions also contain some sodium conductance. The absence of potassium channels at the nodes of Ranvier would result in a loss of the internodal resting potential, leading to a slowed propagation, rather than a block of the impulse. Theoretically, the availability of potassium conductance facilitates repolarization, but at a cost of prolonging the refractory period. In mammalian fibers, the absence of potassium channels at the node of Ranvier, combined with fast inactivation of sodium conductance, allows an increased rate of firing.

Local Current

An action potential initiated at one point on the cell membrane renders the inside of the cell positive in that local region. Intracellular current flows from the active area to the adjacent, negatively charged, inactive region. A return flow through the extracellular fluid from the inactive to active region completes the current.[8] In other words, a current enters the cell at the site of depolarization and passes out in adjacent regions of the polarized membrane (see Fig. 4–3). This local current tends to depolarize inactive regions on both sides of the active area. When the depolarization reaches the threshold, an action potential occurs,

giving rise to a new local current further distally and proximally. Thus, an impulse, once generated in the nerve axon, propagates in both directions from the original site of depolarization, initiating orthodromic as well as antidromic volleys of the action potential (see Chapter 4.3).

Afterpotential

In an extracellular recording, an action potential consists of an initial negative spike about 1 ms in duration and two subsequent phases, negative and positive afterpotentials (Fig. 2–3). An externally negative deflection grafted onto the declining phase of the negative spike (negative afterpotential) represents a supernormal period of excitability. This phase results from extracellular accumulation of potassium ions associated with the generation of an action potential. A prolonged externally positive deflection (positive afterpotential) indicates a subnormal period of excitability. This phase reflects elevated potassium conductance at the end of the action potential and an increased rate of the sodium-potassium pump to counter the internal sodium concentration.

4   VOLUME CONDUCTION

Diphasic Recording of Action Potential

A pair of electrodes placed on the surface of a nerve or muscle at rest register no difference of potential between them. If, in the tissue activated at one end, the propagating action potential reaches the nearest electrode (G_1), then G_1 becomes negative relative to the distant electrode (G_2). This results in an upward deflection of the tracing according to the convention of clinical electrophysiology (although one could also set the oscilloscope to display negativity of G_1 as a downward deflection). With further passage of the action potential, the trace returns to the baseline at the point where the depolarized zone affects G_1 and G_2 equally. When

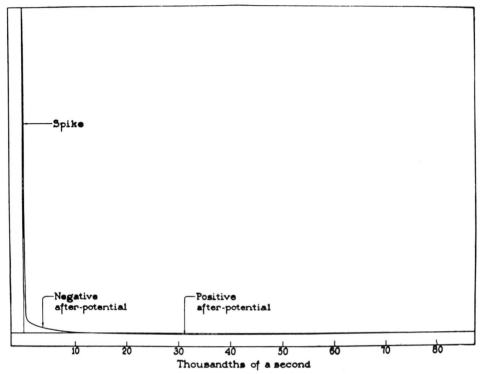

Figure 2–3. Diagrammatic representation of an action potential in A fibers of the cat with the spike and negative and positive after potentials drawn in their correct relative size and true relationships. (From Gasser,[14] with permission.)

the action potential moves further away from G_1 and toward G_2, G_2 becomes negative relative to G_1 (or G_1 becomes positive relative to G_2). Therefore, the trace now shows a downward deflection. It then returns to the baseline as the nerve activity becomes too distant to affect the electrical field near the recording electrodes. This produces a diphasic action potential as shown in Figure 2–4.[36]

Effect of Volume Conduction

The above discussion dealt with the action potential recorded directly with no external conduction medium intervening between the pickup electrodes and the nerve or muscle. During the clinical study, however, connective tissue and interstitial fluid act as a volume conductor surrounding the generator sources.[8,10,15] Here, an electrical field spreads from a source, represented as a dipole; that is, a

pair of positive and negative charges.[3] In a volume conductor, currents move into an infinite number of pathways between the positive and negative ends of the dipole, with the greatest densities (the number of charges passing through a unit area per unit time) along the straight path.

The current flow decreases in proportion to the square of the distance from the generator source. Thus, the effect of the dipole gives rise to voltage difference between an active recording electrode in the area of high current density and a reference electrode at a distance. Whether the electrode records positive or negative potentials depends on its spatial orientation to the opposing charges of the dipole. For example, the active electrode located at a point equidistant from the positive and negative charges registers no potential. The factors that determine the amplitude of the recorded potential include charge density (the net charge per unit area),

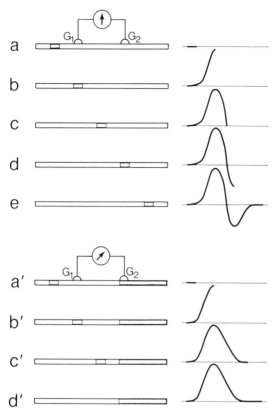

increases in proportion to the size of the polarized membrane viewed by the electrode and decreases with the distance between the electrode and the membrane. Solid angle approximation closely predicts the potential derived from a dipole layer, as schematically shown in Figure 2–5. The propagating action potential, visualized as a positively charged wave front, represents depolarization at the cross-section of the nerve at which the

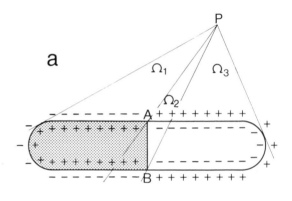

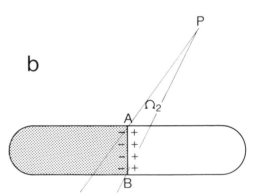

Figure 2–4. Diphasic (**top**) and monophasic recording (**bottom**) of an action potential represented by the *shaded area.* As the impulse propagates from left to right in the top series, the two electrodes see no potential difference in *a, c,* and *e.* Relative to the reference electrode (G_2), the active electrode (G^1) becomes negative in *b* and positive in *d,* resulting in a diphasic potential. In the bottom series, the darkened area on the right indicates a killed end with permanent depolarization, making G_1) positive relative to G_2 in *a', c',* and *d'.* In *b', G_1* and G_2 have no potential difference, causing upward deflection from the positive baseline to 0 potential.

surface area of the dipole, and its proximity to the recording electrode.[11]

The theory of solid angle approximation pertains to the analysis of an action potential recorded through a volume conductor. The resting transmembrane potential consists of a series of dipoles arranged with positive charges on the outer surface and negative charges on the inner surface. The solid angle subtended by an object equals the area of its surface divided by the squared distance from a specific point to the surface.[5,18] Thus, it

Figure 2–5. Potential recorded at *P* from a cell with active (*dark area*) and inactive regions. In **a**, total solid angle consists of Ω_1, Ω_2, and Ω_3. Potential at *P* subtending solid angles Ω_1 and Ω_3 equals zero as, in each, the nearer and farther membranes form a set of dipoles of equal magnitude but opposite polarity. In Ω_2, however, cancellation fails because these two dipoles show the same polarity at the site of depolarization. In **b**, charges of the nearer and farther membranes subtending solid angle Ω_2 are placed on the axial section through a cylindrical cell. A dipole sheet equal in area to the cross-section represents the onset of depolarization traveling along the cell from left to right with positive poles in advance. (Adapted from Patton.[34])

transmembrane potential reverses.[33] A negatively charged wave front follows, signaling repolarization of the activated zone.

Near-Field and Far-Field Potentials

The specific potential recorded under a particular set of conditions depends not only on the location of the recording electrodes relative to the active tissue at any instant in time, but also the physical characteristics of the volume conductor.[9,26,27,32,40] The near-field potential (NFP) and far-field potential (FFP) distinguish two different manifestations of the volume-conducted field.[20,21,39] The NFP represents recording of a potential change near the source, usually by a pair of closely spaced electrodes over the propagating impulses. In contrast, the FFP implies the detection of a voltage step long before the signal arrives at the recording site, usually by a pair of widely separated electrodes located far from the traveling volleys.

A bipolar recording registers primarily, though not exclusively, the NFP from the axonal volley along the course of the nerve, whereas a referential montage usually records the FFP with a variable mixture of the NFP if one of the electrodes lies near the passage of the traveling volleys. A far-field derivation has recently become popular in the study of evoked potentials to detect voltage sources generated at a distance. The original work on short-latency auditory evoked potentials[20,21,39] suggested that synaptically activated neurons in the brainstem give rise to the FFP. Subsequent animal studies[41] emphasized the role of synchronized volleys of action potentials within afferent fiber tracts as the source of the FFP. More recent work with the human peripheral nerve has also clearly documented that stationary peaks can result solely from the propagating impulse.[13,24,26–28,30]

Hence two types of peaks can occur in far-field recording: a volume-conducted potential representing a fixed neural activity such as synaptic discharges and a stationary peak from an advancing front of axonal depolarization. Short sequential segments of the brainstem pathways may summate in far-field recording, resulting in successive peaks of the recorded potentials.[1,2,41] This mechanism by itself, however, does not account for the FFP derived from the propagating volleys along the greater length of the afferent pathways, such as those of somatosensory evoked potentials (SEP) of the median nerve. Instead, a voltage step develops from the traveling impulse at the border of the conducting medium, where the moving volley encounters a sudden geometric change.[27] The designation of potential as junctional or intercompartmental differentiates this type of FFP from that of fixed neural generators and helps specify the source of the voltage step by location.

Here, each volume conductor on the opposite side of the boundary, in effect, acts as a lead connecting any points within the respective compartment to the voltage source at the partition.[9,25] Consequently, the intercompartmental potential remains nearly the same regardless of the distance between G_1 and G_2. Depending on the geometry and the relative size of the volume conductors, the voltage step generated at the partition can give rise to not only intercompartmental but also intracompartmental gradient. Thus, a referential recording with both G_1 and G_2 placed in the same compartment may register a junctional potential generated at the entry; for example, shoulder potential, P_9, recorded between the scalp and the contralateral knee, in the median SEP. Such an uneven distribution of the junctional potential within the volume entered presumably reflects the direction of the impulse crossing the partition (see Chapter 19.3).[12]

Clinical Implications

Analyzing the waveform plays an important role in the assessment of a nerve or muscle action potential. A sequence of potential changes arises as the two sufficiently close wave fronts travel in the volume conductor from left to right

(Fig. 2–6). This results in a positive-negative-positive triphasic wave as the moving fronts of depolarization and repolarization approach, then reach, and finally pass beyond the point of the recording electrode. Thus, an orthodromic sensory action potential from a deeply situated nerve gives rise to a triphasic waveform in surface recording. The potentials originating in the region near the electrode, however, lack the initial positivity in the absence of an approaching volley.

A compound muscle action potential, therefore, appears as a negative-positive diphasic waveform when recorded with the electrode near the end-plate region where the volley initiates. In contrast, a pair of electrodes placed away from the activated muscle registers a positive-negative diphasic potential, indicating that the impulse approaches but does not reach the recording site.

The number of triphasic potentials generated by individual muscle fibers summate to give rise to a motor unit potential in electromyography. The waveform of the recorded potential varies with the location of the recording tip relative to the source of the muscle potential.[6] Thus, the same motor unit shows multiple profiles, depending on the site of the exploring needle. Moving the recording electrode short distances away from the muscle fibers results in obvious reduction in amplitude. Additionally, the duration of the positive-to-negative rising phase (rise time) becomes greater. The rise time gives an important clue in determining the proximity to the generator source. Amplitude may not serve this purpose, because it may decrease with smaller muscle fibers or lower fiber density. The location of the needle also dictates the waveform of spontaneous single fiber discharge; they include initially positive triphasic fibrillation potentials, initially negative biphasic end-plate spikes, and initially positive biphasic positive sharp waves (see Chapter 13.4).

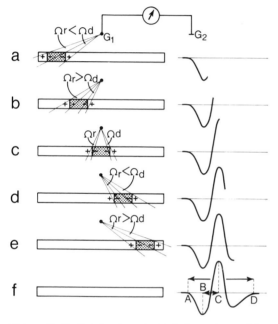

Figure 2–6. Triphasic potential characterized by amplitude, duration ($A - D$), and rise time ($B - C$). A pair of wave fronts of opposite polarity represent depolarization and repolarization. The action potential travels from left to right in a volume conductor with the recording electrode (G_1) near the active region and reference electrode (G_2) on a remote inactive point. As shown in a, G_1 initially sees the positivity of the first dipole, which subtends a greater solid angle (Ω_d) than the second dipole of negative front (Ω_r). In b, this relationship reverses, with gradual diminution of Ω_d, compared with Ω_r, as the active region approaches G_1. In c, the maximal negativity signals the arrival of the impulse directly under G_1, which now sees only negative ends of the dipoles. In d, the negativity declines as G_1 begins to register the positive end of the second dipole. In e, the polarity reverses again as Ω_r exceeds Ω_d. In f, the trace returns to the baseline when the active region moves further away. The last positive phase, though smaller in amplitude, lasts longer than the first, indicating a slower time course of repolarization.

REFERENCES

1. Arezzo, JC, Legatt, AD, and Vaughan, HG Jr: Topography and intracranial sources of somatosensory evoked potentials in the monkey. I. Early components. Electroencephalogr Clin Neurophysiol 46:155–172, 1979.
2. Arezzo, JC, and Vaughan, HG Jr: The contribution of afferent fiber tracts to the somatosensory evoked potentials. Ann NY Acad Sci 388:679–682, 1982.
3. Boyd, DC, Lawrence, PD, and Bratty, P: On modeling the single motor unit action potential. IEEE Trans Biomed Eng 25:236–242, 1978.
4. Brazier, MAB: Electrical Activity of the Nervous System, ed 4. Pitman Medical, Kent, Great Britain, 1977.
5. Brown, BH: Theoretical and experimental

waveform analysis of human compound nerve action potentials using surface electrodes. Med Biol Eng 6:375–386, 1968.

6. Buchthal, F, Guld, C, and Rosenfalck, P: Volume conduction of the spike of the motor unit potential investigated with a new type of multielectrode. Acta Physiol Scand 38:331–354, 1957.

7. Chiu, SY, Ritchie JM, Rogart, RB, and Stagg, D: A quantitative description of membrane currents in rabbit myelinated nerve. J Physiol (Lond) 292:149–166, 1979.

8. Clark, J, and Plonsey, R: The extracellular potential field of the single active nerve fiber in a volume conductor. Biophys J 8:842–864, 1968.

9. Cunningham, K, Halliday, AM, and Jones, SJ: Stationary peaks caused by abrupt changes in volume conductor dimensions: potential field modelling (abstr). Electroencephalogr Clin Neurophysiol (Suppl)61:S100, 1985.

10. Delisa, JA, and Brozovich, FV: AAEE Minimonograph #10, Volume Conduction in Electromyography. American Association of Electromyography and Electrodiagnosis, Rochester, Minnesota, 1979.

11. Delisa, JA, Kraft, GH, and Gans, BM: Clinical electromyography and nerve conduction studies. Orthop Rev 7:75–84, 1978.

12. Desmedt, JE, Huy, NT, and Carmeliet, J: Unexpected latency shifts of the stationary P_9 somatosensory evoked potential far field with changes in shoulder position. Electroencephalogr Clin Neurophysiol 56:623–627, 1983.

13. Eisen, A, Odusote, K, Bozek, C, and Hoirch, M: Far-field potentials from peripheral nerve: Generated at sites of muscle mass change. Neurology 36:815–818, 1986.

14. Gasser, HS: The classification of nerve fibers. Ohio J Sci 41:145–159, 1941.

15. Gath, I, and Stalberg, E: On the volume conduction in human skeletal muscle: In situ measurements. Electroencephalogr Clin Neurophysiol 43:106–110, 1977.

16. Hille, B: Ionic basis of resting and action potentials. In Geiger, SR (ed): Handbook of Physiology, Vol 1, Section 1: The Nervous System. American Physiological Society, Bethesda, 1977, pp 99–136.

17. Hodgkin, AL: Ionic movements and electrical activity in giant nerve fibers. Proc R Soc (Lond) (Series B) 148:1–37, 1958.

18. Hodgkin, AL: The Conduction of the Nervous Impulse. The Sherrington Lectures, Vol 7. Liverpool University Press, Liverpool, 1965.

19. Hodgkin, AL, and Huxley, AF: A quantitative description of membrane current and its application to conduction and excitation in nerve. J Physiol (Lond) 117:500–544, 1952.

20. Jewett, DL: Volume-conducted potentials in response to auditory stimuli as detected by averaging in the cat. Electroencephalogr Clin Neurophysiol 28:609–618, 1970.

21. Jewett, DL, and Williston, JS: Auditory-evoked far fields averaged from the scalp of humans. Brain 94:681–696, 1971.

22. Kandel, E, and Schwartz, JH (eds): Principles of neural science, ed 2. Amsterdam, Elsevier, 1985, pp 13–208.

23. Katz, B: Nerve, Muscle and Synapse. McGraw-Hill, New York, 1966.

24. Kimura, J: Field theory: The origin of stationary peaks from a moving source. In International Symposium on Somatosensory Evoked Potentials. Rochester, Minn., Custom Printing, 1984, pp 39–500.

25. Kimura, J, Kimura, A, Ishida, T, Kudo, Y, Suzuki, S, Machida, M, Matsuoka, H, and Yamada, T: What determines the latency and the amplitude of stationary peaks in far-field recordings? Ann Neurol 19:479–486, 1986.

26. Kimura, J, Mitsudome, A, Beck, DO, Yamada, T, and Dickins, QS: Field distributions of antidromically activated digital nerve potentials: Model for far-field recording. Neurology 33:1164–1169, 1983.

27. Kimura, J, Mitsudome, A, Yamada, T, and Dickins, QS: Stationary peaks from a moving source in far-field recording. Electroencephalogr Clin Neurophysiol 58:351–361, 1984.

28. Kimura, J, Yamada, T, Shivapour, E, and Dickins, QS: Neural pathways of somatosensory evoked potentials: Clinical implication. In Buser, PA, Cobb, WA, and Okuma, T (eds): Kyoto Symposium (EEG Suppl 36). Amsterdam, Elsevier, 1982, pp 328–335.

29. Kocsis, JD, and Waxman, SG: Absence of potassium conductance in central myelinated axons. Nature 287:348–349, 1980.

30. Lin, JT, Phillips, LH II, and Daube, JR: Far-field potentials recorded from peripheral nerves (abstr). Electroencephalogr Clin Neurophysiol 50:174, 1980.

31. Lorente De No, R: Analysis of the distribution of the action currents of nerve in volume conductors. In: Studies from the Rockefeller Institute for Medical Research: A Study of Nerve Physiology, Vol 132. The Rockefeller Institute for Medical Research, New York, 1947, pp 384–482.

32. Nakanishi, T: Origin of action potential recorded by fluid electrodes. Electroencephalogr Clin Neurophysiol 55:114–115, 1983.

33. Patton, HD: Special properties of nerve trunks and tracts. In Ruch, HD, Patton, HD, Woodbury, JW, and Towe, AL (eds): Neurophysiology, ed 2. WB Saunders, Philadelphia, 1965, pp 73–94.

34. Patton, HD: Resting and action potentials of neurons. In Patton, HD, Sundsten, JW, Crill, WE, and Swanson, PD (eds): Introduction to Basic Neurology. WB Saunders, Philadelphia, 1976, pp 49–68.

35. Patton, HD, Sundsten, JW, Crill, WE, and Swanson, PD: Introduction to Basic Neurology. WB Saunders, Philadelphia, 1976.

36. Rosenfalck, P: Intra and extracellular potential fields of active nerve and muscle fibers. Acta Physiol Scand (Suppl) 321:1–168, 1969.

37. Ruch, TC, and Fulton, JF: Medical Physiology and Biophysics, ed 20. WB Saunders, Philadelphia, 1973.

38. Ruch, TC, Patton, HD, Woodbury, JW, and Towe, AL: Neurophysiology, ed 2. WB Saunders, Philadelphia, 1965.

39. Sohmer, H, and Feinmesser, M: Cochlear and cortical audiometry conveniently recorded in the same subject. Isr J Med Sci 6:219–223, 1970.

40. Stegeman, D, Van Oosteron, A, and Colon, E: Simulation of far field stationary potentials due to changes in the volume conductor (abstr). Electroencephalogr Clin Neurophysiol (Suppl) 61:S228, 1985.

41. Vaughan, HG Jr: The nerual origins of human event-related potentials. Ann NY Acad Sci 388:125–138, 1982.

42. Woodbury, JW: Action potential: Properties of excitable membranes. In Ruch, TC, Patton, HD, Woodbury, JW, and Tome, AL (eds): Neurophysiology, ed 2. WB Saunders, Philadelphia, 1965, pp 26–57.

43. Woodbury, JW: The cell membrane: Ionic and potential gradients and active transport. In Ruch, TC, Patton, HD, Woodbury, JW, and Towe, AL (eds): Neurophysiology, ed 2. WB Saunders, Philadelphia, 1965, pp 1–25.

Chapter 3

ELECTRODES AND RECORDING APPARATUS

1 INTRODUCTION

The apparatus used in the performance of routine electrodiagnosis includes electrodes, amplifiers, display, loudspeaker, and data storage devices. Muscle action potentials can be recorded by either a surface electrode placed on the skin over the muscle or a needle electrode inserted into the muscle. Surface electrodes used in conduction studies register a summated electrical activity from many muscle or nerve fibers, whereas needle electrodes in electromyography discriminate individual motor unit potentials discharging within a narrow radius from the recording tip. The electrical and physical characteristics of recording electrodes dictate the amplitude and other aspects of the potentials under study.[24,25] Electromyographers analyze the amplified waveform of action potentials on a visual display and the auditory characteristics of the signals heard through a loudspeaker. The kind of information desired and the type of activities under study determine the optimal amplifier settings. Devices for permanent recordings include photographs with Polaroid films, a fiber-optic system with sensitive papers, a magnetic tape recorder, and digital storage. Amplitude and time calibrations verify the accuracy of the stored signals. This chapter deals with practical aspects of instrumentation, without detailed discussion of electronics (see Appendix 2).

2 ELECTRODES

The signals recorded during voluntary muscle contraction depend to a great extent on the type of recording electrode.[8,21] Surface electrodes placed over the muscle record summated activities from many motor units. The use of a needle electrode allows recording of individual motor unit potentials during mild muscle contraction. With increased effort, synchronous activity in many adjacent motor units precludes the identification of single motor units. For routine purposes, clinical electromyographers use standard concentric, bipolar concentric,[1] or monopolar needles.[30] Single-fiber electrodes have a leading-edge small enough to allow recording of potentials derived from single muscle fibers in isolation.[17] Less commonly used "special purpose" electrodes include the multielectrode, the flexible wire electrode, and the microelectrode placed intracellularly.[7]

Proper Application of Electrodes

The skin preparation for application of surface electrodes consists of cleaning with alcohol, and if necessary, lightly scraping the callous surfaces. These precautions minimize impedance and improve electrical conductivity. Sterilization of needle electrodes in boiling water for at least 20 minutes prevents the transmission of most infectious agents. Commercially available sterilizers bring the water temperature to 100°C and maintain it without excessive boiling. The metal and plastic components of needle electrodes will withstand the time and temperature of steam autoclaving. One must, however, detach nonautoclavable connectors and lead wires before the sterilization procedure. Gas sterilization also suffices, although chemicals used may damage the plastic, causing defects in insulation. Thorough outgassing of electrodes reduces the amount of the agent retained in the plastic material. Electrode manufacturers provide instructions for optimal sterilization methods.

After studying a patient with hepatitis or other contagious disorders, it is wise to discard the needle. Patients with Jakob-Creutzfeldt disease pose special problems because the transmissible agent responsible for the disease may resist conventional sterilization procedures.[20,32] The American Association of Electromyography and Electrodiagnosis recommends discarding needle electrodes after use in all patients with dementia.[23] Before disposal, one must incinerate such needles and blood-contaminated materials or properly sterilize them by autoclaving for 1 hour at 120°C and 15 psi.[2] The same precaution should apply after examining patients with the acquired immunodeficiency syndrome (AIDS). Fiscally, one may dispose monopolar needles much more readily than expensive standard or bipolar concentric electrodes.

The electrical properties of commercial needle electrodes vary considerably. Electrolytic treatment temporarily improves their performance.[15] Periodic examination of needle electrodes ensures their structural integrity. The inner concentric shaft may become corroded. The Teflon coating of monopolar needles may peel off, exposing the insulated portion of the conductor. An increase in recording surface tends to reduce the amplitude and area of the recorded motor unit potential with relatively little effect on its duration.[12] The use of a dissecting microscope with a magnifying factor of 10 can help detect a slight bend in the shaft or a crack in the tip. For testing the needle insulation, one terminal of a battery is connected to the lead of a needle; the other terminal is connected to an ammeter with a small exploring metal hook or moist cotton. A current should flow only if the exploring hook touches the exposed tip of the needle. Any current, if registered while exploring the shaft of the needle, indicates defective insulation. An ammeter should register no current if connected to the battery through the two leads of standard or bipolar concentric needles unless there is a short circuit at the needle tip. A current will flow normally with immersion of the needle tip in water.

Types of Available Electrodes

Figure 3–1 illustrates common electrodes used in electromyography.

SURFACE ELECTRODES

Surface electrodes, square or round metal plates made of platinum or silver, come in different sizes with the average dimension of 1 by 1 cm. An adhesive tape suffices to apply them to the skin, although the use of collodion improves the stability in long-term monitoring. Cleansing the skin with alcohol, scraping the surface, and applying electrolyte cream under the electrode reduce the impedance. Too much paste, however, can form a bridge between the two recording electrodes, canceling the voltage difference. A short circuit between the stimulator and pickup electrodes or ground introduces a large stimulus artifact. Perspiration can act in a similar manner. Electrode offset voltage at the interface is a steady potential not recorded by the amplifier. Nonetheless, it gives rise to an

Figure 3–1. Schematic illustration of standard or coaxial bipolar (a), concentric bipolar (b), monopolar (c), and single fiber needles (d and e). Dimensions vary, but the diameters of the outside cannulas shown resemble 26-gauge hypodermic needles (460 μm) for a, d, and e, a 23-gauge needle (640 μm) for b, and a 28-gauge needle (360 μm) for c. The exposed tip areas measure 150 by 600 μm for a, 150 by 300 μm with spacing between wires of 200 μm center to center for b, 0.14 sq mm for c, and 25 μm in diameter for d and e. A flat-skin electrode completes the circuit with unipolar electrodes shown in c and d. (Modified from Stålberg and Trontelj.[35])

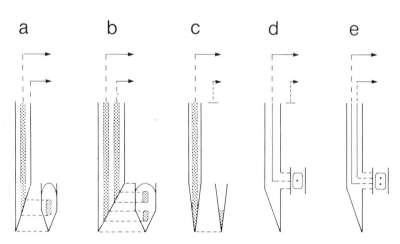

artifact when movement causes a sudden mechanical change in the metal-electrolyte interface. To reduce this type of potential, some surface electrodes allow most movement to occur between electrolyte and skin, rather than at the metal-electrolyte interface.

A surface electrode is used in monitoring voluntary muscle contraction during kinesiologic studies,[4] recording evoked compound nerve or muscle action potentials and stimulating the peripheral nerves. It also serves as a reference or ground lead in conjunction with the monopolar needle. As an active electrode for the study of motor unit potentials, it does not reproduce high frequency components adequately.

STANDARD OR COAXIAL CONCENTRIC NEEDLE

The standard electrode, introduced by Adrian and Bronk,[1] is a stainless steel cannula similar to a hypodermic needle, with a wire in the center of the shaft. The wire, usually made of nichrome, silver, or platinum, is 0.1 mm in diameter or slightly larger, and the external rim of the shaft, 0.3 mm in diameter. The pointed tip of the needle has an oval shape, with an exposed area of about 150 by 600 μm, and an impedance of around 50 kΩ. Both the wire and shaft are bare at the tip. The needle, when near a source of electrical activity, registers the potential difference between the wire and the shaft. A separate surface electrode serves as the ground.

BIPOLAR CONCENTRIC NEEDLE

The cannula of the bipolar concentric needle contains two fine stainless steel or platinum wires. With the same size wires embedded, this electrode is larger in diameter than the standard concentric needle. The electrode registers the potential difference between the two inside wires, with the cannula serving as the ground. The bipolar electrode thus detects potentials from a much smaller volume than the standard needle. The three terminals in the connecting cable consist of two active leads and the ground connection. In this type of recording from a very localized area, only a small number of single muscle fibers contribute as the source for electrical activity. This restricted recording range provides selectivity, but at the risk of disregarding the overall activity of the motor unit.

MONOPOLAR NEEDLE

The monopolar electrode, made of stainless steel for its mechanical properties, has a fine point, insulated except at the distal 0.2 to 0.4 mm. The wire, covered by a sleeve of Teflon, has an average diameter of about 0.8 mm. It requires a surface electrode or a second needle in the subcutaneous tissue as a reference lead. A separate surface electrode placed on the skin serves as a ground. Its sharp tip causes much less pain than other needle electrodes. It is less expensive but less stable electrically, hence noisier than the concentric electrode. The average impedance ranges from 1.4 MΩ at 10 Hz to 6.6 kΩ at 10 kHz.[37] Presoaking the electrodes with a small concentration of a wetting agent in saline solution reduces the impedance by 6- to 20-fold. This pretreatment improves the resolution of low-amplitude signals. A monopolar needle records the voltage changes between the tip of the electrode and the reference. The recording characteristics, of course, differ considerably from one type of monopolar needle to another; but in general, a monopolar needle registers a considerably larger potential, from the same source, than the concentric needle.

SINGLE-FIBER NEEDLE

Single-fiber electromyography requires an electrode with a much smaller leading edge, to record from individual muscle fibers, rather than motor units (see Chapter 15).[16,17] A wire 25 μm in diameter mounted on the side of a needle provides the maximal amplitude discrimination between near and distant muscle fiber potentials.[18,19] As in concentric electrodes, single-fiber needles may contain two or more wires exposed along the shaft, serving as the leading edge. The most commonly used type has one wire inserted into a cannula with its end bent toward the side of the cannula, a few millimeters behind the tip.[35]

MULTIELECTRODES

Multielectrodes contain three or more insulated wires, usually 1 by 1 mm in size, exposed through the side of the cannula.[9] One of the wires serves as the indifferent electrode, whereas the outside cannula of the electrode, 1 mm in diameter, is connected to the ground. The separation between the leads along the side of the multielectrode determines the recording radius. The commonly used distances in measuring the motor unit territory include 0.5 mm for myopathy and 1.0 mm for neuropathy. The single-fiber needle may also contain multiple wires exposed along the shaft.

FLEXIBLE WIRE

The flexible wire electrode permits freedom of movement in kinesiologic examination.[13] The wires are introduced through a hypodermic needle, which is then withdrawn, leaving the wire in the muscle.[6] Some investigators prefer a bipolar electrode made of nylon-coated Evanohm alloy wire, 25 μm in diameter.[5] Although this electrode comes in different sizes, the most commonly used type has insulated platinum wires 50 to 100 μm in diameter with the tip bare. A small hole made in the insulation of the wire may provide smaller lead-off surfaces on the order of 10 to 20 μm.[26] This electrode, however, lacks the rigid standardization required for quantitative studies of action potentials.[35]

GLASS MICROELECTRODES

Glass microelectrodes, used for intracellular recording, consist of fine glass tubing filled with potassium chloride solution. Because of its extreme fragility, one must use a cannula as a carrier to introduce the electrode through the skin and a micromanipulator to insert it into the exposed muscle. The electrode has a very fine tip, less than 1 μm in diameter, and consequently a very high impedance,

on the order of 5 MΩ. Therefore, recording from a glass microelectrode requires amplifiers of exceedingly high input impedance.[7]

3 ELECTRODE AMPLIFIERS

Potentials assessed during electrodiagnostic examinations range in amplitude from microvolts to millivolts. With the oscilloscope display set at 1 V/cm, signals of 1 μV and 1 mV, if amplified 1,000,000 and 1,000 times, respectively, cause a 1-cm deflection. To accomplish this range of sensitivity, the amplifier consists of several stages. One system uses a preamplifier with a gain of 500, followed by several amplifier and attenuator stages, to produce a variable gain of 2 to 2000. This arrangement increases the signal-to-noise ratio by allowing major amplification of the signal near the source prior to the emergence of noise that develops in the following circuits. To achieve this goal, the preamplifier must have a high input impedance, a low noise level, and a large dynamic range.

Differential Amplifiers

During electromyographic examination, a major source of interference comes from the coupled potential of the alternating-current power line. The magnitude of this field can exceed that of biologic potential by a million times. Proper assessment of the signal, therefore, requires its selective amplification without, at the same time, magnifying the noise. This would be impossible if the apparatus amplified any voltage appearing between an input terminal and the ground terminal. The differential amplifiers used in most electromyography, however, amplify only the voltage difference between the two input terminals connected to the recording electrodes. This system effectively rejects "common mode" voltages, which appear between both input terminals and common ground. These include not only the power line interference but also distant muscle action potentials that affect the two recording electrodes equally.

Common Mode Rejection Ratio

Inherent imbalance in the electrical system of the amplifier renders rejection of the common mode voltage less than perfect. The common mode rejection ratio (CMRR) specifies the degree of differential amplification between the signal and the common mode voltage. Good differential amplifiers should have rejection ratios exceeding 100,000, that is, 100,000 times more amplification of the signals than unwanted potentials appearing as a common mode voltage. A very high rejection ratio, however, will not guarantee the complete elimination of external interference caused by undesired distant potentials, for two reasons. First, the electromagnetic interference affects the two recording electrodes almost but not quite equally, depending on their relative positions. Second, the contact impedances inevitably differ between the two recording electrodes, which leads to unequal distribution of the same common mode voltage. A common mode voltage too large to be perfectly balanced overloads the amplifier.

Means of Reducing Interference

Other precautions to minimize electromagnetic interference include reducing and balancing contact impedances of the two electrodes and the use of short, well-shielded electrode cables. The system must effectively ground not only the patient and the bed but also the instrument and, if necessary, the examiner. Major interference may originate from unshielded power cords running to other appliances in the vicinity of the recording instrument. With adequate care, most modern equipment operates well without a shielded room. In the presence of electrical noise uncontrollable by ordinary means, a properly constructed Faraday shield can dramatically reduce the interference. To be effective, it should enclose the examining room as one continuous conductor and be grounded at one point. The 50- or 60-Hz filter available in most instruments reduces power line interference. Such filters also distort electromyographic signals, although special situations, such as portable recording, may warrant their application when all other attempts have failed.

Input Impedance

Analogous to the resistance in a DC circuit, the impedance in an AC circuit determines the current flow for a given alternating voltage source. For recording muscle or sensory action potentials, the tissue and electrode wires add only negligible impedances, compared with those at the needle tip and at the input terminal of the amplifier. In this circuit, the needle tip and the input terminals act as a voltage divider with voltage changes occurring in proportion to the respective impedance. Thus, with the impedance equally divided between these two, only one-half of the original potential will appear across the input terminals. Increasing the input impedance of the amplifier to a level much higher than that of the electrode tip would minimize the loss. The input impedances of most amplifiers range from 100 kΩ to hundreds of megohms. An amplifier with a high input impedance also improves the common mode rejection ratio, because the higher the input impedance, the smaller the effect of electrical asymmetry of the recording electrodes. Higher electrode impedances increase amplifier noise and external interference.

Frequency Response

Most commercially available apparatus have variable high- and low-frequency filters to adjust the frequency response according to the type of potentials under study. Fourier analysis of complex waveforms encountered in electromyography reveals sine waves of different frequencies as their harmonic constituents. The major sine wave frequencies of muscle action potentials, for example, range from 2 Hz to 10 kHz. For clinical electromyography, the frequency band of the amplifier ideally should cover this range.[10,11] However, in the presence of interfering high-frequency noise or low-frequency drift, a frequency band extending from 20 Hz to 5 kHz suffices.

The high-frequency filter, if set too low, reduces the amplitude of high-frequency components disproportionately. Extending the high-frequency response beyond the band required for proper recording results in an unnecessary increase in background noise. The low-frequency filter, if set too high, distorts the recorded potential. Here the new waveform approximates the first derivative (rate of change) of the original signal. Extending the frequency response too low causes instability of the baseline, which then shifts slowly in response to changing biopotentials. The analog filters also affect the peak latency of the recorded response because of phase shift. High-frequency filtering increases, whereas low-frequency filtering reduces the apparent latency of peaks. The use of digital filtering, which introduces zero phase shift, circumvents this problem in clinical assessments.[22,31]

A square wave pulse of known amplitude and duration usually serves as a calibration signal for accurately determining the amplitude and duration of the recorded potentials. The distortion seen in the square pulse results from the effects of high- and low-frequency filters. Its rise time indicates the high-frequency response, and the slope of the flat top, the low frequency response (see Appendix Figs. 2–18 and 2–20). Other calibration signals include sine waves from the power line or discontinuous waveforms of known frequency and amplitude.

4 VISUAL DISPLAYS

After appropriate amplification, the waveforms are displayed for visual analysis. The cathode-ray tube (CRT), with no mechanical limitations in dynamic high-frequency response, provides an optimal means of tracing rapidly changing amplitude against time.

Cathode-Ray Tube

An electron gun discharges an electron beam internally toward the glass screen of the CRT. When struck by a beam of electrons, the phosphor coating on the inside surface of the screen emits light. The adjustable voltage between a pair of horizontal deflection plates determines the horizontal position of this bright spot. Applying a linearly increasing voltage to the plates makes the spot sweep at a constant speed. A pair of vertical deflection plates, connected to the signal voltage from the amplifier, control the vertical position of the electron beam. The waveform displayed on the face of the CRT, therefore, represents changing amplitude of the signal voltage in time. The vertical axis represents response amplitude, whereas the horizontal axis shows units of time. Electromyographic examination usually uses a free-running mode such that, when the spot reaches the end of the screen, it returns rapidly to the beginning to repeat. Many manufacturers now provide digital circuitry to process and store the potentials before displaying them on the CRT, as discussed in Appendix 2.

Delay Line

Instead of being free running, the horizontal sweep may be triggered on command. In this mode of operation, a motor unit potential itself can trigger the sweep. Thus, a given motor unit potential recurs successively at the beginning of each sweep for detailed analysis. Unfortunately, however, the portion of the waveform preceding the trigger point does not appear on the screen. An electronic delay line circumvents this difficulty by storing the recorded motor unit potential for a short period. After a predetermined delay following the onset of a sweep triggered by the real-time potential, the stored signal leaves the delay line for display on the screen. With this arrangement, the potential in question occurs repetitively and in its entirety on the same spot of the screen for precise determination of its amplitude and duration.[34]

Multiple Channel Recording

Some electromyographic instruments have multiple channels to allow simultaneous recording from two or more sets of electrodes. Typically, two or more channels share a beam from a single gun by switching the point vertically between the baselines of different traces as the beam sweeps horizontally across the screen. This electrical switching takes place so fast that each trace in effect appears to be continuous despite the interruption from one trace to the next.

Storage Oscilloscope

Storage oscilloscopes have a different CRT that retains traces on the face of the screen for several hours. One type uses a special phosphor that changes its color when electrons strike it. Application of heat to the phosphor erases the color, a process that takes about 30 seconds. In a more expensive type, the trace is retained as electrostatic charges on a mesh behind the screen. A second electron gun floods the screen to visualize the stored pattern, which may be erased quickly by electrically discharging the mesh. Because of the advent of digital storage techniques, such storage oscilloscopes are now essentially obsolete.

5 OTHER RECORDING APPARATUS

Loudspeaker

The muscle action potentials have certain auditory characteristics when played through a loudspeaker. For clinical analysis, electromyographers depend very heavily on the sounds produced by different kinds of spontaneous or voluntarily activated muscle potentials during needle examination. Acoustic properties also help distinguish a nearby motor unit with a clear, crisp sound, reflecting a short rise time, from distant units with dull sound. In fact, an experienced examiner can detect the difference between near and distant units by sound better than by oscilloscope display. The acoustic cues often guide in properly repositioning the needle close to the source of the discharge.

Magnetic Tape Recorder

Magnetic tape provides one means of storing electrical potentials for later analysis. Amplitude modulation (AM) impresses the signal itself on the tape, whereas frequency modulation (FM) records the signal after converting it to a varying frequency of constant amplitude. The AM recording registers high-frequency potentials well, but low-frequency responses below 10 to 15 Hz are registered poorly. In contrast, the FM method has a better low-frequency response, although it requires a very high tape speed to achieve the high-frequency response required for electromyography. The FM method reproduces the amplitude of potentials more accurately than the AM method.

6 ARTIFACTS

Not all electrical potentials registered during an electromyographic examination originate in skeletal muscle. Any voltage not attributable to the muscle potential sought represents an artifact, which usually causes a unique discharge pattern on the oscilloscope and distinct sounds through the loudspeaker. Some noises, however, mimic muscle activity so closely that even a trained examiner may have difficulty with its identity.

Most artifacts unaffected by the position of the recording electrode originate outside the muscle. These exogenous activities may originate in an event peculiar to the patient, like those induced by the cardiac pacemaker (Fig. 3–2) or transcutaneous stimulator (Fig. 3–3). More commonly, they result from 60 Hz interference caused by the electrostatic or electromagnetic fields of nearby electrical appliances. Different generator sources give rise to characteristic, though not

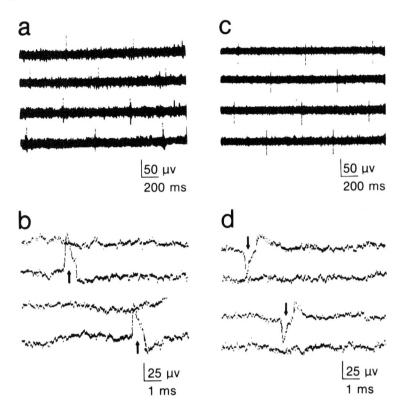

Figure 3–2. Artifacts induced by a cardiac pacemaker recorded by a monopolar needle electrode from the gluteus medius (a and b) and paraspinal muscle (c and d). Note opposite polarity of the sharp discharge in the two recording sites. The interval between the successive impulses of 800 ms corresponds to discharge frequency of 75 impulses per minute. Trains in a and c show continuous recordings from top to bottom, and those in b and d, interrupted tracings from one sweep to the next.

specific, patterns, for easy identification (Fig. 3–4). The artifacts may originate in the recording instruments or a more remote generator, such as a hammer drill (Fig. 3–5). Impedance variability within the muscle tissue may also cause electrical activity, depending on the location of the needle tip. Genuine biologic potentials generated in the muscle include end-plate noises and end-plate spikes (see Chapter 12.4). These artifacts may mimic the intended signals sought during the electromyographic examination (see Figs. 12–3 and 12–4).

Electrode Noise

Potentials may arise from two active metals or the metal-fluid junction at the needle electrode located intramuscularly. With a constant electrode-fluid potential, such a polarized electrode may not receive the signals properly. On the other hand, changing potential will result in electrode noise. A small electrode tip, be-

cause of its high impedance, causes a greater voltage drop during the passage of current. Thus, the smaller the electrode surface, the greater the interference from electrode polarization or electrode noise. Therefore, the type of metal alters the recording characteristics of the needle electrode much more than those of the surface electrode. In fact, an electrode potential from active metals too small to affect skin electrodes could undermine the function of intramuscular electrodes. The use of relatively inert metals, such as stainless steel or platinum, for needle electrodes minimizes such adverse effects.

Amplifier Noise

Electrical noise inherent in an amplifier originates from all components, including resistors, transistors, and integrated circuits. Noise, arising from the thermal agitation of electrons in a resistor, increases with the impedance in the

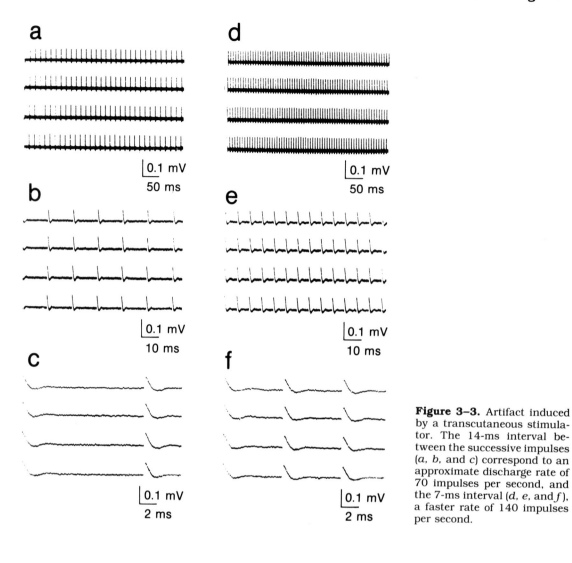

Figure 3–3. Artifact induced by a transcutaneous stimulator. The 14-ms interval between the successive impulses (*a, b,* and *c*) correspond to an approximate discharge rate of 70 impulses per second, and the 7-ms interval (*d, e,* and *f*), a faster rate of 140 impulses per second.

input stage. Microphonic noise results from mechanical vibration of various components. The use of a high-pass filter suppresses low-frequency noise from these and other sources in amplifier circuits. High-frequency noise potentials, however, appear as a thickening of the baseline as it sweeps across the screen, accompanied by a hissing noise on the loudspeaker. The level of amplifier noise as perceived on the oscilloscope increases in proportion to amplifier gain and frequency response. Thus, operating the system at lower gains and with narrower filter band widths substantially reduces this component of noise seen on the screen.

Defective Apparatus

By far the most likely cause of recording problems is a defect in one of the three recording electrodes or its application. A broken wire causes bizarre and sometimes unsuspected artifacts, even when the insulating cover appears intact. A partially severed conductor may cause very deceptive movement-induced artifacts. Other common causes of needle artifacts include defective insulation of the monopolar needle and a concentric needle with a short-circuited tip. A 2-year study on durability revealed the feasibility of reusing monopolar electrodes in 20 to 63 patients.[33] Failure occurred, in

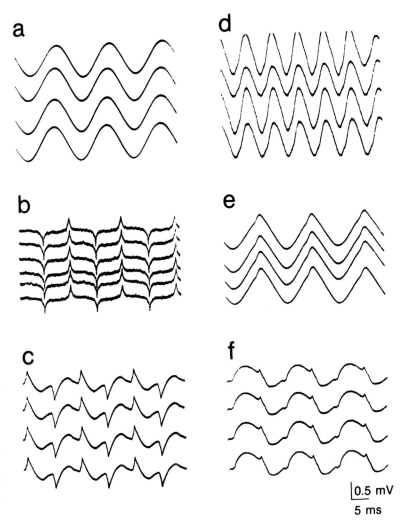

Figure 3–4. Various types of 60 Hz interference induced by nearby electrical appliance. They include a common pattern (*a*), spikes from high impedance of the recording electrode (*b*), interference from a fluorescent light (*c*), 120 Hz interference from a diathermy unit (*d*), and interference from a heat lamp (*e* and *f*).

order of frequency, as the result of Teflon retraction, a dull or burred tip, a break in a wire or pin, electrical artifacts, and bending of the needle shaft. Inadvertent insulation of the electrode tip by blood protein "baked on" in the process of autoclaving can also distort the potential. Careful cleaning of the needle tip prior to autoclaving will alleviate this problem. If necessary, an ultrasonic vibrator can be used to loosen dried material on the needle. Improper or inadequate grounding results in electromagnetic interference from the nearby alternating current source. A loose connection in some parts of the recording circuit may generate electrical activity similar to the muscle action potential.

Movement Artifact

When the patient contracts a muscle, the surface electrode may slide over the skin. This causes a movement artifact primarily because of the change in impedance between the surface electrode and the skin. Movement-induced potentials also result from existing fields near the surface of the skin, particularly those originating from sweat glands.[36] Movement of electrode wires may produce artifacts resembling muscle activity, mostly reflecting changing capacitance. Rubbing the lead of the needle electrode with the finger or cloth sometimes produces friction artifacts from a static charge. Adequate insulation of the needle, ideally

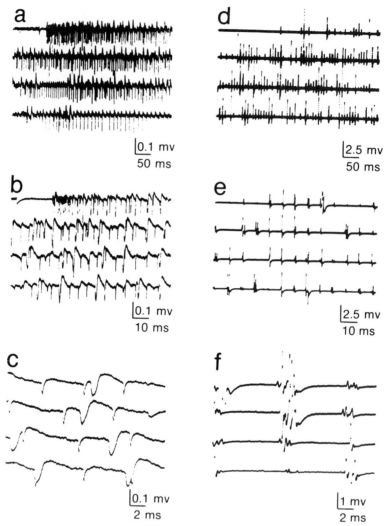

Figure 3–5. Effect of hammer drill operated nearby (*a, b,* and *c*), and oscillation of the amplifier circuits probably induced by an excessively high impedance of the electrode tip (*d, e,* and *f*). Both superficially resemble the complex repetitive discharge, but the recordings with a fast sweep speed (*c* and *f*) uncover a waveform and pattern of recurrence not usually associated with a biologic discharge.

with the use of driven shields, reduces this type of interference.

Electrostatic and Electromagnetic Interference

The sources of 50 or 60 Hz abound (see Fig. 3–4). They include electric fans, lamps, fluorescent lights, CRT screens, electric motors, light dimmers, and even unused power cords plugged into wall outlets. The use of an ungrounded wheelchair or metal examining table enhances this type of artifact. Appliances sharing the same circuit with the electromyographic instrument cause especially noticeable interference. Radio-frequency electromagnetic waves can also "carry" the alternating current. A strong field from a nearby diathermy apparatus produces a characteristic wave pattern. Federal regulations now restrict the amount of interference that such a unit can render to other equipment. Intermittent power-line load causes power-line voltage transient changes, which in turn give rise to artifacts. In the examining room near driveways, auto ignition noise makes a popping sound. If not properly grounded, the examiner acts as an antenna by touching the needle.

Simple but effective measures to reduce electromagnetic interference include relocation of electrode wires relative to the patient or recording apparatus and repositioning the patient and recording apparatus within the room. With power cords near the patient, turning off power to the offending appliance does not necessarily eliminate the artifacts. To avoid interference pickup from a CRT screen, the patient and operator should not come too close to the source. If these simple means fail, adequate control may require removing all the electrical appliances from the room and shielding the examining area.

Radio Interference

High-frequency interference or audio interference may appear on the screen of the oscilloscope from radio broadcasts, television, or radio paging systems. This type of artifact may escape detection because of its transient nature unless the sounds heard through the loudspeaker alert the examiner. Their elimination may require relocation or screening of the electromyographic instrument. The examining room located on the side of the building away from transmitting antennas has the least interference. Screening the noise caused by power wiring may require the use of power-line radio-frequency filters.

7 STIMULATORS

Electrical Stimulation Requirements

Electrical stimulation of the nerve provides a clearly defined, reproducible response for nerve conduction studies. A current of short duration, 50 to 1000 μs, is induced in the fluid surrounding a nerve bundle by applying a potential to electrodes, usually on the skin surface, but sometimes inserted subcutaneously. The stimulating current is directed primarily along the nerve, depolarizing one point and hyperpolarizing another. Increasing the current to obtain a repeatable and maximal recorded response ensures that essentially every nerve in the bundle fired. Surface electrode stimulation requires 100 to 500 V to drive currents of 5 to 75 mA. The higher values are necessary when neuropathy decreases excitability or the stimulating electrodes are distant from the nerve. Subcutaneous needle electrodes, already in good fluid contact and closer to the nerve, require much less voltage and current for adequate stimulation. Just a few volts may elicit a response in this case, requiring much tighter electrical control over stimulus values for consistent and safe stimulation than in the case of surface electrode stimulation.

Effective depolarization displays an inverse relationship between stimulus intensity and duration. Thus, a lower intensity suffices if applied for a longer duration, within limits. Generally, stimulus durations exceeding 1000 μs are not well tolerated. Durations less than 50 μs become ineffective, because tissue capacitances limit the rate of rise, and the stimulus does not reach full amplitude.

The equipment must also provide control and timing of the stimuli for different types of measurements. Some collision techniques require two or three precisely timed stimuli, with adjustable intensities, durations, and latencies, delivered to the same or to different sets of electrodes. Train stimulus techniques have many shocks delivered at rapid, adjustable rates. In the paired-shock facilitation technique, the first shock has reduced intensity to give only a subliminal excitation of the motor neuron pool, and the second shock follows within a few milliseconds on the same electrodes at full intensity. Such complex stimulus generators must have adequate programmability and fail-safe protection features.

Stimulus Isolation

Electrical stimulators are "isolated" from the recording amplifiers and other equipment circuits for artifact reduction and safety. This means that the stimulation circuits have no conductive path except through the patient's body when stimulating and recording electrodes have been applied. Isolation ensures that stimulus current flows only in the loop provided by the two stimulating electrodes. If the stimulator circuit has any connection to the recording circuit, then the stimulus current that is distributed in the body can divide into additional paths, causing a large stimulus artifact, amplifier overload, or even spurious stimulation at unintended sites. Furthermore, under conditions of component failure, these additional paths can conduct hazardous levels of current. Stimulus isolation is usually accomplished by magnetic coupling of energy to the stimulating circuits, although battery-powered stimulators with optical coupling of the control signal have also been used.

Constant-Voltage Versus Constant-Current

Constant-voltage stimulators deliver an adjustable voltage across the stimulating electrodes, essentially independent of stimulus current. One can adjust the voltage to vary the current through the stimulating electrodes to achieve the desired level of stimulation. At a fixed output voltage, changes in stimulating electrode impedance alter the stimulus current level. Constant-current stimulators deliver an adjustable current through the stimulating electrodes, essentially independent of their impedance. The voltage across the stimulating electrodes adjusts dynamically to maintain a constant stimulus current. Constant-current stimulators provide more consistent stimulus control, especially for techniques that require a train of stimuli or response averaging.

Magnetic Coil Stimulation

An alternative means of nerve stimulation, under development at the time of this writing, is magnetic coil stimulation.[3,14,27-29] A rapidly changing magnetic field of high intensity can induce sufficiently localized current in the body fluid to cause nerve excitation. The appara-

tus consists of a hand-held, doughnut-shaped coil and a capacitive-discharge power unit, triggered from conventional electromyography equipment. The advantages of magnetic stimulation include the capability of exciting the brain noninvasively, a lower level of pain associated with the stimulus, and elimination of stimulus electrode application. It might seem that magnetically inducing the stimulus would provide a high degree of isolation and thus reduce stimulus artifact; but in fact, the huge coil currents and high voltages couple substantial artifact into low-level recording circuits. The major disadvantages of magnetic stimulation seem to be a greater uncertainty and variability as to the point of stimulation and the greater expense of the stimulating equipment. The magnetic pulse generation requires a few seconds for recharge between stimuli, eliminating the possibility of closely paired or train stimuli. At this time, the Federal Drug Administration has approved magnetic stimulation for human use only in studies of the peripheral nerve, but some limited research applications are being conducted in the central nervous system.

REFERENCES

1. Adrian, ED, and Bronk, DW: The discharge of impulses in motor nerve fibres. II. The frequency of discharge in reflex and voluntary contractions. J Physiol (Lond) 67:119–151, 1929.
2. Baringer, JR, Gajdusek, DC, Gibbs, CJ Jr, Masters, CL, Stern, WE, and Terry, RD: Transmissible dementias: Current problems in tissue handling. Neurology 30:302–303, 1980.
3. Barker, AT, Freestone, IL, Jalinous, T, Merton, PA, and Morton, HB: Magnetic stimulation of the human brain. J Physiol 369:3P, 1985.
4. Basmajian, JV: Electrodes and electrode connectors. In Desmedt, JE (ed): New Developments in Electromyography and Clinical Neurophysiology, Vol 1. Karger, Basel, 1973, pp 502–510.
5. Basmajian, JV: Muscles Alive: Their Functions Revealed by Electromyography, ed 4. Williams & Wilkins, Baltimore, 1978.
6. Basmajian, JV, and Stecko, G: A new bipolar electrode for electromyography. J Appl Physiol 17:849, 1962.
7. Beranek, R: Intracellular stimulation myography in man. Electroencephalogr Clin Neurophysiol 16:301–304, 1964.
8. Buchthal, F, Guld, C, and Rosenfalck, P: Action potential parameters in normal human muscle

and their dependence on physical variables. Acta Physiol Scand 32:200–218, 1954.

9. Buchthal, F, Guld, C, and Rosenfalck, P: Volume conduction of the spike of the motor unit potential investigated with a new type of multielectrode. Acta Physiol Scand 38:331–354, 1957.

10. Chu, J, and Chan, RC: Changes in motor unit action potential parameters in monopolar recordings related to filter settings of the EMG amplifier. Arch Phys Med Rehabil 66:601–604, 1985.

11. Chu J, Chan, RC, and Bruyninckx, F: Effects of the EMG amplifier filter settings on the motor unit potential parameters recorded with concentric and monopolar needles. Electromyogr Clin Neurophysiol 26:627–639, 1986.

12. Chu J, Chan, RC, and Bruyninckx, F: Progressive Teflon denudation of the monopolar needle: Effects of motor unit potential parameters. Arch Phys Med Rehabil 68:36–40, 1987.

13. Clamann, HP: Activity of single motor units during isometric tension. Neurology 20:254–260, 1970.

14. Day, BL, Dick, JPR, Marsden, CD, and Thompson, PD: Differences between electrical and magnetic stimulation of the human brain. J Physiol 378:36P, 1986.

15. Dorfman, LJ, McGill, KC, and Cummins, KL: Electrical properties of commercial concentric EMG electrodes. Muscle Nerve 8:1–8, 1985.

16. Ekstedt, J: Human single muscle fiber action potentials. Acta Physiol Scand (Suppl 226) 61:1–96, 1964.

17. Ekstedt, J, and Stalberg, E: A method of recording extracellular action potentials of single muscle fibres and measuring their propagation velocity in voluntarily activated human muscle. Bull Am Assoc Electromyogr Electrodiagn 10:16, 1963.

18. Ekstedt, J, and Stalberg, E: How the size of the needle electrode leading-off surface influences the shape of the single muscle fibre action potential in electromyography. Computer Prog Biomed 3:204–212, 1973.

19. Ekstedt, J, and Stalberg, E: Single fibre electromyography for the study of the microphysiology of the human muscle. In Desmedt, JE (ed): New Developments in Electromyography and Clinical Neurophysiology, Vol 1. Karger, Basel, 1973, pp 89–112.

20. Gajdusek, DC, Gibbs, JC Jr, Asher, DM, Brown, P, Diwan, A, Hoffman, P, Nemo, G, Rohwer, R, and White, L: Precautions in medical care of, and in handling materials from, patients with transmissible virus dementia (Creutzfeldt-Jakob disease). N Engl J Med 297:1253–1258, 1977.

21. Geddes, LA, Baker, LE, and McGoodwin, M: The relationship between electrode area and amplifier input impedance in recording muscle action potentials. Med Biol Eng 5:561–569, 1967.

22. Green, JB, Nelson, AV, and Michael, D: Digital zero-phase-shift filtering of short-latency somatosensory evoked potentials. Electroencephalogr Clin Neurophysiol 63:384–388, 1986.

23. Guidelines in Electrodiagnostic Medicine. Professional Standards Committee, American Association of Electromyography and Electrodiagnosis, 1984.

24. Guld, C. Rosenfalck, A, and Willison, RG: Report of the committee on EMG instrumentation: Technical factors in recording electrical activity of muscle and nerve in man. Electroencephalogr Clin Neurophysiol 28:399–413, 1970.

25. Gydikov, A, Gerilovsky, L, Kostov, K, and Gatev, P: Influence of some features of the muscle structure on the potentials of motor units, recorded by means of different types of needle electrodes. Electromyogr Clin Neurophysiol 20:299–321, 1980.

26. Hannerz, J: An electrode for recording single motor unit activity during strong muscle contractions. Electroenceph Clin Neurophysiol 37:179–181, 1974.

27. Hess, CW, Mills, KR, and Murray, WF: Percutaneous stimulation of the human brain: A comparison of electrical and magnetic stimuli. J Physiol 378:35P, 1986.

28. Hess, CW, Mills, KR, and Murray, WF: Methodological considerations for magnetic brain stimulation. In Barber, C, and Blum, T (eds): Evoked Potentials. III. The Third International Evoked Potential Symposium. Butterworths, London, 1987.

29. Hess, CW, Mills, KR, and Murray, MF: Responses in small hand muscles from magnetic stimulation of the human brain. J Physiol 338:397–419, 1987.

30. Jasper, H, and Notman, R: Electromyography in peripheral nerve injuries. National Research Council of Canada, Report C6121, NRC Grant No. Army Med 28 from the Montreal Neurological Institute, Vol IV. McGill University, Montreal, Quebec, 1944.

31. Maccabee, P, Hassan, N, Cracco, R, and Schiff, J: Short latency somatosensory and spinal evoked potentials: Power spectra and comparison between high pass analog and digit filter. Electroencephalogr Clin Neurophysiol 65:177–187, 1986.

32. Manuelidis, EE, Angelo, JN, Gorgacz, EJ, Kim, HA, and Manuelidis, L: Experimental Creutzfeldt-Jakob disease transmitted via the eye with infected cornea. N Engl J Med 296:1334–1336, 1977.

33. Mikolich, LM, and Waylonis, GW: Durability of monopolar Teflon-coated electromyographic needles. Arch Phys Med Rehabil 58:448–451, 1977.

34. Nissen-Petersen, H, Guld, C, and Buchthal, F: A delay line to record random action potentials. Electroencephalogr Clin Neurophysiol 26:100–106, 1969.

35. Stålberg, E, and Trontelj, JV: Single Fibre Electromyography. The Mirvalle Press Limited, Old Woking, Surrey, UK, 1979.

36. Tam, HW, and Webster, JG: Minimizing electrode motion artifact by skin abrasion. IEEE Trans Biomed Eng 24:134–139, 1977.

37. Wiechers, DO, Blood, JR, and Stow, RW: EMG needle electrodes: Electrical impedance. Arch Phys Med Rehabil 60:364–369, 1979.

Part II

NERVE CONDUCTION STUDIES

Chapter 4

ANATOMY AND PHYSIOLOGY OF THE PERIPHERAL NERVE

1 INTRODUCTION

The advent of newer histologic techniques has advanced our understanding of peripheral nerve function in health and disease. The means that helped most defining pathologic processes include the analysis of the fiber diameter spectrum and quantitative assessment of single teased fiber preparations. Electrophysiologic methods have made equally important contributions in elucidating the pathophysiology of peripheral nerve disorders. In particular, in vitro recordings of compound nerve action potentials from the sural nerve have delineated the types of fibers predominantly affected in cer-

55

tain neuropathic processes. These studies also demonstrated the close relationships between histologic and physiologic findings in many disease entities.

Traumatic lesions of the nerve usually, but not always, cause structural changes in the axon with or without separation of its supporting connective tissue sheath. Nontraumatic disorders of the peripheral nerve may primarily affect the nerve cell body, axon, Schwann cell, connective tissue, or vascular supply, singly or in combination. The electrophysiologic abnormalities depend on the kind and degree of nerve damage. Hence, the results of nerve conduction studies closely parallel the structural abnormalities of the nerve. Histologic changes in the nerve and the nature of conduction abnormalities allow subdivision of peripheral nerve lesions into two principal types: axonal degeneration and segmental demyelination.

This chapter will deal with the basic anatomy and physiology of the peripheral nerve and various types of conduction abnormalities. A number of excellent texts provide a more detailed review of the subject for interested readers.[3,12,29,52,69,105,118] Chapters 20 through 23 will deal with the clinical aspects of peripheral nerve disorders.

2 ANATOMY OF PERIPHERAL NERVES

Gross Anatomy

Three kinds of connective tissue—endoneurium, perineurium, and epineurium—surround the axons in the nerve trunks (Fig. 4–1). The endoneurium forms the supporting structure found around individual axons within each fascicle. The perineurium consists of collagenous tissue, which binds each fascicle with elastic fibers and mesothelial cells. This layer serves neither as a connective tissue nor as a simple supporting structure; rather, it provides a diffusion barrier to regulate intrafascicular fluid.[96,125] The epineurium, composed of collagen tissue, elastic fibers, and fatty tissue, tightly binds individual fascicles together. This outermost layer of supporting structure for the peripheral nerve merges in the dura mater of the spinal roots.[46] Paucity of endoneurial collagen at the roots, as compared with the nerve trunk, may explain why some disease processes selectively involve the root. The vasa nervorum, located in the epineurium, branch into arterioles that penetrate the perineurium to form capillary anastomoses in the fascicles. The perineurium probably acts as a blood-nerve barrier, but the elucidation of its detailed function needs further study.[125]

Myelinated and Unmyelinated Fibers

The nerve trunks contain myelinated and unmyelinated fibers. Certain inherent properties of the axon apparently determine whether or not myelination will eventually occur. In myelinated fibers, the surface membrane of a Schwann cell, or axolemma, spirals around the axon to form the myelin sheath (Fig. 4–2). Each myelinated axon has its own Schwann cell, which regulates myelin volume and thereby myelin thickness.[109] The nodes of Ranvier, located at junctions between adjacent Schwann cells, represent uninsulated gaps along the myelinated fiber. In contrast to myelinated fibers, several unmyelinated axons share a single Schwann cell, which gives rise to many separate processes, each surrounding one axon.[38]

The spacing of the Schwann cells at the time of myelination determines the internodal distance. As the nerve grows in length, the internodal distance must increase, because Schwann cells do not proliferate. Thus, the fibers myelinated early achieve larger diameters and wider spacing of the nodes of Ranvier.[128] In myelinated fibers, the action potentials propagate from one node of Ranvier to the next, with the rate of propagation approximately proportional to the fiber diameter. In unmyelinated nerves, conduction velocity varies in proportion to the square root of the fiber diameter. The largest and fastest conducting fibers include the sensory fibers transmitting proprioceptive,

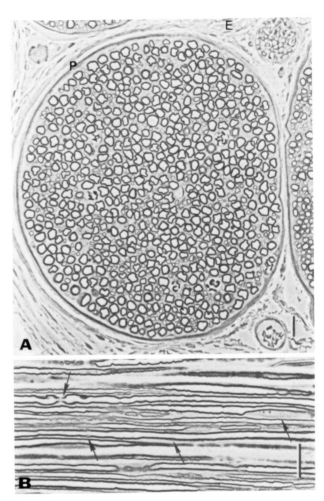

Figure 4–1. Transverse (**A**) and longitudinal (**B**) sections of the sciatic nerve shown at low magnification. Vertical scales at lower right represent 20 μm. In **A**, the epineurium (*E*) contains vessels, fibroblasts, and collagen. The perineurium (*P*) surrounds fascicles of nerve fibers, which are separated by endoneurial connective tissue. The longitudinal section (**B**) includes a node of Ranvier (*upper arrow*), a Schwann cell nucleus (*right arrow*), and Schmidt-Lantermann clefts (*lower arrows*). (From Webster,[134] with permission.)

positional, and touch sensations and the alpha motor neurons. Small myelinated or unmyelinated fibers subserve pain and temperature sense and autonomic functions (see Chapter 4.4).

Axonal Transport

In the peripheral nervous system, the small cell body with the diameter in the order of 50 to 100 μm regulates axons up to 1 m in length. A complicated system of axonal transport provides the metabolic needs of the terminal axon segments. Hence, the axons not only conduct the propagating electrical potential but also actively participate in conveying nutrient and other trophic substances.[45] The velocity of transport varies from several hundred to a few millimeters per day,[13,70] the major flow being centrifugal, although some particles seem to move centripetally.

Axonal transport plays a complex role in maintaining the metabolic integrity of the peripheral nerve. The axonal flow of trophic substances, at least in part, dictates the histochemical and electrophysiologic properties of the muscle fibers. No particles other than acetylcholine (ACh), however, seem to transfer across the neuromuscular junction. Therefore, acetylcholine molecules may have a trophic influence on muscle in addition to its role as a neurotransmitter. Separation of the axon from the cell body first results in failure of neuromuscular junction, fol-

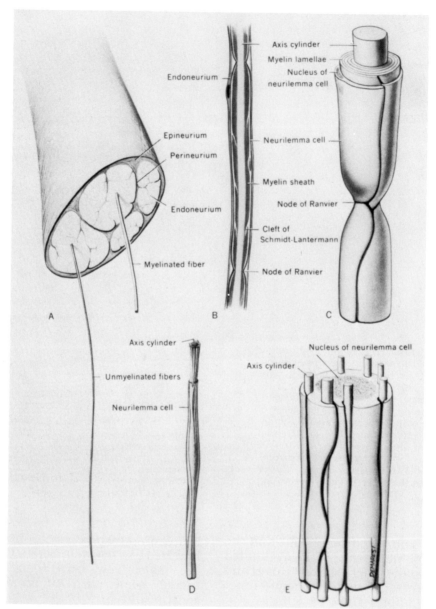

Figure 4–2. Fine structures of the peripheral nerve as visualized with the light microscope (*A*, *B*, and *D*) and as reconstructed from electron micrographs (*C* and *E*). In *A*, the epineurium covers the entire nerve, whereas the perineurium surrounds individual fascicles and endoneurium nerve fibers. In *B*, the myelinated fiber consists of axis cylinder, myelin sheath, and Schwann (neurilemma) cells. The myelin sheath is absent at the node of Ranvier. In *C*, the Schwann cell produces a helically laminated myelin sheath that wraps around an axon individually. In *D*, several unmyelinated nerve fibers share one Schwann cell. In *E*, several axis cylinders of unmyelinated fibers surround the nucleus of the Schwann cell. (From Noback,[80] with permission).

lowed by axonal degeneration and muscle fiber atrophy.[77] Neuromuscular transmission fails and the nerve terminals degenerate faster with distal than with proximal axonotmesis. Similarly, membrane changes in denervated muscle appear more rapidly after nerve injury close to the muscle.[47]

3 PHYSIOLOGY OF NERVE CONDUCTION

Transmembrane Potential

The nerve axons have the electrical properties common to all excitable cells (see Chapter 2.2). Measured transmembrane steady-state potentials vary from about −20 to −100 mV in different tissues, despite the same basic physiologic mechanism underlying the phenomenon. The soma membrane has less polarization (−70 mV) than the axon (−90 mV). The measured membrane potential near the cell body, however, probably represents a partial depolarization from continuous synaptic influences. As in any excitable element, generation of the nerve action potential consists of two steps: graded subliminal excitation caused by any externally applied stimulus and suprathreshold activation as the result of increased sodium conductance. A local subliminal change in the transmembrane potential rapidly diminishes with distance. In contrast, threshold depolarization produces an all-or-none action potential determined by the inherent nature of the cell membrane, irrespective of the type of stimulus applied.

Generation and Propagation of Action Potential

With application of a weak current to a nerve, negative charges from the negative pole (cathode) accumulate outside the axon membrane, making the inside of the cell relatively more positive (cathodal depolarization). Under the positive pole (anode), the negative charges tend to leave the membrane surface, making the inside of the cell relatively more negative (anodal hyperpolarization). The cell plasma resistance and membrane conductance and capacitance limit the subliminal local changes of depolarization or hyperpolarization only within a few millimeters of the point of origin. After about 10 to 30 mV of depolarization, the membrane potential reaches the critical level for degeneration of an action potential with the same maximal response regardless of the kind of stimulus or its magnitude. This energy-dependent process involves complex molecular changes of the cell membrane (see Chapter 2.3).

Nerve excitability change seen after a single nerve impulse has three phases; the initial refractory period of a few milliseconds (see Chapter 7.6), supernormality lasting 30 ms or so, and subnormality extending up to 100 to 200 ms. The supernormal period probably results from partial depolarization of the nerve membrane during negative after-potential. The administration of deltamethrin, known to maintain a proportion of sodium channels open for up to several hundred milliseconds, exaggerates this phase of excitability change.[86,120] Ischemia also increases the supernormal period in human motor and sensory fibers.[113]

An action potential initiated along the course of the axon propagates in both directions from its point of origin (Fig. 4–3). Intracellular current flows from the positively charged active area to the adjacent negatively charged inactive region. An opposing current flows through the extracellular fluid from the inactive to the active region. This local current depolarizes the inactive regions on both sides of the active area. When it attains the critical level, an action potential generated there initiates a new local current further distally or proximally. Hence, the nerve volleys always propagate bidirectionally from the site of external stimulation at one point along the axon. Physiologic impulses originating at the anterior horn cells or sensory terminal travel only orthodromically. In pathologic situations, however, impulses may arise in the midportion of nerve fibers. For example, discharges occur in the middle of the spinal

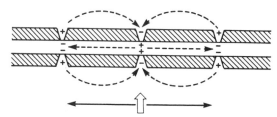

Figure 4–3. Saltatory conduction along the myelinated fiber. The myelin sheath effectively insulates the internodal segment with the bare axon at the node of Ranvier, where the current flows between intracellular and extracellular fluid. A local current (*broken arrows*) produced by an action potential at one node (*open arrow*) depolarizes the axis cylinder at the adjacent nodes on either side, transmitting the impulse in both directions (*solid arrows*). This type of saltatory excitation propagates rapidly as it jumps from one node to the next.

root axons in dystrophic mice, either spontaneously or as a result of ephaptic transmission (cross-talk) from the neighboring fibers.[92]

Strength-Duration Curve

The threshold intensity, just capable of exciting the axons, varies according to the duration of the current; the shorter the duration, the greater the intensity to achieve the same depolarization. The strength-duration curve plots this relationship with a motor point stimulation that elicits a constant muscle response. A shock of long duration excites both nerve and muscle, whereas a stimulus of short duration activates the nerve more effectively than the muscle. The excitability characteristics expressed by this curve, therefore, can differentiate a normally innervated muscle from a partially or totally denervated one.

For the formulation of numeric indices of excitability, rheobase is defined as the minimal current strength below which no response occurs even if the current lasts 300 ms or longer. Chronaxie is the minimal duration of a current required to excite the cell at twice the rheobase strength. The same principle also applies to the study of sensory fibers as a measure of sensory deficit in peripheral neuropathy.[34] Although of historical interest, neither rheobase nor chronaxie has

proven satisfactory as a test in clinical practice. The strength-duration curve itself has fallen into disrepute because of the excessive time required for its determination and the complexity of its interpretation. Nerve conduction studies and electromyography have largely replaced the excitability tests.

Factors Determining Conduction Velocity

Various factors affect the time necessary for generation of action potentials, which in turn determines the conduction velocity of an axon. Rapid propagation results from (1) faster rates of action potential generation, (2) increased current flow along the axons, (3) lower depolarization thresholds of the cell membrane, and (4) higher temperature, which, by increasing sodium conductance, facilitates depolarization. Conduction velocity increases nearly linearly about 5 percent per 1°C from 29°C to 38°C.[48] Thus, the change ranges 2 to 3 m/s per 1°C in a normal nerve conducting at 40 to 60 m/s.[54] Other factors of clinical importance include variation among different nerves and segments and effect of age (see Chapter 5.6).

In the myelinated fibers, action potentials occur only at the nodes of Ranvier. This induces a local current, which, in effect, jumps from one node to the next, producing saltatory conduction (see Fig. 4–3), instead of continuous propagation, as in unmyelinated fibers. An increase in the internodal distance allows a longer jump of the action potential, but at the same time causes greater loss of current through the internodal membrane. Typically, it takes approximately 20 μs for the local current to excite the next node. The conduction velocity would then be 50 m/s, for the internodal distance of 1 mm.

The longitudinal resistance of axoplasm tends to inhibit the flow of the local current. The capacitance and conductance of the internodal membrane also have the same effect by the loss of the current before it reaches the next node. This in turn makes the time required to depolarize the adjacent nodal membrane longer, resulting in slower conduction. Both internodal capacitance and conductance decrease with myelin thickness. Thus, for a fixed axon diameter, conduction velocity increases with myelin thickness up to a point. For a fixed total fiber diameter, an increase in axon diameter reduces myelin thickness, inducing two opposing factors, smaller axoplasmic resistance, on the one hand, and greater membrane conductance and capacitance, on the other.[131] Theoretic considerations indicate that the anatomic characteristics of myelinated fibers fulfill all the conditions required for maximal conduction velocity.

The demyelinated or partially remyelinated segments have an increased internodal capacitance and conductance because of the thin myelin sheath. More local current is then lost to charge the capacitors and by leakage through the internodal membrane before reaching the next node of Ranvier. Failure to activate the next node results in conduction block. If the conduction resolves, impulse propagates slowly, because the dissipated current needs more time to generate an action potential.[65,93,94] Thus, the demyelinated axons characteristically have conduction failure, decreased velocity, and temporal dispersion.[130] After segmental demyelination, smaller diameter fibers may show continuous, rather than saltatory, conduction if there is a sufficient number of sodium channels in the demyelinated region.[11] Reduction in length of the adjacent internodes tends to facilitate conduction past focally demyelinated zones.[132]

Conduction abnormalities do not necessarily imply demyelination, but can result from focal compression. Here, reduced fiber diameter decreases the capacitance of the internodal membrane, which tends to facilitate conduction. Concomitant increases in resistance of the axoplasm, however, more than offset this effect by delaying the flow of the local current to the next node. Most mechanisms known to influence nerve conduction velocity affect the cable properties of the internodal segments. Additionally, altered characteristics of the nodal membrane itself may interfere with generation of the action potential, although this possibility still lacks experimental evidence.[67]

4 TYPES OF NERVE FIBERS AND IN-VITRO RECORDING

Classification of Nerve Fibers

The compound nerve action potential elicited by supramaximal stimulation consists of several peaks, each representing a group of fibers with a different conduction velocity. Erlanger and Gasser,[31] in their original study of the A fibers, designated successive peaks using the Greek letters alpha, beta, gamma, and delta in order of decreasing velocity. Subsequent studies have revealed two additional components with very slow conduction velocity—B and C fibers. The mammalian peripheral nerves contain no B fibers. This designation, therefore, now indicates the preganglionic fibers in mammalian autonomic nerves. There has been some confusion with the terminology for various peaks of the A fibers.[68] The initial peak was referred to as either A-alpha[39] or A-beta.[30] The subsequent peak, originally called A-gamma, was an artifact of recording.[39] Current practice designates the two peaks in the A potential of cutaneous nerves as A-alpha and A-delta.

The A fibers are myelinated somatic axons, either afferent or efferent. The B fibers are myelinated efferent axons that constitute the preganglionic axons of the autonomic nerves. The unmyelinated C fibers consist of the efferent postganglionic axons of autonomic nerves and the small afferent axons of the dorsal root and peripheral nerves. Despite histologic resemblance, physiologic characteristics can differentiate B fibers from small A fibers. For instance, the B fibers lack negative afterpotentials and consequently a supernormal period of excitability after generation of an action potential. The negative spike lasts more than twice as long in B as in A fibers. The B fibers show smooth compound action potentials without discrete peaks, indicating an evenly distributed velocity spectrum. Several C fibers share a single Schwann cell, unlike A and B fibers, which are individually bound. This and the absence of the myelin sheath allow histologic identification of the C fibers. Physiologic features include high thresholds of activation, long spike duration, and slow conduction velocity.

The three types of nerve fibers, A, B, and C, have histologically and electrophysiologically distinctive characteristics:

A fibers: myelinated fibers of somatic nerves
  Muscle nerve
    Afferent
      Group I: 12–21 μm
      Group II: 6–12 μm
      Group III: 1–6 μm
      Group IV: C fiber
    Efferent
      Alpha motor neuron
      Gamma motor neuron
  Cutaneous nerve
    Afferent
      Alpha: 6–17 μm
      Delta: 1–6 μm
B fibers: myelinated preganglionic fibers of autonomic nerve
C fibers: unmyelinated fibers of somatic or autonomic nerve
  sC fibers: efferent postganglionic fibers of autonomic nerve
  drC fibers: afferent fibers of the dorsal root and peripheral nerve

Afferent fibers of the cutaneous nerves show a bimodal diameter distribution, with one component ranging between 6 and 17 μm and the other between 1 and 6 μm, or with the Greek letter designation, A-alpha and A-delta fibers. The A fibers in muscle nerves are either efferent or afferent. The efferent fibers consist of the axons of alpha and gamma motor neurons. In Lloyd's Roman numeral classification, the afferent fibers consist of groups I, II, and III, ranging in diameter from 12 to 21 μm, from 6 to 12 μm, and from 1 to 6 μm, respectively, and group IV, representing small pain fibers. In this designation, the A-alpha fibers of cutaneous nerve correspond in size to groups I

and II, the A-delta fibers to group III, and the C-fibers to group IV. A-alpha and A-delta peaks can be distinguished in direct recording from human nerves, for example, in the intracranially recorded potentials from the sensory root of the trigeminal nerve evoked by electrical stimulation of the different cutaneous branches.[95]

In-Vitro Recording of Sural Nerve Potentials

An in vitro study of the sural nerve action potential complements the quantitative morphometric assessment of the excised nerve.[28] The technique allows comparison between the fiber diameter spectrum and the range of conduction velocities for different components of the sensory nerve action potential. The nerve biopsy consists of dissecting a bundle of several fascicles above the lateral malleolus for a total length of approximately 10 cm. The distal half serves as the specimen for histologic studies, and the proximal half for in vitro electrophysiologic evaluation. The segment used for conduction studies is immediately placed in cool Tyrode's solution and transferred to a sealed chamber filled with 5 percent carbon dioxide in oxygen and saturated with water vapor. Immersing the chamber in a warm water bath helps maintain a constant temperature at 37°C.

A series of silver electrodes support the nerve under slight tension by the pull of a 0.5- to 0.9-g weight attached to each end. Stimulation at the distal end of the nerve allows recording of the compound nerve action potential with a pair of wire electrodes placed 20 to 30 mm proximally. A monophasic waveform is obtained when the nerve is crushed between the recording electrodes after application of 0.1 percent procaine at the distal electrode (see Fig. 2–4). The compound nerve action potential recorded in vitro consists of three distinct peaks: A-alpha, A-delta, and C components with an average conduction velocity of 60 m/s, 20 m/s, and 1 to 2 m/s, respectively (Fig. 4–4). Each component requires a different supramaximal inten-

Figure 4–4. Compound nerve action potential of a normal sural nerve recorded in vitro from an 11-year-old boy who had an above-knee amputation for osteogenic sarcoma. The *arrows* from left to right indicate A-alpha, A-delta, and C components, measuring 2.6 mV, 0.22 mV, and 70 µV in amplitude and 42 m/s, 16 m/s, and 1 m/s in conduction velocity based on the peak latency. (Courtesy of E. Peter Bosch, M.D., Department of Neurology, University of Iowa Hospitals and Clinics.)

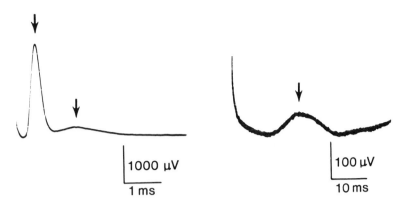

sity for full activation. The gradual onset of A-delta and C peaks precludes accurate calculation of the maximal conduction velocity.

Figure 4–5 shows fiber diameter histograms for the A-alpha and A-delta components. Here, the fiber diameter increases from left to right on the abscissa. Thus, the first peak on the left corresponds to A-delta fibers, and the second smaller peak corresponds to A-alpha fibers. In the opposite arrangement, with decreasing diameter plotted from left to right, fiber groups appear in order of decreasing conduction velocity, as in the tracings of compound action potentials. In normal fiber groups, fiber diameter histograms show a continuous distribution between the large and small myelinated fibers with no clear separation between the two. Similarly, A-alpha and A-delta peaks reflect a high concentration of fibers within the continuous spectrum.[7] The largest fiber, with a diameter close to 12 µm, conducts at an approximate rate of 60 m/s, indicating a 5:1 ratio between the two measurements.

Determining the internodal length spectra in teased fiber preparation also provides quantitative data in elucidating distribution of histologic abnormalities (Fig. 4–6).[53] Statistical analyses show significant correlations between teased fiber changes and conduction abnormalities affecting both motor and sensory nerves in patients with sensory motor polyneuropathies.[9]

Analysis of Compound Nerve Action Potentials

The amplitude of a compound action potential E, recorded over the surface of a nerve, increases in proportion to current flow and external resistance. Ohm's law expresses this as $E = IR$, where I is current and R, resistance. Larger nerves have a greater number of fibers that would collectively generate larger currents, with each fiber contributing an approximately equal amount. Nerves with greater cross-sectional areas, however, have a smaller total resistance. Large nerve size, therefore, may have a negligible overall effect on amplitude. In fact, a whole nerve composed of many fascicles does not necessarily give rise to an action potential larger than the one recorded from a single dissected fascicle.[68]

More current flows as the nerve fibers increase in number, whereas the resistance falls in proportion to the square diameter of the nerve. Thus, fiber density or the number of fibers per unit cross-sectional area determines the amplitude of an action potential. The factors that determine the waveform of a compound action potential include the magnitude of conduction block, diminution of current in individual nerve fibers, and the degree of temporal dispersion. Selective involvement of different groups of fibers results in a major distortion of the recorded potential. In contrast, uniform involvement of all fibers reduces the amplitude with

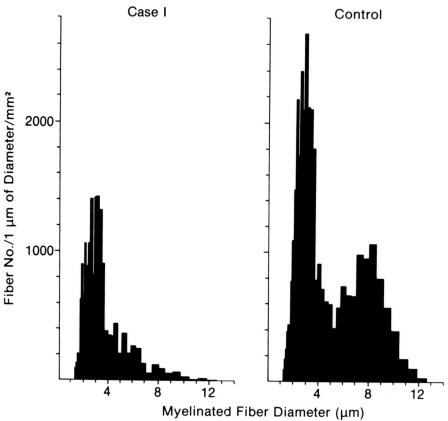

Figure 4–5. Myelinated fiber size-frequency histogram plotting the number of fibers with increasing diameter from left to right. The first large peak on the left corresponds to A-delta and the second smaller peak to A-alpha. Note a bimodal distribution of myelinated fiber diameter in a normal subject (*control*). A patient (*Case 1*) with familial pressure-sensitive neuropathy had an abnormal unimodal pattern with preferential loss of the larger myelinated fibers. (Courtesy of E. Peter Bosch, M.D., Department of Neurology, University of Iowa Hospitals and Clinics.)

relative preservation of all the components. Hence, waveform analysis of compound nerve action potentials provides a means of assessing fiber density and distribution spectrum.

5 CLASSIFICATION OF NERVE INJURIES

Seddon[104,105] defined three degrees of nerve injury: neurapraxia, axonotmesis, and neurotmesis. In neurapraxia, or conduction loss without structural change of the axon, recovery takes place within days or weeks after the removal of the cause. The conduction velocity, if initially slowed because of associated demyelina-

tion, returns to normal with remyelination. In axonotmesis, the axons lose continuity, with subsequent wallerian degeneration along the distal segment. Recovery depends on regeneration of nerve fibers, a process that takes place slowly over months or years at a rate of 1 to 3 mm/day. In neurotmesis, an injury separates the entire nerve, including the supporting connective tissue. Without surgical intervention, regeneration proceeds slowly, resulting in an incomplete and poorly organized repair. This classification was originally proposed for external trauma such as superficial nerve injuries or penetrating wounds in a deep location. The same also applies, however, to compression neuropathies such as the carpal tunnel syndrome, tardy ulnar palsy, and Saturday night palsy.

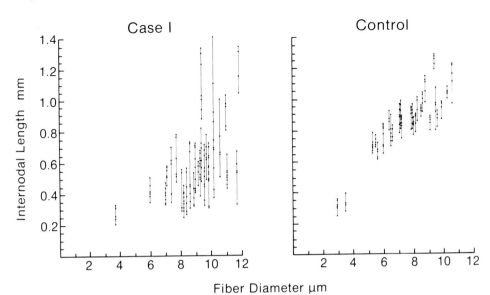

Figure 4–6. Internodal length spectra of the same nerves shown in Figure 4–5. Each vertical line indicates internodal length measured on a given myelinated fiber. The marked variability of internodal length in the patient reflects the effects of chronic demyelination and remyelination. (Courtesy of E. Peter Bosch, M.D., Department of Neurology, University of Iowa Hospitals and Clinics.)

Neurapraxia

The mildest form of nerve block results from local injection of procaine or the transient loss of circulation, for example, with leg crossing. These immediately reversible injuries cause no structural changes in the axon. In rat sciatic nerve, transient conduction block developed within 10 minutes after femoral artery occlusion, reached a nadir at 45 to 60 minutes, and then improved to normal within 24 hours.[88] The fall in amplitude with less than 15 percent slowing of conduction velocity implied the relative sensitivity of slower conducting myelinated fibers to the effect of acute ischemia. During transient paralysis in man experimentally induced by an inflated cuff around the arm, conduction velocity may fall by as much as 30 percent. A complete conduction block usually occurs after 25 to 30 minutes of compression. Serial stimulation along the course of the nerve reveals normal excitability in the segment distal to such a neurapraxic lesion.

These short-term changes in nerve conduction probably result from anoxia secondary to ischemia.[71] Intraneural microelectrode recordings show spontaneous activity in afferent fibers about half a minute after reestablishment of circulation. The perceived paresthesia also suggests ectopic impulses generated along the nerve fibers previously subjected to ischemia.[85] In contrast to short-term effects, chronic nerve ischemia, as may be induced by a bovine shunt, may result in axonal degeneration of sensory fibers initially and of motor fibers later.[8] In experimental animals, partial infarction resulted in degeneration of fibers in the center of the nerve, with no evidence of selective fiber vulnerability.[87]

In most acute compressive neuropathies, such as a Saturday night palsy or crutch palsy of the radial nerve, conduction across the affected segment recovers within a few weeks.[76] A neurapraxia causing incomplete and reversible paralysis could persist for a few months or longer, usually accompanied by demyelination. Similarly, the prolonged application of a tourniquet causes sustained conduction block with paranodal demyelination.[81] Conduction, albeit, markedly slowed, may return immediately after decompression or at times despite continued compression.[4] Complete recovery will ensue with remyelination.

Although demyelination in these cases can result from anoxia secondary to ischemia,[20,21] studies of experimental acute pressure neuropathy have stressed the importance of mechanical factors[37,41,82,83,98] with the initial displacement of axoplasm and myelin in opposite directions under the edges of the compressed region (Figs. 4–7 and 4–8). Part of one myelin segment invaginates the next with occlusions of the nodal gaps. Demyelination of the stretched portions of myelin follows. A patient with documented pneumatic tourniquet paralysis had severe conduction block of sensory and motor fibers localized to the presumed lower margin of the compression.[10,97]

Chronic entrapment states such as carpal tunnel syndrome or tardy ulnar palsy also show focal demyelination.[62,84] Like acute compression, local demyelination in chronic entrapment results from mechanical forces, rather than ischemia, although microscopic findings of single fibers suggest different pathophysiology in the two types. Unexpected abnormalities of nerve conduction studies in asymptomatic subjects suggest a high incidence of subclinical entrapment neuropathy. Routine autopsies in patients without known disease of the peripheral nerve have also documented unpredicted focal anatomic abnormalities.[79]

Patients with demyelinating neuropathy develop paralysis primarily because of conduction block, rather than slowed conduction velocity. The paralyzed muscles may show fibrillation potentials and positive sharp waves after a prolonged lack of neural influence, despite the structural integrity of the axons. In one study,[126] 25 percent of 31 patients had spontaneous discharges solely on the basis of a conduction block lasting more than 14 days. In the remaining 75 percent, spontaneous discharges resulted from axonal degeneration.

Axonotmesis

In this condition, axonal damage results in loss of continuity and wallerian degeneration of the distal segment (Fig. 4–9). Conduction block occurs immediately across the site of nerve injury, followed by irreversible loss of excitability, first at the neuromuscular junction, then at the distal nerve segment. The time course of such degeneration varies among different species but generally does not occur until 4 or 5 days after acute interruption.[44] Available data lack detailed information for precisely characterizing conduction abnormalities during wallerian degeneration in humans. In two cases, serial studies revealed loss of action potentials as early as 185 hours in one case and 168 hours in the other after traumatic transection of the digital nerve. Conduction studies showed a normal velocity during wallerian degeneration prior to the loss of recorded response.[90] During

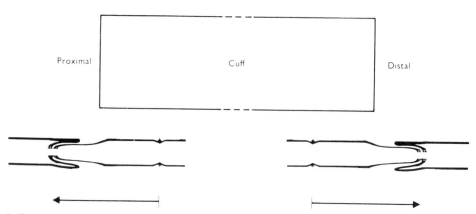

Figure 4–7. Diagram showing the direction of displacement of the nodes of Ranvier in relation to the cuff. Proximal and distal paranodes are invaginated by the adjacent one. (From Ochoa et al.,[82] with permission.)

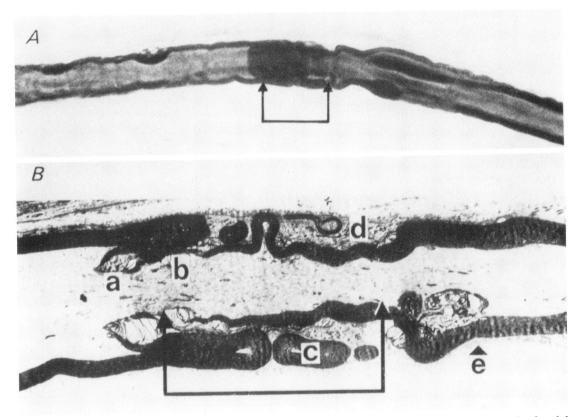

Figure 4–8. A. Part of a single teased fiber showing an abnormal node. **B.** Electron micrograph of nodal region shown in **A.** *a,* terminal myelin loops of ensheathing paranode; *b,* terminal myelin loops of ensheathed paranode; *c,* myelin fold of ensheathing paranode cut tangentially; *d,* Schwann cell cytoplasm; *e,* microvilli indicating site of Schwann cell junction. *Large arrows* show length of ensheathed paranode (~ 20 μm). (From Ochoa et al.,[81] with permission.)

the first few days after nerve injury, therefore, studies of distal nerve excitability fail to distinguish axonotmesis from neurapraxia.

In the baboon, the muscle response to nerve stimulation disappears 4 or 5 days after nerve section, but an ascending nerve action potential may persist in the segment distal to the section for 2 or 3 more days.[43] Preceding conduction failure, there is no change in the maximal conduction velocity of either the descending motor potential or ascending nerve action potential. Histologically, degeneration develops in the terminal portion of the intramuscular nerve at a time when the proximal parts of the same fibers are relatively intact. The central stump of a transected nerve fiber remains excitable, but the nerve action potentials and con-

duction velocity may all be slightly reduced, possibly from retraction of the myelin sheath.[58,123] Transverse section at this level shows a marked reduction in the number of myelinated fibers present.[2]

In an experimental axotomy in cats, sensory fibers degenerated more quickly than did motor fibers of similar diameter, and changes in whole-nerve conduction velocity distribution showed that velocities of fast conducting fibers decreased at the most rapid rate.[78] Permanent axotomy in cats produced by hind limb ablation results in sequential pathologic alteration of myelinated fibers of the proximal nerve stump, namely, axonal atrophy, myelin wrinkling, nodal lengthening, and internodal demyelination and remyelination.[27] This type of interaction between axon and myelin may occur in

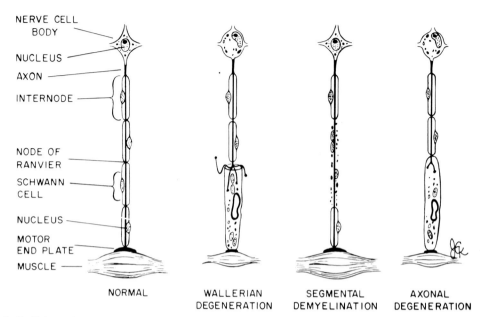

NERVE CELL BODY
NUCLEUS
AXON
INTERNODE
NODE OF RANVIER
SCHWANN CELL
NUCLEUS
MOTOR END PLATE
MUSCLE

NORMAL   WALLERIAN DEGENERATION   SEGMENTAL DEMYELINATION   AXONAL DEGENERATION

Figure 4–9. Schematic representation of nerve axon and myelin sheath. From left to right, normal structures, wallerian degeneration following transection of the fiber, segmental demyelination, and axonal degeneration secondary to disorders of the nerve cell. (From Asbury and Johnson,[3] with permission.)

human uremic neuropathy and other types of systemic atrophy such as Friedrich's atoxia or hereditary motor sensory neuropathy.

Different types of abnormalities coexist in the majority of nerve injuries and neuropathies. Thus, categorizing injuries of a nerve, as opposed to individual nerve fibers, depends on less precise definition. Nonetheless, electrophysiologic studies help elucidate the extent of axonal damage. Nerve stimulation above the site of the lesion reveals reduced amplitude in proportion to the degree of conduction loss but fails to distinguish neurapraxia from axonotmesis. In either condition, unaffected axons, if present, conduct at a normal velocity across the segment in question. Stimulation of the nerve segment distal to the site of the lesion helps differentiate the two entities. Injured axons undergo wallerian degeneration after the first few days of injury. With the loss of distal excitability, the compound action potential elicited by distal stimulation declines steeply. Electromyography shows positive sharp waves 1 to 2 weeks and fibrillation potentials 2 to 3 weeks after axonotmesis.

The process of regeneration accompanies the transport of structural proteins newly synthesized in the cell body to the multiple sprouts derived from the parent axon. Once the axon successfully reaches the periphery and reestablishes the physiologic connections, an orderly sequence of maturation takes place, and fiber diameter progressively increases.[135] The remaining sprouts that fail to make functional reconnection will eventually degenerate. If the Schwann cell basement membrane and the remaining connective tissues are intact, regeneration of the nerve axon occurs in an orderly manner along the intact nerve sheath. The axons regenerate at a rate of approximately 1 to 3 mm/day, nearly restoring the normal number and size of fibers when fully regenerated.

Neurotmesis

Sunderland[118] has proposed three subdivisions of Seddon's neurotmesis. In the first type, the injury damages the axon and surrounding connective tissue, but the perineurium and architecture of the nerve sheath are preserved. Regeneration takes place effectively but less completely than in axonotmesis. Misdirected sprouting leads to innervation of muscle fibers previously not supplied by the nerve. The clinical phenomenon of synkinesis probably indicates an antecedent nerve injury of at least this severity.[64,119] In the second type, which involves the perineurium as well, the nerve barely maintains continuity, although it looks grossly intact on inspection. Some poorly oriented regeneration may occur, often requiring surgical intervention for repair. The last type represents a complete separation of the nerve with loss of continuity. Surgical repair consists of suturing the stumps, usually with a nerve graft to bridge the gap. Regeneration may progress poorly, with frequent formation of neuromas. Spontaneous discharges of nerve impulses probably result from changes in membrane properties. This may be the source of pain associated with neuromas.[129]

After reanastomosis, regenerating nerve fibers do not regain the original number, although their diameter increases gradually over many years. The conduction velocity also increases slowly, reaching 60 percent of the normal value within four years[50] and a mean value of 85 percent after 16 years.[114] Persistent prolongation of the distal latencies suggests the presence of a limited number of fibers distally. Misdirection of motor axons after proximal section of the nerve probably accounts for the loss of orderly recruitment of motor units.[122]

In detailed sequential studies of the median nerve after complete section and suture,[15] the regeneration took place at an average rate of 1.5 to 2.0 mm/day in three patients. The sensory potential, when first recorded 3 to 4 months after the injury, propagated very slowly at a velocity between 10 to 25 percent of normal. The conduction velocity increased 3 percent per month during the first 2 years, and 10 times slower thereafter. The tactile sensibility had returned to normal by 40 months, when the sensory potential showed a normal amplitude but increased duration measuring five times more than normal. Conduction velocity

reached 65 to 75 percent of normal in the adults. In children, the same degree of recovery occurred 13 to 19 months after anastomosis. The sensory potential returned five times faster after a compressive nerve lesion than after section and repair.

Few studies have dealt with the neurophysiologic recovery of human peripheral nerves after repair with an autogenous nerve graft.[5] In one series,[121] motor and sensory nerve conduction velocities showed sustained improvement following sural nerve grafts of the ulnar and median nerves. Two years after surgery motor conduction velocity across the graft itself reached, in most cases, 40 to 50 percent of the normal conduction velocity obtained in the contralateral limb. Sensory nerve action potentials, though greatly reduced in amplitude and conduction velocity, returned in 44 percent of the nerves after 18 months. In another study,[22] based on experience with 67 injured nerves, voluntary motor unit activity returned 7 months after repair and 12 months after grafting. A compound muscle action potential was elicitable by 10 months after suture and 14 months after grafting. Motor unit potentials steadily increased in amplitude with time, but sensory fibers showed poor recovery both clinically and electrophysiologically.

6 INVOLVEMENT OF AXON VERSUS MYELIN IN NEUROPATHIC DISORDERS

The preceding section has dealt with types of conduction abnormalities associated with nerve injuries. These form a basis in assessing electrophysiologic features found in other disease processes, either localized, as in entrapment syndromes or more diffuse, as in polyneuropathies. Histologic[55] and electrophysiologic characteristics indicate the presence of three relatively distinct categories of peripheral nerve disorders (see Fig. 4–9): (1) wallerian degeneration following focal interruption of axons as in vasculitis; (2) axonal degeneration with centripetal or dying-back degeneration from metabolic derangement of the neuron; and (3) segmental demyelination with slowed nerve conduction.[3,59,72,75] Of these, wallerian and axonal degeneration causes denervation and reduces the amplitude of compound action potential, whereas demyelination slows nerve conduction.

Axonal Degeneration

Axons may degenerate in neuropathies, from mechanical compression of the nerve, after application of toxic substances, or after the death of the cell body. This type of abnormality, if mild, affects nerve conduction studies only minimally, especially in diseases that primarily involve the small fibers. More commonly, selective loss of the large, fast conducting fibers results in reduced amplitude and slowing of conduction below the normal range. In these cases, the reduction in size of the compound muscle or sensory action potential is correlated with the degree of slowing in nerve conduction. The small, thinly myelinated regenerating nerves may show much slower conduction velocities. Electromyography reveals normal motor unit potentials that recruit poorly during the acute stage of partial axonal degeneration. Long-duration, high-amplitude polyphasic potentials appear in the chronic phase. Fibrillation potentials and positive sharp waves develop in 2 to 3 weeks after the onset of illness.

Neuropathies with this type of abnormalities include those associated with alcoholism, uremia, polyarteritis nodosa, acute intermittent porphyria, some cases of diabetes and carcinoma, and most toxic and nutritional deficiencies.[40] Most axonal neuropathies affect both sensory and motor fibers, as in uremic neuropathies and amyloidosis. Acute intermittent porphyria and hereditary motor sensory neuropathy type II, or neuronal Charcot-Marie-Tooth disease,[24] however, show prominent motor abnormalities. In contrast, sensory changes predominate in the majority of toxic or metabolic polyneuropathies, Friedreich's ataxia, and some cases of carcinomatous neuropa-

thy. Histologic studies in diabetic rats revealed paranodal axonal swellings and nodal bulgings of the axon during the early metabolic phase of distal symmetric polyneuropathy.[106] These alterations correlate with intra-axonal sodium accumulation and decreased sodium equilibrium potentials, which account for the early nerve conduction defect.

Anterior horn cell diseases can also cause selective loss of the fastest fibers. In poliomyelitis, the motor nerve conduction velocity may fall below the normal value, usually in proportion to the decrease in amplitude.[49] Patients with motor neuron disease also have slightly reduced motor conduction velocities for the same reason. Slower conduction in patients with more severe atrophy may in part reflect the lowered temperature of the wasted extremities.[66]

In neuropathies secondary to chronic alcoholism, carcinoma, and uremia, axonal degeneration initially involves the terminal segment of the longest peripheral nerve fibers. The distal predominance of the disease and its centripetal progression led to the term dying-back neuropathy. Less commonly encountered conditions associated with the dying-back phenomenon include thiamine deficiency,[19] triorthocresyl phosphate neuropathy,[17] acute intermittent porphyria,[18] and experimental acrylamide neuropathies.[36,91,116] In these conditions, impaired axoplasmic flow probably affects the segment of the nerve farthest from the perikaryon initially. Thus, primary involvement of the neurons leads to axonal degeneration in the distal segment most removed from the trophic influence of the nerve cell.[111]

Single-unit recording in dying-back axons has confirmed the earliest failure of impulse generation in the nerve terminal when impulse still propagates normally throughout the remainder of the axon.[115] In acrylamide dying-back neuropathy in rats, a sequential morphometric study of the end-plate region showed the initial enlargement of the nerve terminal area distended by neurofilaments.[127] The postsynaptic regions eventually became denuded with the disappearance of more than half of all nerve terminals.

Unlike the experimental acrylamide neuropathy with a clear dying-back phenomenon,[51,102] not all the peripheral neuropathies with a distal predominance may qualify as truly dying-back in type. Selective loss of the perikarya and axons of the longest and largest fibers can produce the same pattern of abnormality.[14] Distally predominant symptoms do not necessarily indicate a distal pathologic process, according to probabilistic models that reproduce a distal sensory deficit on the basis of randomly distributed axonal lesions.[133] In some neuropathies, studies fail to reveal the exact site of the primary damage responsible for axonal degeneration.

Segmental Demyelination

In the second group, disturbance of the Schwann cells causes segmental demyelination associated with substantial reduction of nerve conduction velocity, commonly, though not always, substantially below the normal range.[16] Axonal degeneration cannot account for this degree of slowing, even with selective loss of the fast conducting fibers, leaving only the slow conducting fibers relatively intact. In animal experiment demyelination blocks the transmission of nerve impulses through the affected zone in some fibers. However, the slowed conduction results primarily from delayed nerve impulses passing through the lesion and not simply from selective block of transmission in the fast-conducting fibers.[73,74] Focal segmental demyelination gives rise to slowed conduction locally across the demyelinated segment but not below the lesion.[94] In addition to various toxins, removal of a small piece of perineurium in amphibian nerve also causes a lesion consistent with demyelination.[89]

Pathophysiologic characteristics and their clinical correlates of demyelination[42] include (1) a raised threshold of conduction block, which results in clinical weakness and sensory loss; (2) increased desynchronization and temporal dispersion of volleys, causing loss of reflexes and vibration sense; (3) a prolonged refractory period with blocking at high frequency, possibly accounting for reduced strength despite maximal voluntary effort; and (4) exaggerated hyperpolarization after the passage of an impulse, giving rise to blocking and the fatigue phenomenon on sustained effort. In addition, a focal demyelination may induce involuntary discharges at steady frequencies or in bursts such as those seen with facial myokymia.[110] These ectopic impulses probably cause the spontaneous and movement-induced paresthesias experienced by patients.

In an experimental study on demyelination induced by diphtheria toxin, conduction velocity began to decline one week after inoculation, reached a plateau during the sixth to eighth week, and recovered to the original level between the 18th and 20th weeks.[56,57] The dose of toxin administered determined the degree of slowing and the severity of paralysis. The amplitude of the compound muscle action potential predicted the loss of strength even more accurately than the conduction velocity. In paranodal demyelination caused by immunodipropionitrite (IDPN)-induced giant axonal swelling, conduction velocity began to slow over the swollen segment between 2 and 4 days after the administration of the toxin. The contact maintained by displaced myelin terminal loops with the axolemma allowed saltatory propagation of impulse to continue without conduction block.[112]

In antiserum-mediated focal demyelination of male Wistar rats, conduction block began between 30 and 60 minutes after injection and peaked within a few hours.[100,117] Paralysis of the foot muscles persisted until about the seventh day, when low-amplitude, long-latency muscle action potentials returned for the first time. Strength gradually recovered thereafter, reaching a normal level by 16 days. Morphologic studies revealed evidence of remyelination with two to eight myelin lamellae around each axon coincident with the onset of clinical and electrophysiologic recovery. Conduction velocities returned to preinjection values by the 37th day, when the myelin layer of remyelinating fibers was only about one-third that of control nerves.

Serial studies of an experimentally produced demyelinating lesion in cat spinal cord[108] revealed the onset of conduction block during the initial phase. Remyelination commenced in the latter part of the second week concomitant with restoration of conduction through the lesion in the affected fibers. Within 3 months the initially prolonged refractory period returned to normal even though the newly formed internodes were still abnormally thin.

Common demyelinating diseases of the peripheral nerve include the Guillain-Barré syndrome, chronic inflammatory demyelinative polyradiculoneuropathies, myelomatous polyneuropathies, hereditary motor sensory neuropathy type I, or hypertrophic Charcot-Marie-Tooth disease,[24] metachromatic leukodystrophy,[35] and Krabbe's leukodystrophy.[23] Some cases of diabetic and carcinomatous neuropathies also belong to this category,[40] although most patients with the paraneoplastic syndrome show axonal degeneration, rather than demyelination. Diphtheritic polyneuritis no longer affects humans very often. Alterations in nerve conduction resemble those seen in animal experiments with marked reduction in conduction velocity diffusely or, in the case of focal demyelination, over a relatively restricted region. Despite the well-established concept of segmental demyelination in experimentally induced chronic lead intoxication, the nerve conduction velocity in human cases is either normal[107] or only mildly slowed.[33] The conventional nerve conduction studies, basically designed for assessment of the distal segments, may fail to elucidate a focal proximal demyelinating lesion in the proximal segment.[61]

In demyelinating polyneuropathy, slowing of nerve conduction often accompanies a reduction of amplitude, indicating localized neurapraxia.[32] Electrophysiologic evidence of conduction block usually, although not always, implies the presence of focal demyelination.[103] Increased range of conduction velocity results if the disease affects smaller fibers exclusively or disproportionately in relation to larger fibers. The evoked action potential broadens because of increased temporal dispersion. Desynchronization of the nerve volley may also result from repetitive discharges at the site of axonal injury after the passage of a single impulse. Unless secondary axonal degeneration is induced by damage of the myelin sheath, electromyography reveals little or no evidence of denervation. The motor unit potentials, though normal in amplitude and waveform, recruit poorly, indicating a conduction block in severely demyelinated fibers.

Types of Abnormalities in the Clinical Domain

In the arbitrary division into axonal and demyelinating neuropathies, few cases fall precisely into one group or the other. A neuropathy with extensive demyelination often accompanies slight axonal degeneration.[25,124] In a study of antiserum-mediated demyelination, the inflammatory reaction could account for axonal degeneration seen in 5 to 15 percent of myelinated fibers.[99] Conversely, axonal atrophy poximal to a neuroma or distal to constriction may cause secondary paranodal demyelination in the presence of healthy Schwann cells. Other conditions that may belong to this category include neuropathies associated with uremia, myeloma, or Friedreich's ataxia, and hereditary motor sensory neuropathy type I.[26,101] Axonal enlargement can also cause axon-triggered demyelination, as in giant axonal neuropathy or hexacarbone intoxication.

Despite the above uncertainty, the electrophysiologic finding of any true axonal or demyelinating component provides an important and major contribution in the differential diagnosis. Certain conduction abnormalities support the diagnosis of a predominantly demyelinating component even when superimposed upon moderate axonal degeneration as demonstrated on needle electromyography. These include reduction of conduction velocity below 70 to 80 percent of the lower limit, prolongation of distal motor or sensory latency and F-wave latency above 120 to 130 percent of the upper limit, and the presence of unequivocal conduction block.[1,60] In contrast, the absence of these criteria does not necessarily preclude an early demyelinating process. In fact, a substantial number of patients with the Gullain-Barré syndrome have no major slowing of conduction along the nerve trunk initially.

Beyond such a broad classification, electrical studies have limited value in distinguishing one variety of neuropathy from another. In particular, conduction studies and electromyography rarely distinguish a specific cause. In some cases, the slight loss of fibers or the mild degree of demyelination demonstrated histologically cannot account for the degree of slowing in the same nerve.[6] Despite these limitations, conduction studies can provide diagnostically pertinent information if used judiciously in appropriate clinical contexts.[63]

REFERENCES

1. Albers, JW, Donofrio, PD, and McGonagle, TK: Sequential electrodiagnostic abnormalities in acute inflammatory demyelinating polyradiculoneuropathy. Muscle Nerve 8:528–539, 1985.
2. Anderson, MH, Fullerton, PM, Gilliatt, RW, and Hern, JEC: Changes in the forearm associated with median nerve compression at the wrist in the guinea-pig. J Neurol Neurosurg Psychiatry 33:70–79, 1970.
3. Asbury, AK, and Johnson, PC: Pathology of Peripheral Nerve, Vol 9. In Bennington, JL (ed): Major Problems in Pathology. WB Saunders, Philadelphia, 1978.
4. Baba, M, and Mastsunaga, M: Recovery from acute demyelinating conduction block in the presence of prolonged distal conduction delay due to peripheral nerve constriction. Electromyogr Clin Neurophysiol 24:611–617, 1984.
5. Ballantyne, JP, and Campbell, MJ: Electrophysiological study after surgical repair of sectioned human peripheral nerves. J Neurol Neurosurg Psychiatry 36:797–805, 1973.
6. Behse, F, and Buchthal, F: Sensory action potentials and biopsy of the sural nerve in neuropathy. Brain 101:473–493, 1978.
7. Bishop, GH: My life among the axons. In Hall, VE (ed): Annual Review of Physiology, Vol 27. Annual Reviews, Palo Alto, California, 1965, pp 1–18.
8. Bolton, CF, Driedger, AA, and Lindsay, RM: Ischaemic neuropathy in uraemic patients caused by bovine arteriovenous shunt. J Neurol Neurosurg Psychiatry 42:810–814, 1979.
9. Bolton, CF, Gilbert, JJ, Girvin, JP, and Hahn, A: Nerve and muscle biopsy: Electrophysiol-

ogy and morphology in polyneuropathy. Neurology (New York) 29:354–362, 1979.
10. Bolton, CF, and McFarlane, RM: Human pneumatic tourniquet paralysis. Neurology 28:787–793, 1978.
11. Bostock, H, and Sears, TA: Continuous conduction in demyelinated mammalian nerve fibres. Nature 263:786–787, 1976.
12. Bradley, WG: Disorders of Peripheral Nerves. Blackwell Scientific Publications, Oxford, 1974.
13. Bradley, WG, Murchison, D, and Day, MJ: The range of velocities of axoplasmic flow: A new approach, and its application to mice with genetically inherited spinal muscular atrophy. Brain Res 35:185–197, 1971.
14. Bradley, WG, and Thomas, PK: The pathology of peripheral nerve disease. In Walton, JN (ed): Disorders of Voluntary Muscle, ed 3. Churchill Livingstone, Edinburgh, 1974, pp 234–273.
15. Buchthal, F, and Kuhl, V: Nerve conduction, tactile sensibility, and the electromyogram after suture or compression of peripheral nerve: A longitudinal study in man. J Neurol Neurosurg Psychiatry 42:436–451, 1979.
16. Buchthal, F, Rosenfalck, A, and Behse, F: Sensory potentials of normal and diseased nerves. In Dyck, PJ, Thomas, PK, Lambert, EH, and Bunge, R (eds): Peripheral Neuropathy, Vol 1. WB Saunders, Philadelphia, 1984, pp 981–1015.
17. Cavanagh, JB: Peripheral nerve changes in ortho-cresyl phosphate poisoning in the cat. J Pathol Bacteriol 87:365–383, 1964.
18. Cavanagh, JB, and Mellick, RS: On the nature of the peripheral nerve lesions associated with acute intermittent porphyria. J Neurol Neurosurg Psychiatry 28:320–327, 1965.
19. Denny-Brown, D: The neurological aspects of thiamine deficiency. Proc Fed Am Soc Exp Biol (Suppl 2) 17:35–39, 1958.
20. Denny-Brown, D, and Brenner, C: Paralysis of nerve induced by direct pressure and by tourniquet. Arch Neurol Psychiatry 51:1–26, 1944.
21. Denny-Brown, D, and Brenner, C: Lesion in peripheral nerve resulting from compression by spring clip. Arch Neurol Psychiatry 52:1–19, 1944.
22. Donoso, RS, Ballantyne, JP, and Hansen, S: Regeneration of sutured human peripheral nerves: An electrophysiological study. J Neurol Neurosurg Psychiatry 42:97–106, 1979.
23. Dunn, HG, Lake, BD, Dolman, CL, and Wilson, J: The neuropathy of Krabbe's infantile cerebral sclerosis (Globoid cell leukodystrophy). Brain 92:329–344, 1969.
24. Dyck, PJ: Inherited neuronal degeneration and atrophy affecting peripheral motor, sensory, and autonomic neurons. In Dyck, PJ, Thomas, PK, Lambert, EH, and Bunge, R (eds): Peripheral Neuropathy, Vol 2. WB Saunders, Philadelphia, 1984, pp 1600–1641.
25. Dyck, PJ, Gutrecht, JA, Bastron, JA, Karnes, WE, and Dale, AJD: Histologic and teased fiber measurements of sural nerve in dis-

orders of lower motor and primary sensory neurons. Mayo Clin Proc 43:81–123, 1968.

26. Dyck, PJ, Johnson, WJ, Lambert, EH, and O'Brien, PC: Segmental demyelination secondary to axonal degeneration in uremic neuropathy. Mayo Clin Proc 46:400–431, 1971.

27. Dyck, PJ, Lais, AC, Karnes, JL, Sparks, M, Hunder, H, Low, PA, and Windebank, AJ: Permanent axotomy: A model of axonal atrophy and secondary segmental demyelination and remyelination. Ann Neurol 9:575–583, 1981.

28. Dyck, PJ, and Lambert, EH: Compound nerve action potentials and morphometry. Electroencephalogr Clin Neurophysiol 36:573–574, 1974.

29. Dyck, PJ, Thomas, PK, Lambert, EH, and Bunge, R (eds): Peripheral Neuropathy, Vols I and II. WB Saunders, Philadelphia, 1984.

30. Erlanger, J: The interpretation of the action potential in cutaneous and muscle nerves. Am J Physiol 82:644–655, 1927.

31. Erlanger, J, and Gasser, HS: Electrical Signs of Nervous Activity. University of Pennsylvania Press, Philadelphia, 1937.

32. Feasby, TE, Brown, WF, Gilbert, JJ, and Hahn, AFD: The pathological basis of conduction block in human neuropathies. J Neurol Neurosurg Psychiatry 48:239–244, 1985.

33. Feldman, RG, Haddow, J, and Chisolm, JJ: Chronic lead intoxication in urban children. In Desmedt, JE (ed): New Developments in Electromyography and Clinical Neurophysiology, Vol 2. Karger, Basel, 1973, pp 313–317.

34. Friedli, WG, and Meyer, M: Strength-duration curve: A measure for assessing sensory deficit in peripheral neuropathy. J Neurol Neurosurg Psychiatry 47:184–189, 1984.

35. Fullerton, PM: Peripheral nerve conduction in metachromatic leucodystrophy (sulphatide lipidosis). J Neurol Neurosurg Psychiatry 27:100–105, 1964.

36. Fullerton, PM, and Barnes, JM: Peripheral neuropathy in rats produced by acrylamide. Br J Industr Med 23:210–221, 1966.

37. Fullerton, PM, and Gilliatt, RW: Median and ulnar neuropathy in the guinea-pig. J Neurol Neurosurg Psychiatry 30:393–402, 1967.

38. Gamble, HJ, and Eames, RA: An electron microscope study of the connective tissues of human peripheral nerve. J Anat 98:655–663, 1964.

39. Gasser, HS: Effect of the method of leading on the recording of the nerve fiber spectrum. J Gen Physiol 43:927–940, 1960.

40. Gilliatt, RW: Recent advances in the pathophysiology of nerve conduction. In Desmedt, JE (ed): New Developments in Electromyography and Clinical Neurophysiology, Vol 2. Karger, Basel, 1973, pp 2–18.

41. Gilliatt, RW: Peripheral nerve compression and entrapment. In Lant, AF (ed): Eleventh Symposium on Advanced Medicine. Pitman Medical, Kent, England, 1975, pp 144–163.

42. Gilliatt, RW: Electrophysiology of peripheral neuropathies. Muscle Nerve (Suppl 5):S108–S116, 1982.

43. Gilliatt, RW, and Hjorth, RJ: Nerve conduction during Wallerian degeneration in the baboon. J Neurol Neurosurg Psychiatry 35:335–341, 1972.

44. Gilliatt, RW, and Taylor, JC: Electrical changes following section of the facial nerve. Proc R Soc Med 52:1080–1083, 1959.

45. Grafstein, B, and Forman, DS: Intracellular transport in neurons. Physiol Rev 60:1167–1283, 1980.

46. Haller, FR, and Low, FN: The fine structure of the peripheral nerve root sheath in the subarachnoid space in the rat and other laboratory animals. Am J Anat 131:1–20, 1971.

47. Harris, JB, and Thesleff, S: Nerve stump length and membrane changes in denervated skeletal muscle. Nature (New Biol) 236:60–61, 1972.

48. Henriksen, JD: Conduction velocity of motor nerves in normal subjects and patients with neuromuscular disorders. Thesis, University of Minnesota, Minneapolis, 1966.

49. Hodes, R: Selective destruction of large motoneurons by poliomyelitis virus I. Conduction velocity of motor nerve fibers of chronic poliomyelitis patients. J Neurophysiol 12:257–266, 1949.

50. Hodes, R, Larrabee, MG, and German, W: The human electromyogram in response to nerve stimulation and the conduction velocity of motor axons: Studies on normal and injured peripheral nerves. Arch Neurol Psychiatry 60:340–365, 1948.

51. Hopkins, AP, and Gilliatt, RW: Motor and sensory nerve conduction velocity in the baboon: Normal values and changes during acrylamide neuropathy. J Neurol Neurosurg Psychiatry 34:415–426, 1971.

52. Hubbard, JI: Neuromuscular transmission—Presynaptic factors. In Hubbard, JI (ed): The Peripheral Nervous System. Plenum Press, New York, 1974, pp 151–180.

53. Jacobs, JM, and Love, S: Qualitative and quantitative morphology of human sural nerve at different ages. Brain 108:897–924, 1986.

54. Johnson, EW, and Olson, KJ: Clinical value of motor nerve conduction velocity determination. JAMA 172:2030–2035, 1960.

55. Johnson, PC, and Asbury, AK: The pathology of peripheral nerve. Muscle and Nerve 3:519–528, 1980.

56. Kaeser, HE: Funktionsprufungen peripherer Nerven bei experimentellen Polyneuritiden und bei der wallerschen Degeneration. Deutsche Z Nervenheilk 183:268–304, 1962.

57. Kaeser, HE: Zur Diagnose des Karpaltunnel-syndroms. Praxis 40:991–995, 1962.

58. Kaeser, HE: Diagnostische Probleme beim Karpaltunnelsyndrom. Deutsche Z Nervenheilk 185:453–470, 1963.

59. Kaeser, HE, and Lambert, EH: Nerve function studies in experimental polyneuritis. Electroencephalogr Clin Neurophysiol Suppl 22:29–35, 1962.

60. Kelly, JJ: The electrodiagnostic findings in pe-

ripheral neuropathy associated with monoclonal gammopathy. Muscle Nerve 6:504–509, 1983.

61. Kimura, J: F-wave velocity in the central segment of the median and ulnar nerves: A study in normal subjects and in patients with Charcot-Marie-Tooth disease. Neurology 24:539–546, 1974.

62. Kimura, J: The carpal tunnel syndrome: Localization of conduction abnormalities within the distal segment of the median nerve. Brain 102:619–635, 1979.

63. Kimura, J: Principles and pitfalls of nerve conduction studies. Ann Neurol 16:415–429, 1984.

64. Kimura, J, Rodnitzky, R, and Okawara, S: Electrophysiologic analysis of aberrant regeneration after facial nerve paralysis. Neurology 25:989–993, 1975.

65. Koles, ZJ, and Rasminsky, M: A computer simulation of conduction in demyelinated nerve fibres. J Physiol (Lond) 227:351–364, 1972.

66. Lambert, EH: Neurophysiological techniques useful in the study of neuromuscular disorders. In Adams, RD, Eaton, LM, and Shy, GM (eds): Neuromuscular Disorders, Vol 38. Williams & Wilkins, Baltimore, 1960, pp 247–273.

67. Lambert, EH: Pathophysiology of focal nerve lesions. American Academy of Neurology (Special Course 15), Clinical Electromyography, 1977.

68. Lambert, EH, and Dyck, PJ: Compound action potentials of sural nerve in vitro in peripheral neuropathy. In Dyck, PJ, Thomas, PK, Lambert, EH, and Bunge, R (eds); Peripheral Neuropathy, Vol 1. WB Saunders, Philadelphia, 1984, pp 1030–1044.

69. Landon, DN (ed): The Peripheral Nerve. Chapman and Hall, London, 1976.

70. Lasek, R: Axoplasmic transport in cat dorsal root ganglion cells: As studied with (3H)-L-leucine. Brain Res 7:360–377, 1968.

71. Lewis, T, Pickering, GW, and Rothschild, P: Centripetal paralysis arising out of arrested bloodflow to the limb, including notes on a form of tingling. Heart 16:1–32, 1931.

72. McDonald, WI: Conduction in muscle afferent fibres during experimental demyelination in cat nerve. Acta Neuropathol 1:425–432, 1962.

73. McDonald, WI: The effects of experimental demyelination on conduction in peripheral nerve: A histological and electrophysiological study. 1. Clinical and histological observations. Brain 86:481–500, 1963.

74. McDonald, WI: The effects of experimental demyelination on conduction in peripheral nerve: A histological and electrophysiological study. II. Electrophysiological observations. Brain 86:501–524, 1963.

75. McLeod, JG, Prineas, JW, and Walsh, JC: The relationship of conduction velocity to pathology in peripheral nerves: A study of sural nerve in 90 patients. In Desmedt, JE (ed): New Developments in Electromyography and Clinical Neurophysiology, Vol 2. Karger, Basel, 1973, pp 248–258.

76. Miller, RG: Acute vs chronic compressive neuropathy. Muscle Nerve 7:427–430, 1984.

77. Miller, RG: Injury to peripheral motor nerves. AAEE Minimonograph 28. Muscle Nerve 10:698–710, 1987.

78. Milner, TE, and Stein, RV: The effects of axotomy on the conduction of action potentials in peripheral, sensory and motor nerve fibers. J Neurol Neurosurg Psychiatry 44:495–496, 1981.

79. Neary, D, Ochoa, J, and Gilliatt, RW: Subclinical entrapment neuropathy in man. J Neurol Sci 24:283–298, 1975.

80. Noback, CR: The Human Nervous System. McGraw-Hill, New York, 1967.

81. Ochoa, J, Danta, G, Fowler, TJ, and Gilliatt, RW: Nature of the nerve lesion caused by a pneumatic tourniquet. Nature 233:265–266, 1971.

82. Ochoa, J, Fowler, TJ, and Gilliatt, RW: Anatomical changes in peripheral nerves compressed by a pneumatic tourniquet. J Anat 113:433–455, 1972.

83. Ochoa, J, Fowler, TJ, and Gilliatt, RW: Changes produced by a pneumatic tourniquet. In Desmedt, JE (ed): New Developments in Electromyography and Clinical Neurophysiology, Vol 2. Karger, Basel, 1973, pp 174–180.

84. Ochoa, J, and Marotte, L: The nature of the nerve lesion caused by chronic entrapment in the guinea-pig. J Neurol Sci 19:491–495, 1973.

85. Ochoa, JL, and Torebjork, HE: Paraesthesiae from ectopic impulse generation in human sensory nerves. Brain 103:835–853, 1980.

86. Parkin, PJ, and Le Quesne, PM: Effect of a synthetic pyrethroid deltamethrin on excitability changes following a nerve impulse. J Neurol Neurosurg Psychiatry 45:337–342, 1982.

87. Parry, GJ, and Brown, MJ: Selective fiber vulnerability in acute ischemic neuropathy. Ann Neurol 11:147–154, 1982.

88. Parry, GJ, Cornblath, DR, and Brown, MJ: Transient conduction block following acute peripheral nerve ischemia. Muscle Nerve 8:490–413, 1985.

89. Pencek, TL, Schauf, CL, Low, PA, Eisenberg, BR, and Davis, FA: Disruption of the perineurium in amphibian peripheral nerve: Morphology and physiology. Neurology 30:593–599, 1980.

90. Pilling, JB: Nerve conduction during wallerian degeneration in man (corres). Muscle Nerve 1:81, 1978.

91. Prineas, J: The pathogenesis of dying-back polyneuropathies. II. An ultrastructural study of experimental acrylamide intoxication in the cat. J Neuropathol Exp Neurol 28:598–621, 1969.

92. Rasminsky, M: Ectopic generation of impulses and cross-talk in spinal nerve roots of "dystrophic" mice. Ann Neurol 3:351–357, 1978.

93. Rasminsky, M: Physiological consequences of demyelination. In Spencer, PS, and Schaum-

burg, HH (eds): Experimental and Clinical Neurotoxicology. Williams & Wilkins, Baltimore, 1980, pp 257–271.

94. Rasminsky, M, and Sears, TA: Saltatory conduction in demyelinated nerve fibres. In Desmedt, JE (ed): New Developments in Electromyography and Clinical Neurophysiology, Vol 2. Karger, Basel, 1973, pp 158–165.

95. Ridderheim, PA, Von Essen, O, Blom, S, and Zetterlund, B: Intracranially recorded compound action potentials from the human trigeminal nerve. Electroencephalogr Clin Neurophysiol 61:138–140, 1985.

96. Ross, MH, and Reith, EJ: Perineurium: Evidence for contractile elements. Science 165:604–606, 1969.

97. Rudge, P: Tourniquet paralysis with prolonged conduction block: An electrophysiological study. J Bone Joint Surg 56B:716–720, 1974.

98. Rudge, P, Ochoa, J, and Gilliatt, RW: Acute peripheral nerve compression in the baboon. J Neurol Sci 23:403–420, 1974.

99. Said, G, Saida, K, Saida, T, and Asbury, AK: Axonal lesions in acute experimental demyelination: A sequential teased nerve fiber study. Neurology 31:413–421, 1981.

100. Saida, K, Sumner, AJ, Saida, T, Brown, MJ, and Silberberg, DH: Antiserum-mediated demyelination: Relationship between remyelination and functional recovery. Ann Neurol 8:12–24, 1980.

101. Schaumburg, HH, and Spencer, PS: Human toxic neuropathy due to industrial agents. In Dyck, PJ, Thomas, PK, Lambert, EH, and Bunge, R (eds): Peripheral Neuropathy, Vol II. WB Saunders, Philadelphia, 1984, pp 2115–2132.

102. Schaumburg, HH, Wisniewski, HM, and Spencer, PS: Ultrastructural studies of the dying-back process. I. Peripheral nerve terminal and axon degeneration in systemic acrylamide intoxication. J Neuropathol Exp Neurol 33:260–284, 1974.

103. Sedal, L, Ghabriel, MN, He, F, Allt, G, Le Quesne, PM, and Harrison, MJG: A combined morphological and electrophysiological study of conduction block in peripheral nerve. J Neurol Sci 60:293–306, 1983.

104. Seddon, HJ: Three types of nerve injury. Brain 66:237–288, 1943.

105. Seddon, H: Surgical Disorders of the Peripheral Nerves, ed 2. Churchill Livingstone, Edinburgh, 1975.

106. Sima, AAF, and Brismar, T: Reversible diabetic nerve dysfunction: Structural correlates to electrophysiological abnormalities. Ann Neurol 18:21–29, 1985.

107. Simpson, JA: Conduction velocity of peripheral nerves in human metabolic disorders. Electroencephalogr Clin Neurophysiol Suppl 22:36–43, 1962.

108. Smith, KJ, Blakemore, WF, and McDonald, WI: The restoration of conduction by central remyelination. Brain 104:383–404, 1981.

109. Smith, KJ, Blakemore, WF, Murray, JA, and Patterson, RC: Internodal myelin volume and axon surface area: A relationship determining myelin thickness? J Neurol Sci 55:231–246, 1982.

110. Smith, KJ, and McDonald, WI: Spontaneous and evoked electrical discharges from a central demyelinating lesion. J Neurol Sci 55:39–47, 1982.

111. Spencer, PS, and Schaumburg, HH: Central-peripheral distal axonopathy: The pathology of dying-back polyneuropathies. In Zimmerman, HM (ed): Progress in Neuropathology, Vol III. Grune & Stratton, New York, 1976, pp 253–295.

112. Stanley, EF, Griffin, JW, and Fahnestock, KE: Effects of IDPN-induced axonal swellings on conduction in motor nerve fibers. J Neurol Sci 69:183–200, 1985.

113. Stohr, M, Gilliatt, RW, and Willison, RG: Supernormal excitability of human sensory fibers after ischemia. Muscle Nerve 4:73–75, 1981.

114. Struppler, A, and Huckauf, H: Propagation velocity in regenerating motor nerve fibres. Electroencephalogr Clin Neurophysiol Suppl 22:58–60, 1962.

115. Sumner, A: Physiology of dying-back neuropathies. In Waxman, SG (ed): Physiology and Pathobiology of Axons. Raven Press, New York, 1978, pp 349–359.

116. Sumner, AJ, and Asbury, AK: Physiological studies of the dying-back phenomenon: Muscle stretch afferents in acrylamide neuropathy. Brain 98:91–100, 1975.

117. Sumner, AJ, Saida, K, Saida, T, Silberberg, DH, and Asbury, AK: Acute conduction block associated with experimental antiserum-mediated demyelination of peripheral nerve. Ann Neurol 11:469–477, 1982.

118. Sunderland, S: Nerves and Nerve Injuries, ed 2. Churchill Livingstone, Edinburgh, 1978.

119. Swift, TR, Leshner, RT, and Gross, JA: Arm-diaphragm synkinesis: Electrodiagnostic studies of aberrant regeneration of phrenic motor neurons. Neurology 30:339–344, 1980.

120. Takahashi M, and Le Quesne, PM: The effects of the pyrethroids deltamethrin and cismethrin on nerve excitability in rats. J Neurol Neurosurg Psychiatry 45:1005–1011, 1982.

121. Tallis, R, Staniforth, P, and Fisher, TR: Neurophysiological studies of autogenous sural nerve grafts. J Neurol Neurosurg Psychiatry 41:677–683, 1978.

122. Thomas, CK, Stein, RB, Gordon, T, Lee RG, and Elleker, MG: Patterns of reinnervation and motor unit recruitment in human hand muscles after complete ulnar and median nerve section and resuture. J Neurol Neurosurg Psychiatry 50:259–268, 1987.

123. Thomas, PK: Motor nerve conduction in the carpal tunnel syndrome. Neurology 10:1045–1050, 1960.

124. Thomas, PK: The morpholigical basis for alterations in nerve conduction in peripheral neuropathy. Proc R Soc Med (Lond) 64:295–298, 1971.

125. Thomas, PK, and Olsson, Y: Microscopic anatomy and function of the connective tissue components of peripheral nerve. In Dyck, PJ,

Thomas, PK, Lambert, EH, and Bunge, R (eds): Peripheral Neuropathy, Vol 1. WB Saunders, Philadelphia, 1984, pp 97–120.

126. Trojaborg, W: Early electrophysiologic changes in conduction block. Muscle Nerve 1:400–403, 1978.

127. Tsujihata, M, Engel, AG, and Lambert, EH: Motor end-plate fine structure in acrylamide dying-back neuropathy: A sequential morphometric study. Neurology 24:849–856, 1974.

128. Vizoso, AD, and Young, JZ: Internode length and fibre diameter in developing and regenerating nerves. J Anat (Lond) 82:110–134, 1948.

129. Wall, PD, and Gutnick, M: Properties of afferent nerve impulses originating from a neuroma. Nature 248:740–743, 1974.

130. Waxman, SG: Conduction in myelinated, unmyelinated, and demyelinated fibers. Arch Neurol 34:585–589, 1977.

131. Waxman, SG: Determinants of conduction velocity in myelinated nerve fibers. Muscle Nerve 3:141–150, 1980.

132. Waxman, SG, and Brill, MH: Conduction through demyelinated plaques in multiple sclerosis: Computer simulations of facilitation by short internodes. J Neurol Neurosurg Psychiatry 41:408–416, 1978.

133. Waxman, SG, Brill, MH, and Geschwind, N: Probability of conduction deficit as related to fiber length in random distribution models of peripheral neuropathies. J Neurol Sci 29:39–53, 1976.

134. Webster, H: Peripheral nerve structure. In Hubbard, JI (ed): The Peripheral Nervous System. Plenum Press, New York, 1974, pp 3–26.

135. Young, JZ: Growth and differentiation of nerve fibres. In Symposia of the Society for Experimental Biology. II. Growth in Relation to Differentiation and Morphogenesis. University Press, Cambridge, 1948, pp 57–74.

Chapter **5**

PRINCIPLES OF NERVE CONDUCTION STUDIES

1 INTRODUCTION

Helmholtz[56] originally recorded the mechanical response of a muscle to measure conduction velocity of motor fibers (see Appendix 1). Piper[97] was the first to use the muscle action potential for this pur-pose, but it was the animal experiment of Berry, Grundfest, and Hinsey[9] and the human study of Hodes, Larrabee, and German[60] that popularized the technique as a clinical test. Eichler[37] successfully recorded nerve potentials percutaneously from mixed nerves in man. Dawson and Scott[29] developed the reliable technique of determining mixed nerve conduction using better resolution, initially by photographic superimposition and later by electrical averaging. Stimulating the digital nerves using ring electrodes, Dawson[28] also recorded pure sensory nerve action potentials through surface electrodes placed over the nerve trunk.

With steady improvement of recording apparatus, nerve conduction studies have become a simple and reliable test of peripheral nerve function. With adequate standardization, the method now provides a means of not only objectifying the lesion but also precisely localizing the site of maximal involvement.[65,72] With this technique, electrical stimulation of the nerve initiates an impulse that travels along motor, sensory, or mixed nerves. The assessment of conduction characteristics depends on the analysis of compound evoked potentials recorded from the muscle in the study of the motor fibers and from the nerve itself in the case of the sensory fibers. The same principles apply in all circumstances, although the anatomic course and pattern of innervation dictates the exact technique used for testing a given nerve. In addition to electrical shocks, used in most clinical studies, tactile stimulation can also elicit nerve action potentials.[7,11,90,100] Assessment of mechanical characteristics may also help delineate contractile properties of the isometric twitch induced by stimulation of the nerve.[87]

2 ELECTRICAL STIMULATION OF THE NERVE

Cathode and Anode

One may use either surface or needle electrodes to stimulate the nerve. Surface electrodes, usually made of silver plate, come in different sizes, commonly in the range of 0.5 to 1.0 cm in diameter. Stimulating electrodes consist of a cathode (negative pole) and an anode (positive pole), so called because they attract cations and anions, respectively. As the current flows between them, negative charges that accumulate under the cath-

ode depolarize the nerve. Conversely, positive charges under the anode hyperpolarize the nerve. In bipolar stimulation, with both electrodes over the nerve trunk, placing the cathode closer to the recording site avoids anodal conduction block of the propagated impulse. Alternatively, locating the anode away from the nerve trunk also prevents its hyperpolarizing effect. Accurate calculation of conduction velocities depends on proper measurements of the distance between the consecutive cathodal points used to stimulate the nerve at multiple sites. One must clearly label the stimulating electrodes to avoid inadvertent surface measurement from the cathode at one stimulus site to the anode at another, which would lead to an erroneous measurement of conduction velocity.

Types of Stimulators

Most commercially available stimulators provide a probe that mounts the cathode and the anode at a fixed distance, usually 2 to 3 cm apart. The intensity control located in the insulated handle, though bulky, simplifies the operation for a single examiner. The ordinary banana plugs connected by shielded cable also serve well as stimulating electrodes. Some electromyographers prefer a monopolar stimulation with a small cathode placed on the nerve trunk and a large anode some distance away in the same extremity. The conduction velocities obtained in this fashion differ slightly but randomly from those determined by bipolar arrangements.[58] Stimulation by a needle electrode inserted subcutaneously close to the nerve requires much less current than surface stimulation to elicit the same response. The anode may be a surface electrode located on the skin nearby or a second needle electrode inserted in the vicinity of the cathode.

Electromyographers use two basically different kinds of electric stimulators in nerve conduction studies (see Chapter 3.7). Of these, constant-voltage stimulators regulate the output in voltage so that the actual current varies inversely with the impedance of the electrode, skin, and subcutaneous tissues. In constant-current units, the voltage changes according to the impedance, so that a specified amount of current reaches the nerve within certain limits of the skin's resistance. Either type suffices for clinical use, provided that the stimulus output has an adequate range to elicit maximal muscle and nerve action potentials in all patients. A constant-current unit provides a better means of serially assessing the level of shock intensity as a measure of nerve excitability.

Stimulus Intensity and Duration

The output impulse provides a square wave of variable duration, ranging from 0.05 to 1.0 ms. Surface stimulation of 0.1 ms duration and 100 to 300 V or 5 to 40 mA intensity usually activate a healthy nerve fully. A study of diseased nerves with decreased excitability may require a maximal output of 400 to 500 V or 60 to 75 mA. Electrical stimulation within the above intensity range causes no particular risk to an ordinary patient. One possible exception is that any current, if delivered near the implantation site, could inhibit a cardiac pacemaker.[101] Special care to safeguard the patient includes proper grounding and placement of the stimulator with sufficient distance from the pacemaker.[1] In patients with indwelling cardiac catheters or central venous pressure lines inserted directly into the heart, all the current may directly reach the cardiac tissue. This possibility makes routine nerve conduction studies contraindicated in these electrically sensitive patients. Electromyographers should always keep in mind this and other problems related to general electrical safety (see Appendix 3).

It is common to qualify electrical stimuli on the basis of the magnitude of the evoked potential. A threshold stimulus barely elicits a response in some, but not all, of the axons contained in the nerve. A maximal stimulus activates the entire group of axons, so that further increase in shock intensity causes no additional increase in the amplitude of the evoked potential. The current required for maximal stimulation varies greatly from one subject to the next and from one nerve to another in the same individual. A supramaximal stimulus has an intensity greater than the maximal stimulus.

If fibers with large diameters have the lowest threshold in man, as in experimental animals,[39,107] then a submaximal stimulus should theoretically suffice for determining the onset latency of the fastest conducting fibers. Although this assumption usually holds, especially with sensory nerves,[99] the exact order of activation also depends on the spatial relationship of various fibers and the stimulating electrode.[43] Further, the length of the axon terminals, which partially determines the latency, varies within a given nerve. Thus, with submaximal stimuli, the onset latency fluctuates considerably from one trial to the next, depending on the excited axons within a nerve. In extreme cases the first axons excited may in fact have the longest latencies.[64] The use of supramaximal stimuli, which activate all of the axons, circumvent this uncertainty.

Most commercial stimulators can provide a pair of stimuli at variable intervals and a train of stimuli of different rates and duration. Ideally, each paired stimulus should have independent controls as to its duration and intensity. A trigger output for the oscilloscope sweep should precede the stimulus by a variable delay, to allow a clear marking of the exact stimulus point on the display.[104]

Stimulus Artifact

The control of a stimulus artifact often poses a major technical challenge in nerve conduction studies. Most electrode amplifiers recover from an overloading input in 5 to 10 ms, depending on the amplifier design and the amount of overload. With the stimulus of sufficient magnitude, an overloading artifact interferes with accurate recording of short-latency responses. Better stimulus isolation from the ground through an isolating transformer serves to reduce excessive shock artifact.[19] Not only does this eliminate amplifier overloading, but it also protects the patient from unexpected current leakage. The use of the transformer, however, makes it difficult to faithfully preserve the waveform of the original stimulus. A radio-frequency isolation also minimizes stimulus artifacts while maintaining the original shape of the stimulus better than the transformer. Unfortunately, high-frequency stimulus isolation units generally fail to deliver adequate intensity for supramaximal stimulation. Finally, the use of a fast-recovery amplifier circumvents the problem of stimulus artifacts.[113] Even then, for optimal recording of short-latency responses, it is important to reduce surface spread of stimulus current, as stated below.

Shock artifacts increase with less separation between stimulus and recording sites and greater distance between the active (G_1) and reference (G_2) electrodes. The stimulator leads, which cannot be shielded, can also cause a large artifact if placed near the recording electrodes. With excessive surface spread, a square pulse of 0.1 ms duration can affect the active electrode for several milliseconds at the signal level of recording with high sensitivity. Thus, reduction in surface spread of stimulus current ensures an optimal recording of short-latency responses. Wiping with alcohol helps dry the moist skin surface with perspiration before the application of the stimulus. Adequate preparation of the stimulating and recording sites reduces the skin resistance. Surface grease will dissolve if cleaned with ether. Callous skin needs gentle abrasion with a dull knife or fine sandpaper. Rubbing the skin with a cream or solvent of high conductance lowers the impedance between the electrode and the underlying tissue. A ground electrode, if placed between the stimulating electrode and the recording electrode, diminishes the stimulus artifact. An alternative location may also suffice, especially with the use of a fast-recovery amplifier.[113]

3 RECORDING OF MUSCLE AND NERVE POTENTIALS

Surface and Needle Electrodes

Surface electrodes, in general, are better than needle electrodes for recording a compound muscle action potential in assessing contributions from all discharging units. Its onset latency indicates the conduction time of the fastest fibers, whereas its amplitude is approximately proportional to the number of available axons. Averaging technique, though not usually required, may help in evaluating markedly atrophic muscles.[6] A needle electrode registers only a small portion of the muscle action potential. With more abrupt onset and less interference from neighboring discharges, the use of a needle electrode improves the recording from small atrophic muscles. It also helps record an action potential from a proximal muscle not excitable in isolation, even after simultaneous stimulation of more than one nerve.

Surface electrodes suffice for recording sensory and mixed nerve action potentials. Some electromyographers, however, prefer needle electrodes placed perpendicular to the nerve to improve the resolution. With this technique, the amplitude of the recorded potential increases, and the noise from the electrode tissue surface decreases by a factor of two to three times.[99] The combination of the two effects enhances the signal-to-noise ratio by about five times and, when averaging, reduces the time required to reach the same resolution considerably. Many laboratories now use ring electrodes to record the antidromic sensory potentials from digital nerves commonly over the proximal and distal interphalangeal joints. Studies in the upper extremities usually require no averaging because individual stimuli give rise to sensory potentials of sufficient amplitude.

Optimal Recording of Intended Signals

The principles of amplification and display used in electromyography also apply to nerve conduction studies (see Chapter 3.3). Instead of continuous runs, a prepulse intermittently triggers the sweep followed, after a short delay, by the stimulus. This arrangement allows precise measurement of the time interval between the stimulus and the onset of the evoked potential. The magnitude of the potential under study dictates the optimal amplifier sensitivity for determination of the amplitude and the latency.

Overamplification results in truncation of the recorded response, whereas underamplification precludes accurate measurements of its takeoff from the baseline.

A 1.0-mV muscle action potential, if amplified 1000 times, causes a 1-cm vertical deflection on the oscilloscope at a display setting of 1 V/cm. A much smaller sensory or mixed nerve action potential, on the order of 10 μV, requires a total amplification of about 100,000 times. With such a high gain, the amplifier must have a very low inherent noise level. The use of low-pass filters helps to further reduce such high-frequency interference. The electrode amplifier should provide differential amplification with a signal-to-noise discrimination ratio close to 100,000:1 and an input impedance greater than 1 MΩ. It should respond to frequencies of wide bandwidth ranging from 2 Hz to 10 kHz without undue distortion.

Averaging Technique

Conventional techniques fail to detect signals within the expected noise level of the system. Interposing a step-up transformer between the recording electrodes and the amplifier improves the signal-to-noise ratio,[12] as does placing the first stage of the amplifier near the electrode site with a remote preamplifier box.[113] The use of digital averaging represents a major improvement over the photographic superimposition[29] and early averager of Dawson,[27] with its motor-driven switch and multiple storage capacitors. The electronic devices now in use sum consecutive samples of waveforms stored digitally after each sweep triggered by repetitive stimulation.

The voltage from noise that randomly changes its temporal relationship to stimulation in successive tracings will average close to zero at each point in time after stimulus onset. In contrast, signals time-locked to the stimulus will sum at a constant latency and appear as an evoked potential, distinct from the background noise. Electrical division of the summated potential by the number of

trials will provide an average value of the signal under consideration. Here, the degree of enhancement increases in proportion to the square root of the trial number. For example, four trials give twice as large a response, whereas nine trials give three times the response. In other words, the signal-to-noise ratio improves by a factor of the square root of 2 every time the number of trials is doubled.

Display and Storage of Recorded Signals

The evoked response, once displayed on an oscilloscope, may be photographed with the use of Polaroid film for latency determination. More conveniently, the use of an oscilloscope with a storage tube allows storing a series of responses with a stepwise vertical shift of the baseline, to facilitate the comparison of successively elicited potentials in waveform and latency. For accurate latency determination, an automatic device digitally displays the value after manual positioning of the marker to the desired spot of the waveform. Modern oscilloscopes provide a very stable time base requiring no marking of calibration signals on the second beam. Consequently, a single channel suffices for most routine nerve conduction studies. Dual channels, however, have a distinct advantage in simultaneous recording of related events. Oscilloscopes with four or more channels are also available for multichannel analysis.

For a permanent record, one may use a 35-mm or Polaroid camera to photograph the display of the storage scope. Alternatively, a magnetic tape recorder can store the original potentials using either frequency (FM) or amplitude modulation (AM). The FM mode has a limited high-frequency response, but can adequately record the frequency range of the compound action potential, including DC changes. Further, in the analysis of evoked muscle or nerve potentials, the FM method preserves the amplitude of the recorded potential very accurately. In contrast, the AM modulation responds well to high-frequency bands but distorts the amplitude of the recorded response.

The AM method preserves the high-frequency components better for recording motor unit potentials with needle electrodes (see Chapter 3.5).

4 MOTOR NERVE CONDUCTION

Stimulation and Recording

For motor conduction studies, the nerve is stimulated at two or more points along its course, with the anode 2 to 3 cm proximal to the cathode. Depolarization under the cathode results in the generation of a nerve action potential, whereas hyperpolarization under the anode tends to block the propagation of the nerve impulse. The pulses of moderate intensity are used to adjust the position of the cathode until further relocation causes no change in the size of the muscle action potential. With the cathode at the best stimulating site, one then defines the maximal intensity that just elicits a maximal potential. Increasing the stimulus further should result in no change in the size of the muscle potential. The use of a 20 to 30 percent supramaximal intensity

guarantees the activation of all the nerve axons innervating the recorded muscle.

Recording action potentials (Fig. 5–1) requires a pair of surface electrodes: an active lead (G_1) placed on the belly of the muscle and an indifferent lead (G_2) placed on the tendon (belly-tendon recording). With this arrangement, the propagating muscle action potential, originating under G_1, located near the motor point, gives rise to a simple biphasic waveform with initial negativity (see Chapter 2.4). A small positive potential may precede the negative peak with inappropriate positioning of the recording electrodes. The usual measurements include amplitude, from the baseline to the negative peak or between negative and positive peaks; duration, from the onset to the negative or positive peak or to the final return to the baseline; and latency, from the stimulus artifact to the onset of the negative response. Electronic integration can provide the area under the waveform, which shows linear correlation to the product of the amplitude and duration measured by conventional means.[42] Latency consists of two components: (1) nerve conduction time, from the stimulus point to the nerve terminal, and (2) neuromuscular trans-

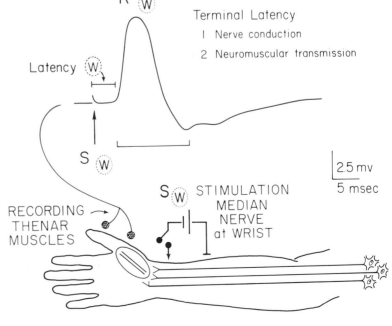

Figure 5–1. Compound muscle action potential recorded from the thenar eminence after stimulation of the median nerve at the wrist. The distal or terminal latency includes (1) nerve conduction from the stimulus point to the axon terminal and (2) neuromuscular transmission, including the time required for generation of the muscle action potential after depolarization of the end plate.

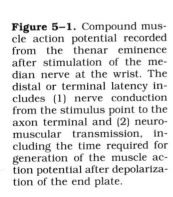

mission time, from the axonal terminal to the motor end plate, including the time required for generation of muscle action potential. Onset latency is a measure of the fastest conducting motor fibers.

Calculation of Conduction Velocity

To measure the motor nerve conduction time, one must eliminate the time for neuromuscular transmission and generation of muscle action potential. The latency difference between the two responses elicited by stimulation at two separate points, in effect, excludes the two components common to both stimuli. Thus, it represents the time necessary for the nerve impulse to travel between the two stimulus points (Fig. 5–2). The conduction velocity is derived as the ratio between the distance from one point of stimulation to the next and the corresponding latency difference. The reliability of results depends on accuracy in determining the length of the nerve segment, estimated with the surface distance along the course of the nerve.[89]

To recapitulate, the nerve conduction velocity equals

$$\frac{D \text{ mm}}{L_p - L_d \text{ ms}} = \frac{D}{L_p - L_d} \text{ m/s},$$

where D is the distance between the two stimulus points in millimeters, and L_p and L_d, the proximal and distal latencies in milliseconds. Stimulation at multiple points along the length of the nerve allows calculation of segmental conduction velocities. Separation of the two stimulation points by at least several, and preferably more than 10, centimeters improves the accuracy of surface measurement and, consequently, determination of conduction velocity. In the case of restricted lesions, as in a compressive neuropathy, however, the inclusion of longer unaffected segments dilutes the effect of slowing and lowers the sensitivity of the test. Here, incremental stimulation across the shorter segment helps isolate the localized abnormality that may otherwise escape detection (see Chapter 6.2).

The latency from the most distal stimulus point to the muscle includes not only the nerve conduction time but also neuro-

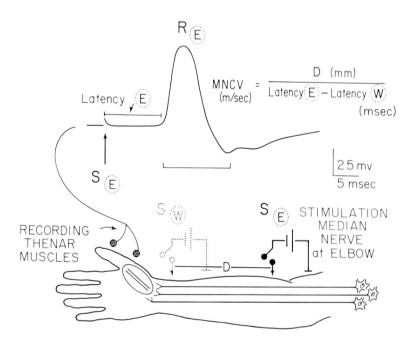

Figure 5–2. Compound muscle action potential recorded from the thenar eminence after stimulation of the median nerve at the elbow. The nerve conduction time from the elbow to the wrist equals the latency difference between the two responses elicited by the distal and proximal stimulation. The motor nerve conduction velocity (MNCV), calculated by dividing the surface distance between the stimulus points by the subtracted times, concerns the fastest fibers.

muscular transmission time. The inclusion of the additional factors precludes calculation of conduction velocity over the most distal segment. Here, meaningful comparison requires the use of either premeasured distance or anatomic landmarks for electrode placement.[91] Both approaches equally improve the accuracy of latency determination.[93] The actual conduction time in the terminal segment (L_d) slightly exceeds the value calculated for the same distance based on the conduction velocity of more proximal segment (L_d'). The difference ($L_d' - L_d$), known as the residual latency, provides a measure of the conduction delay at the nerve terminal and at the neuromuscular junction.[60,66,67,78] The ratio between the calculated and measured latency (L_d'/L_d), referred to as the terminal latency index, also relates to distal conduction delay.[103]

Possible Sources of Error

In normal subjects, shocks of supramaximal intensity elicit almost, but not exactly, identical compound muscle action potentials, depending on the nerve length between the stimulating and recording electrodes. The impulses of the slow conducting fibers lag progressively behind those of the fast conducting fibers over a longer conducting path. Hence, a proximal stimulus gives rise to an evoked potential of slightly longer duration and lower amplitude than a distal shock. This physiologic temporal dispersion does not drastically alter the waveform of the muscle action potentials, as predicted by analysis of duration-dependent phase cancellation (see Fig. 7–11). The evoked potentials of dissimilar shapes preclude accurate calculation of conduction velocity, because the two onset latencies may represent motor fibers of different conduction characteristics.

In diseased nerves with conduction block, the impulse from a proximal site of stimulation fails to propagate in some fibers. More commonly, distorted waveforms result from the use of an inappropriately low stimulus intensity, which activates only part of the nerve fibers.[43] On the other hand, an excessive stimulus intensity can cause an erroneously short latency because the spread of stimulus current depolarizes the nerve a few millimeters away from the cathode.[96] The surface length measured between the two cathodal points under these conditions does not precisely correspond to the conduction distance of the nerve segment under study.[114]

When recorded with a high sensitivity, a small negative peak sometimes precedes the main negative component of the muscle action potential.[13,51,104] This small potential, to be disregarded in latency determination, probably originates from small nerve fibers near the motor point. With awareness of this possibility, one can avoid miscalculation, especially if the nerve potential not seen with stimulation at one point appears at a second point with the use of a higher sensitivity for improved resolution. To avoid this type of error, it is best to use the same amplifier sensitivity for comparison of successively elicited potentials with stimulation along the course of the nerve.

Types of Abnormalities

In general, axonal damage or dysfunction results in loss of amplitude, whereas demyelination leads to prolongation of conduction time (see Chapters 4.5, and 4.6). Assessment of a nerve as a whole, as opposed to individual nerve fibers, usually reveals more complicated features because different types of abnormalities tend to coexist. Nonetheless, three basic types of abnormalities characterize motor nerve conduction studies when stimulating the nerve proximal to the presumed lesion (Fig. 5–3): (1) reduced amplitude with normal or slightly increased latency, (2) increased latency with relatively normal amplitude, and (3) absent response.

In the first variety, a shock below the lesion may elicit a normal compound muscle action potential, even though proximal stimulation above the lesion evokes reduced amplitude (Fig. 5–4). This finding, if seen during the first few days of injury, fails to differentiate a partial nerve lesion causing neurapraxia or early axonotmesis before the onset of distal de-

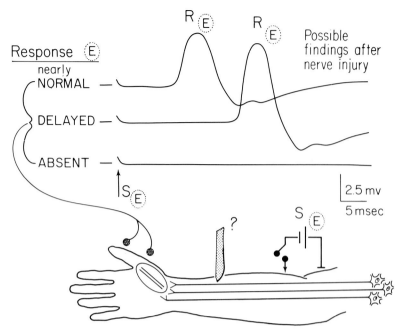

Figure 5–3. Three basic types of alteration in the compound muscle action potential occur after a presumed nerve injury distal to the site of stimulation: mildly reduced amplitude with nearly normal latency (*top*), normal amplitude with substantially increased latency (*middle*) or absent response even with a shock of supramaximal intensity (*bottom*).

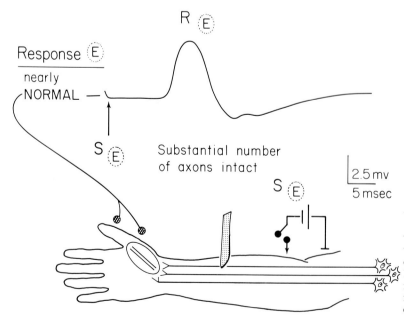

Figure 5–4. Mild reduction in amplitude of the compound muscle action potential with a nearly normal latency. This type of abnormality indicates that a substantial number of axons remain functional. The affected axons, constituting only a small portion of the total population, have either neurapraxia or axonotmesis. The normal latency reflects the surviving axons that conduct normally. Because of inherent individual variability, minor changes in amplitude may escape detection as a sign of abnormality.

generation. Distinction between the two possibilities becomes possible by stimulating the nerve below the lesion several days after the injury, when degenerating axons will have lost their excitability. Partial neurapraxia is indicated when the distally evoked muscle response still exceeds the proximally elicited potential in amplitude. In contrast, stimulation above or below the lesion elicits an equally reduced amplitude in axonotmesis. Because the amplitude of the muscle response varies considerably from one normal subject to another, minor diminution in the recorded potential often escapes detection.

In the second variety, slowed conduction accompanies relatively normal amplitude in stimulation above the lesion (Fig. 5–5). These changes generally imply segmental demyelination affecting a majority of the nerve fibers. As shown in rabbits, incomplete proximal compressive lesions may also give rise to slowed conduction with a reduction in external fiber diameter distal to the site of constriction.[4] The time course of recovery, however, suggests that in these cases, conduction slowing along the distal nerve segment results from distal paranodal demyelination.[5]

A prolonged latency or slowing of the conduction velocity may also result from axonal neuropathy with loss of the fast conducting fibers. A major reduction in amplitude to less than 40 to 50 percent of the mean of the normal value usually accompanies this type of slowing. In fact, if the amplitude remains more than half the control value, a reduction of the conduction velocity to less than 80 to 90 percent of lower limits of normal suggests the presence of demyelination. With a further diminution of amplitude to less than half the mean normal value, the conduction velocity may fall to 70 to 80 percent of the lower limit without demyelination. For the same reason, slowed motor conduction also results from loss of large anterior horn cells in myelopathies. Here, the motor conduction velocity can decrease to 70 percent of the mean normal value with diminution of amplitude to less than 10 percent of normal.[79] Regardless of the amplitude, however, a conduction velocity reduced to less than 60 percent of the mean normal value suggests peripheral nerve disease, rather than myelopathy.[80]

With neurapraxia, proximal stimulation above the lesion gives rise to a smaller compound muscle action poten-

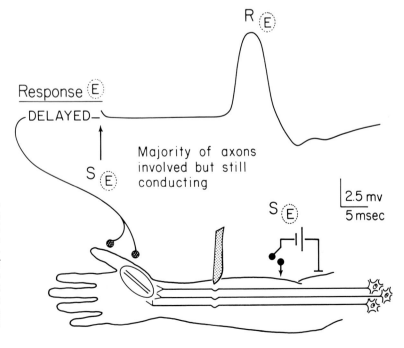

R (E)

Response (E)

DELAYED

Majority of axons involved but still conducting

S (E)

S (E)

2.5 mv
5 msec

Figure 5–5. Increased latency of the compound muscle action potential with normal amplitude. This type of abnormality indicates demyelination affecting the majority of nerve fibers as in a compression neuropathy. Conduction block, if present during acute stages, will also diminish the amplitude of the recorded response.

tial than does a distal stimulation below the lesion (Figs. 5–6 and 5–7). A reduction in size of the compound muscle action potential may also result from phase cancellation between peaks of opposite polarity based on increased temporal dispersion in the absence of a conduction block.[75] Such an excessive temporal dispersion commonly develops in acquired demyelinative neuropathies (Fig. 5–8). If the distal and proximal responses assume dissimilar waveforms, their onset latencies may represent two groups of motor fibers with different conduction characteristics, precluding accurate calculation of velocity.

Absent responses indicate that a majority of the nerve fibers fail to conduct across the site of the presumed lesion (Fig. 5–9). One must then differentiate a neurapraxic lesion from nerve transection. In either case, nerve stimulation distal to the lesion elicits an entirely normal muscle action potential for the first 4 to 7 days. During the second week, how-

ever, the normal excitability of neuromuscular junction and distal nerve segment distinguishes neurapraxic changes from axonal abnormalities. With neurotmesis, stimulation below the point of the lesion produces no muscle action potentials, because of the initial failure at the neuromuscular junction (Fig. 5–10). The loss of nerve excitability follows during subsequent wallerian degeneration.

Serial electrophysiologic studies help delineate progressive recovery from severe axonopathy on the basis of the amplitude of the evoked potential (Fig. 5–11A, B, and C).

5 SENSORY NERVE CONDUCTION

Stimulation and Recording

For sensory conduction studies in the upper extremities, stimulation of the digital nerves elicits an orthodromic sensory

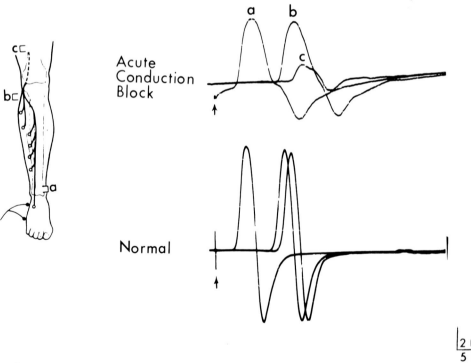

Figure 5–6. A 67-year-old man with an acute onset of footdrop following chemotherapy and radiation treatment of prostate cancer. Although epidural metastasis was suspected clinically because of backache, nerve conduction studies revealed evidence of a conduction block at the knee, indicating a compressive neuropathy. (From Kimura,[73] with permission.)

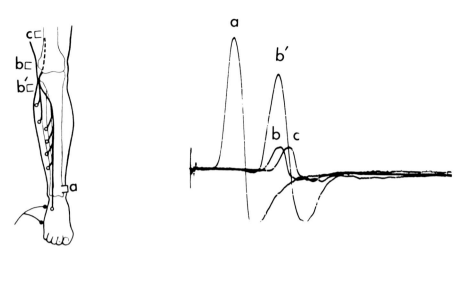

Figure 5–7. A 34-year-old man with selective weakness of foot dorsiflexors and low-back pain radiating to the opposite leg. The nerve conduction studies revealed a major conduction block between the two stimulation sites, *b*, and *b'*, at the knee. The weakness abated promptly when the patient refrained from habitual leg crossing. (From Kimura,[73] with permission.)

potential at a more proximal site. Alternatively, stimulation of the nerve trunk proximally evokes the antidromic digital potential distally and orthodromic potential proximally. For example, shocks applied to the ulnar or median nerve at the wrist give rise to an action potential along the nerve trunk at the elbow. Sensory fibers with large diameters have lower thresholds and conduct faster than motor fibers by about 5 to 10 percent.[28] Thus, mixed nerve potentials allow determination of the fastest sensory nerve conduction velocity. This relationship, however, may not hold in disease states that affect different fibers selectively. Such circumstances would preclude differentiation between the sensory and motor components of mixed nerve potentials.

For routine clinical recordings, surface electrodes provide adequate and reproducible information noninvasively. Some electromyographers prefer needle recording to improve the signal-to-noise ratio, especially in assessing temporal dispersion.[99] Here, a signal averager provides a sensitive measure of early nerve damage by defining small late components that originate from demyelinated, remyelinated, or regenerated fibers.[14,15,45,99]

Amplitude, Duration, and Waveform

With the use of surface electrodes, the antidromic potentials from digits generally have a greater amplitude than the orthodromic response from the nerve trunk, because the digital nerves lie nearer to the surface.[13] The relationship reverses with the use of needle electrodes placed near the nerve. Some motor axons have thresholds similar to those of large myelinated sensory axons.[48] In studying the mixed nerve, therefore, superimposition of action potentials from distal muscles may obscure antidromically recorded sensory potentials. Stimulation distal to the termination of the motor fibers selectively activates sensory fibers of mixed nerves (see Chapter 6.2). Moving more proximally, overlap of muscle action potentials becomes apparent because the waveform of the elicited response shows an abrupt change.[71]

The amplitude of the sensory potential, measured either from the baseline to the negative peak or between the negative and positive peaks, varies substantially among subjects and to a lesser extent between the two sides in the same individ-

Lt Ulnar Nerve   Rt Ulnar Nerve

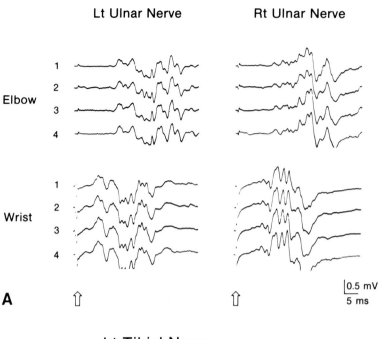

0.5 mV
5 ms

A

Lt Tibial Nerve

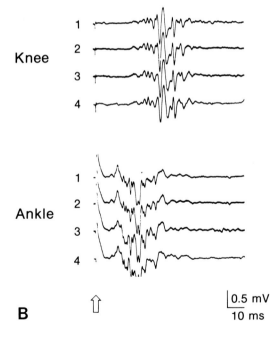

Figure 5–8. A. A 31-year-old man with the Guillain-Barré syndrome. Stimulation of the ulnar nerves at the elbow or wrist elicited delayed temporally dispersed compound muscle action potentials of the abductor digiti minimi bilaterally. Four consecutive trials at each stimulus site confirm the consistency of the evoked potentials. **B.** Compound muscle action potential in the same patient as shown in **A.** Stimulation of the tibial nerve at the knee or ankle elicited delayed and very irregular compound muscle action potentials of the abductor hallucis.

0.5 mV
10 ms

B

ual. The same degree of variability occurs in recording with surface or needle electrodes.[13] The density of sensory innervation in each finger determines the amplitude of the digital nerve potential. It is interesting that left-handers tend to have greater median sensory potentials at the wrist on the right side, and vice versa.[85] Most electromyographers measure the duration of the nerve action potential

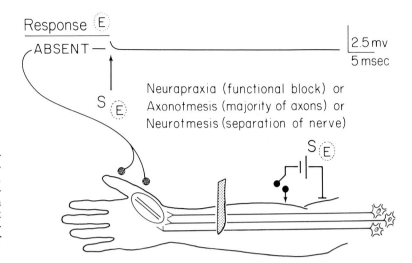

Figure 5–9. No evoked potential with supramaximal stimulation of the nerve proximally. This type of abnormality indicates the loss of conduction in the majority of axons but fails to distinguish neurapraxia from axonotmesis or neurotmesis.

from the initial deflection to the intersection between the descending phase and the baseline. Some use the negative or positive peak as the point of reference, and still others resort to the less definable point where the tracing finally returns to baseline.

The position of the recording electrodes alters the waveform of a sensory nerve action potential.[3] An initially positive triphasic waveform characterizes the orthodromic potential recorded with an active

electrode (G_1) on the nerve and a reference electrode (G_2) at a remote site. A separate late phase may appear in the temporally dispersed response recorded at a more proximal site. Placing G_2 near the nerve at a distance of more than 3 cm from G_1 makes the recorded potential tetraphasic, with addition of the final negativity.[13] A small initial positive phase, clearly seen in the orthodromic potential, is absent in the antidromic digital potential recorded with a pair of ring elec-

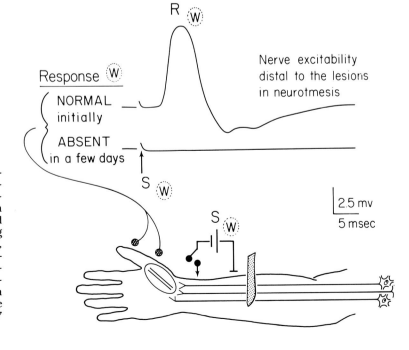

Figure 5–10. Nerve excitability distal to the lesion in neurotmesis or substantial axonotmesis. Distal stimulation elicits a normal compound muscle action potential during the first few days after injury, even with a complete separation of the nerve. Unlike neurapraxia, wallerian degeneration subsequent to transection will render the distal nerve segment inexcitable in 4 to 7 days.

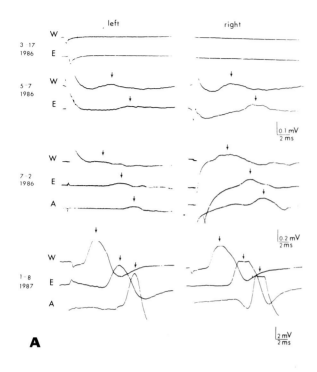

A

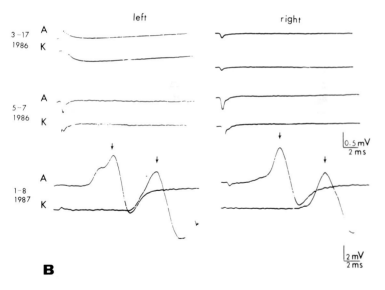

B

Figure 5–11. A 2½-year-old boy with hypothermia-induced axonal polyneuropathy following prolonged exposure to severe freezing weather on a frigid winter night in Iowa. **A.** Compound muscle action potentials recorded over the thenar eminence after stimulation of the median nerve at the wrist (*W*), elbow (*E*), or axilla (*A*). The initial study on March 17, 1986, revealed no response on either side, followed by progressive return in amplitude and latency, with full recovery by January 8, 1987. **B.** Compound muscle action potential recorded from the abductor hallucis after stimulation of the tibial nerve at the ankle (*A*) or knee (*K*). The studies on March 17 and May 7, 1986, revealed no response on either side, with full recovery by January 8, 1987.

trodes. The lack of potential difference between G_1 and G_2 implies the stationary character of the positive phase along the digit, as predicted by the far-field theory (see Chapter 19.3).

The types of abnormalities described for motor conduction apply in principle to sensory conduction as well. Substantial slowing in conduction velocity implies demyelination of the sensory fibers, whereas axonotmesis results in reduced amplitude of the compound nerve action potentials with stimulation distal or proximal to the site of the lesion. The sensory fibers degenerate only with a lesion distal to the sensory ganglion (Fig.

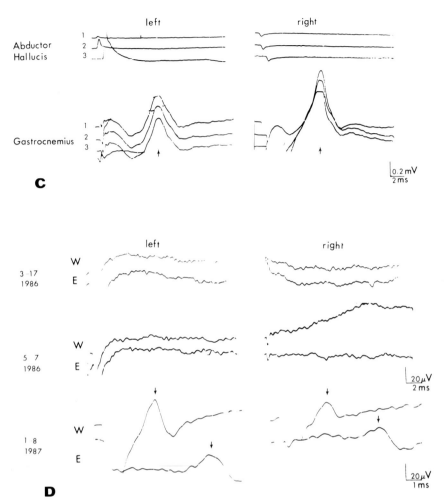

Figure 5–11 (*Cont.*). C. On May 17, 1986, stimulation at the knee elicited no response in the intrinsic foot muscle on either side (*top three tracings*), but a small compound action potential in the gastrocnemius bilaterally (*bottom*) as the result of early reinnervation. **D.** Antidromic sensory nerve action potential recorded from the second digit after stimulation of the median nerve at the wrist (*W*) or elbow (*E*). The studies on March 17 and May 7, 1986, showed no response on either side, with full recovery by January 8, 1987. (From Afifi et al.,[2] with permission.)

5–11D). Thus, the presence of distal sensory potential serves as a criterion for differentiating preganglionic root avulsion from plexopathy.[47] Rare postganglionic root lesions will affect the digital nerve potential as in plexopathy. Distinction between the two then depends on the distribution of sensory involvement. Plexopathy tends to affect multiple digits, whereas radiculopathy will show selective change of the first and second digits by C-6, the third digit by C-7, and the fourth and fifth digits by C-8 root lesions.

This type of assessment must take into account the relative amplitude values of the sensory action potential for each digit.[24,25,94]

Latency and Conduction Velocity

Unlike motor latency, which includes neuromuscular transmission, sensory latency consists only of the nerve conduction time from the stimulus point to the recording electrode. Therefore, stimulation of the nerve at a single site suffices for calculation of conduction velocity. In measuring the latency of the orthodromic sensory potentials, some electromyographers use the initial positive peak and others, the subsequent negative peak as the point of reference.[63] Sensory potentials elicited by stimulation at different sites vary in waveform because of temporal dispersion between fast and slow fibers. The interval between the positive and negative peaks also increases in proportion to the nerve length tested. Therefore, the conduction velocity calculated with the latency to the negative peak cannot necessarily be used as a measure of the fastest conducting sensory fibers.[28]

The measurement to the negative peak circumvents the technical problems of identifying the preceding smaller positive peak, especially in diseased nerves.[49] In this practice, the conduction distance determined to the midpoint of G_1 and G_2, rather than to G_1 itself, compensates for the discrepancy between the arrival of the impulse and the appearance of the negative peak.[34] The use of modern amplifiers with high resolution now makes it feasible in most cases to measure the sensory latency to the initial positive peak. Determining the conduction distance from the stimulus point to G_1 then allows accurate calculation of conduction velocity of the fastest fibers.[13]

With the biphasic digital potential recorded antidromically, the onset latency measured to the initial take-off of the negative peak corresponds to the conduction time of the fastest fibers from the cathode to G_1. The use of the peak latency has little, if any, justification with antidromically recorded digital potentials, which considerably exceed orthodromic potentials in amplitude. In one study, antidromic conduction times, despite identical mean values, showed slightly higher standard deviations than orthodromic measurements.[13] In another study, the orthodromic recording revealed a shorter distal latency than the antidromic method in both median and ulnar nerves.[21]

6 NERVE CONDUCTION IN THE CLINICAL DOMAIN

The validity of the calculated nerve conduction velocity depends on the accuracy in determining the latencies and the conduction distance. Sources of error in measuring latencies include unstable or incorrect triggering of the sweep, poorly defined takeoff of the evoked response, inappropriate stimulus strength, and inaccurate calibration.[80,104] Errors in estimating the conduction distance by surface measurement result from uncertainty as to the exact site of stimulation and the nonlinear course of the nerve segments. Surface determination of the nerve length yields particularly imprecise results when the nerve takes an angulated path, as in the brachial plexus or across the elbow or knee.

Because of these uncontrollable variables, the calculated velocities only approximate the absolute values of nerve conduction. On repeated testing, the results might vary occasionally as much as 10 m/s, because of the limitations inherent in the technique.[88] Strict adherence to the standard procedures minimizes the error and improves the reproducibility. A small range of normal values, then, justifies the use of conduction studies as a clinical diagnostic test. In general, the use of latency instead of conduction velocity requires, for meaningful comparison, a constant distance between the stimulating and recording electrodes. A number of factors listed below can modify the results of motor and sensory conduction studies.

Effect of Temperature

Nerve impulses conduct faster at a higher body temperature,[30] as is seen, for example, after physical activity.[55] The conduction velocity increases almost linearly, by 2.4 m/s, or approximately 5 percent per degree, as the temperature measured near the nerve increases from 29 to 38°C.[58,62] Similarly, distal latencies increase by 0.3 ms per degree for both median and ulnar nerves upon cooling the hand.[17] Lower temperatures augment the amplitude of nerve and muscle potential, as demonstrated in the squid axon,[61] and in human studies.[10,31,81] Cold-induced slowing of Na^+ channel inactivation probably accounts for the increase in amplitude, because a parallel temperature-dependent change occurs in the refractory period.[84] Studies conducted in a warm room with ambient temperature maintained between 21°C and 23°C reduce this type of variability. Although impractical and unnecessary in clinical practice, a warmer room at 26°C to 28°C or even 30°C minimizes the temperature gradient along the course of a nerve.[68]

To check the intramuscular temperature, the insertion of a thermometer through the skin requires an additional puncture for each muscle tested. In practice, the skin temperature measured with a plate thermistor correlates linearly with the subcutaneous and intramuscular temperatures.[53,54] A skin temperature of 34°C or above indicates a muscle temperature close to 37°C.[32] If the measured value falls below this range, one should warm the limbs under study with an infrared heat lamp or by immersion in warm water. Alternatively, one may add 5 percent of the calculated conduction velocity for each degree below 34°C to normalize the result. Such conversion factors, based upon an average of many healthy subjects, however, may provide misleading interpretations in diseases of the peripheral nerve.[16]

Variation Among Different Nerves and Segments

Both motor and sensory fibers conduct substantially more slowly in the legs than in the arms. A small reduction in temperature cannot account for the recorded differences, ranging from 7 to 10 m/s.[76,111] Longer nerves generally conduct more slowly than shorter nerves, as suggested by an inverse relationship between height and nerve conduction velocity.[16] Available data further indicate a good correlation between conduction velocity and estimated axonal length in peroneal and sural nerves, but not in motor or sensory fibers of the median nerve.[105] These findings suggest abrupt distal axonal tapering in the lower extremities, although this view lacks histologic proof. The other factors possibly responsible for the velocity gradient include progressive reduction in axonal diameter, the shorter internodal distances, and lower distal temperatures. Statistical analyses of conduction velocities show no difference between median and ulnar nerves or between tibial and peroneal nerves.

The nerve impulse propagates faster in the proximal than in the distal nerve segments.[50,57] For example, the most proximal motor nerve conduction velocity determined by F-wave latency clearly exceeds the conventionally derived most distal conduction velocity.[23,38,69,74,77] Statistical analyses show no significant difference between cord-to-axilla and axilla-to-elbow segments.[69] Calculation of the F ratio (see Chapter 17.5) allows comparison between motor nerve conduction time from the spinal cord to the stimulus site and that from the remaining nerve segment to the muscle.[70] In healthy subjects, this ratio is close to unity with stimulation at the elbow or at the knee, indicating equal conduction times along the proximal and distal segments from the site of stimulation. Hence, faster proximal conduction must compensate for the difference in length between the cord-to-elbow and elbow-to-muscle segments or between the cord-to-knee and knee-to-muscle segments.[74]

Relative distal slowing of conduction, however, has not been a universal finding. Some authors have reported no difference among various segments, whereas others have noted a slower conduction proximally.[59,106] In the baboon, single motor axons conduct slower in the brachial plexus than in the peripheral nerve trunk.[22]

Effects of Age

Nerve conduction velocities increase rapidly as the process of myelination advances from roughly half the adult value in full-term infants[109] to the adult range at age 3 to 5 years (Fig. 5–12). Conduction

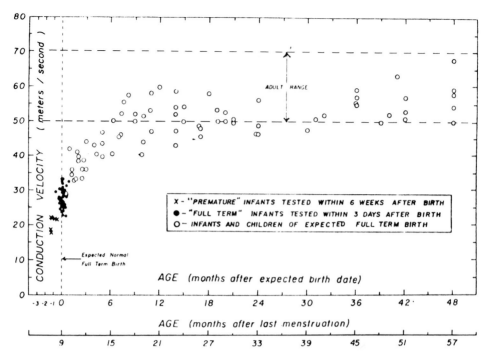

Figure 5–12. Relation of age to conduction velocity of motor fibers in the ulnar nerve between the elbow and wrist. Velocities in normal young adults range from 47 to 73 m/s, with the majority of values between 50 to 70 m/s. Ages plotted indicate the month after the expected birth date based on calculation from the first day of last menstruation. (From Thomas and Lambert,[109] with permission.)

velocity of slower fibers also show a similar time course of maturation.[52] Table 5–1 summarizes the results of one series showing a steep increase in conduction of the peroneal nerve through infancy and a slower maturation of the median nerve during early childhood.[41] Premature infants have even slower conduction velocities, ranging from 17 to 25 m/s in the ulnar nerve and from 14 to 28 m/s in the peroneal nerve.[20] The values at 23 to 24 weeks of fetal life are roughly one-third those of newborns of normal gestational age.[26,92] Fetal nutrition may alter periph-

eral nerve function by influencing myelin formation.[98]

In later childhood and adolescence from age 3 to 19 years, both motor and sensory conduction velocities tend to slightly increase in the upper limb and decrease in the lower limb as a function of age and growth in length.[82] Conduction velocities begin to decline after 30 to 40 years of age, but the values normally change by less than 10 m/s at the 60th year[108,112] or even the 80th year.[95] Table 5–2 summarizes the results of one study[88] in which was shown a reduction

Table 5–1 NORMAL MOTOR NERVE CONDUCTION VELOCITIES (M/S) IN DIFFERENT AGE GROUPS

Age	Ulnar	Median	Peroneal
0–1 week	32 (21–39)	29 (21–38)	29 (19–31)
1 week to 4 months	42 (27–53)	34 (22–42)	36 (23–53)
4 months to 1 year	49 (40–63)	40 (26–58)	48 (31–61)
1–3 years	59 (47–73)	50 (41–62)	54 (44–74)
3–8 years	66 (51–76)	58 (47–72)	57 (46–70)
8–16 years	68 (58–78)	64 (54–72)	57 (45–74)
Adults	63 (52–75)	63 (51–75)	56 (47–63)

From Gamstorp,[41] with permission.

Table 5–2 NORMAL SENSORY AND MOTOR NERVE CONDUCTION VELOCITIES (M/S) IN DIFFERENT AGE GROUPS

Nerve	10–35 Years (30 Cases)		36–50 Years (16 Cases)		51–80 Years (18 Cases)	
	Sensory	Motor	Sensory	Motor	Sensory	Motor
Median nerve						
Digit–wrist	67.5 ± 4.7		65.8 ± 5.7		59.4 ± 4.9	
Wrist–muscle		3.2 ± 0.3*		3.7 ± 0.3*		3.5 ± 0.2*
Wrist–elbow	67.7 ± 4.4	59.3 ± 3.5	65.8 ± 3.1	55.9 ± 2.6	62.8 ± 5.4	54.5 ± 4.0
Elbow–axilla	70.4 ± 4.8	65.9 ± 5.0	70.4 ± 3.4	65.1 ± 4.2	66.2 ± 3.6	63.6 ± 4.4
Ulnar nerve						
Digit–wrist	64.7 ± 3.9		66.5 ± 3.4		57.5 ± 6.6	
Wrist–muscle		2.7 ± 0.3*		2.7 ± 0.3*		3.0 ± 0.35*
Wrist–elbow	64.8 ± 3.8	58.9 ± 2.2	67.1 ± 4.7	57.8 ± 2.1	56.7 ± 3.7	53.3 ± 3.2
Elbow–axilla	69.1 ± 4.3	64.4 ± 2.6	70.6 ± 2.4	63.3 ± 2.0	64.4 ± 3.0	59.9 ± 0.7
Common peroneal nerve						
Ankle–muscle		4.3 ± 0.9*		4.8 ± 0.5*		4.6 ± 0.6*
Ankle–knee	53.0 ± 5.9	49.5 ± 5.6	50.4 ± 1.0	43.6 ± 5.1	46.1 ± 4.0	43.9 ± 4.3
Posterior tibial nerve						
Ankle–muscle		5.9 ± 1.3*		7.3 ± 1.7*		6.0 ± 1.2*
Ankle–knee	56.9 ± 4.4	45.5 ± 3.8	49.0 ± 3.8	42.9 ± 4.9	48.9 ± 2.6	41.8 ± 5.1
H reflex, popliteal fossa		71.0 ± 4.0		64.0 ± 2.1		60.4 ± 5.0
		27.9 ± 2.2*		28.2 ± 1.5*		32.0 ± 2.1*

*Latency in milliseconds.
Values are means ± 1 standard deviation.
From Mayer,[88] with permission.

in the mean conduction rate of about 10 percent at 60 years of age. Aging also causes a diminution in amplitude and changes in the shape of the evoked potential (Table 5–3), especially at the common sites of compression.[24] The latencies of the F wave and somatosensory evoked potentials also gradually increase with advancing age.[33]

Other Factors

Ischemia, induced by a pneumatic tourniquet, alters nerve excitability substantially, with progressive slowing in conduction velocity, decrease in amplitude, and increase in duration of the action potential.[102] This effect affects the median nerve more rapidly in patients with the carpal tunnel syndrome than in normal controls.[40] Conversely, patients with diabetes or uremia or elderly subjects have a greater resistance to ischemia with regard to peripheral nerve function.[18] One study[86] showed no significant difference in the mean conduction veloci-

ties and the range of values before and after a fast of 2 to 3 weeks' duration.

Uses and Limitations

Over the years, nerve conduction studies have made major contributions to the understanding of peripheral nerve function in healthy and disease states.[44,46] Such evaluations play an important role in precisely delineating the extent and distribution of the lesion and in providing an overall distinction between axonal and demyelinating involvement.[110] This dichotomy provides a simple and practical means of correlating conduction abnormalities with major pathologic changes in the nerve fibers. In support of this concept, in vitro recordings from the sural nerve have clearly delineated close relationships between histologic and physiologic findings.[8,35,36]

In addition to such a broad classification, the pattern of nerve conduction abnormalities can often characterize the general nature of the clinical disorder.

**Table 5–3 COMPARISON OF CONDUCTION STUDIES
BETWEEN YOUNGER GROUP (n = 52, 10–40 YEARS) AND
OLDER GROUP (n = 52, 41–84 YEARS)**

Nerve Tested		No. of Nerves	Age 29.7 ± 6.9 Years (Mean ± SD)	No. of Nerves	Age 54.0 ± 10.5 Years (Mean ± SD)	P Value
Peroneal						
M amplitude	(mV)	104	5.4 ± 1.5	98	5.0 ± 1.3	0.03*
M latency	(ms)	104	3.7 ± 0.9	98	3.7 ± 0.7	0.98
MNCV	(m/s)	104	49.5 ± 5.4	98	47.8 ± 3.8	0.01*
F latency	(ms)	44	47.1 ± 5.3	42	47.6 ± 4.9	0.68
FWCV	(m/s)	44	60.6 ± 7.7	42	59.9 ± 7.6	0.66
F number	(#)	10	8.5 ± 1.7	29	9.7 ± 3.1	0.19
Tibial						
M amplitude	(mV)	104	6.7 ± 2.0	100	5.9 ± 1.5	0.001*
M latency	(ms)	104	3.5 ± 0.6	100	3.6 ± 0.6	0.23
MNCV	(m/s)	104	48.6 ± 4.2	100	49.1 ± 4.9	0.52
F latency	(ms)	74	47.9 ± 4.1	74	48.3 ± 4.6	0.63
FWCV	(m/s)	74	58.3 ± 6.2	74	57.5 ± 6.8	0.49
F number	(#)	25	11.6 ± 3.4	27	12.4 ± 2.6	0.39
H amplitude	(mV)	53	1.4 ± 0.8	43	1.2 ± 0.8	0.20
H latency	(ms)	53	29.8 ± 2.3	50	30.7 ± 2.0	0.04*
Sural						
S amplitude	(μV)	53	20.9 ± 8.0	50	17.2 ± 6.7	0.01*
S latency	(ms)	53	2.7 ± 0.3	50	2.8 ± 0.3	0.16
SNCV	(m/s)	53	52.5 ± 5.6	50	51.1 ± 5.9	0.23

MNCV, motor nerve conduction velocity in the distal segment; FWCV, F-wave conduction velocity in the proximal segment; F number, number of responses out of 16 trials.

*Amplitude was significantly reduced in the older group for all the nerves tested, whereas measures of conduction showed no changes except for peroneal MNCV and tibial H-latency.

From Kimura,[73] with permission.

For example, hereditary demyelinating neuropathies commonly show diffuse abnormalities, with little difference from one nerve to another in the same patient and among different members in the same family.[83] In addition, approximately equal involvement of different nerve fibers limits the degree of temporal dispersion despite considerable increases in latency. In contrast, acquired demyelination tends to affect certain segments of the nerve disproportionately,[70,77] giving rise to more asymmetric abnormalities and substantial increases in temporal dispersion.

Optimal application of the nerve conduction study depends on an understanding of the principles and a recognition of the pitfalls of the technique. The conventional methods deal primarily with distal nerve segments in an extremity. Newer techniques allow one to assess nerve segments in less accessible ana-

tomic regions, to improve accuracy in precise localization of a focal lesion, and to increase sensitivity in detecting subclinical abnormalities. Despite certain limitations, these methods can provide diagnostically pertinent information if used judiciously in appropriate clinical contexts.

REFERENCES

1. AAEE: Guidelines in Electrodiagnostic Medicine. Professional Standard Committee, American Association of Electromyography and Electrodiagnosis, Rochester, Minn, 1984.
2. Afifi, AK, Kimura, J, and Bell, WE: Hypothermia-induced reversible polyneuropathy: Electrophysiologic evidence of axonopathy. Pediatr Neurol 4:49–53, 1988.
3. Andersen, K: Surface recording of orthodromic sensory nerve action potentials in median and ulnar nerves in normal subjects. Muscle Nerve 8:402–408, 1985.

4. Baba, M, Fowler, CJ, Jacobs, JM, and Gilliatt, RW: Changes in peripheral nerve fibres distal to a constriction. J Neurol Sci 54:197–208, 1982.

5. Baba, M, Gilliatt, RW, and Jacobs, JM: Recovery of distal changes after nerve constriction by a ligature. J Neurol Sci 60:235–246, 1983.

6. Bamford, CR, Rothrock, JR, and Swenson, M: Average techniques to define the low-amplitude compound motor action potentials. Arch Neurol 41:1307, 1984.

7. Bannister, RG, and Sears, TA: The changes in nerve conduction in acute idiopathic polyneuritis. J Neurol Neurosurg Psychiatry 25:321–328, 1962.

8. Behse, F, and Buchthal, F: Sensory action potentials and biopsy of the sural nerve in neuropathy. Brain 101:473–493, 1978.

9. Berry, CM, Grundfest, H, and Hinsey, JC: The electrical activity of regenerating nerves in the cat. J Neurophysiol 7:103–115, 1944.

10. Bolton, CF, Sawa, GM, and Carter, K: The effects of temperature on human compound action potentials. J Neurol Neurosurg Psychiatry 44:407–413, 1981.

11. Buchthal, F: Action potentials in the sural nerve evoked by tactile stimuli. Mayo Clin Proc 55:223–230, 1980.

12. Buchthal, F, and Rosenfalck, A: Action potentials from sensory nerve in man: Physiology and clinical application. Acta Neurol Scand (Suppl 13) 41:263–266, 1965.

13. Buchthal, F, and Rosenfalck, A: Evoked action potentials and conduction velocity in human sensory nerves. Brain Res 3:1–122, 1966.

14. Buchthal, F, and Rosenfalck, A: Sensory potentials in polyneuropathy. Brain 94:241–262, 1971.

15. Buchthal, F, Rosenfalck, A, and Behse, F: Sensory potentials of normal and diseased nerves. In Dyck, PJ, Thomas, PK, and Lambert, EH (eds): Peripheral Neuropathy, Vol 1. WB Saunders Company, Philadelphia, 1975, pp 442–464.

16. Campbell, WW, Ward, LC, and Swift, TR: Nerve conduction velocity varies inversely with height. Muscle Nerve 3:436–437, 1981.

17. Carpendale, MTF: Conduction time in the terminal portion of the motor fibers of the ulnar, median, and peroneal nerves in healthy subjects and in patients with neuropathy. Thesis, University of Minnesota, Minneapolis, 1956.

18. Caruso, G, Labianca, O, and Ferrannini, E: Effect of ischemia on sensory potentials of normal subjects of different ages. J Neurol Neurosurg Psychiatry 36:455–466, 1973.

19. Casey, EB, and Le Quesne, PM: Digital nerve action potentials in healthy subjects and in carpal tunnel and diabetic patients. J Neurol Neurosurg Psychiatry 35:612–623, 1972.

20. Cerra, D, and Johnson, EW: Motor nerve conduction velocity in premature infants. Arch Phys Med Rehabil 43:160–164, 1962.

21. Chodoroff, G, Tashjian, EA, and Ellenberg, MR: Orthodromic vs antidromic sensory nerve latencies in healthy persons. Arch Phys Med Rehabil 66:589–591, 1985.

22. Clough, JFM, Kernell, D, and Phillips, CG: Conduction velocity in proximal and distal portions of forelimb axons in the baboon. J Physiol (Lond) 198:167–178, 1968.

23. Conrad, B, Aschoff, JC, and Fischler, M: Der Diagnostische Wert der F-Wellen-Latenz. J Neurol 210:151–159, 1975.

24. Cruz Martinez, A, Barrio, M, Perez Conde, MC, and Ferrer, MT: Electrophysiological aspects of sensory conduction velocity in healthy adults. 2. Ratio between the amplitude of sensory evoked potentials at the wrist on stimulating different fingers in both hands. J Neurol Neurosurg Psychiatry 41:1097–1101, 1978.

25. Cruz Martinez, A, Barrio, M, Perez Conde, MC, and Gutierrez, AM: Electrophysiological aspects of sensory conduction velocity in healthy adults. 1. Conduction velocity from digit to palm, from palm to wrist, and across the elbow, as a function of age. J Neurol Neurosurg Psychiatry 41:1092–1096, 1978.

26. Cruz Martinez, A, Ferrer, MT, and Martin, MJ: Motor conduction velocity and H-reflex in prematures with very short gestational age. Electromyogr Clin Neurophysiol 23:13–19, 1983.

27. Dawson, GD: A summation technique for the detection of small evoked potentials. Electroencephalogr Clin Neurophysiol 6:65–84, 1954.

28. Dawson, GD: The relative excitability and conduction velocity of sensory and motor nerve fibres in man. J Physiol (Lond) 131:436–451, 1956.

29. Dawson, GD, and Scott, JW: The recording of nerve action potentials through skin in man. J Neurol Neurosurg Psychiatry 12:259–267, 1949.

30. De Jesus, PV, Hausmanowa-Petrusewicz, I, and Barchi, RL: The effect of cold on nerve conduction of human slow and fast nerve fibers. Neurology 23:1182–1189, 1973.

31. Denys, EH: AAEE Minimonograph #14: The role of temperature in electromyography. American Association of Electromyography and Electrodiagnosis, Rochester, Minn, 1980.

32. Desmedt, JE: The neuromuscular disorder in myasthenia gravis. 1. Electrical and mechanical response to nerve stimulation in hand muscles. In Desmedt, JE (ed); New Developments in Electromyography and Clinical Neurophysiology, Vol 1. Karger, Basel, 1973, pp 241–304.

33. Dorfman, LJ, and Bosley, TM: Age-related changes in peripheral and central nerve conduction in man. Neurology 29:38–44, 1979.

34. Downie, AW, and Newell, DJ: Sensory nerve conduction in patients with diabetes mellitus and controls. Neurology 11:876–882, 1961.

35. Dyck, PJ, Gutrecht, JA, Bastron, JA, Karnes, WE, and Dale, AJD: Histologic and teased fiber measurements of sural nerve in disorders of lower motor and primary sensory neurons. Mayo Clin Proc 43:81–123, 1968.

36. Dyck, PJ, Johnson, WJ, Lambert, EH, and O'Brien, PC: Segmental demyelination sec-

ondary to axonal degeneration in uremic neuropathy. Mayo Clin Proc 46:400–431, 1971.

37. Eichler, W: Uber die Ableitung der Aktionspotentiale vom menschlichen Nerven in situ. Z Biol 98:182–214, 1937.

38. Eisen, A, Schomer, D, and Melmed, C: The application of F-wave measurements in the differentiation of proximal and distal upper limb entrapments. Neurology 27:662–668, 1977.

39. Erlanger, J, and Gasser, HS: Electrical Signs of Nervous Activity. University of Pennsylvania Press, Philadelphia, 1937.

40. Fullerton, PM: The effect of ischaemia on nerve conduction in the carpal tunnel syndrome. J Neurol Neurosurg Psychiatry 26:385–397, 1963.

41. Gamstorp, I: Normal conduction velocity of ulnar, median and peroneal nerves in infancy, childhood and adolescence. Acta Paediatrica Suppl 146: 68–76, 1963.

42. Gans, BM, and Kraft, GH: M-Response quantification: a technique. Arch Phys Med Rehabil 62:376–380, 1981.

43. Gassel, MM: A study of femoral nerve conduction time. Arch Neurol 9:607–614, 1963.

44. Gilliatt, RW: Recent advances in the pathophysiology of nerve conduction. In Desmedt, JE (ed): New Developments in Electromyography and Clinical Neurophysiology, Vol 2, Karger, Basel, 1973, pp 2–18.

45. Gilliatt, RW: Sensory conduction studies in the early recognition of nerve disorders. Muscle Nerve 1:352–359, 1978.

46. Gilliatt, RW: Electrophysiology of peripheral neuropathies-an overview. Muscle Nerve 5:S108–S116, 1982.

47. Gilliatt, RW, Le Quesne, PM, Logue, V, and Sumner, AJ: Wasting of the hand associated with a cervical rib or band. J Neurol Neurosurg Psychiatry 33:615–624, 1970.

48. Gilliatt, RW, Melville, ID, Velate, AS, and Willison, RG: A study of normal nerve action potentials using an averaging technique (barrier grid storage tube). J Neurol Neurosurg Psychiatry 28:191–200, 1965.

49. Gilliatt, RW, and Sears, TA: Sensory nerve action potentials in patients with peripheral nerve lesions. J Neurol Neurosurg Psychiatry 21:109–118, 1958.

50. Gilliatt, RW, and Thomas, PK: Changes in nerve conduction with ulnar lesions at the elbow. J Neurol Neurosurg Psychiatry 23: 312–320, 1960.

51. Gutmann, L: The intramuscular nerve action potential. J Neurol Neurosurg Psychiatry 32:193–196, 1969.

52. Hakamada, S, Kumagai, T, Watanabe, K, Koike, Y, Hara, K, and Miyazaki, S: The conduction velocity of slower and the fastest fibres in infancy and childhood. J Neurol Neurosurg Psychiatry 45:851–853, 1982.

53. Halar, EM, DeLisa, JA, and Brozovich, FV: Nerve conduction velocity: Relationship of skin, subcutaneous and intramuscular temperatures. Arch Phys Med Rehabil 61:199–203, 1980.

54. Halar, EM, DeLisa, JA, and Soine, TL: Nerve conduction studies in upper extremities: Skin temperature corrections. Arch Phys Med Rehabil 64:412–416, 1983.

55. Halar, EM, Hammond, MC and Dirks, S: Physical activity: Its influence on nerve conduction velocity. Arch Phys Med Rehabil 66:605–609, 1985.

56. Helmholtz, H: Vorläufiger Bericht über die Fortpflanzungsgeschwindigkeit der Nervenreizung. Arch Anat Physiol Wiss Med 71–73, 1850.

57. Helmholtz, H, and Baxt, N: Neue Versuche über die Fortpflanzungsgeschwindigkeit der Reizung in den motorischen Nerven der Menschen. Mber Konigl Preus Akad Wiss 184–191, 1870.

58. Henriksen, JD: Conduction velocity of motor nerves in normal subjects and in patients with neuromuscular disorders. Thesis, University of Minnesota, Minneapolis, 1966.

59. Hodes, R: Selective destruction of large motoneurons by poliomyelitis virus. I. Conduction velocity of motor nerve fibers of chronic poliomyelitis patients. J Neurophysiol 12:257–266, 1949.

60. Hodes, R, Larrabee, MG, and German, W: The human electromyogram in response to nerve stimulation and the conduction velocity of motor axons. Arch Neurol Psychiatry 60:340–365, 1948.

61. Hodgkin, AL, and Katz, B: The effect of temperature on the electrical activity of the giant axon of the squid. J Physiol (Lond) 109:240–249, 1949.

62. Johnson, EW, and Olsen, KJ: Clinical value of motor nerve conduction velocity determination. JAMA 172:2030–2035, 1960.

63. Joynt, RL: Calculated nerve conduction velocity dependence upon the method of testing. Arch Phys Med Rehabil 64:212–216, 1983.

64. Kadrie, HA, Yates, SK, Milner-Brown, HS, and Brown, WF: Multiple point electrical stimulation of ulnar and median nerves. J Neurol Neurosurg Psychiatry 39:973–985, 1976.

65. Kaeser, HE: Nerve conduction velocity measurements. In Vinken, PJ, and Bruyn, BW (eds): Handbook of Clinical Neurology, Vol 7. North Holland Publishing Company, Amsterdam, 1970, pp 116–196.

66. Kaplan, PE: Sensory and motor residual latency measurements in healthy patients and patients with neuropathy. Part I. J Neurol Neurosurg Psychiatry 39:338–340, 1976.

67. Kaplan, P, and Sahgal, V: Residual latency: New applications of an old technique. Arch Phys Med Rehabil 59:24–27, 1978.

68. Kato, M: The conduction velocity of the ulnar nerve and the spinal reflex time measured by means of the H-wave in average adults and athletes. Tohoku J Exp Med 73:74–85, 1960.

69. Kimura, J: F-wave velocity in the central segment of the median and ulnar nerves: A study in normal subjects and in patients with Charcot-Marie-Tooth disease. Neurology 24:539–546, 1974.

70. Kimura, J: A method for determining median

nerve conduction velocity across the carpal tunnel. J Neurol Sci 38:1–10, 1978.

71. Kimura, J: The carpal tunnel syndrome. Localization of conduction abnormalities within the distal segment of the median nerve. Brain 102:619–635, 1979.

72. Kimura, J: Principles and pitfalls of nerve conduction studies. Ann Neurol 16:415–429, 1984.

73. Kimura, J: Electromyography and nerve stimulation techniques: Clinical applications (Japanese). Egakushoin, Tokyo, 1989.

74. Kimura, J, Bosch, P, and Lindsay, GM: F-wave conduction velocity in the central segment of the peroneal and tibial nerves. Arch Phys Med Rehabil 56:492–497, 1975.

75. Kimura, J, Machida, M, Ishida, T, Yamada, T, Rodnitzky, R, and Kudo, Y: A relationship between the size of compound sensory and muscle action potentials and the length of the nerve segment under study. Neurology 36:647–652, 1986.

76. Kimura, J, Yamada, T, and Stevland, N: Distal slowing of motor nerve conduction velocity in diabetic polyneuropathy. J Neurol Sci 42:291–302, 1979.

77. King, D, and Ashby, P: Conduction velocity in the proximal segments of a motor nerve in the Guillain-Barré syndrome. J Neurol Neurosurg Psychiatry 39:538–544, 1976.

78. Kraft, GH, and Halvorson, GA: Median nerve residual latency: Normal value and use in diagnosis of carpal tunnel syndrome. Arch Phys Med Rehabil 64:221–226, 1983.

79. Lambert, EH: Neurophysiological techniques useful in the study of neuromuscular disorders. In Adams, RD, Eaton, LM, and Shy, GM (eds): Neuromuscular Disorders. Williams and Wilkins, Baltimore, 1960, pp 247–273.

80. Lambert, EH: Diagnostic value of electrical stimulation of motor nerves. Electroenceph Clin Neurophysiol Suppl 22: 9–16, 1962.

81. Lang, AH, and Puusa, A: Dual influence of temperature on compound nerve action potential. J Neurol Sci 51:81–88, 1981.

82. Lang, HA, Puusa, A, Hynninen, P, Kuusela, V, Jantti, V, and Sillanpaa, M: Evolution of nerve conduction velocity in later childhood and adolescence. Muscle Nerve 8:38–43, 1985.

83. Lewis, RA, and Sumner, AJ: The electrodiagnostic distinctions between chronic familial and acquired demyelinative neuropathies. Neurology 32:592–596, 1982.

84. Louis, AA, and Hotson, JR: Regional cooling of human nerve and slowed NA inactivation. Electroencephalogr Clin Neurophysiol 634:371–375, 1986.

85. Martinez, AC, Perez Conde, MC, Del Campo, F, Mingo, P, and Ferrer, MT: Ratio between the amplitude of sensory evoked potentials at the wrist in both hands of left-handed subjects. J Neurol Neurosurg Psychiatry 43:182–184, 1980.

86. Mattson, RH, and Lecocq, FR: Nerve conduction velocities in fasting patients. Neurology 18:335–339, 1968.

87. Maurer, K, Hopf, HC, and Lowitzsch, K: Iso-metric muscle contraction in endocrine myopathies. Neurology 35:333–337, 1985.

88. Mayer, RF: Nerve conduction studies in man. Neurology 13:1021–1030, 1963.

89. Maynard, FM, and Stolov, WC: Experimental error in determination of nerve conduction velocity. Arch Phys Med Rehabil 53:362–372, 1972.

90. McLeod, JG: Digital nerve conduction in the carpal tunnel syndrome after mechanical stimulation of the finger. J Neurol Neurosurg Psychiatry 29:12–22, 1966.

91. Melvin, JL, Schuchman, JA, and Lanese, RR: Diagnostic specificity of motor and sensory nerve conduction variables in the carpal tunnel. Arch Phys Med Rehab 54:69–74, 1973.

92. Miller, R, and Kuntz, N: Nerve conduction studies in infants and children. J Child Neurol 1:19–26, 1986.

93. Mitz, M, Gokulananda, T, Di Benedetto, M, and Klingbeil, GE: Median nerve determinations: Analysis of two techniques. Arch Phys Med Rehabil 65:191–193, 1984.

94. Newman, M, and Nelson, N: Digital nerve sensory potentials in lesions of cervical roots and brachial plexus. Can J Neurol Sci 10:252–255, 1983.

95. Norris, AH, Shock, NW, and Wagman, IH: Age changes in the maximum conduction velocity of motor fibers of human ulnar nerves. J Appl Physiol 5:589–593, 1953.

96. Pinelli, P: Physical, anatomical and physiological factors in the latency measurement of the M response. Electroencephalogr Clin Neurophysiol 17:86, 1964.

97. Piper, H: Weitere Mitteilungen über die Geschwindigkeit der Erregungsleitung im markhaltigen menschlichen Nerven. Pflugers Arch Ges Physiol 127:474–480, 1909.

98. Robinson, RO, and Robertson, WC, Jr: Fetal nutrition and peripheral nerve conduction velocity. Neurology 31:327–329, 1981.

99. Rosenfalck, A: Early recognition of nerve disorders by near-nerve recording of sensory action potentials. Muscle Nerve 1:360–367, 1978.

100. Rosenfalck, A, and Buchthal, F: Sensory potentials and threshold for electrical and tactile stimuli. In Desmedt, JE (ed): New Developments in Electromyography and Clinical Neurophysiology, Vol 2. Karger, Basel, 1973, pp 45–51.

101. Scranton, PE, Jr, Hasiba, U, and Gorenc, TJ: Intramuscular hemorrhage in hemophiliacs with inhibitors: A medical emergency. JAMA 241:2028–2030, 1979.

102. Seneviratne, KN, and Peiris, OA: The effect of ischaemia on the excitability of human sensory nerve. J Neurol Neurosurg Psychiatry 31:338–347, 1968.

103. Shahani, BT, Young, RR, Potts, F, and Maccabee, P: Terminal latency index (TLI) and late response studies in motor neuron disease (MND), peripheral neuropathies and entrapment syndromes. Acta Neurol Scand (suppl) 73:60, 118, 1979.

104. Simpson, JA: Fact and fallacy in measure-

ment of conduction velocity in motor nerves. J Neurol Neurosurg Psychiatry 27:381–385, 1964.

105. Soudmand, R, Ward, LC, and Swift, TR: Effect of height on nerve conduction velocity. Neurology 32:407–410, 1982.

106. Spiegel, MH, and Johnson, EW: Conduction velocity in the proximal and distal segments of the motor fibers of the ulnar nerve of human beings. Arch Phys Med Rehabil 43:57–61, 1962.

107. Tasaki, I: Electric stimulation and the excitatory process in the nerve fiber. Am J Physiol 125:380–395, 1939.

108. Taylor, PK: Nonlinear effects of age on nerve conduction in adults. J Neurol Sci 66:223–234, 1984.

109. Thomas, JE, and Lambert, EH: Ulnar nerve conduction velocity and H-reflex in infants and children. J Appl Physiol 15:1–9, 1960.

110. Thomas, PK: Morphological basis for alterations in nerve conduction in peripheral neuropathy. Proc Roy Soc Med 64:295–298, 1971.

111. Thomas, PK, Sears, TA, and Gilliatt, RW: The range of conduction velocity in normal motor nerve fibres to the small muscles of the hand and foot. J Neurol Neurosurg Psychiatry 22:175–181, 1959.

112. Wagman, IH, and Lesse, H: Maximum conduction velocities of motor fibers of ulnar nerve in human subjects of various ages and sizes. J Neurophysiol 15:235–242, 1952.

113. Walker, DD, and Kimura, J: A fast-recovery electrode amplifier for electrophysiology. Electroencephalogr Clin Neurophysiol 45:789–792, 1978.

114. Wiederholt, WC: Threshold and conduction velocity in isolated mixed mammalian nerves. Neurology 20:347–352, 1970.

Chapter 6

ASSESSMENT OF INDIVIDUAL NERVES

1 INTRODUCTION

Nerve conduction studies consist of stimulating a nerve and recording the evoked potential either from the nerve it-self or from a muscle innervated by the nerve. The basic principles outlined in the previous section apply to any studies, although the anatomic peculiarities dictate specific approaches to each of the commonly tested individual nerves. This section will describe the usual points of stimulation and recording sites, together with the normal values as reported in the literature or established in our institution. Each laboratory should develop its own normal ranges, using a standardized method, to minimize the bias induced by different techniques.

Most electromyographers assess the latency and conduction velocity against the upper and lower limits of normal, defined as a mean plus or minus two standard deviations (±2 SD), in a healthy population. The same criteria do not apply to amplitude, which distributes in a nongaussian manner. In our experience, most individual measures of amplitude in healthy subjects exceed one-half the mean of the control value, which thus serves as a lower limit of normal. In an alternative approach, one can use a log transformation of the amplitude data to accomplish an equal distribution, and then express the normal range in terms of ±2 SD confidence intervals. Some of the tables cite incomplete data of early studies that often failed to report amplitude values.

The conduction studies are readily available for the motor and sensory fibers of the median, ulnar, and radial nerves, the motor fibers of the peroneal and tibial nerves, the sensory fibers of the sural and superficial peroneal nerves, and the accessory nerves. Less easily accessible structures include the brachial plexus, musculocutaneous and other nerves of the shoulder girdle, dorsal sensory branch of the ulnar nerve, lumbosacral plexus, femoral and sciatic nerves, lateral cutaneous nerve of the forearm, lateral femoral cutaneous nerve, and saphenous nerve.

The ordinary nerve conduction studies provide limited information regarding the central or most proximal nerve segment, such as the radicular portion. Supplemental methods help evaluate the motor and sensory conduction in this region by measuring the F wave, H reflex, or somatosensory evoked potentials (see Chapters 17, 18, and 19). Studies of the facial nerve will be described later because of their technical peculiarities (Chapter 7.3)

and their relationship with the blink reflex testing (Chapter 16.2).

2 COMMONLY TESTED NERVES IN THE UPPER LIMB

Median Nerve—Motor Fibers

The median nerve runs relatively superficially in its entire course from the axilla to the palm (Fig. 6–1A). The conventional sites of stimulation include Erb's point, the axilla, the elbow, and the wrist. Its stimulation at Erb's point or the axilla tends to coactivate other nerves in close proximity.[39] The use of the collision technique circumvents the problem (see Chapter 7.3). In our laboratory, we place the cathode over the brachial pulse near the volar crease at the elbow and 3 cm proximal to the distal crease at the wrist. The anode is located 2 cm proximal to the cathode, with the ground electrode around the forearm between the stimulating and recording electrodes, whenever possible (Fig. 6–2A). Additionally, the nerve is accessible to percutaneous stimulation in the palm.[57,88] Tables 6–1 and 6–2 summarize normal values in our laboratory.

Unlike the sensory axons, the motor axons take a recurrent course along the thenar nerve off the median nerve trunk. Thus, unless one is dealing with the exposed nerve for intraoperative monitoring,[5] palmar stimulation may inadvertently activate unintended portions of the thenar nerve. Specifically, surface stimulation aimed at the origin of the thenar nerve in the palm commonly depolarizes the distal branch near the motor point, which results in an erroneously short latency. An unreasonably large latency increase between the wrist and palm then presents a fallacious impression of the carpal tunnel syndrome. To avoid this error, one must carefully select the origin of the thenar nerve as the point of palmar stimulation. To further compound the problem, the recurrent branch may take an anomalous course in rare instances.[103] Proper stimulation of the

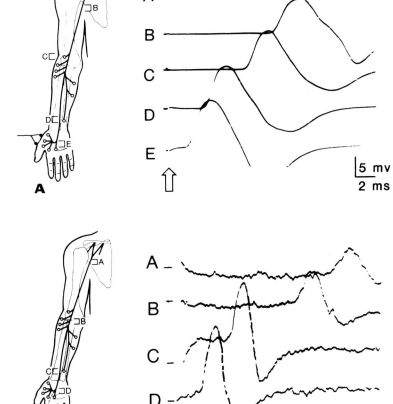

Figure 6–1. A. Motor nerve conduction study of the median nerve. The sites of stimulation include Erb's point (*A*), axilla (*B*), elbow (*C*), wrist (*D*), and palm (*E*). Compound muscle action potentials are recorded with surface electrodes placed on the thenar eminence. **B.** Sensory nerve conduction study of the median nerve. The sites of stimulation include axilla (*A*), elbow (*B*), wrist (*C*), and palm (*D*). Antidromic sensory potentials are recorded with a pair of ring electrodes placed around the second digit.

thenar nerve at the palm[50] requires reversal of the electrode position, with the anode placed distal, rather than proximal, to the cathode. Otherwise, spread of stimulating current will activate the thenar nerve under the anode (see Chapter 7.3).

Recording leads consist of an active electrode (G_1) over the belly of the abductor pollicis brevis and an indifferent electrode (G_2) just distal to the metacarpophalangeal joint (see Fig. 6–2A). Depending upon the electrode positioning, the potentials from other intrinsic hand muscles innervated by the median nerve contribute to the evoked response. In the presence of an anomalous crossover between

the median and ulnar nerve in the forearm, distal and proximal stimulation elicits compound muscle potentials of dissonant wave forms. The latencies of these responses represent two different nerves, precluding their comparison for calculation of the nerve conduction velocity (see Chapter 7.4).

Median Nerve—Sensory Fibers

Stimulation delivered at sites listed for the motor fibers also activate antidromic sensory action potentials of the second or third digits.[75] At the wrist and palm, the cathode is placed 3 cm proximal and 5 cm

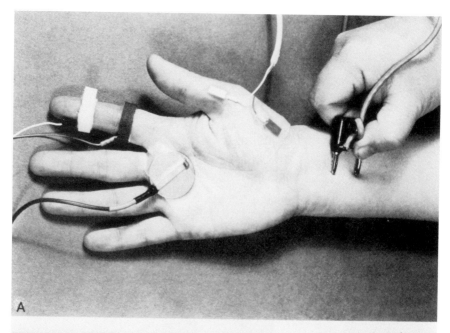

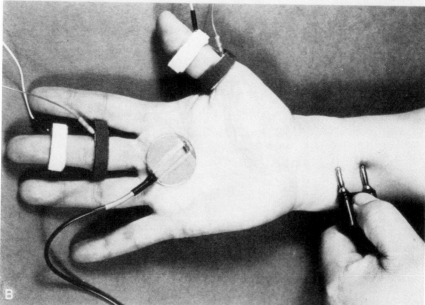

Figure 6–2. A. Motor and sensory conduction studies of the median nerve. Stimulation at the wrist, 3 cm proximal to the distal crease, and recording over the belly (G_1) and tendon (G_2) of the abductor pollicis brevis for motor conduction and around the proximal (G_1) and distal (G_2) interphalangeal joints of the second digit for antidromic sensory conduction. The ground electrode is located in the palm. **B.** Alternative recording sites for sensory conduction study of the median nerve with the ring electrodes placed around the proximal (G_1) and distal (G_2) interphalangeal joints of the third digit or the base (G_1) and the interphalangeal joint (G_2) of the first digit.

Table 6–1 MEDIAN NERVE*

Site of Stimulation	Amplitude†: Motor (mV) Sensory (µV)	Latency‡ to Recording Site (ms)	Difference Between Right and Left (ms)	Conduction Time Between Two Points (ms)	Conduction Velocity (m/s)
Motor fibers					
Palm	6.9 ± 3.2 (3.5)§	1.86 ± 0.28 (2.4)¶	0.19 ± 0.17 (0.5)¶	1.65 ± 0.25 (2.2)¶	48.8 ± 5.3 (38)**
Wrist	7.0 ± 3.0 (3.5)	3.49 ± 0.34 (4.2)	0.24 ± 0.22 (0.7)	3.92 ± 0.49 (4.9)	57.7 ± 4.9 (48)
Elbow	7.0 ± 2.7 (3.5)	7.39 ± 0.69 (8.8)	0.31 ± 0.24 (0.8)	2.42 ± 0.39 (3.2)	63.5 ± 6.2 (51)
Axilla	7.2 ± 2.9 (3.5)	9.81 ± 0.89 (11.6)	0.42 ± 0.33 (1.1)		
Sensory fibers					
Digit	39.0 ± 16.8 (20)	1.37 ± 0.24 (1.9)	0.15 ± 0.11 (0.4)	1.37 ± 0.24 (1.9)	58.8 ± 5.8 (47)
Palm	38.5 ± 15.6 (19)	2.84 ± 0.34 (3.5)	0.18 ± 0.14 (0.5)	1.48 ± 0.18 (1.8)	56.2 ± 5.8 (44)
Wrist	32.0 ± 15.5 (16)	6.46 ± 0.71 (7.9)	0.29 ± 0.21 (0.7)	3.61 ± 0.48 (4.6)	61.9 ± 4.2 (53)
Elbow					

*Mean ± standard deviation (SD) in 122 nerves from 61 patients, 11 to 74 years of age (average, 40), with no apparent disease of the peripheral nerves.
†Amplitude of the evoked response, measured from the baseline to the negative peak.
‡Latency, measured to the onset of the evoked response, with the cathode at the origin of the thenar nerve in the palm.
§Lower limits of normal, based on the distribution of the normative data.
¶Upper limits of normal, calculated as the mean + 2 SD.
**Lower limits of normal, calculated as the mean − 2 SD.

Table 6–2 LATENCY COMPARISON BETWEEN TWO NERVES IN THE SAME LIMB*

Site of Stimulation	Median Nerve (ms)	Ulnar Nerve (ms)	Difference (ms)
Motor fibers			
Wrist	3.34 ± 0.32 (4.0)†	2.56 ± 0.37 (3.3)†	0.79 ± 0.31 (1.4)†
Elbow	7.39 ± 0.72 (8.8)	7.06 ± 0.79 (8.6)	0.59 ± 0.60 (1.8)
Sensory fibers			
Palm	1.33 ± 0.21 (1.8)	1.19 ± 0.22 (1.6)	0.22 ± 0.17 (0.6)
Wrist	2.80 ± 0.32 (3.4)	2.55 ± 0.30 (3.2)	0.29 ± 0.21 (0.7)

*Mean ± standard deviation (SD) in 70 nerves from 35 patients, 14 to 74 years of age (average, 37), with no apparent disease of the peripheral nerve.
†Upper limits of normal, calculated as mean + 2 SD.

distal to the distal crease of the wrist[58] (Fig. 6–1B). Alternative techniques use a fixed distance from the recording electrode, most commonly 12 to 14 cm.[21] Table 6–1 summarizes normal values for the digital potentials recorded with ring electrodes placed around the proximal (G$_1$) and distal (G$_2$) interphalangeal joints of the second digit (see Fig. 6–2A).

The sensory potentials recorded from the first or third digit (Fig. 6–2B) or lateral half of the fourth digit sometimes help identify abnormalities not otherwise detectable. Because of mixed sensory innervation, stimulating the radial or ulnar nerve also elicits a sensory nerve potential over the first or fourth digit, respectively. Thus, inadvertent spread of stimulating current to the other nerves may confuse the issue. Responses recorded from the third digit serve to evaluate C-7 root, middle trunk, and lateral cord, whereas the potentials from the first digit provide assessment of C-6 or C-7 roots, upper or middle trunk, and lateral cord. In contrast to postganglionic lesions, which cause degeneration of the sensory axons, patients with a preganglionic root avulsion have a normal sensory potential over the anesthetic digits.

Unlike the compound muscle action potentials, which maintain nearly the same amplitude irrespective of stimulus sites, the antidromically activated digital potentials diminish substantially with increasing nerve length under study. Indeed, stimulation at Erb's point or the axilla sometimes fails to elicit unequivocal digital potentials without the use of an averaging technique. Here, temporal dispersion between fast and slow conducting fibers results in duration-dependent phase cancellation (see Chapter 7.5).[6,59] In addition, naturally recurring orthodromic sensory impulses may partially extinguish the antidromic impulse by collision. These tendencies favor a proximal, as compared with a more distal, stimulation in proportion to the distance between the stimulating and recording electrodes.

Motor axons have a threshold similar to that of the large myelinated sensory axons. Thus, in studying the mixed nerve, superimposition of action potentials from distal muscles may obscure the antidromically recorded sensory potential. Palmar stimulation distal to the origin of the recurrent motor fibers, however, selectively activates the sensory fibers of the median nerve. This is helpful because muscle action potentials, if elicited with more proximal stimulation, become apparent by a change in the waveform of the evoked response.[58] Recording of the antidromic sensory potentials suffices for routine clinical purposes. Alternatively, digital[7–9,104] or palmar stimulation[15,27] allows recording of the orthodromic sensory potential at the palm, wrist, or elbow with either surface or needle electrodes. This method demands a higher resolution to compensate for a smaller size of the orthodromic potential. The averaging technique offers a distinct advantage in detecting such small nerve potentials, especially in a diseased nerve.

The palmar cutaneous branch of the median nerve usually arises about 5.5 cm proximal to the radial styloid and innervates skin of the thenar eminence. Antidromic stimulation of the median nerve elicits sensory potentials over the mid-thenar eminence. In one series, normal values over 10-cm segments included onset latency of 2.6 ± 0.2 ms (mean $\pm$ SD) and amplitude of 12.0 ± 4.6 μV.[63] This technique may help differentiate the carpal tunnel syndrome, which spares the palmar cutaneous branch, from a more proximal injury.

Multiple Stimulation Across the Carpal Ligament

The use of palmar stimulation provides a simple means of identifying conduction abnormalities of sensory or motor fibers under the transverse carpal ligament or along the most terminal segment.[97] This distinction differentiates the carpal tunnel syndrome from distal neuropathy seen, for example, in digital nerves of diabetics.[13] Stimulation of the median nerve at multiple sites across the wrist (Fig. 6–3A) further localizes the point of maximal conduction delay within the distal segment of the median nerve.[57,58] The sensory axons normally show a predictable latency change of 0.16 to 0.20 ms/cm with series of stimulation from midpalm to distal forearm in 1-cm increments (Fig. 6–3B). A sharply localized latency increase across a 1-cm segment indicates focal abnormalities of the median nerve (Fig. 6–3C and D).

Ulnar Nerve

Like the median nerve, the ulnar nerve takes a relatively superficial course along its entire length. Common sites of stimulation include Erb's point, the axilla, the elbow, the wrist, and the palm (Fig. 6–4). Routine motor conduction studies consist of stimulating the nerve at multiple sites and recording the muscle potential from the hypothenar muscles with surface electrodes placed over the belly of the abductor digiti minimi (G_1) and its tendon

(G_2) 3 cm distally (Fig. 6–5A). Alternatively, one may also record a compound muscle action potential from forearm muscles such as flexor carpi ulnaris or flexor digitorum profundus.[32] For study of the deep palmar branch of the ulnar nerve, the muscle potential is recorded from the first dorsal interosseous or adductor pollicis (Fig. 6–5B). Here, the latency difference between the hypothenar and thenar responses provides a measure of conduction along the deep branch. In one series, the upper limit of the normal range, based on 373 studies, included 4.5 ms for the distal latency to the first dorsal interosseous, 2.0 ms for the latency difference between this muscle and adductor minimi, and 1.3 ms for the latency difference between the two sides.[80] Spread of stimulus current at Erb's point or in the axilla causes less obvious problems in studying the ulnar, as compared with the median nerve, because the hypothenar eminence contains only ulnar-innervated muscles. Nonetheless, coactivation of the median nerve gives rise to volume-conducted potentials from the thenar eminence, unless eliminated by the collision technique.[56] Tables 6–2 and 6–3 show the normal values in our laboratory.

The common sites of stimulation for the motor fibers include sites above and below the elbow, for documenting a tardy ulnar palsy or a cubital tunnel syndrome. A very proximal site of stimulation near the axilla, however, tends to coactivate the median nerve. Segmental stimulation across the elbow in 1-cm increments may detect an abrupt change in the latency and waveform of the compound action potential at the site of localized compression. The conventional studies often fail to uncover such a focal slowing, which induces an insignificant delay when calculated over a longer segment (see Chapter 5.4). The ulnar nerve slides back and forth in the cubital tunnel with flexion and extension of the elbow joint.[42] Holding the arm at either 135 degrees or 90 degrees flexion not only during stimulation of the nerve but also during measurement of the surface distance minimizes the error.[60] Stimulation is applied at a fixed distance from the distal crease of

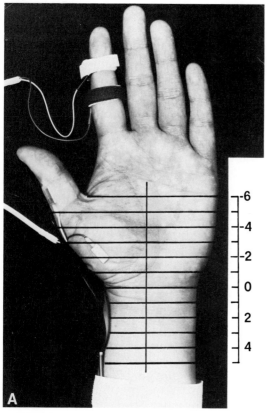

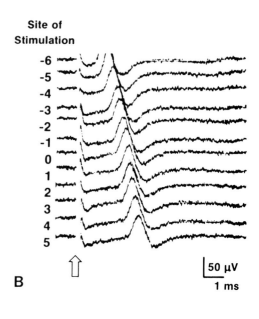

Figure 6–3. A. Twelve sites of stimulation in 1-cm increments along the length of the median nerve. The zero level at the distal crease of the wrist corresponds to the origin of the transverse carpal ligament. The photograph shows a recording arrangement for sensory nerve potentials from the second digit and muscle action potentials from the abductor pollicis brevis. **B.** Sensory nerve potentials in a normal subject recorded after stimulation of the median nerve at multiple points across the wrist. The numbers on the left indicate the site of each stimulus (compare with **A**). The latency increased linearly with stepwise shifts of stimulus site proximally in 1-cm increments. (From Kimura,[58] with permission.)

the wrist or from the recording electrode. This practice allows an accurate comparison of the distal motor latencies between the two sides and among different subjects. In our laboratory, we place the cathode 3 cm proximal to the distal crease of the wrist and the anode 2 cm further, proximally.

Stimulation of the ulnar nerve trunk elicits an antidromic sensory potential of the fourth and fifth digits (see Fig. 6–5A and B). The common cathodal points include sites 3 cm proximal to the distal crease at the wrist and 5 cm distal to the crease in the palm. In either case, the anode is located 2 cm proximal to the cathode. These stimulus sites are compa-

rable to those of the median nerve. Stimulation of the digital nerve with ring electrodes placed around the proximal (cathode) and distal (anode) interphalangeal joints of the fifth finger elicits orthodromic sensory potential at various sites along the course of the nerve. One can also record a mixed nerve potential from the ulnar nerve proximally after stimulation of the nerve at the palm or wrist. These studies help differentiate lesions of C-8 and T-1 roots from those of the lower trunk, medial cord of the brachial plexus, or ulnar nerve. The sensor potential should be normal in preganglionic C-8 and T-1 root avulsion, despite the clinical sensory loss.

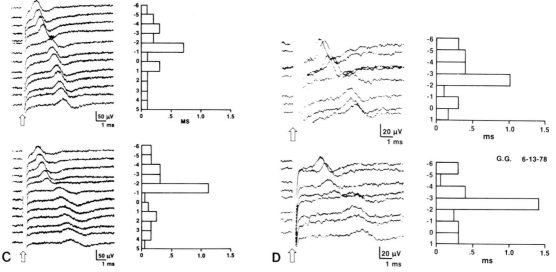

Figure 6–3 (Cont.). C. Sensory nerve potentials in a patient with the carpal tunnel syndrome. Both hands showed a sharply localized slowing from −2 to −1 with the calculated segmental conduction velocity of 14 m/s on the left (*top*) and 9 m/s on the right (*bottom*). Notice the distinct change in waveform of the sensory potential at the point of localized conduction delay. Double humped appearance at −2 on the left suggests sparing of some sensory axons at this level. **D.** Sensory nerve potential in a patient with the carpal tunnel syndrome. Both hands show a sharply localized slowing from −3 to −2 with a segmental conduction velocity of 10 m/s on the left (*top*) and 7 m/s on the right (*bottom*). An abrupt change in waveform of the sensory potential also indicates the point of localized conduction delay. (From Kimura,[58] with permission.)

The dorsal sensory branch leaves the common trunk of the ulnar nerve 5 to 8 cm proximal to the ulnar styloid.[47,54] It becomes superficial between the tendon of the flexor carpi ulnaris and the ulna. Surface stimulation here selectively evokes antidromic sensory potentials over the dorsum of the hand. For recording, the active electrode (G_1) is placed between the fourth and fifth metacarpals and the reference electrode (G_2), at the base of the fifth digit (Fig. 6–5C). Stimulation of the ulnar nerve trunk more proximally elicits a mixed nerve potential that slightly precedes a large muscle action potential from the intrinsic hand muscles.

The normal values of the sensory potential established in one study[47] include latency of 2.0 ± 0.3 ms (mean ± SD) when recorded 8 cm from the point of stimulation, conduction velocity of 60 ± 4.0 m/s between elbow and forearm, and amplitude of 20 ± 6 µV with distal stimulation. This technique is useful for the assessment of the ulnar nerve when a severe lesion at the wrist has precluded the conventional recording from the hypothenar muscles or digits. It also helps in localizing a lesion within the forearm in the segment proximal or distal to the takeoff of this branch.

Radial Nerve

The radial nerve becomes relatively superficial at the supraclavicular fossa, in the axilla near the spinal groove, above the elbow, and in the forearm (Fig. 6. The optimal sites of electrical stim of the motor fibers, therefore, i Erb's point, (2) between th chialis and medial edge about 18 cm proximal condyle, (3) betwee and the tendon o mal to the lat tween the e tensor di of the

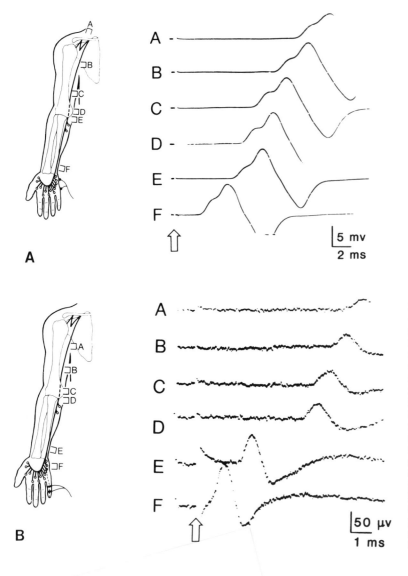

Figure 6–4. **A.** Motor nerve conduction study of the ulnar nerve. The sites of stimulation include Erb's point (*A*), axilla (*B*), above the elbow (*C*), elbow (*D*), below the elbow (*E*), and wrist (*F*). Compound muscle action potentials are recorded with surface electrodes placed on the hypothenar eminence. **B.** Sensory nerve conduction study of the ulnar nerve. The sites of stimulation include axilla (*A*), above the elbow (*B*), elbow (*C*), below the elbow (*D*), wrist (*E*), and palm (*F*). The tracings show antidromic sensory potentials recorded with the ring electrodes placed around the fifth digit.

comparison between the two responses difficult. The use of needle electrodes for stimulation and recording helps circumvent some of these limitations.[29] Needle electrodes also allow relatively selective recording from more proximal muscles such as the anconeus, brachioradialis, or triceps.[37,49]

The sensory fibers cross the extensor pollicis longus at the base of the thumb.[22,23] At this point, one can palpate the nerve and record antidromic sensory potentials with the disk electrode (G$_1$) over the first web space. The reference electrode (G$_2$) is placed near the first dor-

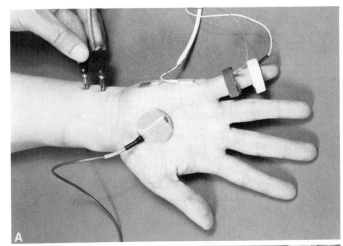

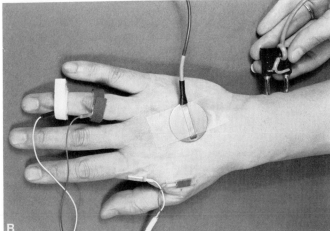

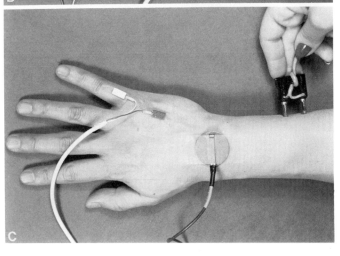

Figure 6–5. A. Motor and sensory conduction study of the ulnar nerve. The photograph shows stimulation at the wrist, 3 cm proximal to the distal crease, and recording over the belly (G_1) and tendon (G_2) of the abductor digiti minimi for motor conduction and around the proximal (G_1) and distal (G_2) interphalangeal joints of the fifth digit for antidromic sensory conduction. **B.** Alternative recording sites for motor and sensory conduction studies of the ulnar nerve with the surface electrodes over the belly (G_1) and tendon (G_2) of the first dorsal interosseous muscle for motor conduction and around the proximal (G_1) and distal (G_2) interphalangeal joints of the fourth digit for antidromic sensory conduction. **C.** Sensory conduction study of the dorsal cutaneous branch of the ulnar nerve. The photograph shows stimulation along the medial aspect of the forearm between the tendon of the flexor carpi ulnaris and the ulna, 14 to 18 cm from the active electrode, and recording over the dorsum of the hand between the fourth and fifth metacarpals (G_1) and the base of the fifth digit (G_2).

Table 6–3 ULNAR NERVE*

Site of Stimulation	Amplitude‡: Motor (mV) Sensory (μV)	Latency† to Recording Site (ms)	Difference Between Right and Left (ms)	Conduction Time Between Two Points (ms)	Conduction Velocity (m/s)
Motor fibers					
Wrist	5.7 ± 2.0 (2.8)§	2.59 ± 0.39 (3.4)¶	0.28 ± 0.27 (0.8)¶		
Below elbow	5.5 ± 2.0 (2.7)	6.10 ± 0.69 (7.5)	0.29 ± 0.27 (0.8)	3.51 ± 0.51 (4.5)¶	58.7 ± 5.1 (49)**
Above elbow	5.5 ± 1.9 (2.7)	8.04 ± 0.76 (9.6)	0.34 ± 0.28 (0.9)	1.94 ± 0.37 (2.7)	61.0 ± 5.5 (50)
Axilla	5.6 ± 2.1 (2.7)	9.90 ± 0.91 (11.7)	0.45 ± 0.39 (1.2)	1.88 ± 0.35 (2.6)	66.5 ± 6.3 (54)
Sensory fibers					
Digit					
Wrist	35.0 ± 14.7 (18)	2.54 ± 0.29 (3.1)	0.18 ± 0.13 (0.4)	2.54 ± 0.29 (3.1)	54.8 ± 5.3 (44)
Below elbow	28.8 ± 12.2 (15)	5.67 ± 0.59 (6.9)	0.26 ± 0.21 (0.5)	3.22 ± 0.42 (4.1)	64.7 ± 5.4 (53)
Above elbow	28.3 ± 11.8 (14)	7.46 ± 0.64 (8.7)	0.28 ± 0.27 (0.8)	1.79 ± 0.30 (2.4)	66.7 ± 6.4 (54)

*Mean ± standard deviation (SD) in 130 nerves from 65 patients, 13 to 74 years of age (average, 39), with no apparent disease of the peripheral nerves.
†Amplitude of the evoked response, measured from the baseline to the negative peak.
‡Latency, measured to the onset of the evoked response, with the cathode 3 cm above the distal crease in the wrist.
§Lower limits of normal, based on the distribution of the normative data.
¶Upper limits of normal, calculated as the mean + 2 SD.
**Lower limits of normal, calculated as the mean − 2 SD.

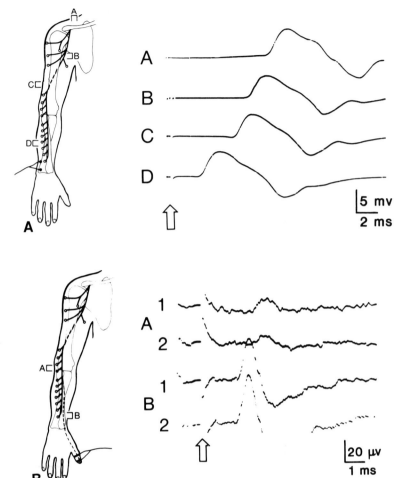

Figure 6–6. A. Motor nerve conduction study of the radial nerve. The sites of stimulation include Erb's point (*A*), axilla (*B*), above the elbow (*C*), and mid forearm (*D*). Compound muscle action potentials are recorded from the extensor indicis with a pair of surface electrodes. **B.** Sensory nerve conduction study of the radial nerve. The sites of stimulation include elbow (*A*) and distal forearm (*B*). Antidromic sensory potentials are recorded with the ring electrodes placed around the first digit.

sal interosseous[65,66] or between the second and third metacarpals.[96] The antidromic sensory potential can also be recorded by a pair of ring electrodes around the thumb (Fig. 6–7B). The sensory nerve becomes readily accessible to percutaneous stimulation at the lateral edge of the radius in the distal forearm 10 to 14 cm proximal to the recording electrodes. An additional stimulation at the elbow under the brachioradialis muscle lateral to the biceps tendon (see Fig. 6–6B) allows determination of conduction velocities in the segments between the elbow and wrist and the wrist and thumb.[31,92,94]

One can also record orthodromic sensory potentials from the wrist, axilla, or elbow after stimulation of the radial nerve at the thumb. With this technique, however, spread of current to the median nerve, which partially supplies the thumb, accounts for 25 percent of the sensory potential recorded at the wrist or elbow and 50 percent at the axilla.[100] Stimulation at the wrist, especially with needle electrodes placed along the nerve, accomplishes more selective activation of the radial nerve. Table 6–4 summarizes the results of one series.[100] The digital nerve potential from the thumb reflects the integrity of the C-6 and C-7 roots, upper and middle trunks, posterior cord of the brachial plexus, and the radial nerve. Preganglionic avulsion of the C-6 and C-7 roots results in a clinical sensory loss associated with no abnormalities of the sensory potentials.

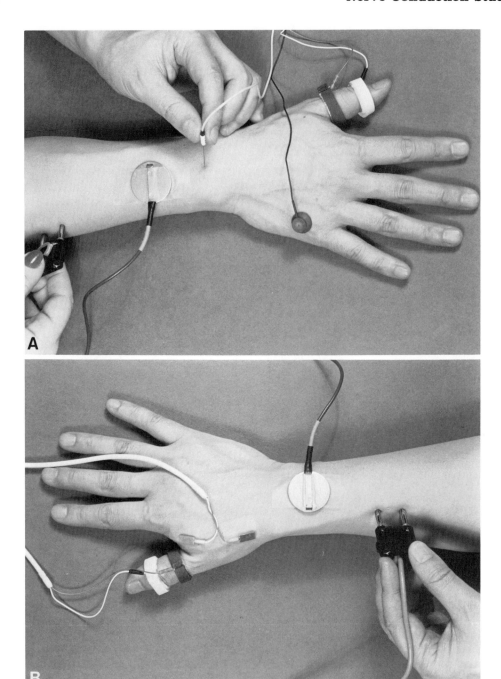

Figure 6–7. A. Motor and sensory conduction studies of the radial nerve. The photo shows stimulation in the forearm with the cathode at the lateral edge of the extensor carpi ulnaris muscle, 8 to 10 cm proximal to the styloid process. The monopolar needle electrode (G_1) is inserted in the extensor indicis with a reference electrode (G_2) over the dorsum of the hand laterally for motor conduction studies. The recording electrodes are placed around the base (G_1) and interphalangeal joint (G_2) of the first digit for antidromic sensory conduction. **B.** Alternative stimulation and recording sites for antidromic sensory nerve conduction study of the radial nerve. The photograph shows the cathode placed at the lateral edge of the radius in the distal forearm, with the anode 2 cm proximally. The recording electrodes are placed either around the base (G_1) and interphalangeal joint (G_2) of the first digit or over the palpable nerve between the first and second metacarpals (G_1) and 2 to 3 cm distally (G_2).

Table 6–4 RADIAL NERVE

Conduction	n	Conduction Velocity (m/s) or Conduction Time (ms)	Amplitude: Motor (mV) Sensory (μV)	Distance (cm)
Motor				
Axilla–elbow	8	69 ± 5.6	11 ± 7.0	15.7 ± 3.3
Elbow–forearm	10	62 ± 5.1	13 ± 8.2	18.1 ± 1.5
Forearm–muscle	10	2.4 ± 0.5	14 ± 8.8	6.2 ± 0.9
Sensory				
Axilla–elbow	16	71 ± 5.2	4 ± 1.4	18.0 ± 0.7
Elbow–wrist	20	69 ± 5.7	5 ± 2.6	20.0 ± 0.5
Wrist–thumb	23	58 ± 6.0	13 ± 7.5	13.8 ± 0.4

From Trojaborg and Sinrup,[100] with permission.

3 OTHER NERVES DERIVED FROM THE CERVICAL OR BRACHIAL PLEXUS

Phrenic Nerve

Conduction studies of the phrenic nerve, though described early,[16,78] have not gained popularity in part because surface stimulation in the cervical area requires shocks of a relatively high intensity. Moreover, some patients do not tolerate the esophageal electrode used for recording the diaphragmatic potentials well.

As an alternative method, some recommend the use of needle electrodes for stimulating the nerve and surface electrodes for recording the response.[68] A standard monopolar needle electrode is inserted medially from the lateral aspect of the neck at the level of the cricoid cartilage (Fig. 6–8). After traversing the posterior margin of the sternocleidomastoid muscle, the needle tip comes to within a few millimeters of the phrenic nerve and adequately distant from the carotid artery anteriorly and the apex of the lung inferiorly. A metal plate placed on the manubrium serves as the anode. With the needle placed appropriately, shocks of very low intensity suffice for selective stimulation of the phrenic nerve. Supramaximal stimulation may coactivate the brachial plexus located posteriorly behind the anterior scalene muscle.

The diaphragmatic action potential gives rise to a strong positivity at the seventh or eighth intercostal space near the costochondral junction and a mild negativity at the xiphoid process.[70] A pair of surface electrodes placed over these recording sites, therefore, register the largest amplitude with summation of out-of-phase activities. Normal ranges es-

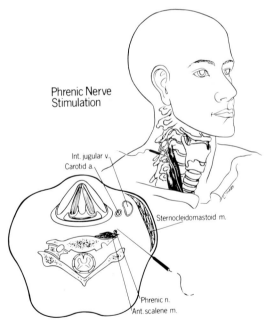

Figure 6–8. Motor conduction study of the phrenic nerve. The diagram shows stimulation with a needle inserted medially through the posterior margin of the sternocleidomastoid at the level of the cricoid cartilage. The recording electrodes are placed on the xiphoid process (G₁) and at the eighth intercostal space near the costochondral junction (G₂). (From MacLean and Mattioni,[68] with permission.)

tablished in 30 healthy subjects using needle stimulation[68] very closely approximate the earlier results obtained by surface stimulation (Table 6–5).[16,78]

Greater Auricular Nerve

This nerve, derived mainly from C-2 and C-3, winds around the posterior border of the sternomastoid and ascends cephalad on the surface of that muscle from the neck to the ear. Stimulation with a pair of surface electrodes firmly placed against the lateral border of the sternocleidomastoid muscle elicits an orthodromic sensory potential easily detectable on the back of the earlobe. Reported values include latency of 1.7 ± 0.2 ms (mean $\pm$ SD) for the distance of 8 cm and conduction velocity of 46.8 ± 6.6 m/s in 20 healthy subjects[81] and latency of 1.9 ± 0.2 ms and amplitude of 22.4 ± 8.9 μV in 32 normal controls.[55]

Brachial Plexus

Stimulation of the brachial plexus at Erb's point (see Fig. 1–7) elicits muscle action potentials in the proximal muscles of the shoulder girdle,[35] including the serratus anterior, via the long thoracic nerve.[51] It also evokes action potentials in the distal muscles, such as those of the thenar and hypothenar eminence. The volume-conducted potentials from a number of coactivated muscles, however, interfere with the accurate recording of the intended signal even with the electrode placed over a specific intrinsic hand muscle. A collision technique circumvents this difficulty by blocking the unwanted impulse with a second stimulus applied distally to the nerve not being tested (see Chapter 7.3). The use of needle electrodes accomplishes more selective stimulation but carries the risk of inducing pneumothorax.[82]

The triceps has the end-plate zone vertically oriented with the distal portion of the muscle innervated by longer nerve branches. Thus, the latency of a recorded response increases with the distance from the stimulus point. The latency changes nonlinearly, reflecting irregularly spaced points of innervation. The biceps and deltoid muscles have one or more horizontally directed end plates, mostly in the middle of the fibers.[71–73,76] The point of recording does not affect the latency of the response in these muscles as much as in the triceps. The same probably applies to the infraspinatus and supraspinatus. The needle electrodes register from a more limited area, providing a reliable measure of latencies even with simultaneous activation of many nerves.[61] Intramuscular recordings, however, fail to reveal the true waveform of the compound muscle action potential because of restricted recording area.

When one is testing a unilateral involvement of the brachial plexus, comparison between the affected and normal

Table 6–5 PHRENIC NERVE

Authors	Stimulation Point	Recording Site	No.	Amplitude (μV)	Duration (ms)	Onset Latency (ms)	Difference Between Sides (ms)
Newsom Davis (1967)[78]			18			7.7 ± 0.80	
Delhez (1965)[16]			30 on right			7.5 ± 0.53	
			30 on left			8.2 ± 0.71	
MacLean and Mattioni (1981)[68]	Needle electrode placed posterior to sterno-cleidomastoid	Xiphoid process	30	8.5 ± 40.5	48.1 ± 12.2	7.4 ± 0.59	0.08 ± 0.42

Table 6–6 NERVE CONDUCTION TIMES FROM ERB'S POINT TO MUSCLE

Muscle	n	Distance (cm)	Latency (ms)
Biceps	19	20	4.6 ± 0.6
	15	24	4.7 ± 0.6
	14	28	5.0 ± 0.5
Deltoid	20	15.5	4.3 ± 0.5
	17	18.5	4.4 ± 0.4
Triceps	16	21.5	4.5 ± 0.4
	23	26.5	4.9 ± 0.5
	16	31.5	5.3 ± 0.5
Supraspinatus	19	8.5	2.6 ± 0.3
	16	10.5	2.7 ± 0.3
Infraspinatus	20	14	3.4 ± 0.4
	15	17	3.4 ± 0.5

Modified from Gassel,[35] with permission.

sides offers the most sensitive indicator (Table 6–6). The standard protocol calls for equalizing the distance between the stimulating and recording electrodes on both sides. One should adhere to this principle in the study of any muscle of the shoulder girdle, but particularly that of the triceps, for the reasons stated previously. Recording from the serratus anterior permits conduction studies of the long thoracic nerve.[84]

Instead of surface stimulation, the use of a needle electrode enables one to deliver a localized stimulus directly to the nerve roots.[4,52,68,69] For this purpose, a standard 50- to 75-mm monopolar needle is inserted perpendicular to the skin surface until the uninsulated tip comes to rest directly on the vertebral transverse process. Joint stimulation of the C-5 and C-6 roots by placing the needle 1 to 2 cm lateral to the C-5 spinous process tests the upper trunk and lateral cord (Fig. 6–9A). Similarly, positioning the needle slightly caudal to the C-7 spinous process stimulates the C-8 and T-1 roots simultaneously for conduction across the lower trunk and medial cord (Fig. 6–9B). The needle inserted between these two points activates the C-6, C-7 and C-8 roots together for evaluation of the posterior cord. A metal plate or disk electrode on the skin surface or a second needle electrode serves as the anode.

Recording from several muscles helps in the evaluation of different portions of the brachial plexus, e.g., biceps for the upper trunk and lateral cord, triceps for the posterior cord, and ulnar-innervated intrinsic hand muscles for the lower trunk and medial cord. Table 6–7 summarizes the conduction time across the brachial plexus, calculated by subtracting the distal latency of the ulnar nerve.[67] The side-to-side difference exceeding 0.6 ms indicates unilateral lesions, making it a more sensitive index than the absolute latency.

Musculocutaneous and Lateral Antebrachial Cutaneous Nerves

Stimulation given at Erb's point[36,61] or the axilla[87] activates the motor fibers for determining conduction characteristics of the proximal segment.[77,99] In the posterior cervical triangle, the stimulus is applied 3 to 6 cm above the clavicle through an insulated stainless steel needle electrode placed just behind the sternocleidomastoid muscle. At the axilla the needle is located between the axillary artery medially and the coracobrachialis muscle laterally. Either surface or needle electrodes suffice for recording the muscle action potentials from the biceps brachii (Table 6–8).

The sensory branch runs superficially at the level of the elbow, just lateral to the tendon of the biceps as it emerges at the lateral cutaneous nerve. Here, one can palpate and stimulate the nerve between the tendon of the biceps medially and the

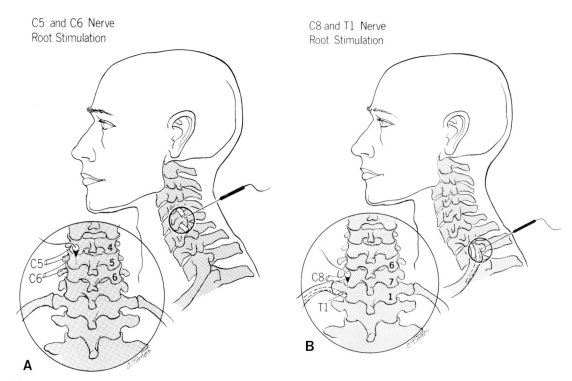

Figure 6–9. A. C-5 and C-6 root stimulation. The diagram shows the needle inserted perpendicular to the skin, 1 to 2 cm lateral to the C-5 spinous process. **B.** C-8 and T-1 root stimulation. The diagram shows the needle inserted slightly caudal to the C-7 spinous process. (From MacLean,[67] with permission.)

brachioradialis laterally. The sensory potentials can be recorded at the posterior cervical triangle and axilla by the same electrodes, positioned to stimulate motor fibers.

The same stimulus also elicits the antidromic sensory potential of the distal branch, the lateral cutaneous nerve of the forearm (Fig. 6–10). The recording electrode is placed 12 cm distally over the course of the nerve in the forearm, along a straight line from the stimulus point to the radial artery at the wrist. Table 6–9 summarizes normal values reported in two studies.[45,95] Like the median sensory potentials recorded from the first digit,

Table 6–7 BRACHIAL PLEXUS LATENCY WITH NERVE ROOT STIMULATION

Plexus	Site of Stimulation	Recording Site	Latency Across Plexus (ms)		
			Range	*Mean*	*SD*
Brachial (upper trunk and lateral cord)	C5 and C6	Biceps brachii	4.8–6.2	5.3	0.4
Brachial (posterior cord)	C6, C7, C8	Triceps brachii	4.4–6.1	5.4	0.4
Brachial (lower trunk and medial cord)	C8 and T1 Ulnar nerve	Abductor digiti quinti	3.7–5.5	4.7	0.5

From MacLean,[67] with permission.

Table 6–8 MUSCULOCUTANEOUS NERVE

| | | Motor Nerve Conduction Between Erb's Point and Axilla | | | Orthodromic Sensory Nerve Conduction Between Erb's Point and Axilla | | | Orthodromic Sensory Nerve Conduction Between Axilla and Elbow | | |
| | | *Range of Conduction Velocity (m/s)* | *Range of Amplitude (μV)* | | | *Range of Conduction Velocity (m/s)* | *Range of Amplitude (μV)* | | *Range of Conduction Velocity (m/s)* | *Range of Amplitude (μV)* |
Age	*n*		*Axilla*	*Erb's Point*	*n*			*n*		
15–24	14	63–78	9–32	7–27	14	59–76	3.5–30	15	61–75	17–75
25–34	6	60–75	8–30	6–26	6	57–74	3–25	8	59–73	16–72
35–44	8	58–73	8–28	6–24	7	54–71	2.5–21	8	57–71	16–69
45–54	10	55–71	7–26	6–22	10	52–69	2–18	13	55–69	15–65
55–64	9	53–68	7–24	5–21	9	49–66	2–15	10	53–67	14–62
65–74	4	50–66	6–22	5–19	4	47–64	1.5–12	6	51–65	13–59

From Trojaborg,[99] with permission.

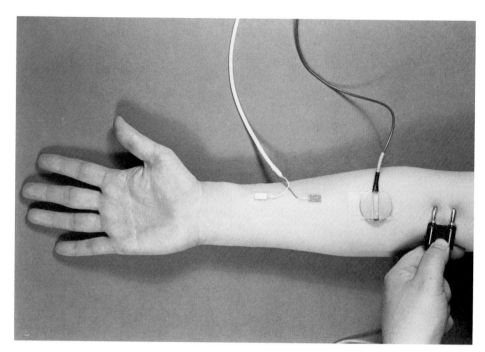

Figure 6–10. Sensory conduction study of the lateral cutaneous nerve of the forearm. The photograph shows stimulation just lateral to the tendon of the biceps and recording from the nerve with the electrodes placed 12 cm distal to the cathode along the straight line to the radial artery (G_1) and 2 to 3 cm further distally (G_2).

Table 6–9 LATERAL AND MEDIAL CUTANEOUS NERVE
(MEAN ± SD)

Authors	Nerve	Number of Patients Seen	Age (mean)	Distance (cm)	Latency Onset (ms)	Latency Peak (ms)	Conduction Velocity (m/s)	Amplitude (μV)
Spindler and Felsenthal (1978)[95]	Lateral cutaneous nerve	30	20–84 (35)	12	1.8 ± 0.1	2.3 ± 0.1	65 ± 4	24.0 ± 7.2
Izzo et al. (1985)[45]	Lateral cutaneous nerve	154	17–80 (45)	14		2.8 ± 0.2	62 ± 4	18.9 ± 9.9
	Medial cutaneous nerve	155	17–80 (45)	14		2.7 ± 0.2	63 ± 5	11.4 ± 5.2
Reddy (1983)[86]	Medial cutaneous nerve	30	23–60 (38)	18	2.7 ± 0.2	3.3 ± 0.2	66 ± 4	15.4 ± 4.1

study of the musculocutaneous nerve provides evaluation of the C-6 root, upper trunk, and lateral cord.

Medial Antebrachial Cutaneous Nerve

This pure sensory branch, derived from the C-8 and T-1 roots subserves the sensation over the medial aspect of the forearm. Surface stimulation in the middle of the arm where the nerve pierces the deep fascia elicits antidromic sensory potentials best recorded over the course of its volar branch. Table 6–9 shows the results of two studies.[45,86] The evaluation of this nerve helps in evaluating the C-8 and T-1 roots, lower trunk, and medial cord of the brachial plexus.

4 COMMONLY TESTED NERVES IN THE LOWER LIMB

Tibial Nerve

For motor conduction studies the tibial nerve is stimulated at the popliteal fossa and at the ankle lateral to the medial malleolus. Tables 6–10 and 6–11 summarize the normal values in our laboratory. The nerve bifurcates into two branches within 1 cm of the malleolar-calcaneal axis in 90% of the feet.[17] One may record the evoked potentials either from the abductor hallucis innervated by the medial plantar nerve or abductor digiti quinti supplied by the lateral plantar nerve (Figs. 6–11 and 6–12). One study reports normal distal latencies (mean ± SD) of 4.9 ± 0.6 ms for medial and 6.0 ± 0.7 ms for lateral plantar nerves over a 12-cm segment.[43] Stimulation of the tibial nerve above and below the medial malleolus determines the conduction characteristics of the motor fibers across the tarsal tunnel. Reported normal values across a 10-cm segment (mean ± SD) include 3.8 ± 0.5 ms for the medial and 3.9 ± 0.5 ms for the lateral plantar nerves.[33]

Sensory conduction studies of the medial and lateral plantar nerves consist of stimulating the first or fifth toes and recording the evoked sensory potentials with surface or needle electrodes placed just below the medial malleolus.[3,89] The orthodromic medial plantar potentials have latencies of 2.4 ± 0.2 ms, 3.2 ± 0.3 ms, 4.0 ± 0.2 ms, and 4.0 ± 0.2 ms for 10-, 14-, and 18-cm segments. The lateral plantar latencies average 3.2 ± 0.3 ms and 4.0 ± 0.3 ms for the distance of 14 and 18 cm. A modification of this method also allows the measurement of orthodromic sensory conduction along the plantar interdigital nerve.[30]

The responses recorded at the knee after stimulation of the tibial nerve at the ankle consist of orthodromic sensory and

Table 6–10 TIBIAL NERVES*

Site of Stimulation	Amplitude† (mV)	Latency† to Recording Site (ms)	Difference Between Two Sides (ms)	Conduction Time Between Two Points (ms)	Conduction Velocity (m/s)
Ankle	5.8 ± 1.9 (2.9)§	3.96 ± 1.00 (6.0)¶	0.66 ± 0.57 (1.8)¶	8.09 ± 1.09 (10.3)¶	48.5 ± 3.6 (41)**
Knee	5.1 ± 2.2 (2.5)	12.05 ± 1.53 (15.1)	0.79 ± 0.61 (2.0)		

*Mean ± standard deviation (SD) in 118 nerves from 59 patients, 11 to 78 years of age (average, 39), with no apparent disease of the peripheral nerves.
†Amplitude of the evoked response, measured from the baseline to the negative peak.
‡Latency, measured to the onset of the evoked response, with a standard distance of 10 cm between the cathode and the recording electrode.
§Lower limits of normal, based on the distribution of the normative data.
¶Upper limits of normal, calculated as the mean + 2 SD.
**Lower limits of normal, calculated as the mean − 2 SD.

**Table 6–11 LATENCY COMPARISON BETWEEN TWO
NERVES IN THE SAME LIMB***

Site of Stimulation	Peroneal Nerve	Tibial Nerve	Difference
Ankle	3.89 ± 0.87 (5.6)†	4.12 ± 1.06 (6.2)†	0.77 ± 0.65 (2.1)†
Knee	12.46 ± 1.38 (15.2)	12.13 ± 1.48 (15.1)	0.88 ± 0.71 (2.3)

*Mean ± standard deviation (SD) in 104 nerves from 52 patients, 17 to 86 years of age (average, 41), with no apparent disease of the peripheral nerve.
†Upper limits of normal, calculated as the mean + 2 SD.

antidromic motor potentials.[74] Stimulation of the tibial nerve below the medial malleolus elicits the antidromic sensory nerve potential of the first and fifth toes.[44] In either case, the use of an averaging technique improves the resolution of small signals that would otherwise escape detection. The study of the plantar nerves helps in evaluating the integrity of the postganglionic sensory fibers derived from the L-4 and L-5 roots in patients with footdrop.[41]

Common and Deep Peroneal Nerve

Stimulation of the nerve above or below the head of the fibula or just above the ankle elicits muscle action potentials in the extensor digitorum brevis (Figs. 6–13 and 6–14). This muscle, primarily supplied by the deep peroneal nerve, may also receive an anomalous innervation from the superficial peroneal nerve. The communicating branch, called the accessory deep peroneal nerve, passes behind the lateral malleolus to reach the lateral portion of the muscle. In the presence of this anomaly, stimulation of the deep peroneal nerve at the ankle gives rise to a much smaller compound muscle action potential than the shocks applied at the knee (see Chapter 7.4).

Series of shocks applied in short increments help localize a focal conduction abnormality across the knee (see Chapter 5.4). In advanced neuropathy, recording from the tibialis anterior, rather than the atrophic extensor digitorum brevis, may facilitate the assessment of this segment.[18] Stimulation of the peroneal nerve at the ankle elicits mixed nerve potentials at the head of the fibula.[38] The use of needle electrode and averaging technique improves resolution in recording this potential which would otherwise escape detection. Tables 6–11 and 6–12 summarize the normal values.

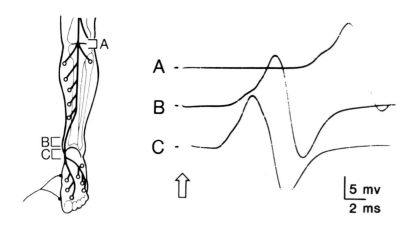

Figure 6–11. Motor nerve conduction study of the tibial nerve. The sites of stimulation include knee (*A*), above the medial malleolus (*B*), and below the medial malleolus (*C*). Compound muscle action potentials are recorded with surface electrodes placed over the abductor hallucis.

5 mv
2 ms

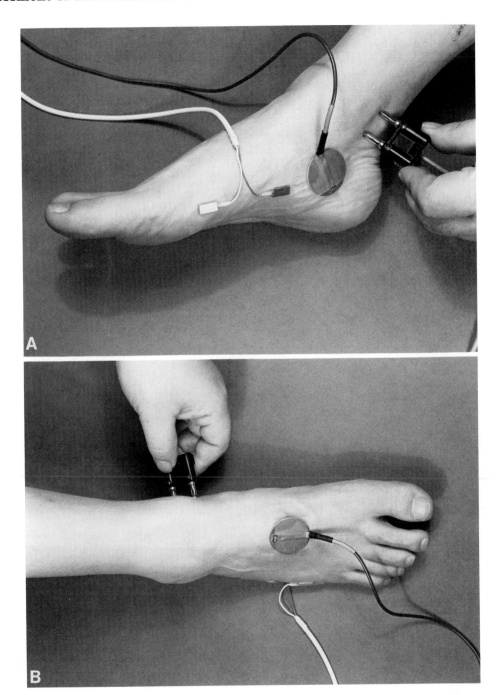

Figure 6–12. A. Motor conduction study of the medial plantar nerve. The photograph shows stimulation of the tibial nerve posterior to the medial malleolus, 10 cm from the recording electrodes placed over the belly (G₁) and tendon (G₂) of the abductor hallucis. **B.** Motor conduction study of the lateral plantar nerve. The photograph shows stimulation posterior to the medial malleolus and recording with surface electrodes placed on the belly (G₁) and tendon (G₂) of the abductor digiti quinti.

Table 6-12 COMMON AND DEEP PERONEAL NERVES*

Site of Stimulation	Amplitude† (mV)	Latency‡ to Recording Site (ms)	Difference Between Right and Left (ms)	Conduction Time Between Two Points (ms)	Conduction Velocity (m/s)
Ankle	5.1 ± 2.3 (2.5)§	3.77 ± 0.86 (5.5)¶	0.62 ± 0.61 (1.8)¶		
Below knee	5.1 ± 2.0 (2.5)	10.79 ± 1.06 (12.9)	0.65 ± 0.65 (2.0)	7.01 ± 0.89 (8.8)¶	48.3 ± 3.9 (40)**
Above knee	5.1 ± 1.9 (2.5)	12.51 ± 1.17 (14.9)	0.65 ± 0.60 (1.9)	1.72 ± 0.40 (2.5)	52.0 ± 6.2 (40)

*Mean ± standard deviation (SD) in 120 nerves from 60 patients, 16 to 86 years of age (average, 41), with no apparent disease of the peripheral nerves.
†Amplitude of the evoked response, measured from the baseline to the negative peak.
‡Latency, measured to the onset of the evoked response, with a standard distance of 7 cm between the cathode and the recording electrode.
§Lower limits of normal, based on the distribution of the normative data.
¶Upper limits of normal, calculated as the mean + 2 SD.
**Lower limits of normal, calculated as the mean − 2 SD.

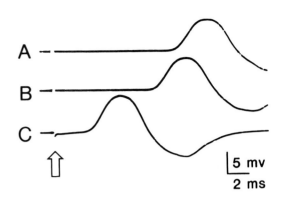

Figure 6–13. Motor nerve conduction study of the common and deep peroneal nerve. The sites of stimulation include above the knee (A), below the knee (B), and ankle (C). Compound muscle action potentials are recorded with surface electrodes over the extensor digitorum brevis.

Superficial Peroneal Nerve

This mixed nerve, derived from the L-5 root, originates below the fibular head as a branch of the common peroneal nerve.

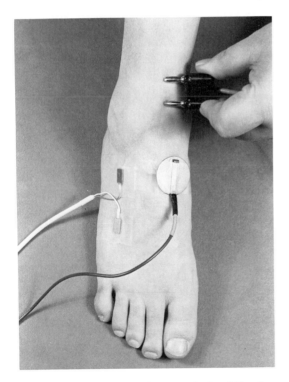

Figure 6–14. Motor conduction study of the deep peroneal nerve. The photograph shows stimulation over the dorsum of the foot near the ankle, 7 cm from the recording electrodes over the belly (G₁) and tendon (G₂) of the extensor digitorum brevis.

It gives rise to two sensory nerves in the lower third of the leg, the medial and intermediate dorsal cutaneous nerves. They innervate the skin of the dorsum of the foot and the anterior and lateral aspects of the leg. The medial dorsal cutaneous nerve pierces the superficial fascia at the anterolateral aspect of the leg about 5 cm above and 2 cm medial to the lateral malleolus.[12,19] Stimulation at this point with the cathode adjusted to produce a sensation radiating into the toes elicits an antidromic sensory potential over the dorsum of the foot medially. The averaging technique helps identify the potential with amplitude approximately half that of the sural nerve, especially in recording from a diseased nerve.

In another method,[46,48] stimulation of the intermediate dorsal cutaneous branch with the cathode placed against the anterior edge of the fibula elicits the antidromic sensory potential at the ankles just medial to the lateral malleolus (Fig. 6–15). Stimulation of the nerve at two points, 12 to 14 cm from the recording electrode and 8 to 9 cm further proximally, allows assessments of the distal and proximal segments. The study of this sensory potential helps distinguish a L-5 radiculopathy from more distal lesions.[48] The near nerve needle recording with signal averaging has also been described for the assessment of sensory action potential from interdigital nerves.[79] Table 6–13 summarizes the normal values.

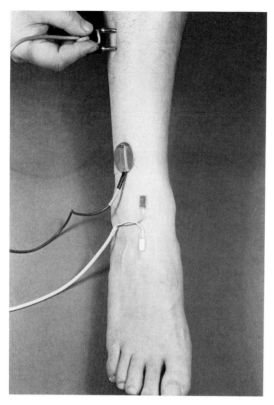

Figure 6–15. Sensory nerve conduction study of the superficial peroneal nerve. The photograph shows stimulation against the anterior edge of the fibula, 12 cm from the active electrode (G_1) located just medial to the lateral malleolus at the ankle with the reference electrode (G_2) placed 2 to 3 cm distally.

Sural Nerve

This sensory nerve, primarily derived from the S-1 root, originates in the popliteal fossa as the medial sural branch of the tibial nerve. It becomes superficial at the junction of the mid and lower third of the leg, where it receives a communicating branch of the common peroneal nerve. In some cases, the peroneal branch contributes more than the main trunk from the tibial nerve. Descending toward the ankle, it turns anterolaterally along the inferior aspect of the lateral malleolus. Its terminal branch, the lateral dorsal cutaneous nerve, supplies the lateral aspect of the dorsum of the foot.

Stimulation of the nerve in the lower third of the leg over the posterior aspect slightly lateral to the midline elicits the antidromic sensory potentials best detectable around the lateral malleolus (Figs. 6–16 and 6–17). Recording sural potentials needs no averaging except perhaps in older populations or patients with diseased nerve.[10,19,20,62,90] Segmental studies dividing the nerve into three contiguous portions of 7 cm each have revealed a smaller mean velocity in the most distal segment than in the middle or proximal segment.[101] It is also possible, with averaging techniques, to measure orthodromic action potentials after stimulation of the nerve over the lateral aspect of the foot.[1,2,53,93] Recording at the popliteal fossa and high at the ankle, 10 to 15 cm proximal to the lateral malleolus, allows comparison between distal and proximal segments of the nerve (Table 6–14).

Sural nerve conduction offers one of the most sensitive means of detecting electrophysiologic abnormalities in various types of neuropathies. In vitro conduction studies and histologic assessments of biopsy specimens[24–26] provide a unique opportunity for direct correlation with in vivo studies (see Chapter 4.4). Preganglionic pathology consistently spares the sensory action potential despite the clinical symptoms. Thus, studies of the sural nerve help distinguish peripheral lesions from S-1 or S-2 radiculopathy or cauda equina involvement.

5   OTHER NERVES DERIVED FROM THE LUMBOSACRAL PLEXUS

Lumbosacral Plexus

The lumbosacral plexus consists of the lumbar plexus with fibers derived from the L-2, L-3, and L-4 roots and the sacral plexus, which arises from the L-5, S-1, and S-2 roots. Neither portion is accessible to percutaneous electrical stimulation. Thus, conventional conduction studies fall short of adequately evaluating their integrity. The use of the F wave and

Table 6–13 SUPERFICIAL PERONEAL NERVE

Stimulation Point	Recording Site	n	Age	Amplitude (μV)	Latency (ms)	Conduction Velocity (m/s)	
5 cm above, 2 cm medial to lateral malleolus	Dorsum of foot	50	1–15	13.0 ± 4.6	1.22 ± 0.40	53.1 ± 5.3	(Distal segment)
		50	Over 15	13.9 ± 4.0	2.24 ± 0.49 (Peak)	47.3 ± 3.4	(Distal segment)
Anterior edge of fibula, 12 cm above the active electrode	Medial border of lateral malleolus	50	3–60	20.5 ± 6.1	2.9 ± 0.3 (Peak)	65.7 ± 3.7	(Proximal segment)
Anterolateral aspect of leg, 14 cm above the active electrode	Medial border of lateral malleolus	80		18.3	2.8 ± 0.3 (Peak)	51.2 ± 5.7	(Proximal segment)
					(Onset)		

Data from Di Benedetto,[14] Jabre,[48] and Izzo et al.[46]

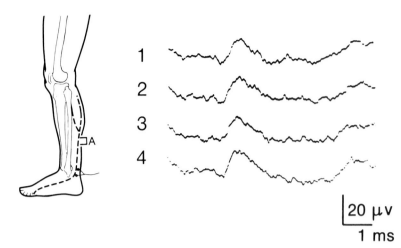

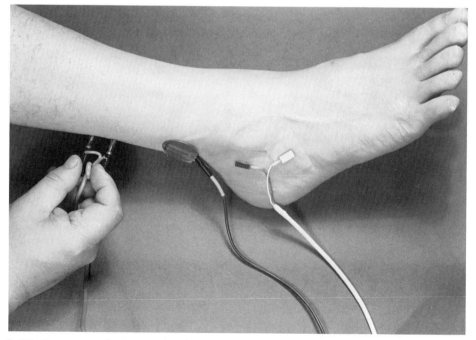

Figure 6–16. Antidromic sensory nerve conduction study of the sural nerve. The diagram shows stimulation on the calf slightly lateral to the midline in the lower third of the leg and recording with surface electrodes placed behind the lateral malleolus.

H reflex provides an indirect measure of impulses propagating across this region (see Chapters 17 and 18). An alternative method[67,83] involves needle stimulation of L-4, L-5, or S-1 root just proximal to the plexus and stimulation of the peripheral nerve just distal to the plexus. Conduction time through the plexus then equals the difference between the distal and proximal latencies.

For the study of the lumbar plexus, the L-4 root is stimulated by a 75-mm standard monopolar needle, placed so as to lie just below the level of the iliac crest. The needle inserted into the paraspinous muscle perpendicular to the skin surface must reach the periosteum of the articular process (Fig. 6–18A). With an optimal needle position, a shock of very low intensity elicits the maximal compound

Figure 6–17. Sensory conduction study of the sural nerve. The photograph shows stimulation along the posterior surface of the leg, slightly lateral to the midline and 7 to 10 cm from the ankle. The active electrode (G$_1$) is placed above or immediately below and behind the lateral malleolus with the reference electrode placed 2 to 3 cm distally along the lateral dorsum of the foot (G$_2$).

Table 6–14 SURAL NERVE

Authors	Stimulation Point	Recording Site	n	Age	Amplitude (μV)	Latency (ms)	Conduction Velocity (m/s)
Shiozawa and Mavor (1969)[93]	Foot	High ankle	40	13–41	6.3 (1.9–17)		44.0 ± 4.7
DiBenedetto (1970)[19]	Lower third of leg	Lateral malleolus	38 / 62	1–15 / Over 15	23.1 ± 4.4 / 23.7 ± 3.8	1.46 ± 0.43 / 2.27 ± 0.43 (Peak)	52.1 ± 5.1 / 46.2 ± 3.3
Behse and Buchthal (1971)[1]	15 cm above lateral malleolus	Dorsal aspect of foot	71	15–30 / 40–65			51.2 ± 4.5 / 48.3 ± 5.3
Schuchmann (1977)[90]	14 cm above lateral malleolus	Lateral malleolus	101	13–66		3.50 ± 0.25 (Peak)	40.1
Wainapel et al. (1978)[102]	Lower third of leg	Lateral malleolus	80	20–79	18.9 ± 6.7	3.7 ± 0.3 (Peak)	41.0 ± 2.5
Truong et al. (1979)[101]	Distal 10 cm / Middle 10 cm / Proximal 10 cm	Lateral malleolus	102 / 102 / 102				33.9 ± 3.25 / 51.0 ± 3.8 / 51.6 ± 3.8
Kimura (Unpublished)	14 cm above lateral malleolus	Lateral malleolus	52	10–40 / 41–84	20.9 ± 8.0 / 17.2 ± 6.7	2.7 ± 0.3 / 2.8 ± 0.3 (Onset)	52.5 ± 5.6 / 51.1 ± 5.9

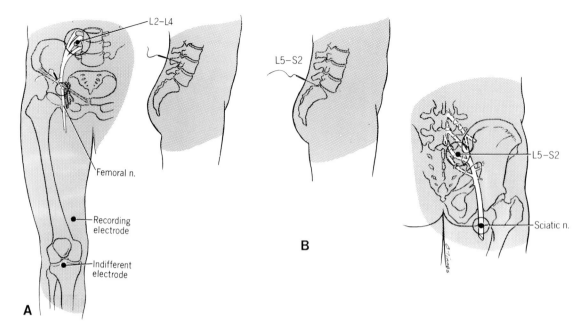

Figure 6–18. A. Motor conduction study of the lumbar plexus. The diagram shows stimulation of the L-4 root with the needle inserted perpendicular to the skin just below the level of the iliac crest and of the femoral nerve distal to the inguinal ligament immediately lateral to the femoral artery. Muscle potentials are recorded with surface electrodes over the vastus medialis (G_1) and patella (G_2). **B.** Motor nerve conduction study of the sacral plexus. The diagram shows stimulation of the S-1 root with the needle inserted at the level of the posterior iliac spine, of the L-5 root halfway in between the L-4 and S-1 roots, and of the sciatic nerve at the level of gluteal skinfold midpoint between the ischial tuberosity and the greater trochanter of the femur. The recording electrodes (not shown) are placed on the belly (G_1) and tendon (G_2) of the tibialis anterior for L-5 and of the abductor hallucis for S-1 root studies. (From MacLean,[67] with permission.)

muscle action potential of the vastus medialis. Stimulation of the femoral nerve just distal to the inguinal ligament, with either a surface or a needle electrode, provides the distal latency. The nerve lies immediately lateral to the readily palpable femoral artery, as discussed later in this section.

For studying the sacral plexus, a needle is placed between the spinous process and posterior iliac spine for the S-1 root and halfway in between the L-4 and S-1 roots for the L-5 root (Fig. 6–18B). At the level of the gluteal skin fold, the sciatic nerve bisects a line drawn between the ischial tuberosity and the greater trochanter of the femur.[105] Needle stimulation here provides the distal latency required for calculation of conduction time across the sacral plexus. With careful adjustment of the needle position, shocks of very low intensity elicit a maximal compound muscle action potential of the ti-

bialis anterior for the study of the L-5 root and of the abductor hallucis for evaluating the S-1 root. Inadvertent activation of the neighboring roots induces volume-conducted potentials from distant muscles. One must carefully avoid such spread of stimulus. Otherwise, the recording electrodes placed over the tibialis anterior, for example, would regularly register a simultaneously activated action potential of the triceps surae. Table 6–15 summarizes the normal value in one series.[67]

Femoral Nerve

Shocks delivered to the femoral nerve above or below the inguinal ligament elicit the response recordable in the rectus femoris muscle at various distances from the point of stimulation. The latency of the response increases pro-

Table 6–15 LUMBOSACRAL PLEXUS

Plexus	Site of Stimulation	Recording Site	Latency Across Plexus (ms)		
			Range	*Mean*	*SD*
Lumbar	L2, L3, L4 Femoral nerve	Vastus medialis	2.0–4.4	3.4	0.6
Sacral	L5 and S1 Sciatic nerve	Abductor hallucis	2.5–4.9	3.9	0.7

From MacLean,[67] with permission.

gressively with the distance reflecting vertical orientation of the end-plate region.[34] The femoral nerve conducts at an average rate of 70 m/s, based on the latency difference between the two responses recorded at 14 and 30 cm from the point of stimulation (Table 6–16). This calculation, however, does not hold unless all branches supplying proximal and distal parts of the muscle have similar and directly comparable electrophysiologic characteristics.

Saphenous Nerve

This largest and longest sensory branch of the femoral nerve lies deep along the medial border of the tibialis anterior tendon (Fig. 6–19). The nerve can be stimulated by surface electrodes pressed firmly between the medial gastrocnemius muscle and tibia, usually 12–14 cm above the ankle. Signal averaging improves the resolution of small antidromic sensory potentials recorded just anterior to the highest prominence of the medial malleolus (Table 6–17). Orthodromic studies[28,91,98] consist of stimulat-ing the nerve at two levels, anterior to the medial malleolus and medial to the knee, and recording the evoked potential with a needle electrode placed near the femoral nerve trunk at the inguinal ligament. The orthodromic potentials are about one-half the size of the antidromic potentials in amplitude. The saphenous nerve may degenerate with postganglionic lesions such as lumbar plexopathy or femoral neuropathy. In contrast, preganglionic L-3 or L-4 radiculopathy spares the distal sensory nerve potentials despite clinical deficits.

Lateral Femoral Cutaneous Nerve

The nerve becomes superficial about 12 cm below the anterior superior iliac spine, where it divides into large anterior and small lateral branches. Surface stimulation at this point elicits the orthodromic sensory potential recordable with a needle electrode inserted 1 cm medial to the lateral end of the inguinal ligament.[64] An alternative technique consists of stimulation at the inguinal ligament with

Table 6–16 FEMORAL NERVE

Stimulation Point	Recording Site	No.	Age	Onset Latency (ms)	Conduction Velocity (m/s)
Just below inguinal ligament	14 cm from stimulus point	42	8–79	3.7 ± 0.45	70 ± 5.5 between the two recording sites
	30 cm from stimulus point	42	8–79	6.0 ± 0.60	

Modified from Gassel,[34] with permission.

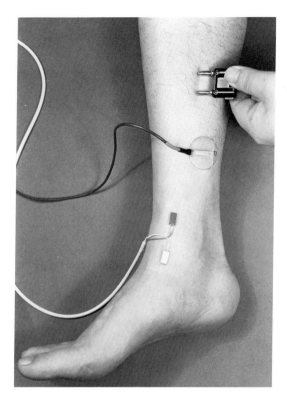

Figure 6–19. Sensory conduction study of the saphenous nerve. The photograph shows stimulation 14 cm above the active electrode (G_1) along the medial surface of the leg between the tibia and gastrocnemius and recording at the ankle 2 to 3 cm above (G_1) and just anterior to the medial malleolus (G_2).

a needle electrode and recording antidromic sensory potentials from the thigh (Fig. 6–20). In one study[11] using a pair of specially constructed 1.2 by 1.9-cm lead strips fastened 4 cm apart, the normal values (mean ± SD) in 25 healthy adults consisted of a latency of 2.6 ± 0.2 ms, an amplitude of 10 to 25 μV, and a calculated velocity of 47.9 ± 3.7 m/s.

6 CRANIAL NERVES

The most commonly tested cranial nerves in electromyographic laboratory include the facial and trigeminal nerves (see Chapters 7.3 and 16.2).

Table 6–17 SAPHENOUS NERVE

Authors	Method	Age	Inguinal Ligament—Knee			Knee—Medial Malleolus		
			Number	Amplitude (μV)	Conduction Velocity	Number	Amplitude (μV)	Conduction Velocity (m/s)
Ertekin (1969)[28]	Orthodromic	17–38	33	4.2 ± 2.3	59.6 ± 2.3	10	4.8 ± 2.4	52.3 ± 2.3
Stöhr et al. (1978)[98]	Orthodromic	<40	28	5.5 ± 2.6	58.9 ± 3.2	22	2.1 ± 1.1	51.2 ± 4.7
		>40	41	5.1 ± 2.7	57.9 ± 4.0	32	1.7 ± 0.8	50.2 ± 5.0
Wainapel et al. (1978)[102]	Antidromic	20–79			Peak latency of 3.6 ± 1.4 for 14 cm	80	9.0 ± 3.4	41.7 ± 3.4
Senden et al. (1981)[91]	Orthodromic	18–56	71					54.8 ± 1.9

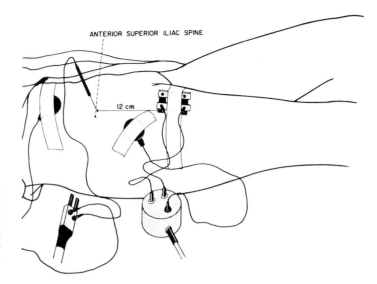

Figure 6–20. Sensory nerve conduction of the lateral femoral cutaneous nerve. The diagram shows stimulation above the inguinal ligament and recording over the thigh 12 cm below the anterior-superior iliac spine (G_1) and 2 to 3 cm distally (G_2). (From Butler et al.,[11] with permission.)

Accessory Nerve

The accessory nerve runs superficially along the posterior border of the sternocleidomastoid muscle. Surface stimulation at this point elicits a compound muscle action potential of the trapezius, usually from the upper portion by an active electrode (G_1) placed at the angle of the neck and shoulder and a reference electrode (G_2) over the tendon near the acromion process. Some electromyographers prefer needle electrodes to stimulate the nerve.[85] In one series of 25 subjects, 10 to 60 years of age, normal latencies to the upper trapezius ranged from 1.8 to 3.0 ms.[14] In another study of 21 nerves, the accessory latency averaged 3.0 ± 0.2 ms (mean $\pm$ SD) to the middle trapezius and 4.6 ± 0.3 ms to the lower trapezius.[40] Changes in amplitude also provide reliable information, with reduction to one-half that of the response on the healthy side, suggesting distal degeneration.

REFERENCES

1. Behse, F, and Buchthal, F: Normal sensory conduction in the nerves of the leg in man. J Neurol Neurosurg Psychiatry 34:404–414, 1971.

2. Behse, F, and Buchthal, F: Sensory action potentials and biopsy of the sural nerve in neuropathy. Brain 101:473–493, 1978.
3. Belen, J: Orthodromic sensory nerve conduction of the medial and lateral plantar nerves, a standardization. Am J Phys Med 64:17–23, 1985.
4. Berger, AR, Busis, NA, Logigian, EL, Wierzbicka, M, and Shahani, BT: Cervical root stimulation in the diagnosis of radiculopathy. Neurology 37:329–332, 1987.
5. Brown, WF, Ferguson, GG, Jones, MW, and Yates, SK: The location of conduction abnormalities in human entrapment neuropathies. Can J Neurol Sci 3:111–122, 1976.
6. Buchthal, F, and Rosenfalck, A: Evoked action potentials and conduction velocity in human sensory nerves. Brain Res 3:1–122, 1966.
7. Buchthal, F, and Rosenfalck, A: Sensory potentials in polyneuropathy. Brain 94:241–262, 1971.
8. Buchthal, F, and Rosenfalck, A: Sensory conduction from digit to palm and from palm to wrist in the carpal tunnel syndrome. J Neurol Neurosurg Psychiatry 34:243–252, 1971.
9. Buchthal, F, Rosenfalck, A, and Trojaborg, W: Electrophysiological findings in entrapment of the median nerve at wrist and elbow. J Neurol Neurosurg Psychiatry 37:340–360, 1974.
10. Burke, D, Skuse, NF, and Lethlean, AK: Sensory conduction of the sural nerve in polyneuropathy. J Neurol Neurosurg Psychiatry 37:647–652, 1974.
11. Butler, ET, Johnson, EW, and Kaye, ZA: Normal conduction velocity in the lateral femoral cutaneous nerve. Arch Phys Med Rehabil 55:31–32, 1974.
12. Cape, CA: Sensory nerve action potentials of the peroneal, sural and tibial nerves. Am J Phys Med 50:220–229, 1971.

13. Casey, EB, and Le Quesne, PM: Digital nerve action potentials in healthy subjects, and in carpal tunnel and diabetic patients. J Neurol Neurosurg Psychiatry 35:612–623, 1972.
14. Cherington, M: Accessory nerve: conduction studies. Arch Neurol 18:708–709, 1968.
15. Daube, JR: Percutaneous palmar median nerve stimulation for carpal tunnel syndrome. Electroencephalogr Clin Neurophysiol 43:139–140, 1977.
16. Delhez, L: Modalites, chez l'homme normal, de la response electrique des piliers du diaphragme a la stimulation electrique des nerfs phreniques par des chocs uniques. Arch Int Physiol Biochim 73:832–839, 1965.
17. Dellon, AL, and Mackinnon SE: Tibial nerve branching in the tarsal tunnel. Arch Neurol 41:645–646, 1984.
18. Devi, S, Lovelace, RE, and Duarte, N: Proximal peroneal nerve conduction velocity: Recording from anterior tibial and peroneus brevis muscles. Ann Neurol 2:116–119, 1977.
19. DiBenedetto, M: Sensory nerve conduction in lower extremities. Arch Phys Med Rehabil 51:253–258, 1970.
20. DiBenedetto, M: Evoked sensory potentials in peripheral neuropathy. Arch Phys Med Rehabil 53:126–131, 1972.
21. DiBenedetto, M, Mitz, M, Klingbeil, G, and Davidoff, DD: New criteria for sensory nerve conduction especially useful in diagnosing carpal tunnel syndrome. Arch Phys Med Rehabil 67:586–589, 1986.
22. Downie, AW, and Scott, TR: Radial nerve conduction studies. Neurology 14:839–843, 1964.
23. Downie, AW, and Scott, TR: An improved technique for radial nerve conduction studies. J Neurol Neurosurg Psychiatry 30:322–336, 1967.
24. Dyck, PJ, and Lambert, EH: Numbers and diameters of nerve fibers and compound action potential of sural nerve: Controls and hereditary neuromuscular disorders. Trans Am Neurol Assoc 91:214–217, 1966.
25. Dyck, PJ, Lambert, EH, and Nichols, PC: Quantitative measurement of sensation related to compound action potential and number and sizes of myelinated and unmyelinated fibers of sural nerve in health, Friedreich's ataxia, hereditary sensory neuropathy, and tabes dorsalis. In Remond, A (ed): Handbook of Electroencephalography and Clinical Neurophysiology, Vol 9. Elsevier, Amsterdam, 1972, pp 83–118.
26. Dyck, PJ, and Lofgren, EP: Method of fascicular biopsy of human peripheral nerve for electrophysiologic and histologic study. Mayo Clin Proc 41:778–784, 1966.
27. Eklund, G: A new electrodiagnostic procedure for measuring sensory nerve conduction across the carpal tunnel. Upsala J Med Sci 80:63–64, 1975.
28. Ertekin, C: Saphenous nerve conduction in man. J Neurol Neurosurg Psychiatry 32:530–540, 1969.
29. Falck, B, and Hurme, M: Conduction velocity of the posterior interosseus nerve across the arcade of Frohse. Electromyogr Clin Neurophysiol 23:567–576, 1983.
30. Falck, B, Hurme, M, Hakkarainen, S, and Aarnio, P: Sensory conduction velocity of plantar digital nerves in Morton's metatarsalgia. Neurology 34:698–701, 1984.
31. Feibel, A, and Foca, FJ: Sensory conduction of radial nerve. Arch Phys Med Rehabil 55:314–316, 1974.
32. Felsenthal, G, Brockman, P, Mondell, D, and Hilton, E: Proximal forearm ulnar nerve conduction techniques. Arch Phys Med Rehabil 67:440–444, 1986.
33. Fu, R, DeLisa, JA, and Kraft, GH: Motor nerve latencies through the tarsal tunnel in normal adult subjects: Standard determinations corrected for temperature and distance. Arch Phys Med Rehabil 61:243–248, 1980.
34. Gassel, MM: A study of femoral nerve conduction time. Arch Neurol 9:57–64, 1963.
35. Gassel, MM: A test of nerve conduction to muscles of the shoulder girdle as an aid in the diagnosis of proximal neurogenic and muscular disease. J Neurol Neurosurg Psychiatry 27:200–205, 1964.
36. Gassel MM: Sources of error in motor nerve conduction studies. Neurology 14:825–835, 1964.
37. Gassel, MM, and Diamantopoulos, E: Pattern of conduction times in the distribution of the radial nerve: A clinical and electrophysiological study. Neurology 14:222–231, 1964.
38. Gilliatt, RW, Goodman, HV, and Willison, RG: The recording of lateral popliteal nerve action potentials in man. J Neurol Neurosurg Psychiatry 24:305–318, 1961.
39. Ginzburg, M, Lee, M, Ginzburg, J, and Alba, A: Median and ulnar nerve conduction determinations in the Erb's point-axilla segment in normal subjects. J Neurol Neurosurg Psychiatry 41:444–448, 1978.
40. Green, RF, and Brien, M: Accessory nerve latency to the middle and lower trapezius. Arch Phys Med Rehabil 66:23–24, 1985.
41. Guiloff, RJ, and Sherratt, RM: Sensory conduction in medial plantar nerve: Normal values, clinical applications, and a comparison with the sural and upper limb sensory nerve action potentials in peripheral neuropathy. J Neurol Neurosurg Psychiatry 40:1168–1181, 1977.
42. Harding, C, and Halar, E: Motor and sensory ulnar nerve conduction velocities: Effect of elbow position. Arch Phys Med Rehabil 64:227–232, 1983.
43. Irnai, KD, Grabois, M, and Harvey, SC: Standardized technique for diagnosis of tarsal tunnel syndrome. Am J Phys Med 61:26–31, 1982.
44. Iyer, KS, Kaplan, E, and Goodgold, J: Sensory nerve action potentials of the medial and lateral plantar nerve. Arch Phys Med Rehabil 65:529–530, 1984.
45. Izzo, KL, Aravabhumi, S, Jafri, A, Sobel, E, and Demopoulous, JT: Medial and lateral antebrachial cutaneous nerves: Standardization of technique, reliability and age effect on

healthy subjects. Arch Phys Med Rehabil 66:592–597, 1985.

46. Izzo, KL, Sridhara, CR, Lemont, H, and Rosenholtz, H: Sensory conduction studies of the branches of the superficial peroneal nerve. Arch Phys Med Rehabil 62:24–27, 1981.

47. Jabre, JF: Ulnar nerve lesions at the wrist: New technique for recording from the sensory dorsal branch of the ulnar nerve. Neurology 30:873–876, 1980.

48. Jabre, JF: The superficial peroneal sensory nerve revisited. Arch Neurol 38:666–667, 1981.

49. Jebsen, RH: Motor conduction velocity in proximal and distal segments of the radial nerve. Arch Phys Med Rehabil 47:597–602, 1966.

50. Johnson, RK, and Shrewsbury, MM: Anatomical course of the thenar branch of the median nerve—usually in a separate tunnel through the transverse carpal ligament (Abstr). J Bone Joint Surg 52A:269–273, 1970.

51. Kaplan, PE: Electrodiagnostic confirmation of long thoracic nerve palsy. J Neurol Neurosurg Psychiatry 43:50–52, 1980.

52. Kaplan, PE: A motor nerve conduction velocity across the upper trunk and the lateral cord of the brachial plexus. Electromyogr Clin Neurophysiol 22:315–320, 1982.

53. Kayed, K, and Rosjo, O: Two-segment sural nerve conduction measurements in polyneuropathy. J Neurol Neurosurg Psychiatry 46:867–870, 1983.

54. Kim, DJ, Kalantri, A, Guha, S, and Wainapel, SF: Dorsal cutaneous ulnar nerve conduction: Diagnostic aid in ulnar neuropathy. Arch Neurol 38:321–322, 1981.

55. Kimura, I, Seki, H, Sasao, S, and Ayyar, DR: The great auricular nerve conduction study: A technique, normative data and clinical usefulness. Electromyogr Clin Neurophysiol 27:39–43, 1987.

56. Kimura, J: Collision technique—Physiological block of nerve impulses in studies of motor nerve conduction velocity. Neurology 26:680–682, 1976.

57. Kimura, J: A method for determining median nerve conduction velocity across the carpal tunnel. J Neurol Sci 38:1–10, 1978.

58. Kimura, J: The carpal tunnel syndrome. Localization of conduction abnormalities within the distal segment of the median nerve. Brain 102:619–635, 1979.

59. Kimura, J, Machida, M, Ishida, T, Yamada, T, Rodnitzky, R, Kudo, Y, and Suzuki, S: Relationship between size of compound sensory or muscle action potentials, and length of nerve segment. Neurology 36:647–652, 1986

60. Kincaid, JC, Phillips, LH, II, and Daube, JR: The evaluation of suspected ulnar neuropathy at the elbow. Arch Neurol 43:44–47, 1986.

61. Kraft, GH: Axillary, musculocutaneous and suprascapular nerve latency studies. Arch Phys Med Rehabil 53:383–387, 1972.

62. Lafratta, CW, and Zalis, AW: Age effects on sural nerve conduction velocity. Arch Phys Med Rehabil 54:475–477, 1973.

63. Lum, PB, and Kanakamedala, RV: Conduction of the palmar cutaneous branch of the median nerve. Arch Phys Med Rehabil 67:805–806, 1986.

64. Lysens, R., Vandendriessche, G, Van Mol, Y, and Rosselle, N: The sensory conduction velocity in the cutaneous femoris lateralis nerve in normal adult subjects and in patients with complaints suggesting meralgia paresthetica. Electromyogr Clin Neurophysiol 21:505–510, 1981.

65. Ma, DM, Kim, SH, Spielholz, N, and Goodgold, J: Sensory conduction study of distal radial nerve. Arch Phys Med Rehabil 62:562–564, 1981.

66. Mackenzie, K, and DeLisa, J: Distal sensory latency measurement of the superficial radial nerve in normal adult subjects. Arch Phys Med Rehabil 62:31–34, 1981.

67. MacLean, IC: Nerve root stimulation to evaluate conduction across the brachial and lumbosacral plexuses. Third Annual Continuing Education Course, American Association of Electromyography and Electrodiagnosis, September 25, 1980, Philadelphia.

68. MacLean, IC, and Mattioni, TA: Phrenic nerve conduction studies: A new technique and its application in quadriplegic patients. Arch Phys Med Rehabil 62:70–73, 1981.

69. MacLean, IC, and Taylor, RS: Nerve root stimulation to evaluate brachial plexus conduction. Abstracts of Communication of the Fifth International Congress of Electromyography, Rochester, Minnesota, 1975.

70. Markand, ON, Kincaid, JC, Pourmand, RA, Moorthy, SS, King, RD, Mahomed, Y, and Brown, JW: Electrophysiologic evaluation of diaphragm by transcutaneous phrenic nerve stimulation. Neurology 34:604–614, 1984.

71. Masuda, T, Miyano, H, and Sadoyama, T: The distribution of myoneural junctions in the biceps brachii investigated by surface electromyography. Electroencephologr Clin Neurophysiol 56:597–603, 1983.

72. Masuda, T, Miyano, H, and Sadoyama, T: A surface electrode array for detecting action potential trains of single motor units. Electroencephalogr Clin Neurophysiol 60:435–443, 1985.

73. Masuda, T, and Sadoyama, T: The propagation of single motor unit action potentials detected by a surface electrode array. Electroencephalogr Clin Neurophysiol 63:590–598, 1986.

74. Mavor, H, and Atcheson, JB: Posterior tibial nerve conduction. Velocity of sensory and motor fibers. Arch Neurol 14:661–669, 1966.

75. Mavor, H, and Shiozawa, R: Antidromic digital and palmar nerve action potentials. Electroencephalogr Clin Neurophysiol 30:210–221, 1971.

76. McComas, AJ, Kereshi, S, and Manzano, G: Multiple innervation of human muscle fibers. J Neurol Sci 64:55–64, 1984.

77. Nelson, RM, and Currier, DP: Motor-nerve conduction velocity of the musculocutaneous nerve. Phys Ther 49:586–590, 1969.

78. Newsom Davis, J: Phrenic nerve conduction in man. J Neurol Neurosurg Psychiatry 30:420–426, 1967.

79. Oh, SJ, Kim, HS, and Ahmad, BK: Electrophysiological diagnosis of interdigital neuropathy of the foot. Muscle Nerve 7:218–225, 1984.

80. Olney, RK, and Wilbourn, AJ: Ulnar nerve conduction study of the first dorsal interosseous muscle. Arch Phys Med Rehabil 66:16–18, 1985.

81. Palliyath, SK: A technique for studying the greater auricular nerve conduction velocity. Muscle Nerve 7:232–234, 1984.

82. Peake, JB, Roth, JL, and Schuchmann, GF: Pneumothorax: A complication of nerve conduction studies using needle stimulation. Arch Phys Med Rehabil 63:187–188, 1982.

83. Peiris, OA: Conduction in the fourth and fifth lumbar and first sacral nerve roots: Preliminary communication. NZ Med J 80:502–503, 1974.

84. Petrera, JE, and Trojaborg, W: Conduction studies of the long thoracic nerve in serratus anterior palsy of different etiology. Neurology 34:1033–1037, 1984.

85. Petrera, JE, and Trojaborg, W: Conduction studies along the accessory nerve and follow-up of patients with trapezius palsy. J Neurol Neurosurg Psychiatry 47:630–636, 1984.

86. Reddy, MP: Conduction studies of the medial cutaneous nerve of the forearm. Arch Phys Med Rehabil 64:209–211, 1983.

87. Redford, JWB: Conduction time in motor fibers of nerves which innervate proximal muscles of the extremities in normal persons and in patients with neuromuscular diseases. Thesis, University of Minnesota, Minneapolis, 1958.

88. Roth, G: Vitesse de conduction motrice du nerf médian dans le canal carpien. Ann Med Phys 13:117–132, 1970.

89. Saeed, MA, and Gatens, PF: Compound nerve action potentials of the medial and lateral plantar nerves through the tarsal tunnel. Arch Phys Med Rehabil 63:304–307, 1982.

90. Schuchmann, JA: Sural nerve conduction: A standardized technique. Arch Phys Med Rehabil 58:166–168, 1977.

91. Senden, R, Van Mulders, J, Ghys, R, and Rosselle, N: Conduction velocity of the distal segment of the saphenous nerve in normal adult subjects. Electromyogr Clin Neurophysiol 21:3–10, 1981.

92. Shahani, B, Goodgold, J, and Spielholz, NI: Sensory nerve action potentials in the radial nerve. Arch Phys Med Rehabil 48:602–605, 1967.

93. Shiozawa, R, and Mavor, H: In vivo human sural nerve action potentials. J Appl Physiol 26:623–629, 1969.

94. Shirali, CS, and Sandler, B: Radial nerve sensory conduction velocity: Measurement by antidromic technique. Arch Phys Med Rehabil 53:457–460, 1972.

95. Spindler, HA, and Felsenthal, G: Sensory conduction in the musculocutaneous nerve. Arch Phys Med Rehabil 59:20–23, 1978.

96. Spindler, HA, and Felsenthal, G: Radial sensory conduction in the hand. Arch Phys Med Rehabil 67:821–823, 1986.

97. Stevens, JC: AAEE Minimonograph #26: The electrodiagnosis of carpal tunnel syndrome. Muscle Nerve 2:99–113, 1987.

98. Stohr, M, Schumm, F, and Ballier, R: Normal sensory conduction in the saphenous nerve in man. Electroenceph Clin Neurophysiol 44:172–178, 1978.

99. Trojaborg, W: Motor and sensory conduction in the musculocutaneous nerve. J Neurol Neurosurg Psychiatry 39:890–899, 1976.

100. Trojaborg, W, and Sindrup, EH: Motor and sensory conduction in different segments of the radial nerve in normal subjects. J Neurol Neurosurg Psychiatry 32:354–359, 1969.

101. Truong, XT, Russo, FI, Vagi, I, and Rippel, DV: Conduction velocity in the proximal sural nerve. Arch Phys Med Rehabil 60:304–308, 1979.

102. Wainapel, SF, Kim, DJ, and Ebel, A: Conduction studies of the saphenous nerve in healthy subjects. Arch Phys Med Rehabil 59:316–319, 1978.

103. Werschkul, JD: Anomalous course of the recurrent motor branch of the median nerve in a patient with carpal tunnel syndrome: Case report. J Neurosurg 47:113–114, 1977.

104. Wiederholt, WC: Median nerve conduction velocity in sensory fibers through carpal tunnel. Arch Phys Med Rehabil 51:328–330, 1970.

105. Yap, CB, and Hirota, T: Sciatic nerve motor conduction velocity study. J Neurol Neurosurg Psychiatry 30:233–239, 1967.

Chapter **7**

FACTS, FALLACIES, AND FANCIES OF NERVE STIMULATION TECHNIQUES

1 INTRODUCTION

Nerve conduction studies help delineate the extent and distribution of the neural lesion and distinguish two major categories of peripheral nerve disease: demyelination and axonal degeneration. With steady improvement and standardization of methods,[33,76] they have become a reliable test not only for precise localization of a lesion but also for accurate characterization of peripheral nerve function.[21,65] In this chapter the fundamental principles and changing concepts of nerve stimulation techniques are reviewed, and their proper application in the differential diagnosis of peripheral nerve disorders is discussed.

Although the method is based on a simple theory, pitfalls abound in practice.[31,66,89,105] Commonly encountered, yet often overlooked, sources of error include intermittent failure in the stimulating or recording system, excessive spread of stimulation current, anomalous innervation, temporal dispersion, and inaccuracy of surface measurement. Unlike a bipolar derivation, which selectively records near-field potentials, a referential recording may give rise to stationary far-field peaks from a moving source (see Chapter 19.3). Lack of awareness of this possibility can be a cause of confusion in the interpretation of the results.

Conventional studies primarily deal with evaluation of the fastest conducting fibers, based on the latency measured to the onset of the evoked potential. In some clinical entities, other aspects of nerve conduction may help. These include conduction velocity of the slower fibers and the time course of the absolute and relative refractory periods. The phenomenon of collision provides a useful means of assessing these features of nerve conduction.[44,49,60–62,71] Here, a second stimulus delivered distally to the nerve blocks the unwanted impulses not under study. Although this is promising as a supplement to the conventional technique, clinical application of newer methods await further clarification.

2 COMMON TECHNICAL ERRORS

Technical problems often account for unexpected observations during routine nerve conduction studies. The failure to appreciate this possibility will lead to an incorrect diagnosis, especially if the findings are consistent with manifestation of the disease. Most properly identified problems can be easily corrected. They include malfunction of the stimulating or recording system.

Stimulating System

Responses may be absent or unexpectedly small in amplitude if the stimulus is inappropriately low in intensity, or misdirected despite adequate current strength. In this situation, electromyographers must relocate the stimulating electrode, press it firmly closer to the nerve, and, if necessary, increase the intensity or duration of the shock. The use of monopolar or concentric needle may help, especially in obese patients. Profuse perspiration or an excessive amount of cream over the skin surface may shunt the cathode and anode, rendering the otherwise sufficient stimulating current ineffective. Inadvertent reversal of the anode and cathode may cause anodal block of the propagating impulse. Regardless of the responsible technical fault, submaximal activation of the nerve proximally may erroneously suggest a conduction block, especially if a distal stimulus elicits a full response.

Recording System

Even optimal stimulation elicits a small response if a faulty connection hampers the recording system. Common problems include inappropriate placement of the pickup electrodes; breaks in the electrode wires; use of a disconnected preamplifier; loss of power supply; and incorrect oscilloscope settings for sensitivity, sweep, or filters. A broken recording electrode may show no change in appearance if the insulating sheath remains intact. With partial damage to the wire, stimulus-induced muscle twitches cause movement-related potentials, which can mimic a compound muscle action potential. To avoid any confusion, one must first test the integrity of the recording system prior to the examination. This is possible simply by asking the patient to contract the muscle with the electrode in position and the recording system turned on. Deficiencies at any step of the recording circuit would prevent a normal display of muscle action potentials on the oscilloscope.

An initial positivity preceding the major negative peak of the compound muscle action potential usually suggests incorrect positioning of the active electrode away from the end-plate region. Alternatively, it may represent a volume conducted potential from distant muscles, activated by anomalous innervation or by accidental spread of stimulation to other nerves. The compound muscle action potential reverses its polarity if the active (G_1) and reference (G_2) electrodes are mistakenly switched. Similarly, any deviation from the standard belly (G_1) and tendon (G_2) placement of recording electrodes distorts the waveform of the compound muscle action potentials.

3   SPREAD OF STIMULATION CURRENT

With an inappropriately high shock intensity, stimulating current can spread to a nerve or muscle not being tested. Failure to recognize this possibility may result in fallacious determination of latencies, because the recording electrodes often register a volume-conducted potential from distant muscles.[31,58] Under these circumstances visual inspection of the contracting muscle helps identify this problem. In some of these cases, the collision technique can, in effect, activate the intended nerve selectively by blocking unwanted nerve stimulation.[60] The use of needle electrodes also restricts the recording to limited areas in the study of innervation by individual motor branches or of patterns of anomaly. This type of recording, however, does not provide any information on the size of compound muscle action potentials, because the electrical activity registered originates from a restricted area in the muscle.

Stimulation of the Facial Nerve

The facial nerve becomes accessible to surface or needle stimulation as it exits from the stylomastoid foramen (see Chapter 16.2) (see Figs. 16–2 and 16–3). The distal segment, tested by stimulating the nerve here and recording compound muscle action potentials from various facial muscles, remains normal for a few days after complete separation of the nerve at a proximal site. The loss of distal excitability by the end of the first week coincides with the onset of nerve degen-eration, which generally implies poor prognosis. With shocks of very high intensity, stimulating current may also activate the motor point of the masseter muscle. A volume-conducted potential then erroneously suggests a favorable prognosis, when in fact the facial nerve has already degenerated (Fig. 7–1). As stated before, visual inspection would verify that the contraction involved the masseter, rather than the facial, muscle. Surface stimulation of the facial nerve may also activate cutaneous fibers of the trigeminal nerve, causing reflexive contraction of the orbicularis oculi (see Chapter 16.2). The reflex response may mimic a late component of the compound muscle action potential or recurrent response from antidromic activation of motor neurons.

Axillary Stimulation and Collision Technique

Stimulation of the median or ulnar nerve at the wrist or elbow ordinarily activates only the nerve in question unless the shock intensity is unusually high.[58] This may not hold with stimulation at the axilla, where the two nerves lie in close proximity.[59] If the current intended for the median nerve spreads to the ulnar nerve, the electrodes placed on the thenar eminence record not only median but also ulnar innervated muscle potential. The measured latency will then indicate normal ulnar conduction even if the median nerve conducts more slowly, as in the carpal tunnel syndrome (Fig. 7–2). In the same case, a stimulus at the elbow activates only the median nerve, revealing a prolonged latency. The calculated conduction time between the axilla and elbow would then suggest an erroneously fast conduction velocity in this segment. In extreme cases, the latency of the median response after stimulation at the elbow exceeds that of the ulnar component elicited with shocks at the axilla.

The reverse discrepancy can occur in a study of tardy ulnar palsy, with spread of axillary stimulation to the median nerve. In this case, the surface electrodes on the hypothenar eminence register a small

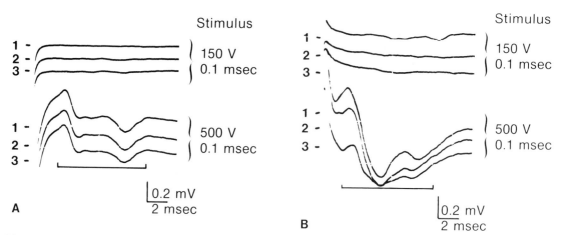

Figure 7–1. Compound muscle action potential from the orbicularis oculi after stimulation of the facial nerve in a patient with traumatic facial diplegia. **A.** Left side. **B.** Right side. Shocks of ordinary intensity (*top three tracings*) elicited no response; but with a much higher intensity, a definite muscle response appeared (*bottom three tracings*). Close observation of the face revealed contraction of the masseter, rather than the orbicularis oculi.

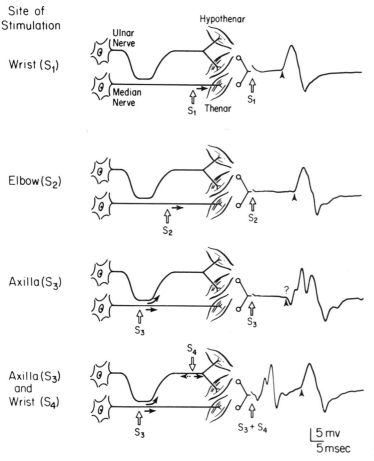

Figure 7–2. A 39-year-old man with carpal tunnel syndrome. The stimulation of the median nerve at the wrist (S_1) or elbow (S_2) elicited a muscle action potential with increased latency in the thenar eminence. Spread of axillary stimulation (S_3) to the ulnar nerve (*third tracing from top*) activated ulnar-innervated thenar muscles with shorter latency. Another stimulus (S_4) applied to the ulnar nerve at the wrist (*bottom tracing*) blocked the proximal impulses by collision. The muscle action potential elicited by S_4 occurred much earlier. The diagnosis on the left shows collision between the orthodromic (*solid arrows*) and antidromic (*open arrows*) impulses. (From Kimura,[60] with permission.)

positive potential of 1 to 5 mV in amplitude and 10 to 20 ms in duration, indicating the volume conducted response from thenar muscles or lumbricals. This positivity, though usually buried in a much larger ulnar response occurring simultaneously, becomes obvious if the ulnar nerve conducts more slowly than the median nerve, as in tardy ulnar palsy (Fig. 7–3). The earlier median component from thenar muscles then obscures the onset of the ulnar response originating from hypothenar muscles. The short latency measured to the onset of the median component fails to correctly reflect a delayed ulnar response. A stimulus at the elbow in the same case activates only the ulnar nerve with a prolonged latency, leading to an erroneously fast conduction velocity from axilla to elbow.

A physiologic nerve block with collision allows selective recording of the median or ulnar component despite coactivation of both nerves proximally.[60] In studying the median nerve, for example, one delivers a distal stimulus to the ulnar nerve (see Fig. 7–2). The antidromic impulse from the wrist collides with the orthodromic impulse from the axilla in the ulnar nerve, so that only the median impulse can reach the muscle. The ulnar response induced by the distal stimulus occurs much earlier without obscuring the median compound muscle action potential under study. If necessary, delivering the stimulus at the wrist a few milliseconds before the proximal stimulation accomplishes a greater separation between these two responses. This time interval should not exceed the conduction

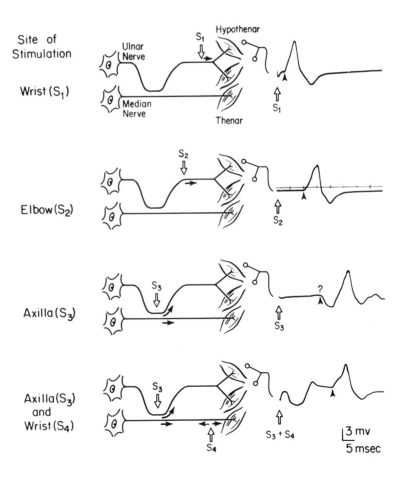

Figure 7–3. A 29-year-old man with tardy ulnar palsy. Stimulation at the wrist (S₁) or elbow (S₂) selectively activated the ulnar nerve, giving rise to an abnormally delayed muscle action potential over the hypothenar eminence. Spread of axillary stimulation (S₃) to the median nerve (*third tracing from top*) elicited an additional short latency median response with initial positivity. This potential, registered through volume conduction, obscured the onset (*arrowhead*) of the muscle response under study. Another stimulus (S₄) applied to the median nerve at the wrist (*bottom tracing*) blocked the proximal impulses by collision. The positive median potential elicited by S₄ clearly preceded the ulnar component under study. (From Kimura,[60] with permission.)

time between the distal and proximal points of stimulation. Otherwise, the antidromic impulse from the distal point passes the proximal site of stimulus without collision. The same principles apply for the use of a distal stimulus to block the median nerve in selective recording of the ulnar response despite coactivation of both nerves at the axilla (see Fig. 7–3).

The collision technique can clarify otherwise confusing results of motor nerve conduction studies in patients with the carpal tunnel syndrome or tardy ulnar palsy. In each of the illustrated cases (see Figs. 7–2 and 7–3), spread of the stimulus caused marked difference in waveform of the evoked potentials elicited by distal and proximal stimuli. To elucidate less apparent discrepancies, one must block unwanted nerve stimulation to uncover the true response from the intended muscle. The collision technique provides a noninvasive means, simpler than the procaine nerve block previously employed, of identifying the origin of the recorded muscle potentials.[31,47] The method improves the accuracy of latency determination even under circumstances that preclude selective stimulation of the median or ulnar nerve at a proximal point. The use of needle electrodes provides an alternative method of obtaining reliable latencies after coactivation of more than one nerve. As stated earlier, however, intramuscular recording from a restricted area does not allow reliable assessment of a compound muscle action potential.

Palmar Stimulation of the Median Nerve

Palmar stimulation provides a unique contribution in evaluating the distal segment of the median nerve, although studies of the motor conduction in this region pose some technical problems.[11,20,63,102,127] With serial stimulation in 1-cm increments from palm to wrist, the sensory latency increases linearly (see Fig. 6–3B). The motor study, however, sometimes shows unexpected latency changes reflecting the recurrent course of the motor fibers. For example, a stimulus directed to the branching point

of the thenar nerve in the palm could accidentally activate a terminal portion near the motor point. If another stimulus, delivered 1 cm proximally, excites only the median nerve trunk, the latency difference between the two stimulus points becomes unreasonably large, erroneously suggesting a focal slowing (Fig. 7–4A). Thus, a disproportionate latency does not automatically indicate a localized pathology. It is valid, however, if serial stimulation shows a linear latency increase in the segment proximal and distal to the site of lesion (Fig. 7–4B).

The same sort of error occurs in the calculation of motor latency over the wrist-to-palm segment unless palmar stimulation activates the median nerve precisely at the origin of the thenar nerve as intended. Incremental stimulation from the wrist toward the digit with the cathode placed distally to the anode can activate the thenar nerve at the anodal point (acting as floating cathode), even when the actual cathode lies clearly distal to the origin of the nerve. A surface distance measured to the cathodal point would then overestimate the nerve length, thereby making the calculated conduction velocity erroneously fast. To circumvent this problem, one must proceed from the distal palm toward the wrist after reversal of the electrode position, that is, cathode proximally to the anode. With this approach, palmar stimulation initially fails to produce a twitch, then causes thumb adduction near the deep branch of the ulnar nerve and finally thumb abduction with the arrival of the cathode just over the origin of the thenar nerve. In most subjects, this point lies 3 to 4 cm from the distal crease of the wrist, near the edge of the transverse carpal ligament.[56]

4 ANOMALIES AS SOURCES OF ERROR

Martin-Gruber Anastomosis

Anatomic studies of Martin (1763)[87] and Gruber (1870)[37] demonstrated frequent communication from the median to the ulnar nerve at the level of the fore-

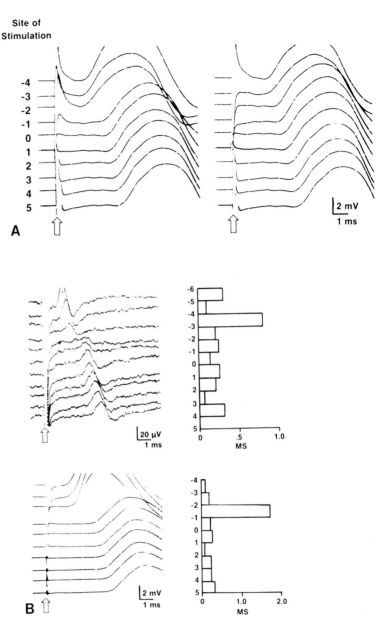

Figure 7–4. A. Compound muscle action potentials in a normal subject recorded after stimulation of the median nerve at multiple points across the wrist. On the initial trial (*left*), the latency decreased, with the cathode inching proximally from −4 to −2, indicating inadvertent spread of stimulating current to a distal portion of the thenar nerve. An apparent steep latency change from −2 to −1 gave an erroneous impression of a focal slowing at this level. A more careful placement of the cathode (*right*) eliminated unintended activation of the thenar nerve. The zero level at the distal crease of the wrist corresponds to the origin of the transverse ligament. (Compare Fig. 6–3A.) **B.** Sensory nerve (*top*) and muscle action potentials (*bottom*) in a symptomatic hand with the carpal tunnel syndrome. Serial stimulation showed a linear motor latency increase from −4 to −2 and from −1 to 5 with a localized slowing between −2 and −1. A temporally dispersed, double-peaked sensory nerve potential indicates the point of localized conduction delay from −4 to −3. (From Kimura,[63] with permission.)

arm. This anastomosis, often originating from the anterior interosseous nerve, predominantly consists of motor axons with rare sensory contribution. The communicating branch usually supplies ordinarily ulnar-innervated intrinsic hand muscles, most notably the first dorsal interosseous, adductor pollicis, and abductor digiti minimi.[85,109,128] The number of axons taking the anomalous course varies widely. A properly adjusted electrical stimulus delivered at the elbow may activate the anomalous fibers, maximally and selectively, without exciting the median nerve proper or vice versa (Fig. 7–5).[71] This observation suggests a grouping of the nerve fibers forming the anastomosis in a separate bundle, rather than a random scattering. The anomaly occurs in 15 to 31 percent of subjects in an unselected population. When present, it tends to involve the arms bilaterally.[71] The high incidence of this anastomosis among congenitally abnormal fetuses in

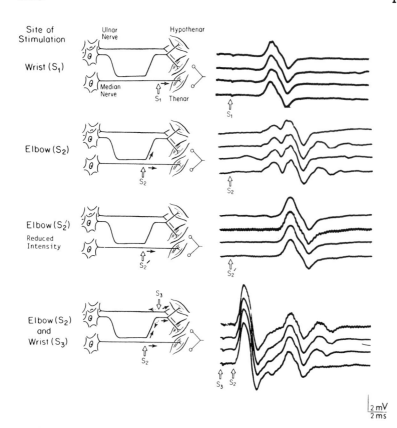

Figure 7-5. A 46-year-old woman with the carpal tunnel syndrome and the Martin-Gruber anomaly. Stimulation at the elbow (S_2) activated not only the median nerve but also communicating fibers, giving rise to a complex compound muscle action potential. With proper adjustment of electrode position and shock intensity, another stimulus at the elbow (S_2') excited the median nerve selectively without activating the anastomosis. Another stimulus (S_3) applied to the ulnar nerve at the wrist (*bottom tracing*) achieved the same effect by blocking the unwanted impulse transmitted through the communicating fibers. (From Kimura,[67] with permission.)

general and those with trisomy 21 in particular indicates its phylogenetic origin.[13,109] The communicating fibers rarely cross from the ulnar to the median nerve in the forearm.[114]

Careful analysis of the compound muscle action potentials readily reveals the presence of a Martin-Gruber anomaly during routine nerve conduction studies. Stimulation of the median nerve at the elbow evokes a response not only from the median-innervated thenar muscle but also from thenar and hypothenar muscles anomalously innervated by fibers crossing to the ulnar nerve in the forearm. In contrast, stimulation of the median nerve at the wrist elicits a smaller response lacking the ulnar component. Studies of the ulnar nerve show a reverse discrepancy in the amplitude of thenar or hypothenar compound muscle action potentials elicited via proximal and distal stimulation. Here stimulation at the wrist activates the additional anomalous fibers, giving rise to a full response, whereas stimulation at the elbow

spares the communicating branch still attached to the median nerve.

Evaluation of equivocal cases requires selective recording of action potentials from the first dorsal interosseous, adductor pollicis, or hypothenar muscles after stimulation of the median nerve at the elbow. In this case distant potentials from median-innervated muscles may mimic an anomalously activated response.[12,31,60,76,105] Volume-conducted potentials with initial positivity appear regardless of the site of stimulus, whereas an anomalous response occurs only with proximal stimulation. In difficult cases, the use of a needle electrode may clarify the origin of the recorded response, although intramuscular recordings do not entirely eliminate distant activities.

Again, a collision technique provides selective blocking of unwanted impulses transmitted via the communicating fibers (Fig. 7-6). Normally, antidromically directed impulses from the distal stimulation will block the orthodromic impulses

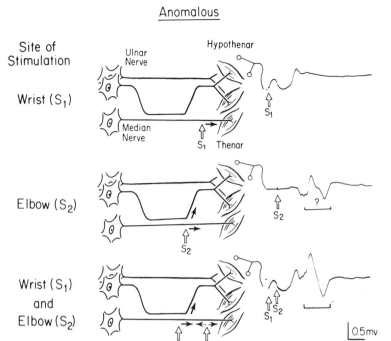

Figure 7–6. Muscle action potentials recorded from the hypothenar eminence after stimulation of the median nerve at the wrist (S_1) or elbow (S_2). The *top tracing* shows a volume conducted potential from thenar muscles (U-shaped wave of positive polarity). The *middle tracing* reveals a small negative potential superimposed upon the thenar component. In the *bottom tracing*, the collision technique clearly separated the anomalous response (*bracket*), with S_1 preceding S_2 by 4 ms. (From Kimura,[67] with permission.)

from the proximal stimulation in the same nerve.[44,121] The orthodromic impulses traveling through an anastomotic branch to the ulnar nerve would bypass the antidromic impulses and escape collision.[60] The technique helps not only in calculating conduction velocity accurately but also in characterizing the anomalous response. Even with collision, a thenar, as opposed to hypothenar, response elicited via anastomosis tends to overlap with a large median potential elicited distally (Fig. 7–7). Delivering the distal stimulus a few milliseconds before the proximal stimulation usually achieves satisfactory separation of the two responses. The time interval must not exceed the conduction time between the two stimulus sites. Otherwise, the orthodromic impulse produced by proximal stimulation would escape antidromic collision even in the absence of an anomalous route of transmission.

If this anastomosis accompanies the carpal tunnel syndrome, stimulation of the median nerve at the elbow evokes two temporally dispersed potentials, a normal ulnar component and a delayed median component. The latency of the initial ulnar response erroneously suggests the presence of normal conducting median fibers. In contrast, stimulation of the median nerve at the wrist evokes a delayed response without an ulnar component.[55,60,76] The discrepancy between proximal and distal stimulation would lead to an unreasonably fast conduction velocity from the elbow to the wrist.[12,60,76] The anomalously innervated ulnar muscles usually lie at some distance from the recording electrodes placed on the thenar eminence. Thus, the initial ulnar component commonly, though not always, displays an initial positive deflection.[39,40] As mentioned earlier, a collision technique can block impulses in the anomalous fibers without affecting those transmitted along the median nerve proper (Fig. 7–8).

Severence or substantial injury of the ulnar nerve at the elbow ordinarily results in wallerian degeneration and inexcitability of the distal segment. In the presence of this anomaly, however, stimulation at the wrist will excite the communicating fibers that bypass the lesion to evoke a small but otherwise normal muscle action potential. In extreme cases, separation of the ulnar nerve at

Anomalous

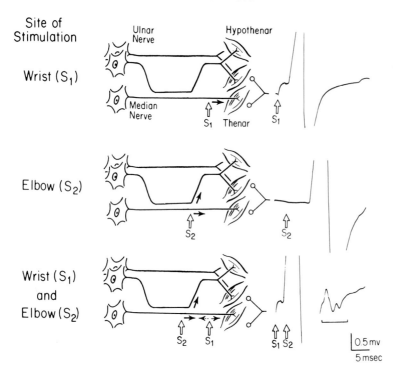

Figure 7–7. Muscle action potentials recorded from the thenar eminence after stimulation of the median nerve at the wrist (S_1) or elbow (S_2) as in Figure 7–5. In the *middle tracing*, a large compound action potential buried a small anomalous response mediated by the anastomosis. In the *bottom tracing*, a collision technique separated the anomalous response (*bracket*), with S_1 preceding S_2 by 4 ms. (From Kimura,[67] with permission.)

Anomalous Communication

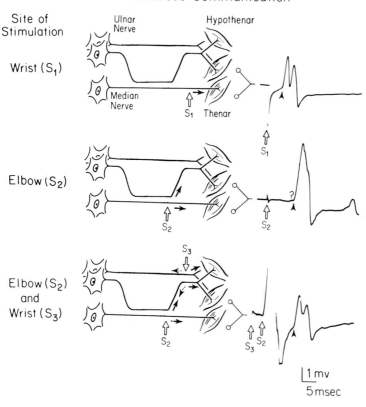

Figure 7–8. A 55-year-old man with the carpal tunnel syndrome and the Martin-Gruber anastomosis. Stimulation at the elbow (S_2) spread to the ulnar nerve through the anomalous communication (*middle tracing*). Another stimulus (S_3) applied to the ulnar nerve at the wrist (*bottom tracing*) blocked the impulses transmitted through the communicating fibers. In the *bottom tracing*, S_3 preceded S_2 by 4 ms to avoid the overlap of the muscle responses. (From Kimura,[67] with permission.)

the elbow may not appreciably affect the intrinsic hand muscles. In this rare condition, called all-median hand, the intrinsic hand muscles, ordinarily supplied by the ulnar nerve, receive innervation via the communicating fibers.[86] Electromyography may reveal normal motor unit potentials in the ulnar-innervated muscles, despite severe damage to the ulnar nerve at the elbow. Conversely, an injury to the median nerve at the elbow could lead to the appearance of spontaneous discharges in the ulnar-innervated intrinsic hand muscles. Hence, an anomaly of this type, if undetected, gives rise to considerable confusion in the interpretation of electrophysiologic findings.

Anomalies of the Hand

Common anomalies of the peripheral nerves include variations in innervation of the intrinsic hand muscles. Although not as widely recognized as the median-to-ulnar communication, they constitute sources of error in the evaluation of nerve conduction velocity and electromyography. Electrophysiologic techniques often hint at the presence of such anastomoses, although anatomic studies are necessary for precise characterization and delineation of the extent of the anomaly.[116] Various communications may exist between motor branches of the median and ulnar nerves in the lateral portion of the hand.[14,100] Variations may occur in any of the intrinsic hand muscles but in particular involve the flexor pollicis brevis, which may receive median, ulnar, or dual innervation.[103] In a small percentage of cases, thenar muscles, including the adductor pollicis, may derive their supply exclusively from the median or ulnar nerve. In addition to neural anastomoses, skeletal anomalies of the upper limb may cause confusing clinical pictures. The congenital absence of thenar muscles, for example, may suggest a false diagnosis of the carpal tunnel syndrome.[15]

Accessory Deep Peroneal Nerve

The most frequent anomaly of the lower extremity involves the innervation of the extensor digitorum brevis, the muscle commonly used in conduction studies of the peroneal nerve. This muscle usually derives its supply from the deep peroneal nerve, a major branch of the common peroneal nerve. In 20 to 28 percent of an unselected population, however, the superficial peroneal nerve also contributes via a communicating fiber. This branch, called the accessory deep peroneal nerve (Fig. 7–9), descends on the lateral aspect of the leg after arising from the superficial peroneal nerve, then passes behind the lateral malleolus and proceeds anteriorly to innervate the lateral portion of the extensor digitorum brevis.[51,77,129] Occasionally the extensor digitorum brevis may receive exclusive supply from this communication.[92] The anomaly, when inherited, shows a dominant trait.[16]

In patients with the anastomosis, stimulation of the deep peroneal nerve at the ankle elicits a smaller compound muscle action potential than stimulation of the common peroneal nerve at the knee. Stimulation of the accessory deep peroneal nerve behind the lateral malleolus activates the anomalously innervated lateral portion of the muscle. Injury to the deep peroneal nerve ordinarily causes weakness of the tibialis anterior, extensor digitorum longus, extensor hallucis longus, and extensor digitorum brevis. In the presence of the anastomosis, however, such a lesion would spare the lateral portion of the extensor digitorum brevis. Overlooking this possibility would, therefore, lead to an erroneous interpretation.[38]

5 FAST VERSUS SLOW CONDUCTING FIBERS

Temporal Dispersion of the Compound Action Potential

In nerve conduction studies, latency measure of the fastest fibers allows calculation of the maximal motor or sensory velocities. In addition, waveform analyses of compound muscle and sensory nerve action potentials help estimate the range of the functional units.[21,66,94] This aspect of the study provides an equally, if

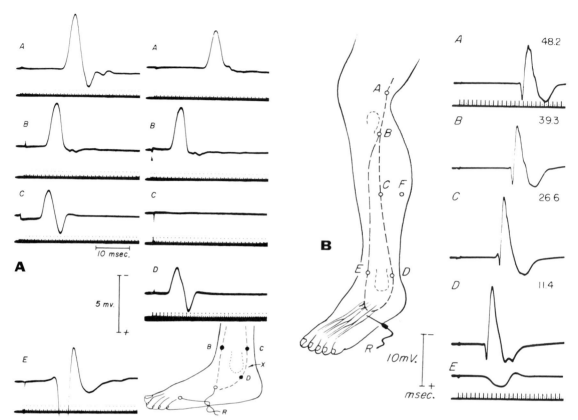

Figure 7–9. A. Compound muscle action potentials recorded from surface electrodes over the extensor digitorum brevis after a maximal stimulus to the common peroneal nerve at the knee (A), deep peroneal nerve on the dorsum of the ankle (B), accessory deep peroneal nerve posterior to the lateral malleolus (C and D), and posterior tibial nerve posterior to the medial malleolus (E) at the ankle. Left and right panels show responses before and after block of the accessory deep peroneal nerve with 2% lidocaine posterior to the lateral malleolus. Diagram of the foot indicates the site of block (X) and the points of stimulation (B, C, and D) and recording (R). **B.** Course of the accessory deep peroneal nerve and action potentials recorded with coaxial needle electrode (R) in the lateral belly of the extensor digitorum brevis muscle following stimulation of the common peroneal nerve at the knee (A), just below the head of fibula (B), superficial peroneal nerve (C), accessory deep peroneal nerve posterior to the lateral malleolus (D), and deep peroneal nerve on the dorsum of the ankle (E). The volume-conducted potential from the medial bellies of the extensor digitorum brevis (E) reduces amplitude of action potential of the lateral belly with simultaneous stimulation of the common peroneal nerve at A or B. (From Lambert,[77] with permission.)

not more, important assessment, especially in the study of peripheral neuropathies with segmental block, in which surviving axons may conduct normally.[32,33,42,80,90,115,120] In clinical tests of motor and sensory conduction, the size of the recorded response approximately parallels the number of excitable fibers. Any discrepancy between responses to proximal and distal shocks, however, does not necessarily imply the presence of a conduction block. The impulses of slow conducting fibers lag increasingly behind those of fast conducting fibers over a long conduction path.[10,19,76] Thus, the size of the recorded response depends to a great extent on the site of stimulation.

With increasing distance between stimulating and pickup electrodes, the recorded potentials become smaller in amplitude and longer in duration; and, contrary to the common belief, the area under the waveform also diminishes.

Thus, stimulation proximally in the axilla or Erb's point may give rise to a small or inconsistent digital potential, in contrast to a large response elicited by stimulation at the wrist or palm.[65,93,126] For the same number of conducting fibers activated by the stimulus, the size of sensory potentials change almost linearly with the length of the nerve segment.[68] A physiologic reduction both in amplitude and the area under the waveform may erroneously suggest a conduction block between the proximal and the distal sites of stimulation.

Physiologic temporal dispersion affects the sensory action potential more than the muscle response, perhaps because of the difference in duration of individual unit discharges between nerve and muscle (Fig. 7–10). With short-duration diphasic sensory spikes, a slight latency difference could line up the positive peaks of the fast fibers with the negative peaks of the slow fibers, canceling both (Fig. 7–11). In motor unit potentials of longer duration, however, the same latency shift would superimpose peaks of

opposite polarity only partially, resulting in little cancellation (Fig. 7–12). In support of this view, the duration change of the sensory potential, expressed as a percentage of the respective baseline values, far exceeds that of the muscle response.[68]

To further test the hypothesis, it is possible to study the effect of desynchronized inputs using a simple model. A shock applied to the median (S_1) or ulnar (S_2) nerve at the wrist evokes a sensory potential of the fourth digit and a muscle potential over the thenar eminence. Hence, a concomitant application of S_1 and S_2 with varying interstimulus intervals simulates the effect of desynchronized inputs (Figs. 7–13 and 7–14). In 10 hands, an interstimulus interval on the order of 1 ms between S_1 and S_2 caused a major reduction in sensory potential by as much as 30 to 40 percent but little change in muscle action potential. With further separation of S_1 and S_2, the muscle response began to decrease in amplitude and area, reaching a minimal size at interstimulus intervals of 5 to 6 ms. A latency difference slightly less than one-half the total duration of

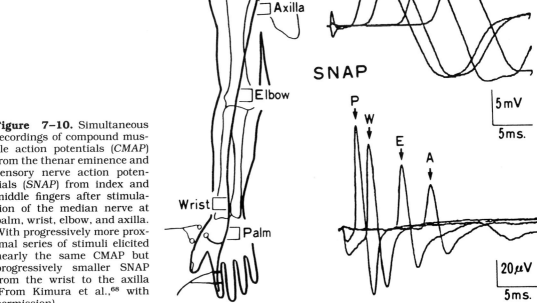

Figure 7–10. Simultaneous recordings of compound muscle action potentials (*CMAP*) from the thenar eminence and sensory nerve action potentials (*SNAP*) from index and middle fingers after stimulation of the median nerve at palm, wrist, elbow, and axilla. With progressively more proximal series of stimuli elicited nearly the same CMAP but progressively smaller SNAP from the wrist to the axilla (From Kimura et al.,[68] with permission).

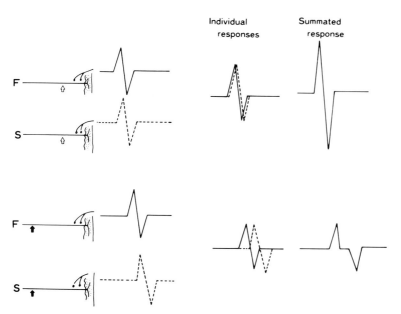

Figure 7–11. Sensory action potentials. A model for phase cancellation between fast (*F*) and slow (*S*) conducting sensory fibers. With distal stimulation, two unit discharges summate in phase to produce a sensory action potential twice as large. With proximal stimulation, a delay of the slow fiber causes phase cancellation between the negative peak of the fast fiber and positive peak of the slow fiber, resulting in a 50% reduction in size of the summated response. (From Kimura et al.,[68] with permission.)

unit discharge maximized the phase cancellation between the two components and consequently the loss of area under the waveform.

The overlap between peaks of opposite polarity varies, depending on the degree of separation between G_1 and G_2, which dictates the duration and waveform of unit discharges.[10] In our experience, the cancellation effect is maximized when a waveform contains negative and positive phases of comparable size. In a triphasic orthodromic sensory potential, as compared with biphasic antidromic digital potentials, the initial positivity provides an additional probability for phase can-

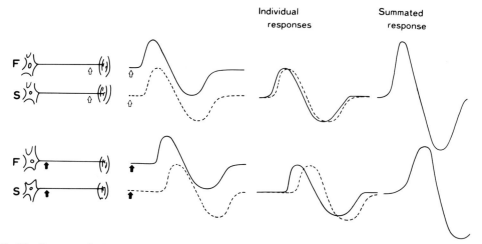

Figure 7–12. Compound muscle action potentials. Same arrangements as in Figure 7–11 to show the relationship between fast (*F*) and slow (*S*) conducting motor fibers. With distal stimulation, two unit discharges representing motor unit potentials summate to produce a muscle action potential twice as large. With proximal stimulation, motor unit potentials of long duration still superimpose nearly in phase despite the same latency shift of the slow motor fiber as the sensory fiber shown in Figure 7–11. Thus, a physiologic temporal dispersion alters the size of the muscle action potential only minimally, if at all. (From Kimura et al.,[68] with permission.)

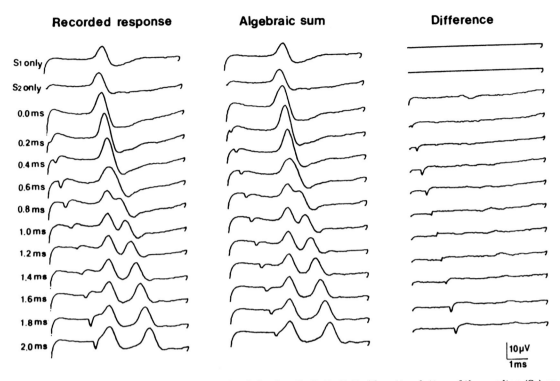

Figure 7-13. Antidromic sensory potentials of the fourth digit elicited by stimulation of the median (S_1) or ulnar (S_2) nerve (top two tracings), or by both S_1 and S_2 at interstimulus intervals ranging from 0 to 2.0 ms (*left*). Algebraic sums of the two top tracings (*middle*) closely matched the actual recording at each interval, as evidenced by small difference shown in computer subtraction (*right*). The area under the negative peak reached a minimal value at 0.8 ms in actual recordings as well as in the calculated waveforms. (From Kimura et al.,[72] with permission.)

cellation. Therefore, the equations for the best fit lines between nerve length and other measurements in one study may not necessarily apply to another unless the recording technique conforms to the particular specifications.

A major reduction in size of the compound sensory action potential can result solely from physiologic phase cancellation. In contrast, the same degree of temporal dispersion has little effect on compound muscle action potential.[29,93,126] This may not hold if the latency difference between fast and slow conducting motor fibers increases substantially as predicted by computer simulation, with a broader spectrum of motor nerve conduction velocities.[78] Thus, in a demyelinating neuropathy, a substantial phase cancellation of the muscle action potential could give rise to a false impression of motor conduction block. This phenomenon also explains an occasionally encountered

discrepancy between severe reduction in amplitude of the compound muscle action potential, on the one hand, and relatively normal recruitment of the motor units and preserved strength, on the other. As an inference, sustained reduction in size of compound muscle action potential does not necessarily imply a prolonged neurapraxia.[90]

In summary, temporal dispersion can effectively reduce the area of diphasic or triphasic evoked potentials recorded in bipolar derivation. The loss of area under the waveform seen in the absence of conduction block implies a duration-dependent phase cancellation of unit discharges within the compound action potential. An awareness of this possibility helps in properly analyzing dispersed action potentials in identifying various patterns of neuropathic processes.[66] Referential derivation of a monophasic waveform in a "killed-end" arrangement

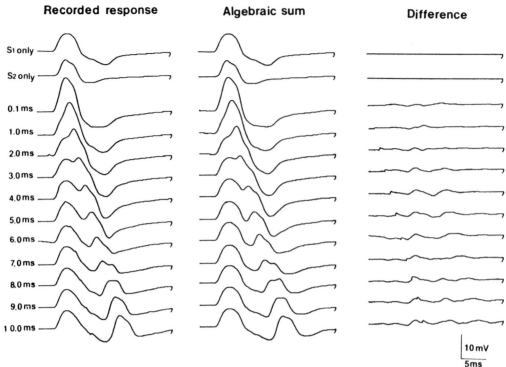

Figure 7–14. Compound muscle action potentials from the thenar eminence elicited by stimulation of the median (S_1) and ulnar (S_2) nerve (*top two tracings*), or by both S_1 and S_2 at interstimulus intervals ranging from 0 to 10 ms (*left*). Algebraic sums of the top two tracings almost, but not exactly, equaled the actual recordings as shown by computer subtraction at each interstimulus interval (*right*). The area under the negative peak reached a minimal value at 5 ms in actual recordings as well as in the calculated waveform. (From Kimura et al.,[72] with permission.)

conserves the area irrespective of stimulus sites and circumvents the ambiguity. This type of recording, however, may register a stationary far-field potential generated by the propagating impulse crossing the partition of the volume conductor.[69,70] Such a steady potential could, in turn, distort the waveform of the near-field activity (see Chapter 19.3).

Waveform Analysis and Distribution of Conduction Velocities

Whereas the onset latency of the action potential relates only to the fastest conducting fibers, its waveform reveals the functional status of the remaining slower conducting fibers. With the loss of nerve fibers a smaller range of conduction velocity reduces the duration of the compound action potential. Conversely,

disproportionate slowing of slower conducting fibers will result in increased temporal dispersion. The greater the range between the fastest and slowest nerve fibers, the longer the duration of the evoked potential. Temporal dispersion also increases with more proximal stimulation in proportion to the distance to the recording site.[29,54,68]

A few recent publications have dealt with mathematical models for studying the waveform.[19,79,91,110–112] The method allows estimation of conduction velocity distribution in a nerve bundle based on a detailed model of the compound muscle action potential as a weighted sum of asynchronous single-fiber action potentials.[17–19,27] In one study,[25,26] nerve conduction velocity of the large myelinated axons, which contribute to the surface recorded response, varied by as much as 25 m/s between fast and slow sensory fibers but over a much narrower range of

11 m/s for motor fibers. This observation, although not universally accepted,[24] would in part explain the different effect of temporal dispersion on sensory and motor fibers for a given length of nerve segment.[93]

Careful attention to the waveform of each evoked potential improves the accuracy of interpretation in any electrophysiologic study. If the responses have dissimilar shapes when elicited by distal and proximal stimuli, the onset latencies probably represent fibers of different conduction characteristics. This discrepancy results, for example, from the use of a submaximal stimulus at one point and a supramaximal stimulus at a second site. In diseased nerves the impulse from a proximal site of stimulation may fail to propagate in some fibers because of conduction block even with an adequate shock intensity. In addition, apparently supramaximal stimuli may not activate a bundle of regenerating or severely demyelinated axons if fibrosis or other local structural changes effectively prevent the current from reaching the nerve. The impulses, once generated voluntarily or reflexively at a proximal site, however, may propagate along these fibers, giving rise to a confusing set of electrophysiologic findings. Any of these circumstances preclude the calculation of conduction velocity with the conventional formula.

Collision Technique to Block Fast or Slow Fibers

The duration of the compound action potentials, although useful as an indirect estimate, falls short of providing a precise measure of slow fibers. Different methods devised for a more quantitative assessment commonly employ the principle of

collision.[121] A distal stimulus of submaximal intensity initially excites the large-diameter, fast fibers with low thresholds. A shock of supramaximal intensity given simultaneously at a proximal site, then, allows selective passage of impulses in the slower fibers, because antidromic activity from the distal stimulation blocks the fast fibers. This assumption, however, does not always hold, because the order of activation with threshold stimulation depends in part on the position of the stimulating electrode in relation to the different fascicles.[30,57]

An alternative method utilizes paired shocks of supramaximal intensity.[34,41,44,52,101] This technique, in essence, consists of incremental delay of proximal shock after distal stimulation without varying stimulus intensity. Shocks applied simultaneously cause collision to occur in all fibers. With increasing intervals between the two stimuli, the fastest fibers escape collision before the slow fibers. Measurement of the minimal interstimulus interval sufficient to produce a full muscle action potential provides an indirect assessment of the slowest conduction (Table 7–1).

Direct latency determination of the slowest fibers requires blocking of the fast conducting fibers, leaving the activity in the slower fibers unaffected. The use of two sets of stimulating electrodes, one placed at the axilla and the other at the wrist, allows delivering two stimuli, S(A$_1$) and S(A$_2$), through the proximal electrodes and another shock, S(W), through the distal electrodes. The antidromic impulse of S(W) blocks the orthodromic impulse of S(A$_1$) provided the distal shock precedes the arrival of the proximal impulse. With an appropriate adjustment of the interstimulus interval between S(A$_1$) and S(W), the collision

Table 7–1 RANGE OF CONDUCTION VELOCITY IN
MOTOR FIBERS OF THE ULNAR NERVE

Authors	Fastest Fibers	Slowest Fibers	Range
Thomas et al. (1959)[121]			30–40%
Poloni and Sala (1962)[98]			35–39%
Hopf (1962)[44] (1963)[45]	60.0 ± 3.2		4–7 m/s
Skorpil (1965)[106]	61.1 ± 4.5	37.7 ± 7.1	22.4 m/s

takes place only in the slow fibers, sparing the antidromic activity from S(W) in the fast fibers. Thus, the impulse of the subsequent proximal stimuli, S(A₂), collides with the antidromic activity only in the fast fibers. In this way, the muscle action potential elicited by S(A₂) corresponds to the remaining slow conducting fibers that selectively transmit the orthodromic impulses (Fig. 7–15).

This technique allows direct determination of the amplitude and latency of the slowest conducting fibers. The muscle action potential elicited by S(A₂) shows progressive diminution of amplitude as the antidromic impulse of S(W) eliminates an increasing number of fast con-

ducting fibers. The latency changes, however, do not always coincide exactly with the values expected from the time interval between S(A₁) and S(W), presumably because the impulses in the slowest conducting fibers do not necessarily arrive at the motor end-plate last. The conduction time must depend not only on the speed of the propagated impulse but also, and perhaps more importantly, on the length of fine terminal branches.[97] Even though they vary in length only on the order of a few millimeters, this degree of difference can still give rise to a substantial latency change at this level, where the impulse normally conducts at a very slow rate.

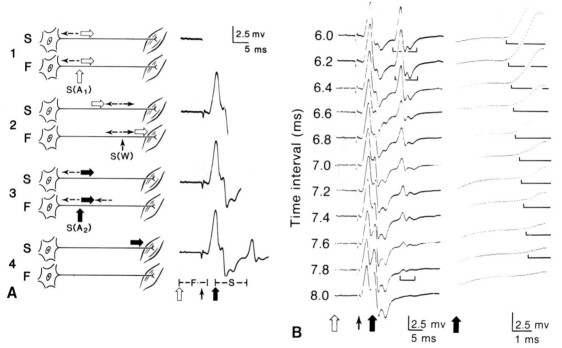

Figure 7–15. A. Compound muscle action potential recorded by surface electrode placed over the abductor digiti minimi after stimulation of the ulnar nerve. The diagrams on the left show orthodromic (*solid line*) and antidromic (*dotted line*) impulses generated by three stimuli, S(A₁), S(W), and S(A₂), delivered at the axilla, wrist, and axilla, respectively. Notice the collision between the orthodromic impulse from S(A₁) and antidromic impulse of S(W) in slow conduction fibers (S) and between the orthodromic impulse of S(A₂) and antidromic impulse of S(W) in the fast conducting fibers (F). The orthodromic impulse of S(A₂) propagates along the slow conducting fibers and elicits the second compound muscle action potential. **B.** Paired axillary shocks of supramaximal intensity combined with a single shock at the wrist. (Compare bottom tracing in Figure 7–15A.) The first axillary stimulation, S(A₁), preceded the wrist stimulation, S(W) by intervals ranging from 6.0 to 8.0 ms in increments of 0.2 ms. Adjusting the second axillary shock, S(A₂), to recur always 6.0 ms after S(W) automatically determined the interstimulus interval between S(A₁) and S(A₂). The figures on the left show the entire tracing with a slow sweep triggered by S(A₁) for amplitude measurement. The figures on the right illustrate latency determination with a fast sweep triggered by S(A₂) and displayed after a predetermined delay of 6.0 ms.

Other Techniques of Clinical Interest

The use of needle, instead of surface, electrodes improves the selectivity of recording in sampling different motor units within a given muscle. By this means, a wide range of motor fibers shows different conduction characteristics. The technique has limited clinical application, because it requires multiple needle insertions for isolation of the slowest conducting fibers.[4,5,7,124]

In the conventional nerve conduction studies, unmyelinated fibers do not contribute to the surface recorded responses. Recording sympathetic skin responses provides a means of testing these axons.[104] Randomly timed electrical stimuli over the median nerve elicit a biphasic potential with the initial negativity over the palmar surface of the hand and the plantar surface of the foot. In one study of 30 healthy subjects, normal values (mean ±SD) consisted of an onset latency of 1.52 ± 0.135 and 2.07 ± 0.165 seconds and an amplitude of 479 ± 105 and 101 ± 40 μV for palmar and plantar responses, respectively.[74] Iontophoresis of atropine into the skin under the recording site abolishes the response. Patients with diabetes or sympathectomy have absent or reduced response on the affected limbs.

Microneurographic techniques also allow recording neural activity in single C fibers[122,123] or autonomic fibers.[113] Despite theoretic interest in correlating cutaneous pain with neural discharges and vasoconstriction with sympathetic activity, the technique has only limited clinical value.

6 ASSESSMENT OF THE REFRACTORY PERIOD

Physiologic Basis

After passage of an impulse, an axon becomes totally inexcitable for a fraction of a millisecond (absolute refractory period), then gradually recovers its prestimulus excitability within the ensu-

ing few milliseconds (relative refractory period). Direct measurement of the nerve action potentials in experimental animals[1,3,36,43,119] substantiates the results in human studies, mostly tested in the sensory or mixed nerves.[2,10,35,48,50,84,117,118] Modified paired-shock techniques enable one to study the refractory characteristics of the motor fibers.[49,73,75,99] Although a considerable amount of data has accumulated during the past decade, their clinical value and limitations await further clarification.[64]

The physiologic mechanism underlying the refractory period centers on inactivation of sodium (Na^+) conductance. Following the passage of an impulse, sodium channels will close to initiate repolarization. Once closed or inactivated, they cannot open immediately, regardless of the magnitude of depolarization by a subsequent impulse. This constitutes the absolute refractory period, lasting 0.5 to 1.0 ms. During the subsequent relative refractory period lasting 3 to 5 ms, only an excessive depolarization, far beyond the ordinary range, can reactivate sodium conductance. Here, the impulse propagates more slowly than usual, because it takes longer to reach the elevated critical level required to generate the action potential. The refractory period is prolonged with low temperature,[10,95,96] advanced age,[23] slow conduction velocity,[95,96] and after experimental demyelination.[22,81,107,108]

A brief suprathreshold current, if applied repetitively, causes conduction block at the site of cathodal stimulation after generation of one or more action potentials. This phenomenon, called cathodal block, occurs even when the interstimulus interval exceeds the absolute refractory period. Like the refractory period, it results from inactivation of sodium conductance. A hyperpolarizing current relieves this type of block by reducing depolarization and reactivating sodium conductance. In contrast, a depolarizing current that brings the transmembrane potential closer to the normal resting value relieves anodal block induced by hyperpolarization of the membrane.

Whereas a suprathreshold current

always induces a refractory period or conduction block, the effect of subthreshold current on the excitability of the cell membrane depends on the duration of the stimulus. A brief current usually increases the excitability by depolarizing the cell membrane. Hence, the cell may fire by summation with a second subthreshold stimulus if applied to the already partly depolarized membrane. A prolonged subthreshold stimulus may decrease, rather than increase, the excitability, because inactivation of sodium conductance more than compensates for the effect of partial depolarization. Thus, a slowly rising current may inactivate sodium conductance before the depolarization reaches the threshold. In this case, conduction ceases at the cathodal point prior to the generation of any action potentials. The term accommodation indicates such reduction of nerve excitability following prolonged subthreshold stimulation.

Paired-Shock Technique

A second shock delivered at a varying time interval after the first reveals excitability changes induced by a preceding impulse. In this method, called the paired-shock technique or the conditioning and testing technique, the first shock conditions the nerve, and the second impulse tests the resulting effect. The test stimulus, given during the refractory period of the conditioning stimulus, elicits no response. During the relative refractory period that ensues, the test response shows reduced amplitude and increased latency. After extensive investigation in experimental animals,[3,119] the paired-shock technique has found its way to the study of human sensory[10,117,118] and mixed nerve potentials.[35,84]

Studies of Motor Fibers

In testing the motor fibers with the short interstimulus interval required for the study of the refractory period, the muscle responses elicited by the first and second stimuli overlap. A computerized subtraction technique circumvents this problem by separating the test from the conditioning muscle response.[2,75] The size of the test response measured thereby, however, still depends on the excitability change of not only the motor axons but also the neuromuscular junction and muscle fibers.[9] Therefore, this technique, based on successively evoked muscle responses, fails to measure the nerve refractory period per se. This difficulty can be avoided if one determines the refractory period of antidromic motor impulses by paired distal stimuli given at various intervals, followed by a single proximal stimulus.[49]

Alternatively, paired proximal stimuli, combined with a single distal stimulus, also allows assessment of neural motor impulses without the muscle effect.[61] In this arrangement, the orthodromic impulse generated by the first of the paired axillary shocks, S(A$_1$), eliminates the antidromic impulse from the distal shock at the wrist, S(W). The orthodromic impulse of the second axillary stimulus, S(A$_2$), will propagate distally along the motor fibers cleared of antidromic activity (Fig. 7–16). Its magnitude and speed depends solely on the neural excitability after the passage of the conditioning stimulus, S(A$_1$). The S(A$_1$)-to-S(W) interval dictates the point of collision and consequently the length of the nerve segment made refractory by S(A$_1$), before it is eliminated by the antidromic activity of S(W). Changing the S(A$_1$)-to-S(A$_2$) interval defines the range of the absolute refractory periods of the different motor fiber by demonstrating the serial recovery of the test response amplitude (Fig. 7–17A). In contrast, the latency of the test response elucidates the duration of the relative refractory period of the most excitable fibers (Fig. 7–17B). Table 7–2 summarizes the results in 20 ulnar nerves from 10 healthy subjects studied in our laboratory.[73]

Changes in Amplitude Versus Latency

The amplitude changes of the test response obtained with shocks of maximal intensity follow nearly an identical

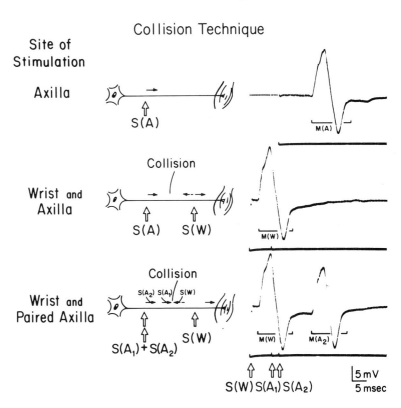

Figure 7–16. Compound muscle action potentials recorded by surface electrodes placed over the abductor digiti minimi after stimulation of the ulnar nerve. The diagram on the left shows the collision between orthodromic (*solid arrows*) and antidromic (*dotted arrows*) impulses. Axillary stimulation, S(A), given 6.0 ms after the stimulus at the wrist, S(W), triggered sweeps on the oscilloscope. With single stimulation at the axilla and at the wrist (*middle tracing*), the orthodromic impulse elicited by S(A) collided with the antidromic impulse of S(W) from the wrist. With paired shocks at the axilla (*bottom tracing*), M(A$_2$) appeared because the first axillary stimulus, S(A$_1$), cleared the path for the second stimulus, S(A$_2$). (From Kimura et al.,[73] with permission.)

course irrespective of the length of the refractory segment (Fig. 7–18). Therefore, reduction in amplitude of the test response must result from failure of nerve activation at the site of stimulation, rather than cessation of propagation along the course of the nerve. The impulse conducts at a slower speed than normal, if transmitted at all, during the relative refractory period, showing the greatest delay near the absolute refractory period (Fig. 7–19). Thereafter, the conduction progressively recovers to normal as the interstimulus interval between the conditioning and test stimuli increases. The length of the refractory segment, which hardly influences the recovery of the amplitude, substantially alters the time course of the latency. The longer the refractory segment, the greater the change in latency of the test response. The delay, however, does not increase linearly in proportion to the length of the refractory segment. In fact, a change in latency per unit length decreases for a longer conduction distance. Therefore, the average conduction veloc-

ity improves as the refractory segment increases (Fig. 7–20).

These findings confirm the results of an animal study[119] that indicated (1) that a delay of the test impulse during the refractory period allows an increasing interval between conditioning and test impulses as they travel further distally and (2) that an increasingly longer interval between the two impulses, in turn, leads to progressive recovery of the test impulse conduction velocity. Because of this regressive process, the test impulse conducts at a relatively normal speed by the time it reaches the end of the refractory segment, especially when one is studying a longer nerve.[61] Electrophysiologic studies of human sensory fibers[10] as well as computer simulation have shown the same relationship between the refractory period and the length of the nerve segment.[125]

Human studies of the refractory period suffer from technical limitation in precisely measuring the amplitude and latency of the test response. Specific problems include small signals, unstable

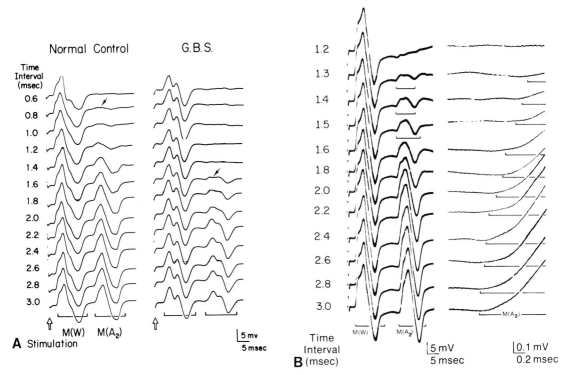

Figure 7–17. A. Paired axillary shocks, S(A$_1$) and S(A$_2$), of just maximal intensity combined with a single shock at the wrist, S(W). (Compare bottom tracing in Figure 7–16.) Interstimulus intervals between S(A$_1$) and S(A$_2$) ranged from 0.6 to 3.0 ms. S(A$_2$) always occurred 5.0 ms after S(W), which triggered sweeps on the oscilloscope. In the normal subject, M(A$_2$) first appeared (*small arrows*) at an interstimulus interval of 0.8 ms and recovered completely by 3.0 ms. The patient with the Guillain-Barré syndrome showed delayed and incomplete recovery. (From Kimura,[61] with permission.) **B.** Paired axillary shocks, S(A$_1$) and S(A$_2$), of just maximal intensity combined with a single shock at the wrist, S(W). (Compare bottom tracing in Figure 7–16.) Delivering S(A$_1$) 6.0 ms after S(W) allowed collision to occur 1.5 ms after S(A$_1$). The interstimulus intervals between S(A$_1$) and S(A$_2$) ranged from 1.2 to 3.0 ms in increments of 0.2 ms. The figures on the *left* show amplitude measurements with a slow sweep triggered by S(W). The figures on the *right* illustrate latency determination with a fast sweep triggered by S(A$_2$) and displayed after a predetermined delay of 11.0 ms. (From Kimura et al.,[73] with permission.)

Table 7–2 INTERSTIMULUS INTERVALS OF THE PAIRED SHOCKS AND CONDUCTION VELOCITY OF THE TEST RESPONSE (MEAN ± SD)

Length of Refractory Segment	Initial Recovery in Amplitude (Test Response >5% of Unconditioned Response)		Full Recovery in Amplitude (Test Response >95% of Unconditioned Response)		Full Recovery in Velocity (>95%)
	Interstimulus Interval Between Paired Shocks (ms)	*Conduction Velocity of Test Impulse (% of normal)*	*Interstimulus Interval Between Paired Shocks (ms)*	*Conduction Velocity of Test Impulse (% of normal)*	*Interstimulus Interval Between Paired Shocks (ms)*
A distance normally covered in 0.5 ms	1.16 ± 0.18	55.3 ± 19.2	2.11 ± 0.50	81.2 ± 17.4	2.65 ± 0.65
A distance normally covered in 1.5 ms	1.18 ± 0.16	70.3 ± 13.5	2.16 ± 0.52	87.3 ± 14.2	2.36 ± 0.45

From Kimura et al.,[73] with permission.

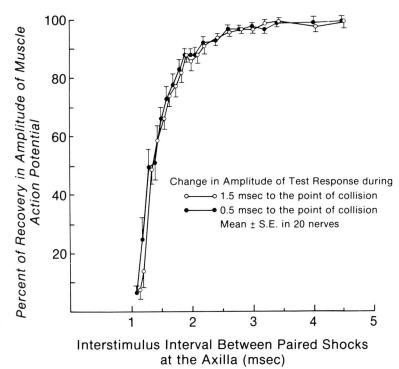

Figure 7–18. The pattern of recovery in amplitude of M(A₂) during the refractory period in 10 healthy subjects. The return of M(A₂) followed the identical time course with the passage of impulse along the shorter (0.5 ms) or longer refractory segment (1.5 ms). The gradual increase of M(A₂) indicates the range of the absolute refractory periods of different motor fibers. (From Kimura et al.,[73] with permission.)

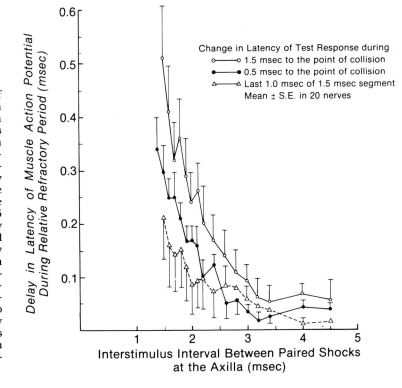

Figure 7–19. The pattern of recovery in latency of M(A₂) in the same subjects as shown in Figure 7–18. The curve shows the latency difference between the response to a single axillary shock M(A) and the response to the second axillary shock M(A₂) of the pair. The passage of impulse across the longer refractory segment (1.5 ms) showed significantly slower recovery as compared with the shorter refractory segment (0.5 ms). The bottom curve (*triangles*) plots the difference in delay of latency between 1.5 ms and 0.5 ms segments. The values so calculated represent the delay attributable to the last 1.0 ms of the 1.5 ms segment. (From Kimura et al.,[73] with permission.)

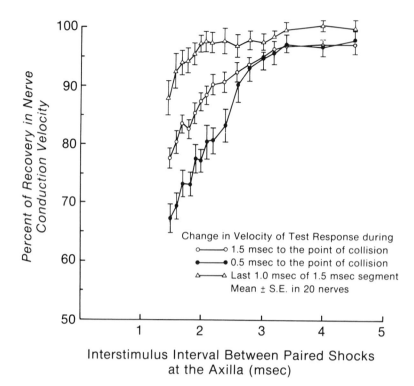

Change in Velocity of Test Response during
○———○ 1.5 msec to the point of collision
●———● 0.5 msec to the point of collision
△———△ Last 1.0 msec of 1.5 msec segment
Mean ± S.E. in 20 nerves

Interstimulus Interval Between Paired Shocks
at the Axilla (msec)

Figure 7–20. The time course of recovery in conduction velocity of M(A_2) in the same subjects shown in Figures 7–18 and 7–19. The conduction velocities were calculated on the assumption that the delay of M(A_2) occurred primarily in the segment proximal to the point of collision. In contrast to the pattern of recovery in latency (compare Figure 7–19), the conduction velocity returned significantly faster for the passage of impulse across the longer (1.5 ms) than the shorter (1.0 ms) refractory segment. The *top curve* (*triangles*) shows the estimated velocity of M(A_2) over the last 1.0 ms of the 1.5-ms segment. (From Kimura et al.,[73] with permission.)

baseline, gradual onset of the evoked response, and partial overlap of the test response with the preceding events, despite the use of a collision technique. A computerized cross-correlation analysis helps improve numeric quantification of the compound muscle potential in shape and latency.[28] In this method, the height of the peak in the correlation curve gives a shape-weighted measure of the size of the test response, and the time lag of the peak indicates the delay of the test response as compared with an averaged unconditioned muscle response. Another technique, called the double collision method, alleviates the transient changes in nerve and muscle fiber conduction that can distort test muscle responses.[53]

Clinical Value and Limitations

A number of studies have shown prolongation of the refractory period of sensory and mixed nerve fibers in diseases of the peripheral nerve.[83,84,117,118] In patients with chronic renal failure, the initially abnormal relative refractory period reverted to normal after hemodialysis.[82] An increased refractory period of median nerve sensory fibers in patients with multiple sclerosis suggests the possible involvement of peripheral nerve fibers in this disorder.[46] Conversely, hypokalemia of various origins shortens the relative refractory period.[88]

Most previous studies in man have dealt with the sensory or mixed nerve fibers, but similar alterations probably occur in the refractory characteristics of motor fibers. In fact, the absolute and relative refractory periods affect motor,[73] sensory, and mixed fibers[35] alike. For example, full recovery in the amplitude of the test response precedes full recovery of the conduction velocity, regardless of the type of the nerve fibers tested.[10,35,48,73]

Determining the refractory period of individual motor fibers requires recording of single motor unit potentials following paired stimuli to the nerve.[5–6,8] Studies of the whole nerve, however, lack precision because fibers with different conduction characteristics contribute to the absolute and relative refractory period. Furthermore, in contrast to amplitude, which follows a predictable time course, small, often variable changes in latency provide limited value in clinical assessment. Further work is necessary to define the clinical value and limitations of the measurement of the refractory period in diagnosing diseases of the motor fibers and in elucidating their pathophysiology.

REFERENCES

1. Bergmans, J: Physiological observations on single human nerve fibres. In Desmedt, JE (ed): New Developments in Electromyography and Clinical Neurophysiology, Vol 2. Karger, Basel, 1973, pp 89–127.
2. Betts, RP, Johnston, DM, and Brown, BH: Nerve fibre velocity and refractory period distributions in nerve trunks. J Neurol Neurosurg Psychiatry 39:694–700, 1976.
3. Bishop, GH, and Heinbecker, P: Differentiation of axon types in visceral nerves by means of the potential record. Am J Physiol 94:170–200, 1930.
4. Blackstock, E, Rushworth, G, and Gath, D: Electrophysiological studies in alcoholism. J Neurol Neurosurg Psychiatry 35:326–334, 1972.
5. Borg, J: Axonal refractory period of single short toe extensor motor units in neuropathies and neuromuscular diseases. J Neurol Neurosurg Psychiat 44:1136–1140, 1981.
6. Borg, J: Effects of prior activity on the conduction in single motor units in man. J Neurol Neurosurg Psychiatry 46:317–321, 1983.
7. Borg, J: Conduction velocity and refractory period of single motor nerve fibres in motor neuron disease. J Neurol Neurosurg Psychiatry 47:349–353, 1984.
8. Borg, J: Refractory period of single motor nerve fibres in man. J Neurol Neurosurg Psychiatry 47:344–348, 1984.
9. Buchthal, F, and Engbaek, L: Refractory period and conduction velocity of the striated muscle fibre. Acta Physiol Scand 59:199–220, 1963.
10. Buchthal, F, and Rosenfalck, A: Evoked action potentials and conduction velocity in human sensory nerves. Brain Res 3:1–122, 1966.
11. Buchthal, F, and Rosenfalck, A: Sensory potentials in polyneuropathy. Brain 94:241–262, 1971.
12. Buchthal, F, Rosenfalck, A, and Trojaborg, W: Electrophysiological findings in entrapment of the median nerve at wrist and elbow. J Neurol Neurosurg Psychiatry 37:340–360, 1974.
13. Bunnels, S: Surgery of the Hand, ed 4. JB Lippincott, Philadelphia, 1964.
14. Cannieu, JMA: Note sur une anastomose entre la branche profonde du cubital et le median. Bull Soc D'Anat Physiol Bordeaux 18:339–340, 1897.
15. Cavanagh, NPC, Yates, DAH, and Sutcliffe, J: Thenar hypoplasia with associated radiologic abnormalities. Muscle Nerve 2:431–436, 1979.
16. Crutchfield, CA, and Gutmann, L: Hereditary aspects of accessory deep peroneal nerve. J Neurol Neurosurg Psychiatry 36:989–990, 1973.
17. Cummins, KL, and Dorfman, LJ: Nerve fiber conduction velocity distributions: Studies of normal and diabetic human nerves. Ann Neurol 9:67–74, 1981.
18. Cummins, KL, Dorfman, LJ, and Perkel, DH: Nerve fiber conduction-velocity distributions. II. Estimation based on two compound action potentials. Electroencephalogr Clin Neurophysiol 46:647–658, 1979.
19. Cummins, KL, Perkel, DH, and Dorfman, LJ: Nerve fiber conduction-velocity distributions. I. Estimation based on the single-fiber and compound action potentials. Electroencephalogr Clin Neurophysiol 46:634–646, 1979.
20. Daube, JR: Percutaneous palmar median nerve stimulation for carpal tunnel syndrome. Electroencephalogr Clin Neurophysiol 43:139–140, 1977.
21. Daube, JR: Nerve Conduction Studies. Churchill Livingstone, New York, 1980, pp 229–264.
22. Davis, FA: Impairment of repetitive impulse conduction in experimentally demyelinated and pressure-injured nerves. J Neurol Neurosurg Psychiatry 35:537–544, 1972.
23. Delbeke, J, Kopec, J, and McComas, AJ: The effects of age, temperature, and disease on the refractoriness of human nerve and muscle. J Neurol Neurosurg Psychiatry 41:65–71, 1978.
24. Dominque, J, Shahani, BT, and Young, RR: Conduction velocity in different diameter ulnar sensory and motor nerve fibers. Electroencephalogr Clin Neurophysiol 50:239P–245P, 1980.
25. Dorfman, LJ: The distribution of conduction velocities (DCV) in peripheral nerves: A review. Muscle Nerve 7:2–11, 1984.
26. Dorfman, LJ, Cummins, KL, and Abraham, GS: Conduction velocity distributions of the human median nerve: Comparison of methods. Muscle Nerve 5:S148–S153, 1982.
27. Dorfman, LJ, Cummins, KL, Reaven, GM, Ceranski, J, Greenfield, MS, and Doberne, L: Studies of diabetic polyneuropathy using conduction velocity distribution (DCV) analysis. Neurology 33:773–779, 1983.
28. Faisst, S, and Meyer, M: A non-invasive computerized measurement of motor neurone refractory period and subnormal conduction in man. Electroencephalogr Clin Neurophysiol 51:548–558, 1981.
29. Felsenthal, G, and Teng, CS: Changes in duration and amplitude of the evoked muscle action potential (EMAP) over distance in peroneal, median, and ulnar nerves. Am J Phys Med 62:123–134, 1983.
30. Gassel, MM: A study of femoral nerve conduction time. Arch Neurol 9:607–614, 1963.
31. Gassel, MM: Sources of error in motor nerve

conduction studies. Neurology 14:825–835, 1964.

32. Gilliatt, RW: Acute compression block. In Sumner, A (ed): The Physiology of Peripheral Nerve Disease. WB Saunders, Philadelphia, 1980, pp 287–315.

33. Gilliatt, RW: Electrophysiology of peripheral neuropathies: An overview. Muscle Nerve 5:S108–S116, 1982.

34. Gilliatt, RW, Hopf, HC, Rudge, P, and Baraitser, M: Axonal velocities of motor units in the hand and foot muscles of the baboon. J Neurol Sci 29:249–258, 1976.

35. Gilliatt, RW, and Willison, RG: The refractory and supernormal periods of the human median nerve. J Neurol Neurosurg Psychiatry 26:136–147, 1963.

36. Graham, HT: The subnormal period of nerve response. Am J Physiol 11:452–465, 1935.

37. Gruber, W: Ueber die Verbindung des Nervus medianus mit dem Nervus ulnaris am Unterarme des Menschen und der Saugethiere. Arch Anat Physiol Med 1870, pp 501–522.

38. Gutmann, L: Atypical deep peroneal neuropathy in presence of accessory deep peroneal nerve. J Neurol Neurosurg Psychiatry 33:453–456, 1970.

39. Gutmann, L: Important anomalous innervations of the extremities. Meeting of the American Association of Electromyography and Electrodiagnosis, 1977.

40. Gutmann, L: Median-ulnar nerve communications and carpal tunnel syndrome. J Neurol Neurosurg Psychiatry 40:982–986, 1977.

41. Hakamada, S, Kumagai, T, Watanabe, K, Koike, Y, Hara, K, and Miyazaki, S: The conduction velocity of slower and the fastest fibres in infancy and childhood. J Neurol Neurosurg Psychiatry 45:851–853, 1982.

42. Harrison, MJG: Pressure palsy of the ulnar nerve with prolonged conduction block. J Neurol Neurosurg Psychiatry 39:96–99, 1976.

43. Hodgkin, AL: The Conduction of the Nervous Impulse. The Sherrington Lectures VII, Liverpool University Press, Liverpool, 1965.

44. Hopf, HC: Untersuchungen über die Unterschiede in der leitgeschwindigkeit motorischer Nervenfasern beim Menschen. Deutsche Zeitschrift Für Nervenheilkunde 183:579–588, 1962.

45. Hopf, HC: Electromyographic study on so-called mononeuritis. Arch Neurol 9:307–312, 1963.

46. Hopf, HC, and Eysholdt, M: Impaired refractory periods of peripheral sensory nerves in multiple sclerosis. Ann Neurol 4:499–501, 1978.

47. Hopf, HC, and Hense, W: Anomalien der motorischen Innervation an der Hand. Z EEG-EMG 5:220–224, 1974.

48. Hopf, HC, Le Quesne, PM, and Willison, RG: Refractory periods and lower limiting frequencies of sensory fibres of the hand. In Kunze, K, and Desmedt, JE (eds): Studies on Neuromuscular Diseases. Proceedings of the International Symposium (Giessen). Karger, Basel, 1975, pp 258–263.

49. Hopf, HC, and Lowitzsch, K: Relative refractory periods of motor nerve fibres. In Kunze, K, and Desmedt, JE (eds): Studies on Neuromuscular Diseases. Proceedings of the International Symposium (Giessen). Karger, Basel, 1975, pp 264–267.

50. Hopf, HC, Lowitzsch, K, and Galland, J: Conduction velocity during the supernormal and late subnormal periods in human nerve fibres. J Neurol 211:293–296, 1976.

51. Infante, E, and Kennedy, WR: Anomalous branch of the peroneal nerve detected by electromyography. Arch Neurol 22:162–165, 1970.

52. Ingram, DA, Davis, GR, and Swash, M: The double collision technique: a new method for measurement of the motor nerve refractory period distribution in man. Electroencephalogr Clin Neurophysiol 66:225–234, 1987.

53. Ingram, DA, Davis, GR, and Swash, M: Motor conduction velocity distributions in man: Results of a new computer-based collision technique. Electroencephalogr Clin Neurophysiol 66:235–243, 1987.

54. Isch, F, Isch-Treussard, C, Buchheit, F, Delgado, V, and Kirchner, JP: Measurement of conduction velocity of motor nerve fibres in polyneuritis and polyradiculoneuritis (abstr) Electroencephalogr Clin Neurophysiol 16:416, 1964.

55. Iyer, V, and Fenichel, GM: Normal median nerve proximal latency in carpal tunnel syndrome: A clue to coexisting Martin-Gruber anastomosis. J Neurol Neurosurg Psychiatry 39:449–452, 1976.

56. Johnson, RK, and Shrewsbury, MM: Anatomical course of the thenar branch of the median nerve—usually in a separate tunnel through the transverse carpal ligament. J Bone Joint Surg 52A:269–273, 1970.

57. Kadrie, HA, Yates, SK, Milner-Brown, HS, and Brown, WF: Multiple point electrical stimulation of ulnar and median nerves. J Neurol Neurosurg Psychiatry 39:973–985, 1976.

58. Kaeser, HE: Nerve conduction velocity measurements. In Vinken, PJ, and Bruyn, BW (eds): Handbook of Clinical Neurology, Vol 7. North Holland, Amsterdam, 1970, pp 116–196.

59. Kimura, J: F-wave velocity in the central segment of the median and ulnar nerves: A study in normal subjects and in patients with Charcot-Marie-Tooth disease. Neurology 24:539–546, 1974.

60. Kimura, J: Collision technique: Physiologic block of nerve impulses in studies of motor nerve conduction velocity. Neurology 26:680–682, 1976.

61. Kimura, J: A method for estimating the refractory period of motor fibers in the human peripheral nerve. J Neurol Sci 28:485–490, 1976.

62. Kimura, J: Electrical activity in voluntarily contracting muscle. Arch Neurol 34:85–88, 1977.

63. Kimura, J: The carpal tunnel syndrome: Localization of conduction abnormalities within the distal segment of the median nerve. Brain 102:619–635, 1979.

64. Kimura, J: Refractory period measurement in the clinical domain. In Waxman, SA, and Ritchie, JM (eds): Demyelinating Disease: Basic and Clinical Electrophysiology. Raven Press, New York, 1981, pp 239–265.

65. Kimura, J: Electrodiagnosis in Diseases of Nerve and Muscle: Principles and Practices. FA Davis, Philadelphia, 1983.

66. Kimura, J: Principles and pitfalls of nerve conduction studies. Ann Neurol 16:415–429, 1984.

67. Kimura, J: Electromyography and nerve stimulation techniques: Clinical applications (Japanese). Egakushoin, Tokyo, 1989.

68. Kimura, J, Machida, M, Ishida, T, Yamada, T, Rodnitzky, RL, Kudo, Y, and Suzuki, S: Relation between size of compound sensory or muscle action potentials and length of nerve segment. Neurology 36:647–652, 1986.

69. Kimura, J, Mitsudome, A, Beck, DO, Yamada, T, and Dickins, QS: Field distribution of antidromically activated digital nerve potentials: model for far-field recording. Neurology 33:1164–1169, 1983.

70. Kimura, J, Mitsudome, A, Yamada, T, and Dickins, QS: Stationary peaks from a moving source in far-field recording. Electroencephalogr Clin Neurophysiol 58:351–361, 1984.

71. Kimura, J, Murphy, JM, and Varda, DJ: Electrophysiological study of anomalous innervation of intrinsic hand muscles. Arch Neurol 33:842–844, 1976.

72. Kimura, J, Sakimura, Y, Machida, M, Fuchigami, Y, Ishida, T, Claus, D, Kameyama, S, Nakazumi, Y, Wang, J, and Yamada, T: Effect of desynchronized inputs on compound sensory and muscle action potentials. Muscle Nerve 11:694–702, 1988.

73. Kimura, J, Yamada, T, and Rodnitzky, RL: Refractory period of human motor nerve fibres. J Neurol Neurosurg Psychiatry 41:784–790, 1978.

74. Knezevic, W, and Bajada, S: Peripheral autonomic surface potential: A quantitative technique for recording sympathetic conduction in man. J Neurol 67:239–251, 1985.

75. Kopec, J, Delbecke, J, and McComas, AJ: Refractory period studies in a human neuromuscular preparation. J Neurol Neurosurg Psychiatry 41:54–64, 1978.

76. Lambert, EH: Diagnostic value of electrical stimulation of motor nerves. Electroencephalogr Clin Neurophysiol (Suppl 22):9–16, 1962.

77. Lambert, EH: The accessory deep peroneal nerve: A common variation in innervation of extensor digitorum brevis. Neurology 19:1169–1176, 1969.

78. Lee, R, Ashby, P, White D, and Aguayo, A: Analysis of motor conduction velocity in human median nerve by computer simulation of compound action potentials. Electroencephalogr Clin Neurophysiol 39:225–237, 1975.

79. Leifer, LJ: Nerve-fiber conduction velocity distributions of the human median nerve: Comparison of methods. Alan R Liss, 1981, pp 233–263.

80. Lewis, RA, Sumner, AJ, Brown, MJ, and Asbury, AK: Multifocal demyelinating neuropathy with persistent conduction block. Neurology 32:958–964, 1982.

81. Low, PA, and McLeod, JG: Refractory period, conduction of trains of impulses, and effect of temperature on conduction in chronic hypertrophic neuropathy: Electrophysiological studies on the trembler mouse. J Neurol Neurosurg Psychiatry 40:434–447, 1977.

82. Lowitzsch, K, Gohring, U, Hecking, E, and Kohler, H: Refractory period, sensory conduction velocity and visual evoked potentials before and after haemodialysis. J Neurol Neurosurg Psychiatry 44:121–128, 1981.

83. Lowitzsch, K, and Hopf, HC: Refractory periods and propagation of repetitive mixed nerve action potentials in severe and mild neuropathy. In Hausmanowa-Petrusewicz, I, and Jedrzejowska, H (eds): Structure and Function of Normal and Diseased Muscle and Peripheral Nerve. Polish Medical Publisher, 1972.

84. Lowitzsch, K, Hopf, HC, and Schlegel, HJ: Conduction of two or more impulses in relation to the fibre spectrum in the mixed human peripheral nerve. In Desmedt, JE (ed): New Developments in Electromyography and Clinical Neurophysiology, Vol 2. Karger, Basel, 1973, pp 272–278.

85. Mannerfelt, L: Studies on the hand in ulnar nerve paralysis: A clinical-experimental investigation in normal and anomalous innervation. Acta Orthopaedica Scand (Suppl 87): 23–176, 1966.

86. Marinacci, AA: Diagnosis of "all median hand." Bull LA Neurol Soc 29:191–197, 1964.

87. Martin, R: Tal om Nervers allmanna egenskapper i manniskans kropp. L Salvius, Stockholm, 1763.

88. Maurer, K, Hopf, HC, and Lowitzsch, K: Hypokalemia shortens relative refractory period of peripheral sensory nerves in man. J Neurol 216:67–71, 1977.

89. Maynard, FM, and Stolov, WC: Experimental error in determination of nerve conduction velocity. Arch Phys Med Rehabil 53:362–372, 1972.

90. Miller, RG, and Olney, RK: Persistent conduction block in compression neuropathy. Muscle Nerve (Suppl)5:S154–S156, 1982.

91. Milner, TE, Stein, RV, Gillespie, J, and Hanley, B: Improved estimates of conduction velocity distributions using single unit action potentials. J Neurol Neurosurg Psychiatry 44:476–484, 1981.

92. Neundorfer, B, and Seiberth, R: The accessory deep peroneal nerve. J Neurol 209:125–129, 1975.

93. Olney, RK, and Miller, RG: Pseudo-conduction block in normal nerves. Muscle Nerve 6:530, 1983.

94. Olney, RK, and Miller, RG: Conduction block in compression neuropathy: Recognition and quantification. Muscle Nerve 7:662–667, 1984.

95. Paintal, AS: Block of conduction in mammalian myelinated nerve fibres by low temperatures. J Physiol (Lond) 180:1–19, 1965.

96. Paintal, AS: Effects of temperature on conduction in single vagal and saphenous myelinated

nerve fibres of the cat. J Physiol (Lond) 180:20–49, 1965.

97. Pinelli, P: Physical, anatomical and physiological factors in the latency measurement of the M response (abstr). Electroencephalogr Clin Neurophysiol 17:86, 1964.

98. Poloni, AE, and Sala, E: The conduction velocity of the ulnar and median nerves stimulated through a twin-needle electrode. Electroencephalogr Clin Neurophysiol (Suppl 22):17–19, 1962.

99. Reitter, BF, and Johannsen, S: Neuromuscular reaction to paired stimuli. Muscle Nerve 5:593–603, 1982.

100. Richie, P: Le nerf cubital et les muscles de l'eminence thenar. Bull Mem Soc Anat Paris (Series 5)72:251–252, 1897.

101. Rossi, B, Sartucci, F, and Stefanini, A: Measurement of motor conduction velocity with Hopf's technique in the diagnosis of mild peripheral neuropathies. J Neurol Neurosurg Psychiatry 44:168–170, 1981.

102. Roth, G: Vitesse de conduction motrice du nerf median dans le cana carpien. Ann Med Physique 13:117–132, 1970.

103. Seddon, H: Surgical Disorders of the Peripheral Nerves, ed 2. Churchill Livingstone, Edinburgh, 1975, pp 203–211.

104. Shahani, BT, Halperin, JJ, Boulu, P, and Cohen, J: Sympathetic skin response: A method of assessing unmyelinated axon dysfunction in peripheral neuropathies. J Neurol Neurosurg Psychiatry 47:536–542, 1984.

105. Simpson, JA: Fact and fallacy in measurement of conduction velocity in motor nerves. J Neurol Neurosurg Psychiatry 27:381–385, 1964.

106. Skorpil, V: Conduction velocity of human nerve structures. Roszpr Cesk Akad Ved 75:1–103, 1965.

107. Smith, KJ: A sensitive method for detection and quantification of conduction deficits in nerve. J Neurol Sci 48:191–199, 1980.

108. Smith, KJ, and Hall, SM: Nerve conduction during peripheral demyelination and remyelination. J Neurol Sci 48:201–219, 1980.

109. Srinivasan, R, and Rhodes, J: The median-ulnar anastomosis (Martin-Gruber) in normal and congenitally abnormal fetuses. Arch Neurol 38:418–419, 1981.

110. Stegeman, DE, and De Weerd, JPC: Modelling compound action potentials of peripheral nerves in situ. I. Model description: Evidence for a non-linear relation between fibre diameter and velocity. Electroencephalogr Clin Neurophysiol 54:436–448, 1982.

111. Stegeman, DE, and De Weerd, JPC: Modelling compound action potentials of peripheral nerves in situ. II. A study of the influence of temperature. Electroencephalogr Clin Neurophysiol 54:516–529, 1982.

112. Stegeman, DF, De Weerd, JPC, and Notermans, SLH: Modelling compound action potentials of peripheral nerves in situ. III. Nerve propagation in the refractory period. Electroencephalogr Clin Neurophysiol 55:668–679, 1983.

113. Stjernberg, L, Blumberg, H, and Wallin, B: Sympathetic activity in man after spinal cord injury: outflow to muscle below the lesion. Brain 109:695–715, 1986.

114. Streib, EW: Ulnar-to-median nerve anastomosis in the forearm: Electromyographic studies. Neurology 29:1534–1537, 1979.

115. Sumner, AJ: The physiological basis for symptoms in Guillain-Barré syndrome. Ann Neurol (Suppl)9:28–30, 1981.

116. Sunderland, S: Nerves and Nerve Injuries. ed 2. Churchill Livingstone, Edinburgh, 1978.

117. Tackmann, W, and Lehmann, HJ: Refractory period in human sensory nerve fibres. Eur Neurol 12:277–292, 1974.

118. Tackmann, W, and Lehmann, HJ: Relative refractory period of median nerve sensory fibres in the carpal tunnel syndrome. Eur Neurol 12:309–316, 1974.

119. Tasaki, I: Nervous Transmission. Charles C Thomas, Springfield, 1953.

120. Thomas, PK: The morphological basis for alterations in nerve conduction in peripheral neuropathy. Proc R Soc Med 645:295–298, 1971.

121. Thomas, PK, Sears, TA, and Gilliatt, RW: The range of conduction velocity in normal motor nerve fibres to the small muscles of the hand and foot. J Neurol Neurosurg Psychiatry 22:175–181, 1959.

122. Torebjork, HE, Lamotte, RH, and Robinson, CJ: Peripheral neural correlates of magnitude of cutaneous pain and hyperalgesia: Simultaneous recordings in humans of sensory judgments of pain and evoked responses in nociceptors with C-fibers. J Neurophysiol 51:325–339, 1984.

123. Torebjork, HE, Schady, W, and Ochoa, JL: A new method for demonstration of central effects of analgesic agents in man. J Neurol Neurosurg Psychiatry 47:862–869, 1984.

124. Van der Most Van Spijk, D, Hoogland, RA, and Dijkstra, S: Conduction velocities compared and related to degrees of renal insufficiency. In Desmedt, JE (eds): New Developments in Electromyography and Clinical Neurophysiology, Vol 2. Karger, Basel, 1973, pp 381–389.

125. Waxman, SG, Kocsis, JD, Brill, MH, and Swadlow, HA: Dependence of refractory period measurements on conduction distance: A computer simulation analysis. Electroencephalogr Clin Neurophysiol 47:717–724, 1979.

126. Wiechers, D, and Fatehi, M: Changes in the evoked potential area of normal median nerves with computer techniques. Muscle Nerve 6:532, 1983.

127. Wiederholt, WC: Median nerve conduction velocity in sensory fibers through carpal tunnel. Arch Phys Med Rehabil 51:328–330,1970.

128. Wilbourn, AJ, and Lambert, EH: The forearm median-to-ulnar nerve communication: Electrodiagnostic aspects. Neurology 26:368, 1976.

129. Winckler, G: Le nerf péronier accessoire profond: Étude d'anatomie comparée. Arch Anat Histol Embryol 18:181–219, 1934.

Part III

ASSESSMENT OF NEUROMUSCULAR TRANSMISSION

Chapter 8

ANATOMY AND PHYSIOLOGY OF THE NEUROMUSCULAR JUNCTION

1 INTRODUCTION

The neuromuscular junction is a synaptic structure consisting of the motor nerve terminal, junctional cleft, and muscle end plate. Its chemical mode of transmission has fundamentally differ-ent properties, compared with electrical propagation of impulses along the nerve and muscle. For example, the release of acetylcholine (ACh) ensures unidirectional conduction from the axon terminal to the muscle end plate. The same basic principle applies to synaptic transmission in a sequence of neurons. In con-

169

trast, the nerve axons and muscle fibers conduct an impulse bidirectionally.

There are other physiologic characteristics common to the nerve synapse and neuromuscular junction. Analogous to synaptic delay, transmission at the muscle end plate requires a fraction of a millisecond. As in a synapse, its mobilization store must continuously replenish the liberated transmitters. Otherwise, the neuromuscular junction would fail, with depletion of immediately available molecules. Neither synaptic nor end-plate potentials (EPPs) propagate. These local potentials cause no refractoriness, unlike the all-or-none response of the nerve or muscle action potential. The nonpropagating, graded response allows temporal as well as spatial summation of subliminal stimuli, thereby providing greater flexibility and adaptability.

A simplified overview of the complex physiology is in order in preparation for a subsequent more detailed discussion. The presynaptic ending contains many minute vesicles, each with up to 10,000 ACh molecules. At rest, these vesicles randomly migrate into the junctional cleft. At the muscle end plate, they produce small depolarizations of the postsynaptic membrane. These miniature EPPs (MEPPs) do not attain the critical level for generation of a muscle action potential. Depolarization of the presynaptic ending at the axon terminal triggers an influx of calcium (Ca^{2+}), initiating the calcium-dependent release of immediately available vesicles into the junctional cleft. The greatly enhanced and synchronized ACh activity gives rise to a nonpropagated EPP from summation of multiple MEPPs. When the EPP exceeds the excitability threshold of the muscle cell, an action potential ensues. Propagation of the muscle potential activates the contractile elements through excitation-contraction coupling.

2 ANATOMY OF THE NEUROMUSCULAR JUNCTION

End Plate

The name motor end plate originally implied the specialized efferent endings that terminate on a striated muscle as a whole. Most authors, however, now use the term to describe the postsynaptic membrane of the muscle alone. Each muscle fiber usually has only one end plate, and each branch of a motor axon innervates one end plate. The motor nerve fiber looses the myelin sheath at the nerve terminals. Distal to the myelin sheath, therefore, only the Schwann cells separate the nerve terminals from the surrounding tissue. Thus, the neuromuscular junction consists of the motor nerve ending, Schwann cell, and muscle end plate (Fig. 8–1). At the junctional region between the nerve ending and end plate, Schwann cells are absent. Here, the nerve ending forms a flattened plate lying within a surface depression of the end plate. This indentation of the muscle fiber, called a synaptic gutter or a primary synaptic cleft, is about 200 to 500 Å deep. The thickened postsynaptic membrane in this region has narrow infoldings called junctional folds or secondary clefts. A large number of mitochondria, nuclei, and small granules accumulate close to the secondary clefts. Many mitochondria and synaptic vesicles also lie in the axon terminals, just proximal to the presynaptic membrane.

Electron-microscopic studies have delineated the ultrastructural features of the end plates in human external intercostal muscles.[23] A nerve terminal that occupies an area close to 4 μm^2 contains approximately 50 synaptic vesicles per square micrometer. The postsynaptic membrane is 10 times longer than the presynaptic membrane. The postsynaptic folds covers the area about two and a half times that of the terminal itself. Diseases of neuromuscular transmission alters the end-plate profile (Fig. 8–2). In myasthenia gravis, the area occupied by the terminal is reduced, with a simplified appearance of the postsynaptic folds. Conversely, in the myasthenic syndrome or Lambert-Eaton syndrome,[13,39] the terminal, though normal in area, contains elongated and sometimes markedly hypertrophic postsynaptic membrane. Neither disease shows significant alteration in the mean synaptic vesicle diameter or mean synaptic vesicle count per unit nerve terminal area. Clinically unaffected limb muscles may show the ultrastruc-

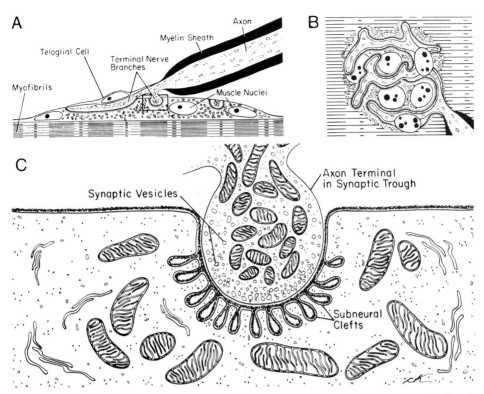

Figure 8–1. Motor end plate as seen in histologic sections in the long axis of the muscle fiber (**A**) and in surface view (**B**) under the light microscope and a section through the motor end plate (*area in the rectangle in* **A**) under the electron microscope (**C**). The myeline sheath ends at the junction at which the axon terminal fits into the synaptic cleft. The Schwann (teloglial) cells cover the remaining portion without extending into the primary cleft. The plasma membrane of axon (axolemma) forms the presynaptic membrane and that of muscle fiber (sarcolemma), the postsynaptic membrane of the end plate. Interdigitation of the sarcolemma gives rise to the subneural or secondary clefts. The axon terminal contains synaptic vesicles and mitochondria. (From Bloom and Fawcett,[4] with permission.)

Figure 8–2. Schematic representation of the motor end plates in control, myasthenia gravis, and the myasthenic syndrome drawn to the scale of the mean figure. The diagram shows an oversimplification of the postsynaptic membrane in myasthenia gravis and marked hypertrophy in the myasthenic syndrome. (From Engel and Santa,[22] with permission.)

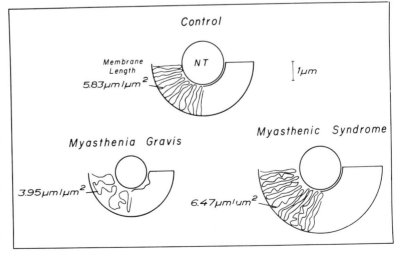

tural changes of the motor end plates in patients with ocular myasthenia gravis.[55]

Synaptic Vesicles

Minute intracellular structures, 300 to 500 Å in diameter, encapsulate ACh inside the presynaptic axoplasm. In addition to the synaptic vesicles, the nerve endings contain high concentrations of choline acetyl transferase, which synthesizes ACh, and acetylcholinesterase, which hydrolyzes ACh. The proximal portions of neurons also have the neurotransmitter and the two enzymes, although to a much lesser extent. This finding suggests that enzymatic synthesis takes place in the cell body before the transport to the nerve terminals.[43] Each vesicle contains 5000 to 10,000 molecules of ACh or a quantum.[31] Some quanta (about 1000) located adjacent to the cell membrane are immediately available for release. Many more quanta (10,000), contained in the mobilization store, move toward the membrane to continuously replace liberated ACh. The remaining and largest portion of quanta (300,000) forms the main store as a reserve supply for the mobilization store.

3 ELECTRICAL ACTIVITY AT THE END PLATE

Miniature End-Plate Potential

Many resting muscle fibers show a spontaneous subliminal electrical activity, MEPP. It represents a small depolarization of the postsynaptic membrane induced by sustained but random release of a single quantum of ACh from the nerve terminal.[26] An ordinary needle electrode placed near the end plate of the muscle fibers can record the MEPP (see Chapter 12.4). More quantitative analysis requires a microelectrode inserted directly into the end-plate region to achieve a higher resolution. Each ACh quantum liberated from the nerve terminal contains a nearly equal number of ACh molecules, irrespective of external factors such as temperature or ionic concentration. In contrast, the frequency of the MEPP varies over a wide range. It increases with elevated temperatures and upon depolarization of the motor nerve terminals. It decreases with deficiency of calcium (Ca^{2+}), the ion known to enhance quantal release by increasing fusion of the ACh vesicles with the membrane of the nerve terminal.

The factors that dictate the amplitude of the MEPP or quantum size include the number of ACh molecules in a vesicle, diffusion properties of the liberated molecules, structural characteristics of the end plate, and sensitivity of the ACh receptors. In normal human intercostal muscles, an MEPP recurs roughly every 5 seconds and is approximately 1 mV in amplitude when recorded intracellularly.[16] Hence, the MEPP is much below the excitability threshold of the muscle fiber and is about 1 percent of the normal EPP generated by a volley of nerve impulses. A small dose of curare greatly reduces the amplitude of the MEPP, whereas neostigmine (Prostigmin) increases it.[36] The MEPP ceases after denervation or nerve anesthesia. In myasthenia gravis, receptor insensitivity results in reduced amplitude of the MEPP, despite normal discharge frequency. Conversely, defective release of ACh reduces the rate of firing in the myasthenic syndrome and in botulism, although the MEPP is normal in amplitude.

Events Related to Nerve Action Potential

In the resting state, the interior of the muscle fibers is negative relative to the exterior by about 90 mV. This transmembrane potential primarily results from an unequal distribution of inorganic ions across the membrane with a high concentration of potassium (K^+) intracellularly and of sodium (Na^+) and chloride (Cl^-) extracellularly (see Chapter 2.2). It also depends on differential permeability across the muscle membrane with a high conductance for potassium and chloride and low conductance for sodium. The energy-dependent sodium-potassium pump compensates for a slight inward movement of sodium and outward movement of potassium at steady state to maintain the electrochemical potential equilibrium (see Fig. 2–1).

As mentioned earlier, spontaneous release of a single quantum of ACh induces an MEPP that falls far below the critical level necessary for generation of a muscle action potential. With the arrival of a nerve impulse, depolarization of the motor nerve ending initiates an influx of calcium into the motor axons. The increased amount of calcium accelerates fusion of the vesicle membrane with the nerve terminal membrane, thereby producing a large increase in the rate of quantal release. Massive synchronized release of ACh triggered by the arrival of a nerve action potential results in summation of many MEPPs, giving rise to a localized EPP. Thus, the number of immediately available ACh quanta and the voltage-dependent concentration of calcium within the axon terminal determine the size of the EPP. The number of quanta emitted per nerve impulse, or quantum content, is estimated to be about 100, based on the amplitude ratio, EPP/MEPP.

End-Plate Potential

Like MEPPs, EPPs are attributable to the depolarization of the motor end plate by ACh. The synaptic transmitter increases the conductance of various diffusible ions, but principally those of sodium and potassium. Therefore, these ions move freely down their electrochemical gradients, resulting in depolarization of the motor end plate. The rise time, amplitude, and duration characterize this non-propagated local response that declines rapidly with distance from the end plate. It normally begins about 0.5 ms after the release of ACh, reaches its peak in about 0.8 ms, and decreases exponentially with a half decay time of about 3.0 ms. The EPP, a graded, rather than all-or-none, response, increases in proportion to the number of ACh quanta liberated from the nerve terminal. The sensitivity of the end

plate to the depolarizing action of ACh also affects the degree of depolarization. Like the excitatory postsynaptic potential (EPSP), two or more subthreshold EPPs generated in near synchrony can summate to cause a depolarization exceeding the critical level for generation of an action potential.

4 EXCITATION-CONTRACTION COUPLING

Generation of Muscle Action Potential

An EPP exceeding the critical level of depolarization, called threshold, generates an all-or-none muscle action potential. A molecular change of the depolarized membrane results in selective increase of sodium (Na^+) conductance, followed by an increase in potassium (K^+) conductance. This phenomenon, inherent to the muscle membrane, occurs irrespective of the nature of the stimulus, as long as depolarization reaches the critical value. Once generated at the end plate, the action potential propagates bidirectionally to the remaining parts of the fiber. The impulse conducts only in the range of 3 to 5 m/s along the muscle membrane, compared with 60 m/s over the nerve. A neuromuscular block results when the EPP fails to reach the critical level. A subliminal EPP may imply insufficient liberation of ACh from the axon terminal or reduced sensitivity of the muscle end plate. In contrast to the all-or-none generation of a muscle action potential in each muscle fiber, the compound muscle action potential shows a graded response in proportion to the number of activated muscle fibers.

Transverse and Longitudinal Tubules and Triad

The spread of action potential from the motor end plate to the transverse tubules initiates muscle contraction. This process, called excitation-contraction coupling, links electrical and mechanical activity. Electrical activity of a muscle fiber consists of two temporally separate components attributable to different structures within the fiber.[9] The first portion originates at the motor end plate and spreads along the outer surface of the muscle fiber. The second part occurs within a complex tubular system that surrounds and interpenetrates the muscle fiber. The network, called the transverse tubules, because of its orientation relative to the axis of the muscle fiber, lies at the junctions of the A and I bands in humans (see Fig. 11–1). These tridimensional tubules, though structurally internal to the cell, contain extracellular fluid. Consequently, the inside of the tubule is electropositive relative to the outside, surrounded by intracellular fluid.

Muscle action potentials propagate along the tubules into the depth of the muscle.

A second tubular system, called the longitudinal tubule or sarcoplasmic reticulum, surrounds the myofibrils of a muscle fiber (Fig. 8–3). These tubules have longitudinal orientation with respect to the myofibrillar axis and, unlike transverse tubules, form a closed system devoid of continuity with either extracellular fluids or sarcoplasm. They appear as fenestrated sacs surrounding the myofibrils. The longitudinal tubules expand to form bulbous terminal cisterns on both sides of the transverse tubules where they come into close contact. The two terminal cisterns and one interposed transverse tubule form a triad in longitudinal sections of the muscle.

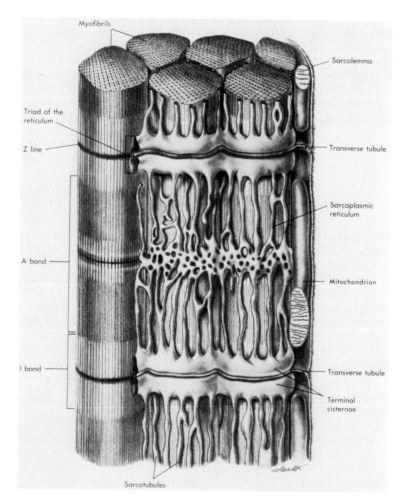

Figure 8–3. Anatomic relationship between the perpendicularly oriented longitudinal and transverse tubules. Propagating muscle action potentials initiate electromechanical coupling at the triad of the reticulum, which consists of two terminal cisterns of the longitudinal tubules and one transverse tubule between them. (From Bloom and Fawcett,[4] with permission.)

Role of Calcium Ions

Propagated action potentials invade the muscle fibers along the transverse tubules to come into contact with the terminal cisterns of the longitudinal tubules at the triad. This coupling process to the sarcoplasmic reticulum gives rise to a small electrical potential referred to as intramembranous charge movement.[8] The action potential crossing the terminal cistern initiates the release of calcium (Ca^{2+}) from the longitudinal tubules into the sarcoplasm that surrounds the myofilaments. The presence of calcium here in turn triggers a chemical interaction that leads to the formation of bridges between thin and thick filaments. Sliding of thin filaments against thick filaments results in contraction of the myofibril (see Chapter 11.2). At the end of the muscle action potential, rapid resequestering of calcium into the longitudinal tubules lowers its concentration in the sarcoplasm. The myofibers relax as adenosine triphosphate breaks the existing bridges between filaments.

5 POSTSYNAPTIC ABNORMALITIES

Pathophysiology of Myasthenia Gravis

In this disorder, intracellular recordings from the intercostal muscles have revealed reduced amplitude of MEPP or small quantum size but normal or nearly normal discharge frequency.[16] Consequently, the EPP elicited by a nerve impulse is also reduced in amplitude, despite a normal number of ACh quanta liberated by a single volley or normal EPP quantum content. On repetitive stimulation, the number of quanta released falls gradually, as it does in normal muscle, causing a further decrease in the amplitude of the initially small EPP. With successive stimuli, the EPP becomes insufficient to bring the membrane potential to the critical level in a progressively greater number of fibers, thus causing reduction in amplitude of compound muscle action

potential. Neuromuscular transmission fails first in small motor units, perhaps because they have a lower margin of safety than the large motor units.[33,34]

Reduction in amplitude of MEPP suggests (1) decreased numbers of ACh molecules per quantum; (2) diffusional loss of ACh within the synaptic cleft; or (3) reduced sensitivity of the ACh receptor. In early studies, postsynaptic sensitivity to carbachol and decamethonium added to the bath solution appeared to be normal.[17] A presynaptic abnormality proposed on the basis of this finding, however, has subsequently received neither morphologic nor electrophysiologic confirmation. Indeed, microiontophoretic application of ACh at the end-plate region has since disclosed impaired postsynaptic sensitivity to ACh.[1] The observed electrophysiologic changes may also imply diffusional ACh loss resulting from alterations in postsynaptic membrane structure.

Morphologic and Immunologic Changes

Ultrastructural histometric studies in myasthenic intercostal muscles have shown a distinct end-plate profile, indicating postsynaptic membrane abnormalities.[23] Another experiment has revealed three types of neuromuscular junctions in the surface fibers of internal and external intercostal muscles of myasthenics.[1] One group with mild morphologic alterations had EPPs of sufficiently large amplitude to trigger an action potential. A second group with a grossly altered postjunctional membrane showed marked reduction not only in amplitude but also in frequency of the MEPP and in amplitude of the EPP. The last group had the totally degenerated end plate showing neither MEPPs nor EPPs.

Not every myasthenic end plate shows morphologic alterations, despite diminished MEPP amplitude demonstrated uniformly. Therefore, changes in end-plate geometry per se may not totally explain the physiologic defect. Myasthenic muscles have decreased functional receptor sites detected by radioactively labeled alpha-bungarotoxin, a snake poison that binds to the ACh receptor.[15,25,28] Further, the number of functional ACh receptors when counted by this technique, shows positive correlation with the mean amplitude of the MEPP.[32] These findings indicate the presence of an ACh receptor abnormality in myasthenia gravis. Partial blocking of the ACh receptors with curare produces a similar physiologic defect.

Studies using plasma exchange have revealed an inverse relationship between clinical muscle strength and antibody titers. This finding supports the view that the antibody is the most important factor impairing neuromuscular transmission.[45] The underlying mechanisms, however, must be complex, because, despite a precipitous drop in antibody titers, electrophysiologic findings usually improve, with a delay of at least 7 days from the start of plasmapheresis.[7] Antibodies mediate obstruction of the ACh receptor, presumably by binding with complement to the receptor zone of the postsynaptic membrane.[21] Intercostal muscle biopsies show reduced numbers of ACh receptors and binding of antibodies to many of the remaining receptors in patients with myasthenia gravis.[41]

Experimental Models

Experimental autoimmune myasthenia gravis shares the same morphologic and physiologic abnormalities with human disease.[24,47,53] Studies in rats showed reduced receptor content and increased receptor-bound antibody. Thus, defective neuromuscular transmission seems to result from a reduced number of fully active receptors.[41] Typical histologic and electrophysiologic myasthenic features develop in mice after passive transfer of human serum fractions obtained from patients with myasthenia gravis.[54] Decamethonium causes paralysis by persistent depolarization of the end-plate region in normal muscle.[44] Reduction of postsynaptic sensitivity to depolarization renders myasthenic muscles resistant to this type of neuromuscular blocking. Antibodies to the ACh receptor do not impair

the ionophore, an ion conductance modulator protein thought to control the permeability change following a reaction of ACh with ACh receptor.[50]

Intracellular recordings from muscle end plates of immunized rabbits show reduced amplitude of MEPPs but a normal number of ACh quanta released per nerve impulse.[19] Rats with chronic experimental myasthenia have reduced amplitude of MEPPs despite normal ACh output at rest and during stimulation.[37] After passive transfer of human myasthenia gravis to rats, reduction of MEPP amplitude does not develop immediately, but occurs after the first 24 hours, reaching minimum levels by six days.[30] The delayed development of reduced MEPP amplitude suggests a more complex mechanism than a simple block, like that caused by curare, of ACh receptors by IgG antibodies.[30]

6 PRESYNAPTIC ABNORMALITIES

Pathophysiology of the Myasthenic Syndromes

In contrast to the receptor insensitivity of myasthenia gravis, defective release of ACh quanta characterizes the myasthenic syndrome.[13] Microelectrode recordings from excised intercostal muscles reveal no abnormality in amplitude of the MEPPs; that is, quantum size is normal. The sensitivity of the muscle end plate to ACh is also normal. The discharge frequency of the MEPP, however, does not increase as expected in response to depolarization of the motor nerve terminal.[39] Thus, a single nerve impulse releases a smaller number of ACh quanta than normal; that is, there is decreased quantum content. The EPP then fails to trigger an action potential in some muscle fibers, which leads to a reduced amplitude of the compound muscle action potential.[40]

The defect improves immediately with various maneuvers to prime the nerve terminals.[18] For example, the EPP augments progressively with repetitive stimulation of the nerve. An increase of external calcium (Ca^{2+}) or the addition of quanidine also enhances the EPP. These findings suggest a normal number of quanta available in the presynaptic store despite a low probability of quantum release at the nerve terminal. Indeed, ultrastructural studies have revealed no alteration in the mean nerve terminal area or in the synaptic vesicle count per unit.[23]

A congenital myasthenic syndrome may result from different types of presynaptic abnormalities.[20] Typical patients have such features as deficient muscle acetylcholinesterase, decreased frequency but normal amplitude of the MEPP, decreased number of quanta liberated per nerve impulse, small nerve terminals, and focal degeneration of the postsynaptic membrane (see Chapter 24.4). In some types, a low number of quanta per EPP primarily reflects a reduced store, rather than a low probability of release, as in the case of the classic myasthenic syndrome. A congenital defect in the molecular assembly of acetylcholinesterase or its attachment to the postsynaptic membrane might represent the basic abnormality. A familial congenital myasthenic syndrome shows deficient synthesis of ACh.[29]

Effect of Botulinum and Chemicals

Abnormalities in calcium-dependent ACh release also reduce the amplitude of the EPP in a number of other conditions, including a neuromuscular block by botulinum toxin. Here, the reduced frequency of the MEPP, not affected by the addition of calcium, recovers after the administration of a spider venom known to neutralize the toxin. Thus, the neuromuscular insufficiency in botulism results neither from blockage of calcium entry into the nerve nor from reduced storage of ACh vesicles. The toxin may interfere with the ACh release process itself, possibly by blocking exocytosis at the release sites.[35]

High concentrations of magnesium block neuromuscular transmission.[5,48] Lowering the temperature increases transmitter release and reactivates previously paralyzed muscle in botulinum paralysis but not in normal muscle blocked by high magnesium concentration.[42] Experimental evidence also indicates an inhibitory effect of manganese on transmitter release at the neuromuscular junction.[2] Aminoglycoside antibiotics such as neomycin and kanamycin not only interfere with ACh release directly but also inhibit the transmission by postsynaptic block.[10] Experimental autoimmune myasthenia gravis improves by administration of dantrolene sodium, which induces accumulation of free calcium in the subcellular store.[51]

7 RECOVERY CYCLES OF NEUROMUSCULAR TRANSMISSION

Release of Acetylcholine Molecules

The amount of ACh in the immediately available store and the concentration of calcium (Ca^{2+}) at the nerve terminal determine the number of ACh molecules released by a nerve action potential. Paired or repetitive stimulation affects the release of ACh and the EPP in two opposing manners. The first shock utilizes a portion of the store, partially depleting the amount of ACh available for subsequent stimuli, until the mobilization store has refilled the loss. After each shock, however, calcium accumulates in the nerve terminal, enhancing ACh release. These two competing phenomena, though initiated by the same stimulus, follow different time courses.[12]

The partially depleted ACh store recovers exponentially in 5 to 10 seconds through the slow reloading of ACh ejection sites. Influx of calcium into the terminal axons takes place immediately after depolarization of the nerve, but the ion diffuses out of the axon over the next 100 to 200 ms. Hence, paired or repetitive stimulation with a shorter interstimulus interval causes accumulation of calcium. Such fast rates of stimulation, therefore, tend to facilitate release of ACh, despite concomitant reduction of its immediately available store. In contrast, slower rates of repetition result in suppression because the negligible electrosecretory facilitation can no longer compensate for the loss of ACh stores. The dichotomy between the fast and slow rates of stimulation, however, does not always hold. For example, even at high rates of stimulation, ACh depletion far exceeding its mobilization will lead to reduced release of the transmitter.

Neuromuscular Depression and Facilitation

Reduction in the number of ACh quanta released by the second nerve impulse results in a smaller EPP, which no longer reaches the threshold in some muscle fibers. The amplitude of the second compound muscle action potential decreases or shows a decrement, compared with the first response. Conversely, an increase in the number of quanta released by the second nerve impulse gives rise to a larger EPP. Such a true facilitation is based on the neurosecretory potentiation, rather than summation of two EPPs elicited by paired shocks with a very short interstimulus interval.[12]

Both facilitation and summation result in larger compound muscle action potentials through recruitment, provided that the initial stimulus failed to activate all the muscle fibers. The greater amplitude and area under the waveform in recruitment imply the discharge of additional muscle fibers. An increased amplitude may also result from better synchronization of different muscle fibers without recruitment. In this phenomenon, called pseudofacilitation, the area under the waveform, which approximates the number of active muscle fibers, shows no major changes.

Normal Recovery Cycle

Studies of the recovery cycle consist in recording the muscle action potentials after delivering paired stimuli to the nerve at various interstimulus intervals.

A second shock delivered a few milliseconds after the first falls in the refractory periods of the muscle and nerve (Fig. 8–4). For the intervals of 10 to 15 ms, an overlap between the first and second muscle responses precludes accurate measurement of the individual potentials. Thereafter, the second compound muscle potential recovers to the size of the first in the normal muscle. This finding, however, does not necessarily imply that the first and second stimuli elicit the same EPPs.

At interstimulus intervals of 100 to 200 ms, the second shock may normally evoke a greater EPP than the first through neurosecretory potentiation. If the EPP by the first stimulus exceeds the threshold of excitation in all muscle fibers, however, enhanced EPP by the second stimulus recruits no additional fibers. A slow rate of stimulation depresses the number of ACh quanta released successively even in normal muscles. Because of a large margin of safety, however, the decreased amount of ACh suffices to cause an EPP well above the critical level of excitation in all muscle fibers. In normal muscles, therefore, changes in the amount of ACh do not alter the size of compound muscle action potential elicited by the second or subsequent stimuli.

Effects of Disease States

Partially curarized mammalian muscle, with reduced margin of safety, serves as good model in studying the recovery cycle of the EPP.[6,11,14] With paired stimuli, the second muscle response equals or exceeds the first for interstimulus intervals of 100 to 200 ms that accompany calcium-dependent neurosecretory facilitation.[3,11,52] At longer intervals, the second response falls below the first, because depleted stores of available ACh quanta can no longer overcome the receptor insensitivity. The maximal depression at interstimulus intervals ranging from 300 to 600 ms is followed by a slow recovery. Full return to the control value in about 10 seconds implies restoration of releasable ACh through replenishment of the

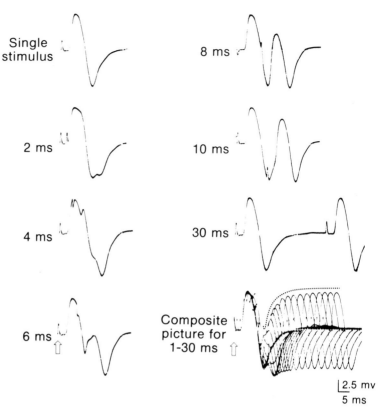

Single stimulus

8 ms

2 ms

10 ms

4 ms

30 ms

6 ms

Composite picture for 1-30 ms

2.5 mv
5 ms

Figure 8–4. Compound action potentials from the thenar muscles elicited by paired shocks delivered to the median nerve at the wrist. Time intervals ranged from 2 to 30 ms between conditioning (*arrow*) and test stimuli. The top tracing on the left shows a response to a single stimulus. The bottom tracing on the right is a composite picture superimposing 20 paired responses. The conditioning response of each pair appeared in the same spot, whereas the test responses shifted to the right in proportion to the interstimulus interval. An imaginary line connecting the peaks of the sequential test responses represents the time course of neuromuscular excitability change following the conditioning stimulus. (From Kimura,[38] with permission.)

stores. In myasthenia gravis, a reduced amount of ACh also fails to activate some muscle fibers with receptor insensitivity. Hence, the recovery cycle of the muscle action potential shows a great resemblance to that of curarized muscle (Figs. 8–5, 8–6, 8–7, and 8–8). In either case, repetitive stimulation at the rate of two to three per second induces the maximal depression because it is fast enough for the depletion of ACh but slow enough for the diffusion of calcium out of the axon.

In the myasthenic syndrome with a defective release of ACh, the EPP elicited by a single stimulus falls short of activating many muscle fibers. With the second stimulus given in less than a few milliseconds, the summated EPPs will recruit additional muscle fibers. With stimuli delivered at a longer interval of 100 to 200 ms, the electrosecretory facilitation partially overcomes the defective release of ACh (Fig. 8–8). Increased EPPs will in turn recruit some of the muscle fibers not acti-

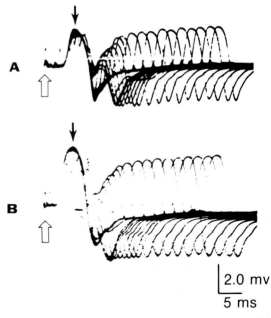

A

B

2.0 mv
5 ms

Figure 8–5. Composite pictures superimposing 20 paired responses from the thenar muscles (compare bottom tracing on right in Figure 8–4.) A patient with myasthenia gravis (**A**) and a normal control (**B**) showed the same recovery course for the interstimulus intervals ranging from 1 to 30 ms.

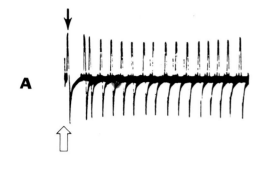

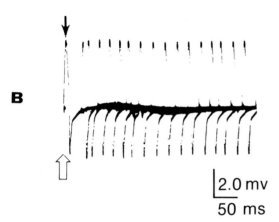

Figure 8–6. Composite pictures of 16 paired responses from the thenar muscles arranged in the same manner as in Figure 8–5. The interstimulus intervals of paired shocks ranged from 30 to 400 ms. The conditioning response of each pair appeared in the same spot of each tracing (*arrows pointing down*), whereas the test responses shifted to the right successively. The test response showed a mild but definite reduction in amplitude at the interstimulus intervals of 150 to 250 ms in the myasthenic muscle (**A**), but not in the normal muscle (**B**).

tributable to an increased number of quanta released by the second impulse at interstimulus intervals of 100 to 200 ms. Summation of the EPP also augments the second response at intervals of less than 10 ms. This, however, is a less specific phenomenon, seen whenever the first stimulus evokes less than the maximal response, as seen in some cases of myasthenia gravis (see Fig. 8–7).[49] As expected, paired shocks of longer intervals may cause depression, though not as consistently as in myasthenia gravis.

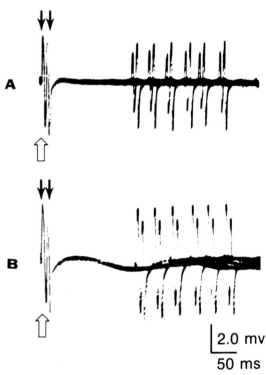

Figure 8–7. Composite pictures similar to those shown in Figures 8–5 and 8–6. Unlike the previous tracings, both conditioning and test stimuli consist of paired shocks with interstimulus intervals of 10 ms. The paired test stimuli followed the paired conditioning stimuli (*open arrow*) by an interval of 200 to 400 ms. The double peaked conditioning responses appeared in the same spot of each tracing (*paired arrows*). The second peak of the pair, though displaced downwards, had the same amplitude as the first. In myasthenia gravis (**A**), depletion of ACh by the conditioning stimuli reduced the first peak of each test response. The second peak of each test response, elicited 10 ms after the first, recovered to a normal level, which indicated the summation of the two EPPs. In each test response of the normal muscle, the maximal size of the first peak precluded any amplitude increase of the second peak.

vated by the first stimulus, which leads to an increase in amplitude of the second compound muscle action potential.[12] At a slower rate separated by more than 200 ms, the second EPP diminishes because calcium no longer accumulates to compensate for depletion of available ACh stores. Limited release of ACh by the first stimulus, however, may preclude major decremental muscle responses in most patients.

Defective release of ACh also underlines the electrophysiologic abnormality in botulism. Paired stimuli usually, though not always, show facilitation at-

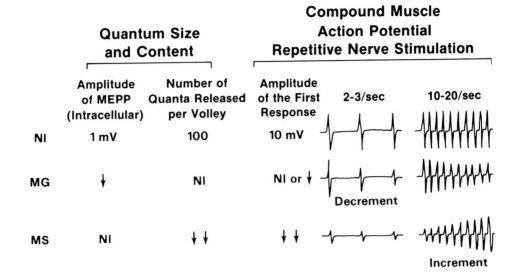

Quantum Size and Content		Compound Muscle Action Potential Repetitive Nerve Stimulation		
Amplitude of MEPP (Intracellular)	Number of Quanta Released per Volley	Amplitude of the First Response	2-3/sec	10-20/sec
NI 1 mV	100	10 mV		
MG ↓	NI	NI or ↓	Decrement	
MS NI	↓↓	↓↓		Increment

Decrement: Reduced Quantum Content Below Safety Margin.

Increment: Neurosecretory Potentiation (Ca++ dependent?)

Figure 8–8. Typical changes in quantum size and quantum content as determined by intracellular recordings in myasthenia gravis (*MG*) and the myasthenic syndrome (*MS*). The compound muscle action potential shows a decrement to repetitive nerve stimulation with dropout of individual muscle fibers and an increment with recruitment of additional fibers.

Posttetanic Potentiation and Exhaustion

With prolonged repetitive stimulation or after a sustained voluntary muscle contraction, the immediately available store of ACh may increase as a result of a greater mobilization rate. This, coupled with the accumulation of calcium in the axon, potentiates the release of ACh and the EPP for 1 to 2 minutes (posttetanic potentiation). Subsequent stimuli release fewer ACh quanta for up to 15 minutes, probably because of metabolic changes in the nerve terminal (posttetanic exhaustion). This finding resembles the experimentally induced block by hemicholinium, which interferes with ACh synthesis.[11]

Rabbits exposed to neomycin also have prominent posttetanic exhaustion. In rat muscles, the same drug causes a reduction in both the amplitude of the MEPP and the number of quanta released per single nerve impulse. These findings suggest a combined presynaptic and postsynaptic action.[10] The posttetanic exhaustion affects the fast twitch muscles more than the slow ones in the rat. Such differences in sensitivity to a blocking agent suggest variability in structural characteristics of the end plate[46] or pattern of blood flow.[27]

REFERENCES

1. Albuquerque, EX, Rash, JE, Mayer, RF, and Satterfield, JR: An electrophysiological and morphological study of the neuromuscular junction in patients with myasthenia gravis. Exp Neurol 51:536–563, 1976.
2. Balnave, RJ, and Gage, PW: The inhibitory effect of manganese on transmitter release at the neuromuscular junction of the toad. Br J Pharmacol 47:339–352, 1973.
3. Betz, WJ: Depression of transmitter release at the neuromuscular junction of the frog. J Physiol (Lond) 206:629–644, 1970.
4. Bloom, W, and Fawcett, DW: A Textbook of Histology, ed 10. WB Saunders, Philadelphia, 1975.
5. Branisteanu, DD, Miyamoto, MD, and Volle, RL: Effects of physiologic alterations on binomial transmitter release at magnesium-depressed neuromuscular junctions. J Physiol (Lond) 254:19–37, 1976.

6. Brooks, VB, and Thies, RE: Reduction of quantum content during neuromuscular transmission. J Physiol (Lond) 162:298–310, 1962.

7. Campbell, WW, Jr, Leshner, RT, and Swift, TR: Plasma exchange in myasthenia gravis: Electrophysiologic studies. Ann Neurol 8:584–589, 1980.

8. Chandler, WK, Rakowski, RF, and Schneider, MF: A non-linear voltage dependent charge movement in frog skeletal muscle. J Physiol (Lond) 254:245–283, 1976.

9. Costantin, LL: The role of sodium current in the radial spread of contraction in frog muscle fibers. J Gen Physiol 55:703–715, 1970.

10. Daube, JR, and Lambert, EH: Post-activation exhaustion in rat muscle. In Desmedt, JE (ed): New Developments in Electromyography and Clinical Neurophysiology, Vol 1. Karger, Basel, 1973, pp 343–349.

11. Desmedt, JE: Presynaptic mechanisms in myasthenia gravis. Ann NY Acad Sci 135:209–246, 1966.

12. Desmedt, JE: The neuromuscular disorder in myasthenia gravis. 1. Electrical and mechanical response to nerve stimulation in hand muscles. In Desmedt, JE (ed): New Developments in Electromyography and Clinical Neurophysiology, Vol. 1. Karger, Basel, 1973, pp 241–304.

13. Eaton, LM, and Lambert, E: Electromyography and electric stimulation of nerves in diseases of motor units: Observations on myasthenic syndrome associated with malignant tumors. JAMA 163:117–1124, 1957.

14. Eccles, JC, Katz, B, and Kuffler, SW: Nature of the "endplate potential" in curarized muscle. J Neurophysiol 4:362–387, 1941.

15. Elias, SB, and Appel, SH: Acetylcholine receptor in myasthenia gravis: Increased affinity for α-bungarotoxin. Ann Neurol 4:250–252, 1978.

16. Elmqvist, D: Neuromuscular transmission defects. In Desmedt, JE (ed): New Developments in Electromyography and Clinical Neurophysiology, Vol 1. Karger, Basel, 1973, pp 229–240.

17. Elmqvist, D, Hofmann, WW, Kugelberg, J, and Quastel, DMJ: An electrophysiological investigation of neuromuscular transmission in myasthenia gravis. J Physiol (Lond) 174:417–434, 1964.

18. Elmqvist, D, and Lambert, EH: Detailed analysis of neuromuscular transmission in a patient with the myasthenic syndrome sometimes associated with bronchogenic carcinoma. Mayo Clin Proc 43:689–713, 1968.

19. Elmqvist, D, Mattsson, C, Heilbronn, E, Londh, H, and Libelius, R: Acetylcholine receptor protein: Neuromuscular transmission in immunized rabbits. Arch Neurol 34:7–11, 1977.

20. Engel, AG, Lambert, EH, and Gomez, MR: A new myasthenic syndrome with end-plate acetylcholinesterase deficiency, small nerve terminals, and reduced acetylcholine release. Ann Neurol 1:315–330, 1977.

21. Engel, AG, Lambert, EH, and Howard, FM, Jr: Immune complexes (IgG and C3) at the motor end-plate in myasthenia gravis. Mayo Clin Proc 52:267–280, 1977.

22. Engel, AG, and Santa, T: Histometric analysis of the ultrastructure of the neuromuscular junction in myasthenia gravis and in the myasthenic syndrome. Ann NY Acad Sci 183:46–63, 1971.

23. Engel, AG, and Santa, T: Motor endplate fine structure. In Desmedt, JE (ed): New Developments in Electromyography and Clinical Neurophysiology, Vol 1. Karger, Basel, 1973, pp 196–228.

24. Engel, AG, Tsujihata, M, Lambert, EH, Lindstrom, JM, and Lennon, VA: Experimental autoimmune myasthenia gravis: A sequential and quantitative study of the neuromuscular junction ultrastructure and electrophysiologic correlations. J Neuropathol Exp Neurol 35:569–587, 1976.

25. Fambrough, DM, Drachman, DB, and Satyamurti, S: Neuromuscular junction in myasthenia gravis: Decreased acetylcholine receptors. Science 182:293–295, 1973.

26. Fatt, P. and Katz, B: Spontaneous subthreshold activity at motor nerve endings. J Physiol (Lond) 117:109–128, 1952.

27. Friess, SL, Durant, RC, Martin, HL, Hudak, WV, and Weems, H: Blockade in simultaneously perfused soleus and gastrocnemius muscles in the cat. Toxicol Appl Pharmacol 18:133–140, 1971.

28. Green, DPL, Miledi, R, Perez de la Mora, M, and Vincent, A: Acetylcholine receptors. Phil Trans R Soc (Lond) B 270:551–559, 1975.

29. Hart, ZH, Sahashi, K, Lambert, EH, Engel, AG, and Lindstrom, JM: A congenital familial myasthenic syndrome caused by a presynaptic defect of transmitter resynthesis or mobilization. Neurology 29:556–557, 1979.

30. Howard, JF, Jr, and Saunders, DB: Passive transfer of human myasthenia gravis to rats. 1. Electrophysiology of the developing neuromuscular block. Neurology 30:760–764, 1980.

31. Hubbard, JI: Neuromuscular transmission: Presynaptic factors. In Hubbard, JI (ed): The Peripheral Nervous System. Plenum Press, New York, 1974, pp 151–180.

32. Ito, Y, Miledi, R, Vincent, A, and Newsom Davis, J: Acetylcholine receptors and end-plate electrophysiology in myasthenia gravis. Brain 101:345–368, 1978.

33. Kadrie, HA, and Brown, WF: Neuromuscular transmission in human single motor units. J Neurol Neurosurg Psychiatry 41:193–204, 1978.

34. Kadrie, HA, and Brown, WF: Neuromuscular transmission in myasthenic single motor units. J Neurol Neurosurg Psychiatry 41:205–214, 1978.

35. Kao, I, Drachman, DB, and Price, DL: Botulinum toxin: Mechanism of presynaptic blockade. Science 193:1256–1258, 1976.

36. Katz, B: Microphysiology of the neuro-muscular junction: A physiological "quantum of action" at the myoneural junction. Bull Johns Hopkins Hosp 102:275–312, 1958.

37. Kelly, JJ, Jr, Lambert, EH, and Lennon, VA: Acetylcholine release in diaphragm of rats with chronic experimental autoimmune myasthenia gravis. Ann Neurol 4:67–72, 1978.

38. Kimura, J: Electrodiagnostic study of pesticide toxicity. In Xintaras, C, Johnson, BL, and De Groot, I (eds): Behavioral Toxicology. U.S. Department of Health, Education and Welfare, U.S. Government Printing Office, Washington, D.C., 1974, pp 174–181.

39. Lambert, EH, and Elmqvist, D: Quantal components of end-plate potentials in the myasthenic syndrome. Ann NY Acad Sci 183:183–199, 1971.

40. Lambert, EH, Okihiro, M, and Rooke, ED: Clinical physiology of the neuromuscular junction. In Paul, WM, Daniel, EE, Kay, CM, and Monckton, G (eds): Muscle. Proceedings of the Symposium, The Faculty of Medicine, University of Alberta, Pergamon Press, London, 1965, pp 487–499.

41. Lindstrom, JM, and Lambert, EH: Content of acetylcholine receptor and antibodies bound to receptor in myasthenia gravis, experimental autoimmune myasthenia gravis, and Eaton-Lambert syndrome. Neurology 28:130–138, 1978.

42. Lundh, H: Antagonism of botulinum toxin paralysis by low temperature. Muscle Nerve 6:56–60, 1983.

43. MacIntosh, FC: Formation, storage, and release of acetylcholine at nerve endings. Can J Biochem Physiol 37:343–356, 1959.

44. Meadows, JC, Ross-Russell, RW, and Wise, RP: A re-evaluation of the decamethonium test for myasthenia gravis. Acta Neurol Scand 50:248–256, 1974.

45. Newsom Davis, J, Pinching, AJ, Vincent, A, and Wilson, SG: Function of circulating antibody to acetylcholine receptor in myasthenia gravis: Investigation by plasma exchange. Neurology 28:266–272, 1978.

46. Nystrom, B: Histochemical studies of end-plate bound esterases in "slow-red" and "fast-white" cat muscles during postnatal development. Acta Neurol Scand 44:259–318, 1968.

47. Satyamurti, S, Drachman, DB, and Slone, F: Blockade of acetylcholine receptors: A model of myasthenia gravis. Science 187:955–957, 1975.

48. Swift, TR: Weakness from magnesium-containing cathartics: Electrophysiologic studies. Muscle Nerve 2:295–298, 1979.

49. Takamori, M, and Gutmann, L: Intermittent defect of acetylcholine release in myasthenia gravis. Neurology 21:47–54, 1971.

50. Takamori, M, Ide, Y, and Kasai, M: Neuromuscular defect after suppression of ion conductance. Neurology 29:772–779, 1979.

51. Takamori, M, Sakato, S, Matsubara, S, and Okumura, S: Therapeutic approach to experimental autoimmune myasthenia gravis by dantrolene sodium. J Neurol Sci 58:17–24, 1983.

52. Takeuchi, A: The long-lasting depression in neuromuscular transmission of frog. Jpn J Physiol 8:102–113, 1958.

53. Thornell, LE, Sjostrom, M, Mattsson, CH, and Heilbronn, E: Morphological observations on motor end-plates in rabbits with experimental myasthenia. J Neurol Sci 29:389–410, 1976.

54. Toyka, KV, Drachman, DB, Griffin, DE, Pestronk, A, Winkelstein, JA, Fischbeck, KH, Jr, and Kao, I: Myasthenia gravis: Study of humoral immune mechanisms by passive transfer to mice. N Engl J Med 296:125–131, 1977.

55. Tsujihata, M, Hazama, R, Ishii, N, Ide, Y, Mori, M, and Takamori, M: Limb muscle endplates in ocular myasthenia gravis: Quantitative ultrastructural study. Neurology 29:654–661, 1979.

Chapter 9

TECHNIQUES OF REPETITIVE STIMULATION

1 INTRODUCTION

Nerve stimulation techniques as tests for neuromuscular transmission began with Jolly,[39] who applied faradic current repeatedly at short intervals. Using a kymographic recording and visual inspection of skin displacement, he found that the size of the muscle response deteriorated rapidly in patients with myasthenia gravis during the faradization. Faradic current failed to elicit a response in the volitionally fatigued muscle prior to testing. Conversely, after faradization, muscle responded poorly to subsequent volitional contraction. On the basis of these findings, Jolly concluded that the myasthenics had motor failure of the peripheral, rather than central, nervous system. His insight was remarkable, considering the limitations of his equipment, which consisted of a double-coil stimulator capable of eliciting only submaximal responses and a mechanical, rather than electrical, recorder.

The use of supermaximal stimulation and the recording of the muscle action potential have increased the reliability and sensitivity of nerve stimulation techniques considerably. Harvey and Masland[35] noted that in myasthenia a single muscle response induced a prolonged depression, during which a second maximal motor nerve stimulus excited a reduced number of muscle fibers, and that a train of impulses resulted in a progressive decline in amplitude of compound muscle potential. Later studies have established optimal frequency of stimulation, proper control of temperature, appropriate selection of muscles, and various activation procedures to enhance an equivocal neuromuscular block.[22]

Microelectrode studies provide direct recording of end-plate potentials from muscle in vitro.[2,28] All other electrophysiologic methods assess the neuromuscular junction only indirectly. Nonetheless, such an approach allows quantitation of the motor response to paired stimuli, tetanic contraction, or repetitive stimulation at fast and slow rates.[19,46] Transmission defects affect a variety of disease states, such as myasthenia gravis, the myasthenic syndrome, botulism, amyotrophic lateral sclerosis, poliomyelitis, and multiple sclerosis. This section deals with the physiologic significance of decremental or incremental responses in the differential diagnosis of clinical disorders (see Chapter 24).

2 METHODS AND TECHNICAL FACTORS

Belly-Tendon Recording

The method[45] consists of stimulating the nerve with supramaximal intensity and recording the muscle action potential with the active electrode (G_1) placed over the motor point and the reference electrode (G_2), on the tendon (belly-tendon response). The initially negative potential thus recorded represents the summated electrical activity from the entire muscle fiber population, discharging relatively synchronously. The area under the negative phase changes primarily with the number of muscle fibers activated. The magnitude of the unit discharge from individual muscle fibers also alters the size of the compound muscle potential, especially in myogenic disorders. In clinical studies, measurement of the amplitude suffices in a train of responses that shows the same duration and waveform.

Movement-Induced Artifacts

Movement-related artifacts abound during repetitive stimulation of the nerve. The recording electrode may continuously slide away from the muscle belly, or the stimulating electrodes may gradually slip from the nerve, causing subthreshold activation. In either case, a progressively smaller amplitude of a train mimics a decremental response. Shortening of the muscle during an isotonic contraction also changes the spatial relationship of muscle and recording electrodes, leading to a misleading alteration in amplitude of the recorded response (Fig. 9–1). Firm immobilization of the extremity and the muscle under study

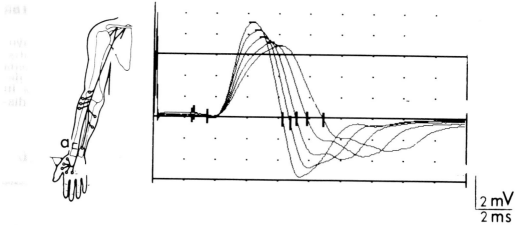

Figure 9–1. A train of responses recorded from the thenar muscle with stimuli delivered one per second to the median nerve at the wrist in a healthy subject. Intentional stepwise alteration in thumb position from abduction to adduction after each shock gave rise to a smooth reduction in amplitude with concomitant increase in duration of successive potentials. The area under the waveform showed relatively little change from the first to the fifth response.

minimizes these movement-induced changes.

In most instances, technical problems cause abrupt, irregular changes in the amplitude or shape of the evoked response. Some movement artifacts, however, induce a smooth, progressive alteration indistinguishable from the myasthenic process. Changing waveforms probably result from artifacts if visual inspection of the contracting muscle detects excessive movement. Repeated trials help establish the reproducibility of the finding, increasing the reliability of the results. Intertrial intervals should be 30 seconds or longer to avoid the effect of subnormality of neuromuscular transmission that lasts for a few seconds after a single stimulus and a greater time period after repetitive impulses.

Temperature and Other Factors

Exposure to warm sunlight may precipitate ptosis and diplopia in myasthenic patients.[5,63] Similarly, electrophysiologic abnormalities of weak muscles may appear only after local warming. Four physiologic mechanisms account for the improved neuromuscular transmission with cooling: (1) facilitation of transmitter replacement in the presynaptic termi-nal[36,37]; (2) reduction in the amount of transmitter released at the neuromuscular junction by the first of a train of impulses, leaving more quanta available for subsequent stimuli[20]; (3) decreased hydrolysis of acetylcholine (ACh) by acetylcholinesterase, allowing sustained action of the transmitter already released from the axon terminal[30]; and (4) increased postsynaptic receptor sensitivity to ACh.[34]

Lowering the intramuscular temperature from 35°C to 28°C increases the amplitude of the compound muscle action potential and enhances the force of the isometric twitch and tetanic contraction.[6] Patients with the myasthenic syndrome also experience distinct improvement after cooling,[58,67] as do those with amyotrophic lateral sclerosis,[21] botulism,[68] or tick paralysis.[16] Cooling reduces the decrement to repetitive nerve stimulation (Fig. 9–2). It is, therefore, important to maintain the skin temperature over the tested muscle, above 34°C for diagnostic application. This is best achieved by prior immersion of the limb in warm water or the use of an infrared heat lamp. Paradoxically, brief stimulation at high rates may produce a decremental response in normal muscles cooled below 32°C.[47]

The effect of cholinesterase inhibitors also influences the results of repetitive

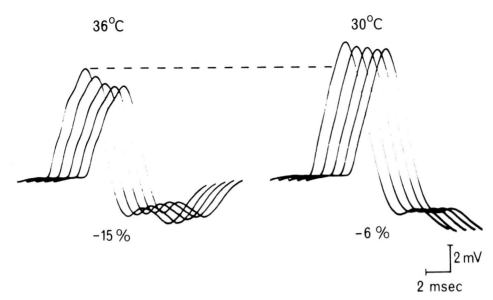

Figure 9–2. Decremental response of the hypothenar muscle with stimulation of the ulnar nerve at two per second in a patient with myasthenia gravis, on the left, at 36°C, and on the right, following cooling of the hand to 30°C. Note the reduction in the decrement from 15 to 6 percent and the increase in amplitude following cooling of the hand. (From Denys,[20] with permission.)

stimulation. Administration of anticholinesterase drugs within a few hours before the test reduces the probability of obtaining a decremental response. Discontinuance of the short-acting medication for several hours improves the sensitivity of the test. The patient must withhold a long-acting medication for a longer period, if clinically feasible. With an overdose of anticholinesterase drugs, a single nerve impulse may cause a repetitive muscle response, and repetitive stimuli at a high rate give rise to a decremental response.

3   COMMONLY USED NERVES AND MUSCLES

Distal Versus Proximal Muscle

Patients with myasthenia gravis rarely have a decremental response in clinically unaffected muscle. Thus, weak proximal or facial muscles show a higher incidence of electrical abnormality than stronger distal muscles. In a series of experiments, electrical and mechanical responses to repetitive stimuli revealed substantially greater decrement and posttetanic potentiation in the platysma than in the adductor pollicis.[43,44] Electrophysiologic findings in botulism may also be limited to weak muscles in the clinically affected limbs. In contrast, patients with the myasthenic syndrome usually have prominent abnormalities not only in the proximal muscles but also to a lesser degree in distal muscles.[10]

In principle, the method consists of applying repetitive stimulation to a motor or mixed nerve and recording a train of responses from the innervated muscle. Although less sensitive, studies of the distal musculature provide technically more reliable results than those of more proximal muscles in the limb or facial muscles. With stimulation of the ulnar nerve at the elbow, it is possible to make simultaneous recordings from one proximal and three distal muscles, the flexor carpi ulnaris, abductor digiti quinti, first dorsal interosseous, and adductor pollicis.[23] A negative result with distal muscles should prompt the examination of the proximal muscles, such as the deltoid, biceps, and upper trapezius. Stimulation of the brachial plexus at the supraclavicular fossa tends to activate many muscles simultaneously. In contrast, selective stimulation of the accessory nerve allows recording from the trapezius without contamination from other muscles.[52,61] Although the general approach is the same, wise choice of the nerve and muscle based on distribution of weakness determines the test sensitivity.

Upper Limb and Shoulder Girdle

HYPOTHENAR MUSCLES

The ulnar nerve is stimulated at the wrist with G_1 over the belly of abductor digiti quinti and G_2 on the tendon. Binding the four fingers together with a bandage or Velcro strap prevents interference from movement. A restraining metal bar is also used to hold the hand, palm down, flat on the examining table. The patient exercises by abducting the fifth digit against the restraint.

THENAR MUSCLES

The median nerve is stimulated at the wrist with G_1 on the belly of the abductor pollicis brevis and G_2 2 cm distally. The hand, held palm up by the restraining metal bars, lies flat on the board with the thumb in the adducted position. The patient exercises the muscle by abducting the thumb against the bar.

BICEPS

The musculocutaneous nerve is stimulated at the axilla with G_1 on the belly of the biceps and G_2 over the tendon. The position of the arm depends on the type of mechanical board available. A handlebar attached under a solid table can serve as an excellent restraint. The patient, upright in a chair, holds on to the handle from below with the arm flexed approximately 45 degrees in the adducted and supinated position. Pulling up against the handlebar with flexion at the elbow exercises the muscle.

DELTOID

The brachial plexus is stimulated at Erb's point with G_1 on the belly of the muscle and G_2 on the acromion. The arm is adducted, flexed at the elbow, and internally rotated to place the hand in front of abdomen for self-restraint by the opposite hand. The patient exercises by abducting the arm against his own resistance. For the weak or uncooperative patient, a Velcro strap is applied firmly against the trunk with the arm adducted at the side.

TRAPEZIUS

The spinal accessory nerve is stimulated along the posterior border of the sternocleidomastoid muscle, with G_1 on the upper trapezius muscle at the angle of neck and shoulder and G_2 over the tendon, near the acromion process. The patient, upright in a chair, with the arms adducted and extended, holds on to the bottom of the chair and exercises, shrugging the shoulders against his own resistance.

Lower Limb

ANTERIOR TIBIAL

The peroneal nerve is stimulated at the fibular head with G_1 on the belly of the muscle and G_2 a few centimeters distally. The patient sits in a chair with the thigh restrained firmly by Velcro straps and exercises by dorsiflexing the foot held on a restraining foot board.

QUADRICEPS

The femoral nerve is stimulated at the groin, just lateral to the femoral artery, with G_1 on the rectus femoris and G_2 on the patellar tendon. The patient sits in a chair with the thigh and the leg fastened to the chair with Velcro straps and exercises by extending the leg against the restraint. The patient may also lie supine with the thigh bound to the bed by a Velcro strap and exercise by lifting the foot off the bed.

Face

ORBICULARIS OCULI, ORBICULARIS ORIS, AND NASALIS

The facial nerve is stimulated in front of the ear, with G_1 on the belly of the muscle and G_2 on the opposite side or the bridge of the nose. The patient, lying supine, exercises by contracting the muscle as vigorously as possible without the benefit of a restraining device to immobilize facial muscles.

4 PAIRED STIMULATION

Short Interstimulus Intervals

The paired stimuli are used to plot the time course of recovery of neuromuscular transmission quantitatively (see Chapter 8.7). In normal muscles, the first supramaximal stimulus activates the entire group of muscle fibers. A second stimulus delivered within a few milliseconds evokes a smaller response, indicating refractoriness of the nerve and muscle (see Fig. 8–4). The second potential then recovers, although the two responses overlap at intervals of less than 15 ms.

In typical cases of myasthenia gravis, the first stimulus elicits a maximal or near maximal muscle response. The recovery curve also follows a normal pattern. The curve deviates from the normal in the myasthenic syndrome, where the first stimulus elicits a submaximal response. A second shock given at very short interstimulus intervals evokes a larger response with the amplitude one and a half to two times that of the first (Fig. 9–3). The increment represents recruitment, based on summation of two

end-plate potentials (EPPs), of those fibers activated only subliminally by the first stimulus. This phenomenon is also observed in most patients with botulism and in occasional cases of myasthenia gravis.[14]

Long Interstimulus Intervals

Two EPPs no longer summate at interstimulus intervals exceeding 15 ms. Potentiation of the second response here represents true facilitation, resulting from an increased number of quanta liberated by the second stimulus. Despite the release of a greater amount of ACh, the second muscle potential normally shows no increment from the already maximal first response. The change may also be minimal at this interstimulus range in most patients with myasthenia gravis or botulism. In contrast, an increment regularly occurs at intervals ranging from 15 to 100 ms in the myasthenic syndrome. Indeed, this is one of the most characteristic electrophysiologic features of this syndrome.

The decremental response in myasthenia gravis begins at intervals of about 20 ms but becomes more definite at between 100 to 700 ms. It slowly returns to the baseline in about 10 seconds. The response reaches the trough at an interstimulus interval of about 300 to 500 ms (see Fig. 8–6). At shorter intervals, concomitant facilitation attributable to the electrosecretory mechanism obscures the

Figure 9–3. The effect of paired shocks given at interstimulus intervals of 2.5 ms (*top*), 15 ms (*middle*) and 25 ms (*bottom*). On the top tracings, the reduced test response in a healthy subject (**A**) indicates the effect of the refractory period, whereas the increased test response in a patient with botulism (**B**) suggests summation of two closely elicited EPPs. (From Cherington,[12] with permission.)

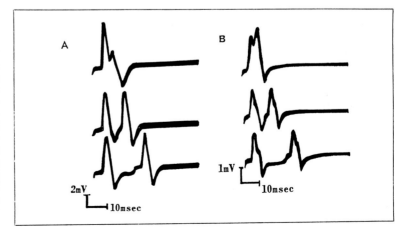

depression. The results of paired stimuli predict that a train of stimuli produces the maximal decrement at the rate of two to three per second.[22]

5 REPETITIVE STIMULATION AT SLOW RATES

Method and Normal Values

Repetitive stimulation at a rate of one to five per second depletes the immediately available ACh store, without superimposed facilitation from neurosecretory mechanisms (see Fig. 8–8). At slow rates of stimulation, movement-related artifacts are minimal because the muscle returns close to its original relaxed position before the next stimulus. Most patients tolerate a train at faster rates poorly. Moreover, continuous muscle contraction alters the geometry of the volume conductor, which in turn affects the waveform of the successive responses.

Random or irregular variations in amplitude or waveform suggest artifacts. In contrast, technically satisfactory trains with smooth, reproducible changes suggest abnormality of neuromuscular transmission. Most modern equipment automatically calculates the percentage reduction for the smallest of the initial five to seven responses, compared with the first in the same train. Before accepting the computed results, however, one must verify the waveform stored on the oscilloscope or photographed with Polaroid film. In normal muscles, decrement at stimulation of two to three per second, if present, does not exceed 5 to 8 percent.[7,64] In fact, an optimal train comprises practically identical responses from the first to the last. Thus, the presence of any reproducible decrement should raise suspicion in a tracing free of any technical problems.

Myasthenia Gravis

In myasthenia gravis, a maximal drop in amplitude occurs between the first and second responses of a train, followed by

further decline up to the fourth or fifth potential (Fig. 9–4). Subsequent responses in the series then level off or, more typically, reverse the course by regaining some of the lost amplitude. Occasionally, the recovery may even exceed the original value by 10 to 20 percent, especially after several seconds of repetitive stimulation. More characteristically, continued stimulation induces a long, slow decline following a transient increment.[38] To avoid a false-positive result, most electromyographers use a conservative criterion of abnormality, i.e., reproducible decrement of 10 percent or more between the first response and the smallest of the next four to six responses. In addition to the changes in amplitude, the latency may progressively increase in some myasthenic muscles. In equivocal cases, sampling several muscles improves the chance of documenting localized myasthenic weakness. In particular, a negative result in the distal limb muscles by no means precludes electrical abnormalities detectable in the proximal or facial musculature (Fig. 9–5).

The administration of edrophonium (Tensilon) or neostigmine (Prostigmin) helps further delineate the characteristics of defective neuromuscular transmission. These agents potentiate the action of ACh by blocking acetylcholinesterase in patients with postjunctional abnormalities. Therefore, a partial or complete reversal of the decrement by anticholinesterase agents tends to confirm the diagnosis of myasthenia gravis.

Other Neuromuscular Disorders

A train of stimuli at a slow rate causes decrementing responses not only in myasthenia gravis but also in a number of other conditions with reduced margins of safety. These include the myasthenic syndrome (Figs. 9–6 and 9–7), botulism, multiple sclerosis,[27] motor neuron disease,[4,54] and regenerating nerve.[31] A partially curarized muscle will also develop a similar decrement to a train of stimuli. In the patient with the myasthenic syndrome or botulism, single stimuli typi-

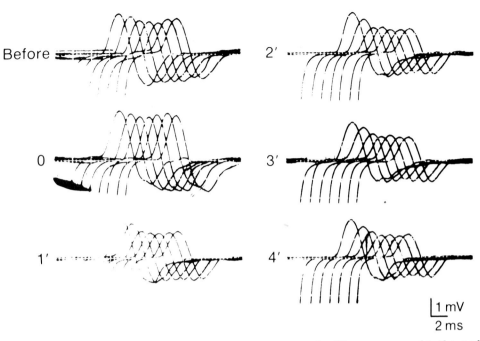

Figure 9–4. Thenar muscle potentials elicited by a train of stimuli of three per second to the median nerve before and after 1 minute of exercise in a patient with generalized myasthenia gravis. Amplitude comparison between the first and fifth responses revealed a decrement of 25 percent at rest, 12 percent immediately after exercise, and 50 percent 4 minutes later.

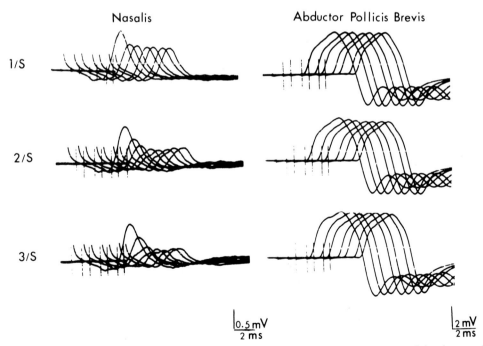

Figure 9–5. A 25-year-old woman with double vision of 1½-month duration. A train of shocks one, two, and three per second to the median nerve revealed no detectable abnormalities in the abductor pollicis brevis. Stimulation of the facial nerve elicited decrementing responses in the nasalis. Note greater change within the train as the rate of stimulation increased from one to three per second. (From Kimura,[41] with permission.)

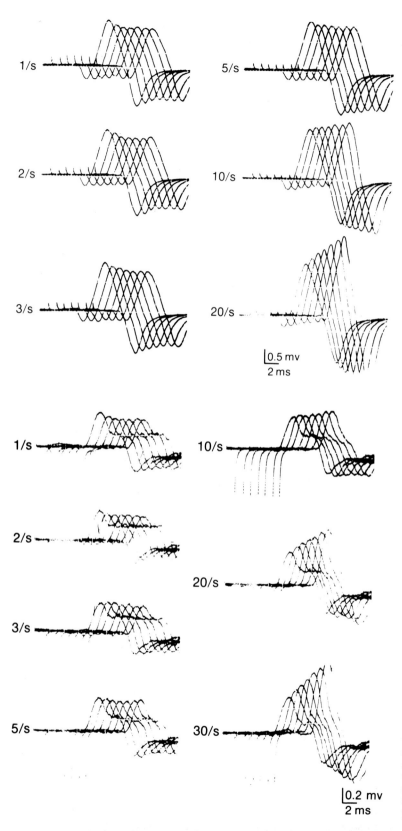

Figure 9–6. Thenar muscle potential elicited by a train of stimuli 1 through 20 per second to the median nerve in a patient with the myasthenic syndrome. Note decremental responses to slow rates of stimulation up to 5 per second and incremental responses to faster rates of stimulation at 10 per second and to a much greater degree at 20 per second.

Figure 9–7. A repeat study in the same patient as in Figure 9–6 using the same recording arrangements. Notice further diminution in amplitude of the compound muscle action potentials, compared with the earlier study, slight decrement at slow rates of stimulation up to 5 per second, and progressively more prominent increment at faster rates of 10 to 30 per second.

cally elicit very small muscle action potentials. A decremental tendency with a slow rate of repetitive stimulation, though present in most cases, does not constitute an essential feature of these disorders, characterized by defective release of ACh.

6 REPETITIVE STIMULATION AT FAST RATES

Normal Muscles

Supramaximal stimulation normally activates all muscle fibers innervated by the nerve. This precludes any increment with subsequent stimuli even though greater amounts of acetylcholine (ACh) are released. The recruitment of muscle fibers not activated by the first stimulus underlines the incremental tendency seen in the myasthenic syndrome, botulism, and occasionally myasthenia gravis.

In normal adults, muscle action potentials remain stable during repetitive stimulation at a rate of up to 20 to 30 per second.[55] Some healthy infants, however, may show progressive decline in amplitude at this rate.[15] In adults, trains of 50 per second may cause apparent decremental or incremental responses. At such a fast rate, however, inherent movement artifacts render the measurement unreliable. Muscles stimulated repetitively at a high rate tend to discharge with increased synchrony without recruitment of additional muscle fibers. The compound muscle action potential may then increase in amplitude, although not in area under the waveform.

Myasthenic Syndrome and Botulism

In the myasthenic syndrome[26] single stimuli typically elicit a strikingly small compound muscle action potential (Fig. 9–8). The amplitude varies over a wide range among different subjects. Thus, a decrease by as much as 50 percent of the maximal response in some individuals may still remain above the lower limit of a population norm. An apparent lack of reduction in amplitude, therefore, does not necessarily rule out the syndrome. A marked potentiation following a brief voluntary exercise would disclose the subnormality of the initial amplitude and confirm the diagnosis.

A train of stimulation at high rates reveals an interesting result from theoretic point of view, although its clinical application is limited because most patients tolerate the procedure poorly (see Fig. 8–8). Repetitive stimulation given at 20 to 50 per second produces a remarkable increment of successive muscle action potentials to a normal or near normal level (see Figs. 9–6 and 9–7). A slight initial decrement may precede the increment, but the last response of a train at the end of 1 minute usually exceeds the first response several times.[49] The electrophysiologic abnormalities often normalize in parallel to clinical improvement following the administration of guanidine.

Patients with botulism may have entirely normal electrical responses in early stages of illness or have a small muscle potential in response to a single stimulus.[12,13] An initially small response usually potentiates after voluntary exercise or with a train of stimuli (Fig. 9–9). Incrementing responses, though smaller in range, resemble those found in the myasthenic syndrome.[33,56] The most characteristic abnormality of infantile botulism is tetanic and posttetanic facilitation, which persists for a number of minutes.[17,29]

Other Neuromuscular Disorders

An incremental response, though characteristic of the myasthenic syndrome and botulism, by no means excludes other disorders of the neuromuscular junction (see Chapter 24.6). Patients with myasthenia gravis not infrequently show such a pattern, either during a progressive phase of the disease or during steroid therapy.[18,53,62,66] In contrast to the marked potentiation in the myasthenic

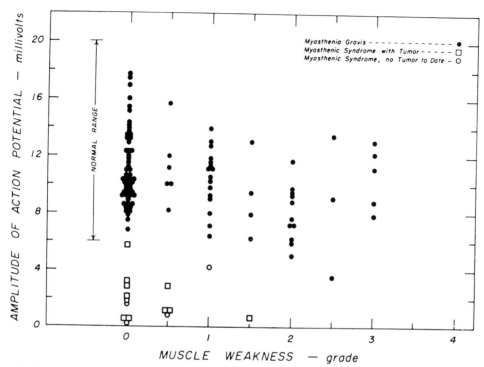

Figure 9–8. Relationship between the clinical estimate of weakness and the amplitude of muscle action potential in patients with myasthenia gravis and the myasthenic syndrome. The histogram plots the amplitude of the hypothenar muscle potential elicited by single maximal stimuli to the ulnar nerve. The scale on the abscissa denotes normal strength (0), 75 percent (1), 50 percent (2), 25 percent (3), and complete paralysis (4). (From Lambert et al.,[50] with permission.)

syndrome, changes in myasthenia gravis do not usually exceed the initial value by more than 40 percent at the end of 1 minute. Other disorders associated with depressed neuromuscular transmission and incremental tendency by a train of stimuli include antibiotic toxicity,[57] hypocalcemia, hypermagnesemia,[8,65] and snake poisoning.[40] Again, a limited degree of potentiation seen in these conditions stands in sharp contrast to the multifold increase characteristic of the myasthenic syndrome.

Use of Prolonged Stimulation

A short train of several shocks at slow rate suffices for routine evaluation of neuromuscular transmission. Prolonged stimulation at a rapid rate adds diagnostic information in the evaluation of infantile botulism. Otherwise, clinical yields are too little to justify subjecting the patient to this painful procedure. Besides, sustained muscle contraction causes excessive movement artifacts that often interfere with accurate assessments of the

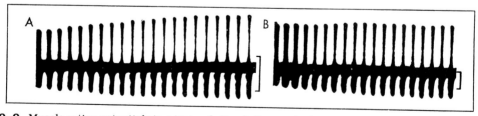

Figure 9–9. Muscle action potentials to a train of stimulation applied to the motor nerve at 50 per second in a patient with botulism. Note incremental responses when the patient received a 7 mg/kg daily dose of guanidine (**A**) and electrophysiologic recovery after the dosage was increased to 35 mg/kg (**B**). Vertical calibration is 2 mV. (From Cherington,[12] with permission.)

waveform. As a research tool, however, a long train helps elucidate the time course of the mechanical force of contraction. The force of muscle twitch increases during prolonged stimulation in healthy subjects but not in patients with myasthenia gravis. This phenomenon, called a positive staircase, however, lacks diagnostic specificity as a clinical test.[24,64] Whatever the purpose, clinicians must resort to a train of rapid stimulation judiciously to avoid inflicting unnecessary pain.

7   EFFECT OF TETANIC CONTRACTION

To induce tetany electrically, it is necessary to apply a 20- to 30-second train at 50 per second or a continuous run for a few minutes at 3 per second. Most subjects tolerate these procedures poorly. Fortunately, voluntary muscle contrac-

tion accomplishes the same effect painlessly, with motor fibers discharging up to 50 impulses per second during maximal effort. A typical postactivation cycle following voluntary or involuntary tetanic contraction consists of two phases: posttetanic potentiation,[38] lasting for about 2 minutes, and posttetanic exhaustion,[22] lasting up to 15 minutes.

Posttetanic Potentiation

Tetanic contraction not only causes calcium (Ca^{2+}) to accumulate inside the axon but also mobilizes ACh vescicles from the main store. Subsequent nerve stimulation gives rise to a larger EPP, thus recruiting additional muscle fibers not previously activated in the myasthenic syndrome or related disorders with defective release of ACh (Figs. 9–10 and 9–11).

In practice, a simple procedure for test-

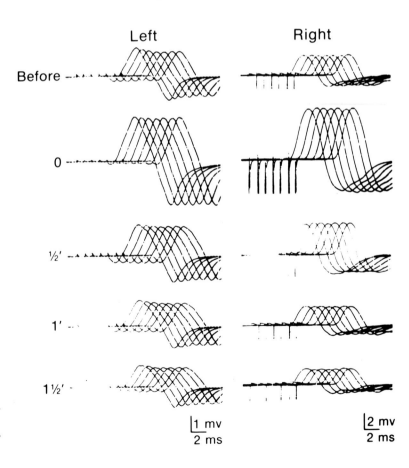

Left   Right

Before

0

½'

1'

1½'

|1 mv
2 ms

|2 mv
2 ms

Figure 9–10. Thenar muscle potential elicited by a train of stimuli three per second to the median nerve before and after 10 seconds of exercise in a patient with the myasthenic syndrome. Notice the posttetanic potentiation of 70 percent on the left and 160 percent on the right immediately after the exercise and posttetanic exhaustion 1½ minutes later.

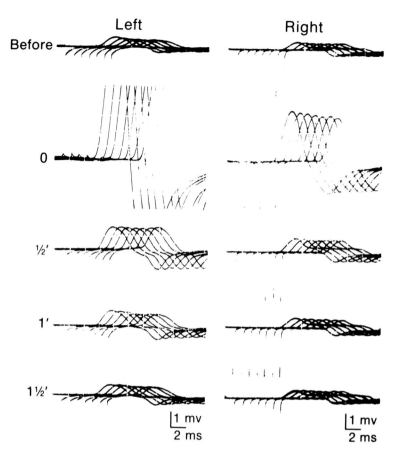

Figure 9–11. A repeat study in the same patient as in Figure 9–10 with the same recording arrangements. Compared with the earlier study, the patient had further diminution in amplitude of the compound muscle action potentials and a greater posttetanic potentiation on both sides.

ing this phenomenon consists of delivering single shocks of supramaximal intensity to the nerve and comparing the size of the muscle response measured before and after exercise. A striking increase in amplitude, usually reaching a level more than twice the baseline value, indicates a presynaptic defect of neuromuscular transmission.[50] Duration of exercise should not exceed 10 seconds, so that depletion of ACh during voluntary contraction is minimized. In general, a posttetanic potentiation greater than twice the preactivation response suggests the diagnosis of the myasthenic syndrome. The magnitude of potentiation, however, varies considerably from one subject to another and during the course of the illness within the same patient (Fig. 9–12). A lesser degree of facilitation also implies a presynaptic disturbance, which may be seen not only in the myasthenic syndrome but also in botulism or occasional cases of myasthenia gravis.

The use of a train of stimuli at three per second instead of a single shock allows simultaneous evaluation of the decremental trends. The same train is repeated before and immediately after the exercise and then every 30 seconds thereafter for a few minutes. In this arrangement, posttetanic potentiation partially compensates for depletion of ACh during each train, repairing the deficit caused by the slow rate of stimulation (see Fig. 9–4). Thus, the characteristic decrement seen within a train in myasthenia gravis tends to normalize immediately after the tetanic stimulation.

Posttetanic Exhaustion

After exercise, a transient potentiation is followed by decreased excitability of the neuromuscular junction in 2 to 4 minutes after exercise. The underlying physiologic mechanism for this phenom-

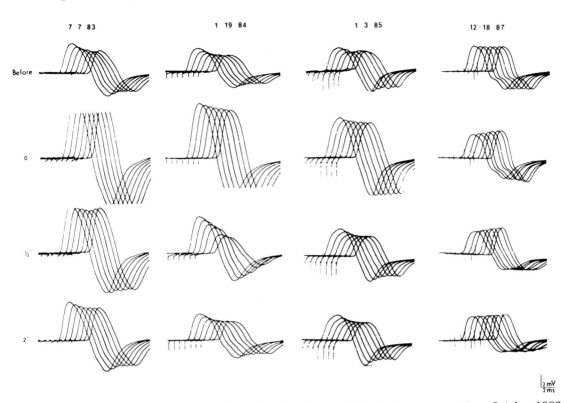

Figure 9–12. A 63-year-old woman with proximal weakness of all four extremities since October 1982. Thenar muscle potentials were elicited by stimuli applied to the median nerve at the wrist at three per second before and after 15 seconds of exercise. Notice the gradual reduction in the magnitude of posttetanic potentiation from 1983 through 1987. In the last study, the exercise induced only an incrementing tendency within the train, rather than the absolute increase in amplitude usually considered mandatory for the diagnosis of the myasthenic syndrome. (From Kimura,[41] with permission.)

enon awaits further clarification, but prolonged contraction will deplete the immediately available store of ACh, despite an increased rate of ACh mobilization. In normals, because of a large margin of safety, the reduced amount of ACh released during posttetanic exhaustion will still generate an adequate EPP in each individual muscle fiber. In premature infants and some newborns with limited neuromuscular reserve,[42] however, the amplitude of the compound muscle action potential progressively declines at high rates of stimulation.

In myasthenia gravis, neuromuscular block worsens during posttetanic exhaustion, indicating a reduced margin of safety. Some patients with an equivocal decrement at three per second in resting muscle may develop a definite abnormality after exercise (see Fig. 9–4). In the myasthenic syndrome, a reduced EPP after exercise results in further diminution of the originally small compound muscle action potential (see Figs. 9–8 and 9–9). Thus, the use of exercise increases the sensitivity of the nerve stimulation technique as a test of neuromuscular transmission. In the evaluation of posttetanic exhaustion, a 1-minute period of voluntary contraction is optimal to deplete the ACh store. In contrast, a shorter exercise, for a period ranging from 10 to 15 seconds, is recommended for assessment of the posttetanic potentiation to avoid exercise depletion of ACh.

8 REPETITIVE STIMULATION IN MYOGENIC DISORDERS

A train of stimuli causes an apparent decrement of the compound muscle action potentials in a number of myogenic disorders, such as myotonia, McArdle's disease, and periodic paralysis.

Myotonia

In myotonic muscles repetitive nerve stimulation produces commonly but not invariably decrementing responses.[3,9,48,51] In myotonia, unlike myasthenia gravis, however, responses in a train do not level off after the fourth or fifth stimulus. Instead, progressive decline continues for the initial few seconds; then gradual recovery ensues during subsequent stimulation for many seconds. In general, the higher the rate of stimulation, the greater the change in amplitude and the shorter the latent periods. The presence of clinical weakness also favors the possibility of finding prominent electrical decrement. The change occurs at a lower stimulation frequency in myotonia congenita than in myotonia dystrophica. In paramyotonia congenita, cooling potentiates both the weakness and electrical findings.

The decremental changes in myotonia may result from prolonged afterdepolarization, induced by accumulation of potassium (K^+) in the transverse tubules.[1] Direct stimulation of the muscle also evokes decreasing response, suggesting an excitability change of the muscle, rather than the neuromuscular junction.[9] Intracellular recording of a myotonic discharge also shows a progressive decline in amplitude.[60] Myotonic bursts may render some of the muscle fibers refractory to subsequent stimuli. In contrast to muscles in myasthenia gravis, myotonic muscles show neither posttetanic potentiation nor exhaustion. Indeed, the amplitude of the muscle response is less than the baseline value immediately after exercise. The decremental tendency also worsens after exercise, gradually restoring the resting value in about 2 to 3 minutes.

McArdle's Disease

In McArdle's disease, painful muscle contracture develops after exercise. The stiff muscle is electrically silent. With rapid repetitive stimulation of a motor nerve the amplitude of the compound muscle action potential progressively declines, eventually leading to the development of contracture.[25,59]

Periodic Paralysis

In periodic paralysis, a single stimulus elicits a small muscle during a paralytic episode. The amplitude may progressively increase with repetitive stimulation of the nerve at high rates[32] or during intermittent repetitive stimulation,[11] although it falls again during rest.

REFERENCES

1. Adrian, RH, and Bryant, SH: On the repetitive discharge in myotonic muscle fibers. J Physiol (Lond) 240:505–515, 1974.
2. Albuquerque, EX, Rash, JE, Mayer, RF, and Satterfield, JR: An electrophysiological and morphological study of the neuromuscular junction in patients with myasthenia gravis. Exp Neurol 51:536–563, 1976
3. Aminoff, MJ, Layzer, RB, Satya-Murti, S, and Faden, AI: The declining electrical response of muscle to repetitive nerve stimulation in myotonia. Neurology 27:812–816, 1977
4. Bernstein, LP, and Antel, JP: Motor neuron disease: Decremental responses to repetitive nerve stimulation. Neurology 31:202–204, 1981.
5. Borenstein, S, and Desmedt, JE: Temperature and weather correlates of myasthenic fatigue. Lancet 2:63–66, 1974.
6. Borenstein, S, and Desmedt, JE: Local cooling in myasthenia. Improvement of neuromuscular failure. Arch Neurol 32:152–157, 1975.
7. Botelho, SY, Deaterly, CF, Austin, S, and Comroe, JH, Jr: Evaluation of the electromyogram of patients with myasthenia gravis. Arch Neurol Psychiatry 67:441–450, 1952.
8. Branisteanu, DD, Miyamoto, MD, and Volle, RL: Effects of physiologic alterations on binomial transmitter release at magnesium-depressed neuromuscular junctions. J Physiol (Lond) 254:19–37, 1976.
9. Brown, JC: Muscle weakness after rest in myotonic disorders: An electrophysiological study. J Neurol Neurosurg Psychiatry 37:1336–1342, 1974.
10. Brown, JC, and Johns, RJ: Diagnostic difficulties encountered in the myasthenic syndrome

sometimes associated with carcinoma. J Neurol Neurosurg Psychiatry 37:1214–1224, 1974.

11. Campa, JF, and Sanders, DB: Familial hypokalemic periodic paralysis: Local recovery after nerve stimulation. Arch Neurol 31:110–115, 1974.

12. Cherington, M: Botulism: Electrophysiologic and therapeutic observations. In Desmedt, JE (ed): New Developments in Electromyography and Clinical Neurophysiology, Vol 1. Karger, Basel, 1973, pp 375–379.

13. Cherington, M: Botulism: Ten-year experience. Arch Neurol 30:432–437, 1974.

14. Cherington, M, and Ginsberg, S: Type B botulism: Neurophysiologic studies. Neurology 21:43–46, 1971.

15. Churchill-Davidson, HC, and Wise, RP: Neuromuscular transmission in the newborn infant. Anesthesiology 24:271–278, 1963.

16. Cooper, BJ, and Spence, I: Temperature-dependent inhibition of evoked acetylcholine release in tick paralysis. Nature 263:693–695, 1976.

17. Cornblath, DR, Sladky, JT, and Sumner, AJ: Clinical electrophysiology of infantile botulism. Muscle Nerve 6:448–452, 1983.

18. Dahl, DS, and Sato, S: Unusual myasthenic state in a teen-age boy. Neurology 24:897–901, 1974.

19. Daube, JR: Minimonograph #8: Electrophysiologic testing for disorders of the neuromuscular junction. American Association of Electromyography and Electrodiagnosis, Rochester, Minn, 1978.

20. Denys, EH: Minimonograph 14: The role of temperature in electromyography. American Association of Electromyography and Electrodiagnosis, Rochester, Minn, 1980.

21. Denys, EH, and Norris, FH, Jr: Amyotrophic lateral sclerosis: Impairment of neuromuscular transmission. Arch Neurol 36:202–205, 1979.

22. Desmedt, JE: The neuromuscular disorder in myasthenia gravis. 1. Electrical and mechanical response to nerve stimulation in hand muscles. In Desmedt, JE (ed): New Developments in Electromyography and Clinical Neurophysiology, Vol 1. Karger, Basel, 1973, pp 241–304.

23. Desmedt, JE, and Borenstein, S: Diagnosis of myasthenia gravis by nerve stimulation. Ann NY Acad Sci 274:174–188, 1976.

24. Desmedt, JE, Emeryk, B, Hainaut, K, Reinhold, H, and Borenstein, S: Muscular dystrophy and myasthenia gravis: Muscle contraction properties studied by the staircase phenomenon. In Desmedt, JE (ed): New Developments in Electromyography and Clinical Neurophysiology, Vol 1. Karger, Basel, 1973, pp 380–399.

25. Dyken, ML, Smith, DM, and Peake, RL: An electromyographic diagnostic screening test in McArdle's disease and a case report. Neurology 17:45–50, 1967.

26. Eaton, LM, and Lambert, EH: Electromyography and electric stimulation of nerves in diseases of motor unit: Observations on myasthenic syndrome associated with malignant tumors. JAMA 163:1117–1124, 1957.

27. Eisen, A, Yufe, R, Trop, D, and Campbell, I: Reduced neuromuscular transmission safety factor in multiple sclerosis. Neurology 28:598–602, 1978.

28. Elmqvist, D: Neuromuscular transmission defects. In Desmedt, JE (ed): New Developments in Electromyography and Clinical Neurophysiology, Vol 1. Karger, Basel, 1973, pp 229–240.

29. Fakadej, AV, and Gutmann, L: Prolongation of post-tetanic facilitation in infant botulism. Muscle Nerve 5:727–729, 1982.

30. Foldes, FF, Kuze, S, Vizi, ES, and Derry, A: The influence of temperature on neuromuscular performance. J Neurol Transm 43:27–45, 1978.

31. Gilliatt, RW: Nerve conduction in human and experimental neuropathies. Proc R Soc Med (Lond) 59:989–993, 1966.

32. Grob, D, Johns, RJ, and Liljestrand, A: Potassium movement in patients with familial periodic paralysis: Relationship to the defect in muscle function. Am J Med 23:356–375, 1957.

33. Gutmann, L, and Pratt, L: Pathophysiologic aspects of human botulism. Arch Neurol 33:175–179, 1976.

34. Harris, JB, and Leach, GDH: The effect of temperature on end-plate depolarization of the rat diaphragm produced by suxamethonium and acetylcholine. J Pharm Pharmacol 20:194–198, 1968.

35. Harvey, AM, and Masland, RL: The electromyogram in myasthenia gravis. Bull Johns Hopkins Hosp 48:1–13, 1941.

36. Hofmann, WW, Parsons, RL, and Feigen, GA: Effects of temperature and drugs on mammalian motor nerve terminals. Am J Physiol 211:135–140, 1966.

37. Hubbard, JI, Jones, SF, and Landau, EM: The effect of temperature change upon transmitter release, facilitation and post-tetanic potentiation. J Physiol (Lond) 216:591–609, 1971.

38. Johns, RJ, Grob, D, and Harvey, AM: Studies in neuromuscular function. 2. Effects of nerve stimulation in normal subjects and in patients with myasthenia gravis. Bull Johns Hopkins Hosp 99:125–135, 1956.

39. Jolly, F: Myasthenia gravis pseudoparalytica. Berliner Klinische Wochenschrift 32:33–34, 1895.

40. Kamenskaya, MA, and Thesleff, S: The neuromuscular blocking action of an isolated toxin from the elapid (oxyuranus scutellactus). Acta Physiol Scand 90:716–724, 1974.

41. Kimura, J: Electromyography and nerve stimulation techniques: Clinical applications (Japanese). Egakushoin, Tokyo, 1989.

42. Koenigsberger, MR, Patten, B, and Lovelace, RE: Studies of neuromuscular function in the newborn. 1. A comparison of myoneural function in the full term and the premature infant. Neuropadriatrie 4:350–361, 1973.

43. Krarup, C: Electrical and mechanical responses in the platysma and in the adductor pollicis muscle: In patients with myasthenia gravis. J Neurol Neurosurg Psychiatry 40:241–249, 1977.

44. Krarup, C: Electrical and mechanical responses in the platysma and in the adductor pollicis

muscle: In normal subjects. J Neurol Neurosurg Psychiatry 40:234–240, 1977.

45. Lambert, EH: Electromyography and electric stimulation of peripheral nerves and muscle. In Clinical Examinations in Neurology, ed 4. Departments of Neurology, Physiology, and Biophysics, Mayo Clinic and Mayo Foundation, WB Saunders, Philadelphia, 1976, pp 298–329.

46. Lambert, EH: Electromyographic responses to repetitive stimulation of nerve. American Academy of Neurology, Special Course #16, April 23–28, 1979.

47. Lambert, EH: Neuromuscular transmission studies. American Association of Electromyography and Electrodiagnosis, Third Annual Continuing Education Course, September 25, 1980.

48. Lambert, EH, Millikan, CH, and Eaton, LM: Stage of neuromuscular paralysis in myotonia (abstr). Am J Physiol 171:741, 1952.

49. Lambert, EH, and Rooke, ED: Myasthenic state and lung cancer. In Brain, WR, and Norris, FH, Jr (eds): Contemporary Neurology Symposia, Vol 1, The Remote Effects of Cancer on the Nervous System. Grune & Stratton, New York, 1965, pp 67–80.

50. Lambert, EH, Rooke, ED, Eaton, LM, and Hodgson, CH: Myasthenic syndrome occasionally associated with bronchial neoplasm: Neurophysiologic studies. In Viets, HR (ed): Myasthenia Gravis, The Second International Symposium Proceedings. Charles C Thomas, Springfield, Ill, 1961, pp 362–410.

51. Lundberg, PO, Stalberg, E, and Thiele, B: Paralysis periodica paramyotonica: A clinical and neurophysiological study. J Neurol Sci 21:309–321, 1974.

52. Ma, DM, Wasserman, EJL, and Giebfried, J: Repetitive stimulation of the trapezius muscle: Its value in myasthenic testing. Muscle Nerve 3:439–440, 1980.

53. Mayer, RF, and Williams, IR: Incrementing responses in myasthenia gravis. Arch Neurol 31:24–26, 1974.

54. Mulder, DW, Lambert, EH, and Eaton, LM: Myasthenic syndrome in patients with amyotrophic lateral sclerosis. Neurology 9:627–631, 1959.

55. Ozdemir, C, and Young, RR: The results to be expected from electrical testing in the diagnosis of myasthenia gravis. Ann NY Acad Sci 274:203–225, 1976.

56. Pickett, J, Berg, B, Chaplin, E, and Brunstetter-Shafer, MA: Syndrome of botulism in infancy: Clinical and electrophysiological study. N Engl J Med 295:770–772, 1976.

57. Pittinger, C, and Adamson, R: Antibiotic blockade of neuromuscular function. Ann Rev Pharmacol 12:169–184, 1972.

58. Ricker, K, Hertel, G, and Stodieck, S: The influence of local cooling on neuromuscular transmission in the myasthenic syndrome of Eaton and Lambert. J Neurol 217:95–102, 1977.

59. Ricker, K, and Mertens, HG: Myasthenic reaction in primary muscle fibre disease. Electroencephalogr Clin Neurophysiol 25:413–414, 1968.

60. Rudel, R, and Keller, M: Intracellular recording of myotonic runs in dantrolene-blocked myotonic muscle fibres. In Bradley, WG, Gardner-Medwin, D, and Walton, JN (eds): Recent Advances in Myology. Proceedings of the Third International Congress on Muscle Diseases, Excerpta Medica, Amsterdam, 1975, pp 441–445.

61. Schumm, F, and Stohr, M: Accessory nerve stimulation in the assessment of myasthenia gravis. Muscle Nerve 7:147–151, 1984.

62. Schwartz, MS, and Stalberg, E: Myasthenia gravis with features of the myasthenic syndrome: An investigation with electrophysiologic methods including single-fiber electromyography. Neurology 25:80–84, 1975.

63. Simpson, JA: Myasthenia gravis: A new hypothesis. Scott Med J 5:419–436, 1960.

64. Slomic, A, Rosenfalck, A, and Buchthal, F: Electrical and mechanical responses of normal and myasthenic muscle with particular reference to the staircase phenomenon. Brain Res 10:1–78, 1968.

65. Swift, TR: Weakness from magnesium-containing cathartics: Electrophysiologic studies. Muscle Nerve 2:295–298, 1979.

66. Takamori, M, and Gutmann, L: Intermittent defect of acetylcholine release in myasthenia gravis. Neurology (Minn) 21:47–54, 1971.

67. Ward, CD, and Murray, NMF: Effect of temperature on neuromuscular transmission in the Eaton-Lambert syndrome. J Neurol Neurosurg Psychiatry 42:247–249, 1979.

68. Wright, GP: The neurotoxins of clostridium botulinum and clostridium tetani. Pharmacol Rev 7:413–465, 1955.

Chapter 10

ACTIVATION PROCEDURES AND OTHER METHODS

1 INTRODUCTION

Not all muscles show electrophysiologic abnormalities in diseases of the neuromuscular junction. Thus, the conventional nerve stimulation techniques may fail to substantiate the clinical diagnosis. Imposing a metabolic stress reduces the safety factor of the system, thus increasing the detection ratio of neuromuscular transmission abnormalities. Obviously, such activation maneuvers used to improve the diagnostic yields must avoid false-positive results. The ischemic test[10] and regional curare test[22] affect the normal muscles very little but cause a transmission defect in overt as well as some occult myasthenic muscles.

A number of electrodiagnostic methods are available to supplement the technique of paired or repetitive electrical stimuli. Measuring the fatigability of the muscle force serves to document a decremental tendency, although an accurate recording of mechanical activity is possible in only a few muscles. Electromyographic analyses of motor unit potentials can also elucidate the stability of neuromuscular transmission. In particular, assessment of the extraocular muscles provides useful information in patients with ocular myasthenia. Single fiber electromyography has added the calculation of jitter as a very sensitive and most useful measure of neuromuscular transmission. The in vitro analysis of miniature end-plate potentials (MEPPs) of intercostal muscles allows a direct quantitative measure of neuromuscular transmission.[11]

This technique, however, is not feasible as a routine clinical test.

Other laboratory methods of diagnostic value include tonography and infrared optokinetic nystagmography for ocular myasthenia[39,43] and stapedius reflex for facial weakness.[2] Electrophysiologic methods also help in quantitating edrophonium (Tensilon) and neostigmine (Prostigmin) tests. Measurement of serum antibodies against the muscle end plate detects immunologic abnormalities in most patients with myasthenia gravis.[1,25,26] As a part of clinical evaluation, pulmonary function tests also provide useful objective criteria in documenting neuromuscular fatigue.[16,30]

2   ISCHEMIC TEST

A double-step test increases the diagnostic sensitivity of nerve stimulation techniques in some patients with mild, generalized or ocular myasthenia gravis.[4,9] The first step consists of continuous supramaximal stimulation of the ulnar nerve at a rate of three per second for 4 minutes and recording the compound muscle action potential from the ulnar-innervated intrinsic hand muscles or the flexor carpi ulnaris in the forearm. Thereafter, a train of three-per-second stimulation for several shocks at 30-second intervals determines the amount of decrement within each three-per-second trial. In cases of a negative or equivocal result, the second step consists of the same procedure under ischemia induced by a cuff inflated above arterial pressure proximal to the stimulation site.

Normal Response

In normal subjects, the amplitude of the compound muscle action potential may fall as much as 50 percent after 2 or 3 minutes of continuous three-per-second stimulation under ischemia.[10] Subsequent three-per-second tests, however, elicit no decrement within each train in the intrinsic hand muscles unless ischemic exercise has lasted 7 to 15 minutes. Such substantial reduction in amplitude without an intratrial decrement suggests conduction block of the nerve during ischemia. The amplitude returns to normal in about 3 minutes after the release of the cuff. For unknown reasons, the three-per-second series may show a decrement in the flexor carpi ulnaris following only 4 minutes of ischemic exercise.

Myasthenic Response

In myasthenic hand muscles, the ischemic exercise at three-per-second stimulation for 4 minutes generally elicits marked exhaustion. The amplitude of the muscle response decreases markedly 2 to 3 minutes after the beginning of the continuous three-per-second stimulation. A train of three-per-second stimulation following ischemic exercise shows increased block in the overtly myasthenic muscle and, more importantly, in muscles with subclinical involvement. Muscles with subclinical myasthenia develop decrement only when sensitized by both ischemia and sustained exercise. Cooling ischemic muscles by $1\,^{\circ}C$ to $2\,^{\circ}C$ improves neuromuscular transmission, increasing the incidence of false-negative results.[5]

The ischemic exercise should last at least 2 minutes to unmask the reduced safety factor but should not exceed 5 minutes, so that one can avoid the decrementing response seen in normal muscles. Thus, a series of three-per-second stimulation for 4 minutes before and 4 minutes during ischemia provides optimal results. Various muscles are affected to different degrees in the same patient with myasthenia gravis; neuromuscular block affects some muscles in the resting state, others after exercise, and the rest not at all, even after considerable stress. The double-step test has helped elucidate different degrees of myasthenic involvement in the same patient. How much additional help this procedure provides in the early diagnosis of myasthenia gravis awaits further clarification.[32]

3 REGIONAL CURARE TEST

The amount of acetylcholine (ACh) released with each nerve impulse normally produces an EPP that substantially exceeds the critical level for excitation of all the muscle fibers. This margin of safety protects the neuromuscular transmission with a latent deficit, rendering clinical and electrophysiologic evaluation difficult. Curare causes a nondepolarizing block by competing with ACh for the endplate receptors. Its administration, therefore, reduces or eliminates the functional reserve and elucidates the defect of neuromuscular transmission that otherwise escapes detection.

Principles and Procedures

A small dose of curare regionally injected at the wrist with a pressure tourniquet occluding the forearm circulation alters voluntary grip strength[13] and force of evoked muscle contraction.[12] Changes in amplitude of the compound muscle action potentials provide an optimal means of analyzing the effect of curare in the upper limb under ischemia.[6,7,19–22]

One twentieth of the usual curarizing dose suffices to produce neuromuscular block restricted to one hand. The recommended steps[18] consist of insertion of a scalp vein needle, pointing distally, into a superficial forearm vein, elevation of the arm for 1 minute for draining venous circulation, inflation of the cuff to at least 100 mm Hg above systolic pressure, and rapid injection of 0.2 mg d-tubocurarine in 20 ml of 0.9 percent sodium chloride with the arm lowered. Repetitive stimulation at three per second is applied 5 minutes and again 6 minutes after the start of the injection, with the arm still ischemic. Deflation of the cuff then reestablishes circulation before the next trains of three per second at 8, 9, and 10 minutes after injection. If tracing reveals equivocal decrement, three-per-second stimulus trains are given at 1-minute intervals for 10 minutes after 1-minute voluntary contraction, for evaluation of posttetanic exhaustion.

Clinical Value and Limitations

The regional curare test, as a measure of last resort, supplements the conventional nerve stimulation techniques in difficult cases. Patients with established myasthenia gravis should not undergo the procedure to avoid the possible risk. In one series of 600 patients the repetitive stimulation at the wrist and shoulder verified the diagnosis in 320 (53 percent).[22] In the remaining 280 patients, the regional curare test revealed abnormality in 72 (26 percent), including 52 (74 percent) of 70 patients with definite generalized myasthenia gravis, 13 (10 percent) of 136 patients with possible generalized myasthenia gravis, and 7 (14 percent) of 49 with ocular myasthenia.

The concentration of curare that reaches the muscle depends on diffusion through the volume of tissue, which probably varies from one case to the next. Thus, titrated dosages of curare do not always differentiate normal and pathologic responses.[17] Studies have revealed undue sensitivity to curare not only in myasthenia gravis[36] but also in amyotrophic lateral sclerosis[29] and muscular weakness after administration of antibiotics.[34] Therefore, an abnormal regional curare test indicates a defect in neuromuscular transmission but not necessarily myasthenia gravis. In one series, the conventional stimulation technique without the use of curare confirmed the diagnosis in 95% of patients with myasthenia gravis.[32] Provocative techniques are contraindicated in this group of patients.

4 ELECTROMYOGRAPHY

In normal muscles, a motor unit firing repetitively under voluntary control gives rise to identical potentials in waveform and amplitude every time it discharges, as long as the needle electrode remains in the same location relative to the generator source. This does not hold in myasthenic muscles, because nerve impulses may not always depolarize individual muscle fibers to the critical level, as the result of a reduced margin of safety. Intermittent failure of some muscle fibers innervated by the same axon in response to successive nerve impulse causes amplitude variability of the recurring motor unit potentials. Blocking at the neuromuscular junction also explains diminished mean amplitude and duration of motor unit potentials in myasthenic muscles.[31]

Varying Motor Unit Potentials

Electromyography can assess the stability of isolated motor unit potentials by slowly advancing or retracting the needle for optimal display of the repetitive discharges. During minimal contraction of the muscle, the amplitude variability of an isolated potential can be heard over the loudspeaker, alerting the examiner to search for unstable motor unit discharges. This method does not necessarily allow accurate quantitative assessment of neuromuscular transmission. The needle examination, however, is applicable to any muscles including those not tested by the stimulation technique. As an added advantage, the needle study requires no immobilization of the muscle. For example, electromyography helps establish the diagnosis of ocular myasthenia (see Chapter 14.3).

The administration of anticholinesterase reverses the abnormalities of motor unit potentials in patients with myasthenia gravis. Thus, the injection of edrophonium (Tensilon) increases motor unit potentials recorded in the extensor digitorum communis by 30 to 130 percent in amplitude and 10 to 25 percent in duration.[33] In patients with ocular myasthenia gravis, a progressive decrease in amplitude and frequency during a prolonged period of voluntary contraction partially reverses after intravenous administration of edrophonium.[23,37]

Jitter and Blocking

The single fiber recording has proven useful in early detection of neuromuscular disturbances. It allows assessment of single fiber neuromuscular transmission

as an important adjunct technique in the evaluation of myasthenia gravis.[40,41] As described in detail later (see Chapter 15) the method consists of recording a pair of single fiber potentials simultaneously and measuring fluctuation of the neuromuscular transmission by the stability of the interpeak intervals. Either blocking or increased jitter characterizes neuromuscular disturbances.

The EPPs generated by voluntary contraction normally reach the threshold in all muscle fibers. If the EPP falls short of this critical level at some neuromuscular junction, those muscle fibers fail to discharge. This blocking affecting only some fibers of a motor unit reduces the size of the motor unit potential on standard needle recordings. If an EPP barely reaches the necessary level, its rate of rise falls below the normal range, delaying, rather than blocking, the action potential. This abnormality escapes detection in the usual electromyography because unlike blocking, delayed discharge of muscle fibers alters the motor unit potential very little.

On single-fiber recording of a pair of potentials, an intermittent delay of the action potential in either fiber increases the jitter or the variability of interpotential interval. This finding, as the first sign of neuromuscular instability, precedes blocking of transmission. Thus, increased jitter heralds variation of motor unit potentials or decrementing response to repetitive stimulation of the nerve. The practice of single-fiber electromyography has added a new dimension to the assessment of neuromuscular transmission, although it requires additional training. Most commercially available instruments have the capability of computerizing the method, which surpasses the manual calculation of the recorded responses.

5 OTHER TECHNIQUES

Microelectrode Recording

Microelectrode recordings from single intercostal muscles provide the only means of measuring the size and the number of the ACh quanta released per volley of nerve impulses. These determinations in turn allow precise characterization of the abnormality of neuromuscular transmission. The method helps elucidate the specific pathophysiology underlying the deficits in production or mobilization of ACh. It also measures the sensitivity of the motor end plate by quantitative assessment. This method is beyond the scope of routine clinical tests because it requires biopsy of the intercostal muscles. In selected cases that pose a diagnostic challenge, however, it helps differentiate myasthenia gravis, the myasthenic syndrome, and other disorders involving the neuromuscular junction with certainty.

Tonography

Other techniques not ordinarily used in an electromyographic laboratory include edrophonium (Tensilon) tonography. The intraocular pressure results in part from contraction of the extraocular muscles.[23] Thus, measurements of the pressure with an electronic tonometer reveal the effect of the anticholinesterase on the ocular motility.[14,15] Some investigators advocate simultaneous recording of muscle action potentials with needle electrodes placed in the extraocular muscles.

In normal subjects, intraocular pressure may fall on the average 1.6 to 1.8 mm Hg over a 1-minute period, after an intravenous injection of edrophonium up to 10 mg.[8] Intraocular pressure is low in patients with decreased extraocular tone, as in ocular myasthenia. The administration of edrophonium, then, produces changes in tonography coincident with a moderate increase in electrical activity in the extraocular muscles. A sudden increase in extraocular muscle tone alters intraocular pressure by a mean of 1.6 mm Hg within 35 seconds of injection. This phenomenon, however, does not necessarily imply ocular myasthenia, as it occurs, for example, in ocular myositis without other features of myasthenia gravis.[43] Intraocular pressure may also rise with the Valsalva maneuver. In this case, a control injection of saline can identify a false-positive result.

Stapedius Reflex

The stapedius muscles contract bilaterally in response to unilateral sound stimulation. This contraction in turn dampens the acoustic sensitivity of the middle ear and prevents hyperacusis. Thus, impedance audiometry can be used for measuring the function of the stapedius muscle. In normal subjects, a sound stimulus 70 to 100 dB above the hearing threshold elicits the stapedius reflex. It shows no decay during sustained contraction for up to 1 minute with stimulus frequencies of 250 to 1000 Hz.

In patients with myasthenia gravis, weakened stapedius muscles enhance transmission of sound in the 1- to 4-kHz range, resulting in hyperacusis. Here, only high-intensity sound can induce the acoustic reflex.[28] In addition, reflex contraction of the stapedius muscle shows a rapid decrement, analogous to the similar response of the limb muscles to repetitive electrical stimulation of the nerve.[2] The administration of edrophonium enhances the acoustic reflex, diminishes hyperacusis, and improves the decay of the stapedius reflex in response to repetitive sound stimulation. In some patients with myasthenia gravis, testing the stapedius reflex may provide the only electrophysiologic abnormality.[42] In one study, stapedial reflex showed clear abnormalities in 84 percent of the patients with myasthenia gravis as compared with 56 percent by repetitive stimulation and 91 percent by single-fiber electromyography.[24]

Tests for Oculomotor Function

Electronystagmography provides quantitative measurements of amplitude, velocity, and frequency of optokinetic nystagmus to document fatigue of extraocular muscles. In patients with ocular myasthenia, edrophonium (Tensilon) induces an increase in previously reduced oculomotor function.[3] In one series, electrooculography revealed neuromuscular fatigue in 50 percent of myasthenic patients.[8] Infrared reflection technique improves the sensitivity of the test with the use of numeric criteria in grading neuropharmacologic effect on oculomotor fatigue.[38,39] For example, velocity of saccad measured by this means increases following administration of edrophonium.[27] The Lancaster red-green test also detects oculomotor fatigue, which improves after rest or with administration of edrophonium in patients with myasthenia gravis.[35]

REFERENCES

1. Almon, RR, Andrew, CG, and Appel, SH: Serum globulin in myasthenia gravis: Inhibition of a-bungarotoxin binding to acetylcholine receptors. Science 186:55–57, 1974.
2. Blom, S, and Zakrisson, JE: The stapedius reflex in the diagnosis of myasthenia gravis. J Neurol Sci 21:71–76, 1974.
3. Blomberg, LH, and Persson, T: A new test for myasthenia gravis. Preliminary report. Acta Neurol Scand 41 (Suppl 13):363–364, 1965.
4. Borenstein, S, and Desmedt, JE: New diagnostic procedures in myasthenia gravis. In Desmedt, JE (ed): New Developments in Electromyography and Clinical Neurophysiology, Vol 1. Karger, Basel, 1973, pp 350–374.
5. Borenstein, S, and Desmedt, JE: Local cooling in myasthenia: Improvement of neuromuscular failure. Arch Neurol 32:152–157, 1975.
6. Brown, JC, and Charlton, JE: A study of sensitivity to curare in myasthenic disorders using a regional technique. J Neurol Neurosurg Psychiatry 38:27–33, 1975.
7. Brown, JC, Charlton, JE, and White, DJK: A regional technique for the study of sensitivity to curare in human muscle. J Neurol Neurosurg Psychiatry 38:18–26, 1975.
8. Campbell, MJ, Simpson E, Crombie, AL, and Walton, JN: Ocular myasthenia: Evaluation of Tensilon tonography and electronystagmography as diagnostic tests. J Neurol Neurosurg Psychiatry 33:639–646, 1970.
9. Desmedt, JE: The neuromuscular disorder in myasthenia gravis. 1. Electrical and mechanical response to nerve stimulation in hand muscles. In Desmedt, JE (ed): New Developments in Electromyography and Clinical Neurophysiology, Vol 1. Karger, Basel, 1973, pp 241–304.
10. Desmedt, JE, and Borenstein, S: Double-step nerve stimulation test for myasthenic block: Sensitization of postactivation exhaustion by ischemia. Ann Neurol 1:55–64, 1977.
11. Elmqvist, D: Neuromuscular transmission defects. In Desmedt, JE (ed): New Developments in Electromyography and Clinical Neurophysiology, Vol 1. Karger, Basel, 1973, pp 229–240.
12. Feldman, SA, and Tyrrell, MF: A new theory of the termination of action of the muscle relaxants. Proc R Soc Med 63:692–695, 1970.
13. Foldes, FF, Klonymus, DH, Maisel, W, and Osserman, KE: A new curare test for the diagnosis of myasthenia gravis. JAMA 203:649–653, 1968.
14. Glaser, JS: Tensilon tonography in the diagnosis of myasthenia gravis. Invest Opthalmol 6:135–140, 1967.
15. Glaser, JS, Miller, GR, and Gass, JDM: The edrophonium tonogram test in myasthenia gravis. Arch Ophthalmol 76:368–373, 1966.
16. Griggs, RC, Donohoe, KM, Utell, MJ, Goldblatt, D, and Moxley, RT, III: Evaluation of pulmonary function in neuromuscular disease. Arch Neurol 38:9–12, 1981.
17. Hertel, G, Ricker, K, and Hirsch, A: The regional curare test in myasthenia gravis. J Neurol 214:257–265, 1977.
18. Horowitz, SH: The regional curare test and electrophysiologic diagnosis of myasthenia gravis. American Academy of Neurology, Special Course #16, April 23–28, 1979.
19. Horowitz, SH, Genkins, G, Kornfeld, P, and Papatestas, AE: Regional curare test in evaluation of ocular myasthenia. Arch Neurol 32:84–88, 1975.
20. Horowitz, SH, Genkins, G, Kornfeld, P, and Papatestas, AE: Electrophysiologic diagnosis of myasthenia gravis and the regional curare test. Neurology 26:410–417, 1976.
21. Horowitz, SH, and Krarup, C: A new regional curare test of the elbow flexors in myasthenia gravis. Muscle Nerve 2:478–490, 1979.
22. Horowitz, SH, and Sivak, M: The regional curare test and electrophysiologic diagnosis of myasthenia gravis: Further studies. Muscle Nerve 1:432–434, 1978.
23. Kornblueth, W, Jampolsky, A, Tamler, E, and Marg, E: Contraction of the oculorotary muscles and intraocular pressure: A tonographic and electromyographic study of the effect of edrophonium chloride (Tensilon) and succinylcholine (Anectine) on the intraocular pressure. Am J Ophthalmol 49:1381–1387, 1960.
24. Kramer, LD, Ruth, RA, Johns, ME, and Sanders, DB: A comparison of stapedial reflex fatigue with repetitive stimulation and single fiber EMG in myasthenia gravis. Ann Neurol 9:531–536, 1981.
25. Lefvert, AK, Bergstrom, K, Matell, G, Osterman, PO, and Pirskanen, R: Determination of acetylcholine receptor antibody in myasthenia gravis: Clinical usefulness and pathogenetic implications. J Neurol Neurosurg Psychiatry 41:394–403, 1978.
26. Lindstrom, JM, Lennon, VA, Seybold, ME, and Whittingham, S: Experimental autoimmune myasthenia gravis and myasthenia gravis: Biochemical and immunochemical aspects. Ann NY Acad Sci 274:254–274, 1976.
27. Metz, HS, Scott, AB, and O'Meara, DM: Saccadic eye movements in myasthenia gravis. Arch Ophthalmol 88:9–11, 1972.
28. Morioka, WT, Neff, PA, Boisseranc, TE, Hartman, PW, and Cantrell, RW: Audiotympanometric findings in myasthenia gravis. Arch Otolaryngol 102:211–213, 1976.
29. Mulder, DW, Lambert, EH, and Eaton, LM: Myasthenic syndrome in patients with amyotrophic lateral sclerosis. Neurology 9:627–631, 1959.
30. O'Donohue, WJ, Jr, Baker, JP, Bell, GM,

Muren, O, Parker, CL, and Patterson, JL, Jr: Respiratory failure in neuromuscular disease: Management in a respiratory intensive care unit. JAMA 235:733–735, 1976.

31. Oosterhuis, HJGH, Hootsman, WJM, Veenhuyzen, HB, and Van Zadelhoff, I: The mean duration of motor unit action potentials in patients with myasthenia gravis. Electroencephalogr Clin Neurophysiol 32:697–700, 1972.

32. Ozdemir, C, and Young, RR: The results to be expected from electrical testing in the diagnosis of myasthenia gravis. Ann NY Acad Sci 274:203–235, 1976.

33. Pinelli, P: The effect of anticholinesterases on motor unit potentials in myasthenia gravis. Muscle Nerve 1:438–441, 1978.

34. Pittinger, C, and Adamson, R: Antibiotic blockade of neuromuscular function. Ann Rev Pharmacol 12:169–184, 1972.

35. Retzlaff, JA, Kearns, TP, Howard, FM, Jr, and Cronin, ML: Lancaster red-green test in evaluation of edrophonium effect in myasthenia gravis. Am J Ophthalmol 67:13–21, 1969.

36. Rowland, LP, Aranow, H, Jr, and Hoefer, PFA: Observations on the curare test in the differential diagnosis of myasthenia gravis. In Viets, HR (ed): Myasthenia Gravis, The Second International Symposium Proceedings. Charles C Thomas, Springfield, Ill, 1961, pp 411–434.

37. Sears, ML, Walsh, FB, and Teasdall, RD: The electromyogram from ocular muscles in myasthenia gravis. Arch Ophthalmol 63:791–798, 1960.

38. Spector, RH, and Daroff, RB: Edrophonium infrared optokinetic nystagmography in the diagnosis of myasthenia gravis. Ann NY Acad Sci 274:642–651, 1976.

39. Spector, RH, Daroff, RB, and Birkett, JE: Edrophonium infrared optokinetic nystagmography in the diagnosis of myasthenia gravis. Neurology 25:317–321, 1975.

40. Stalberg, E, Ekstedt, J, and Broman, A: Neuromuscular transmission in myasthenia gravis studied with single fibre electromyography. J Neurol Neurosurg Psychiatry 37:540–547, 1974.

41. Stalberg, E, Trontelj, JV, and Schwartz, MS: Single-muscle-fiber recording of the jitter phenomenon in patients with myasthenia gravis and in members of their families. Ann NY Acad Sci 274:189–202, 1976.

42. Warren, WR, Gutmann, L, Cody, RC, Flowers, P, and Segat, AT: Stapedius reflex decay in myasthenia gravis. Arch Neurol 34:496–497, 1977.

43. Wray, SH, and Pavan-Langston, D: A reevaluation of edrophonium chloride (Tensilon) tonography in the diagnosis of myasthenia gravis: With observations on some other defects of neuromuscular transmission. Neurology 21:586–593, 1971.

Part IV

ELECTROMYOGRAPHY

Chapter 11

ANATOMY AND PHYSIOLOGY OF THE SKELETAL MUSCLE

1 INTRODUCTION

The skeletal muscles comprise the extrafusal and intrafusal fibers, which are anatomically and physiologically distinct. The alpha motor neurons innervate the extrafusal fibers that occupy the bulk of muscle mass as contractile elements. The gamma motor neurons subserve the stretch-sensitive intrafusal fibers that constitute the muscle spindles found in parallel with the extrafusal fibers. The Golgi tendon organs, aligned in series with the tendon of the extrafusal fibers, also respond to stretch. The spindles and Golgi tendon organ continuously monitor and regulate the tonus of the reflexive or volitional muscle contraction. The motor unit, the smallest contractile element, consists of a single motor neuron and all the muscle fibers innervated by its axon.

A nerve impulse initiates muscle contraction in two distinct steps: neuromuscular transmission and electromechanical coupling (see Chapter 8.3 and 8.4). Acetylcholine (ACh), released at the neuromuscular junction, depolarizes the end-plate region and generates an action potential, which then propagates along the muscle membrane. As the impulse reaches the triad, ionized calcium (Ca^{2+}) is released into the sarcoplasm in response to depolarization of the transverse tubules. The interaction between calcium and the thin filaments triggers electromechanical coupling, leading to the formation of bridges between the thin and thick filaments. The sliding of thin filaments between the thick filaments results in a shortening of the muscle fibers.

This section will present an anatomic description of the contractile elements, the mechanism underlying the shortening of the muscle fibers, and the anatomy and physiology of motor units.

2   FUNCTIONAL ANATOMY

Gross Anatomy of Muscle

A connective tissue called epimysium covers the surface of each muscle. Inside this sheath are many fascicles bound by the coarse sleeves of the connective tissue perimysium (Fig. 11–1). Individual fascicles contain many muscle fibers, each surrounded by a delicate network of fine connective tissue, called endomysium. A muscle fiber is the smallest anatomic unit capable of contraction, averaging in diameter 10 μm in a newborn and 50 μm in an adult.[18] Individual muscle fibers range from 2 to 12 cm in length, some extending the entire length of the muscle and others only through a short segment of the total length.

The sarcolemma on the surface membrane of a muscle fiber contains multiple nuclei distributed beneath the thin sheath. The membrane has functional properties of excitability and conductivity similar to those of an axon. A myoelectric signal, originating from a neuromuscular junction, propagates in both the proximal

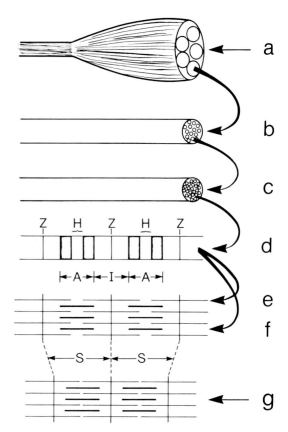

Figure 11–1. Anatomic composition of the skeletal muscle. The epimysium surrounds the entire muscle (*a*), which consists of many fascicles bound by perimysium (*b*). Individual muscle fibers (*c*) in the fascicle are covered by endomysium. Each muscle fiber contains many bundles of myofibrils (*d*), which in turn consist of thin (*e*) and thick (*f*) myofilaments. Thin actin filaments slide past thick myosin filaments during muscle contraction (*d*).

and distal directions.[75] The muscle fibers conduct considerably more slowly than nerve axons, with an estimated rate of 3 to 5 m/s.[18,47,64,97,109,121,128]

Myofibrils and Myofilaments

The semifluid intracellular content of a muscle fiber, called sarcoplasm, contains many bundles of cylindrical myofibrils. They appear as a thin, threadlike substance with light and dark bands of striations under the light microscope. Myofibrils consist of two types of myofilaments, which represent the basic substrates for the contraction of muscle fibers. The transverse striations seen by light microscopy result from their specific arrangements. The structural subunit, called the sarcomere, extends between two adjacent Z lines. The center of the sarcomere contains the longitudinally oriented thick myosin myofilaments. The thin actin filaments extend from either side of the Z line into the two adjacent sarcomeres to interdigitate with the myosin filaments.

The thick filaments consist of only myosin molecules, which form parallel elongated rods. The thin filaments contain not only actin molecules but also two other proteins, troponin and tropomyosin, both intimately bound to actin molecules. Globular-shaped troponins are attached to each end of the elongated tropomyosin molecule and several actin molecules are found along its interwoven course (Fig. 11–2). During muscle fiber contraction, actin filaments slide relative to the myosin filaments. This brings the adjacent Z lines closer together, shortening the sarcomere, rather than individual filament.[21]

Mechanism of Contraction

The mechanism of the sliding begins with the formation of calcium (Ca^{2+})-dependent bridges that link the actin and myosin filaments. At rest, tropomyosin physically blocks the formation of bridges between myosin and actin. The propagation of the action potential into the sarcoplasmic reticulum via the transverse tubules releases calcium from the terminal cistern of the longitudinal tubules. The free calcium binds to troponin, the only calcium-receptive protein in the contractile system. This interaction shifts the position of tropomyosin relative to the actin molecule, allowing the globular heads of myosin to gain access to the actin molecules. Myosin-actin cross-bridges pull the actin filaments past the myosin filaments. The tension develops in proportion to the number of cross-bridges formed by this chemical interaction. The dissociation of actin and myosin by adenosine triphosphate (ATP) shears old bridges to allow further sliding with new bridges.

Without a sustained muscle action potential, ATP-dependent active transport sequesters calcium into the sarcoplasmic reticulum. The removal of calcium from troponin allows tropomyosin to return to the resting position, and the muscle relaxes.[46,125] In McArdle's disease, characterized by deficiency of muscle phosphorylase, this initial step of muscle relaxation does not occur, presumably because of an insufficient amount of ATP. Failure of relaxation results in persistent shortening of the muscle in the absence of ongoing muscle action potentials. This condition, called contracture, typically develops when patients exercise under ischemic conditions (Fig. 11–3).

3 TYPES OF MUSCLE FIBERS

The subdivision of muscle fibers depends on their histologic and physiologic profiles. Important determining factors include enzymatic properties demonstrated by histochemical reactions; rate of rise in twitch tension, regulating the

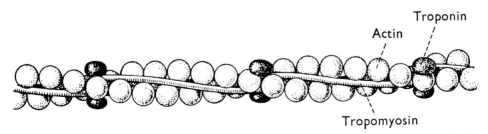

Figure 11–2. Fine structure of the thin actin filament with actin molecules attached to globular-shaped troponin and rod-shaped tropomyosin in an orderly arrangement. (From Ebashi et al.,[46] with permission.)

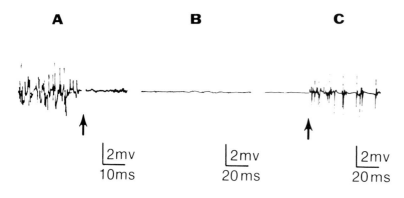

A   **B**   **C**

$\begin{array}{c}2mv\\10ms\end{array}$ $\begin{array}{c}2mv\\20ms\end{array}$ $\begin{array}{c}2mv\\20ms\end{array}$

Figure 11–3. Contracture during ischemic exercise in a 66-year-old man with McArdle's disease. The patient exercised the forearm flexors with an inflated pressure tourniquet placed around the arm and a concentric needle electrode inserted in the flexor digitorum profundus. Contracture began 45 seconds after the start of ischemic exercise (*arrow* in **A**), and persisted (**B**). Electrical activity returned 15 minutes after the release of the cuff (*arrow* in **C**). (Courtesy of E. Peter Bosch, M.D., Department of Neurology, University of Iowa Hospitals and Clinics.)

speed of contraction; degree of fatigability; and the nature of motor innervation. Table 11–1 summarizes the commonly used classification of muscle fibers into type I and type II, according to histochemical reactions[14,45,51]; slow (S), fast resistant (FR), and fast fatiguing (FF), based on twitch and fatigue characteristics[30]; or slow oxidative (SO), fast oxidative glycolytic (FOG), and fast glycolytic (FG), by twitch and enzymatic properties.[102]

Type I and Type II Fibers

Histochemical reactions (Fig. 11–4) reveal two types of human muscle fibers. Type I fibers react strongly to oxidative enzymes such as nicotinamide adenine dinucleotide dehydrogenase (NADH) and reduced diphosphopyridine nucleotide (DPNH) and weakly to both phosphorylase and myofibrillar adenosine triphosphatase (ATPase). Type II fibers show the reverse reactivity.[45] Three subtypes of type II fibers, IIA, IIB, and IIC, emerge according to their ATPase reactions (see Table 11–1) after preincubation at different pH values.[13,14,51] Type IIC fibers constitute fetal precursor cells, rarely seen in adult muscles.

The myosine ATPase content dictates the speed of contraction,[4] which forms the basis for physiologic subdivision of muscle fibers. Thus, in general, physio-

Table 11–1 TYPES OF MUSCLE FIBERS

Commonly used designations			
Fiber types[14]	Type I	Type II A	Type II B
Twitch and fatigue characteristics[30]	Slow (S)	Fast resistant (FR)	Fast fatigue (FF)
Twitch and enzymatic properties[102]	Slow oxidative (SO)	Fast oxidative-glycolytic (FOG)	Fast glycolytic (FG)
Properties of muscle fibers			
Resistance to fatigue	High	High	Low
Oxidative enzymes	High	High	Low
Phosphorylase (glycolytic)	Low	High	High
Adenosine triphosphate	Low	High	High
Twitch speed	Low	High	High
Twitch tension	Low	High	High
Characteristics of motor units			
Size of cell body	Small	Large	Large
Size of motor unit	Small	Large	Large
Diameter of axons	Small	Large	Large
Conduction velocity	Low	High	High
Threshold for recruitment	Low	High	High
Firing frequency	Low	High	High
Frequency of miniature end-plate potentials	Low	High	High

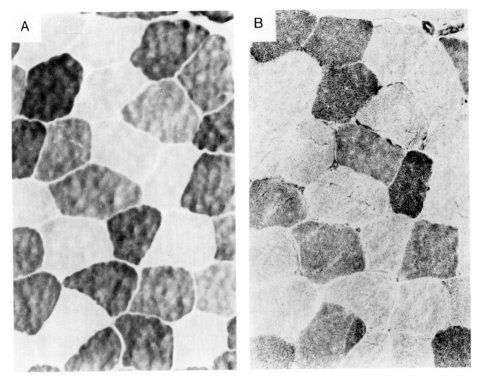

Figure 11–4. Cross-section of a normal skeletal muscle stained with ATPase at pH 9.4 in **A**, and with nicotinamide adenine dinucleotide dehydrogenase (NADH) in **B**. The darker fibers represent type II in **A** and type I in **B**. (Courtesy of Linda Ansbacher, M.D., and Michael N. Hart, M.D., Department of Pathology, University of Iowa Hospitals and Clinics.)

logic data correlate the slow twitch fibers to histochemical type I, and fast twitch fibers, to type II.[31,72] This association between histochemical types and physiologic characteristics of muscle, however, is not always predictable. For example, histochemically mixed extensor digitorum longus of the rat contains only fast fibers,[35] slow soleus muscle of cats shows greater myosin ATPase activity than fast gastrocnemius muscle. Therefore, the intensity of histochemical ATPase reaction cannot serve as the sole criterion in distinguishing fast and slow twitch fibers.[32]

Fast and Slow Twitch Fibers

Muscle fibers differ in their contraction time, force-velocity curves, and rates of decay.[102] Slow fibers (S) with high oxidative properties (SO) resist fatigue. Fast resistant (FR) fibers with high oxidative and glycolytic properties (FOG) also resist fatigue, whereas the fast fatigue (FF) fibers with high glycolytic activity but low oxidative enzyme (FG) fatigue easily.[30] These findings suggest that glycolytic capacity generally relates to twitch characteristics, and oxidative capability dictates fatigability. Intracellular recordings have shown that, compared with slow fibers, fast glycolytic fibers have greater resting membrane potential, a larger amplitude of the action potential, higher maximum rates of depolarization and repolarization, and a more variable shape of the repolarization phase.[122]

Fast and Slow Muscles

In animals, most muscles consist mainly of one muscle fiber type. Slow muscles appear deeper red in color, reflecting a higher myoglobin content, whereas fast muscles tend to show a whitish hue. Functionally, slow muscles have a tonic postural role, like that of the soleus in the cat: whereas fast muscles provide willed phasic movements, like those of the wing muscles of a chicken. This distinction, however, blurs in man because most human limb muscles consist of slow and fast twitch motor units in various combinations.[26] For example, the slow fibers with contraction times longer than 60 ms constitute a majority in triceps surae, one half in tibialis anterior, one third in biceps brachii, and a small percentage in triceps brachii.[23] Slow oxidative fibers occupy 38 and 44 percent of superficial and deep areas in the vastus lateralis and 47 and 61 percent in the vastus medialis.[66] Further, fibers of the same types do not necessarily share the same contractile speed in different muscles.[106]

Effect of Innervation

After the transplantation of the nerve normally innervating a fast fiber to a slow fiber, the originally slow muscle fiber will acquire the properties of a fast muscle fiber.[27,67,104,129] Such a relationship between the type of innervation and muscle activity also determines the mechanical characteristics of contraction in some fibers[24] but not others.[123] The driving forces for this regulation, though not yet elucidated, probably include the discharge pattern of the motor neuron and the axoplasmic transplantation of trophic substances from the nerve to the muscle. For example, motor neurons innervating fast twitch muscles have shorter afterhyperpolarization than those supplying slow twitch muscles.[48]

In animal experiments, the rate of stimulation dictates the contractile characteristics of muscle fibers.[74,103] Athletes engaged in endurance training have a greater number of slow fibers,[55] whereas weight lifters have more fast fibers.[119] Brachial plexus palsy at birth alters isometric contraction time and half relaxation time of the affected muscles.[114] The finding suggests that denervation during infancy impairs normal development of muscle contractile properties. In patients with chronic neuromuscular diseases,

normal muscle fiber histochemistry persists as long as motor neuron differentiation remains. In patients with long-term spastic hemiplegia, some motor units show greater fatigability and longer twitch contraction times than normal. Thus, the dynamic properties of the muscle seem to change even in upper motor neuron lesions.[130]

Alterations in histochemical properties reflect the firing pattern and axonal conduction velocity of the motor neurons.[8] Exercise training alone, however, induces little change in basic muscle contractility in man.[2,55,119] Hence, motor neuron activity does not suffice in itself to alter the distribution of fast and slow fibers in a muscle. The findings in favor of additional neurotrophic influences[58] include effects of neurons on muscle in tissue cultures[131] and the inverse relationship of nerve length on the time interval before the development of muscle membrane changes after nerve section.[33]

4 STRETCH-SENSITIVE RECEPTORS

Anatomy of Muscle Spindles

Muscle spindles consist of small specialized muscle fibers encapsulated by connective tissue. The intrafusal fibers measure only 4 to 10 mm in length and 0.2 to 0.35 mm in diameter, in contrast to the much larger extrafusal fibers of striated muscle.[68] The connective tissue capsule surrounding the intrafusal fiber joins the sarcolemma of the extrafusal fibers attached to the origin and insertion of the muscle. The muscle spindles lie parallel to the striated muscle fibers. The nuclear arrangement in their equatorial region distinguishes two types of intrafusal fibers, nuclear bag and nuclear chain (Fig. 11–5). Both dynamic and static bag fibers expand near their midpoint over a short length of about 100 μm by a collection of some 50 nuclei. The smaller nuclear chain fibers contain a linear array of nuclei along the center of the fiber.

The afferent and efferent nerves that supply muscle spindles each have two

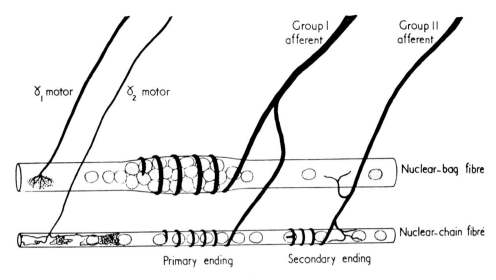

Figure 11–5. Simplified diagram of the central region (about 1 mm) of the nuclear bag fiber (*top*) and nuclear chain fiber (*bottom*) showing two types of motor endings, two types of afferent fibers, and two types of gamma motor neurons. (From Matthews,[76] with permission.)

different kinds of endings: primary (annulospiral) and secondary (flower-spray) sensory endings; and plate (single, discrete) and trail (multiple, diffuse) motor endings. The primary sensory ending spirals around the center of the bag and chain fibers. In contrast, the secondary ending terminates more peripherally and chiefly on nuclear chain fibers. The large diameter fast conducting group IA afferent nerve fibers from the primary endings subserve the monosynaptic stretch reflex. In contrast, the secondary ending gives rise to group II afferent nerve fibers that terminate on the interneurons in the spinal cord. Although both kinds of motor endings can innervate either type of intrafusal fibers, the plate endings tend to supply preferentially the nuclear bag and the trail endings, the chain fibers.

Function of Muscle Spindles

The dynamic afferent fibers respond to the velocity of the actively stretching spindles. The static afferent fibers detect a sustained change in the length. The primary ending has both dynamic and static function, but the secondary ending mainly mediates static changes (Fig. 11–6). The dynamic and static axons of the fusimotor system influence the dynamic and static muscle spindle.[25,76] The trail endings mediate static changes, whereas the plate endings primarily control dynamic changes.[78] The bag and chain fibers receive a sufficiently distinctive motor innervation to subserve preferentially, dynamic and static fusimotor effects.[9,10] Muscle spindles using the contraction-dependent discharge pattern monitor activity of motor units in the vicinity.[85] Receptor feedback, however, has a negligible effect on the motor neuron pool, compared with the excitatory drive during voluntary contraction.[86]

Table 11–2 provides a simplified summary of sensory endings found in muscle spindles. The basic structural elements comprise two types of intrafusal fibers, nuclear bag and nuclear chain; two types of sensory receptor endings, primary and secondary, giving rise to group IA and group II afferent fibers; and two types of fusimotor endings, plate and trail, which preferentially subserve dynamic and static function. The dynamic bag fibers, receive innervation from the primary sensory endings and the fusimotor fibers

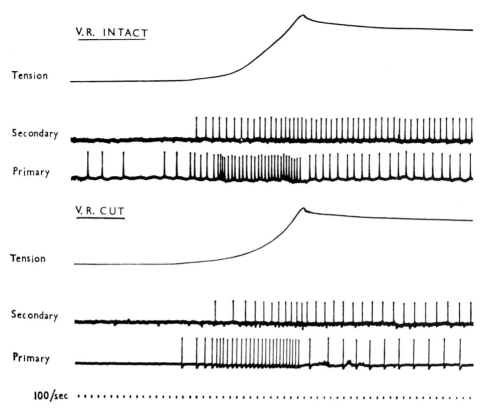

Figure 11–6. Responses of primary and secondary endings to a rapidly applied stretch before (*top*) and after (*bottom*) cutting the ventral root. Spontaneous fusimotor discharge maintained a steady intrafusal contraction in the decerebrate cat. The primary endings show a greater sensitivity to stretch than the secondary endings, but both types respond equally to changes in muscle length. (From Matthews,[77] with permission.)

with plate endings, and modulate dynamic function. The static bag fibers and chain fibers, innervated mainly by fusimotor fibers with trail endings, give rise to both types of sensory afferents and regulate static muscle length. Muscle receptors play a role in proprioception, as evidenced by sensory effects of pulling or vibrating exposed tendons in man,[79] although cutaneous afferents may also provide the dominant input.[95]

Golgi Tendon Organ

The Golgi tendon organ, arranged in series with extrafusal striated muscle fibers, monitors not only active muscle contraction but also passive stretch. The group IB afferent fibers originating herein subserve disynaptic inhibition of the motor neurons that innervate the stretched muscle. According to the traditional view, this inhibitory mechanism

Table 11–2 SENSORY ENDINGS OF MUSCLE SPINDLES

	Primary Sensory Ending	Secondary Sensory Ending
Location	Both bag and chain fibers	Mainly chain fibers
Sensitivity	Both length and velocity	Mainly length
Fusimotor system	Both dynamic and static	Mainly static
Form of ending	Half rings and annulospirals	Spirals and flower sprays
Length of ending	About 300 μm	About 400 μm
Type of afferent fiber	Group IA	Group II
Diameter of afferent fiber	12–20 μm	6–12 μm

provides a safety function to prevent an excessive muscle tension when motor neuron firing reaches a certain level. It is now known that the threshold tension much less than previously believed excites the tendon organ, especially during active stretch.[65] The activation of group IB afferent fibers during mild tension helps continuously monitor and adjust the magnitude of muscle activity for smooth contraction even at a low level of tension.

5 ANATOMY OF THE MOTOR UNIT

As defined by Liddell and Sherrington,[73] the motor unit consists of a motor neuron and the few hundred muscle fibers that it supplies. A single discharge of a motor neuron gives rise to synchronous contraction of all muscle fibers innervated by the axon. Hence, even though individual muscle fibers represent the anatomic substrate, the motor unit constitutes the smallest functional element of contraction.[108]

Motor Unit Count

Progressively stronger electrical shocks applied to a nerve result in stepwise increase in amplitude of the evoked compound muscle action potential. If each addition represents the recruitment of one motor unit, the ratio between the full response elicited by maximal stimulation and the average increment gives an estimate of the motor unit count.[15,38,39,84,107] The technique provides a valid count only if each increment in amplitude in fact corresponds to a single motor unit and those initially recruited truly represent an average size. These assumptions may not always hold, because the thresholds of two or more axons frequently overlap. A number of motor units may then discharge simultaneously with stimulation at a threshold intensity. Moreover, the smaller motor units, which are initially activated according to the size principle, contribute less than average evoked po-

tential. Thus, even modified techniques using computer-assisted analysis tend to underestimate the average size of motor units.[16] Despite such shortcomings, the method offers the only noninvasive means currently available for motor unit count.

The mean number of motor units calculated by this technique ranges from 200 for extensor digitorum brevis[3,81] to 250 or 340 for the thenar muscles.[15,80] Clinical studies have shown subnormal numbers of functioning motor units in several disorders, including Duchenne muscular dystrophy.[82,83]

Innervation Ratio

The innervation ratio relates to the average size of a motor unit expressed as a ratio between the total number of extrafusal fibers and the number of innervating motor axons. Depending on the type of muscle, the ratio ranges from 3:1 in extrinsic eye muscles, which require fine gradations of movement, to 30:1 to 120:1 in some limb muscles subserving only coarse movement.[117] Table 11–3 summarizes the results of one study.[52] Table 11–4 shows the territory of motor units estimated histologically[36] or electrically.[17]

Distribution of Muscle Fibers

Muscle fibers of a given motor unit have identical histologic characteristics. Therefore, the apparent random distribution of different histologic fiber types seen in muscle cross-sections indicates considerable overlap in the territories of adjacent motor units.[66] Single fiber electromyography[50,112,113] and electrophysiologic cross-section analysis[111] have also demonstrated the scattering of muscle fibers belonging to a given motor unit.

Another mapping technique has also substantiated motor unit overlap.[12,44,49] Repetitive stimulation of an isolated single ventral root nerve fiber exhausts glycogen storage in all the muscle fibers belonging to the motor unit of the stimulated axon. The muscle excised im-

Table 11–3 SUMMARY OF INNERVATION STUDY

Material	Muscle	Number of Large Nerve Fibers	Number of Muscle Fibers	Calculated Number of Motor Units	Mean Number of Fibers per Motor Unit	Mean Diameter of Muscle Fibers (μm)	Cross-sectional Area of Motor Units (mm³)
♂22	Platysma	1,826	27,100	1,096	25	20	0.008
♂40	Brachio-radialis	Right 525	>129,200	315	>410	34	
		Left 584		350			
♂22	First dorsal interosseous	199	40,500	119	340	26	0.18
♂54	First lumbrical	155	10,038	93	108	19	0.031
♀29		164	10,500	98	107	21	0.037
♂40	Anterior tibial	742	250,200	445	562		
♂22			292,500		657	57	1.7
♂28	Gastrocnemius medial head	965	1,120,000	579	1,934		
♂22			946,000		1,634	54	3.4

From Feinstein et al.,[52] with permission.

mediately after tetanic stimulation and stained for glycogen in a frozen section shows a scattered distribution of unstained muscle fibers. This method not only confirms the territorial overlapping of adjacent motor units but also the histochemical uniformity of a given motor unit.

Indeed, a muscle fiber of a single motor unit rarely makes direct contact with other fibers of the same unit. One study even refutes a random arrangement of mammalian muscle fibers but argues for more orderly disposition at certain stages of development to minimize adjacencies of individual muscle fibers of the same motor unit.[126] Such specification may have the functional advantage of maximizing muscle action potential dispersal for smooth muscle contraction and in compensating for lost motor units.

Histologic findings in partially denervated muscle once prompted some investigators[19,20] to propose that the fibers of each motor unit might consist of many subunits, each containing an average of 10 to 30 fibers. According to this theory, the motor unit potential recorded during routine electromyography results from completely synchronized firing of all

Table 11–4 MEAN VALUES OF MOTOR UNIT TERRITORY AND MAXIMUM VOLTAGE IN NORMAL MUSCLES

Muscle	Number of Muscles	Number of Motor Units	Spike Level (μV)	Territory at Spike Level (mm)	Standard Deviation (mm)	Maximum Voltage (μV)	Standard Deviation (μV)
Biceps brachii	24	129	100	5.1 ± 0.2	2.4	370 ± 17	190
Deltoid	7	52	100	6.7 ± 0.4	3.0	450 ± 27	190
Extensor digitorum communis	11	43	100	5.5 ± 0.3	2.1	800 ± 59	390
Opponens pollicis	10	34	150	7.4 ± 0.4	2.6	1,000 ± 83	500
Rectus femoris	9	65	100	10.0 ± 0.6	4.6	550 ± 38	300
Biceps femoris	5	35	150	8.8 ± 0.7	4.1	900 ± 67	400
Tibialis anterior	8	47	100	7.0 ± 0.4	3.0	620 ± 43	300
Extensor digitorum brevis	5	25	200	11.3 ± 0.8	4.1	3,000 ± 300	1,500

From Buchthal et al.,[17] with permission.

fibers belonging to a subunit. Electrophysiologic studies in rat phrenic-hemi-diaphragm preparation[71] and in rat peroneus longus muscle,[98] however, failed to substantiate this concept. Histochemical studies showed no groupings of fibers within the motor unit in rat or cat muscle.[11,44,49] Human studies with the single-fiber needle revealed no evidence of muscle fiber grouping within a motor unit in normal extensor digitorum communis or biceps brachii muscles.[110] Moreover, high amplitude spikes do not necessarily imply a synchronized discharge from a subunit, because a single muscle fiber can give rise to such a potential.[50,105] These findings have led most electromyographers to abandon the concept of the subunit in normal human muscle.[22]

6 PHYSIOLOGY OF THE MOTOR UNIT

Table 11–1 summarizes types of muscle fibers, described earlier in this chapter. The same criteria apply to the classification of motor units, because all the muscle fibers of a given motor unit have identical histologic and physiologic properties. The animal and human data briefly reviewed below pertain to the understanding of motor unit potentials in clinical electromyography.

Animal Experiments

Series of animal experiments have clearly established the close relationship between the fundamental physiologic properties of motor units and the size of the motor neuron (see Table 11–1). The large motor neurons have fast conducting axons of large diameter[69,87] and a higher innervation ratio, i.e., a greater number of muscle fibers supplied by one axon.[61,88,127] Larger motor units have, in turn, greater twitch tensions, faster twitch contractions, and a greater tendency to fatigue.[29,62,63] According to the size principle of Henneman, the motor neurons recruit not at random but in an orderly manner determined by the fixed central drive that preferentially activates small motor neurons first.[61,69,70]

In brief, the larger the cell body, the greater the conduction velocity, stronger the twitch tension, faster the twitch contraction, and, in general, greater the tendency to fatigue. Smaller motor neurons, innervating smaller motor units, discharge initially with minimal effort, before a greater effort of contraction activates larger motor neurons.

Recruitment

Most findings in animal studies also apply to man (see Table 11–1). In the first dorsal interosseous muscle, the motor units activated early at low threshold have lower twitch tensions and slower twitch contractions, compared with those units recruited at higher levels of effort.[28,92–94] Factors correlated with motor neuron excitability include axon diameter,[54] conduction velocity,[7,53,60] and motor unit size.[92–94] High and low threshold motor units also differ histochemically.[124] Earlier studies hinted the distinction between tonic and phasic motor units on the basis of their firing pattern and the order of recruitment.[120] More recent studies, however, have shown a relatively continuous rather than bimodal pattern of recruitment.[53,59,93,94,99]

Despite certain exceptions documented under some experimental circumstances,[5,56,57] the size principle generally applies to any voluntary activation of motor units, including rapid ramp or ballistic contractions.[40,42,100,101] In one study,[41] the same rule governed the order of presynaptic inhibition after activation of group IA afferent fibers by tonic vibration. Neuropathy or motor neuron disease does not alter the size principle, but a previously transected peripheral nerve may show a random pattern of recruitment.[91] Misdirection of motor axons account for the absence of orderly recruitment following complete ulnar or above-elbow median nerve sections. The size principle of recruitment can be reestablished after section in humans, if motor axons reinnervate their original muscles or ones with closely synergistic function, as seen after complete median nerve section at the wrist.[118]

Twitch Characteristics

Different human muscles contain either fast or slow units whose twitch contraction approximates the contraction time of the whole muscle.[107] An averaging technique, using repetitive discharges from a single muscle fiber as a trigger, can provide a selective summation of the muscle twitch attributable to that motor unit.[92–94,96,115] Twitch tensions analyzed by this means range from 0.1 to 1.0 g, with contraction times varying between 20 and 100 ms.

As in animals, the twitch tension generated by a motor unit in man increases in proportion to its action potential amplitude and the voluntary force required for its activation. The units recruited with slight contraction have smaller twitch tensions, slower contraction times, and greater resistance to fatigue, compared with the units that appear with stronger contraction.[115] Partially denervated muscles generally have a prolonged contraction time and reduced twitch tension.[89] It is not clear, however, whether the twitch tension of individual motor units becomes larger[84] or smaller[91] after denervation.

Rate Coding

The muscle force is augmented either by recruitment of previously inactive motor neurons or through more rapid firing of already active units. In early studies, discharge frequency appeared to stabilize over a wide range of forces, although the firing rate ranged from several to 30 impulses per second during early phases of voluntary contraction.[6,34,37,116] These findings suggested to some that rate coding mostly regulated fine control at the beginning of contraction and during maximal effort. More recent work in humans,[94,96] however, has emphasized the importance of rate coding for increasing force, as originally suggested by Adrian and Bronk.[1]

Recruitment must play an important role at low levels of contraction, when all units fire at about the same rate, ranging from 5 to 15 impulses per second.[40,53,94]

After the activation of most units, additional increases in force must result from faster firing of individual motor units. In strong or ballistic contractions, instantaneous firing may reach 60 to 120 impulses per second at the onset.[40] To maintain the same twitch tensions, muscle fibers tend to fire at a higher rate in myopathy and at a lower rate in neuropathy, compared with controls,[43] although only severely weak muscles show a significant difference.[90]

REFERENCES

 1. Adrian, ED, and Bronk, DW: The discharge of impulses in motor nerve fibers. Part II. The frequency of discharge in reflex and voluntary contractions. J Physiol (Lond) 67:119–151, 1929.
 2. Andersen, P, and Henriksson, J: Capillary supply of the quadriceps femoris muscle of man: Adaptive response to exercise. J Physiol (Lond) 270:677–690, 1977.
 3. Ballantyne, JP, and Hansen, S: New method for the estimation of the number of motor units in a muscle. 2. Duchenne, limb-girdle and facioscapulohumeral, and myotonic muscular dystrophies. J Neurol Neurosurg Psychiatry 37:1195–1201, 1974.
 4. Barany, M, and Close, RI: The transformation of myosin in cross-innervated rat muscles. J Physiol (Lond) 213:455–474, 1971.
 5. Basmajian, JV: Control and training of individual motor units. Science 141:440–441, 1963.
 6. Bigland, B, and Lippold, OCJ: Motor unit activity in the voluntary contraction of human muscle. J Physiol (Lond) 125:322–335, 1954.
 7. Borg, J, Grimby, L, and Hannerz, J: Axonal conduction velocity and voluntary discharge properties of individual short toe extensor motor units in man. J Physiol (Lond) 277:143–152, 1978.
 8. Borg, J, Grimby, L, and Hannerz, J: Motor neuron firing range, axonal conduction velocity, and muscle fiber histochemistry in neuromuscular diseases. Muscle Nerve 2:423–430, 1979.
 9. Boyd, IA: The structure and innervation of the nuclear bag muscle fibre system and the nuclear chain muscle fibre system in mammalian muscle spindles. Philos Trans R Soc Lond 245:81–136, 1962.
10. Boyd, IA: Muscle spindles and stretch reflexes. In Swash, M, and Kennard, C (eds): The Scientific Basis of Clinical Neurology, Churchill Livingstone, London, 1985.
11. Brandstater, ME, and Lambert, EH: A histological study of the spatial arrangement of muscle fibers in single motor units within rat tibialis anterior muscle (abstr). Bulletin of the

American Association of Electromyography and Electrodiagnosis 16:82, 1969.

12. Brandstater, ME, and Lambert, EH: Motor unit anatomy. In Desmedt, JE (ed): New Developments in Electromyography and Clinical Neurophysiology, Vol 1. Karger, Basel, 1973, pp 14–22.

13. Brooke, MH, and Kaiser, KK: The use and abuse of muscle histochemistry. Ann NY Acad Sci 228:121–144, 1974.

14. Brooke, MH, Williamson, E, and Kaiser, KK: The behavior of four fiber types in developing and reinnervated muscle. Arch Neurol 25:360–366, 1971.

15. Brown, WF: A method for estimating the number of motor units in thenar muscles and the changes in motor unit count with ageing. J Neurol Neurosurg Psychiatry 35:845–852, 1972.

16. Brown, WF, and Milner-Brown, HS: Some electrical properties of motor units and their effects on the methods of estimating motor unit numbers. J Neurol Neurosurg Psychiatry 39:249–257, 1976.

17. Buchthal, F, Erminio, F, and Rosenfalck, P: Motor unit territory in different human muscles. Acta Physiol Scand 45:72–87, 1959.

18. Buchthal, F, Guld, C, and Rosenfalck, P: Propagation velocity in electrically activated muscle fibres in man. Acta Physiol Scand 34:75–89, 1955.

19. Buchthal, F, Guld, C, and Rosenfalck, P: Volume conduction of the spike of the motor unit potential investigated with a new type of multielectrode. Acta Physiol Scand 38:331–354, 1957.

20. Buchthal, F, Guld, C, and Rosenfalck, P: Multielectrode study of the territory of a motor unit. Acta Physiol Scand 39:83–104, 1957.

21. Buchthal, F, and Kaiser, E: The rheology of the cross striated muscle fibre with particular reference to isotonic conditions. Dan Biol Med 21:1–318, 1951.

22. Buchthal, F, and Rosenfalck, P: On the structure of motor units. In Desmedt, JE (ed): New Developments in Electromyography and Clinical Neurophysiology, Vol 1. Karger, Basel, 1973, pp 71–85.

23. Buchthal, F, and Schmalbruch, H: Contraction time and fibre types in intact human muscle. Acta Physiol Scand 79:435–452, 1970.

24. Buchthal, F, Schmalbruch, H, and Kamieniecka, Z: Contraction times and fiber types in neurogenic paresis. Neurology 21:58–67, 1971.

25. Buller, AJ: The motor unit in reflex action. In Creese, R (ed): Recent Advances in Physiology, ed 8. J & A Churchill, London, 1963.

26. Buller, AJ: The physiology of the motor unit. In Walton, JN (ed): Disorders of Voluntary Muscle, ed 3. Churchill Livingstone, London, 1974.

27. Buller, AJ, Eccles, JC, and Eccles, RM: Interactions between motoneurons and muscles in respect of the characteristic speeds of their responses. J Physiol (Lond) 150:417–439, 1960.

28. Burke, D, Skuse, NF, and Lethlean, AK: Isometric contraction of the abductor digiti minimi muscle in man. J Neurol Neurosurg Psychiatry 37:825–834, 1974.

29. Burke, RE, Levine, DN, Tsairis, P, and Zajac, FE, III: Physiological types and histochemical profiles in motor units of the cat gastrocnemius. J Physiol (Lond) 234:723–748, 1973.

30. Burke, RE, Levine, DN, Zajac, FE, III, Tsairis, P, and Engel, WK: Mammalian motor units: Physiological-histochemical correlation in three types in cat gastrocnemius. Science 174:709–712, 1971.

31. Burke, RE, and Tsairis, P: The correlation of physiological properties with histochemical characteristics in single muscle units. Ann NY Acad Sci 228:145–159, 1974.

32. Burke, RE, Tsairis, P, Levine, DN, Zajac, FE, III, and Engel WK: Direct correlation of physiological and histochemical characteristics in motor units of cat triceps surae muscle. In Desmedt, JE (ed): New Developments in Electromyography and Clinical Neurophysiology, Vol 1. Karger, Basel, 1973, pp 23–30.

33. Card, DJ: Denervation: Sequence of neuromuscular degenerative changes in rats and the effect of stimulation. Exp Neurol 54:251–265, 1977.

34. Clamann, HP: Activity of single motor units during isometric tension. Neurology 2:254–260, 1970.

35. Close, R: Properties of motor units in fast and slow skeletal muscles of the rat. J Physiol (Lond) 193:45–55, 1967.

36. Coers, C, and Woolf, AL: The Innervation of Muscle: A Biopsy Study. Charles C Thomas, Springfield, Ill, 1959.

37. Dasgupta, A, and Simpson, JA: Relation between firing frequency of motor units and muscle tension in the human. Electromyography 2:117–128, 1962.

38. Defaria, CR, and Toyonaga, K: Motor unit estimation in a muscle supplied by the radial nerve. J Neurol Neurosurg Psychiatry 41:794–797, 1978.

39. Delbeke, J: Reliability of the motor unit count in the facial muscles. Electromyogr Clin Neurophysiol 22:277–290, 1982.

40. Desmedt, JE, and Godaux, E: Fast motor units are not preferentially activated in rapid voluntary contractions in man. Nature 267:717–719, 1977.

41. Desmedt, JE, and Godaux, E: Mechanism of the vibration paradox: Excitatory and inhibitory effects of tendon vibration on single soleus muscle motor units in man. J Physiol (Lond) 285:197–207, 1978.

42. Desmedt, JE, and Godaux, E: Voluntary motor commands in human ballistic movements. Ann Neurol 5:415–421, 1979.

43. Dietz, V, Budingen, HJ, Hillesheimer, W, and Freund, HJ: Discharge characteristics of single motor fibres of hand muscles in lower motoneurone diseases and myopathies. In Kunze, K, and Desmedt, JE (ed): Studies on Neuro-

muscular Diseases. Proceedings of the International Symposium of the German Neurological Society. Karger, Basel, 1975, pp 122–127.

44. Doyle, AM, and Mayer, RF: Studies of the motor unit in the cat: A preliminary report. Bull School Med Univ Maryland 54:11–17, 1969.

45. Dubowitz, V: Histochemical aspects of muscle disease. In Walton, JN (ed): Disorders of Voluntary Muscle, ed 3. Churchill Livingstone, London, 1974.

46. Ebashi, S, Endo, M, and Ohtsuki, I: Control of muscle contraction. Q Rev Biophys 2:351–384, 1969.

47. Eberstein, A, and Goodgold, J: Slow and fast twitch fibers in human skeletal muscle. Am J Physiol 215:535–541, 1968.

48. Eccles, JC: Specificity or neural influence on speed of muscle contraction. In Gutmann, E, and Hink, P (ed): The Effect of Use and Disuse on Neuromuscular Functions. Elsevier, Amsterdam, 1963, pp 111–128.

49. Edstrom, L, and Kugelberg, E: Histochemical composition, distribution of fibres and fatiguability of single motor units. J Neurol Neurosurg Psychiatry 31:424–433, 1968.

50. Ekstedt, J: Human single muscle fiber action potentials. Acta Physiol Scand 61(Suppl 226):1–96, 1964.

51. Engel, WK: Selective and nonselective susceptibility of muscle fiber types. Arch Neurol 22:97–117, 1970.

52. Feinstein, B, Lindegard, B, Nyman, E, and Wohlfart, G: Morphologic studies of motor units in normal human muscles. Acta Anat (Basel) 23:127–142, 1955.

53. Freund, HJ, Budingen, HJ, and Dietz, V: Activity of single motor units from human forearm muscles during voluntary isometric contractions. J Neurophysiol 38:933–946, 1975.

54. Freund, HJ, Dietz, V, Wita, CW, and Kapp, H: Discharge characteristics of single motor units in normal subjects and patients with supraspinal motor disturbances. In Desmedt, JE (ed): New Developments in Electromyography and Clinical Neurophysiology, Vol 3. Karger, Basel, 1973, pp 242–250.

55. Gollnick, PD, Armstrong, RB, Saltin, B, Saubert, CW, IV, Sembrowich, WL, and Shepherd, RE: Effect of training on enzyme activity and fiber composition of human skeletal muscle. J Appl Physiol 34:107–111, 1973.

56. Grimby, L, and Hannerz, J: Recruitment order of motor units on voluntary contraction: Changes induced by proprioceptive afferent activity. J Neurol Neurosurg Psychiatry 31:565–573, 1968.

57. Grimby, L, and Hannerz, J: Disturbances in voluntary recruitment order of low and high frequency motor units on blockades of proprioceptive afferent activity. Acta Physiol Scand 96:207–216, 1976.

58. Gutmann, E: Considerations of neurotrophic relations in the central and peripheral nervous system. Acta Neurobiol Exp 35:841–851, 1975.

59. Hannerz, J: Discharge properties of motor units in relation to recruitment order in voluntary contraction. Acta Physiol Scand 91:374–384, 1974.

60. Hannerz, J, and Grimby, L: The afferent influence on the voluntary firing range of individual motor units in man. Muscle Nerve 2:414–422, 1979.

61. Henneman, E: Relation between size of neurons and their susceptibility to discharge. Science 126:1345–1347, 1957.

62. Henneman, E, Somjen, G, and Carpenter, DO: Functional significance of cell size in spinal motoneurons. J Neurophysiol 28:560–580, 1965.

63. Henneman, E, Somjen, G, and Carpenter, DO: Excitability and inhibitibility of motoneurons of different sizes. J Neurophysiol 28:599–620, 1965.

64. Hilfiker, P, and Meyer, M: Normal and myopathic propagation of surface motor unit action potentials. Electroencephalogr Clin Neurophysiol 57:21–31, 1984.

65. Houk, J, and Henneman, E: Responses of Golgi tendon organs to active contractions of the soleus muscle of the cat. J Neurophysiol 30:466–481, 1967.

66. Johnson, MA, Polgar, J, Weightman, D, and Appleton, D: Data on the distribution of fibre types in thirty-six human muscles: An autopsy study. J Neurol Sci 18:111–129, 1973.

67. Karpati, G, and Engel, WK: "Type grouping" in skeletal muscles after experimental reinnervation. Neurology 18:447–455, 1968.

68. Kennedy, WR: Innervation of normal human muscle spindles. Neurology 20:463–475, 1970.

69. Kernell, D: Input resistance, electrical excitability, and size of ventral horn cells in cat spinal cord. Science 152:1637–1640, 1966.

70. Kernell, D, and Sjoholm, H: Recruitment and firing rate modulation of motor unit tension in a small muscle of the cat's foot. Brain Res 98:57–72, 1975.

71. Krnjevic, K, and Miledi, R: Motor units in the rat diaphragm. J Physiol (Lond) 140:427–439, 1958.

72. Kugelberg, E: Properties of the rat hind-limb motor units. In Desmedt, JE (ed): New Developments in Electromyography and Clinical Neurophysiology, Vol 1. Karger, Basel, 1973, pp 2–13.

73. Liddell, EGT, and Sherrington, CS: Recruitment and some other features of reflex inhibition. Proc R Soc Lond [Biol] 97:488–518, 1925.

74. Lomo, T, Westgaard, RH, and Dahl, HA: Contractile properties of muscle: Control by pattern of muscle activity in the rat. Proc R Soc Lond [Biol] 187:99–103, 1974.

75. Masuda, T, Miyano, H, and Sadoyama, T: The propagation of motor unit action potential and the location of neuromuscular junction investigated by surface electrode arrays. Electroencephalogr Clin Neurophysiol 55:594–600, 1983.

76. Matthews, PBC: Muscle spindles and their motor control. Physiol Rev 44:219–288, 1964.

77. Matthews, PBC: Mammalian Muscle Receptors and Their Central Actions. Edward Arnold, London, 1972.

78. Matthews, PBC: The advances of the last decade of animal experimentation upon muscle spindles. In Desmedt, JE (ed): New Developments in Electromyography and Clinical Neurophysiology, Vol 3. Karger, Basel, 1973, pp 95–125.

79. McCloskey, DI, Cross, MJ, Honner, R, and Potter, EK: Sensory effects of pulling or vibrating exposed tendons in man. Brain 106:21–27, 1983.

80. McComas, AJ: Neuromuscular Function and Disorders. Butterworth, London, 1977.

81. McComas, AJ, Fawcett, PRW, Campbell, MJ, and Sics, REP: Electrophysiological estimation of the number of motor units within a human muscle. J Neurol Neurosurg Psychiatry 34:121–131, 1971.

82. McComas, AJ, Preswick, G, and Garner, S: The sick motoneurone hypothesis of muscular dystrophy. Prog Neurobiol 30:309–331, 1988.

83. McComas, AJ, Sica, REP, and Brandstater, ME: Further motor unit studies in Duchenne muscular dystrophy. J Neurosurg Psychiatry 40:1147–1151, 1977.

84. McComas, AJ, Sica, REP, Campbell, MJ, and Upton, ARM: Functional compensation in partially denervated muscles. J Neurol Neurosurg Psychiatry 34:453–460, 1971.

85. McKeon, B, and Burke, D: Muscle spindle discharge in response to contraction of single motor units. J Neurophysiol 49:291–302, 1983.

86. McKeon, B, Gandevia, S, and Burke, D: Absence of somatotopic projection of muscle afferents onto motoneurons of same muscle. J Neurophysiol 51:185–193, 1984.

87. McLeod, JG, and Wray, SH: Conduction velocity and fibre diameter of the median and ulnar nerves of the baboon. J Neurol Neurosurg Psychiatry 30:240–247, 1967.

88. McPhedran, AM, Wuerker, RB, and Henneman, E: Properties of motor units in a homogenous red muscle (soleus) of the cat. J Neurophysiol 28:71–84, 1965.

89. Miller, RG: Dynamic properties of partially denervated muscle. Ann Neurol 6:51–55, 1979.

90. Miller, RG, and Sherratt, M: Firing rates of human motor units in partially denervated muscle. Neurology 28:1241–1248, 1978.

91. Milner-Brown, HS, Stein, RB, and Lee, RG: Contractile and electrical properties of human motor units in neuropathies and motor neurone disease. J Neurol Neurosurg Psychiatry 37:670–676, 1974.

92. Milner-Brown, HS, Stein, RB, and Yemm, R: The contractile properties of human motor units during voluntary isometric contractions. J Physiol (Lond) 228:285–306, 1973.

93. Milner-Brown, HS, Stein, RB, and Yemm, R: The orderly recruitment of human motor units during voluntary isometric contractions. J Physiol (Lond) 230:359–370, 1973.

94. Milner-Brown, HS, Stein, RB, and Yemm, R: Changes in firing rate of human motor units during linearly changing voluntary contractions. J Physiol (Lond) 230:371–390, 1973.

95. Moberg, E: The role of cutaneous afferents in position sense, kinaesthesia, and motor function of the hand. Brain 106:1–19, 1983.

96. Monster, AW, and Chan, H: Isometric force production by motor units of extensor digitorum communis muscle in man. J Neurophysiol 40:1432–1443, 1977.

97. Nishizono, H, Saito, Y, and Miyashita, M: The estimation of conduction velocity in human skeletal muscle in situ with surface electrodes. Electroencephalogr Clin Neurophysiol 46:659–664, 1979.

98. Norris, FH, Jr, and Irwin, RL: Motor unit area in a rat muscle. Am J Physiol 200:944–946, 1961.

99. Person, RS, and Kudina, LP: Discharge frequency and discharge pattern of human motor units during voluntary contraction of muscle. Electroencephalogr Clin Neurophysiol 32:471–483, 1972.

100. Petajan, JH: Clinical electromyographic studies of diseases of the motor unit. Electroencephalogr Clin Neurophysiol 36:395–401, 1974.

101. Petajan, JH, and Philip, BA: Frequency control of motor unit action potentials. Electroencephalogr Clin Neurophysiol 27:66–72, 1969.

102. Peter, JB, Barnard, RJ, Edgerton, VR, Gillespie, CA, and Stempel, KE: Metabolic profiles of three different types of skeletal muscle in guinea pigs and rabbits. Biochemistry 11:2627–2633, 1972.

103. Pette, D, Smith, ME, Staudte, HW, and Vrbova, G: Effects of long-term electrical stimulation on some contractile and metabolic characteristics of fast rabbit muscles. Pflugers Arch 338:257–272, 1973.

104. Romanul, FCA, and Van Der Meulen, JP: Slow and fast muscles after cross innervation: Enzymatic and physiological changes. Arch Neurol 17:387–402, 1967.

105. Rosenfalck, P: Intra- and extracellular potential fields of active nerve and muscle fibres. Acta Physiol Scand (Suppl 321):1–168, 1969.

106. Round, JM, Jones, DA, Chapman, SJ, Edwards, RHT, Ward, PS, and Fodden, DL: The anatomy and fibre type composition of the human adductor pollicis in relation to its contractile properties. J Neurol Sci 66:263–293, 1984.

107. Sica, REP, McComas, RJ, Upton, ARM, and Longmire, D: Motor unit estimations in small muscles of the hand. J Neurol Neurosurg Psychiatry 37:55–67, 1974.

108. Sissons, H: Anatomy of the motor unit. In Walton, JN (ed): Disorders of Voluntary Muscle, ed 3. Churchill Livingstone, London, 1974.

109. Sollie, G, Hermens, JH, Boon, KL, Wallinga-De Jonge, W, and Zilvold, G: The measurement of the conduction velocity of muscle fibres with surface EMG according to the cross-correlation method. Electromyogr Clin Neurophysiol 25:193–204, 1985.

110. Stalberg, E: Single fibre electromyography for

motor unit study in man. In Shahani, M (ed): The Motor System: Neurophysiology and Muscle Mechanisms. Elsevier, Amsterdam, 1976.

111. Stalberg, E, and Antoni, L: Electrophysiological cross section of the motor unit. J Neurol Neurosurg Psychiatry 43:469–474, 1980.

112. Stalberg, E, and Ekstedt, J: Single fibre EMG and microphysiology of the motor unit in normal and diseased human muscle. In Desmedt, JE (ed): New Developments in Electromyography and Clinical Neurophysiology, Vol 1. Karger, Basel, 1973, pp 113–129.

113. Stalberg, E, Schiller, HH, and Schwartz, MS: Safety factor in single human motor endplates studied in vivo with single fibre electromyography. J Neurol Neurosurg Psychiatry 38:799–804, 1975.

114. Stefanova-Uzunova, M, Stamatova, L, and Gatev, V: Dynamic properties of partially denervated muscle in children with brachial plexus birth palsy. J Neurol Neurosurg Psychiatry 44:497–502, 1981.

115. Stephens, JA, and Usherwood, TP: The mechanical properties of human motor units with special reference to their fatiguability and recruitment threshold. Brain Res 125:91–97, 1977.

116. Tanji, J, and Kato, M: Recruitment of motor units in voluntary contraction of a finger muscle in man. Exp Neurol 40:759–770, 1973.

117. Tergast, P: Ueber das Verhaltniss von Nerve und Muskel. Arch Mikr Anat 9:36–46, 1873.

118. Thomas, CK, Stein, RB, Gordon, T, Lee, RG, and Elleker, MG: Patterns of reinnervation and motor unit recruitment in human hand muscles after complete ulnar and median nerve section and resuture. J Neurol Neurosurg Psychiatry 50:259–268, 1987.

119. Thorstensson, A: Muscle strength, fibre types and enzyme activities in man. Acta Physiol Scand (Suppl 433):3–45, 1976.

120. Tokizane, T, and Shimazu, H: Functional Differentiation of Human Skeletal Muscle. Charles C Thomas, Springfield, Ill, 1964.

121. Troni, W, Cantello, R, and Rainero, I: Conduc-

tion velocity along human muscle fibers in situ. Neurology 33:1453–1459, 1983.

122. Wallinga-De Jonge, W, Gielen, FLH, Wirtz, P, De Jong, P, and Broenink, J: The different intracellular action potentials of fast and slow muscle fibres. Electroencephalogr Clin Neurophysiol 60:539–547, 1985.

123. Ward, KM, Manning, W, and Wareham, AC: Effects of denervation and immobilization during development upon [^{3}H]ouabain binding by slow-and fast-twitch muscle of the rat. J Neurol Sci 78:213–224, 1987.

124. Warmolts, JR, and Engel, WK: Open-biopsy electromyography. I. Correlation of motor unit behavior with histochemical muscle fiber type in human limb muscle. Arch Neurol 27:512–517, 1972.

125. Weber, A, and Murray, JM: Molecular control mechanisms in muscle contraction. Physiol Rev 53:612–673, 1973.

126. Willison, RG: Arrangement of muscle fibers of a single motor unit in mammalian muscles (letter to the editor). Muscle Nerve 3:360–361, 1980.

127. Wuerker, RB, McPhedran, AM, and Henneman, E: Properties of motor units in a heterogeneous pale muscle (m. gastrocnemius) of the cat. J Neurophysiol 28:85–99, 1965.

128. Yaar, I, Shapiro, MB, Mitz, AR, and Pottala, EW: New technique for measuring fiber conduction velocities in full interference patterns. Electroencephalogr Clin Neurophysiol 57:427–434, 1984.

129. Yellin, H: Neural regulation of enzymes in muscle fibers of red and white muscle. Exp Neurol 19:92–103, 1967.

130. Young, JL, and Mayer, RF: Physiological alterations of motor units in hemiplegia. J Neurol Sci 54:401–412, 1982.

131. Younkin, SG, Brett, RS, Davey, B, and Younkin, LH: Substances moved by axonal transport and released by nerve stimulation have an innervation-like effect on muscle. Science 200:1292–1295, 1978.

Chapter 12

TECHNIQUES AND NORMAL FINDINGS

1   INTRODUCTION

Electromyography tests the integrity of the entire motor system, which consists of upper and lower motor neurons, neuromuscular junction, and muscle. Further subdivision in each category reveals seven possible sites of involvement that may cause muscle weakness (Fig. 12–1). One must first learn physiologic mechanisms underlying normal muscle contraction to understand various abnormalities that characterize disorders of the motor system in each step. Electromyographers must also consider multiple factors that can significantly affect the outcome of recordings. These include the age of patients and the particular properties of the muscle under study, in addition to the electrical specifications of the needle electrodes and recording apparatus, as discussed earlier.

Electromyography serves best if performed as an extension of the physical examination, rather than a laboratory procedure. The clinical symptoms and signs guide the optimal selection of specific muscle groups. An adequate study calls for multiple sampling of each muscle at rest and during different degrees of voluntary contraction. The findings in the initially tested muscles dictate the course of subsequent exploration. Thus, no rigid protocol suffices for a routine electromyographic examination. Certain basic principles apply, but a flexible approach best fulfills the needs of individual patients.

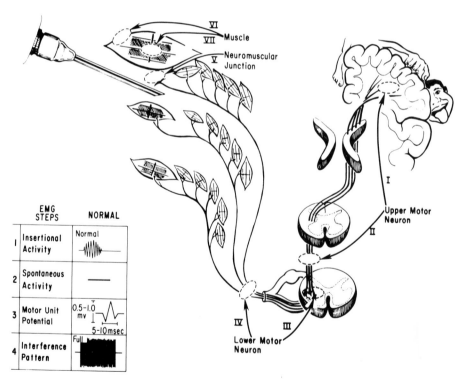

Figure 12–1. Schematic view of the motor system with seven anatomic levels. They include (1) upper motor neuron from the cortex (*I*) to the spinal cord (*II*); (2) lower motor neuron with the anterior horn cell (*III*) and nerve axon (*IV*); (3) neuromuscular junction (*V*); and (4) muscle membrane (*VI*) and contractile elements (*VII*). The inset illustrates diagrammatically four steps of electromyographic examination and normal findings. The figure of cortical representation is adapted from Netter.[73]

2 PRINCIPLES OF ELECTROMYOGRAPHY

Recording of Muscle Action Potential

The electrical properties of the cells (see Chapter 2.2) form the basis of clinical electromyography. Extracellular recording of the muscle action potential through the volume conductor reveals an initially positive triphasic waveform as the impulse approaches, arrives at, and departs from the active electrode. If the muscle fiber is traumatized by the needle, a negative spike cannot be generated at the damaged membrane. In this case, a low-amplitude, slow negativity follows a large initial positivity.

The size of an action potential detected in the external field varies, depending on the spacial relationship between the cell and the tip of the needle electrode. For example, when recorded by an electrode with a small lead-off surface, the amplitude falls off to less than 10 percent at a distance of 1 mm from the generator source. Normally, neural impulses give rise to synchronous discharges of all muscle fibers of a motor unit, producing a motor unit potential. In an unstable denervated muscle, individual fibers fire independently in the absence of neural control. The detection of these spontaneous single fiber potentials constitute one of the most important findings in electromyography.

Contraindications for Electromyography

Two possibilities deserve special mention in screening patients for electromyographic examination: bleeding tendencies and unusual susceptibility to recurrent systemic infections. Specific inquiry in this regard often reveals pertinent information that the patient may not volunteer. To prevent unnecessary complications, the electromyographer should consult with the referring physician to weigh the diagnostic benefits against the risks. Bleeding, partial thromboplastin, and prothrombin times must be tested in patients on anticoagulants or those with any coagulopathy, including hemophillia[82] and thrombocytopenia. Unless the platelet count falls below 20,000/mm,[4] however, local pressure can adequately counter the minimal bleeding tendency to accomplish hemostasis. Transient bacteremia following needle examination could cause endocarditis in the presence of valvular disease or prosthetic valves. Although these patients must avoid needle studies, unless clearly indicated, few recommend prophylactic administration of antibiotics for the procedure.[1]

Electromyography, if conducted prematurely, could interfere with the interpretation of subsequent histologic or biochemical findings that supplement clinical evaluation. Repeated trauma during insertion and movement of the needle electrode consistently induces localized inflammation and, less frequently, focal myopathic changes. These abnormalities may preclude the confirmation of a clinical diagnosis, which often requires a muscle biopsy. With the anticipated need for pathologic exploration, needle examination must spare the muscle under consideration.

Serum creatine kinase (CK) increases in certain muscle diseases, such as muscular dystrophy and polymyositis,[45] and in other conditions, including cardiac ischemia, hypothyroidism, and sustained athletic participation. The level may also rise considerably in normal muscles by the combination of electromyography, diurnal variation, and prolonged exercise.[5,16,64,75] Needle examination by itself, however, should not elevate the enzyme to a misleading level in normal persons. In one series no significant changes occurred within 2 hours after electromyographic studies.[17] The value reached a peak of one and one-half times baseline in 6 hours and returned to baseline 48 hours after the needle examination. To avoid any confusion, one should test enzymes prior to needle examination, but a sufficient elevation of CK activities indicates abnormality, even for the serum drawn after the procedure.

Recording Techniques

Electromyographic examination of skeletal muscle consists of four steps:

1. Insertional activity caused by movement of a needle electrode in the muscle
2. Spontaneous activity recorded with the muscle at rest, that is, with the needle stationary in a relaxed muscle
3. Motor unit potentials evoked by isolated discharges of motor neurons during mild voluntary contraction
4. Recruitment and interference pattern evaluated by the change in electrical activity during progressively increasing levels of contraction to a maximum level

Routine oscilloscope settings consist of a sweep speed ranging from 2 to 20 ms/cm and an optimal gain to maximize the recorded potentials without truncating the peaks. The sensitivity varies from 50 to 500 μV/cm for insertional and spontaneous activities and from 100 μV to 1 mV/cm for motor unit potentials. Obviously, a lower amplification suffices for the study of larger potentials. Most investigators use the low-frequency filter of 10 to 20 Hz and high-frequency filter of 10 kHz, but some prefer lowering the lower limit to 2 Hz or less when determining the waveform of motor unit potentials.

The needle electrode registers muscle action potentials only from a restricted area of the muscle. For an adequate survey, therefore, frequent needle repositioning is necessary for multiple sampling in small steps. Exploration in four directions from a single puncture site minimizes the patient's discomfort. Studies of larger muscles require additional insertions in proximal, central, and distal portions.

3 INSERTIONAL ACTIVITY

Origin and Characteristics

Insertion of a needle electrode into the muscle normally gives rise to brief bursts of electrical activity.[54,96] The same discharges also occur with each repositioning. The insertional activity on the average lasts a few hundred milliseconds, slightly exceeding the movement of the needle (Fig. 12–2). It appears as positive or negative high-frequency spikes in a cluster,[99] accompanied by a crisp static sound over the loudspeaker. As implied by the commonly used term, injury potential, the discharges originate from muscle fibers injured or mechanically stimulated by the penetrating needle.

Clinical Significance

The waveforms seen on the oscilloscope and, perhaps more importantly, the sounds over the loudspeaker allow a somewhat loose but useful categorization of the insertional activity into normal, decreased, and increased patterns. The level of response depends, among other things, on the magnitude and speed of needle movement. Nonetheless, semiquantitative analysis provides an important measure of muscle excitability, being typically reduced in fibroses and exaggerated in denervation or inflammatory processes. Such findings often reveal the first clue to the nature of the lesion, directing the electromyographer toward the proper course of subsequent examination. As mentioned earlier, a complete study consists of sampling the activities at several locations in each muscle by shifting the electrode from one point to another. Otherwise, patchy areas of hyperexcitability, if present, may remain undetected.

In denervated muscles, insertion of the exploring needle may provoke positive sharp waves and, less frequently, fibrillation potentials.[96,97] These early abnormalities of denervation resemble a normal insertional activity that may also take the form of positive sharp potentials. In a quantitative analysis using a mechanical electrode inserter, one or two isolated positive waves commonly appeared in normal muscles at the end of the insertional activity.[98] None of these potentials, however, fired repetitively or in a train, or in a reproducible fashion with further insertions. Their audio characteristics lacked the typical pitch of pos-

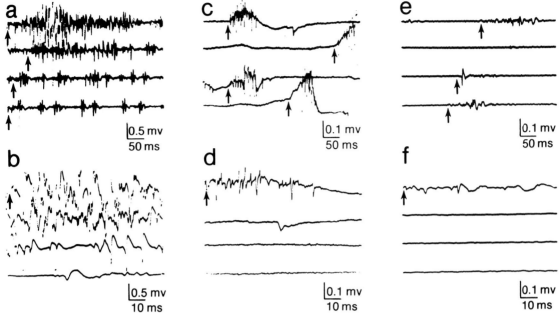

Figure 12–2. Increased (*a* and *b*), normal (*c* and *d*), and decreased (*e* and *f*) insertional activities (*arrows*) from the first dorsal interosseus in tardy ulnar palsy, tibialis anterior in a control, and fibrotic deltoid in severe dermatomyositis.

itive sharp waves associated with denervation. These findings suggest the nonspecificity of isolated positive waves induced by insertion, unless they give rise to reproducible trains with characteristic audio displays reminiscent of the spontaneous discharge.

4   END-PLATE ACTIVITIES

If the needle is held stationary, normal resting muscles show no electrical activity except at the end-plate region. Here, irritation of the small intramuscular nerve terminals by the tip of the electrode causes end-plate activities that consist of two components: low-amplitude, undulating end-plate noise (Fig. 12–3) and high-amplitude intermittent spikes (Fig. 12–4). These two types of potentials occur conjointly or independently. The patient usually experiences a dull pain, which dissipates with slight withdrawal of the needle. End-plate activities, although physiologic in nature, tend to become excessive in denervated muscles.

End-Plate Noise

The background activity in the end-plate region consists of frequently recurring irregular negative potentials, 10 to 50 μV in amplitude and 1 to 2 ms in duration, producing over the loudspeaker a characteristic sound much like a seashell held to the ear. It represents extracellularly recorded miniature end-plate potentials (MEPP), that is, nonpropagating depolarizations caused by spontaneous release of acetylcholine (ACh) quanta.[14,100] The corresponding potentials recorded intracellularly with microelectrodes show monophasic positivity, about 1 mV in amplitude, that is, opposite in polarity and much greater in amplitude.[26]

End-Plate Spike

Intermittent spikes, 100 to 200 μV in amplitude and 3 to 4 ms in duration, fire irregularly at 5 to 50 impulses per second. The typical pattern with an initial negativity indicates that the spikes origi-

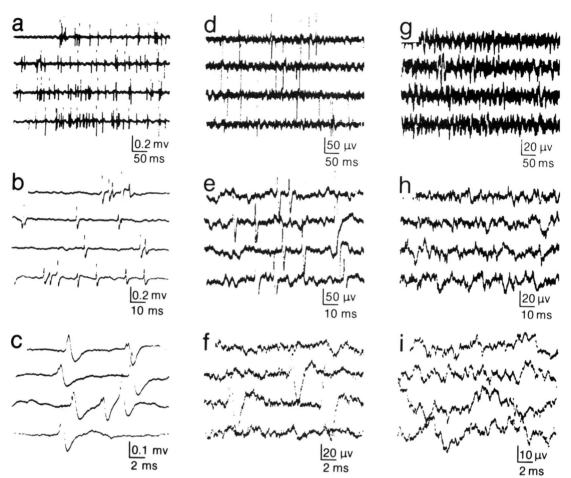

Figure 12–3. End-plate activities recorded from the tibialis anterior in a healthy subject. Two types of potentials shown represent the initially negative, high-amplitude end-plate spikes (a, b, and c) and low amplitude end-plate noise (g, h, and i). The spikes and end-plate noise usually, though not necessarily, appear together (d, e, and f).

nate at the tip of the recording electrode. In contrast, fibrillation potentials, recorded elsewhere, have a small positive phase preceding the major negative spike. Although sometimes referred to as nerve potentials, the end-plate spikes probably result from discharges of single muscle fibers excited by the needle.[6,14,40] In fact, end-plate spikes are indistinguishable in waveform from fibrillation potentials, which also show an initial negativity when recorded at the end-plate region. Similarity in firing patterns between discharges of muscle spindle afferents and the end-plate spikes led some investigators to postulate that the poten-tials originated in the intrafusal muscle fibers.[74]

Repositioning of the recording needle may injure the cell membrane at the end-plate region. Slight relocation of the needle tip near the source of discharge may then reverse the polarity of the ordinarily negative end-plate spikes. Small irregularly occurring positive potentials also appear in the end-plate region when recorded with a concentric needle electrode. Here, the positive discharges probably represent cannula-recorded end-plate spikes, hence reversed in polarity and reduced in amplitude.[78] These positive potentials favor the more distal

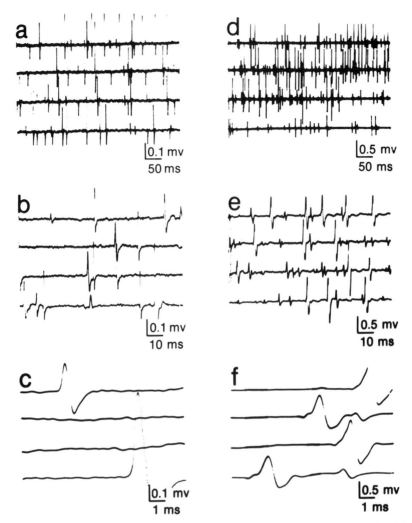

Figure 12-4. End-plate spikes recorded from the abductor pollicis brevis in a normal subject (*a, b,* and *c*) and in a patient with the carpal tunnel syndrome (*d, e,* and *f*). An unusual prominence of end-plate activity in denervated muscle, although common, carries uncertain diagnostic value.

muscles, perhaps because of their higher innervation ratios.[79] The irregular pattern of firing and shorter duration distinguish the physiologic positive discharges at the end plate from positive sharp waves seen in denervation or other pathologic conditions.

5 MOTOR UNIT POTENTIAL

The motor unit consists of a group of muscle fibers innervated by a single anterior horn cell (Fig. 12-5). It has anatomic and physiologic properties based on the innervation ratio, fiber density, propagation velocity, and integrity of neuromuscular transmission. These factors vary not only from one muscle group to another but also with age for a given muscle. Isolated potentials attributed to an individual motor unit represent the sum of all single muscle fiber spikes that occur nearly synchronously within the recording radius of the electrode.

Motor Unit Profile

The shape of motor unit potentials reflects, in addition to the inherent proper-

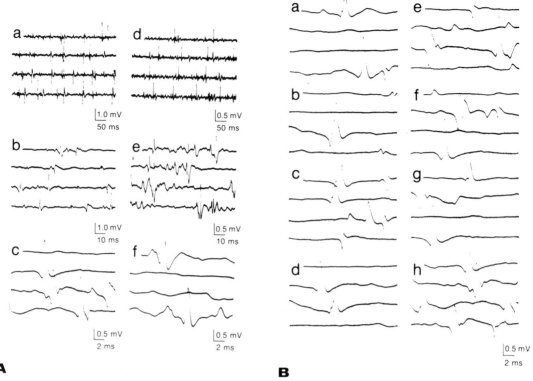

Figure 12–5. A. Normal motor unit potentials from minimally contracted biceps in a 40-year-old healthy man (*a*, *b*, and *c*) and maximally contracted tibialis anterior in a 31-year-old woman with hysterical weakness (*d*, *e*, and *f*). In both, low firing frequency indicates weak voluntary effort. **B.** Normal variations of motor unit potentials from the same motor unit in the normal biceps. Tracings *a* through *h* represent eight slightly different sites of recording with the subject maintaining isolated discharges of a single motor unit.

ties of the motor unit itself, many other factors. Of these, the spatial relationships between the needle and individual muscle fibers play the most important role in determining the waveform.[8–11] Thus, slight repositioning of the electrode, altering the spacial orientation, introduces a new profile for the same motor unit.

Other important variables include the resistance and capacitance of the intervening tissue and intramuscular temperature. Cooling from 37°C to 30°C, for example, causes the duration to increase by 10 to 30 percent, but the amplitude decreases by 2 to 5 percent per 1°C. The number of polyphasic potentials increases as much as tenfold with a 10°C

decrease.[13] The amplitude decreases slightly with hypothermia, despite the local facilitatory effect on the muscle membrane, because differential slowing and desynchronization more than counter the anticipated change.

Finally, a number of nonphysiologic factors influence the configuration of the recorded potentials. These include the type of needle electrode, size of the recording surface or lead-off area, electrical properties of the amplifier, choice of oscilloscope sensitivity, sweep or filters, and the methods of storage and display. These factors together dictate the amplitude, rise time, duration, number of phases, and other characteristics of a motor unit potential.[20]

Amplitude

Although all the individual muscle fibers in a motor unit discharge in near synchrony, a limited number of muscle fibers located near the tip of the recording electrode determine the amplitude of a motor unit potential (Fig. 12–6). Single muscle fiber potentials fall off to less than 50 percent at a distance of 200 to 300 μM from the source and to less than 1 percent a few millimeters away.[10,11,25,34,60] Fewer than 20 muscle fibers lying within a 1-mm radius of the electrode tip contribute to the high voltage spike of the motor unit potential.[91] Therefore, the same motor unit can give rise to many motor unit potentials of different ampli-

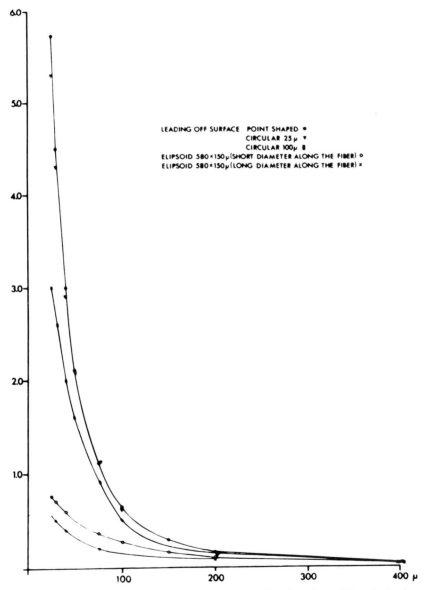

LEADING OFF SURFACE POINT SHAPED •
CIRCULAR 25 μ ▾
CIRCULAR 100μ ▮
ELIPSOID 580×150μ (SHORT DIAMETER ALONG THE FIBER) ○
ELIPSOID 580×150μ (LONG DIAMETER ALONG THE FIBER) ×

Figure 12–6. Reduction in amplitude of recorded response with relocation of the electrode away from the source. The needle with a large leading-off surface registers a low amplitude even near the spike generator, showing only minor reduction as the distance between the electrode and the source increases. In contrast, amplitude declines per unit distance steeply with a smaller leading-off surface. (See Figure 15–1.) (From Ekstedt and Stalberg,[25] with permission.)

tudes at different recording sites. The amplitude normally ranges from several hundred microvolts to a few millivolts with a concentric needle and is substantially greater with a monopolar needle.

Rise Time

The rise time, measured as a time lag from the initial positive peak to the subsequent negative peak (see Fig. 2–6), helps estimate the distance between the recording tip of the electrode and the discharging motor unit. A distant unit has a greater rise time because the resistance and capacitance of the intervening tissue act as a high-frequency filter. A unit accepted for quantitative measurement should have a rise time less than 500 μs, preferably 100 to 200 μs. Such a motor unit produces a sharp, crisp sound over the loudspeaker, which provides an important clue for the proximity of the unit to the electrode. A distant discharge accompanies a dull sound, indicating the need to reposition the electrode closer to the source. The measurement of the rise time confirms the suitability of the recorded potential for quantitative assessment of the amplitude.

Duration

Duration is measured from the initial takeoff to the return to the baseline (Table 12–1). It indicates the degree of synchrony among many individual muscle fibers with variable length, conduction velocity, and membrane excitability. Generators more than 1 mm away from the electrode also contribute to the initial and terminal low-amplitude portions of the motor unit potential. Thus, unlike the spike amplitude, exclusively determined by a small number of muscle fibers near the electrode, the duration of a motor unit potential reflects the activity from all the muscle fibers. Therefore, a slight shift in the needle position influences the duration much less than the amplitude. The duration normally varies from 5 to 15 ms, depending on the age of the subject. In one study,[7] the values measured at the

ages of 3 and 75 were 7.3 and 12.8 ms in biceps brachii, 9.2 and 15.9 ms in tibialis anterior, and 4.3 and 7.5 ms in the facial muscles.

Phases

A phase is defined as that portion of a waveform between the departure from and return to the baseline. The number of phases, determined by counting negative and positive peaks to and from the baseline, equals the number of baseline crossings plus one. Normally, motor unit potentials have four or fewer phases. Polyphasic motor unit potentials with more than four phases suggest desynchronized discharge or drop-off of individual fibers. These potentials do not exceed 5 to 15 percent of the total population in a healthy muscle, if recorded with a concentric needle electrode. Polyphasic activities occur more commonly with the use of a monopolar needle, although no studies have established the exact incidence. Some action potentials show several "turns" or directional changes without crossing the baseline. These serrated action potentials or, less appropriately, complex or pseudopolyphasic potentials, also indicate desynchronization among discharging muscle fibers.

6 QUANTITATIVE MEASUREMENTS

Conventional Assessment

In clinical tests, electromyographers assess motor unit parameters by oscilloscope displays of waveforms and their audio characteristics. Using these simple means, an experienced examiner can detect abnormalities with reasonable certainty. Such subjective assessment, though satisfactory for the detection of unequivocal abnormalities, may not suffice to delineate less obvious deviations or mixed patterns of abnormalities. These ambiguous circumstances call for quantitative measurement of motor unit potentials.[12] With the use of a standardized

Table 12–1 MEAN ACTION POTENTIAL DURATION (IN MILLISECONDS) IN VARIOUS MUSCLES AT DIFFERENT AGES (CONCENTRIC ELECTRODES)

Age in Years	Arm Muscles						Leg Muscles					Facial Muscles
	Deltoideus	Biceps Brachii	Triceps Brachii	Extensor Digitorum Communis	Opponens Pollicis; Interosseus	Abductor Digiti Quinti	Biceps Femoris; Quadriceps	Gastrocnemius	Tibialis Anterior	Peroneus Longus	Extensor Digitorum Brevis	Orbicularis Oris Superior; Triangularis; Frontalis
0	8.8	7.1	8.1	6.6	7.9	9.2	8.0	7.1	8.9	6.5	7.0	4.2
3	9.0	7.3	8.3	6.8	8.1	9.5	8.2	7.3	9.2	6.7	7.2	4.3
5	9.2	7.5	8.5	6.9	8.3	9.7	8.4	7.5	9.4	6.8	7.4	4.4
8	9.4	7.7	8.6	7.1	8.5	9.9	8.6	7.7	9.6	6.9	7.6	4.5
10	9.6	7.8	8.7	7.2	8.6	10.0	8.7	7.8	9.7	7.0	7.7	4.6
13	9.9	8.0	9.0	7.4	8.9	10.3	9.0	8.0	10.0	7.2	7.9	4.7
15	10.1	8.2	9.2	7.5	9.1	10.5	9.2	8.2	10.2	7.4	8.1	4.8
18	10.4	8.5	9.6	7.8	9.4	10.9	9.5	8.5	10.5	7.6	8.4	5.0
20	10.7	8.7	9.9	8.1	9.7	11.2	9.8	8.7	10.8	7.8	8.6	5.1
25	11.4	9.2	10.4	8.5	10.2	11.9	10.3	9.2	11.5	8.3	9.1	5.4
30	12.2	9.9	11.2	9.2	11.0	12.8	11.1	9.9	12.3	8.9	9.8	5.8
35	13.0	10.6	12.0	9.8	11.7	13.6	11.8	10.6	13.2	9.5	10.5	6.2
40	13.4	10.9	12.4	10.1	12.1	14.1	12.2	10.9	13.6	9.8	10.8	6.4
45	13.8	11.2	12.7	10.3	12.5	14.5	12.5	11.2	13.9	10.1	11.1	6.6
50	14.3	11.6	13.2	10.7	12.9	15.0	13.0	11.6	14.4	10.5	11.5	6.8
55	14.8	12.0	13.6	11.1	13.3	15.5	13.4	12.0	14.9	10.8	11.9	7.0
60	15.1	12.3	13.9	11.3	13.6	15.8	13.7	12.3	15.2	11.0	12.2	7.1
65	15.3	12.5	14.1	11.5	13.9	16.1	14.0	12.5	15.5	11.2	12.4	7.3
70	15.5	12.6	14.3	11.6	14.0	16.3	14.1	12.6	15.7	11.4	12.5	7.4
75	15.7	12.8	14.4	11.8	14.2	16.5	14.3	12.8	15.9	11.5	12.7	7.5

The values given in the table are mean values from different subjects without evidence of neuromuscular disease. The standard deviation of each value is 15 percent (20 potentials for each muscle). Therefore, deviations up to 20 percent are considered within the normal range when comparing measurements in a given muscle with the values of the table.

From Buchthal,[7] with permission.

method, objective descriptions also allow meaningful comparison of test results obtained sequentially or in different laboratories. Physiologic properties that characterize the motor unit quantitatively include duration, spike amplitude, spike area, phases, turns, number of satellites, and degree of waveform variability.[87]

Selection and Analysis

In quantitative analysis,[3] most investigators use the standard concentric needle electrode with a lead-off surface of about 0.07 mm^2. The optimal recording requires amplifier frequency range of 10 Hz to 10 kHz and standard sensitivity of 100 to 500 μV/cm. It is important to raise the skin temperature above 34°C to maintain muscle temperature at about 37°C.[23] The motor unit potentials selected for assessment must have a rise time less than 500 μs. A storage oscilloscope with a delay line offers a distinct advantage for quick identification of such potentials. Recorded waveforms vary a great deal from one motor unit to another and within the same unit, depending upon the relative position of the needle tip to the source of discharge. For an ideal quantification, one must count at least 20 different motor units in each muscle using multiple needle insertions.[7]

Table 12–1 summarizes the duration of motor unit potentials recorded with a concentric needle in normal subjects of different ages.[7] These values were measured from the point of takeoff to the return to baseline. They exclude late or satellite components seen as a separate peak after return to the baseline.[59] The normal ranges depend on many factors other than simply the characteristics of the motor unit itself, as discussed earlier. Hence, each laboratory should construct its own table of normal values to avoid indiscriminate application of published data.

Automated Methods

Different investigators have explored the possibility of automatically analyzing motor unit potentials using analog[72] or digital techniques.[80] Such a system converts a motor unit potential to a digital equivalent for analysis by computer. The usual measurements include duration, amplitude, polarity, number of phases, and integrated area under the waveform.[56] One of the inherent difficulties with this approach centers on the selection of the signals. In early methods, the examiner screened the motor unit potentials by visual inspection, using a monitor scope, before processing them for automated analysis.[80] With another technique, motor unit potentials qualified automatically if their peak-to-peak amplitudes exceeded 100 μV.[51] This system measures the duration of the discharge at 20 μV above the baseline and counts the number of phases as a deflection exceeding 40 μV. Some investigators advocate lowering the cutoff level to less than 50 μV so that a greater number of motor unit potentials can be analyzed.[42]

Most studies have shown that there is no major discrepancy between the results of time-consuming manual quantification and quick automatic analysis.[49,50,52,53,55,61] Indeed, the computer can accurately and efficiently discriminate typical neuropathic and myopathic changes.[43,83,104] It is not known, however, whether these techniques also help resolve borderline cases in which conventional methods fail to provide useful information. Using automatic analysis, one cannot separate female relatives of patients with Duchenne dystrophy from healthy subjects individually, even though the results are statistically different between the two as groups.[92]

Frequency Spectrum

The waveform of any action potential can be decomposed into many sine waves of different frequencies. Thus, a frequency spectrum provides another objective means of characterizing motor unit potentials. This type of analysis reveals that the shorter the duration of the motor unit potential, the greater are the high-frequency components. Several investigators have studied frequency spectra in normal and diseased muscles.[46] The highest peak seen during maximal contraction falls between 100 and 200 Hz in normal subjects.[94] This peak to a higher frequency in subjects with myopathy,[81] and to a lower frequency in subjects with anterior horn cell lesions.[27] Frequency analysis has also been used to characterize fatigue trends in normal subjects and those with myasthenia gravis.[103] The clear difference seen in typical cases, however, does not imply its practical value as a diagnostic test. For clinical use, it is important to control the variables, such as needle position or level of muscle contraction, that appreciably influence the results.[19]

7 DISCHARGE PATTERN OF MOTOR UNITS

Recruitment

Motor units are activated according to the physiologic rule called the size principle (see Chapter 11.6). Mild voluntary contraction induces isolated discharges of one or a few motor units at a rate of five to seven impulses per second. These discharges are typically semirhythmic, with slowly increasing, then decreasing interspike intervals, during constant contraction. Motor units may fire irregularly in basal ganglia disorders such as parkinsonism or chorea.[22] Greater muscle force brings about two separate but related changes in the pattern of motor unit discharge: (1) recruitment of previously inactive units and (2) more rapid firing of already active units (Fig. 12–7). Which of the two plays a greater role is not known, but both mechanisms operate simultaneously. A normal recruitment pattern implies the discharge of an appropriate number of motor units for the effort (Fig. 12–8A). A reduced or increased pattern indicates a fewer or greater number of discharging units than expected (Fig. 12–8B and C). For accurate assessment, the examiner must know how many motor units should normally fire for a given force being exerted.

A healthy subject can initially excite only one or two motor units before re-cruiting additional units in a fixed order.[41] The units activated early consist primarily of small, type I muscle fibers according to the size principle. Large, type II units participate later, during strong voluntary contraction.[95] The recruitment frequency is a measure of motor unit discharge defined as the firing frequency at the time an additional unit is recruited in the vicinity. Normal values measured during mild contraction average 5 to 10 impulses per second, depending on the types of motor units under study.[18,77] The reported ranges show a considerable overlap between healthy subjects and patients with neuromuscular disorders.[33] Some electromyographers prefer the ratio of the average firing rate to the number of active units. Normal subjects should have a ratio less than 5 with, for example, three units firing less than 15 impulses per second each.[21] If two units are firing over 20 impulses per second, the ratio exceeds 10, indicating a loss of motor units.

Interference Pattern

With greater contraction, many motor units begin to fire very rapidly (Fig. 12–9). Simultaneous activation of different units preclude recognition of individual motor unit potentials; hence the name interference pattern. The spike density and the average amplitude of the summated response are determined by a number of factors: descending input from the cortex, number of motor neurons capable of discharging, firing frequency of each motor unit, waveform of individual potentials, and probability of phase cancellation. Despite such complexity, its analysis provides a simple quantitative means of evaluating the relationship between the number of firing units and the muscle force exerted with maximal effort.

Automatic Analysis

Examination of individual motor unit potentials during weak voluntary effort only relates to low-threshold type I muscle fibers. Studies of the interference pat

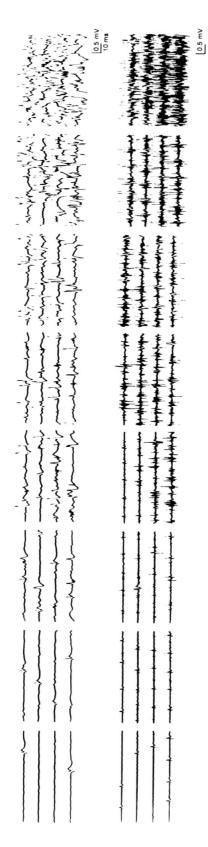

Figure 12–7. Normal recruitment and full interference pattern with increasing strength in the same healthy subject shown in Figure 12–5. The tracings show the same activity recorded with fast (*top*) and slow (*bottom*) sweep.

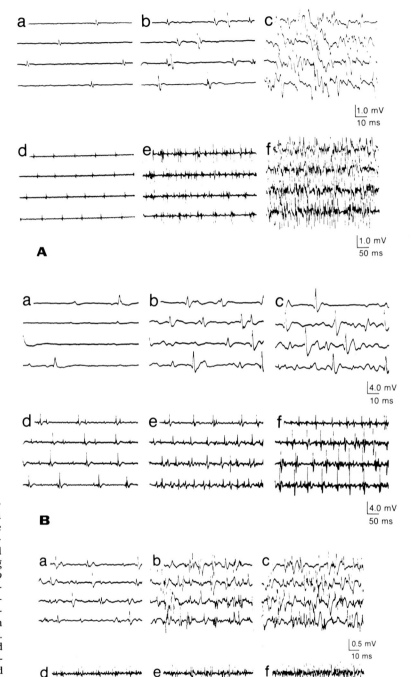

Figure 12–8. A. Normal recruitment in the triceps of a 44-year-old healthy man. The tracings show the same activity recorded with fast (*top*) and slow sweep (*bottom*) during minimal (*a* and *d*), moderate (*b* and *e*), and maximal contraction (*c* and *f*). **B.** Reduced recruitment in the tibialis anterior of a 44-year-old man with amyotrophic lateral sclerosis. A single motor unit discharged rapidly during strong contraction. **C.** Early recruitment and full interference pattern in the quadriceps of a 20-year-old patient with limb-girdle dystrophy. The tracings show an excessive number of motor units for the amount of muscle force exerted during weak contraction.

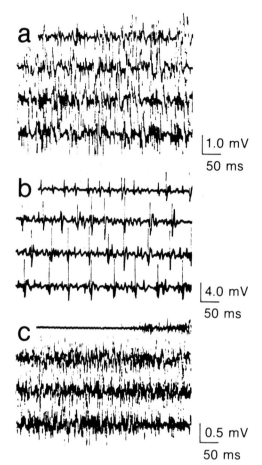

Figure 12–9. Interference patterns seen in the triceps of a 44-year-old healthy man (*a*), tibialis anterior of a 52-year-old man with amyotrophic lateral sclerosis (*b*), and quadriceps of a 20-year-old man with limb-girdle dystrophy (*c*). Discrete single motor unit discharge in *b* stands in good contrast to abundant motor unit potentials with reduced amplitude in *c*.

tern induced by strong muscle contraction allows quantitative assessment over a wider range. One such analysis utilizes an automated technique designed to count the number of "turns" or directional changes of a waveform that exceeds a minimum excursion without necessarily crossing the baseline.[101] In this method (Fig. 12–10), one measures the amplitude from a point of change in direction to the next, not from baseline to peak, selecting potentials greater than 100 μV to avoid contamination from noise.[37,38] During a fixed time epoch, the

subject must maintain constant levels of muscle contraction.

After automatic analysis, a special-purpose digital computer displays the total number of reversals, histograms of the intervals between potential reversals, and cumulative amplitude of all potentials during a fixed time period.[24] In one study[29] the number of turns and mean amplitude had 10 to 25 percent variability on repeated trials. Interindividual differences diminished with the use of a relative, rather than absolute, force in each subject. Diagnostic yield reached an acceptable level at muscle contraction producing 21 to 50 percent of the maximum force, with the best reproducibility at 10 to 30 percent.[36] The ratio of turns to mean amplitude increased in myopathy, especially at 10 to 20 percent of maximum force, and decreased in neurogenic disorders, mainly at a force of 20 to 30 percent.[28,30] The use of a calculated index independent of force may improve the diagnostic sensitivity.[15] The method has also been used to differentiate primary muscle disease from neurogenic lesions in infants and young children.[85,86]

Quantitative measurements of recruitment patterns complement studies of single motor units. Evaluation of individual potentials allows precise description of normal and abnormal motor units and their temporal stability. Analysis of recruitment reveals an overall muscle performance by demonstrating the number and discharge pattern of all the motor units. These methods, though not widely used at this time, hold great promise as supplements to routine electromyography.[3,28,31,32,35,39,65–68,88]

8 ELECTRICAL POTENTIAL AND MUSCLE FORCE

During maximal effort, motor units discharging at frequencies up to 50 impulses per second give rise to a tetanic contraction. Despite intermittency of electrical impulses, the accompanying mechanical response fuses at high discharge frequencies to produce a relatively smooth tension. In contrast, unfused twitches of

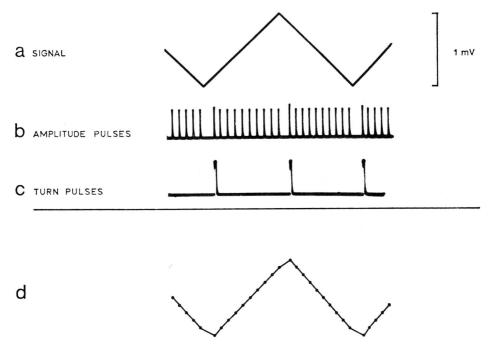

Figure 12–10. Conversion of calibration waveform (*a*) into two serial pulse trains: amplitude (*b*) and turns (*c*). The outputs of these two pulse generators characterize the original input accurately, as evidenced by graphic reconstruction of the waveform (*d*) from *b* and *c*. (From Hayward and Willison,[38] with permission.)

intermittently firing motor units induce a force tremor during isometric contraction. Spectral analysis of muscle force, therefore, can be used to estimate overall motor unit activity.[44] Isokinetic measurements of muscle strength reveal the level of consistency in motor performance, which in turn aids in the identification of unusual patterns, for example, those seen in hysterical paresis.[48] A tetanic contraction generated by a high degree of fusion produces more than twice the tension of a single twitch. Smooth contraction of the whole muscle also results from asynchronous firing of different motor units. The power spectrum of the electromyogram is known to shift during fatigue, a phenomenon best explained by accumulation of extracellular potassium.[69]

Rectification and Integration

Waveform integration helps correlate the muscle force and the electrical activity, although with repeated trials, the result may vary considerably.[84] For determining the total area, a process called full-wave rectification is necessary to reverse the polarity of all positive peaks. The tracing now consists only of negative deflections, allowing one to measure the area between the waveform and baseline without phase cancellation.[2] The integral of a waveform increases in proportion to the amplitude, frequency, and duration of the original potential. The integrated electromyogram usually relates linearly to the isometric tension up to the maximal contraction.[4,57,58,62,63,70,89,93] Muscles of mixed fiber composition, however, may show a nonlinear relationship.[102]

Collision Technique

A collision technique helps determine the relationship between the electrical potential and force produced by voluntary contraction of the first dorsal interosseous muscle. Shocks of supramaximal intensity, delivered at either the wrist or axilla, evoke nearly identical compound muscle action potentials, M(W) or M(A) (Fig. 12–11). Shocks applied simultaneously at the wrist and axilla with the subject at rest elicit M(W) but not M(A) because the orthodromic impulse from the axilla collides with the antidromic impulse from the wrist. During muscle contraction, antidromic impulses from the wrist first collide with voluntary impulses. Therefore, the distal shock cannot completely block the impulse evoked by axillary stimulation. The fraction of M(A) so recorded, termed M(V), represents the magnitude of the voluntary impulse. This technique produces, in effect, a synchronized equivalent to the asynchronous motor neuron activity associated with voluntary contraction.[47] The

amplitude M(V) relates linearly to the force of contraction under isometric conditions (Fig. 12–12).

Needle study during weak voluntary contraction best characterizes the recruitment and discharge pattern of individual motor units.[71,76,77,90] Strong muscle contraction interferes with the identification of single motor units in the presence of a large shower of spikes from many different units. Moreover, the few motor units selected for observation do not necessarily reveal the behavior of the total population of motor neurons. The collision technique provides a direct means of elucidating the relationship between the discharge pattern of the motor neuron pool and muscle force over a wide range of voluntary contraction.

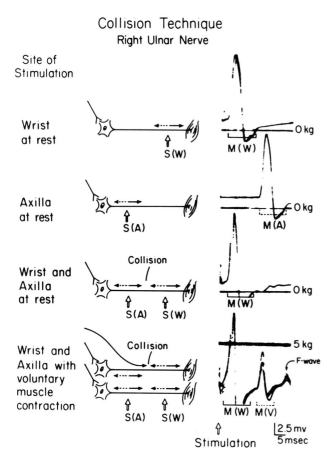

Figure 12–11. Compound muscle action potentials, M(W) and M(A), from the first dorsal interosseous and muscle force (*straight line*). At rest, the antidromic impulse from stimulation at the wrist eliminated the orthodromic impulse from the axilla by collision. With muscle contraction (*bottom tracing*), M(V) appeared in proportion to the number of axons in which the voluntary impulse first collided with the antidromic impulse from the wrist. (From Kimura,[47] with permission.)

Right Ulnar Nerve

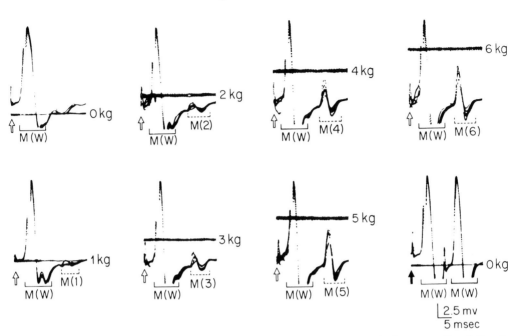

Figure 12–12. Correlation between muscle force and electrical activity, with the same stimulation (*open arrow*) and recording as in the bottom tracing of Figure 12–11. Muscle force ranged from 0 to 6.0 kg (*straight line*). In the last tracing, paired stimuli (*closed arrow*) delivered at the wrist elicited the second M(W) to appear with the same time delay as M(V). The second M(W) equaled the first in amplitude, indicating the integrity of the neuromuscular excitability. (From Kimura,[47] with permission.)

REFERENCES

1. AAEE: Guidelines in Electrodiagnostic Medicine. Professional Standards Committee, American Association of Electromyography and Electrodiagnosis, Rochester, Minnesota, 1984.
2. Bak, MJ, and Loeb, GE: A pulsed integrator for EMG analysis. Electroencephalogr Clin Neurophysiol 47:738–741, 1979.
3. Bergmans, J: Clinical applications: Applications to EMG. In Remond, A (ed): EEG Informatics: A Didactic Review of Methods and Applications of EEG Data Processing. Elsevier, Amsterdam, 1977.
4. Bouisset, S: EMG and muscle force in normal motor activities. In Desmedt, JE (ed): New Developments in Electromyography and Clinical Neurophysiology, Vol 1. Karger, Basel, 1973, pp 547–583.
5. Brooke, MH, Carroll, JE, Davis, JE, and Hagberg, JM: The prolonged exercise test. Neurology 29:636–643, 1979.
6. Brown, WF, and Varkey, GP: The origin of spontaneous electrical activity at the endplate zone. Ann Neurol 10:557–560, 1981.
7. Buchthal, F: An Introduction to Electromyography. Scandinavian University Books, Copenhagen, 1957.
8. Buchthal, F: The general concept of the motor unit. In Adams, RD, Eaton, LM, and Shy, GM (eds): Neuromuscular Disorders. Williams and Wilkins, Baltimore, 1960.
9. Buchthal, F, Guld, C, and Rosenfalck, P: Action potential parameters in normal human muscle and their dependence on physical variables. Acta Physiol Scand 32:200–218, 1954.
10. Buchthal, F, Guld, C, and Rosenfalck, P: Volume conduction of the spike of the motor unit potential investigated with a new type of multielectrode. Acta Physiol Scand 38:331–354, 1957.
11. Buchthal, F, Guld, C, and Rosenfalck, P: Multielectrode study of the territory of a motor unit. Acta Physiol Scand 39:83–104, 1957.
12. Buchthal, F, and Kamieniecka, Z: The diagnostic yield of quantified electromyography and quantified muscle biopsy in neuromuscular disorders. Muscle Nerve 5:265–280, 1982.
13. Buchthal, F, Pinelli, P, and Rosenfalck, P: Action potential parameters in normal human muscle and their physiological determinants. Acta Physiol Scand 32:219–229, 1954.

14. Buchthal, F, and Rosenfalck, P: Spontaneous electrical activity of human muscle. Electroencephalogr Clin Neurophysiol 20:321–336, 1966.

15. Cenkovich, F, Hsu, SF, and Gersten, JW: A quantitative electromyographic index that is independent of the force of contraction. Electroencephalogr Clin Neurophysiol 54:79–86, 1982.

16. Cherington, M, Lewin, E, and McCrimmon, A: Serum creatine phosphokinase changes following needle electromyographic studies. Neurology 18:271–272, 1968.

17. Chrissian, SA, Stolov, WC, and Hongladarom, T: Needle electromyography: Its effect on serum creatine phosphokinase activity. Arch Phys Med Rehabil 57:114–119, 1976.

18. Clamann, HP: Activity of single motor units during isometric tension. Neurology 20:254–260, 1970.

19. Cosi, V, and Mazzella, GL: Frequency analysis in clinical electromyography: A preliminary report (abstr). Electroencephalogr Clin Neurophysiol 27:100, 1969.

20. Daube, JR: The description of motor unit potentials in electromyography. Neurology 28:623–625, 1978.

21. Daube, JR: Needle Examination in Electromyography. American Association of Electromyography and Electrodiagnosis, Mayo Clinic, Rochester, Minnesota, 1979.

22. Dengler, R, Wolf, W, Schubert, M, and Struppler, A: Discharge pattern of single motor units in basal ganglia disorders. Neurology 36:1061–1066, 1986.

23. Desmedt, JE: The neuromuscular disorder in myasthenia gravis. 1. Electrical and mechanical response to nerve stimulation in hand muscles. In Desmedt, JE (ed): New Developments in Electromyography and Clinical Neurophysiology, Vol 1. Karger, Basel, 1973.

24. Dowling, MH, Fitch, P, and Willison, RG: A special purpose digital computer (Biomac 500) used in the analysis of the human electromyogram. Electroencephalogr Clin Neurophysiol 25:570–573, 1968.

25. Ekstedt, J, and Stalberg, E: How the size of the needle electrode leading-off surface influences the shape of the single muscle fibre action potential in electromyography. Computer Prog Biomed 3:204–212, 1973.

26. Fatt, P, and Katz, B: An analysis of the end-plate potential recorded with an intra-cellular electrode. J Physiol (Lond) 115:320–370, 1951.

27. Fex, J, and Krakau, CET: Some experiences with Walton's frequency analysis of the electromyogram. J Neurol Neurosurg Psychiatry 20:178–184, 1957.

28. Fuglsang-Frederiksen, A, Dahl, K, and Monaco, ML: Electrical muscle activity during a gradual increase in force in patients with neuromuscular diseases. Electroencephalogr Clin Neurophysiol 57:320–329, 1984.

29. Fuglsang-Frederiksen, A, and Mansson, A: Analysis of electrical activity of normal muscle in man at different degrees of voluntary effort. J Neurol Neurosurg Psychiatry 38:683–694, 1975.

30. Fuglsang-Frederiksen, A, Monaco, ML, and Dahl, K: Turns analysis (peak ratio) in EMG using the mean amplitude as a substitute of force measurement. Electroencephalogr Clin Neurophysiol 60:225–227, 1985.

31. Fuglsang-Frederiksen, A, Scheel, U, and Buchthal, F: Diagnostic yield of analysis of the pattern of electrical activity and of individual motor unit potentials in myopathy. J Neurol Neurosurg Psychiatry 39:742–750, 1976.

32. Fuglsang-Frederiksen, A, Scheel, U, and Buchthal, F: Diagnostic yield of the analysis of the pattern of electrical activity of muscle and of individual motor unit potentials in neurogenic involvement. J Neurol Neurosurg Psychiatry 40:544–554, 1977.

33. Fuglsang-Frederiksen, A, Smith, T, and Hogenhaven, H: Motor unit firing intervals and other parameters of electrical activity in normal and pathological muscle. J Neurol Sci 78:51–62, 1987.

34. Griep, PAM, Gielen, FLH, Boom, HBK, Boon, KL, Hoogstraten, LLW, Pool, CW, and Wallinga-DeJonge, W: Calculation and registration of the same motor unit action potential. Electroencephalogr Clin Neurophysiol 53:388–404, 1982.

35. Hausmanowa-Petrusewicz, I, and Jozwik, A: The application of the nearest neighbor decision rule in the evaluation of electromyogram in spinal muscular atrophy (SMA) of childhood. Electromyogr Clin Neurophysiol 26:689–703, 1986.

36. Hausmanowa-Petrusewicz, I, and Kopec, J: EMG parameters changes in the effort pattern at various load in dystrophic muscle. Electromyogr Clin Neurophysiol 24:121–136, 1984.

37. Hayward, M: Automatic analysis of the electromyogram in healthy subjects of different ages. J Neurol Sci 33:397–413, 1977.

38. Hayward, M, and Willison, RG: The recognition of myogenic and neurogenic lesions by quantitative EMG. In Desmedt, JE (ed): New Developments in Electromyography and Clinical Neurophysiology, Vol 2. Karger, Basel, 1973, pp 448–453.

39. Hayward, M, and Willison, RG: Automatic analysis of the electromyogram in patients with chronic partial denervation. J Neurol Sci 33:415–423, 1977.

40. Heckmann, R, and Ludin, HP: Differentiation of spontaneous activity from normal and denervated skeletal muscle. J Neurol Neurosurg Psychiatry 45:331–336, 1982.

41. Henneman, E: Relation between size of neurons and their susceptibility to discharge. Science 126:1245–1247, 1957.

42. Hirose, K, and Uono, M: Noise in quantitative electromyography. Electromyogr Clin Neurophysiol 25:341–352, 1985.

43. Hirose, K, Uono, M, and Sobue, I: Quantitative electromyography comparison between manual values and computer ones on normal subjects. Electromyogr Clin Neurophysiol 14:315–320, 1974.

44. Homberg, V, Reiners, K, Hefter, H and Freund, HJ: The muscle activity spectrum: Spectral analysis of muscle force as an estimator of overall motor unit activity. Electroencephalogr Clin Neurophysiol 63:209–222, 1986.

45. Hughes, RC, Park, DC, Parson, ME, and O'Brien, MD: Serum creatine kinase studies in the detection of carriers of Duchenne dystrophy. J Neurol Neurosurg Psychiatry 34: 527–530, 1971.

46. Kaiser, E, and Petersen, I: Muscle action potentials studied by frequency analysis and duration measurement. Acta Neurol Scand 41(Suppl 13):213–236, 1965.

47. Kimura, J: Electrical activity in voluntarily contracting muscle. Arch Neurol 34:85–88, 1977.

48. Knutsson, E, and Martensson, A: Isokinetic measurements of muscle strength in hysterical paresis. Electroencephalogr Clin Neurophysiol 61:370–374, 1985.

49. Kopec, J: Two new descriptions for complete EMG evaluation, applied in automatic analysis. Electromyogr Clin Neurophysiol 24:321–330, 1984.

50. Kopec, J, and Hausmanowa-Petrusewicz, I: Histogram of muscle potentials recorded automatically with the aid of the averaging computer "ANOPS." Electromyography 9:371–381, 1969.

51. Kopec, J, and Hausmanowa-Petrusewicz, I: Application of automatic analysis of electromyograms in clinical diagnosis. Electroenceph Clin Neurophysiol 36:575–576, 1974.

52. Kopec, J, and Hausmanowa-Petrusewicz, I: On-line computer application in clinical quantitative electromyography. Electromyogr Clin Neurophysiol 16:49–64, 1976.

53. Kopec, J, Hausmanowa-Petrusewicz, I, Rawski, M, and Wolynski, M: Automatic analysis in electromyography. In Desmedt, JE (ed): New Developments in Electromyography and Clinical Neurophysiology, Vol 2. Karger, Basel, 1973.

54. Kugelberg, E, and Petersen, I: "Insertion activity" in electromyography: With notes on denervated muscle response to constant current. J Neurol Neurosurg Psychiatry 12: 268–273, 1949.

55. Kunze, K: Quantitative EMG analysis in myogenic and neurogenic muscle diseases. In Desmedt, JE (ed): New Developments in Electromyography and Clinical Neurophysiology, Vol 2. Karger, Basel, 1973, pp 469–476.

56. Kunze, K, and Erbsloh, F: Automatic EMG analysis, a new approach. Electroencephalogr Clin Neurophysiol 25:402, 1968.

57. Kuroda, E, Klissouras, V, and Milsum, JH: Electrical and metabolic activities and fatigue in human isometric contraction. J Appl Physiol 29:358–367, 1970.

58. Lam, HS, Morgan, DL, and Lampard, DG: Derivation of reliable electromyograms and their relation to tension in mammalian skeletal muscles during synchronous stimulation. Electroencephalogr Clin Neurophysiol 46:72–80, 1979.

59. Lang, AH, and Partanen, VSJ: "Satellite potentials" and the duration of motor unit potentials in normal, neuropathic and myopathic muscles. J Neurol Sci 27:513–524, 1976.

60. Lang, AH, and Tuomola, H: The time parameters of motor unit potentials recorded with multi-electrodes and the summation technique. Electromyogr Clin Neurophysiol 14: 513–525, 1974.

61. Lee, RG, and White, DG: Computer analysis of motor unit action potentials in routine clinical electromyography. In Desmedt, JE (ed): New Developments in Electromyography and Clinical Neurophysiology, Vol 2. Karger, Basel, 1973, pp 454–461.

62. Lenman, JAR: A clinical and experimental study of the effects of exercise on motor weakness in neurological disease. J Neurol Neurosurg Psychiatry 22:182–194, 1959.

63. Lippold, OCJ: The relation between integrated action potentials in a human muscle and its isometric tension. J Physiol (Lond) 117:492–499, 1952.

64. Maeyens, E, Jr, and Pitner, SE: Effect of electromyography on CPK and aldolase levels. Arch Neurol 19:538–539, 1968.

65. Mambrito, B, and DeLuca, CJ: A technique for the detection, decomposition and analysis of the EMG signal. Electroencephalogr Clin Neurophysiol 58:175–188, 1984.

66. Martinez, AC, Ferrer, MT, and Perez Conde, MC: Automatic analysis of the electromyogram. 2. Studies in patients with primary muscle disease and neurogenic involvement: Comparison of diagnostic yields versus individual motor unit potential parameters. Electromyogr Clin Neurophysiol 24:17–38, 1984.

67. McComas, AJ, and Sica, REP: Automatic quantitative analysis of the electromyogram in partially denervated distal muscles: Comparison with motor unit counting. Can J Neurol Sci 5:377–383, 1978.

68. McGill, KC, and Dorfman, LJ: Automatic decomposition electromyography (ADEMG): Validation and normative data in brachial biceps. Electroencephalogr Clin Neurophysiol 61: 453–461, 1985.

69. Mills, KR, and Edwards, RHT: Muscle fatigue in myophosphorylase deficiency: Power spectral analysis of the electromyogram. Electroencephalogr Clin Neurophysiol 57:330–335, 1984.

70. Milner-Brown, HS, and Stein, RB: The relation between the surface electromyogram and muscular force. J Physiol (Lond) 246:549–569, 1975.

71. Milner-Brown, HS, Stein, RB, and Yemm, R: Changes in firing rate of human motor units during linearly changing voluntary contractions. J Physiol (Lond) 230:371–390, 1973.

72. Moosa, A, and Brown, BH: Quantitative electromyography: A new analogue technique for detecting changes in action potential duration. J Neurol Neurosurg Psychiatry 35:216–220, 1972.

73. Netter, FH: The Ciba Collection of Medical Il-

lustrations. Vol 1, Nervous System. Part 1. Anatomy and Physiology. Ciba Pharmaceutical Co., Medical Education Division, Summit, N.J., 1983.

74. Partanen, JV, and Nousiainen, U: End-plate spikes in electromyography are fusimotor unit potentials. Neurology 33:1039–1043, 1983.

75. Pedinoff, S, and Sandhu, RS: Electromyographic effect on serum creatine phosphokinase in normal individuals. Arch Phys Med Rehabil 59:27–29, 1978.

76. Person, RS: Rhythmic activity of a group of human motoneurones during voluntary contraction of a muscle. Electroencephalogr Clin Neurophysiol 36:585–595, 1974.

77. Petajan, JH: Clinical electromyographic studies of diseases of the motor unit. Electroencephalogr Clin Neurophysiol 36:395–401, 1974.

78. Pickett, JB: Small sputtering positive waves: Cannula recorded "nerve" potentials (abstr). Electroencephalogr Clin Neurophysiol 45:178, 1978.

79. Pickett, JB, and Schmidley, JW: Sputtering positive potentials in the EMG: An artifact resembling positive waves. Neurology 30:215–218, 1980.

80. Rathjen, R, Simons, DG, and Peterson, CR: Computer analysis of the duration of motor unit potentials. Arch Phys Med Rehabil 49:524–527, 1968.

81. Richardson, AT: Electromyography in myasthenia gravis and the other myopathies. Am J Phys Med 38:118–124, 1959.

82. Scranton, PE, Hasiba, U, and Gorenc, TJ: Intramuscular hemorrhage in hemophiliacs with inhibitors. JAMA 241:2028–2030, 1979.

83. Sica, REP, McComas, AJ, and Ferreira, JCD: Evaluation of an automated method for analysing the electromyogram. Can J Neurol Sci 5:275–281, 1978.

84. Siegler, S, Hillstrom, HJ, Freedman, W, and Moskowitz, G: The effect of myoelectric signal processing on the relationship between muscle force and processed EMG. Am J Phys Med 64:130–149, 1985.

85. Smyth, DPL: Quantitative electromyography in babies and young children with primary muscle disease and neurogenic lesions. J Neurol Sci 56:199–207, 1982.

86. Smyth, DPL, and Willison, RG: Quantitative electromyography in babies and young children with no evidence of neuromuscular disease. J Neurol Sci 56:209–217, 1982.

87. Stalberg, E, Andreassen, S, Falck, B, Lang, H, Rosenfalck, A, and Trojaborg, W: Quantitative analysis of individual motor unit potentials: A proposition for standardized terminology and criteria for measurement. J Clin Neurophysiol 3:313–348, 1986.

88. Stalberg, E, Chu, J, Bril, V, Nandedkar, S, Stalberg, S, and Ericsson, M: Automatic analysis of the EMG interference pattern. Electroencephalogr Clin Neurophysiol 56:672–681, 1983.

89. Stephens, JA, and Taylor, A: The relationship between integrated electrical activity and force in normal and fatiguing human voluntary muscle contractions. In Desmedt, JE (ed): New Developments in Electromyography and Clinical Neurophysiology, Vol 1. Karger, Basel, 1973, pp 623–627.

90. Tanji, J, and Kato, M: Recruitment of motor units in voluntary contractions of a finger muscle in man. Exp Neurol 40:759–770, 1973.

91. Thiele, B, and Bohle, A: Number of spike-components contributing to the motor unit potential. Z EEG-EMG 9:125–130, 1978.

92. Toulouse, P, Coatrieux, JL, and LeMarec, B: An attempt to differentiate female relative of Duchenne type dystrophy from healthy subjects using an automatic EMG analysis. J Neurol Sci 67:45–55, 1985.

93. Vredenbregt, J, and Rau, G: Surface electromyography in relation to force, muscle length and endurance. In Desmedt, JE (ed): New Developments in Electromyography and Clinical Neurophysiology, Vol 1. Karger, Basel, 1973, pp 607–622.

94. Walton, JN: The electromyogram in myopathy: Analysis with the audio-frequency spectrometer. J Neurol Neurosurg Psychiatry 15:219–226, 1952.

95. Warmolts, JR, and Engel, WK: Open-biopsy electromyography. I. Correlation of motor unit behavior with histochemical muscle fiber type in human limb muscle. Arch Neurol 27:512–517, 1972.

96. Weddell, G, Feinstein, B, and Pattle, RE: The electrical activity of voluntary muscle in man under normal and pathological conditions. Brain 67:178–257, 1944.

97. Wiechers, DO: Mechanically provoked insertional activity before and after nerve section in rats. Arch Phys Med Rehabil 58:402–405, 1977.

98. Wiechers, DO: Electromyographic insertional activity in normal limb muscles. Arch Phys Med Rehabil 60:359–363, 1979.

99. Wiechers, DO, Stow, R, and Johnson, EW: Electromyographic insertional activity mechanically provoked in biceps brachii. Arch Phys Med Rehabil 58:573–578, 1977.

100. Wiederholt, WC: "End-plate noise" in electromyography. Neurology 20:214–224, 1970.

101. Willison, RG: Analysis of electrical activity in healthy and dystrophic muscle in man. J Neurol Neurosurg Psychiatry 27:386–394, 1964.

102. Woods, JJ, and Bigland-Ritchie, B: Linear and non-linear surface EMG/force relationships in human muscles. Am J Phys Med 62:287–299, 1983.

103. Yaar, I, Mitz, AR, and Pottala, EW: Fatigue trends in and the diagnosis of myasthenia gravis by frequency analysis of EMG interference patterns. Muscle Nerve 8:328–335, 1985.

104. Yu, YL, and Murray, NMF: A comparison of concentric needle electromyography, quantitative EMG and single fibre EMG in the diagnosis of neuromuscular diseases. Electroencephalogr Clin Neurophysiol 58:220–225, 1984.

Chapter 13

TYPES OF ABNORMALITY

1 INTRODUCTION

Electromyographic studies analyze the propagating muscle action potentials extracellularly (see Chapter 12.2). Except for the end-plate activities and brief injury potentials coincident with the inserting of the needle, a relaxed muscle is electrically silent. Several types of spontaneous discharges seen at rest, therefore, all signal diseases of the nerve or muscle, although they do not necessarily carry the same clinical implications. Both fibrillation potentials and positive sharp waves result from excitation of individual muscle fibers, whereas complex repetitive discharges comprise high-frequency spikes derived from multiple muscle fibers.

A motor unit is the smallest functional element of volitional contraction. In conventional electromyography, isolated discharges of this element give rise to motor unit potentials. Diseases of the nerve or muscle cause structural or functional disturbances of the motor unit, which in turn lead to alterations in the waveform and discharge patterns of their electrical signals. Because certain characteristics of such abnormalities suggest a particular pathologic process, the study of mus-

249

cle action potentials provides information useful in elucidating the nature of the disease.

Electromyography serves as a clinical tool only if the examiner interprets the findings in light of the patient's history, physical examination, and other diagnostic studies. In fact, the study should be regarded as an extension of physical examination, rather than a laboratory test. The four steps of electromyography (see Fig. 12–1) help categorize motor dysfunction into upper and lower motor neuron disorders and myogenic lesions. Each entity has typical electromyographic findings, as shown in Figures 13–1, 2, and 3 and summarized in Figure 13–4. As a means of introduction, the illustrations emphasize the basic principles at the risk of oversimplification. The description in the text will amplify these points and clarify certain variations and exceptions not apparent in the diagrams.

2 INSERTIONAL ACTIVITY

Decreased Versus Prolonged Activity

A marked diminution or absence of insertional activity usually indicates a reduced number of healthy muscle fibers in fibrotic or severely atrophied muscles (see Fig. 12–2). Functionally inexcitable muscle fibers will also show the same abnormality during attacks of familial periodic paralysis. One must of course first exclude technical problems such as a broken lead wire, a faulty needle, or an inadvertent exploration of the subcutaneous fat instead of muscle tissue by the examiner (misjudging the depth of obesity, for example). An abnormally prolonged insertional activity outlasting the cessation of needle movement indicates irritability of the muscle or, more specifically, insta-

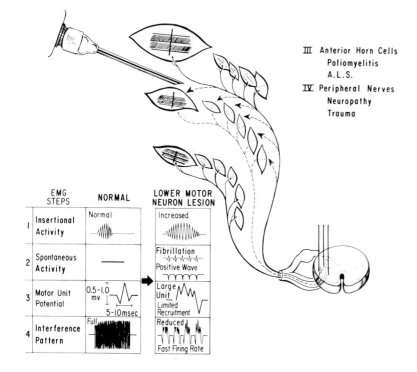

III Anterior Horn Cells
   Poliomyelitis
   A.L.S.
IV Peripheral Nerves
   Neuropathy
   Trauma

Figure 13–1. Typical findings in lower motor neuron lesions. They include (1) prolonged insertional activity; (2) spontaneous activities in the form of fibrillation potentials and positive sharp waves; (3) large amplitude, long duration polyphasic motor unit potentials; and (4) discrete single unit activity firing rapidly during maximal effort of contraction. The diagram depicts reinnervation of muscle fibers supplied by a diseased axon. (Compare with Fig. 12–1.) Although not apparent in this illustration, the sprouting axon respects the anatomic constraint, incorporating only those muscle fibers found within its own boundary. Thus, regeneration increases muscle fiber density but not necessarily motor unit territory.

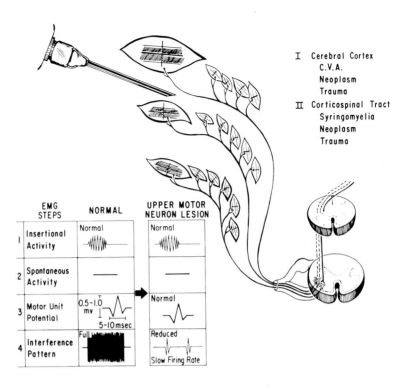

Figure 13–2. Typical findings in upper motor neuron lesions. They include (1) normal insertional activity; (2) no spontaneous activity; (3) normal motor unit potential if detected in an incomplete paralysis; and (4) reduced interference pattern with slow rates of firing of individual motor unit potentials. The diagram illustrates degeneration of the corticospinal tract, resulting in a reduced number of descending impulses reaching the anterior horn cells, which in turn activate a small number of motor unit potentials.

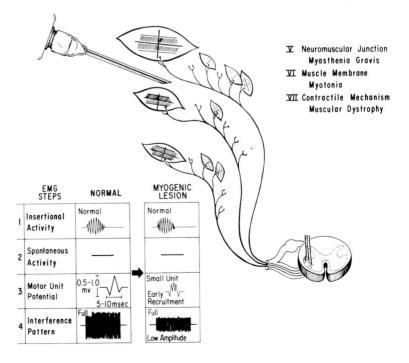

Figure 13–3. Typical findings in myogenic lesions. They include (1) normal insertional activity; (2) no spontaneous activity, with some notable exceptions; (3) low-amplitude, short-duration, polyphasic motor unit potentials; and (4) early recruitment leading to a low-amplitude full interference pattern at less than a maximal effort of contraction. The diagram illustrates a random loss of individual muscle fibers, resulting in a reduced number of fibers per motor unit.

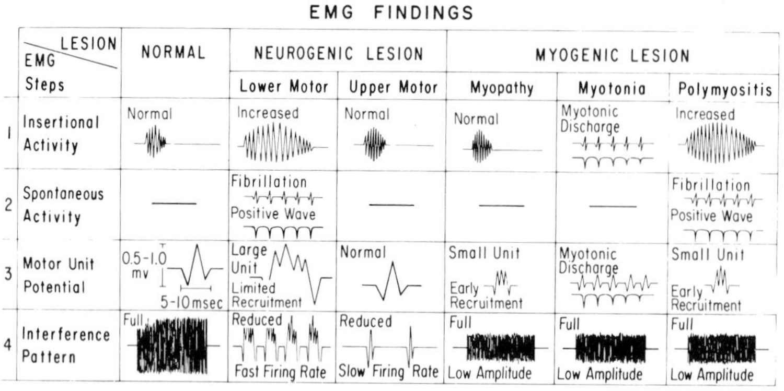

EMG FINDINGS

LESION / EMG Steps	NORMAL	NEUROGENIC LESION		MYOGENIC LESION		
		Lower Motor	Upper Motor	Myopathy	Myotonia	Polymyositis
1. Insertional Activity	Normal	Increased	Normal	Normal	Myotonic Discharge	Increased
2. Spontaneous Activity		Fibrillation / Positive Wave				Fibrillation / Positive Wave
3. Motor Unit Potential	0.5-1.0 mv / 5-10 msec	Large Unit / Limited Recruitment	Normal	Small Unit / Early Recruitment	Myotonic Discharge	Small Unit / Early Recruitment
4. Interference Pattern	Full	Reduced / Fast Firing Rate	Reduced / Slow Firing Rate	Full / Low Amplitude	Full / Low Amplitude	Full / Low Amplitude

Figure 13–4. Typical findings in lower and upper motor neuron disorders and myogenic lesions as shown in Figures 13–1 through 13–3. Myotonia shares many features common to myopathy in general in addition to myotonic discharges triggered by insertion of the needle or with voluntary effort to contract the muscle. Polymyositis shows combined features of myopathy and neuropathy, including (1) prolonged insertional activity; (2) abundant spontaneous discharges; (3) low-amplitude, short-duration, polyphasic motor unit potentials; and (4) early recruitment leading to a low-amplitude full interference pattern.

bility of the muscle membrane (see Fig. 12–2). This type of activity often develops in conjunction with frank denervation, myotonic disorders, or certain other myogenic disorders such as myositis.[55] In addition, some healthy individuals may have one or two isolated positive potentials at the end of the discharge.[101,102] The lack of reproducibility distinguishes this variant of normal insertional activity from qualitatively similar insertional positive waves described below.

Insertional Positive Waves

A briefly sustained run of positive waves may follow insertional activity, lasting several seconds to minutes after cessation of the needle movement. Less frequently, a train of negative spikes with or without initial positivity may develop instead of positive sharp waves. These discharges, ranging from 3 to 30 impulses per second in firing frequency, closely resemble the spontaneous discharges recorded from frankly denervated muscles at rest. In fact, abnormal insertional activity of this type commonly appears in the early stages of denervation, 10 days to 2 weeks after nerve injury, before the appearance of spontaneous activity. It may also occur in chronically denervated muscles or in association with rapidly progressive degeneration of muscle fibers in acute polymyo-

sitis. In these cases, positive sharp waves also appear spontaneously—not initiated by needle movement (Fig. 13–5). By definition, insertional activity immediately follows the mechanical stimulus by the needle, even if it continues after cessation of needle movement; whereas true spontaneous activities occur without apparent triggering mechanisms. Needle movement also enhances spontaneous activity, making the differentiation between insertional and noninsertional activities somewhat arbitrary.

Insertional positive waves seen in denervated muscles have an abrupt onset and termination without the waxing and waning quality. Nonetheless, a few positive waves in the first seconds after insertion of the needle may mimic a mild form of myotonic discharge; hence the use of the now discarded term pseudomyotonia. On the other hand, an abortive form of myotonic discharge, seen immediately after prolonged exercise, resembles positive waves of early denervation. This finding in otherwise asymptomatic subjects possibly suggests a forme fruste of myotonia congenita of autosomal dominant inheritance.[102]

3 MYOTONIC DISCHARGE

In myotonia, a sustained contraction of the muscle follows voluntary movement on electrical or mechanical stimulation.

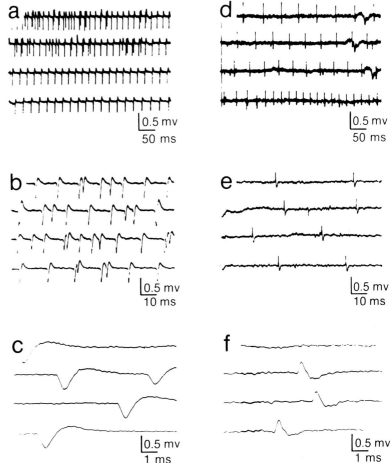

Figure 13–5. Spontaneous single fiber discharges from the right paraspinal muscle in a 62-year-old woman with polymyositis. The tracings show two types of discharges: trains of positive sharp waves (a, b, and c) and negative spikes (d, e, and f) initiated by insertion of the needle electrode. The lack of initial positivity indicates the recording of the negative spikes near the end-plate region, although their rhythmic pattern speaks against the physiologic end-plate spikes. Note progressively declining amplitude (a and d) in the absence of the waxing and waning quality typically associated with myotonic potentials. (Compare Fig. 13–6.)

This phenomenon is seen in myotonia congenita, myotonia dystrophica, paramyotonia congenita, and hyperkalemic periodic paralysis.[18,64] The electromyographic correlates of clinical myotonia consist of rhythmic discharges triggered by insertion of the needle electrode, but outlasting the external source of excitation. Myotonic discharges do not necessarily accompany clinical myotonia when seen in polymyositis, type II glycogen storage disease with acid maltase deficiency,[45] or other disorders with chronic denervation.

Positive Versus Negative Discharge

Myotonic discharges take two forms, probably depending on the spatial relationship between the recording surface of the needle electrode and the discharging muscle fibers. One type of myotonic discharge occurs as a sustained run of sharp positive waves, each followed by a slow negative component of much longer duration (Fig. 13–6). These waveforms, like those of denervation, probably represent recurring single fiber potentials recorded from an injured area of the muscle membrane. A second type of myotonic discharge consists of a sustained run of negative spikes with a small initial positivity. They resemble fibrillation potentials seen in denervation. In contrast to the positive sharp waves usually initiated by needle insertion, the negative spikes tend to occur at the beginning of slight volitional contraction. Both positive sharp waves and negative spikes typically wax and

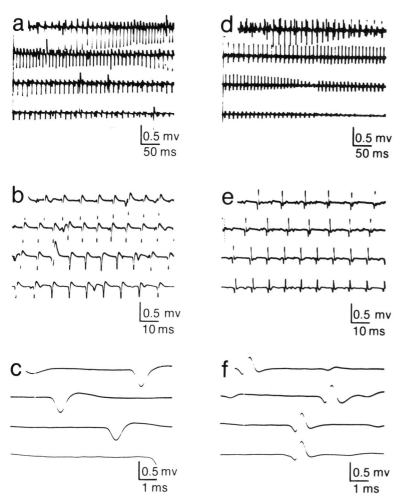

Figure 13–6. Myotonic discharges from the right anterior tibialis in a 39-year-old man with myotonic dystrophy. The tracings show two types of discharges: trains of positive sharp waves (*a, b,* and *c*) and negative spikes with initial positivity (*d, e,* and *f*). The discharges in *a* and *d* reveal waxing and waning. (Compare Fig. 13–5.)

wane in amplitude over the range of 10 μV to 1 mV, varying inversely to the rate of firing. Their frequency may increase or decrease within the range of 50 to 100 impulses per second, giving rise to a characteristic noise over the loudspeaker, reminiscent of an accelerating or decelerating motorcycle or chain saw. Despite common belief, a myotonic discharge does not closely simulate the sound of a dive-bomber, judged from my extensive personal experience (with dive-bombers).

Pathophysiology

The pathophysiology of the myotonic discharge, although yet unestablished in man, centers on a decrease in resting chloride (Cl^-) conductance. This results in repetitive electrical activity in isolated frog[36] and mammalian skeletal muscles.[59] Electrophysiologic studies show abnormalities attributable to decreased density of chloride channels in hereditary myotonia of goats.[58] In normal fibers the presence of chloride conductance stabilizes the membrane potential by shunting the depolarizing current and dampening its effect.[47] Conversely, the absence of chloride conductance in effect raises the resistance of the membrane. According to Ohm's law, $E = IR$, increased resistance, R, will reduce the amount of current, I, necessary to initiate a threshold depolarization, E.[36]

The critical level of depolarization opens the sodium (Na^+) channel with a rapid change in sodium conductance, which in turn initiates an action potential. The action potential falls with inactivation of sodium conductance and delayed activation of potassium (K^+) conductance, which tends to hyperpolarize the membrane. As potassium conductance slowly returns to its resting value, the cell becomes slightly depolarized, with accumulation of potassium in the transverse tubule system. In an unstable membrane without chloride shunting the current, this slow change may trigger another action potential, and the cycle repeats itself.[12] Thus, the process of depolarization begins as soon as repolarization ends, leading to a series of repetitive action potentials. The explanation of myotonic phenomena based on low chloride conductance, however, does not seem to apply to all human myotonic dystrophy.[56]

Animal studies also suggest possible relationships between the increase in muscle peroxisomal enzymes and the onset of myotonic discharges.[4] Pharmacologic blocking of acetylcholine receptor or atropine binding site effectively silences fibrillation potentials, but not myotonic discharges.[11] The excitability of myotonic muscle changes in proportion to the plasma level of infused potassium. The membrane characteristics can be reliably assessed by the duration of the electromyographic relaxation time after maximal voluntary effort. Myotonic afterdischarges, however, vary too much to provide such a measure.[31] Cultured muscle cells from patients with myotonic dystrophy show the same electrophysiologic properties as the control cell.[86]

4 SPONTANEOUS ACTIVITY

Origin and Clinical Significance

In the first 2 weeks after denervation, the sensitivity of a muscle fiber to acetylcholine (ACh) increases by as much as 100-fold.[10,60,90] This phenomenon, known as denervation hypersensitivity, may explain spontaneous discharges of denervated muscle fibers in response to small quantities of circulating ACh.[28,88] The disappearance of fibrillation potentials following artificially induced ischemia[44] and in isolated muscle fibers[88] also supports the presence of some circulating substance. In rats, fibrillation potentials cease after application of alpha-bungarotoxin or atropine sulfate.[11] Therefore, the receptor molecules for these agents must play an essential part in the production of spontaneous activity.

Experimental data have been marshaled against the ACh hypersensitivity hypothesis: (1) The large amount of circulating ACh reaching the end-plate combines with acetylcholinesterase concentrated in this region. This results in continuous hydrolysis of ACh to choline and acetate. (2) Denervation hypersensitivity reflects the development of many highly reactive sites along the entire length of the denervated muscle fiber,[92] rather than a specific change localized to the end-plate region.[3] Spontaneous activity, however, seems to originate only in the end-plate zone and not elsewhere along the nonjunctional membrane.[5] Further, the infusion of curare blocks the end-plate receptors but fails to abolish spontaneous discharges. (3) Denervation of frog muscle may cause increased sensitivity to ACh but produces no spontaneous activity.[62] These findings suggest that ACh hypersensitivity alone cannot explain the generation of spontaneous activity. One alternative hypothesis invokes slowly changing membrane potentials of metabolic origin that may periodically reach the critical level and evoke propagated spikes.[88]

Spontaneous activity, if reproducible in at least two muscle sites, provides an unequivocal sign of abnormality and one of the most useful findings in clinical electromyography. It is usually seen in denervated muscles, but also in certain primary muscle diseases, as discussed later in the section on fibrillation potential. Because of the latent period of 2 to 3 weeks, its absence does not preclude denervation during the early weeks of nerve injury. When found in disorders of the lower motor neuron, the distribution of spontaneous potentials can aid in lo-

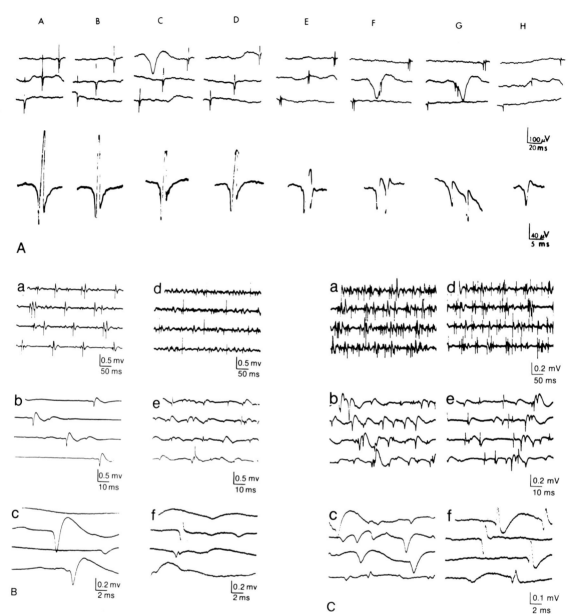

Figure 13–7. A. Single fiber discharges recorded from the denervated tibialis anterior in a 67-year-old man with acute onset of a footdrop (compare Fig. 5–6). Note gradual alteration of the waveform from a triphasic spike with major negativity to paired positive potentials and finally to single positive sharp wave over the time course of some 8 seconds without movement of the needle. This fortuitous recording provides direct evidence that the same single fiber discharge can be recorded either as fibrillation potentials or positive sharp waves. Long duration positive deflections seen in c, f, and g represent a pulse artifact. (From Kimura,[52] with permission.) **B.** Spontaneous single fiber activity of the anterior tibialis in a 68-year-old woman with amyotrophic lateral sclerosis. The tracings show two types of discharges: positive sharp waves (a, b, and c) and fibrillation potentials (d, e, and f).

calizing lesions of the spinal cord, root, plexus, or peripheral nerve.

Spontaneous discharges also occur, though not consistently, in otherwise uninvolved paretic limbs between 6 weeks and 3 months after the onset of acute upper motor neuron lesions.[24,50] This phenomenon may be analogous to hemiplegic atrophy, which also follows a stroke.[79] Some, however, argue that posi-

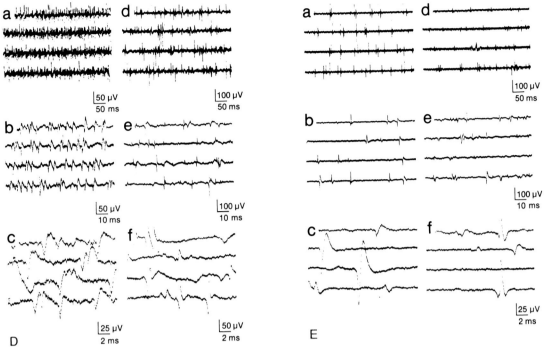

Figure 13–7 *(Cont).* C. Spontaneous single fiber activity of the paraspinal muscle in a 40-year-old man with radiculopathy, consisting of positive sharp waves (*a, b,* and *c*) and fibrillation potentials (*d, e,* and *f*). **D.** Spontaneous single fiber activity of the deltoid (*a, b,* and *c*) and tibialis anterior (*d, e,* and *f*) in a 9-year-old boy with a 6-week history of dermatomyositis, showing two types of discharges: positive sharp waves (*a, b,* and *c*) and fibrillation potentials (*d, e,* and *f*). **E.** Spontaneous single fiber activity of the tibialis anterior in a 7-year-old boy with Duchenne type dystrophy, showing positive sharp waves (*a, b,* and *c*) and fibrillation potentials (*d, e,* and *f*).

tive sharp waves and fibrillation potentials seen in hemiplegic patients reflect secondary disease of the lower motor neurons.[23] Spontaneous activity may also appear in the paraspinous muscles, after myelography or lumbar puncture, developing by the first day after the procedure and resolving by the second or fourth day.[25,99] As a rule, no spontaneous activity develops in disuse atrophy.

Basic types of spontaneous activity comprise fibrillation potentials, positive sharp waves, fasciculation potentials, myokymic discharges, and complex repetitive discharges. Isolated visible muscle twitches over a localized area may accompany fasciculation potentials and complex repetitive discharges but neither fibrillation potentials nor positive sharp waves. Myokymic discharges seen in cramp syndromes cause sustained contraction (see Chapter 26.6).

Both fibrillation potentials and positive sharp waves represent single fiber acti-

vation.[26] Physical relationship between the generator and the recording electrode dictate the waveform of the potential (Fig. 13–7A). If the tip of the needle damages the membrane, the sustained standing depolarization here precludes the generation of a negative spike at this point. Thus, propagating action potential that approaches the site of injury gives rise to a sharp positive discharge, followed by low-amplitude negative deflection. Fasciculation potentials are isolated spontaneous discharges of a motor unit. In contrast, myokymic discharges represent repetitive firing of a motor unit, as the name grouped fasciculation indicates. Complex repetitive discharges result from rapid firing of many muscle fibers in sequence, driven ephaptically at a point of lateral contact.[32,85] A spontaneously activated single fiber serving as a pacemaker regulates the frequency and pattern of discharge by two different, usually independent, mechanisms: rate of rhythmic depolarization of the denervated muscle fiber and circus movements of currents among muscle fibers.[48]

A numeric grading serves to semiquantitate fibrillation potentials and positive sharp waves in a clinical study:

+1—Transient but reproducible runs of positive discharges after moving the needle electrode (i.e., insertional positive waves).

+2—Occasional spontaneous potentials at rest in more than two different sites.

+3—Spontaneous activities present at rest, regardless of the position of the needle electrode.

+4—Abundant spontaneous potentials nearly filling the screen of the oscilloscope.

Fibrillation Potentials

Fibrillation potentials range from 1 to 5 ms in duration and 20 to 200 μV in amplitude when recorded with a concentric needle electrode.[21] They usually show diphasic or triphasic waveforms with initial positivity (Fig. 13–7A to E), although the needle electrode placed near the end-plate zone registers an initial negativity. To differentiate from physiologic end-plate spikes, therefore, one must analyze spontaneous activity away from the end-plate region. Over the loudspeaker, the fibrillation potentials produce a crisp clicking noise reminiscent of the sound caused by wrinkling tissue paper. The discharges increase after warming the muscle or with administration of cholinesterase inhibitors, such as edrophonium (Tensilon) or neostigmine (Prostigmin), and decrease after moderate cooling of the muscle or hypoxia. Thus, warming the muscle under study enhances the chance of detecting spontaneous activity during electromyography.

Fibrillation potentials triggered by spontaneous oscillations in the membrane potential typically fire in a regular pattern at a rate of 1 to 30 impulses per second, with an average frequency of 13 impulses per second.[21,88] The decreased resting membrane potential in the dener-

vated muscle plays a critical role as the cause of the oscillations.[89] A very irregular firing pattern usually represents discharges from more than one fiber. Fibrillation potentials originating from the same muscle fiber, however, may also fire irregularly in the range of 0.1 to 25 impulses per second.[17,63] These potentials result from random, discrete, spontaneous depolarization of nearly constant amplitude.[17] A new class of sodium channels that develops after denervation may cause reduced sodium inactivation. Increased sodium conductance presumably accounts for progressive lowering of the firing threshold, giving rise to cyclical activities.[71]

Fibrillation potentials and voluntarily activated single fiber potentials have the same shape and amplitude distribution when studied with single-fiber electromyography (SFEMG).[84] Close scrutiny of a train reveals no change in shape between the first and the last discharges. These findings indicate that fibrillation potentials originate from single muscle fibers, a view consistent with the observation that they represent the smallest unit recorded by the needle electrode.[28,49] The now abandoned concept of the subunit led to the earlier erroneous belief that 10 to 30 muscle fibers must discharge to generate a single potential.[21,22]

Fibrillation potentials, although usually regarded as pathognomonic of denervation, may occasionally appear in otherwise healthy muscles. The isolated incidence, therefore, cannot serve as absolute evidence of abnormality. Nonetheless, the presence of reproducible discharges in at least two different areas of the muscle usually suggests lower motor neuron disorders. These include diseases of the anterior horn cells, radiculopathies, plexopathies, and axonal mono- or polyneuropathies. Spontaneous discharges are commonly seen in certain myopathic processes such as muscular dystrophy, dermatomyositis, or polymyositis. Less consistently, diseases of neuromuscular junction also give rise to fibrillation potentials, as do many other disorders,[39,69] such as facioscapulohumeral dystrophy, limb girdle dystrophy, oculopharyngeal dystrophy,[42] myotubular (centronuclear) myopathy,[83] and trichinosis.[97,98]

Fibrillation potentials found in 25 percent of patients with progressive muscular dystrophy[21] result at least in part from denervation secondary to muscle necrosis.[29] Spontaneous activity in polymyositis suggests increased membrane irritability,[7] inflammation of intramuscular nerve fibers,[74] or focal degeneration separating a part of the muscle fiber from the end-plate region.[81] In support of postulated functional denervation, SFEMG and histochemical techniques have revealed evidence of reinnervation in the terminal innervation pattern.[43]

Positive Sharp Waves

These discharges have a saw tooth appearance with the initial positivity and a subsequent slow negativity, much lower in amplitude but longer in duration. They often follow insertion of the needle but also fire spontaneously, at regular intervals (see Fig. 13–7). The absence of a negative spike implies recording near the damaged part of the muscle fiber. Although usually seen together following nerve section, the appearance of fibrillation potential often lags behind that of positive sharp waves, which can be triggered by the insertion of a needle.[101]

Like fibrillation potentials, positive sharp waves are seen not only in denervated muscles but also in a variety of myogenic conditions. The latter group includes polymyositis, dermatomyositis, trichinosis, ischemic myositis, and progressive muscular dystrophy. As discussed earlier, positive sharp waves may form part of myotonic discharges, triggered by insertion of the needle or mild voluntary contraction. Despite close resemblance in waveform, myotonic discharges, which characteristically wax and wave, do not appear spontaneously.

Fasciculation Potentials and Myokymic Discharges

Clinicians once referred to visible twitching of muscle bundles as fibrillation, a term now reserved for the electro-myographic description of spontaneously firing single muscle fibers. To avoid confusion, Denny-Brown and Pennybacker[28] proposed the term fasciculation to describe the spontaneous contraction of motor units. Fasciculation potentials result from spontaneous discharges of a group of muscle fibers representing either a whole or possibly part of a motor unit (Fig. 13–8). Motor unit potentials in the depth of the muscle may not necessarily induce visible twitches. In these instances, electromyography allows detection of this spontaneous activity, which would otherwise remain unrecognized.

Unlike voluntary motor unit potentials, fasciculation potentials may undergo slight changes in amplitude and waveform from time to time. Mild voluntary contraction of agonistic or antagonistic muscles fails to alter the firing rate or discharge pattern. The generator source remains unknown, although the existing evidence suggests that the neural discharge may originate in the spinal cord or anywhere along the length of the peripheral nerve.[100] In one study using a collision method and F-wave analysis, nearly all fasciculations had an axonal origin.[76] Fasciculation potentials may sometimes persist despite distal nerve block. After the total removal of the nerve supply to the muscle, fasciculations remain for about 4 days and then disappear.[37]

In contrast to isolated discharges of one motor unit, more complex bursts of repetitive discharges cause vermicular movements of the skin, called myokymia.[82] Repetitive firing of the same motor units (see Chapter 26.6), usually occur in bursts at regular intervals of 0.1 to 10 seconds, with 2 to 10 spikes discharging at 30 to 40 impulses per second in each burst (Fig. 13–9). Myokymic discharges commonly, though not specifically, involve facial muscles in patients with brainstem glioma or multiple sclerosis. Myokymic discharges also favor certain chronic neuropathic processes, such as Guillain-Barré syndrome[61] and radiation plexopathies.[1,2,27] Hyperventilation induces hypocalcemia, which in turn amplifies axonal excitability and myokymic bursts, generated ectopically in demyelinated motor fibers.[9]

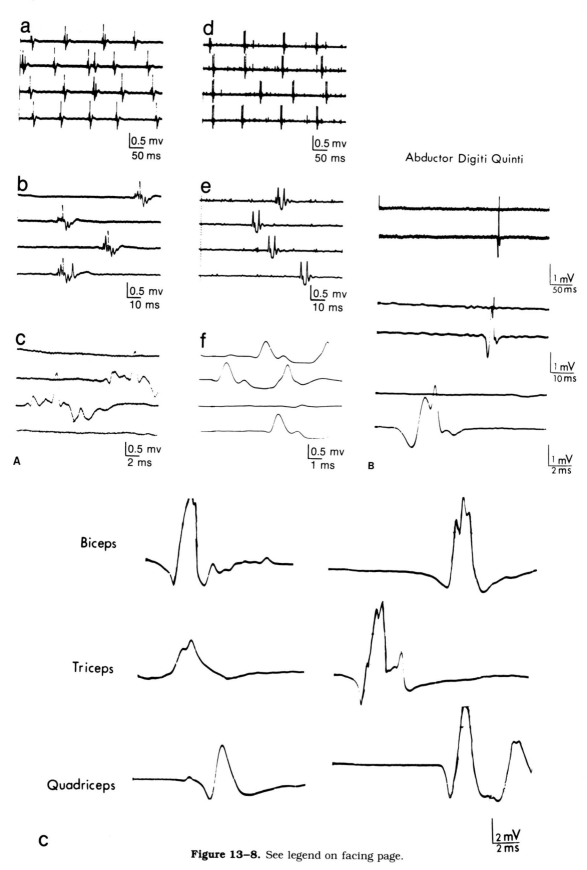

Abductor Digiti Quinti

Biceps

Triceps

Quadriceps

Figure 13–8. See legend on facing page.

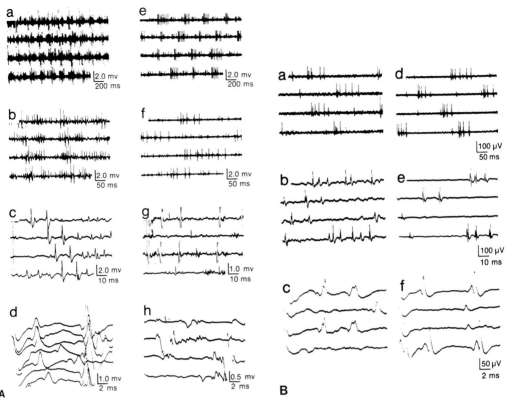

Figure 13–9. A. Myokymic discharges in a 21-year-old woman with multiple sclerosis. The patient had visible undulating movement of the facial muscles on the right associated with characteristic bursts of spontaneous activity recorded from the orbicularis oris (*a, b, c,* and *d*) as well as orbicularis oculi (*e, f, g,* and *h*). In *d*, each sweep, triggered by a recurring spontaneous potential, shows a repetitive but not exactly time-locked pattern of the waveform. **B.** Myokymic discharges in a 57-year-old man with a 2-week history of the Guillain-Barré syndrome and a nearly complete peripheral facial palsy. Despite the absence of visible undulating movement, rhythmically recurring spontaneous discharges appeared in the upper (*a, b,* and *c*) and lower (*d, e,* and *f*) portions of the left orbicularis oris. In *c* and *f*, each sweep, triggered by a recurring spontaneous potential, shows the repetitive pattern.

Fasciculation potentials, although typically associated with diseases of anterior horn cells, are also seen in radiculopathy, entrapment neuropathy, and the muscular pain-fasciculation syndrome.[46] In patients with cervical spondylotic myelopathy, fasciculation potentials may appear in the lower extremities, presumably secondary to loss of inhibition, vascular insufficiency, cord traction, or denervation. Although these hypotheses lack anatomic or physiologic evidence, spontaneous discharges do abate after cervical decompression.[51,53] Fasciculation poten-

Figure 13–8. A. Fasciculation potentials from the tibialis anterior in two patients with polyneuropathy, displayed with a continuous run from top to bottom in *a, b, d,* and *e* and interrupted sweep in *c* and *f*. Both the polyphasic, long-duration potential (*a, b,* and *c*) and a doublet (*d, e,* and *f*) are indistinguishable in shape from voluntarily activated motor unit potentials, especially when they fire at high frequency, as shown in these examples. **B.** A 59-year-old man with pain in the posterior calf after he fell from a ladder, landing on his feet. Electromyography showed fibrillation potentials and sharp positive waves in the abductor hallucis and only fasciculation potentials in the abductor digiti quinti. (From Kimura,[52] with permission.) **C.** A 58-year-old man with muscular pain fasciculation syndrome of 1 year's duration. He came to the hospital for evaluation of "muscle twitching," which began in the right arm but soon became generalized. Electromyography showed fasciculation potentials in most muscles tested with no evidence of other spontaneous discharges such as fibrillation potentials or sharp positive waves. (From Kimura,[52] with permission.)

tials also appear in some metabolic derangements such as tetany, thyrotoxicosis, and overdoses of anticholinesterase medication.[26] Grouped fasciculation potentials from multiple units tend to imply an ominous prognosis because of their frequent association with either amyotrophic lateral sclerosis or progressive spinal muscular atrophy. They are also seen in other degenerative diseases of the anterior horn cells, including poliomyelitis and syringomyelia.

Either single or grouped spontaneous discharges may occur in otherwise normal muscle, sometimes, but not always, causing cramps.[73] Data obtained from a questionnaire survey of a group of 539 healthy medical personnel indicate that 70 percent have experienced some type of muscle twitch.[72] Because of the serious implications, a number of investigators have sought to differentiate this form of fasciculation potential from that associated with motor neuron disease, but in vain. No single method reliably distinguishes one type from the other on the basis of waveform characteristics, such as amplitude, duration, and number of phases.[73,91] The frequency of discharge, however, may possibly separate the two categories because the discharges seen in motor neuron disease fire irregularly at an average interval of 3.5 seconds, compared with 0.8 seconds in asymptomatic individuals.[91]

In conclusion, fasciculation potentials by themselves cannot provide absolute proof of abnormality, unless accompanied by either fibrillation potentials or positive sharp waves. Excluding those seen in healthy subjects, they suggest disease of the lower motor neuron with the origin at any level from the anterior horn cells to axon terminals. Electrophysiologic studies fail to offer reliable means of distinguishing between "benign" forms seen in otherwise normal muscle and "malignant" forms associated with motor neuron disease.[73] The dichotomy, therefore, serves no useful purpose in the clinical domain. For description, one must characterize a recorded discharge by its waveform, amplitude, duration, firing pattern, and frequency of occurrence.

Complex Repetitive Discharges

The repetitive discharges range from 50 μV to 1 mV in amplitude and up to 50 to 100 ms in duration, representing a group of muscle fibers firing in near synchrony (Figs. 13–10 and 13–11). The entire sequence repeats itself at slow or fast rates, usually in the range of 5 to 100 impulses per second. The polyphasic and complex waveform remains uniform from one discharge to another, although the shape may change suddenly. These discharges typically begin suddenly, maintain a constant rate of firing for a short period, and cease as abruptly as they started. Over the loudspeaker, they mimic the sound of a machine gun. The unique repetitive pattern once prompted the use of a now discarded term, bizarre high frequency discharges. Superficial similarities to myotonic sound led to the even less appropriate term pseudomyotonia, despite the lack of waxing and waning. The waveforms seen in complex repetitive discharges, myokymia, neuromyotonia, and cramp syndromes are similar to one another, although each differs in rate of repetition and firing pattern (see Chapter 26).

In single-fiber recordings,[85] complex repetitive discharges often consist of 10 or more distinct unit potentials separated by intervals ranging from less than 0.5 ms to more than 200 ms. The individual spikes within the complex fire in the same order, as the discharge recurs repetitively. One fiber in the complex, serving as a pacemaker, initiates the burst, driving one or several other fibers ephaptically.[85,93] In successive cycles, one of the remaining fibers, activated late in the previous cycle, reexcites the principal pacemaker to repeat the whole cycle until the pacemaker fibers eventually become blocked. The electrical field associated with this repetitive pattern must effectively induce ephaptic activation of neighboring muscle fibers. Thus, complex repetitive discharges often give rise to high-amplitude spikes, compared with fibrillation potentials.

This discharge is seen in some myopathies, such as muscular dystrophy or polymyositis, and a wide variety of

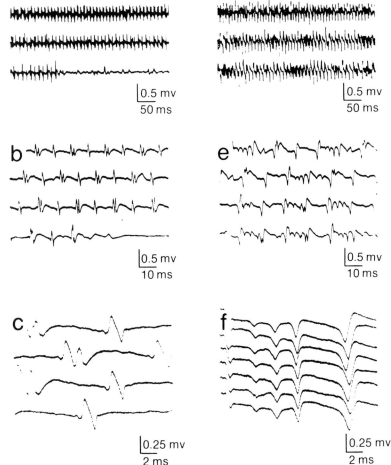

Figure 13–10. Complex repetitive discharges of the left quadriceps in a 58-year-old man with herniated lumbar disk. The tracings show two types of discharges: trains of single- or double-peaked negative spikes (a, b, and c) and complex positive sharp waves (d, e, and f). In f, each sweep, triggered by a recurring motor unit potential, shows remarkable reproducibility of the waveform within a given train.

chronic denervating conditions, such as motor neuron disease, radiculopathy, chronic polyneuropathy, myxedema, and the Schwarz-Jampel syndrome. In a large series,[32] an overall analysis of the prevalence revealed its highest incidence in Duchenne muscular dystrophy, spinal muscular atrophy, and Charcot-Marie-Tooth disease. Profuse activity of this type has been found in the striated muscle of the urethral sphincter associated with urinary retention in women.[38] The complex repetitive discharges are occasionally seen in apparently healthy subjects. These foci of a clinically silent irritative process tend to involve deeper muscles in general and the iliopsoas in particular.

5 MOTOR UNIT POTENTIALS

A motor unit potential can be defined by amplitude, rise time, duration, phases, stability, and territory. A wide range of neuromuscular disorders affects these measures in different combinations. Hence, such abnormalities per se fail to establish a specific diagnosis, although they help distinguish primary muscle diseases from lower motor neuron disorders. In myopathies, motor unit potentials decrease in spike duration and amplitude, reflecting random loss of individual fibers.[20] In neuropathies or anterior horn cell diseases, a loss of axons results in a reduced number of units, al-

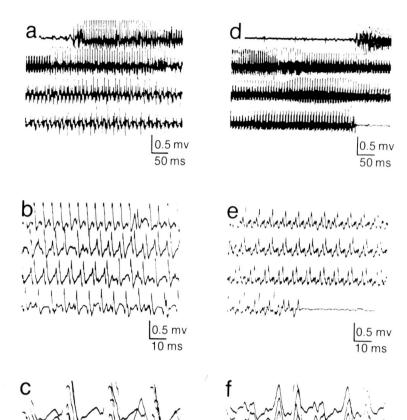

Figure 13–11. Complex repetitive discharges with trains of negative spikes from the same muscle shown in Figure 13–10. Note gradual decline of discharge frequency in one train (*a*, *b*, and *c*) but not in the other (*d*, *e*, and *f*) and characteristically abrupt onset and cessation (*a*, *d*, and *e*). In *c* and *f*, each sweep, triggered by a recurring motor unit potential, shows a detailed waveform of the repetitive patterns.

though surviving fibers with sprouting give rise to a larger potential than normal. Thus, taken together with abnormalities of insertional and spontaneous activities, changes in the size and recruitment pattern of the motor unit potential play an essential role in the classification of weakness in diseases of the nerve and muscle.[41]

Abnormalities of Motor Unit Potentials

The following discussion will deal with the contrasting features of the motor unit potential seen in myopathies and lower motor neuron disorders. Each type of abnormality is common to a number of disease categories, as listed here and described in greater detail later from clinical points of view (see Chapters 20 through 26).

The recorded amplitude varies greatly with the position of the needle electrode relative to the discharging unit. One must, therefore, select a motor unit potential with a short rise time of 500 μs or less, which guarantees its proximity to the recording surface. The number of single muscle fibers at the tip of the needle determines the size of the negative spike. The higher the amplitude is, the closer together are the muscle fibers near the recording surface. Hence, in general, the amplitude aids in determining the muscle fiber density, not the motor unit territory. Distant units, not contributing to the amplitude of the negative spike, add to the motor unit duration, increasing the time of the initial and terminal positivity. Thus, the duration of the motor unit potential serves as a good index of the motor unit territory. For meaningful assessment, one must compare the measured value with the normal range established in the same muscle for the same age group by the same technique.[13]

Diphasic or triphasic motor unit potentials abound in normal muscles, with only 5 to 15 percent having four or more phases. The number of polyphasic units increases either in myopathy, neuropathy, or motor neuron disease (Fig. 13–12). Polyphasia indicates temporal dispersion of muscle fiber potentials within a motor unit. Excessive temporal dispersion, in

turn, results from differences in conduction time along the terminal branch of the nerve or over the muscle fiber membrane. During neurapraxia or an acute stage of axonotmesis, motor unit potentials, if recorded at all, show normal waveforms, indicating the integrity of the surviving axons (Fig. 13–13).

Motor units normally discharge semi-rhythmically, with successive potentials showing nearly identical configuration. Fatigue causes irregularity and reduction in the firing rate, without altering the waveform of the motor unit potential. In patients with defective neuromuscular transmission, the amplitude of a repetitively firing unit may fluctuate or diminish steadily. This finding suggests intermittent blocking of individual muscle fibers within the unit as recurring discharges deplete the store of immediately available ACh. Waveform variability of a repetitively firing motor unit potential serves to document deficient neuromuscular transmission. This observation is especially useful in muscles not accessible by conventional nerve stimulation techniques (see Chapter 10.4). Such an instability of motor unit potential, however, is consistent not only with myasthenia gravis, myasthenic syndrome, and botulism but also with motor neuron disease, poliomyelitis, and syringomyelia and during the early stages of reinnervation. In myotonia congenita, a characteristic decline in amplitude of the successive discharges typically recovers during continued contraction.

In another abnormality called doublets or triplets (see Fig. 13–8A), a motor unit fires twice or three times at very short intervals. In doublets, or double discharges, two action potentials maintain the same relationship to one another at intervals of 2 to 20 ms. The term paired discharges is used to describe a set of spikes with longer intervals, ranging from 20 to 80 ms. In triplets, the middle spike discharges closer to the first than to the third, although both intervals range from 2 to 20 ms. The physiologic origin and clinical significance of multiple discharges remain to be elucidated. They are characteristic of latent tetani, hyperventilation, and other metabolic states asso-

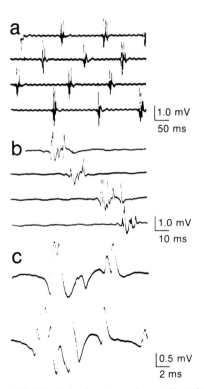

a

1.0 mV
50 ms

b

1.0 mV
10 ms

c

0.5 mV
2 ms

Figure 13–12. Polyphasic motor unit potentials from the anterior tibialis in a 52-year-old man with amyotrophic lateral sclerosis. Temporal variability of repetitive discharges in waveform suggests intermittent blocking of some axon terminals.

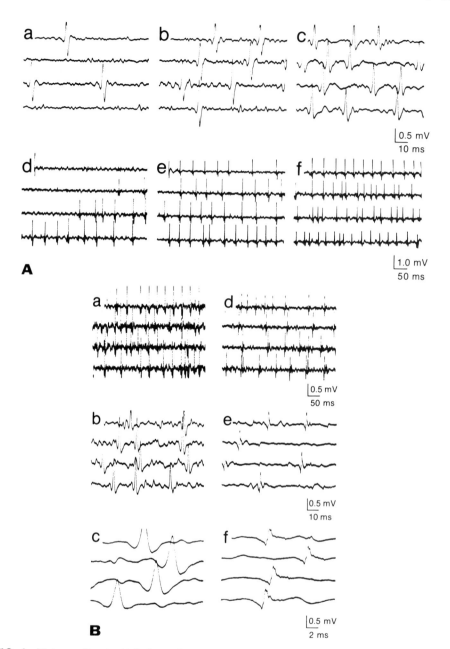

Figure 13–13. A. Motor unit potentials from the extensor digitorum communis in a 20-year-old man with partial radial nerve palsy. Minimal (a and d), moderate (b and e), and maximal voluntary contraction (c and f) recruited only a single motor unit, which discharged at progressively higher rates. **B.** Motor unit potentials from the extensor carpi ulnaris (a, b, and c) and extensor carpi radialis longus (d, e, and f) in the same subject. Maximal voluntary contraction recruited only a single motor unit firing at a high discharge rate.

ciated with hyperexcitability of the motor neuron pool.[78] Other possibilities include poliomyelitis,[94] motor neuron disease,[87] Guillain-Barré syndrome,[75] radiculopathy, and myotonic dystrophy.[66] Doublets may also be observed at the begin- ning and termination of voluntary contraction in normal muscles.[68] Patients with fasciculation potentials do not necessarily have a higher incidence of double discharges during voluntary activation of motor unit potentials.[67]

Lower Motor Neuron Versus Myopathic Disorders

Increased amplitude and duration (Fig. 13–14) generally suggest disorders of the lower motor neuron, such as motor neuron disease, poliomyelitis, and syringomyelia, or diseases of the peripheral nerve, such as chronic neuropathy and reinnervation after nerve injury. In these disorders, the increased size of motor unit potential indicates anatomic reorganization of denervated muscle fibers by means of axon sprouting. Alternatively, two or more motor units may discharge simultaneously with abnormal synchronization at the cord level or with

ephaptic activation near the terminal axons.[77] Increased duration probably results from usual variability in length and conduction time of regenerating axon terminals, rather than enlarged territory. In contrast, increased amplitude implies greater muscle fiber density, with incorporation of denervated fibers within the territory of the surviving axon (see Fig. 13–1).

Studies on the time course of reinnervation[8] have revealed characteristic changes of motor unit potentials following traumatic nerve lesions. Complete nerve transection leads to increased polyphasicity and temporal instability, with intermittent segmental conduction

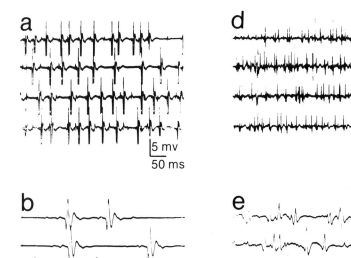

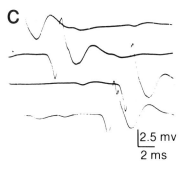

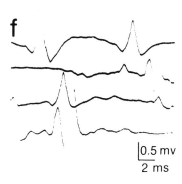

Figure 13–14. High-amplitude, long-duration motor unit potentials from the first dorsal interosseous (*a*, *b*, and *c*), compared with relatively normal motor unit potentials from the orbicularis oculi (*d*, *e*, and *f*) in a patient with polyneuropathy. Note a discrete single unit interference pattern in both muscles during maximal voluntary contraction.

block of regenerating motor axons. After a partial nerve lesion, healthy motor axons give rise to extensive collaterals for reinnervation of the denervated muscle fibers. Late potentials linked to the main unit will substantially increase the total duration. These long-latency components, easily overlooked in free-running modes, become apparent if recorded with the use of an internal trigger. Here, a recurring motor unit potential itself initiates the sweep, but a delay line allows the display of the potential in its entirety.

In general, reduction in amplitude and duration of the motor unit potential (Fig. 13–15) suggests primary myopathic disorders such as muscular dystrophy, congenital or other myopathies, periodic paralysis, myositis, and disorders of neuromuscular transmission, including myasthenia gravis, myasthenic syndrome, and botulism. All these entities have in common the random loss of functional muscle fibers from each motor unit, caused by muscle degeneration, inflammation, metabolic changes, or failure of neuromuscular activation. In extreme cases, voluntary contraction activates only a single muscle fiber, displaying a motor unit potential indistinguishable from a fibrillation potential. The short spikes, 1 to 2 ms in duration, produce a high-frequency sound over the loudspeaker, reminiscent of spontaneously discharging fibrillation potentials. Unlike some inherited disorders of muscle, metabolic or toxic myopathies may cause reversible changes.[14] Mild metabolic and endocrine myopathies characteristically show little or no alteration in duration or amplitude of the motor unit potential.

Contrasting changes in amplitude and duration of the motor unit potential generally help differentiate myopathies from lower motor neuron disorders.[15,16,20,40] Electromyography and histochemical findings from muscle biopsies have an overall concordance of 90 percent or greater,[6,15,16,41] although the distinction may not always be unequivocal.[33] Sick axon terminals in distal neuropathy, for example, may result in random loss of muscle fibers within a motor unit. Similarly, during early reinnervation, immature motor units consist of only a few muscle fibers. Motor unit potentials may then become polyphasic, of low amplitude and of short duration. In both instances a neuropathic process will produce changes classically regarded as consistent with a myopathy.[65]

Conversely, motor unit potentials may be of long duration in myopathies with regenerating muscle fibers, erroneously suggesting a neuropathic process.[29,57,70] These potentials commonly appear quite distinct from the main unit, thus giving rise to the terms satellite or parasite potentials, now discarded in favor of the more descriptive name late component. In addition, if the fiber density increases during regeneration, so does the amplitude, to a range much greater than ordinarily expected in myopathy. Hence, the oversimplified dichotomy between myopathy and neuropathy may not necessarily hold in interpreting abnormalities of motor unit potentials and correlating them with clinical diagnoses.[34,35,95,96]

Despite these uncertainties, the elec-

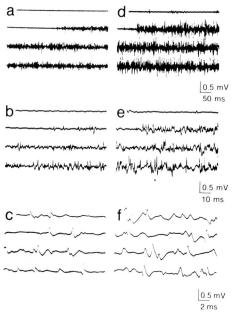

Figure 13–15. Low-amplitude, short-duration motor unit potentials from the biceps (*a, b,* and *c*) and tibialis anterior (*d, e,* and *f*) in a 7-year-old boy with Duchenne type dystrophy. (Compare with Fig. 13–7E.) Minimal voluntary contraction recruited an excessive number of motor units in both muscles.

tromyographic studies allow division between myogenic and lower motor neuron involvements in most patients with definite weakness.[19] Findings often vary among different muscles in the same patient or even from one site to another in a given muscle. An adequate study consists of exploration in different parts of the limb, sampling each muscle in several areas. In some disease states, muscles with minimal dysfunction may show no abnormality, whereas very severely diseased muscles may reveal only nonspecific end-stage changes. Optimal evaluations, therefore, should include those moderately affected but not totally destroyed by the disease process.

6 RECRUITMENT PATTERN

Lower and Upper Motor Neuron Disorders

The number and the average force contributed by each functional motor unit dictates the recruitment pattern. In disorders of the motor neuron, root, or peripheral nerve with reduced numbers of excitable motor units, recruitment is necessarily limited despite increased effort to contract the muscle. To maintain a certain force, surviving motor neurons must fire at an inappropriately rapid rate to compensate for the loss in number. In extreme instances, a single motor unit potential produces the discrete "picket fence" interference pattern at maximal effort (Fig. 13–16).

In late recruitment caused by failure of descending impulses, the excited motor units discharge more slowly than expected for normal maximal contraction (Fig. 13–17). Thus, a lower motor neuron weakness with a rapid rate of discharge stands in sharp contrast to an upper motor neuron or hysterical paralysis with a slow rate of discharge, even though both show a reduced interference pattern. In addition, hysterical weakness or poor cooperation often produces irregular tremulous firing of motor units, not seen in a genuine paresis. Thus, isokinetic measurements of muscle strength reveal increased variability of tonus in repeated tests and other signs of inconsistency and contradictory motor performance.[54] Electromyography also has value in quantitatively assessing paresis caused by upper motor neuron lesions. For example, it may reveal the presence of surviving fibers that traverse the injured portion of the spinal cord even in patients diagnosed as having a complete transection.[30]

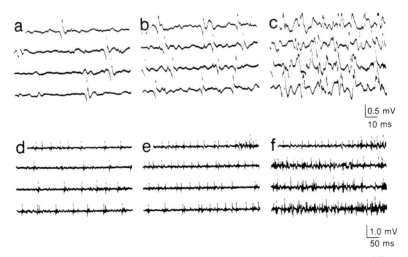

Figure 13–16. Reduced recruitment and incomplete interference pattern of the mildly paretic extensor carpi radialis in a 20-year-old man with partial radial nerve palsy. The rate of firing, rather than the number of discharging motor units, increased during minimal (a and d), moderate (b and e), and maximal voluntary contraction (c and f). (Compare Fig. 13–13.)

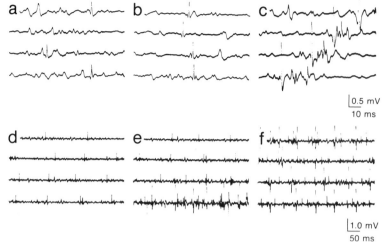

Figure 13–17. Reduced recruitment and incomplete interference pattern of the tibialis anterior in a 31-year-old woman with hysterical weakness. Minimal (*a* and *d*), moderate (*b* and *e*), and maximal (*c* and *f*) effort to contract the muscle altered neither the rate of firing nor the number of discharging motor units appreciably.

Myopathy

Motor unit potentials, if low in amplitude and short in duration, must recruit early to functionally compensate in quantity for the smaller force per unit. The number of units required to maintain a given force increases in proportion to the inefficiency of unit discharge. Thus, with slight voluntary effort, many axons begin to fire almost instantaneously in advanced myopathy (Fig. 13–18). A full interference pattern develops at less than maximal contraction, although its amplitude remains low, reflecting the decreased fiber density of individual motor units. For the same reason, the motor units also recruit early, reaching full interference prematurely in diseases of neuromuscular transmission. This general rule may not apply in advanced myogenic disorders with loss of whole motor units, rather than individual muscle fibers. Here, a limited recruitment leads to an incomplete interference pattern, mimicking a neuropathic change.

Involuntary Movement

Electromyographic findings in movement disorders with involuntary motor symptoms may resemble changes seen in lower motor neuron disease (see Chapter 26). Tremor shows characteristic bursts of motor unit potentials repeating at a fairly constant rate. Although many motor units fire in a group during each burst, no fixed temporal or spatial relationships emerge among them. Thus, successive bursts vary in amplitude, duration, waveform, and number of motor unit potentials. A subclinical tremor burst could masquerade a polyphasic motor unit potential of long duration, despite the varying appearance and rhythmic pattern. Electromyographic recordings help characterize different types of tremor on the basis of their rate, rhythm, and distribution.[80]

In most other movement disorders, normal motor unit potentials recruit involuntarily. These include the stiff-man syndrome, common cramps, tetanus, tetani, and hemifacial spasm (see Chapter 26). Synkinesis seen in hemifacial spasm or following aberrant regeneration gives rise to unintended activation of motor units in the muscles not under voluntary contraction (see Fig. 14–1). Simultaneous recording from multiple muscles confirms the presence of time-locked discharge of aberrant motor unit potentials, thus differentiating associated voluntary activity from involuntary synkinetic discharges.

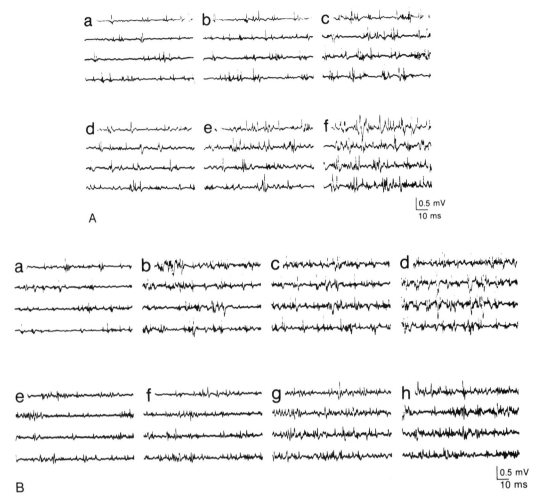

Figure 13–18. A. Early recruitment of the deltoid (*a, b,* and *c*) and tibialis anterior (*d, e,* and *f*) in a 9-year-old boy with a 6-week history of dermatomyositis. (Compare Fig. 13–7D.) Note abundant motor units discharging with increasing effort from *a* through *c* and *d* through *f* during minimal muscle contraction. **B.** Early recruitment of the biceps (*a, b, c,* and *d*) and tibialis anterior (*e, f, g,* and *h*) in a 7-year-old boy with Duchenne type dystrophy. (Compare Fig. 13–7E.) An excessive number of motor unit potentials appeared during minimal (*a* and *e*), mild (*b* and *f*), moderate (*c* and *g*), and maximal contraction (*d* and *h*).

REFERENCES

1. Aho, K, and Sainio, K: Late irradiation-induced lesions of the lumbosacral plexus. Neurology 33:953–955, 1983.
2. Albers, JW, Allen, AA, Bastron, JA, and Daube, JR: Limb myokymia. Muscle Nerve 4:494–504, 1981.
3. Axelsson, J, and Thesleff, S: A study of supersensitivity of denervated mammalian skeletal muscle. J Physiol (Lond) 149:178–193, 1957.
4. Behrens, MI, Soza, MA, and Inestrosa, NC: Increase of muscle peroxisomal enzymes and myotonia induced by nafenopin, a hypolipidemic drug. Muscle Nerve 6:154–159, 1983.
5. Belmar, J, and Eyzaguirre, C: Pacemaker site of fibrillation potentials in denervated mammalian muscle. J Neurophysiol 29:425–441, 1966.
6. Black, JT, Bhatt, GP, DeJesus, PV, Schotland, DL, and Rowland, LP: Diagnostic accuracy of clinical data, quantitative electromyography and histochemistry in neuromuscular disease: A study of 105 cases. J Neurol Sci 21:59–70, 1974.
7. Bohan, A, and Peter, JB: Polymyositis and dermatomyositis. N Engl J Med 292:344–347, 1975.
8. Borenstein, S, and Desmedt, JE: Range of variations in motor unit potentials during reinnervation after traumatic nerve lesions in humans. Ann Neurol 8:460–467, 1980.
9. Brick, JF, Gutmann, L, and McComas, CF:

Calcium effect on generation and amplification of myokymic discharges. Neurology 32:618–622, 1982.

10. Brown, GL: The actions of acetylcholine on denervated mammalian and frog's muscle. J Physiol (Lond) 89:438–461, 1937.

11. Brumback, RA, Bertorini, TE, Engel, WEK, Trotter, JL, Oliver, KL, and Zirow, GC: The effect of pharmacologic acetylcholine receptor on fibrillation and myotonia in rat skeletal muscle. Arch Neurol 35:8–10, 1978.

12. Bryant, SH: The electrophysiology of myotonia, with a review of congenital myotonia of goats. In Desmedt, JE (ed): New Developments in Electromyography and Clinical Neurophysiology, Vol 1. Karger, Basel, 1973, pp 420–450.

13. Buchthal, F: An Introduction to Electromyography. Scandinavian University Books, Copenhagen, 1957.

14. Buchthal, F: Electrophysiologic abnormalities in metabolic myopathies and neuropathies. Acta Neurol Scand (Suppl 43)46:129–176, 1970.

15. Buchthal, F: Diagnostic significance of the myopathic EMG. In Rowland, LP (ed): Pathogenesis of Human Muscular Dystrophies. Proceedings of the Fifth International Scientific Conference of the Muscular Dystrophy Association, Durango, Colorado, June 1976. Excerpta Medica, Amsterdam, 1977.

16. Buchthal, F: Electrophysiological signs of myopathy as related with muscle biopsy. Acta Neurol (Napoli) 32:1–29, 1977.

17. Buchthal, F: Fibrillations: Clinical electrophysiology. In Culp, WJ, and Ochoa, J (eds): Abnormal Nerves and Muscles as Impulse Generators. Oxford University Press, Oxford, pp 632–662, 1982.

18. Buchthal, F, Engbaek, L, and Gamstorp, I: Paresis and hyperexcitability in adynamia episodica hereditaria. Neurology 8:347–351, 1958.

19. Buchthal, F, and Kamieniecka, Z: The diagnostic yield of quantified electromyography and quantified muscle biopsy in neuromuscular disorders. Muscle Nerve 5:265–280, 1982.

20. Buchthal, F, and Rosenfalck, P: Electrophysiological aspects of myopathy with particular reference to progressive muscular dystrophy. In Bourne, GH, and Golarz, MN (ed): Muscular Dystrophy in Man and Animals. Hafner Publishing Company, New York, 1963.

21. Buchthal, F, and Rosenfalck, P: Spontaneous electrical activity of human muscle. Electroenceph Clin Neurophysiol 20:321–336, 1966.

22. Buchthal, F, and Rosenfalck, P: On the structure of motor units. In Desmedt, JE (ed): New Developments in Electromyography and Clinical Neurophysiology, Vol 1. Karger, Basel, 1973, pp 71–85.

23. Chokroverty, S, and Medina, J: Electrophysiological study of hemiplegia. Arch Neurol 35:360–363, 1978.

24. Cruz Martinez, A: Electrophysiological study in hemiparetic patients: Electromyography, motor conduction velocity, and response to repetitive nerve stimulation. Electromyogr Clin Neurophysiol 23:139–148, 1983.

25. Danner, R: Occurrence of transient positive sharp wave-like activity in the paraspinal muscles following lumbar puncture. Electromyogr Clin Neurophysiol 22:149–154, 1982.

26. Daube, JR: Needle Examination in Electromyography. American Association of Electromyography and Electrodiagnosis, Minimonograph #11, Rochester, Minn, 1979.

27. Daube, JR, Kelly, JJ, Jr, and Martin, RA: Facial myokymia with polyradiculoneuropathy. Neurology 29:662–669, 1979.

28. Denny-Brown, D, and Pennybacker, JB: Fibrillation and fasciculation in voluntary muscle. Brain 61:311–332, 1938.

29. Desmedt, JE, and Borenstein, S: Regeneration in Duchenne muscular dystrophy: Electromyographic evidence. Arch Neurol 33:642–650, 1976.

30. Dimitrijevic, MR, Dimitrijevic, MM, Faganel, J, and Sherwood, AM: Suprasegmentally induced motor unit activity in paralyzed muscles of patients with established spinal cord injury. Ann Neurol 16:216–221, 1984.

31. Durelli, L, Mutani, R, Piredda, S, Fassio, F, and Delsedime, M: The quantification of myotonia. J Neurol Sci 59:167–173, 1983.

32. Emeryk, B, Hausmanowa-Petrusewicz, I, and Nowak, T: Spontaneous volleys of bizarre high frequency potentials (b.h.f.p.) in neuro-muscular diseases. Part I. Occurrence of spontaneous volleys of b.h.f.p. in neuro-muscular diseases. Part II. An analysis of the morphology of spontaneous volleys of b.h.f.p. in neuromuscular diseases. Electromyogr Clin Neurophysiol (Part I) 14:303–312, 1974; (Part II) 14:339–354, 1974.

33. Engel, WK: Brief, small, abundant motor-unit action potentials: A further critique of electromyographic interpretation. Neurology 25:173–176, 1975.

34. Engel, WK, and Warmolts, JR: Multiplicity of muscle changes postulated from motoneuron abnormalities. In Cassens, RG (ed): Muscle Biology. Marcel Dekker, New York, 1972.

35. Engel, WK, and Warmolts, JR: The motor unit. Diseases affecting it in toto or in portio. In Desmedt, JE (ed): New Developments in Electromyography and Clinical Neurophysiology, Vol 1. Karger, Basel, 1973, pp 141–177.

36. Falk, G, and Landa, JF: Effects of potassium on frog skeletal muscle in a chloride-deficient medium. Am J Physiol 198:1225–1231, 1960.

37. Forster, FM, Borkowski, WJ, and Alpers, BJ: Effects of denervation on fasciculations in human muscle: Relation of fibrillations to fasciculations. Arch Neurol Psychiatry 56:276–283, 1946.

38. Fowler, CJ, Kirby, RS, and Harrison, MJG: Decelerating burst and complex repetitive discharges in the striated muscle of the urethral sphincter, associated with urinary retention in women. J Neurol Neurosurg Psychiatry 48:1004–1009, 1985.

39. Fusfeld, RD: Electromyographic abnormalities in a case of botulism. Bulletin of the Los An-

geles Neurological Societies 35:164–168, 1970.

40. Hausmanowa-Petrusewicz, I, Emeryk, B, Wasowicz, B, and Kopec, A: Electromyography in neuro-muscular diagnostics. Electromyography 7:203–225, 1967.

41. Hausmanowa-Petrusewicz, I, and Jedrzejowska, H: Correlation between electromyographic findings and muscle biopsy in cases of neuromuscular disease (abstr). J Neurol Sci 13:85–106, 1971.

42. Heffernan, LP, Rewcastle, NB, and Humphrey, JG: The spectrum of rod myopathies. Arch Neurol 18:529–542, 1968.

43. Henriksson, KG, and Stalberg, E: The terminal innervation pattern in polymyositis: A histochemical and SFEMG study. Muscle Nerve 1:3–13, 1978.

44. Hnik, P, and Skorpil, V: Fibrillation activity in denervated muscle. In Gutmann, E (ed): The Denervated Muscle. Czech Academy of Science, Prague, 1962, pp 135–150.

45. Hudgson, P, Gardner-Medwin D, Worsfold, M, Pennington, RJT, and Walton, JN: Adult myopathy from glycogen storage disease due to acid maltase deficiency. Brain 91:435–462, 1968.

46. Hudson, AJ, Brown, WF, and Gilbert, JJ: The muscular pain-fasciculation syndrome. Neurology 28:1105–1109, 1978.

47. Hutter, OF, and Noble, D: The chloride conductance of frog skeletal muscle. J Physiol (Lond) 151:89–102, 1960.

48. Jablecki, C, and Knoll, D: Fibrillation potentials and complex repetitive discharges (abstr). Electroencephalogr Clin Neurophysiol 50:242, 1980.

49. Jasper, H, and Ballem, G: Unipolar electromyograms of normal and denervated human muscle. J Neurophysiol 12:231–244, 1949.

50. Johnson, EW, Denny, ST, and Kelley, JP: Sequence of electromyographic abnormalities in stroke syndrome. Arch Phys Med Rehabil 56:468–473, 1975.

51. Kadson, DL: Cervical spondylotic myelopathy with reversible fasciculations in the lower extremities. Arch Neurol 34:774–776, 1977.

52. Kimura, J: Electromyography and nerve stimulation techniques: Clinical applications (Japanese). Egakushoin, Tokyo, 1989.

53. King, RB, and Stoops, WL: Cervical myelopathy with fasciculations in the lower extremities. J Neurosurg 20:948–952, 1963.

54. Knutsson, E, and Martensson, A: Isokinetic measurements of muscle strength in hysterical paresis. Electroencephalogr Clin Neurophysiol 61:370–374, 1985.

55. Kugelberg, E, and Petersén, I: "Insertion activity" in electromyography: With notes on denervated muscle response to constant current. J Neurol Neurosurg Psychiatry 12:268–273, 1949.

56. Kuhn, E: Myotonia: The clinical evidence. In Desmedt, JE (ed): New Developments in Electromyography and Clinical Neurophysiology, Vol 1. Karger, Basel, 1973, pp 413–419.

57. Lang, AH, and Partanen, VSJ: "Satellite po-

tentials" and the duration of motor unit potentials in normal, neuropathic and myopathic muscles. J Neurol Sci 27:513–524, 1976.

58. Lipicky, RJ, and Bryant, SH: A biophysical study of the human myotonias. In Desmedt, JE (ed): New Developments in Electromyography and Clinical Neurophysiology, Vol 1. Karger, Basel, 1973, pp 451–463.

59. Lullmann, H: Das Verhalten normaler und denervierter Skelemuskulatur in chloridfreiem Medium (Methylsulfat-Tyrodelosung). Naunyn Schmiedebergs Arch Exp Pathol Pharmacol 240:351–360, 1961.

60. Maillis, AG, and Johnstone, BM: Observations on the development of muscle hypersensitivity following chronic nerve conduction blockage and recovery. J Neurol Sci 38:145–161, 1978.

61. Mateer, JE, Gutmann, L, and McComas, CF: Myokymia in Guillain-Barré syndrome. Neurology 33:374–376, 1983.

62. Miledi, R: The acetylcholine sensitivity of frog muscle fibres after complete or partial denervation. J Physiol (Lond) 151:1–23, 1960.

63. Miller, RG: AAEE Minimonograph #28: Injury to peripheral motor nerves. Muscle Nerve 10:698–710, 1987.

64. Morrison, JB: The electromyographic changes in hyperkalaemic familial periodic paralysis. Ann Phys Med 5:153–155, 1960.

65. Nakashima, K, Tabuchi, Y, and Takahashi, K: The diagnostic significance of large action potentials in myopathy. J Neurol Sci 61:161–170, 1983.

66. Partanen, VSJ: Double discharges in neuromuscular diseases. J Neurol Sci 36:377–382, 1978.

67. Partanen, VSJ: Lack of correlation between spontaneous fasciculations and double discharges of voluntarily activated motor units. J Neurol Sci 42:261–266, 1979.

68. Partanen, VSJ, and Lang, AH: An analysis of double discharges in the human electromyogram. J Neurol Sci 36:363–375, 1978.

69. Petersen, I, and Broman, AM: Elektromyografiska fynd från ett fall av botulism. Nordisk Med 65:259–261, 1961.

70. Pickett, JB: Late components of motor unit potentials in a patient with myoglobinuria. Ann Neurol 3:461–464, 1978.

71. Purves, D, and Sakmann, B: Membrane properties underlying spontaneous activity of denervated muscle fibers. J Physiol (Lond) 239:125–153, 1974.

72. Reed, DM, and Kurland, LT: Muscle fasciculations in a healthy population. Arch Neurol 9:363–367, 1963.

73. Richardson, AT: Muscle fasciculation. Arch Phys Med Rehabil 35:281–286, 1954.

74. Richardson, AT: Clinical and electromyographic aspects of polymyositis. Proc R Soc Med 49:111–114, 1956.

75. Roth, G: Réflexe d'axone moteur. Arch Suisses Neurol Neurochir Psychiatr 109:73–97, 1971.

76. Roth, G: Fasciculations and their F-response

localization of their axonal origin. J Neurol Sci 63:299–306, 1984.

77. Roth, G, and Magistris, MD: Ephapse between two motor units in chronically denervated muscle. Electromyogr Clin Neurophysiol 25:331–339, 1985.

78. Scherrer, J, and Metral, S: Electromyography. In Vinken, PJ, and Bruyn, GW (eds): Handbook of Clinical Neurology, Vol 1, Disturbances of Nervous Function. North-Holland, Amsterdam, 1969.

79. Segura, RP, and Sahgal, V: Hemiplegic atrophy: electrophysiological and morphological studies. Muscle Nerve 4:246–248, 1981.

80. Shahani, BT, and Young, RR: The blink, H, and tendon vibration reflexes. In Goodgold, J, and Eberstein, A (eds): Electrodiagnosis of Neuromuscular Diseases, ed 2. Williams and Wilkins, Baltimore, 1977, pp 245–263.

81. Simpson, JA: Handbook of Electromyography and Clinical Neurophysiology, Vol 16, Neuromuscular Diseases. Elsevier, Amsterdam, 1973.

82. Sindermann, F, Conrad, B, Jacobi, HM, and Prochazka, VJ: Unusual properties of repetitive fasciculations. Electroencephalogr Clin Neurophysiol 35:173–179, 1973.

83. Spiro, AJ, Shy, GM, and Gonatas, NK: Myotubular myopathy: Persistence of fetal muscle in an adolescent boy. Arch Neurol 14:1–14, 1966.

84. Stalberg, E, and Ekstedt, J: Single fibre EMG and microphysiology of the motor unit in normal and diseased human muscle. In Desmedt, JE (ed): New Developments in Electromyography and Clinical Neurophysiology, Vol 1. Karger, Basel, 1973, pp 113–129.

85. Stalberg, E, and Trontelj, JV: Single Fiber Electromyography. The Mirvalle Press Limited, Old Woking, Surrey, UK, 1979.

86. Tahmoush, AJ, Askanas, V, Nelson, PG, and Engel, WK: Electrophysiologic properties of aneurally cultured muscle from patients with myotonic muscular atrophy. Neurology 33:311–316, 1983.

87. Taraschi, G, and Lanzi, G: Decharges multiples d'une unite motrice, durant l'activite volontaire dans un cas de sclerose laterale amyotrophique. Electroencephalogr Clin Neurophysiol (Suppl 22):146–148, 1962.

88. Thesleff, S: Fibrillation in denervated mammalian skeletal muscle. In Culp, WJ, and Ochoa, J (eds): Abnormal Nerves and Muscle as Impulse Generators. Oxford University Press, Oxford, 1982, pp 678–694.

89. Thesleff, S, and Sellin, LC: Denervation supersensitivity. Trends Neurosci 4:122–126, 1980.

90. Trojaborg, W: Early electrophysiologic changes in conduction block. Muscle Nerve 1:400–403, 1978.

91. Trojaborg, W, and Buchthal, F: Malignant and benign fasciculations. Acta Neurol Scand (Suppl 13)41:251–254, 1965.

92. Trontelj, J, and Stålberg, E: Responses to electrical stimulation of denervated human muscle fibers recorded with single fibre EMG. J Neurol Neurosurg Psychiatry 46:305–309, 1983.

93. Trontelj, J, and Stålberg, E: Bizarre repetitive discharges recorded with single fibre EMG. J Neurol Neurosurg Psychiatry 46:310–316, 1983.

94. Valle, M: Decharges multiples dans la poliomyelite par mecanismes fonctionnels nonreconductibles a cryptotetanie. Electroenceph Clin Neurophysiol (Suppl 22):144–146, 1962.

95. Warmolts, JR, and Engel, WK: Open-biopsy electromyography. I. Correlation of motor unit behavior with histochemical muscle fiber type in human limb muscle. Arch Neurol 27:512–517, 1972.

96. Warmolts, JR, and Mendell, JR: Open-biopsy electromyography: Direct correlation of a pattern of excessively recruited, pathologically small motor unit potentials with histologic evidence of neuropathy. Arch Neurol 36:406–409, 1979.

97. Waylonis, GW, and Johnson, EW: The electromyogram in acute trichinosis: Report of four cases. Arch Phys Med Rehabil 45:177–183, 1964.

98. Waylonis, GW, and Johnson, EW: Electromyographic findings in induced trichinosis. Arch Phys Med Rehabil 46:615–625, 1965.

99. Weber, RJ, and Weingarden, SI: Electromyographic abnormalities following myelography. Arch Neurol 36:588–589, 1979.

100. Wettstein, A: The origin of fasciculations in motoneuron disease. Ann Neurol 5:295–300, 1979.

101. Wiechers, DO: Mechanically provoked insertional activity before and after nerve section in rats. Arch Phys Med Rehabil 58:402–495, 1977.

102. Wiechers, DO, and Johnson, EW: Diffuse abnormal electromyographic insertional activity: A preliminary report. Arch Phys Med Rehabil 60:419–422, 1979.

Chapter 14

EXAMINATION OF NONLIMB MUSCLES

1 INTRODUCTION

The muscles of mastication, face, soft palate, and tongue are as readily accessible to the needle electrode as the skeletal muscles in the limbs. Electromyographic evaluation of laryngeal muscles requires the assistance of an otolaryngologist for proper placement of the needle electrode. Examination of the extraocular muscles also poses technical difficulty, but ophthalmologists with the special skill and knowledge can place the electrode safely in the intended muscles. These muscles have the same physiologic and pharmacologic properties as the peripheral skeletal muscles.[19]

The same technique applies to the truncal musculature and the muscles of the extremities. The abdominal muscles are innervated by the intercostal nerves derived from the anterior rami of the spinal nerve, whereas the paraspinal muscles are supplied by the posterior rami. The study of these muscles and the external anal sphincter requires no special instrumentation. Full evaluation of the paraurethral muscles depends to a great extent on cystometry and other urodynamic procedures, which are beyond the scope of this book.[16,26,40,54,63]

2 MUSCLES OF THE FACE, LARYNX, AND NECK

The ordinary techniques used for the skeletal muscles also apply in studies of most voluntary muscles innervated by the cranial nerves. The exceptions include the laryngeal and extraocular muscles as discussed below. The most commonly tested muscles in the face and neck are the masseter, temporalis, orbicularis oculi, orbicularis oris, tongue, trapezius, and sternocleidomastoid. The insertion of a needle electrode can be facilitated if the examiner holds the belly of the muscles between the index finger and thumb for better immobilization.

Facial Muscles

Because of anatomic proximity, the needle electrode placed in the orbicularis oris or oculi may detect distant potentials generated in the masseter or temporalis muscle. To avoid this interference, the patient should open the mouth slightly and relax the jaw. In the mimetic muscles

of the face, motor unit potentials are not only low in amplitude but also short in duration, although the reported values vary widely from 2.28 ± 0.3 ms (mean $\pm$ standard deviation [SD])[51] to 5 or 6 ms.[25] The orbicularis oris contains some muscle fibers crossing from one side to the other. In the case of unilateral denervation, therefore, activity of muscle fibers innervated by the normal facial nerve on the unaffected side may confuse the issue. Anesthetic block on the healthy side can establish a complete loss of innervation on the side of the lesion.[23]

After nerve injury, fibrillation potentials appear slightly earlier in the face than in the limb. Detection of spontaneous activity helps differentiate structural damage to the axon from functional block in patients with peripheral facial palsy. The brevity and small amplitude of normal motor unit potentials can mimic fibrillation potentials in waveform (Fig. 14–1). Accurate assessment of spontaneous potentials, therefore, calls for complete relaxation of the muscle under study. As in the skeletal muscles, the appearance of nascent units precedes the clinical return of voluntary movement, as

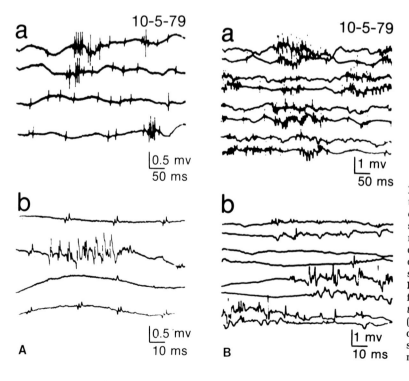

Figure 14–1. Motor unit potentials recorded in a 54-year-old woman with hemifacial spasm. **A.** Recurrent spontaneous bursts of high frequency discharges from the orbicularis oris shown at a slow (*a*) and fast (*b*) sweep. **B.** Simultaneous recording from the orbicularis oculi (*top tracing in each pair*) and oris (*bottom*). The patient blinked quickly several times to show synkinesis involving the two muscles.

the electrical evidence of reinnervation. Aberrant regeneration is the rule, rather than the exception, after the degeneration of the nerve from the proximal trunk (see Fig. 16–9).[44] Random misdirection may involve two branches of the facial nerve or two distinct but anatomically close nerves, such as the facial and trigeminal nerves.[60] In these cases, simultaneous recording from the affected muscles substantiate the presence of synkinesis.

Laryngeal and Nuchal Muscles

The glossopharyngeal nerve and the recurrent branches of the vagal nerve subserve the same motor function in the larynx. Electromyographic studies can characterize the paralytic involvement of the vocal cord, palate, and pharyngeal and laryngeal muscles. These muscles are not routinely tested in an ordinary electromyographic laboratory. Studies of these anatomic structures need a flexible wire electrode usually inserted with the help of an otolaryngologist.

In contrast, the conventional needle electrode suffices for examination of the tongue. The needle can be inserted with little pain from the bottom through the under surface of the mandible, 2 to 3 cm posterior to the tip of the chin. Alternatively, it is also possible to place the needle in the lateral portion of the tongue without much discomfort. To study spontaneous activity, the patient withdraws the tongue on the floor of the mouth with the electrode in place. Deviation of the tongue away from the needle generates the motor unit potentials. Conversely, muscle potentials subside with deviation toward the needle. Its protrusion in the midline requires simultaneous contraction on both sides. The innervation ratio of these muscles probably falls between those of the extraocular and limb muscles.

The spinal accessory nerve supplies two muscles, the sternocleidomastoid and the trapezius. Both muscles can be tested with ease using a regular electromyographic needle. The sternocleido-mastoid has unique ipsilateral supranuclear control, unlike most other muscles, which receive crossed input from the contralateral cerebral hemisphere.[5] Unilateral activation turns the head away from the contracting muscle. The muscle on the opposite side receives reciprocal inhibition in healthy subjects but not in patients with torticollis (Fig. 14–2). Bilateral contraction flexes the head forward. The activation of the trapezius causes the patient to shrug the shoulders upward toward the ears.

Diaphragm

The sternal origin of the diaphragm arises from the xiphoid process. Here, the muscle is easily accessible to the needle electrode inserted behind the bone slightly off the midline to either side.[36]

3   EXTRAOCULAR MUSCLES

Bjork,[8] Bjork and Kugelberg,[10] and Breinin[19] provided detailed descriptions of electromyography in the extraocular muscles. More recent reviews[21,45,56] indicate its usefulness in differentiating causes of paralytic squint, such as denervation, ocular myopathy, and myasthenia gravis. Ocular electromyography also helps detect abnormalities of eye movements attributable to mechanical limitations, such as dislocation of the globe, anomalies in tendon attachment, presence of fascial bands connecting one muscle with another, and fibrous tissue partly replacing the extraocular muscles. Assessment of electrical activity of the extraocular muscles reveals no abnormality in most patients with mechanical strabismus.

Recording Technique

Monopolar needle electrodes currently in use have an insulated shaft about 0.25 mm in diameter with a bare tip.[10,11] Recording requires an indifferent electrode placed either on the tip of the nose

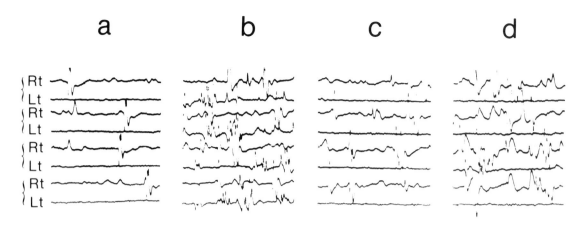

Figure 14–2. Torticollis on the right in a 30-year-old woman. Each pair of recordings shows muscle action potentials registered simultaneously from right (*upper tracing*) and left (*lower tracing*) sternocleidomastoid. During the sequential recordings, the patient either faced straight ahead (*a* and *c*) or turned the head to the right (*b*) or left (*d*). The muscle on the right continuously discharged regardless of the head position, whereas the muscle on the left fired only when the subject turned the head in the opposite direction (*b*).

or a blepharostat attached to the eyelid. Some investigators prefer a fine concentric electrode, 1 to 1.5 inches long and similar to a 30-gauge hypodermic needle in diameter. The needles come in different sizes, ranging from 0.25 to 0.5 mm in external diameter with a leading area varying from 0.005 to 0.015 mm^2.[2,19,32] Simultaneous recording from a second needle electrode placed in an agonist or antagonist muscle allows studies of synergistic actions or reciprocal inhibition.

The patient lies supine on the examining table for placement of the needle electrodes through the skin of the lid following application of a topical anesthetic to the eye. To evaluate voluntary eye movements, the subject must be awake during the examination and cooperate fully. This requirement precludes the use of any form of general anesthesia. Electrical activity decreases during general, retrobulbar, or local anesthesia as the level deepens, leading to complete electrical silence with the eyes assuming a position of divergence.[17–19]

An ophthalmologist familiar with the extraocular muscles can easily reach the inferior oblique, and with some search any of the remaining extraocular muscles. The study of the least accessible superior oblique requires a considerably longer needle. Monitoring the waveform and sound of motor unit discharges helps adjust the position of the needle inserted subconjunctivally into the belly of a muscle along its long axis. Most patients tolerate the procedure well, with minimal discomfort. Rare complications include ecchymoses of the conjunctiva, subcapsular hemorrhage, and exposure keratitis, all of which clear without sequelae.[10,19,47] Inadvertent perforation of the globe can occur, especially in the presence of undetected glaucoma.

Unique Properties of Extraocular Muscles

The eyes move rapidly and accurately. Complex coactivation of synergistic muscles and relaxation of the antagonists achieves precisely controlled movements of a constant load. Sherrington first described this principle of reciprocal inhibition based on studies of the extraocular muscles. The eye muscles can discharge at a rate of up to 200 impulses per second.[10] This stands in sharp contrast to the usual rates of firing of less than 50 impulses per second in the skeletal muscles. Extraocular muscle fibers are very thin, ranging from 10 to 50 μm in diameter.[20] The motor units consist of a small number of muscle fibers, averaging 23 in one study[30] and 6 to 12 in others.[15,59] These values fall considerably below the corresponding value in the limb muscles, which varies from 100 to 2000.[22,34,55] The low innervation ratio and other physiologic characteristics of the fast twitch fibers permit rapid and very finely graded eye movements. The extraocular muscles also contain slow twitch fibers, which show characteristic monophasic, low-amplitude potential near the surface layer of the extraocular muscle.[20,21]

Despite electromyographic similarity to skeletal muscles, a certain electrophysiologic peculiarity characterizes the extraocular muscles. Placement of the needle electrode causes a brief insertional activity, presumably representing an injury potential. Unlike the limb musculature, however, the extraocular muscles show constant electrical discharges following the cessation of needle movement. The tonic activity of the muscles is required to maintain the eyes in the primary position during the alert state. With ocular movement, motor unit discharge increases in the contracting muscles and decreases in the others (Fig. 14–3). The antagonist develops complete electrical silence only through reciprocal inhibition during fast eye movements.[11]

The motor unit potentials in the extraocular muscles are lower in amplitude and shorter in duration than in the limb muscles, reflecting the smaller diameter of the muscle fibers and lower innervation ratio. Reported normal values (mean $\pm$ SD) include amplitude of 108 $\pm$ 9.2 μV and a duration ranging from 1.60 $\pm$ 0.06 ms[10] to 2.8 $\pm$ 0.1 ms.[32] Another study[21] showed the normal amplitude to be between 20 and 600 μV, averaging 200 μV in the primary position, and the normal duration, 1 to 2 ms, with an average of 1.5 ms. As in the limb muscles, individual potentials show mostly triphasic and occasionally polyphasic waveforms. With maximal effort of contraction they discharge at a rate of up to 200 impulses per second.[10]

Neurogenic Extraocular Palsy

A neurogenic extraocular palsy results from lesions of the third, fourth, or sixth nerve. Electromyographic findings are in

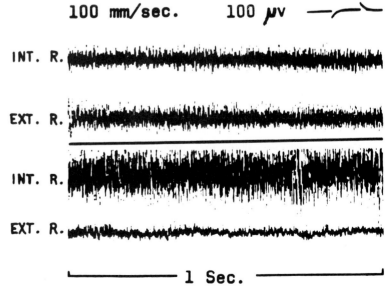

100 mm/sec.   100 μV

INT. R.

EXT. R.

INT. R.

EXT. R.

1 Sec.

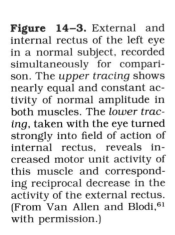

Figure 14–3. External and internal rectus of the left eye in a normal subject, recorded simultaneously for comparison. The *upper tracing* shows nearly equal and constant activity of normal amplitude in both muscles. The *lower tracing*, taken with the eye turned strongly into field of action of internal rectus, reveals increased motor unit activity of this muscle and corresponding reciprocal decrease in the activity of the external rectus. (From Van Allen and Blodi,[61] with permission.)

principle the same as those in denervated limb muscles. In the extraocular muscle, however, physiologic tonic discharge with the eyes in the primary position may obscure pathologic discharges. To compound the problem, the normally brief motor unit potentials resemble fibrillation potentials.[24] Electromyography can still confirm denervation with certainty in a paretic muscle where spontaneous activities occur independently of any attempted contraction.[9] After reinnervation, studies reveal high-amplitude motor unit potentials of long duration with increased polyphasic activities, but to a lesser extent than in skeletal muscles. Large motor unit potentials frequently accompany aberrant regeneration of oculomotor nerves.[21]

As in limb muscles, slow recruitment of motor unit potentials suggests neurogenic weakness of the extraocular muscles with the reduction of the interference pattern approximately in proportion to the degree of paresis. Examination shows no motor unit potentials in a totally paretic muscle with attempted maximal contraction, although complete electrical silence is rare, even in severe palsies. The interference pattern may consist of repetitive discharges from a single motor unit in severe, but incomplete, paralysis. Patients without definite limitation of rotation may have abundant electrical activity in the remaining normal units despite mild palsies. This finding will mimic those found in patients with ordinary strabismus, who also recruit motor unit normally on attempted rotation into the field of action.

Myopathy and Myasthenia Gravis

Electromyography in ocular myopathy, unlike in neurogenic paralysis, shows the preservation of a normal interference pattern with no evidence of denervation.[12,49] The abundance of brief, low-amplitude motor unit potentials suggests random loss of individual muscle fibers without major loss in the number of functional motor units.[33] The conditions known to have such findings include pro-

gressive external ophthalmoplegia, thyroid ophthalmopathy, and pseudotumor of the orbit.[21] Except in advanced cases, myopathic features may escape detection, because normal extraocular muscles show a similar pattern. Automatic frequency analysis reveals a higher firing frequency in the patients with ocular myopathy than in healthy subjects. Unfortunately, this approach provides only limited success in detecting abnormalities from individual patients against the established normal range.[19]

Myasthenia gravis affects the ocular muscles early, causing diplopia and abnormal fatigue of eye movements. Thus, electromyography of the extraocular muscles may help establish the diagnosis in patients with normal extremity muscles. In myasthenia gravis, the amplitude of a motor unit potential fluctuates or steadily declines during sustained contraction. Progressive decrease in the number of discharging motor units results in a reduced interference pattern, which may return to normal immediately after injection of edrophonium (Tensilon).[17] In patients with ptosis as the presenting sign, studies of the levator palpebrae may reveal abnormality, despite the technical difficulty in localizing the muscle. Overmedication with anticholinesterase may cause clinical and electrical abnormalities of cholinergic block. In the event of secondary myopathic changes of the extraocular muscle, certain electrical abnormalities may persist after the administration of anticholinesterase.[49]

Other Types of Gaze Palsy

Musculofascial anomalies generally result in limitation of gaze in one direction, either vertically or horizontally. During contraction of an apparently paretic muscle for mechanical reasons, electromyography reveals normal motor unit potentials and a complete interference pattern disproportionate to the failure of rotation. In a blow-out fracture of the orbit, for example, incarceration of the extraocular muscle in the fracture line may prevent the globe from normal rotation. In such a case, ocular electromyography establishes the presence of normally innervated muscle by demonstrating abundant activity with the effort to rotate the eye. Conversely, the detection of electromyographic abnormalities in patients with limited rotation suggests a direct injury to the nerve or muscle.

In Duane's syndrome, a deficiency of ocular motility results from congenital absence of the sixth nucleus with aberrant innervation of the lateral rectus by the third nerve.[48] The syndrome typically consists of impaired abduction of one eye and retraction and ptosis on attempted adduction of the same eye. A fibrotic lateral rectus presumably neither contracts on abduction nor relaxes on adduction. Thus, in electromyography, the muscle shows a reduced number of motor unit potentials when activated and fails to produce electrical silence when reciprocally inhibited.[18,53] In addition, a central

supranuclear lesion may disrupt normal reciprocal inhibition (Fig. 14–4), contributing to the ophthalmoplegia in this condition.[13]

Limitation of eye movements may occur on a central basis, as in internuclear ophthalmoplegia.[3,46] In this syndrome, caused by a lesion of the medial longitudinal fasciculus, the eye on the side of the lesion fails to adduct, despite the integrity of the peripheral motor system. In such supranuclear disturbances, extraocular electromyography reveals an altered pattern of innervation. The medial rectus has normal electrical discharges with the eyes in the primary position but shows neither an increase in activity on attempted adduction nor reciprocal inhibition on attempted abduction.

The Mobius syndrome consists of facial diplegia and restriction of horizontal eye movements. Electromyographic studies

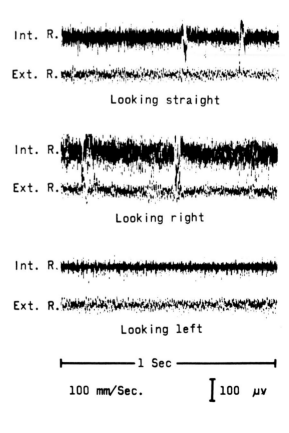

Figure 14–4. External and internal rectus of the left eye, recorded simultaneously in a patient with Duane's syndrome. The tracings show a normal innervation pattern of the internal rectus but neither increment nor decrement of the external rectus on attempted gaze to the left or to the right. Note the normal electrical activity of this muscle in the primary position. (From Blodi et al., 1964, with permission.)

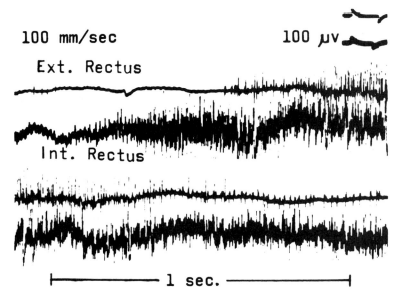

100 mm/sec   100 μV

Ext. Rectus

Int. Rectus

├─────── 1 sec. ───────┤

Figure 14–5. External and internal rectus of the left eye, recorded simultaneously in a patient with the Mobius syndrome. Note the spontaneous volley in external rectus with simultaneous waxing of activity in internal rectus, indicating the lack of physiologic reciprocal innervation. (From Van Allen and Blodi,[61] with permission.)

show synchronous bursts of activity from the medial and lateral rectus, rather than the normally expected reciprocal pattern of innervation (Fig. 14–5). These findings suggest supranuclear abnormalities responsible for the abnormal ocular motility in these patients, despite the designation of the syndrome as congenital nuclear agenesis of the sixth and seventh nerves.

A pair of electrodes inserted into the extraocular muscles can elucidate the pattern of various types of nystagmus.[21] Electromyographic techniques also help explore the reciprocal relationship between orbicularis oculi and levator palpebrae.[62]

4 TRUNCAL MUSCULATURE

Abdominal Muscles

The anterior rami of the cervical spinal nerves supply the upper limb muscles and those of the lumbosacral spinal nerves, the lower limb muscles. Similarly, 12 pairs of intercostal nerves derived from the anterior rami of the thoracic spinal nerves innervate intercostal and abdominal muscles. Involvement of the intercostal nerve results in segmental

paralysis of the abdominal muscles and weak respiration. In this condition, the abdomen would protrude on coughing and the umbilicus would deviate to the unaffected side by unopposed action of the normal muscle. The electromyographic study of the abdominal muscles helps detect a lesion at the thoracic levels, which do not have appendicular representation in the extremities. The considerable overlap in segmental representation, however, precludes the exact localization of the involved cord level. Each segmental level receives at least two adjoining intercostal nerves in both thoracic and abdominal regions.

Electromyographers can study the abdominal musculature with a needle just as easily as the muscles of the four extremities. The external oblique is tested at the anterior axillary line 5 to 10 cm above the anterior superior spine of the iliac crest. The needle, if inserted obliquely, can sample the electrical activities along the course of the muscle fibers, which run medially and downward. The needle, if placed too deep, may reach the internal oblique or transverse abdominis (or the abdominal cavity!). Even with the patient completely relaxed, the intercostal muscles fire rhythmically with respiration. Volume-conducted potentials from this source may mimic spontaneous discharges, but the time relationship to the breathing cycle differentiates the two. For analysis of motor unit potentials, the patient contracts the muscle by bending the upper trunk forward.

The abdominal rectus is tested between the linea alba, which connects the xiphoid and umbilicus in the midline, and the linea semilunaris, which forms the lateral margin of the rectus. The needle insertion into the muscle must avoid the three transverse tendinous bands located at the xiphoid, the umbilicus, and halfway in between.[36] The patient bends forward against a resistance to contract the muscle for the assessment of motor unit potentials.

Paraspinal Muscles

In contrast to the muscles of the extremities and abdominal musculature, innervated by the anterior ramus of the spinal nerve, the posterior ramus supplies the paraspinal muscles at respective segmental levels. Documentation of electromyographic abnormalities in this region thus identifies a radicular lesion that affects the spinal nerve at a point proximal to its bifurcation into the posterior and anterior rami (see Figs. 1–6 and 13–7C). A more distally located lesion at the level of the plexus or the peripheral nerve would entirely spare the paraspinal muscles without the involvement of the posterior rami. Hence, the examination of paraspinal muscles plays a critical role in the investigation of cervical or lumbar disk herniation.[37,41,42] In fact, patients in early stages of radiculopathy within 1 to 2 weeks after the onset may have electrical abnormalities limited to this region. Some systemic disorders, most notably polymyositis, may affect the paraspinal muscles preferentially and sometimes exclusively.[1,57] Relatively selective denervation in this region also develops in degenerative joint disease, arachnoiditis, diabetic polyradiculopathy, and rare local metastasis to the muscles.

The erector spinae consists of two portions, short spinal muscles, also called multifidus, and long spinal muscles, known as longissimus dorsi. The short

spinal muscles, located deep, immediately posterior to the transverse process, receive a fairly discrete nerve supply from corresponding posterior rami. To study this portion, one must insert the needle deeply, just lateral to the spinal process, toward the transverse process. The long spinal muscles, located more superficially, extend several centimeters to either side of the spinous process and ligamentum nuchae.[36] Their nerve supply overlaps at least one to two segments caudally and rostrally. This portion of the muscle can be reached quite superficially with the needle inserted 2 to 3 cm lateral to the spinous process at either the cervical or the lumbar level.

To achieve complete relaxation, the subject lies in the prone position with pillows under the neck for cervical examination and under the abdomen for lumbar studies. For relaxation of the lumbar paraspinal muscles, the patient raises the hips slightly toward the ceiling. The cervical paraspinal muscles usually relax if the patient bends the neck forward, pressing the forehead against the table. In some subjects, lung tissue extends above the clavicle with a distance from skin surface of approximately 3.3 cm.[38] Thus, one should direct the exploring needle in a direction perpendicular to the spine or slightly upward to minimize the risk of inducing pneumothorax, especially in patients with a long neck.

Motor unit potentials of low amplitude and short duration seen in the cervical region mimic fibrillation potentials. Transient positive sharp waves may appear in the paraspinal muscles by the end of the first day and last up to 4 days after myelography[65] or lumbar puncture.[31]

5 ANAL SPHINCTER

Indications and Technique

The anal sphincter receives the innervation of the pudendal nerve, which derives from the anterior division of the S-3, S-4, and occasionally also S-2 spinal nerves. Interdigitation of muscle fascicles across the midline results in substantial overlap of innervation between the two sides. This enables partial reinnervation from the contralateral side following unilateral pudendal neurectomy.[66] The anal sphincter, which normally receives volitional control, shares similar physiologic properties with the skeletal muscles of the limbs. Since the initial attempt by Beck,[7] electromyography has long contributed to kinesiologic studies of the normal anal sphincter at rest and during defecation. Surface recordings from the sphincter have shown increased activity during coughing, speaking, and body movements of the trunk and decreased activity in sleep.[35] Other studies have used dental needles with a 25-μm wire in the center for kinesiologic studies[43] and steel needles for recording reflex contraction induced by digital stretching of the sphincter.[58] The conventional concentric or monopolar needle suffices for routine clinical use.

Electromyographic studies quantitate sphincter dysfunction in neurologic disorders.[64] They help establish or rule out the possibility of agenesis of the striate sphincter in the preoperative assessment of the newborn with an imperforate anus. Electrical studies not only localize the sphincter precisely, if it is present, but also determine its functional capacity.[2] The anal sphincter may sustain traumatic injury during parturition, prostatectomy, or rectal surgery for repair of an anal fistula or prolapse. Electromyography helps determine the extent of damage in such cases. The anal and external urethral sphincters share a common segmental derivation. Thus, confirming the integrity of the anal sphincter provides an important, albeit indirect, guide in ilioconduit surgery for prominent urologic dysfunction. Electromyography of the urethral sphincter ideally requires a laboratory equipped with tools for urodynamic investigations.

For studies of the anal sphincter, adults and older children usually prefer the lateral decubitus position.[26,40] The patient may assume the knee-chest position or modified lithotomy position, which allows the best examination in infants. After digital examination of the sphincter tone, a gloved finger, still in place, can be used to guide the needle, inserted through the perianal skin adjacent to the mucocutaneous junction. The tip of the electrode should enter perpendicular to the skin surface close to the anal orifice, 0.5 to 1 cm from the ring. The ring of the anal orifice has four parts, anterior and posterior quadrants on both sides. A complete study consists of exploration of the four quadrants with the anal sphincter at rest and while contracted voluntarily or reflexively.

Resting and Voluntary Activities

Unlike peripheral skeletal muscles, the anal sphincter maintains a certain tonus without volitional effort. Thus, the subject at rest maintains sustained firing of isolated motor unit potentials at a low rate. This activity varies considerably with changes in subject position. The activity continues during sleep, although the discharge rate drops substantially, compared with that during wakefulness. Sphincter activity ceases completely only during attempted defecation. Conversely, volitional contraction of the anal sphincter inhibits rectal motility based on reciprocal innervation between the rectal musculature and the striated muscle of the anal sphincter. The presence of physiologic tonic activity at rest makes detection of abnormal spontaneous potentials difficult in a partially denervated muscle. In the paretic sphincter one can recognize fibrillation potentials and positive sharp waves with certainty, as in any extremity muscle.

To test voluntary activity, the patient contracts the sphincter as though attempting to hold a bowel movement. The motor unit potentials range from 5.5 to 7.5 ms in duration and 0.2 to 0.5 μV in amplitude.[26,50] In one study,[6] patients with fecal incontinence exhibited prolongation of mean motor unit potential duration, compared with matched controls. Digital examination of the anus, coughing, or crying elicits reflex activity of the sphincter.[35,58] A full interference pattern is expected to accompany a normal maximal contraction, whether induced voluntarily or reflexively. Reliability of grading the degree of such discharge, as in the skeletal muscles of the limb, depends on patient cooperation.[4,64] Some subjects can neither relax nor contract the sphincter during the test, as instructed by the examiner. In these cases, an appraisal of sphincteric tone by the interference pattern might erroneously suggest a central lesion. Experienced electromyographers, however, can usually correlate electrical activity and sphincter tone with reasonable accuracy.

Central Versus Peripheral Paralysis

Paralysis of the striated sphincter may result from a pure central, pure peripheral, or mixed lesion. Central paralysis causes reduction in voluntary discharges with preservation of reflexive activation. The interference pattern is incomplete with motor unit potentials of normal amplitude discharging at low frequency.

With a complete loss of voluntary activity, the low frequency discharge normally seen at rest continues during maximal effort of contraction.

Peripheral paralysis of the anal sphincter usually suggests lesions in the cauda equina or in the sacral or pudendal plexus. In an incomplete paralysis, volitional effort recruits a few motor units that fire at a high frequency. In contrast to central paralysis, the surviving units show a polyphasic waveform and a long duration. In an acute cauda equina syndrome, the initial paralysis may result from a functional block. Axonal degeneration, if present, gives rise to fibrillation potentials, positive sharp waves, and complex repetitive discharges (Fig. 14–6). Amyotrophic lateral sclerosis typically spares the sphincter muscle, even when the limb muscles show evidence of conspicuous denervation.[52]

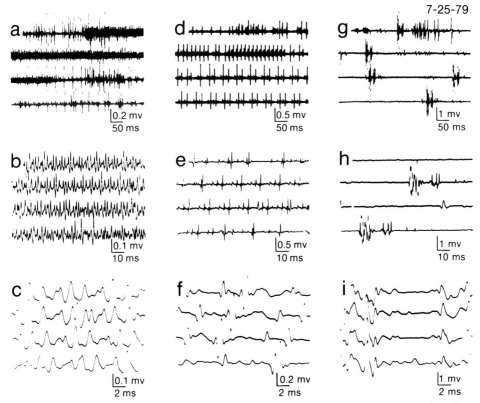

Figure 14–6. Recording from anal sphincter in a 16-year-old girl with incontinence. Tracings include continuous discharge at high frequency, resembling very prominent end-plate noise (*a, b,* and *c*), complex repetitive discharges (*d, e,* and *f*), and very polyphasic fasciculation potentials (*g, h,* and *i*), all recorded in a small localized area of the sphincter with the patient completely at rest. In *i,* each sweep triggered by a recurring fasciculation potential, shows a consistent late component following the main discharge.

Patients often have a mixture of central and peripheral paresis in congenital malformation, vascular disease, or traumatic injury of the conus medullaris.[14,29,39] Spina bifida with meningomyelocele characteristically affects both upper and lower motor neurons.[27] Electromyography of the anal sphincter in these cases reveals absent or markedly reduced voluntary activity. Reflexive contraction, if present, shows isolated high-frequency discharges of a few motor units. With complete damage to the sacral segment of the conus medullaris, no sphincter response is possible either voluntarily or reflexively. Spontaneous potentials recorded in these cases indicate the involvement of the anterior horn cells.[28]

REFERENCES

1. Albers, JW, Mitz, M, Sulaiman, AR, and Chang, GJ: Spontaneous electrical activity and muscle biopsy abnormalities in polymyositis and dermatomyositis (corres). Muscle Nerve 2:503, 1979.
2. Archibald, KC, and Goldsmith, EI: Sphincter electromyography. Arch Phys Med Rehabil 48:387–392, 1967.
3. Bach-y-Rita, P, and Collins, CC (eds): The Control of Eye Movements. Academic Press, New York, 1971.
4. Bailey, JA, Powers, JJ, and Waylonis, GW: A clinical evaluation of electromyography of the anal sphincter. Arch Phys Med Rehabil 51:403–408, 1970.
5. Balagura, S, and Katz, RG: Undecussated innervation to the sternocleidomastoid muscle: A reinstatement. Ann Neurol 7:84–85, 1980.
6. Bartolo, DCC, Jarratt, JA, and Read, NW: The use of conventional electromyography to assess external sphincter neuropathy in man. J Neurol Neurosurg Psychiatry 46:1115–1118, 1983.
7. Beck, A: Elektromyographische Untersuchungen Sphincter ani. Arch Physiol 224:278–292, 1930.
8. Bjork, A: Electrical activity of human extrinsic eye muscles. Experientia 8:226–227, 1952.
9. Bjork, A: Electromyographic study of conditions involving limited mobility of the eye, chiefly due to neurogenic pareses. Br J Ophthalmol 38:528–544, 1954.
10. Bjork, A, and Kugelberg, E: Motor unit activity in the human extraocular muscles. Electroencephalogr Clin Neurophysiol 5:271–278, 1953.
11. Bjork, A, and Kugelberg, E: The electrical activity of the muscles of the eye and eyelids in various positions and during movement. Electroencephalogr Clin Neurophysiol 5:595–602, 1953.
12. Blodi, FC, and Van Allen, MW: Electromyographic studies in some neuro-ophthalmologic disorders. XVIII Concilium Ophthalmologicum, Vol 2. (Belgica), 1958, pp 1621–1627.
13. Blodi, FC, Van Allen, MW, and Yarbrough, JC: Duane's syndrome: A brain stem lesion. Arch Ophthalmol 72:171–177, 1964.
14. Bonnal, J, Stevenaert, A, and Chantraine, A: Bilan électromyographique pré et post opératoire du spina bifida avec troubles neurologiques. Neuro-Chir (Paris) 15:299–306, 1969.
15. Bors, E: Uber das Zahlenverhaltnis zwischen Nervenund Muskelfasern. Anat Anz 60:415–416, 1926.
16. Bradley, WE, Timm, GW, Rockswold, GL, and Scott, FB: Detrusor and urethral electromyography. J Urol 114:891–894, 1975.
17. Breinin, GM: Electromyography—A tool in ocular and neurologic diagnosis. I. Myasthenia gravis. Arch Ophthalmol 57:161–164, 1957.
18. Breinin, GM: Electromyography—A tool in ocular and neurologic diagnosis. II. Muscle palsies. Arch Ophthalmol 57:165–175, 1957.
19. Breinin, GM: The Electrophysiology of Extraocular Muscle with Special Reference to Electromyography. University of Toronto Press, Toronto, 1962.
20. Breinin, GM: The structure and function of extraocular muscle—An appraisal of the duality concept. Am J Ophthalmol 72:1–9, 1971.
21. Breinin, GM: Ocular electromyography. In Goodgold, J, and Eberstein, A (eds): Electrodiagnosis of Neuromuscular Diseases, ed 2. Williams and Wilkins, Baltimore, 1977.
22. Buchthal, F: The general concept of the motor unit. In Adams, R, Eaton, L, and Shy, GM (eds): Neuromuscular Disorders, Vol 38. Williams and Wilkins, Baltimore, 1960.
23. Buchthal, F: Electromyography in paralysis of the facial nerve. Arch Otolaryngol (Stockh) 81:463–469, 1965.
24. Buchthal, F: Electrophysiological abnormalities in metabolic myopathies and neuropathies. Acta Neurol Scand (Suppl 43)46:129–176, 1970.
25. Buchthal, F, and Rosenfalck, P: Action potential parameters in different human muscles. Acta Psychiatr Neurol Scand 30:125–131, 1955.
26. Chantraine, A: EMG examination of the anal and urethral sphincters. In Desmedt, JE (ed): New Developments in Electromyography and Clinical Neurophysiology, Vol 2. Karger, Basel, 1973, pp 421–432.
27. Chantraine, A, Lloyd, K, and Swinyard, CA: The sphincter ani externus in spina bifida and myelomeningocele. J Urol (Baltimore) 95:250–256, 1966.
28. Chantraine, A, Stevenaert, A, Carlier, G, and Bonnal, J: Evolution électromyographique du bilan pré et postopératoire du spina bifida avec troubles neurologiques. Acta Paediat Belg 22:127–140, 1968.
29. Chantraine, A, Stevenaert, A, and Timmermans, L: Electromyographic study before and after operation in spina bifida with myelomeningocele: A preliminary report. Dev Med Child Neurol (Suppl 13):136–137, 1967.

30. Christensen, E: Topography of terminal motor innervation in striated muscles from stillborn infants. Am J Phys Med 38:65–78, 1959.

31. Danner, R: Occurrence of transient positive sharp wave-like activity in the paraspinal muscles following lumbar puncture. Electromyogr Clin Neurophysiol 22:149–154, 1982.

32. Faurschou Jensen, S: The normal electromyogram from the external ocular muscles. Acta Ophthalmol 49:615–626, 1971.

33. Faurschou Jensen, S: Endocrine ophthalmoplegia: Is it due to myopathy or to mechanical immobilization? Acta Ophthalmol 49:679–684, 1971.

34. Feinstein, B, Lindegard, B, Nyman, E, and Wohlfart, G: Morphologic studies of motor units in normal human muscles. Acta Anat 23:127–142, 1955.

35. Floyd, WF, and Walls, EW: Electromyography of the sphincter ani externus in man. J Physiol (Lond) 122:599–609, 1953.

36. Goodgold, J: Anatomical Correlates of Clinical Electromyography. Williams and Wilkins, Baltimore, 1974.

37. Gough, JG, and Koepke, GH: Electromyographic determination of motor root levels in erector spinae muscles. Arch Phys Med Rehabil 47:9–11, 1966.

38. Honet, JE, Honet, JC, and Cascade, P: Pneumothorax after electromyographic electrode insertion in the paracervical muscles: Case report and radiographic analysis. Arch Phys Med Rehabil 67:601–603, 1986.

39. Ingberg, HO, and Johnson, EW: Electromyographic evaluation of infants with lumbar meningomyelocele. Arch Phys Med Rehabil 44:86–92, 1963.

40. Jesel, M, Isch-Treussard, C, and Isch, F: Electromyography of striated muscle of anal and urethral sphincters. In Desmedt, JE (ed): New Developments in Electromyography and Clinical Neurophysiology, Vol 2. Karger, Basel, 1973, pp 406–420.

41. Johnson, EW, and Melvin, JL: The value of electromyography in the management of lumbar radiculopathy. Arch Phys Med Rehabil 50:720, 1969.

42. Jonsson, B: Morphology, innervation, and electromyographic study of the erector spinae. Arch Phys Med Rehabil 50:638–641, 1969.

43. Kawakami, M: Electro-myographic investigation on the human external sphincter muscle of anus. Jpn J Physiol 4:196–204, 1954.

44. Kimura, J, Rodnitzky, RL, and Okawara, S: Electrophysiologic analysis of aberrant regeneration after facial nerve paralysis. Neurology 25:989–993, 1975.

45. Lenman, JAR, and Ritchie, AE: Clinical Electromyography, ed 2. JB Lippincott, Philadelphia, 1977.

46. Loeffler, JD, Hoyt, WF, and Slatt, B: Motor excitation and inhibition in internuclear palsy. Arch Neurol 15:664–671, 1966.

47. Marg, E, Jampolsky, A, and Tamler, E: Elements of human extraocular electromyography. Arch Ophthalmol 61:258–269, 1959.

48. Miller, NR, Kiel, SM, Areen, WR, and Clark, AW: Unilateral Duane's Retraction syndrome (Type I). Arch Opthalmol 100:1468–1472, 1982.

49. Papst, W, Esslen, E, and Mertens, HG: Klinische Erfahrungen mit der Elektromyographie bei ocularen Myopathien. Zusamm Deutsch Ophth Ges, 1957.

50. Petersen, I, and Franksson, C: Electromyographic study of the striated muscles of the male urethra. Br J Urol 27:148–153, 1955.

51. Petersen, I, and Kugelberg, E: Duration and form of action potential in the normal human muscle. J Neurol Neurosurg Psychiatry 12:124–128, 1949.

52. Sakuta, M, Nakanishi, T, and Toyokura, Y: Anal muscle electromyograms differ in amyotrophic lateral sclerosis and Shy-Drager syndrome. Neurology 28:1289–1293, 1978.

53. Sato, S: Electromyographic study on retraction syndrome. Jpn J Ophthalmol 4:57–66, 1960.

54. Scott, FB, Quesada, EM, and Cardus, D: The use of combined uroflowimetry, cystometry and electromyography in evaluation of neurogenic bladder dysfunction. In Boyarsky, S (ed): Neurogenic Bladder. Williams and Wilkins, Baltimore, 1967, pp 106–114.

55. Sissons, HA: Anatomy of the motor unit. In Walton, JN (ed): Disorders of Voluntary Muscle, ed 3. Churchill-Livingstone, London, 1974.

56. Strachan, IM: Clinical electromyography of the extra-ocular muscles. Br Orthopt J 26:60–67, 1969.

57. Streib, EW, Wilbourn, AJ, and Mitsumoto, H: Spontaneous electrical muscle fiber activity in polymyositis and dermatomyositis. Muscle Nerve 2:14–18, 1979.

58. Taverner, D, and Smiddy, FG: An electromyographic study of the normal function of the external anal sphincter and pelvic diaphragm. Dis Colon Rectum 2:153–160, 1959.

59. Torre, M: Nombre et dimensions des unités motrices dans les muscles extrinsèques de l'oeil et, en général, dans les muscles squélettiques reliés à des organes de sens. Arch Suisses Neurol Psychiatr 72:362–376, 1953.

60. Trojaborg, W, and Siemssen, SO: Reinnervation after resection of the facial nerve. Arch Neurol 26:17–24, 1972.

61. Van Allen, MW, and Blodi, FC: Neurologic aspects of the Mobius syndrome: A case study with electromyography of the extraocular and facial muscles. Neurology 10:252–259, 1960.

62. Van Allen, MW, and Blodi, FC: Electromyography study of reciprocal innervation in blinking. Neurology 12:371–377, 1962.

63. Vereecken, RL, and Verduyn, H: The electrical activity of the paraurethral and perineal muscles in normal and pathological conditions. Br J Urol 42:457–463, 1970.

64. Waylonis, GW, and Krueger, KC: Anal sphincter electromyography in adults. Arch Phys Med Rehabil 51:409–412, 1970.

65. Weber, RJ, and Weingarden, SI: Electromyographic abnormalities following myelography. Arch Neurol 36:588–589, 1979.

66. Wunderlich, M, and Swash, M: The overlapping innervation of the two sides of the external anal sphincter by the pudendal nerves. J Neurol Sci 59:97–109, 1983.

Chapter 15

SINGLE-FIBER AND MACRO ELECTROMYOGRAPHY

1 INTRODUCTION

The concentric needle electrode[1] and other bipolar or monopolar needles record single motor unit potentials that represent the smallest functional unit of muscle activation. However, they fail to discriminate potentials from different muscle fibers within a motor unit, all of which fire more or less synchronously. A multielectrode, introduced for study of muscle fiber physiology,[3] primarily measures the territory of a motor unit.

In contrast, the single-fiber needle[9] allows extracellular recording of individual muscle fiber action potentials during voluntary contraction. Termed single-fiber electromyography (SFEMG), this technique has contributed substantially to the understanding of muscle physiology and pathophysiology. In the clinical domain, the SFEMG supplements conventional electromyography by determining (1) fiber density—the number of single-fiber action potentials within the recording radius of the electrode—and (2) electromyographic jitter—the variability of the interpotential interval between two or more single muscle fibers belonging to the same motor unit.[57]

2 RECORDING APPARATUS

Electrode Characteristics

A small leading-off surface of the single-fiber needle electrode lies close to fewer muscle fibers than the larger tip of the conventional needle that commands a wider territory.[7] In addition, a smaller pickup area causes little shunting and consequently less distortion of the electrical field (Fig. 15–1). This type of needle, therefore, helps establish selective recording from the generator under study. Here, the action potential decreases almost exponentially as the recording electrode moves away from the origin.[19] Thus, the recorded amplitude declines very steeply as the distance between the electrode and the source increases (see Fig. 12–6). This in turn results in sharp discrimination of single-fiber potentials with minimal interference from action potentials of neighboring muscle fibers.

A recording surface diameter of 25 to 30 μm has proven to be optimal for this purpose, considering the average muscle fiber diameter of 50 μm.[11] Most SFEMG needles have an active recording surface, 25 μm in diameter, located 3 mm from the needle tip along the side port. This arrangement minimizes the chance of recording from fibers damaged by needle penetration. A system such as this allows an uptake area approximately 300 μm from the needle and consequently records signals from only one or two fibers. The

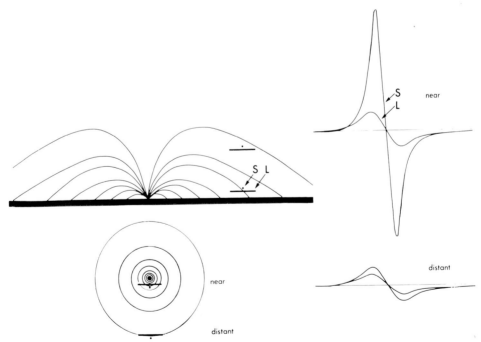

Figure 15–1. Electrical field around a muscle fiber recorded with a small (S) and a large (L) leading-off surface. The size of the recording area primarily determines the magnitude of shunting across the high-density isopotential lines near the generator source, and the low-density isopotential lines further in the periphery. This, in turn, dictates the relationship between the amplitude recorded and the electrode distance from the source. Note a much steeper decline in potential per unit radius with a smaller leading-off surface (see Figure 12–6). (From Stålberg and Trontelj,[57] with permission.)

selectivity of single-fiber recording can be further improved by "bipolar" derivation with two small electrodes separated by a short interelectrode distance, as opposed to a "monopolar" arrangement with a reference electrode outside the muscle (see Fig. 3–1). The suggested interelectrode distance of 200 μm provides enough separation for optimal amplification of a single discharge but precludes recording from two independent sources.[57]

In comparison, the conventional needle electrode has a leading-off surface of about 150 by 600 μm, which records from an area within a 1-mm radius (see Fig. 3–1). This larger leading-off surface induces prominent shunting across the electrical fields that becomes disproportionately greater near the source, where the isopotential lines gather in high density (see Fig. 15–1). Thus, the ordinary electrode registers comparatively less amplitude near the potential generator. Farther from the source, the shunting effect diminishes with either type of electrode, because the larger radius of the isopotential lines give rise to lower gradient of the electrical field. With large leading-off surfaces, therefore, the action potential does not decrease exponentially with increasing recording distance.[17] Consequently, potentials derived from near and distant fibers show relatively little difference in amplitude.

Amplifier Settings

Because of the small leading-off surface, a single-fiber electrode has a much higher electrical impedance than a conventional monopolar or concentric needle. Impedances range on the order of megaohms (MΩ) at 1 kHz for a platinum needle but vary for different metals. To maintain a high signal-to-noise ratio, therefore, the amplifier must have a very high input impedance on the order of 100 MΩ. This helps maintain an adequate common mode rejection ratio or differential amplification between the signal and the interference potential.[57] The initial amplifier settings include a sensitivity of 0.2 to 1 mV/cm and a sweep speed of 0.5 to 1 ms/cm.

Short-duration, high-amplitude single-fiber action potentials, recorded near the generator, consist mainly of high-frequency components. In contrast, distant potentials have a larger proportion of low-frequency discharges because the intervening muscle tissue tends to filter out high-frequency components. Thus, the use of a low-frequency cutoff of 500 Hz, for example, selectively attenuates volume-conducted background activity. The action potential from fibers close to the electrode also decreases by about 10 percent.[18,19] This slight change in shape of the single-fiber potential barely affects the measurements of propagation velocity, fiber density, or jitter. In the analysis of waveforms, however, one must lower the high-pass (low-frequency) filter setting to about 2 Hz. A high-frequency cutoff of 35 kHz, though ideal, adds little in practice, because a low-pass (high-frequency) filter of 10 kHz can substantially maintain the amplitude and shape of the original spike.

3 SINGLE-FIBER POTENTIAL

This is a biphasic spike with a rise time of 75 to 200 μs and total duration of about 1 ms, when recorded with an optimally placed single-fiber electrode.[6] The peak-to-peak amplitude varies widely from a low of 200 μV to a high of 20 mV, but usually within the range of 1 to 7 mV. The recorded amplitude attenuates exponentially as the distance between the electrode and the discharging muscle fiber increases.[19] With a time resolution of 5 to 10 μs, the shape of the potential remains nearly constant for successive discharges. The frequency spectrum ranges from 100 Hz to 10 kHz, with a peak at 1.61 ± 0.30 kHz.[17]

Recording Procedures

Either electrical or voluntary activation can be used to generate motor unit potentials for SFEMG. Surface stimulation of the motor fibers evokes many motor units simultaneously, making single-fiber recording difficult. In contrast, stimulation of an endplate zone with a bipolar needle electrode can excite only a few terminal twigs of a motor neuron. The activated terminal twigs conduct the action potential first antidromically to the branching point, then orthodromically to the remaining nerve twigs of the entire motor unit.[43,56] This allows recording of the SFEMG from a single motor unit firing in response to electrical stimulation. In cooperative subjects, slight, steady voluntary muscle contraction also reliably generates isolated motor unit potentials. This is the preferred method of studying SFEMG.

The recommended recording procedure[12,57] calls for amplifier sensitivity of 0.2 to 1 mV and sweep of 0.5 to 1 ms/cm for initial exploration. The needle is inserted into the slightly contracting muscle with the subject comfortably lying down or seated. For optimal acquisition of single-fiber potentials, it is important for the examiner to maintain the needle at the critical area with a steady hand. Small shifts in position result in radical changes in the waveform and amplitude of the recorded response. The clear, high-pitched sound of a single-fiber discharge, audible over the loudspeaker, indicates a suitable site for further study. Careful rotation and advancement or retraction of the needle then maximizes the potential on the oscilloscope. The trigger level set on the initial positive deflection of the action potential allows consecutive discharges to superimpose on Polaroid film or a storage scope screen using a new sweep of 20 µs/cm. A constant waveform of the successive tracings confirms a single muscle fiber discharge, whereas varying waveforms indicate a composite action potential not suitable for analysis.

Recommended Criteria

The criteria for accepting a potential as generated by a single muscle fiber near the needle include peak-to-peak amplitude exceeding 200 µV; rise time from the positive to the negative peak of less than 300 µs; and successive discharges with a constant waveform, assessed with a time

resolution of 10 µs or better. The amplitude of a single-fiber discharge decreases to less than 200 µV at a distance greater than 300 µm. Thus, counting the spike discharge fulfilling the above criteria reveals all the muscle fibers of a motor unit located within this radius. Commercially available SFEMG systems may provide different time resolution of the amplifier and other particulars, which dictate the accuracy of analysis. Each laboratory should establish its own normal values.

It is possible to record single-fiber potentials with the monopolar or concentric needle by low-frequency attenuation. The use of a high-pass filter set at 500 Hz eliminates most low-frequency responses that represent volume-conducted potentials from distant muscle fibers.[2] In this situation, even regular needle electrodes register the activity selectively from a few muscle fibers located nearby. Although this type of recording does not accurately distinguish single-fiber responses from summated potentials of more than one fiber, it sometimes reveals abnormal complexity and instability of the motor unit not otherwise appreciated. This approach may bridge the gap between the SFEMG and conventional electromyography.[33,62]

4 FIBER DENSITY

Definition and Clinical Significance

The single-fiber electrode randomly inserted into a slightly contracting normal muscle generally records activities derived from only one muscle fiber. The electrode may occasionally lie close to two or more muscle fibers of the same motor unit. The recorded activity then consists of multiple single fiber potentials discharging synchronously within the recording radius of the single fiber electrode (Fig. 15–2). Repeated counting of such spikes with amplitude greater than 200 µV determines the electromyographic fiber density defined as the mean number of associated single fiber potentials that fire almost synchronously with the ini-

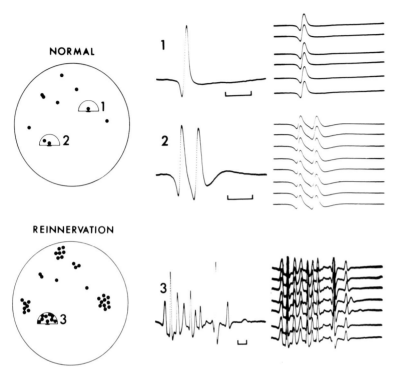

Figure 15–2. Fiber density in normal and reinnervated muscles. All muscle fibers belonging to one motor unit (*small closed circles*) discharge synchronously, but the recording radius of the single fiber electrode (*half circle*) normally contains only one (*1*) or two (*2*) muscle fibers. Following reinnervation, however, a large number of fibers (*3*) cluster within the same radius, reflecting an increase in fiber density. Time calibration is 1 ms. (From Stålberg and Trontelj,[57] with permission.)

tially identified potential.[55] All potentials greater than 200 μV originate within a 300-μm radius of the recording surface in the normal adult.[57] Thus, the motor unit fiber density indicates the average number of single muscle fibers belonging to the same motor unit within this radius.

Fiber density provides a measure of muscle fiber clustering, rather than the total number of muscle fibers within a motor unit. Random loss of muscle fibers generally escapes detection by this technique, because, by definition, the lowest possible value is 1.0. However, a local concentration of action potentials or an increase in fiber density usually indicates the presence of collateral sprouting.[53] Fiber density rivals histochemical fiber grouping in identifying rearrangements within the motor unit.[15,57] Studies have shown a slightly higher density in the frontalis and lower values in the biceps brachii. Subjects under the age of 10 years and over the age of 60, in general, have slightly higher counts (Table 15–1). Fiber density increases slowly during life,

Table 15–1 FIBER DENSITY IN NORMAL SUBJECTS*

| | Ages | | | | | | | | | | | |
| | *10–25 Years* | | | *26–50 Years* | | | *51–75 Years* | | | *Above 75 Years* | | |
Muscles	*Mean*	*SD*	*n*	*Mean*	*SD*	*n*	*Mean*	*SD*	*n*	*Mean*	*SD*	*n*
Frontalis	1.61	0.21	11	1.72	0.21	15						
Deltoid	1.36	0.16	20	1.40	0.11	10						
Biceps	1.25	0.09	20	1.33	0.07	17						
Extensor digitorum communis	1.47	0.16	61	1.49	0.16	98	1.57	0.17	59	2.13	0.41	21
First dorsal interosseous	1.33	0.13	14	1.45	0.12	6						
Rectus femoris	1.43	0.18	11	1.57	0.23	14						
Tibialis anterior	1.57	0.22	18	1.56	0.22	21	1.77	0.12	4	3.8		1
Extensor digitorum brevis	2.07	0.42	16	2.62	0.30	11						

*Fiber density in different muscles of normal subjects arranged in four age groups. n, number of subjects. From Stålberg and Trontelj,[57] with permission.

with faster progression after the age of 70 years, perhaps indicating degeneration of motor neurons with aging, compensated for by reinnervation.[55]

Determination of Fiber Density

To determine the fiber density, one must first record a single-fiber potential with the leading-off surface of the electrode optimally positioned close to the identified fiber. In practice, moving the needle tip back and forth and rotating it will achieve the maximal amplitude of the identified potential with the trigger level of the oscilloscope set at 200 μV. After stabilization of the first action potential, one counts the number of simultaneously firing single muscle fibers for a time interval of at least 5 ms after the triggering spike. For inclusions, an action potential must have an amplitude exceeding 200 μV and a rise time shorter than 300 μs with a high-pass filter set at 500 Hz. The needle is then further advanced to identify another single muscle fiber potential.

This procedure, repeated at 20 to 30 different sites in the muscle, allows calculation of the fiber density as the average number of simultaneously firing single muscle fibers within the recording radius of the single-fiber electrode. For example, isolated discharges of a single muscle fiber in 15 different recording sites and two fiber discharges in 15 other insertions yield an average fiber density of 1.5. In some disease states, a complex pattern of discharges may preclude counting the number of associated spikes. In this situation, one may report the percentage of needle insertions that encounter only one single-fiber potential without associated discharges. Isolated discharges of one single fiber occur in 65 to 70 percent of random insertions in the normal extensor digitorum communis muscle. Only two fibers discharge in the remaining 30 to 35 percent, and triple potentials occur in 5 percent or less.[55]

Duration and Mean Interspike Intervals

The duration of the action potential complex determined during the fiber density search provides an additional means of characterizing the motor unit. This value, defined as the time difference between the first and last single-fiber potentials of the same motor unit recorded at each random insertion, reflects the difference in nerve terminal conduction, neuromuscular transmission, and muscle fiber conduction times within the recording radius of the needle. In practice, one measures the interval from the baseline intersection of the first potential to the corresponding point of the last potential and calculates the average value of at least 20 such measurements. The duration normally falls short of 4 ms in over 95 percent of all multiple-potential recordings in the extensor digitorum communis. In contrast, values may reach as high as 40 to 50 ms in some pathologic conditions.

Dividing the total duration by the number of intervals (number of spikes minus one) yields another measure, called the mean interspike interval (MISI). The normal MISI in the extensor digitorum communis ranges from 0.3 to 0.7 ms. This value increases in muscular dystrophy, polymyositis, and early reinnervation.[57]

5 JITTER

Definition and Basic Physiology

A series of single-fiber potentials recorded after repetitive stimulation of the nerve show almost, but not exactly, the same latencies with each stimulus.[6] This latency variability, on the order of tens of microseconds, represents electromyographic jitter (Fig. 15–3), the term previously used in the engineering literature to denote instability of a timebase generator.[10]

Instead of nerve stimulation used in comatose or uncooperative patients, voluntary activation of muscle suffices for jitter measurements in cooperative patients. For this purpose, one isolates a pair of single-fiber potentials simultaneously recorded from two muscle fibers innervated by adjacent terminal branches of the same axon (Fig. 15–4). The patient slightly activates the muscle under study, and the examiner moves and rotates the

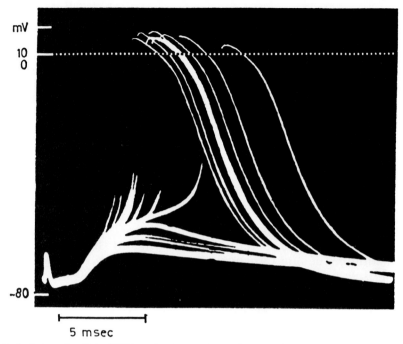

Figure 15–3. End-plate potentials (EPP) and action potentials recorded intracellularly from the end-plate region of a human muscle fiber. The inconsistency of neuromuscular transmission time (jitter) results, primarily because amplitude and slope of EPP vary from one discharge to the next. (From Elmqvist,[14] with permission.)

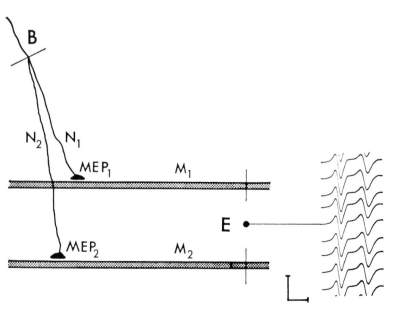

Figure 15–4. Determination of jitter by simultaneous recording from two muscle fibers, M_1 and M_2, innervated by two nerve branches, N_1 and N_2, terminating at two motor end plates, MEP_1 and MEP_2, within the same motor unit. The potential from M_1 triggers the sweep, although the use of a delay line allows its display from the onset. The potential from M_2 appears after a short interpotential interval determined by the difference in conduction time from the common branching point (B) to the recording electrode (E). The variability of the interpotential interval (jitter) mainly occurs at the motor end plates, with some contribution from changes in propagation time along the terminal axons and muscle fibers. Calibration in the strip recording: 2 mV and 500 μs. (From Dahlbäck et al.,[4] with permission.)

needle until at least two time-locked single potentials appear. Skillful use of triggering mechanisms, coupled with delay lines, allows stable repetition of those discharges on the screen. The interpotential interval (IPI), then, represents the difference in conduction time from the common branching point to each fiber within the same motor unit.

In this type of recording, electromyographic jitter is defined as the degree of variability in the IPI, that is, the combined variability of the two responses, measured with one of the two discharges taken as a time of reference. Thus, any factor influencing the conduction of any component will affect jitter. For example, jitter may result from variability in the conduction of impulses along the nerve and muscle fibers. These factors contribute little unless the paired potentials show an excessive IPI or very rapid firing, as discussed below. Thus, the motor end plate constitutes the main source of jitter in normal muscles.[6] A slight change in the rising slope of the end-plate potential (see Fig. 15–3) and fluctuation in the threshold of the muscle membrane necessary for generation of an action potential probably account for most of the variability in transmission time at the neuromuscular junction.

Sometimes after increases in the IPI, the second potential fails to appear. This phenomenon, referred to as "blocking," occurs more commonly in pathologic conductions such as in myasthenia gravis, or to a lesser extent in normal subjects over age 50.[13]

Determination of Jitter

Jitter measurement uses the same techniques as those described for fiber density assessments, except for the need to identify paired single-fiber potentials fulfilling the criteria. If the first of the paired responses triggers the oscilloscope sweep, then the changing delay of the second potential of the pair indicates the variability in the IPI. Jitter may increase erroneously unless the examiner strictly adheres to the recommended criteria to analyze only potentials greater than 200

μV in amplitude with rise time shorter than 300 μs. Other sources of error include use of an unstable trigger, measurement of a potential pair separated by less than 150 μs, and determination of jitter in potentials on the descending phase of the triggering potential.

Most investigators[8] express electromyographic jitter as the mean value of consecutive differences (MCD), rather than the standard deviation (SD) about the mean interpotential interval (MIPI). The SD reflects not only the short-term random variability but also the slow changes in MIPI that result from fluctuation in muscle fiber propagation velocity. Superimposed slow latency shifts will cause an increase in the overall SD even though actual jitter between potentials on sequential firing remains the same. In contrast, the MCD measures only the short-term variation by comparison of sequential discharges. A series of consecutive differences has the additional advantage of being more easily computed. Jitter values expressed by the MCD remain the same during continuous activity lasting up to 1 hour.[8]

Most digital instruments offer software for automatic analysis of jitter and display of the results by numeric or graphic means. Without such a program, manual determination of jitter depends on photographic superimposition of 50 sweeps in groups of 5 or 10 discharges with a sweep speed of 200 μs/cm or faster and conversion of variability measured in IPI into an MCD (Fig. 15–5). If the first potential triggers the oscilloscope, then the jitter equals the variability of a series of second potential, which follows within approximately 1 ms. After superimposition of 10 paired discharges, the latency difference between the baseline intersection points of the earliest and latest second potentials provides the time range of 10 discharges (R_{10}) that represents the variability of the IPI. The average of R_{10}s from five different sampling sites in the same muscle gives the mean range of 10 discharges (MR_{10}). To convert MR_{10} to an estimated MCD, one must multiply it by a factor of 0.37[8]; that is, estimated MCD = $MR_{10} \times$ 0.37. Similarly, for the mean range of 5 discharges (MR_5) from 10 different sites,

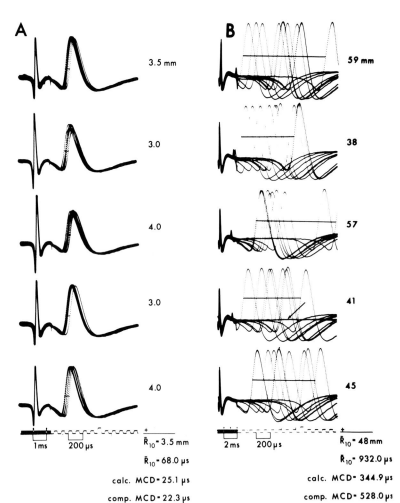

Figure 15–5. Manual calculation of jitter in a normal (*A*) and abnormal (*B*) action potential pair. The tracings show five groups of 10 superimpositions to measure the ranges of variability for the IPI in each group. The mean value of the IPI variability multiplied by 0.37 equals the calculated MCD (*calc. MCD*), which approximates very closely the result determined by a computer (*comp. MCD*). (From Stålberg and Trontelj,[57] with permission.)

estimated MCD = $MR_5 \times 0.49$. The value obtained by these formulas gives a good approximation of the actual MCD calculated by a computer program.[31,34]

Muscle fiber propagation slows substantially upon rapid firing because successive action potentials occur in the relative refractory period of the muscle. This delay may differentially affect the activation of two muscle fibers, depending on the lengths of their respective axon terminals. In general, a long IPI and rapid firing rates tend to increase jitter because these factors induce physiologic slowing that influences two muscle fibers differently. As a rule, the firing rate affects jitter if the IPI exceeds 4 ms. Here, a computer can sort the IPI on the basis of the interdischarge interval (IDI) to calculate the corrected MCD, termed mean sorted

interval difference (MSD). If firing rate has not affected jitter, MCD/MSD = 1. If the ratio exceeds 1.25, one must use MSD instead of MCD, because the firing rate has influenced jitter. On the other hand, a ratio less than 0.8, indicating slow trends, favors the use of MCD, not MSD. In calculating jitter without a computer, selection of IPI less than 4 ms precludes the effect of IDI, and the use of MCD avoids slow trends.

Normal and Abnormal Jitter Values

Table 15–2 summarizes the jitter values for various muscles determined at a very fast sweep speed with the time resolution of 0.3 μs.[57] Jitter measurements

Table 15–2 JITTER IN NORMAL SUBJECTS*

Muscles	Number of Potential Pairs	MCD—Pooled Data		SD of MCD Values from Individual Subjects		Upper Normal Limit Close to Mean + 3 SD
		Mean	*SD*	*Mean*	*SD*	
Frontalis (range of individual means	258	20.4 (15.7–29.2)	8.8	6.2 (5.5–8.7)	2.3	45
Biceps	125	15.6	5.9			35
Extensor digitorum communis (range of individual means)	759	24.6 (16.5–32.0)	10.6	8.3 (2.3–12.4)	3.2	55 (65)†
Rectus femoris	73	31.0	12.6			60 (75)†
Tibialis anterior	153	32.1	15.0			60
Extensor digitorum brevis	29	85.3	68.6			None

*Jitter (MCD) measured with voluntary activation in normal subjects aged 10 to 70 years.
†Because of some extreme high values, the data deviate from a gaussian distribution. Thus, a more appropriate upper normal limit is 60 μs. In no one normal subject was there more than one value exceeding this limit.
From Stålberg and Trontelj,[57] with permission.

may show a different range and higher mean value than those listed if recorded with less time resolution. Jitter value varies, depending on the subject's age and the individual muscles tested, although blocking in more than one fiber or jitter values exceeding 55 μs constitute an abnormality in any muscles. The jitter remains relatively constant below the age of 70 in the extensor digitorum communis. It increases around the age of 50 in the tibialis anterior, probably secondary to neurogenic change.[56] Normal muscles show the same jitter regardless of the innervation rates or the recording site relative to the end-plate zone. Neuromuscular jitter may increase during continuous voluntary activation in patients with myasthenia gravis, spinal muscular atrophy, or motor neuron disease, but not in normal subjects.[27]

In normal muscles, jitter values are independent of the IPI. In most recordings showing an interval of less than 1 ms, changes in conduction time by prior discharge largely cancel out between the two potentials. Thus, jitter results primarily from variability in neuromuscular transmission time. To further support this view, nonparalytic doses of tubocurarine, known to block end-plate depolarization, cause jitter to increase without changing the shape and amplitude of the single muscle fiber potentials.[10] In pathologic conductions where the IPI may reach many milliseconds, however, variability in the propagation velocity may contribute to the jitter. In fact, jitter changing with firing rate may reflect this type of underlying pathology. In myasthenia gravis characterized by postsynaptic defect, the rapid firing rate increases jitter even with an IPI of less than 4 ms. In presynaptic disorders such as myasthenic syndrome and botulism, jitter increases at slow firing rates and decreases at fast rates.

Jitter increases 2 to 3 μs per degree centigrade as the temperature of the muscle falls from 36°C to 32°C, followed by a more rapid change of about 7.5 μs per degree centigrade thereafter.[50] Despite an increase in the jitter value, cooling reduces decrement of the compound muscle action potentials to a train of stimuli. A number of factors may contribute to the apparent discrepancy. Defective release of transmitters at low temperatures would explain increased jitter and paradoxically reduced decrement; fewer quanta released by the first impulse leave more quanta available for subsequent release. Increases in temperature between 35°C and 38°C do not normally change jitter value.

In normal muscles, jitter may increase during ischemia or following administration of curare. Abnormal jitter occurs not only in diseases of neuromuscular transmission[13] but also in many other conditions associated with conduction defects of nerve and muscle.[28,49] It may also result from unusually low end-plate potentials or from a high threshold of the muscle fiber membrane. In general, an increase in jitter values, typically beyond 80 to 100 μs, precedes the transmission block. Chronic muscular activity also leads to increased jitter and other minor SFEMG abnormalities, presumably as the result of mild denervation and reinnervation of nerve terminals.[41]

6 MACRO EMG

Compared with the single-fiber electrode that covers the radius of some 300 μm, the concentric or monopolar needle records action potentials from a much wider zone with about a 1-mm radius. Motor unit territories, however, extend much further, varying in size from 5 to 10 mm. To capture the total electrical activity generated by a motor unit, the electrode must have a much greater recording surface. Such an electrode, however, registers activities from a number of motor units because muscle fibers from different units intermingle within the recording zone. An averaging technique with a special needle circumvents this difficulty and is termed macro EMG.[47,48]

The needle used for macro EMG consists of two recording surfaces, one capable of recording SFEMG from a side port and the other dedicated for territorial pickup with a bare cannula 15 mm in length (Fig. 15–6). A two-channel system provides SFEMG recording with a 500-Hz low-frequency filter in one channel, and macro EMG recording at standard EMG setting in the other. The SFEMG side port referenced to the cannula produces single-fiber signals that trigger the oscilloscope sweep. The active cannula electrode with reference to a skin or distant electrode registers electrical activities along its entire length but only if time-locked to the SFEMG trigger. Simultaneous discharges from neighboring motor units do not time-lock with the trigger and therefore cancel as background noise during signal averaging.

The resultant response differs substantially from the ordinary motor unit potential. A monopolar or concentric needle registers the activities generated by only a few muscle fibers within a 1-mm radius. In contrast, macro EMG motor unit potentials receive contribution from many more muscle fibers located outside the range of the conventional recording.[32] Thus, macro EMG signals give information about the whole motor unit, in contrast to the regional electrical activity measured in conventional electromyography, or focal pickup in single-fiber EMG (Fig. 15–7).

In summary, this technique extracts the contribution from all muscle fibers belonging to a motor unit by recording the electrical activity obtained by the electrode shaft during voluntary muscle contraction.[47] With a modified single-fiber EMG electrode, single-fiber action potentials provide the trigger for selective averaging of the intended motor unit potential. The factors that determine the characteristics of macro EMG include number of fibers, fiber diameter, endplate scatter, pattern of nerve branching, motor unit territory, and electrode position. Table 15–3 summarizes the suggested normal data for the different age groups. Macro motor unit potentials increase in size after the age of 60, in part reflecting reinnervation following physiologic loss of anterior horn cells with age.[52]

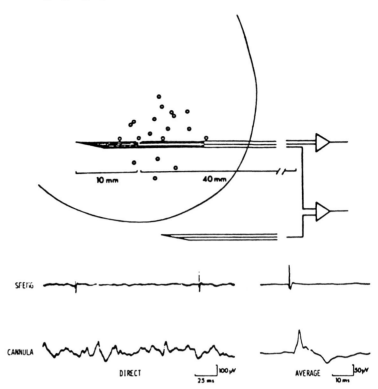

Figure 15–6. Principles for macro EMG. Single-fiber action potentials recorded by the small lead-off surface provide triggers to average cannula activities time-locked to the discharge from one muscle fiber. (From Stålberg,[47] with permission).

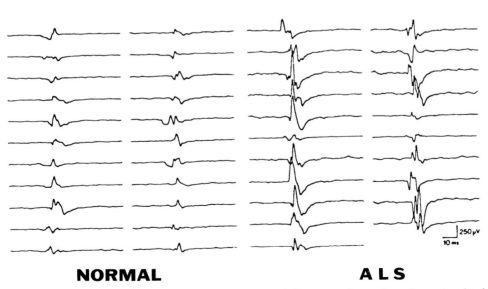

Figure 15–7. Examples of macro motor unit potentials recorded in normal muscle and amyotrophic lateral sclerosis (ALS). (From Stålberg,[48] with permission.)

Table 15–3 MACRO EMG IN NORMAL SUBJECTS

	Suggested Amplitude Limits (μV)											
	Biceps				Vastus Lateralis				Tibialis Anterior			
	Median		Individual Macro-MUP		Median		Individual Macro-MUP		Median		Individual Macro-MUP	
Age	Min	Max	Min	Max	Min	Max	Min	Max	Min	Max	Min	Max
10–19	65	100	30	350	70	150	20	350	65	200	30	350
20–29	65	140	30	350	70	240	20	525	65	250	30	450
30–39	65	180	30	400	70	240	20	550	65	260	30	450
40–49	65	180	30	500	70	250	20	575	65	330	30	575
50–59	65	180	30	500	70	260	20	575	65	375	40	700
60–69	65	250	30	650	80	370	20	1250	120	375	45	700
70–79	65	250	30	650	90	600	20	1250	120	620	65	800

From Stålberg,[48] with permission.

7 CLINICAL APPLICATION

SFEMG has become available in most laboratories with a dedicated system or with minor modification of the conventional units of electromyography. A computer-assisted method has rendered the technique simple enough to conduct as part of routine electromyography with a little extra training.[57] The method has clinical and research applications for lower motor neuron disorders and diseases of muscle in general. It has proven most useful, from an electrodiagnostic point of view, as a test for neuromuscular transmission.[13,46,51]

Motor Neuron Disease

Disorders associated with abnormal SFEMG include degenerative processes affecting the anterior horn cell[53] and tetanus.[16] The chronic processes with marked collateral sprouting, such as spinal muscular atrophy, show the highest fiber density among motor neuron diseases. Clinical studies have revealed an inverse relationship between muscle strength and fiber density.[59] The increased duration of the action potential found in this entity suggests a mixture of hypertrophic and atrophic fibers and slowly conducting, regenerated nerve sprouts, forming collateral reinnervation but not as effectively as in other neuro-genic disorders.[59] In contrast, rapidly progressive diseases such as amyotrophic lateral sclerosis show increased jitter and blocking. The SFEMG characterizes the functional status of the motor unit and may help establish the diagnosis and prognosis. Detecting abnormalities not apparent clinically or with conventional electromyography provides early evidence of motor neuron involvement.

Peripheral Neuropathy

Disorders of the peripheral nerves also show increased jitter, occasional blocking, and increased fiber density.[49] These findings become particularly prominent during the process of reinnervation, having been observed, for example, up to 1 year after autogenous facial muscle transplants.[20] Abnormalities of the SFEMG in part result from reinnervation when seen in patients with polyneuropathies and motor neuron disease, and perhaps also in those with polymyositis and muscular dystrophy. Not all the polyneuropathies show similar changes, however. In studies of the extensor digitorum communis,[60] fiber density, duration, jitter, and blocking were abnormal in patients with alcoholic neuropathy, but not in those with diabetic or uremic neuropathy. Persistent changes may reflect ongoing reinnervation in certain types of neuropathy, but the exact reason for the discrepancy is not known. Jitter values in chronic renal failure improve after intermittent hemodialysis.[29] Jitter is also increased in some patients with multiple sclerosis[61] and idiopathic fecal incontinence.[42]

Disorders of Neuromuscular Transmission

Normal muscles show increased jitter in 1 of 20 recorded pairs of potentials.[50] Increased jitter or blocking, if found in 2 or more of 20 pairs of potentials, can be regarded as evidence of defective neuromuscular transmission. A patient with myasthenia gravis may have normal or increased jitter values within any one muscle. If the jitter exceeds 100 μs, intermittent blocking may occur.[58] In generalized myasthenia gravis, more than 30 percent of recorded potential pairs show abnormalities in the extensor digitorum communis. Patients with ocular myasthenia may have such findings only in the facial muscles and not necessarily in the muscles of the extremities.[36,58] About 25 percent of patients with myasthenia gravis have slightly increased fiber density above the normal range. Statistical analyses showed no correlation between this abnormality and disease severity or duration, but the patients treated with cholinesterase inhibitors had a significantly greater increase in jitter value.[25]

The SFEMG can detect disturbances of neuromuscular transmission before the appearance of clinical symptoms. Thus, normal jitter in a clinically weak muscle tends to exclude the diagnosis of myasthenia gravis.[35] In one series,[58] SFEMG showed increased jitter or blocking in the hypothenar muscles in all 40 patients with mild to moderate generalized myasthenia gravis, even though the repetitive nerve stimulation technique revealed equivocal results in 40 percent of these cases. In another series,[36] 127 of 131 patients demonstrated defective neuromuscular transmission by SFEMG, whereas less than 50 percent of these patients had an abnormality by the conventional nerve stimulation technique. The SFEMG of the extensor digitorum communis showed abnormality in 8 of 24 first-degree relatives of 12 patients with juvenile myasthenia gravis. In this asymptomatic group, increased jitter occurred, on the average, in 5 of 20 potential pairs. Hence only 25 percent of all recordings showed abnormalities, in contrast to 75 percent in clinically symptomatic patients.[58]

The SFEMG abnormalities correlated well with the clinical course in serial studies of individual patients. Administration of edrophonium (Tensilon) shortens abnormal jitter and decreases the incidence of blocking, although it does not affect initially normal jitter. A therapeutic dosage of anticholinesterase medication may correct jitter in myasthenia. In some cases, recovery from blocking in a number of fibers may give an apparent increase in jitter values after treatment. In a healthy subject, anticholinesterase has no effect on the jitter value. Indeed, the jitter value remained normal in a patient who had received the medication for years with an incorrect diagnosis of myasthenia gravis.[57] The SFEMG may occasionally return to normal during spontaneous remission or after thymectomy, but most of these patients still have increased jitter without blocking.[57]

In the myasthenic syndrome a slight increase in fiber density[57] probably results from type II fiber grouping.[22] Blocking tends to occur with higher jitter values than in myasthenia gravis. In fact, jitter may reach as high as 500 μs, with the interval between the first and last of the second potentials reaching 2 ms for 50 discharges.[38,39] The degree of blocking and jitter values decrease as the stimulation rate increases, and the transmission worsens after rest. These findings stand in sharp contrast to those typically seen in patients with myasthenia gravis.

In four patients who received periocular injections of botulinum toxin for blepharospasm, abnormal neuromuscular transmission of the arm muscles was demonstrated by SFEMG.[37] The time course as well as the inverse relationship between jitter and the firing rate in the affected muscle indicated that the toxin spread remotely from the site of injection.

Myopathy

Dystrophic muscle in general shows increase in fiber density and jitter, although abnormally low jitter also occurs in some recordings. Macro EMG study shows normal diameter of motor units with no signs of abnormal volume conduction.[23,24] These findings suggest a remodeling of the motor unit as the result of fiber loss, fiber regeneration, and reinnervation. The increased fiber density probably reflects a localized abnormality in the distribution of muscle fibers within each motor unit.

In one series,[46] patients with Duchenne dystrophy had markedly increased fiber density, averaging 3.5 initially and less in the late stage, although still above the normal value of 1.45. Another series showed increased jitter in about 30 percent of the recordings in each muscle and occasional blocking in 10 percent of the recordings with increased jitter.[54] Interestingly, some pairs had jitter values below the normal range, suggesting that the potential originated from split muscle fibers sharing a common innervation zone.[26] In support of this view, pairs with reduced jitter always block simultaneously when subjected to tubocurarine or other agents that inhibit neuromuscular transmission. Ordinary potential pairs would show clear dissociation with this type of inhibition.

Fiber density also increases in limb girdle dystrophy, but to a lesser degree than in Duchenne dystrophy. In one series,[45] patients with limb girdle dystrophy had increased jitter in 54 percent of the recordings in clinically weak muscles and blocking in less than 10 percent. In another series of 20 patients with either myopathic limb girdle syndrome or chronic spinal muscular atrophy, SFEMG confirmed the original diagnosis in 16 unequivocal cases, and helped differentiate the four indeterminate cases into myopathic and neurogenic categories.[40] Studies in patients with facioscapulohumeral dystrophy showed findings similar to those reported in limb-girdle dystrophy.[45] Although the pathophysiology underlying these SFEMG abnormalities remains unclear, increased jitter may result from altered propagation time in the muscle fibers.[43] Increased fiber density may indicate reinnervation of a portion of the muscle fiber separated from the endplate by transverse lesions, as shown in Duchenne dystrophy.[5] Alternatively, it may reflect new innervation of regenerating muscle fibers or splitting of muscle fibers.[44]

In polymyositis, a segmental degeneration separates a portion of the affected muscle fiber from its motor end plate. Collateral sprouts then reinnervate the denervated portion of the muscle fiber. This probably accounts for the presence of fibrillation potentials, increased fiber density, and increased jitter and blocking.[21]

In myotonic dystrophy, high-frequency discharges recorded in SFEMG progressively decrease in amplitude and increase in rise time. In one series,[30] fiber density was increased in 84 percent and jitter exceeded the normal range in 20 percent of the measurements.

REFERENCES

1. Adrian, ED, and Bronk, DW: The discharge of impulses in motor nerve fibres. Part II. The frequency of discharge in reflex and voluntary contractions. J Physiol (Lond) 67:119–151, 1929.
2. Borenstein, S, and Desmedt, JE: Local cooling in myasthenia: Improvement of neuromuscular failure. Arch Neurol 32:152–157, 1975.
3. Buchthal, F, Guld, C, and Rosenfalck, P: Multielectrode study of the territory of a motor unit. Acta Physiol Scand 39:83–104, 1957.
4. Dahlback, LO, Ekstedt, J, and Stålberg, E: Ischemic effects of impulse transmission to muscle fibers in man. Electroenceph Clin Neurophysiol 29:579–591, 1970.
5. Desmedt, JE, and Borenstein, S: Regeneration in Duchenne muscular dystrophy: Electromyographic evidence. Arch Neurol 33:642–650, 1976.
6. Ekstedt, J: Human single muscle fiber action potentials. Acta Physiol Scand (Suppl 226) 61:1–96, 1964.
7. Ekstedt, J, Haggqvist, P, and Stålberg, E: The construction of needle multi-electrodes for single fiber electromyography. Electroencephalogr Clin Neurophysiol 27:540–543, 1969.
8. Ekstedt, J, Nilsson, G, and Stålberg, E: Calculation of the electromyographic jitter. J Neurol Neurosurg Psychiatry 37:526–539, 1974.
9. Ekstedt, J, and Stålberg, E: A method of recording extracellular action potentials of single

muscle fibres and measuring their propagation velocity in voluntarily activated human muscle. Bulletin of the American Association of Electromyography and Electrodiagnosis 10:16, 1963.

10. Ekstedt, J, and Stålberg, E: The effect of nonparalytic doses of D-tubocurarine on individual motor end-plates in man, studied with a new electrophysiological method. Electroencephalogr Clin Neurophysiol 27:557–562, 1969.

11. Ekstedt, J, and Stålberg, E: How the size of the needle electrode leading-off surface influences the shape of the single muscle fibre action potential in electromyography. Computer Prog Biomed 3:204–212, 1973.

12. Ekstedt, J, and Stålberg, E: Single fibre electromyography for the study of the microphysiology of the human muscle. In Desmedt, JE (ed): New Developments in Electromyography and Clinical Neurophysiology, Vol 1. Karger, Basel, 1973, pp 89–112.

13. Ekstedt, J, and Stålberg, E: Single muscle fibre electromyography in myasthenia gravis. In Kunze, K, and Desmedt, JE (eds): Studies in Neuromuscular Diseases. Proceedings of the International Symposium (Giessen). Karger, Basel, 1975, pp 157–161.

14. Elmqvist, D. Hofmann, WW, Kugelberg, J, and Quastel, DMJ: An electrophysiological investigation of neuromuscular transmission in myasthenia gravis. J Physiol (Lond) 174:417–434, 1964.

15. Fawcett, PRW, Johnson, MA, and Schofield, IS: Comparison of electrophysiological and histochemical methods for assessing the spatial distribution of muscle fibres of a motor unit within muscle. J Neurol Sci 69:67–79, 1985.

16. Fernandez, JM, Ferrandiz, M, Larrea, L, Ramio, R, and Boada, M: Cephalic tetanus studied with single fibre EMG. J Neurol Neurosurg Psychiatry 46:862–866, 1983.

17. Gath, I, and Stålberg, E: Frequency and time domain characteristics of single muscle fibre action potentials. Electroencephalogr Clin Neurophysiol 39:371–376, 1975.

18. Gath, I, and Stålberg, E: On the volume conduction in human skeletal muscle: In situ measurements. Electroencephalogr Clin Neurophysiol 43:106–110, 1977.

19. Gath, I, and Stålberg, E: The calculated radial decline of the extracellular action potential compared with in situ measurements in the human brachial biceps. Electroencephalogr Clin Neurophysiol 44:547–552, 1978.

20. Hakelius, L, and Stålberg, E: Electromyographical studies of free autogenous muscle transplants in man. Scand J Plast Reconstr Surg 8:211–219, 1974.

21. Henriksson, KG, and Stålberg, E: The terminal innervation pattern in polymyositis: A histochemical and SFEMG study. Muscle Nerve 1:3–13, 1978.

22. Henriksson, KG, Nilsson, O, Rosen, I, and Schiller, HH: Clinical, neurophysiological and morphological findings in Eaton Lambert syndrome. Acta Neurol Scand 56:117–140, 1977.

23. Hilton-Brown, P, and Stålberg, E: The motor unit in muscular dystrophy, a single fibre EMG and scanning EMG study. J Neurol Neurosurg Psychiatry 46:981–995, 1983.

24. Hilton-Brown, P, and Stålberg, E: Motor unit size in muscular dystrophy, a macro EMG and scanning EMG study. J Neurol Neurosurg Psychiatry 46:996–1005, 1983.

25. Hilton-Brown, P, Stålberg, E, and Osterman, PO: Signs of reinnervation in myasthenia gravis. Muscle Nerve 5:215–221, 1982.

26. Hilton-Brown, P, Stålberg, E, Trontelj, J, and Mihelin, M: Causes of the increased fiber density in muscular dystrophies studied with single fiber EMG during electrical stimulation. Muscle Nerve 8:383–388, 1985.

27. Ingram, DA, Davis, GR, Schwartz, MS, and Swash, M: The effect of continuous voluntary activation of neuromuscular transmission: An SFEMG study of myasthenia gravis and anterior horn cell disorders. Electroencephalogr Clin Neurophysiol 60:207–213, 1985.

28. Jamal, GA, and Hansen, S: Electrophysiological studies in the post-viral fatigue syndrome. J Neurol Neurosurg Psychiatry 48:691–694, 1985.

29. Konishi, T, Nishitani, H, and Motomura, S: Single fiber electromyography in chronic renal failure. Muscle Nerve 5:458–461, 1982.

30. Martinez, AC, Ferrer, MT, and Conde, MCP: Electrophysiological studies in myotonic dystrophy. 2. Single fibre EMG. Electromyogr Clin Neurophysiol 24:537–546, 1984.

31. Mihelin, M, Trontelj, JV, and Trontelj, JK: Automatic measurement of random interpotential intervals in single fibre electromyography. Int J Biomed Comput 6:181–191, 1975.

32. Nandedkar, S, and Stålberg, E: Simulation of macro EMG motor unit potentials. Electroencephalogr Clin Neurophysiol 56:52–62, 1983.

33. Payan, J: The blanket principle: A technical note. Muscle Nerve 1:423–426, 1978.

34. Salmi, T: A duration matching method for the measurement of jitter in single fibre EMG. Electroencephalogr Clin Neurophysiol 56:515–520, 1983.

35. Sanders, DB, and Howard, JI: AAEE Minimonograph #25: Single fiber electromyography in myasthenia gravis. Muscle Nerve 9:809–819, 1986.

36. Sanders, DB, Howard, JF, Jr, and Johns, TR: Single-fiber electromyography in myasthenia gravis. Neurology 29:68–76, 1979.

37. Sanders, DB, Massey, EW, and Buckley, EG: Botulinum toxin for blepharospasm: Single-fiber EMG studies. Neurology 36:545–547, 1986.

38. Schwartz, MS, and Stålberg, E: Myasthenia gravis with features of the myasthenic syndrome: An investigation with electrophysiologic methods including single-fiber electromyography. Neurology 25:80–84, 1975.

39. Schwartz, MS, and Stålberg, E: Myasthenic syndrome studied with single fiber electromyography. Arch Neurol 32:815–817, 1975.

40. Shields, RW, Jr: Single fiber electromyography in the differential diagnosis of myopathic limb girdle syndromes and chronic spinal muscular atrophy. Muscle Nerve 7:265–272, 1984.

41. Shields, RW, Robbins, N, and Verrilli, AA, III: The effects of chronic muscular activity on age-related changes in single fiber electromyography. Muscle Nerve 7:273–277, 1984.

42. Snooks, SJ, Barnes, PRH, and Swash, M: Damage to the innervation of the voluntary anal and periurethral sphincter musculature in incontinence: An electrophysiological study. J Neurol Neurosurg Psychiatry 47:1269–1273, 1984.

43. Stålberg, E: Propagation velocity in human muscle fibers in situ. Acta Physiol Scand (Suppl 287) 70:1–112, 1966.

44. Stålberg, E: Single fibre electromyography for motor unit study in man. In Shahani, M (ed): The Motor System: Neurophysiology and Muscle Mechanisms. Elsevier, Amsterdam, 1976, pp 79–92.

45. Stålberg, E: Electrogenesis in human dystrophic muscle. In Rowland, LP (ed): Pathogenesis of Human Muscular Dystrophies. Excerpta Medica, Amsterdam, 1977, pp 570–587.

46. Stålberg, E: Neuromuscular transmission studied with single fibre electromyography. Acta Anaesthesiol Scand Suppl 70:112–117, 1978.

47. Stålberg, E: Macro EMG, a new recording technique. J Neurol Neurosurg Psychiatry 43:475–482, 1980.

48. Stålberg, E: AAEE Minimonograph #20, Macro EMG. Muscle Nerve 6:619–630, 1983.

49. Stålberg, E, and Ekstedt, J: Single fibre EMG and microphysiology of the motor unit in normal and diseased human muscle. In Desmedt, JE (ed): New Developments in Electromyography and Clinical Neurophysiology, Vol 1. Karger, Basel, 1973, pp 113–129.

50. Stålberg, E, Ekstedt, J, and Broman, A: The electromyographic jitter in normal human muscles. Electroencephalogr Clin Neurophysiol 31:429–438, 1971.

51. Stålberg, E, Ekstedt, J, and Broman, A: Neuromuscular transmission in myasthenia gravis studied with single fibre electromyography. J Neurol Neurosurg Psychiatry 37:540–547, 1974.

52. Stålberg, E, and Fawcett, PRW: Macro EMG in healthy subjects of different ages. J Neurol Neurosurg Psychiat 45:870–878, 1982.

53. Stålberg, E. Schwartz, MS, and Trontelj, JV: Single fibre electromyography in various processes affecting the anterior horn cell. J Neurol Sci 24:403–415, 1975.

54. Stålberg, E, and Thiele, B: Transmission block in terminal nerve twigs: A single fibre electromyographic finding in man. J Neurol Neurosurg Psychiatry 35:52–59, 1972.

55. Stålberg, E, and Thiele, B: Motor unit fibre density in the extensor digitorum communis muscle: Single fibre electromyographic study in normal subjects at different ages. J Neurol Neurosurg Psychiatry 38:874–880, 1975.

56. Stålberg, E, and Trontelj, JV: Demonstration of axon reflexes in human motor nerve fibres. J Neurol Neurosurg Psychiatry 33:571–579, 1970.

57. Stålberg, E, and Trontelj, J: Single Fibre Electromyography. The Miraville Press Limited, Old Woking, Surrey, UK, 1979.

58. Stålberg, E, Trontelj, JV, and Schwartz, MS: Single-muscle-fiber recording of the jitter phenomenon in patients with myasthenia gravis and in members of their families. Ann NY Acad Sci 274:189–202, 1976.

59. Swash, M: Vulnerability of lower brachial myotomes in motor neurone disease: A clinical and single fibre EMG study. J Neurol Sci 47:59–68, 1980.

60. Thiele, B, and Stålberg, E: Single fibre EMG findings in polyneuropathies of different etiology. J Neurol Neurosurg Psychiatry 38:881–887, 1975.

61. Weir, A, Hansen, S, and Ballantyne, JP: Single fibre electromyographic jitter in multiple sclerosis. J Neurol Neurosurg Psychiatry 42:1146–1150, 1979.

62. Wiechers, DO: Single fiber electromyography with a standard monopolar electrode. Arch Phys Med Rehabil 66:47–48, 1985.

Part V

TESTS FOR LESS ACCESSIBLE REGIONS OF THE NERVOUS SYSTEM

Chapter 16

THE BLINK REFLEX

1 INTRODUCTION

The mechanically or electrically elicited blink reflex bears resemblance to the corneal reflex tested in clinical practice.[8,21,23,54,59,74,83,92,93] The use of an oscilloscope display allows quantitative analysis for more meaningful assessment of the reflex responses.[1,65,76,97] Stimulation of the supraorbital nerve elicits two temporally separate responses of the orbicularis oculi, an early (R_1) component and a late (R_2) component (Fig. 16–1A and B). Of the two, R_1 is evoked only on the side of stimulation via a pontine pathway,[38,92] but an inconsistent crossed response may also appear with priming.[104] In contrast, unilateral stimulation elicits R_2 bilaterally, presumably relayed through a more complex route, including the pons and lateral medulla.[16,33,52,75,98]

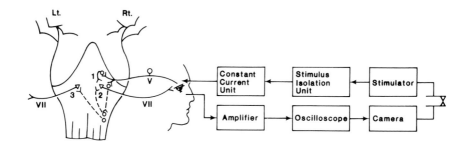

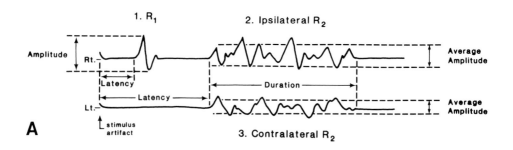

1. R₁   2. Ipsilateral R₂

Amplitude

Rt.

Latency

Latency

Duration

Average Amplitude

Lt.

Average Amplitude

stimulus artifact

A

3. Contralateral R₂

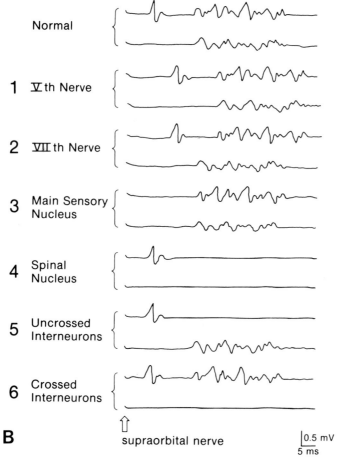

Normal

1   Ⅴ th Nerve

2   Ⅶ th Nerve

3   Main Sensory Nucleus

4   Spinal Nucleus

5   Uncrossed Interneurons

6   Crossed Interneurons

B

supraorbital nerve

0.5 mV

5 ms

Figure 16–1. See facing page for legend.

The more reproducible R_1 is a better measure of nerve conduction along the reflex pathways. Analysis of R_2 helps localize the lesion to the trigeminal nerve, the facial nerve, or the brainstem.[31,52] Involvement of the trigeminal nerve causes an afferent pattern with delays or diminution of R_2 bilaterally after stimulation of the affected side. In diseases of the facial nerve, the pattern indicates an efferent abnormality with alteration of R_2 only on the affected side regardless of the side of stimulation. (Afferent and efferent delays of R_2 are analogous to the two types of abnormality in the pupillary light reflex.)

2 DIRECT VERSUS REFLEX RESPONSES

Stimulation of the Facial Nerve

Nerve excitability is tested by applying shocks of increasing intensity and inspecting the visible contraction of the facial muscles. The normal threshold ranges from 3.0 to 8.0 mA, depending on skin resistance, skin temperature, and the anatomic course of the facial nerve. Comparisons with the unaffected nerve on the opposite side reduce the number of variables to a minimum. In healthy subjects, differences between left and right should not exceed 2.0 mA. After complete section of the nerve at a proximal site, distal excitability remains normal up to 4 days, followed by complete loss by the end of the first week, with the emergence of wallerian degeneration. Hence, a nor-mal distal response during the first week after onset suggests a good prognosis for recovery.[30]

As opposed to nerve excitability testing by visual inspection of contracting muscle, electrical recording of muscle action potentials provides a quantitative assessment. Stimulating the facial nerve just below the ear and anterior to the mastoid process[102] or directly over the stylomastoid foramen[96] elicits compound muscle action potentials in the facial muscles (Fig. 16–2). This potential is designated as direct or M response to distinguish it from the reflex activation of the orbicularis oculi by stimulation of the trigeminal nerve. The amplitude of the direct response varies with the number of functional motor axons, whereas the onset latency reveals the distal conduction of the fastest fibers. Determination of nerve excitability, although no longer widely practiced, may provide additional measures that supplement studies of the direct response.[22]

For surface recording, the active electrode (G_1) is usually placed over the orbicularis oculi, orbicularis oris, quadratus labii, or nasalis, and the indifferent electrode (G_2) over the same muscle on the opposite side or on the nose. When necessary, selective stimulation of a given branch of the facial nerve elicits an isolated response from any muscles of the face, including the posterior auricular muscle.[18] Some investigators prefer recording from a needle placed in the orbicularis oris just superior to the corner of the mouth or orbicularis oculi at the lateral epicanthus. The coaxial needle gives a slightly better endpoint than the mono-

---

Figure 16–1. A. *Top,* stimulation and recording arrangement for the blink reflex, with the presumed pathway of R_1 through the pons (*1*) and ipsilateral and contralateral R_2 through the pons and lateral medulla (*2* and *3*). The schematic illustration shows the primary afferents of R_1 and R_2 shown as one fiber. This and other details of central connections of these reflexes await further clarification. *Bottom,* a typical oscilloscope recording of the blink reflex after right-sided stimulation. Note an ipsilateral R_1 response and bilateral simultaneous R_2 responses. (Modified from Kimura.[44]) **B.** Five basic types of blink reflex abnormalities. From *top* to *bottom,* the finding suggests the conduction abnormality of the afferent pathway along the trigeminal nerve (*1*); the efferent pathway along the facial nerve (*2*); the main sensory nucleus or pontine interneurons relaying to the ipsilateral facial nucleus (*3*) (*1* of **A**); the spinal tract and nucleus or medullary interneuronal pathways to the facial nuclei on both sides (*4*); uncrossed medullary interneurons to the ipsilateral facial nucleus (*5*) (*2* of **A**); and crossed medullary interneurons to the contralateral facial nucleus (*6*) (*3* of **A**). Increased latencies of R_1 usually indicate the involvement of the reflex arc itself, whereas the loss or diminution of R_1 or R_2 may result not only from lesions directly affecting the reflex pathway but also from those indirectly influencing the excitability of the interneurons or motor neurons.

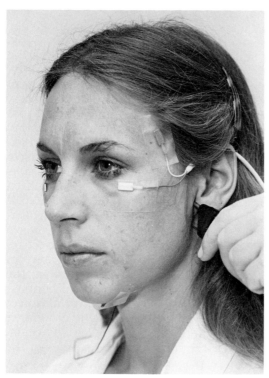

Figure 16–2. Technique for recording the direct response. Stimulation of the facial nerve trunk with the cathode placed just anterior to the mastoid process elicits compound action potentials in all muscles of the face ipsilaterally. The test performed in conjunction with the blink reflex uses the active electrode (G_1) placed on the orbicularis oculi and the reference electrode (G_2) on the temple or the side of the nose. Recording from the ipsilateral nasalis (not shown) often gives rise to a more discrete compound muscle action potential.

Table 16–1 FACIAL NERVE LATENCY IN 78 SUBJECTS DIVIDED INTO DIFFERENT AGE GROUPS

Age	Mean (ms)	Range (ms)
0–1 month	10.1	6.4–12.0
1–12 months	7.0	5.0–10.0
1–2 years	5.1	3.5–6.3
2–3 years	3.9	3.8–4.5
3–4 years	3.7	3.4–4.0
4–5 years	4.1	3.5–5.0
5–7 years	3.9	3.2–5.0
7–16 years	4.0	3.0–5.0

From Waylonis and Johnson,[102] with permission.

polar needle. The use of a monopolar needle requires a reference electrode usually placed on the side of the nose and the ground placed on the forehead or under the chin.

Reported normal values for facial nerve latencies (mean ± 1 standard deviation [SD]) in adults range from 3.4 ± 0.8 ms[102] to 4.0 ± 0.5 ms.[96] Table 16–1 summarizes the normal values measured to the onset of the negative deflection of the evoked potential in 78 subjects divided into different age groups.[102] For the assessment of a proximal lesion in Bell's palsy, the latency of the direct response rarely provides useful information. Even with substantial axonal degeneration, the remaining axons tend to show a normal or only slightly increased onset latency. In contrast, the amplitude of the direct response determines the degree of axonal loss for accurate assessment of prognosis. Comparison between the sides in the same individual provides a more sensitive measure than the absolute value, which varies substantially from one subject to the next. An amplitude reduction to one half that of the response on the normal side suggests distal degeneration.

More importantly, serial determinations reveal progressive amplitude changes as an increasing number of axons degenerate (Fig. 16–3). Distal stimulation elicits a normal response for a few days even after complete separation of the nerve at a proximal site. By the end of the first week, however, the amplitude drops abruptly, coincident with the onset of nerve degeneration. Thus, normal direct response during the first week after injury generally speaks for a good prognosis. With shocks of very high intensity, stimulating current may inadvertently activate the masseter muscle at its motor point (see Fig. 7–1). A volume conducted potential from this muscle can erroneously suggest a favorable prognosis when in fact the facial nerve has already degenerated (see Chapter 7.3).

Stimulation of the Trigeminal Nerve

Stimulation of the trigeminal nerve elicits reflex contraction of the orbicularis oculi. In contrast to the direct response, which provides a measure of distal nerve excitability, the blink reflex

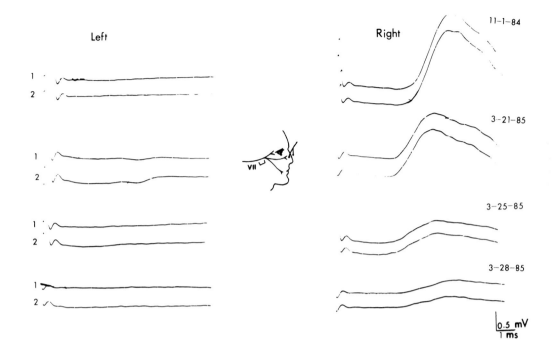

Figure 16–3. A 63-year-old man with acute facial palsy on the left in November 1984 and on the right in March 1985. Stimulation of the left facial nerve elicited no response in the nasalis at the initial evaluation, with no recovery thereafter. Stimulation on the right evoked a normal response in November but progressive reduction in amplitude of the compound muscle action potential in March. This finding indicates axonal degeneration during the first few days after the onset of illness. (From Kimura,[47] with permission.)

reflects the integrity of the afferent and efferent pathways including the proximal segment of the facial nerve. As mentioned earlier, a single shock to the supraorbital nerve evokes two separate contractile responses of the orbicularis oculi. The latency of R_1 represents the conduction time along the trigeminal and facial nerves and pontine relay. R_2 is less reliable for this purpose, because of inherent latency variability from one trial to the next. Furthermore, the latency of R_2 reflects the excitability of interneurons and the delay for synaptic transmission in addition to the axonal conduction time.

The subject lies supine on a bed in a warm room with the eyes open or gently closed. Surface electrodes suffice for stimulation of the nerve and recording of the evoked muscle action potentials.[54] The recording leads consist of an active electrode (G_1) on the upper or lower lateral aspect of the orbicularis oculi and a reference electrode (G_2) on the temple or the lateral surface of the nose, with a ground electrode under the chin or around the arm (Fig. 16–4). The supraorbital, infraorbital, or mental nerve is stimulated with the cathode placed over the respective foramen on one side. A two-channel oscilloscope allows simultaneous recording from the orbicularis oculi on both sides. Assessment of facial synkinesis described later in this chapter requires two pairs of recording electrodes on the same side of the face, one pair over the orbicularis oculi and the other over the orbicularis oris or platysma.[3,55]

Shocks of optimal intensity elicit a maximum and nearly stable response with repeated trials. The reflex latency of R_1, measured from the stimulus artifact to the initial deflection of the evoked potential, corresponds to the maximal conduction time of the reflex pathway. Eight or more trials on each side ensure the recording of the shortest latency response. The latency ratios of R_1 to the direct response (R/D ratio) provide a measure for comparison of the conduction through the distal segment of the facial nerve with that of the entire reflex arc,

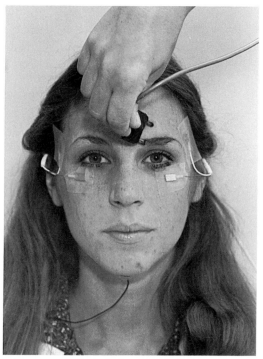

Figure 16–4. Technique for recording the blink reflex. Unilateral stimulation of the supraorbital nerve with the cathode placed at the supraorbital foramen elicits R_1 ipsilaterally and R_2 bilaterally in the orbicularis oculi muscles. Recording leads consist of the active electrode (G_1) placed over either the inferior or superior (not shown) portion of the orbicularis oculi near the outer canthus and the reference electrode (G_2) on the temple or the side of the nose (not shown). Rotation of the anode around the cathode helps establish the optimal position of the stimulating electrodes to minimize the shock artifact.

which includes the trigeminal nerve and the proximal segment of the facial nerve.

In 5 to 10 percent of healthy subjects, single shocks of appropriate intensities may fail to elicit a stable R_1, usually regardless of the side of stimulation. In these cases, paired stimuli with an interstimulus interval of 3 to 5 ms facilitate the response for accurate determination of the shortest latency. A pair of stimuli ideally comprises a subthreshold conditioning shock to subliminally excite the motor neurons and a supramaximal test stimulus (Figs. 16–5 and 16–6). Because the second, and not the first, stimulus elicits the recorded response, the latency is measured from the second shock artifact.

Because G_1 and G_2 lie only a few centimeters away from the cathode, R_1 tends to overlap the stimulus artifact, which can last more than 10 ms. Usual care in reducing surface spread of stimulus current helps accomplish optimal recording of this short-latency response. A specially designed amplifier with a short blocking time (0.1 ms) and low internal noise (0.5 μV RMS at a bandwidth of 2 kHz) also minimizes the problem of stimulus artifact.[99] Frequency response in the range of 20 Hz to 10 kHz suffices for recording either the R_1 or R_2 component.

In addition to electrical stimulation of the supraorbital nerve, mechanical visual or auditory stimuli also elicit the blink reflex. A mechanical tap over the glabella[28,59,92] is given by a specially constructed reflex hammer with a built-in microswitch that closes on impact and triggers a sweep. Other pressure-sensitive devices are also available commercially. A gentle tap over the glabella causes a cutaneous, rather than a stretch, reflex, probably relayed by the same polysynaptic reflex pathways as the electrically elicited blink reflex. In contrast to unilateral electrical stimulation, the glabellar tap elicits the R_1 component bilaterally, allowing instantaneous comparison between the two sides (see Fig. 16–6C). A glabellar tap elicits R_1 with the latency 2 to 3 ms greater than the electrically evoked response. The additional length of the afferent arc from the glabella to the supraorbital foramen seems to account for the discrepancy only partially.

The R_2 component elicited by a glabellar tap provides confirmation of an afferent or efferent abnormality of the electrically elicited R_2. A glabellar tap stimulates the right and left trigeminal nerves simultaneously, each of which activates the facial nuclei on both sides to elicit bilateral R_2 responses. A consistent latency or amplitude difference between simultaneously recorded right- and left-sided R_2 indicates a delay or block in the facial nerve or in the final common path. A lesion affecting the afferent arc unilaterally does not alter R_2 on either side, because the crossed afferent input from the unaffected side compensates for the

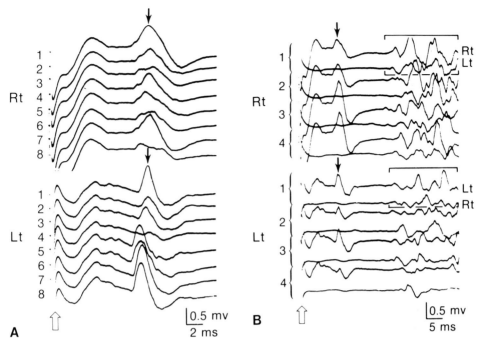

Figure 16–5. A. R_1 components recorded from the orbicularis oculi after stimulation of the supraorbital nerve by single supramaximal stimuli (*top four trials on each side*) or by paired stimuli with an interstimulus interval of 5 ms (*bottom four trials on each side*). The paired stimuli consist of the first shock of subthreshold intensity that subliminally primes the motor neuron pool and the second shock of supramaximal intensity that activates the reflex and triggers the oscilloscope sweep. (From Kimura,[45] with permission).
B. Simultaneous recording from ipsilateral (*upper tracing in each frame*) and contralateral (*lower tracing*) orbicularis oculi after unilateral stimulation of the supraorbital nerve either with single shocks (*top two trials on each side*) or with paired shocks (*bottom two trials on each side*). The paired stimuli consist of the first shock, of subthreshold intensity, and the second shock, of supramaximal intensity. The latter triggers the oscilloscope sweep. Note unilateral R_1 (*arrows*) recorded only in the upper tracing in each frame and bilateral R_2 (*brackets*) in both upper and lower tracings. (From Kimura,[45] with permission.)

loss (see Fig. 16–6C). A glabellar tap inflicts less discomfort to the patients and causes no shock artifacts. In our experience, however, electrical stimulation of the supraorbital nerve generally provides more precise information.

3 NORMAL VALUES IN ADULTS AND INFANTS

Latencies of the Direct and Reflex Responses

Table 16–2 shows the normal latency range of the direct response, R_1, the R/D ratio, and R_2 elicited by stimulation of the supraorbital nerve in 83 healthy subjects 7 to 86 years of age (average age, 37) and in 30 full-term neonates[44,48] and R_1 elicited by a midline glabellar tap in another group of 21 healthy adult subjects.[57] R_1 was present in all but 3 (infants) of the 113 subjects. Despite a considerably shorter relex arc, the neonates had significantly greater latency than the adults. Stimulation of the supraorbital nerve elicited R_2 bilaterally in all adults but only in two thirds of neonates, mostly on the side ipsilateral to the stimulus.[9,13,34,48] Both direct and reflex responses vary considerably in amplitude from one individual to the next. In 60 nerves from 30 healthy subjects, 7 to 67 years of age, the mean values were 1.21 mV for direct response, 0.38 mV for R_1, 0.53 mV for ipsilateral R_2, and 0.49 mV for contralateral R_2.[54]

In 50 other healthy subjects, 12 to 77

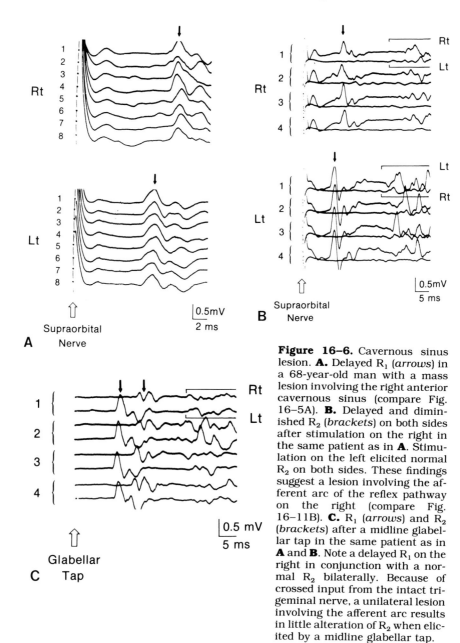

Figure 16–6. Cavernous sinus lesion. **A.** Delayed R_1 (*arrows*) in a 68-year-old man with a mass lesion involving the right anterior cavernous sinus (compare Fig. 16–5A). **B.** Delayed and diminished R_2 (*brackets*) on both sides after stimulation on the right in the same patient as in **A**. Stimulation on the left elicited normal R_2 on both sides. These findings suggest a lesion involving the afferent arc of the reflex pathway on the right (compare Fig. 16–11B). **C.** R_1 (*arrows*) and R_2 (*brackets*) after a midline glabellar tap in the same patient as in **A** and **B**. Note a delayed R_1 on the right in conjunction with a normal R_2 bilaterally. Because of crossed input from the intact trigeminal nerve, a unilateral lesion involving the afferent arc results in little alteration of R_2 when elicited by a midline glabellar tap.

years of age (average age, 40), stimulation of the supraorbital nerve elicited both R_1 and R_2 regularly, whereas that of the infraorbital nerve evoked R_1 in some and R_2 in all. Both R_1 and R_2 had similar latencies regardless of the nerve tested. Shocks applied to the mental nerve elicited R_1 rarely and R_2 inconsistently, showing considerably prolonged latency.

Upper and Lower Limits of Normal Values

The upper limits of normal, defined as the mean latency plus three standard deviations, are 4.1 ms for direct response, 13.0 ms for electrically elicited R_2, and 16.7 ms for mechanically evoked R_1. Additionally, latency difference between the

Table 16-2 BLINK REFLEX ELICITED BY ELECTRICAL STIMULATION OF SUPRAORBITAL NERVE IN NORMAL SUBJECTS AND PATIENTS WITH BILATERAL NEUROLOGIC DISEASES (MEAN ± SD)

Category	Number of Patients	Direct Response Right and Left Combined			R_1 Right and Left Combined			Direct Response (ms)	R_1 (ms)	R/D Ratio	Ipsilateral R_2 (ms)	Contralateral R_2 (ms)
		Abs	Delay	Nl	Abs	Delay	Nl					
Normal	83 (Glabellar tap 21)*	0	0	166	0	0	166	2.9 ± 0.4	10.5 ± 0.8 (12.5 ± 1.4)*	3.6 ± 0.5	30.5 ± 3.4	30.5 ± 4.4
Guillain-Barré syndrome	90	12	63	105	20	78	82	4.2 ± 2.1	15.1 ± 5.9	3.9 ± 1.3	37.4 ± 8.9	37.7 ± 8.4
Chronic inflammatory polyneuropathy	14	4	13	11	7	13	8	5.8 ± 2.6	16.4 ± 6.4	3.1 ± 0.5	39.5 ± 9.4	42.0 ± 10.3
Fisher syndrome	4	0	0	8	0	1	7	2.7 ± 0.2	10.7 ± 0.8	3.9 ± 0.4	31.8 ± 1.3	31.4 ± 1.9
Hereditary motor sensory neuropathy type I	62	9	88	27	0	105	19	6.7 ± 2.7	17.0 ± 3.7	2.8 ± 0.9	39.5 ± 5.7	39.3 ± 6.4
Hereditary motor sensory neuropathy type II	17	0	0	34	1	0	33	2.9 ± 0.4	10.1 ± 0.6	3.6 ± 0.6	30.1 ± 3.8	30.1 ± 3.7
Diabetic polyneuropathy	86	2	20	150	1	17	154	3.4 ± 0.6	11.4 ± 1.2	3.4 ± 0.5	33.7 ± 4.6	34.8 ± 5.3
Multiple sclerosis	62	0	0	124	1	44	79	2.9 ± 0.5	12.3 ± 2.7	4.3 ± 0.9	35.8 ± 8.4	37.7 ± 8.0

*R_1 elicited bilaterally by a midline glabellar tap in another group of 21 healthy subjects.

two sides should not exceed 0.6 ms for direct response, 1.2 ms for electrically elicited R_1, and 1.6 ms for mechanically evoked R_1 in one subject. The R/D latency ratio should not fall outside the range of 2.6 to 4.6, 2 SD above and below the mean in normals.

With stimulation of the supraorbital nerve, R_2 latency should not exceed 40 ms on the side of stimulus and 41 ms on the contralateral side. In addition, the ipsilateral and the contralateral R_2 simultaneously evoked by stimulation on one side should not vary more than 5 ms in latency. A latency difference between R_2 evoked by right-sided stimulation and corresponding R_2 evoked by left-sided stimulation may show a slightly greater value but not more than 7 ms. With stimulation of the infraorbital nerve, the upper limit is 41 ms on the side of stimulus and 42 ms on the contralateral side. Studies of the mental nerve provide less consistent results but R_2 response rarely exceeds 50 ms in latency.

4   NEUROLOGIC DISORDERS WITH ABNORMAL BLINK REFLEX

Tables 16–2 through 16–4 summarize a 10-year experience with the blink reflex

in our laboratory.[39,41,42,44,48,49,52,56,63] A brief summary of each category follows.

Lesions of the Trigeminal Nerve

Only 7 of 93 patients with trigeminal neuralgia had absent or slowed R_1 (see Table 16–3). Excluding three patients who had undergone nerve avulsion prior to the test, only four patients had abnormalities attributable to the disease. These findings suggest that the impulse conducts normally along the first division of the trigeminal nerve in most patients with this disorder. Usual sparing of the first division and minimal compression of the nerve, if any, probably account for this finding. Conduction abnormalities however, may appear following surgery.[15]

In contrast, 10 of 17 patients with tumor, infection, or other demonstrable causes for facial pain showed significant delay of R_1 on the affected side (see Fig. 16–6A). In these patients, reproducible delay of R_2 bilaterally with stimulation on the affected side indicated involvement of the afferent arc of the blink reflex (see Fig. 16–6B). The R/D ratio increased, reflecting normal conduction along the distal segment of the facial nerve, combined with a delay along the trigeminal nerve.

Table 16–3 BLINK REFLEX ELICITED BY ELECTRICAL STIMULATION OF SUPRAORBITAL NERVE ON THE AFFECTED AND NORMAL SIDES IN PATIENTS WITH UNILATERAL NEUROLOGIC DISEASES (MEAN ± SD)

Category and Side of Stimulation	Number of Patients	Direct Response (ms)	R_1 (ms)	R/D Ratio	Ipsilateral R_2 (ms)	Contralateral R_2 (ms)
Trigeminal neuralgia						
Affected side	89	2.9 ± 0.4	10.6 ± 1.0	3.7 ± 0.6	30.4 ± 4.4	31.6 ± 4.5
Normal side	89	2.9 ± 0.5	10.5 ± 0.9	3.7 ± 0.6	30.5 ± 4.2	31.1 ± 4.7
Compressive lesion of the trigeminal nerve						
Affected side	17	3.1 ± 0.5	11.9 ± 1.8	3.9 ± 1.0	36.0 ± 5.5	37.2 ± 5.7
Normal side	17	3.2 ± 0.6	10.3 ± 1.1	3.4 ± 0.6	33.7 ± 3.5	34.8 ± 4.1
Bell's palsy						
Affected side	100	2.9 ± 0.6	12.8 ± 1.6	4.4 ± 0.9	33.9 ± 4.9	30.5 ± 4.9
Normal side	100	2.8 ± 0.4	10.2 ± 1.0	3.7 ± 0.6	30.5 ± 4.3	34.0 ± 5.4
Acoustic neuroma						
Affected side	26	3.2 ± 0.7	14.0 ± 2.7	4.6 ± 1.7	38.2 ± 8.2	36.6 ± 8.2
Normal side	26	2.9 ± 0.4	10.9 ± 0.9	3.8 ± 0.5	33.1 ± 3.5	35.3 ± 4.5
Wallenberg syndrome						
Affected side	23	3.2 ± 0.6	10.9 ± 0.7	3.6 ± 0.6	40.7 ± 4.6	38.4 ± 7.1
Normal side	23	3.2 ± 0.4	10.7 ± 0.5	3.4 ± 0.4	34.0 ± 5.7	35.1 ± 5.8

Table 16–4 DIRECT RESPONSE AND R₁ AND R₂ OF THE BLINK REFLEX

Disorders	Direct Response	R_1	R_2
Trigeminal neuralgia	Normal	Normal (95%)	Normal
Compressive lesion of the trigeminal nerve	Normal	Abnormal on the affected side (59%)	Abnormal on both sides when affected side stimulated (afferent type)
Bell's palsy	Normal unless distal segment degenerated	Abnormal on the affected side (99%)	Abnormal on the affected side regardless of the side of stimulus (efferent type)
Acoustic neuroma	Normal unless distal segment degenerated	Abnormal on the affected side (85%)	Afferent and/or efferent type
Guillain-Barré syndrome	Abnormal (42%)	Abnormal (54%)	Afferent and/or efferent type
Hereditary motor sensory neuropathy type I	Abnormal (78%)	Abnormal (85%)	Afferent and/or efferent type
Diabetic polyneuropathy	Abnormal (13%)	Abnormal (10%)	Afferent and/or efferent type
Multiple sclerosis	Normal	Abnormal with pontine lesion, variable incidence determined by patient's selection	Afferent and/or efferent type
Wallenberg syndrome	Normal	Normal or borderline	Afferent type
Facial hypesthesia	Normal	Abnormal with lesions of the trigeminal nerve or pons	Afferent type
Comatose state, akinetic mutism, locked-in syndrome	Normal	Abnormal with pontine lesion; reduced excitability in acute supratentorial lesion	Absent on both sides regardless of side of stimulus

Bell's Palsy

Blink reflex latencies reflect the conduction along the entire length of the facial nerve, including the interosseous portion involved in Bell's palsy.[49,54,78,79,87] All 144 patients showed either block or slowing of R_1 during the first week of Bell's palsy, although the abnormalities did not necessarily emerge at the onset. Delayed or absent R_2 on the paretic side, regardless of the side of stimulation, indicated an efferent involvement. Two patients not included in this study had normal blink reflex despite minimal unilateral facial weakness lasting 1 to 2 days, perhaps representing an unusually mild form of Bell's palsy.

In 100 of 127 patients tested serially, the previously absent R_1 or R_2 returned, with preservation of the direct response throughout the course. This finding implied recovery of conduction across the involved segment without substantial distal degeneration (Fig. 16–7). These patients generally showed good clinical recovery within a few months after onset. The latency of R_1, initially delayed by more than 2 ms on the average, decreased during the second month and returned to normal during the third or fourth months (Fig. 16–8). The magnitude of latency change at onset and subsequent time course of recovery indicate a demyelinative nature of the responsible lesion. The R/D ratios increased as expected in conduction abnormalities involving the proximal segment of the facial nerve. In the remaining 27 patients, marked diminution of the direct response

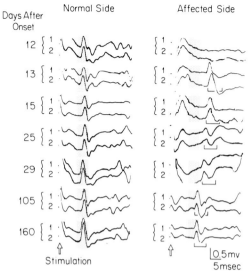

Figure 16–7. Serial changes of R_1 in a 16-year-old girl with Bell's palsy on the right. Two consecutive tracings recorded on each side show consistency of R_1 on a given day. On the affected side, delayed R_1 first appeared on the 13th day of onset, recovering progressively thereafter. *Shaded areas* indicate normal range (mean $\pm$ 3 SD in 83 subjects). (From Kimura,[40] with permission.)

without return of the reflex response during the first 2 weeks indicated axonal degeneration.[55] This group of patients had slow and usually incomplete recovery with evidence of synkinesis.

Synkinesis of Facial Muscles

Both R_1 and R_2 components of the blink reflex normally involve the orbicularis oculi alone and only rarely, if at all, other facial muscles.[29,92] During aberrant axon regeneration, however, the fibers that originally innervated the orbicularis oculi may, by misdirection, supply other facial muscles.[55] Under such circumstances, the blink reflex elicited in muscles other than the orbicularis oculi serves as a sign of aberrant reinnervation (Fig. 16–9).

Recording an aberrant blink reflex helps identify time-locked discharges involving two independent muscles showing synkinetic movements. This stands in contrast to volitional, associated movements that clinically mimic synkinesis

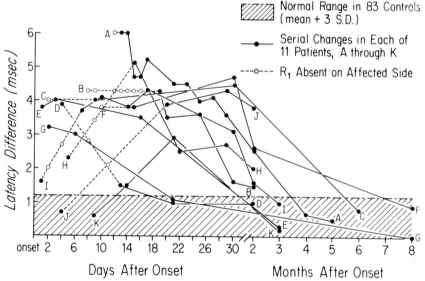

Figure 16–8. Serial changes in latency difference of R_1 between normal and paretic sides in 11 patients recovering without nerve degeneration (*A* through *K*). *Shaded area* indicates the normal range (mean $\pm$ 3 SD in 83 subjects). The response, if present at onset, showed relatively normal latencies but rapidly deteriorated during the first few days. Delayed R_1 usually returned during the second week, plateaued for 2 to 4 weeks, and then progressively recovered in latency within the next few months. (From Kimura et al.,[49] with permission.)

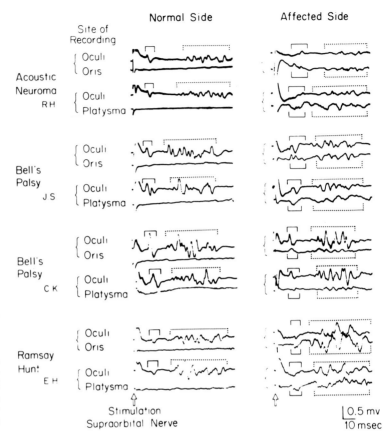

Figure 16–9. The blink reflex in the orbicularis oris and platysma in four patients following various diseases of the facial nerve. Stimulation on the affected side of the face elicited both R_1 (*small brackets*) and R_2 (*dotted brackets*) not only in the orbicularis oris but also in the platysma, indicating widespread synkinesis. The blink reflex, elicited only in the orbicularis oculi on the normal side of the face, served as a control in each patient. (From Kimura et al.,[55] with permission.)

but lack the exact temporal relationship between the two co-contracting muscles. Measuring the size of the blink reflex elicited in muscles other than orbicularis oculi also elucidates the extent of aberrant reinnervation. In one series, the blink reflex confirmed synkinesis involving the orbicularis oris or platysma in 26 of 29 patients tested at least 4 months after total facial nerve degeneration.[55] One of the remaining three had injury only to a peripheral branch of the facial nerve and experienced return of function with no evidence of synkinesis. In the other two, the affected side of the face showed total paralysis with no evidence of any regeneration. These findings suggest that synkinetic movements ultimately occur in nearly all cases after degeneration of the facial nerve unless the lesion involves a distal branch or the facial nerve fails to regenerate.

Hemifacial Spasm

Patients with hemifacial spasm (see Chapter 26.7) also exhibit clinical and electrical evidence of synkinetic movements.[3,11,26,44,66,100] In these cases the appearance of the blink reflex in muscles other than the orbicularis oculi may indicate hyperexcitability at the facial nucleus, ephaptic activation of motor axons not normally involved in blinking, or aberrant regeneration of the facial nerve fibers. Unlike the constant responses seen after peripheral facial paresis,[55] however, successive responses in hemifacial spasm vary in latency and waveform, a finding in favor of ephaptic transmission.[3] The blink reflex reveals no evidence of synkinesis in essential blepharospasm, focal seizures, or facial myokymia. Inhalation of anesthetics during surgery completely suppresses R_1 or

R_2 in normal subjects but not on the affected side of patients with hemifacial spasm.[71]

Acoustic Neuroma

A cerebellopontine angle tumor frequently compresses the trigeminal nerve, facial nerve, or brainstem. With possible involvement of the afferent, efferent, or central pathways,[6,23,53,63,82,91] the blink reflex provided unique diagnostic value in 33 patients studied. Stimulation of the facial nerve elicited no direct response in seven, including five tested only after surgical sacrifice of the facial nerve. In the remaining 26 patients, R_1 on the affected side was absent in 5, delayed in 17, and normal in 4. Analyses of R_2 revealed six efferent, six afferent, and seven mixed patterns and seven normal responses.

Polyneuropathy

Facial or trigeminal nerve involvement in various polyneuropathies affects the blink reflex (Fig. 16–10A). Although the two components of blink reflex are clearly separated normally, R_1 tends to merge with R_2 in a demyelinative neuropathy (Fig. 16–10B). On bilateral recording, however, the R_1 can still be recognized on the basis of the onset of the contralateral R_2, which should approximately coincide with the ipsilateral R_2 (Fig. 16–10C).

Different category of neuropathy shows distinct abnormalities as briefly described below.[46] Most patients have either absent or delayed direct and R_1 responses in the Guillain-Barré syndrome (GBS), chronic inflammatory demyelinative polyradiculoneuropathy (CIDP), and hereditary motor and sensory neuropathy (HMSN) type I or the hypertrophic type of Charcot-Marie-Tooth disease. The patients with diabetic polyneuropathy have a considerably lower incidence of abnormality. The Fisher syndrome does not regularly affect the blink reflex, but one of four patients with peripheral facial palsy had a delayed R_1 on the affected side. The blink reflex is usually normal in

HMSN type II or the neuronal type of Charcot-Marie-Tooth disease. Patients with chronic renal failure have an abnormal blink reflex, which improves after hemodialysis in some.[95]

Statistical analyses of the direct response and R_1 latencies revealed a marked increase in GBS, CIDP, and HMSN type I, a much lesser degree of slowing in diabetic polyneuropathy, and no change in the Fisher syndrome and HMSN type II (see Table 16–2). The latency ratio of R_1 to the direct response showed a mild increase in the GBS, a moderate decrease in HMSN type I and CIDP, and a mild decrease in diabetic polyneuropathy. The latencies of R_2, although commonly within the normal range when analyzed individually, had a significantly greater average value in the neuropathies than in the controls.

Multiple Sclerosis

Alterations of the electrically elicited blink reflex may result from disorders of the central reflex pathways. Of various lesions affecting the brain stem, multiple sclerosis causes a most conspicuous delay of R_1,[37,38,44,60,64,73,77] as expected from the effect of demyelination on impulse propagation.[67,88,101] The incidence of blink reflex abnormality varies greatly, depending on the selection of patients. In general, the longer the history of clinical symptoms is, the higher is the rate of abnormality.

An earlier study of 260 patients with long-standing disease[44] showed a delayed R_1 in 96 of 145 patients (66 percent) who clinically had disseminated lesions as well as episodes of remission and exacerbation (Fig. 16–11). The study was abnormal in 32 of 57 patients (56 percent) who had either multiple sites of involvement without relapse or a history of recurrence of a localized lesion. The test revealed alteration of R_1 in 17 of the remaining 58 patients (29 percent) in whom the diagnosis was suspected but not clinically established. In the same 260 patients, R_1 was abnormal in 49 of 63 patients with clinical evidence of pontine lesions (78 percent), 50 of 104 with

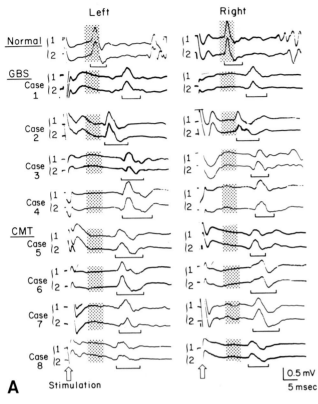

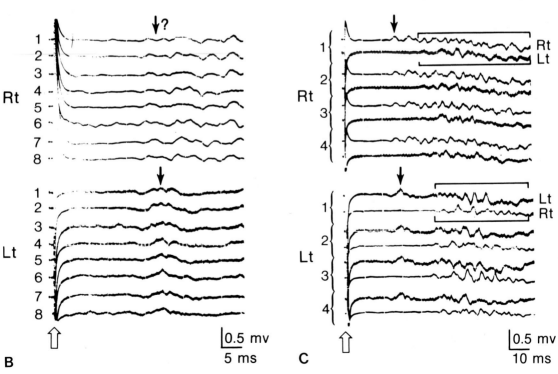

Figure 16–10. A. Bilateral delay of R_1 in four patients with the Guillain-Barré syndrome (*GBS*) and four patients with hereditary motor sensory neuropathy type I (*CMT*). Two tracings recorded on each side in each subject show consistency. The *top tracings* from a healthy subject serve as a control, with *shaded areas* indicating the normal range. (From Kimura,[46] with permission.) **B.** R_1 in a 55-year-old woman with chronic peripheral neuropathy and a monoclonal gammopathy (compare Figure 16–5A). Note a substantially delayed and temporally dispersed R_1 recorded by a slower sweep of 5 ms/cm instead of 2 ms/cm, normally used for this response. **C.** R_1 and R_2 in the same patient shown in Figure 16–10B. Note delayed R_2 recorded by a slower sweep of 10 ms/cm instead of the usual 5 ms/cm. The continuity between R_1 and R_2 precluded accurate latency determination of R_2 on the right. Nonetheless, the contralateral R_2 recorded simultaneously allows approximate separation of R_1 and R_2 ipsilateral to the side of stimulation.

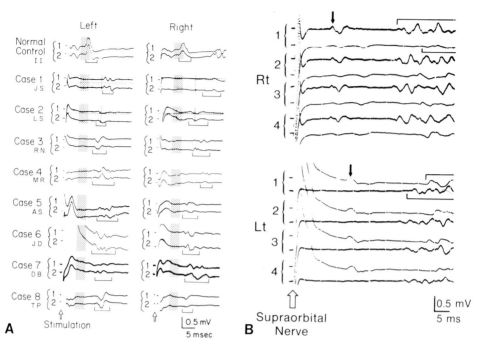

Figure 16–11. A. Delayed R_1 on both sides in multiple sclerosis. Two tracings recorded on each side in each subject show consistency of the R_1 response. The top tracings from a healthy subject serve as a control, with *shaded areas* indicating the normal range (mean ± 3 SD in 83 subjects). In addition to increased latency, R_1 obtained in the patients shows temporal dispersion and very irregular waveforms, compared with the normal response. None of these patients had unequivocal pontine signs clinically, except for mild horizontal nystagmus seen in cases 1, 2, 5, 6, and 7. (From Kimura,[44] with permission.) **B.** R_1 and R_2 in a 35-year-old woman with multiple sclerosis and mild facial and abducens paresis on the left (compare Fig. 16–5B). Stimulation on the right elicited normal R_1 and R_2 ipsilaterally and delayed R_2 contralaterally, whereas stimulation on left evoked delayed R_1 and delayed R_2 ipsilaterally. This finding suggests a lesion involving the efferent arc of the reflex on the left, i.e., the intrapontine portion of the facial nerve (compare Fig. 16–6B). (From Kimura,[46] with permission.)

other brainstem lesions (57 percent), and 37 of 93 with neither brainstem signs nor symptoms (40 percent).

In the 63 patients with clinical signs of pontine lesions, the average latency of R_1 substantially exceeded the normal value but fell short of the delay seen in the Guillain-Barré syndrome or Charcot-Marie-Tooth disease (Fig. 16–12). The normal direct response, combined with delayed R_1, markedly increased the R/D ratio. Hyperthermia did not induce significant changes in mean reflex latency, amplitude, or duration, even in those with unequivocal blink reflex abnormalities before warming.[81]

Subsequent studies of multiple sclerosis showed comparable results.[60,64,73,77,84] A more recent study revealed a delayed R_1 in 41 percent of patients with definite diagnosis and 18 percent in those with

possible diagnoses.[37] Other investigators report similar rates of abnormality in a series of patients referred for electrophysiologic testing soon after the onset of their symptoms.[36,85] The blink reflex detects only those lesions that affect the short pontine pathway. Thus, the incidence of abnormalities is less, compared with visual, somatosensory, or brainstem auditory evoked potentials. A delayed R_1, however, helps in localizing a lesion to the pons and establishing subclinical dissemination.

Wallenberg Syndrome

Patients with the Wallenberg syndrome have selective alteration of R_2 as expected from lesions affecting the lateral medulla.[52,75] Unless the infarct extends to

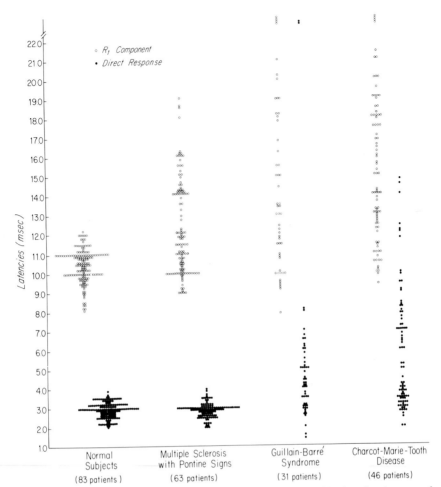

Figure 16–12. Latency distribution of the direct response and R_1 of the blink reflex in normal subjects and in patients with central or peripheral demyelination of the reflex pathways. The histogram shows delayed direct response in Charcot-Marie-Tooth disease and to a slightly lesser extent in the Guillain-Barré syndrome and normal response in multiple sclerosis. The R_1 response is delayed equally in the two polyneuropathies but to a lesser degree in multiple sclerosis. (From Kimura,[46] with permission.)

the pons, the latency of R_1 falls within the normal range; but when analyzed individually, the values on the affected side may slightly exceed those on the normal side (see Table 16–3). In a series of 23 typical cases, stimulation on the affected side of the face elicited no R_2 on either side in 7, low-amplitude R_2 in 6, and delayed R_2 in 10 (Fig. 16–13). In contrast, stimulation on the normal side of the face evoked normal R_2 bilaterally in 20 of 23 patients. The remaining three patients showed normal R_2 only on the side of stimulation. Stimulation of the infraorbital nerve or mental nerve gives rise to the same pattern of abnormality.

Facial Hypoesthesia

Patients with contralateral hemispheric lesions also develop an afferent delay of R_2 indistinguishable from that seen in the Wallenberg syndrome.[17,28,43,70] This type of abnormality commonly, although not exclusively, accompanies sensory disturbances of the face. Thus, the electrically elicited blink reflex provides a means of quantitating facial sensation.[2] In equivocal cases, repetitive stimulation of the right and left sides of the face alternately every 5 to 10 seconds reveals consistent asymmetry beyond random variations that follow no predictable pattern.

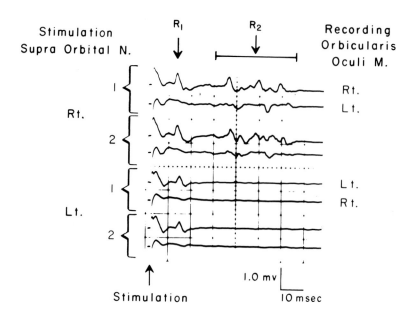

Figure 16–13. Left lateral medullary syndrome. Two successive stimuli given on the right (*top two pairs*) elicited a normal R_1 and R_2 bilaterally. Two successive stimuli on the left (*bottom two pairs*) evoked normal R_1 but absent R_2 bilaterally (compare Fig. 16–17). (From Kimura and Lyon,[52] with permission.)

In six patients with bilateral trigeminal neuropathy, blink reflex studies revealed slowed or absent R_1 bilaterally in three and delayed or diminished R_2 regardless of the side of stimulation in four. In 19 patients with unilateral disease of either the trigeminal nerve[14] or brainstem,[13] R_1 was absent on the affected side in 6, delayed in 7, and normal in the others. Stimulation on the affected side of the face elicited a small or no R_2 bilaterally in four, a delayed R_2 in seven, and a normal response in the rest (Fig. 16–14). Generally, a smaller response indicated more complete sensory loss, and stimulation of an anesthetic part of the face failed to elicit any response at all.

5 ANALYSIS OF THE R_1 COMPONENT

Direct Involvement of the Reflex Arc

R_1 of the blink reflex shows increased latency in diseases associated with demyelination either centrally[38,41,44,60,72,73] or peripherally affecting the trigeminal nerve,[32,56,74] facial nerve,[49,54,78,79,87] or both.[6,23,39,53,63] Posterior fossa tumors may affect R_1 either by compressing the

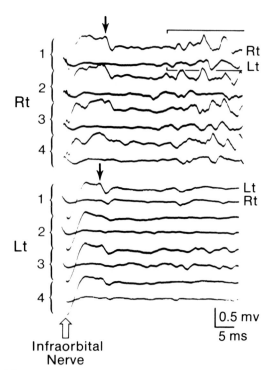

Figure 16–14. R_1 and R_2 elicited by stimulation of the infraorbital nerve in a 39-year-old woman with syringobulbia and facial numbness on the left (compare Fig. 16–5B). Stimulation of the right side of the face elicited normal R_1 and R_2 bilaterally, but stimulation on the left evoked only the R_1 component.

cranial nerves extra-axially or by involvement of the brainstem itself.[14,40,53]

Effect of Lesions Outside the Reflex Pathway

Alteration of R_1 does not necessarily indicate a pathologic process of the reflex arc itself, because edema or lesions outside the brainstem can also cause conduction abnormalities.[44] A reversible block of R_1 seen in comatose patients usually results from acute supratentorial lesions or massive drug intoxication.[62] The latency of R_1 elicited by a glabellar tap shows a mild increase in patients with acute hemispheric strokes but recovers almost completely within a few days.[28] In contrast, electrically elicited R_1 has a normal latency even during acute stages of hemispheric disease, especially with strong or paired stimuli or with other facilitatory maneuvers[80] to compensate for reduced excitability.[17,35,43]

In some patients with acute supranuclear lesions, single electric shocks elicit R_1 only partially or not at all when given contralateral to the hemispheric lesion. An apparent increase in the latency of R_1 results if such a stimulus fails to activate the fastest conducting fibers. In this instance, paired stimuli with an interstimulus interval of 3 to 5 ms usually elicit a maximal R_1 with normal latency.[40,43,44] Thus, the latency of a fully activated R_1 indicates the conduction characteristics of the reflex arc itself—that is, a delay of R_1 by several milliseconds implies a lesion directly involving the pathway, rather than a remote process altering excitability. In these cases, smaller, slower conducting fibers may mediate the reflex response following the conduction block of the larger myelinated fibers, or all axons may have slowed conduction across the demyelinated area.

Degree of Slowing

In multiple sclerosis, central demyelination increases the latency of R_1 to 12.3 ± 2.7 ms (mean $\pm$ SD) as compared with 15.1 ± 5.9 ms in GBS and 17.0 ± 3.7 ms in HMSN type 1. The degree of latency prolongation presumably reflects the difference in length of the demyelinated segment in the pons and along the peripheral reflex arc. In support of this view, the latency of R_1 increases only to 12.8 ± 1.6 ms in Bell's palsy with focal involvement of the facial nerve. In contrast to patients with trigeminal neuralgia with normal reflex latency, those with compressive lesions of the trigeminal nerve often have an unequivocal delay of R_1.

Conduction abnormalities affect HMSN type I and GBS to the same degree (see Figs. 16–10A and 16–12). A decreased R/D ratio found in HMSN type I suggests distal slowing of facial nerve conduction, whereas a slightly increased R/D ratio in GBS indicates proximal involvement of the facial nerve, if the trigeminal nerve conducts normally. The R/D ratio is also increased in patients with multiple sclerosis or compressive lesions of the trigeminal nerve or in Bell's palsy without distal degeneration of the facial nerve.

6 ANALYSIS OF THE R_2 COMPONENT

Direct and Remote Effect on Polysynaptic Pathways

As mentioned earlier, the reflex abnormality can be categorized into either afferent or efferent types on the basis of analysis of the R_2 component. However, some brainstem lesions may give rise to a more complex pattern of reflex change (see Fig. 16–1B). Stimulation on one side may reveal unilateral abnormality of R_2 either ipsilateral or contralateral to the stimulus, whereas stimulation on the opposite side shows R_2 to be normal, absent or delayed bilaterally, or affected unilaterally but not on the same side as implicated by the contralateral stimulation.

Like R_1, changes of R_2 may imply lesions directly affecting the reflex pathways per se, as in the case of the Wallenberg syndrome, or lesions elsewhere indirectly influencing the excitability of the polysynaptic connections (Fig. 16–15).[20,52,75] For example, R_2 is absent or markedly diminished in size in

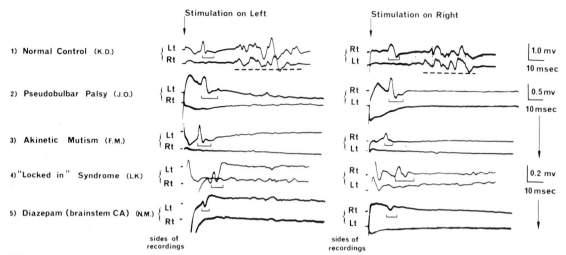

Figure 16–15. Various neurologic disorders associated with absent R_2 after stimulation of the supraorbital nerve. Shock intensity was slowly advanced up to 40 mA and 0.5 ms duration. Note virtual absence of R_2 regardless of the side of stimulation in cases 2 through 5, with normal R_1 in cases 2, 3, and 5, and delayed R_1 in case 4. (From Kimura,[41] with permission.)

any comatose state (Fig. 16–16), regardless of the site of lesion.[12,62,69] A hemispheric lesion (Fig. 16–17) also suppresses R_2, producing either an afferent or an efferent pattern of abnormality, perhaps based on the site of involvement.[7,17,19,28,35,43,57,70]

Level of Consciousness and Perception of Pain

The state of arousal alters the excitability of R_2 and, to a lesser extent, R_1.[5,10,21,27,50,51,89,90,94] Analysis of R_2 during sleep has shown marked reduction in

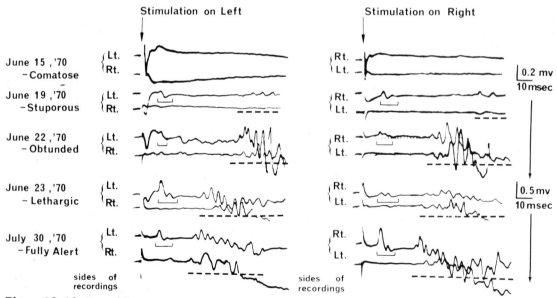

Figure 16–16. R_1 and R_2 in a patient recovering from herpes simplex encephalitis. The stimulus delivered to the supraorbital nerve elicited neither R_1 nor R_2 on June 9 (not shown) and on June 15 with the patient in a coma. A repeat study on June 19 showed a normal R_1 but markedly delayed and diminished R_2. Note the progressive recovery in amplitude and latency of R_2 contemporaneous with her improvement to full alertness in July. (From Kimura,[41] with permission.)

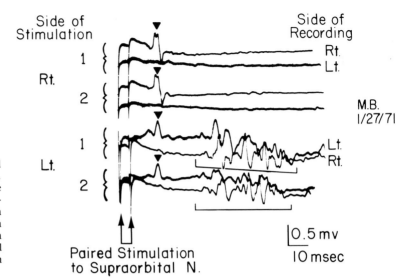

Side of
Stimulation

1 {

Rt.

2 {

1 {

Lt.

2 {

Side of
Recording

Rt.
Lt.

M.B.
1/27/71

Lt.
Rt.

| 0.5 mv

10 msec

Paired Stimulation
to Supraorbital N.

Figure 16–17. Left cerebral stroke (compare Fig. 16–13). Paired stimuli delivered to the right supraorbital nerve elicited normal R_1 but no R_2 on either side. Stimulation on the left, however, evoked an ipsilateral R_1 and a bilateral R_2. (From Kimura,[43] with permission.)

stages II, III, and IV and substantial recovery during rapid eye movement (REM) sleep, when the excitability approaches but does not quite reach that of full wakefulness. Blink reflex studies may show absent R_2 with normal or nearly normal R_1 in some alert but immobile patients with features of the locked-in syndrome, in alert and ambulatory patients with pseudobulbar palsy, and in alert patients given therapeutic dosages of diazepam (Valium), which presumably blocks the multisynaptic reflex arc.[41] Complex psychologic events may also selectively affect different reflex pathways.[86]

Stimulation on a hypesthetic area of the face elicits a smaller R_2 than that evoked by a shock of the same intensity applied to the corresponding area on the normal side. Sensory deficits of the face often cause alteration of R_2; the reverse, however, does not hold, because similar reduction of R_2 occurs in pure motor hemiplegia.[4,17,28,43] In these cases, clinical evaluation may have failed to detect minor sensory deficits. Certain supratentorial lesions outside the somatosensory pathways, however, can also inhibit or disfacilitate the reflex excitability.

Altered Excitability of Interneurons

Finally, R_2 habituates readily in normals but not in patients with Parkinson's disease, whether tested clinically as the glabellar sign or electromyographically.[58,61,68,79,83] Similarly, the blink reflex fails to show physiologic habituation in nocturnal myoclonus, a syndrome associated with additional reflex components after R_2. These findings suggest a disorder of the central nervous system producing increased excitability of segmental reflexes.[103]

The paired-shock technique reveals the effect of a single cutaneous conditioning stimulus on this reflex (Fig. 16–18). Dissociation between the recovery curves of the oligosynaptic R_1 and polysynaptic R_2 presumably results from excitability changes at the interneuron level.[42] Healthy subjects show a greater suppression of R_2 than R_1 following a conditioning stimulus. This finding, then, indicates physiologic inhibition at the interneuron, rather than at the motor neuron, that constitutes the final common path of both reflexes. In normal subjects, the conditioning stimuli delivered anywhere on the face or neck suppress R_2 elicited by a subsequent test stimulus applied to the same or different, ipsilateral or contralateral trigeminal cutaneous fields.[41]

In Parkinson's disease, the recovery of R_1 follows a normal time course, whereas the physiologic suppression of R_2 by a conditioning stimulus lasts substantially less than the normal range. The recovery

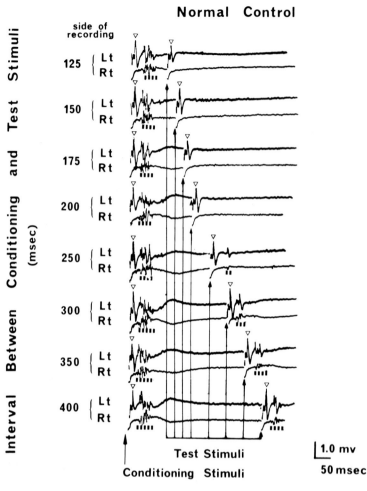

Normal Control

side of recording

Test Stimuli

Conditioning Stimuli

Stimulation - Left Supra Orbital Nerve

|1.0 mv

50 msec

Figure 16–18. Normal responses to paired shocks (*arrows*) delivered to the left supraorbital nerve with time intervals ranging from 125 to 400 ms between test and conditioning stimuli. R_1 of the test response, though slightly suppressed at time intervals of 125 to 175 ms, remained relatively constant thereafter, with the amplitude equal to the conditioning response. The test stimuli failed to elicit R_2 up to the time interval of 200 ms, with gradual recovery subsequently. (From Kimura,[42] with permission.)

curve of R_2 indicates that, unlike in normal subjects, a cutaneous conditioning stimulus fails to inhibit interneurons in this disease. Additional evidence of a change in excitability includes an abnormally short latency of R_2 in response to a single maximal stimulus in advanced cases. These findings imply facilitation or distribution of the interneurons, rather than the motor neurons, as the primary cause of motor dysfunction in this disease. In contrast, diminution of the R_2 component in Huntington's chorea represents the opposite extreme and probably reflects a decreased interneuronal activity.[24,25]

REFERENCES

1. Accornero, N, Berardelli, A, Bini, G, Cruccu, G, and Manfredi, M: Corneal reflex elicited by electrical stimulation of the human cornea. Neurology 30:782–785, 1980.
2. Ashworth, B, and Tait, GBW: Trigeminal neuropathy in connective tissue disease. Neurology 21:609–614, 1971.
3. Auger, RG: Hemifacial spasm: Clinical and electrophysiologic observations. Neurology 29:1261–1272, 1979.
4. Barron, SA, Heffner, RR, Jr, and Zwirecki, R: A familial mitochondrial myopathy with central defect in neural transmission. Arch Neurol 36:553–556, 1979.
5. Beck, U, Schenck, E, and Ischinger, TH: Spinale und bulbäre Reflexe im Schlaf beim Mens-

chen. Arch Psychiatr Nervenkr 217:157–168, 1973.

6. Bender, LF, Maynard, FM, and Hastings, SV: The blink reflex as a diagnostic procedure. Arch Phys Med Rehabil 50:27–31, 1969.

7. Berardelli, A, Accornero, N, Cruccu, G, Fabiano, F, Guerrisi, V, and Manfredi, M: The orbicularis oculi response after hemispheral damage. J Neurol Neurosurg Psychiatry 46:837–843, 1983.

8. Berardelli, A, Cruccu, G, Manfredi, M, Rothwell, JC, Day, BL, and Marsdens, CD: The corneal reflex and the R_2 component of the blink reflex. Neurology 35:797–801, 1985.

9. Blank, A, Ferber, I, Shapira, Y, and Fast, A: Electrically elicited blink reflex in children. Arch Phys Med Rehabil 64:558–559, 1983.

10. Boelhouwer, AJW, and Brunia, CHM: Blink reflexes and the state of arousal. J Neurol Neurosurg Psychiatry 40:58–63, 1977.

11. Bohnert, B, and Stohr, M: Beitrag zum Spasmus facialis. Arch Psychiatr Nervenkr 224:11–21, 1977.

12. Buonaguidi, R, Rossi, B, Sartucci, F, and Ravelli, V: Blink reflexes in severe traumatic coma. J Neurol Neurosurg Psychiatry 42:470–474, 1979.

13. Clay, SA, and Ramseyer, JC: The orbicularis oculi reflex in infancy and childhood: Establishment of normal values. Neurology 26:521–524, 1976.

14. Clay, SA, and Ramseyer, JC: The orbicularis oculi reflex: Pathologic studies in childhood. Neurology 27:892–895, 1977.

15. Cruccu, G, Inghilleri, M, Fraioli, B, Guidetti, B, and Manfredi, M: Neurophysiologic assessment of trigeminal function after surgery for trigeminal neuralgia. Neurology 37:631–638, 1987.

16. Csecsei, G: Facial afferent fibers in the blink reflex of man. Brain Res 161:347–350, 1979.

17. Dehen, H, Willer, JC, Bathien, N, and Cambier, J: Blink reflex in hemiplegia. Electroencephalogr Clin Neurophysiol 40:393–400, 1976.

18. De Meirsman, J, Claes, G, and Geerdens, J: Normal latency value of the facial nerve with detection in the posterior auricular muscle and normal amplitude value of the evoked action potential. Electromyogr Clin Neurophysiol 20:481–485, 1980.

19. Dengler, R, Kossev, A, Gippner, C, and Struppler, A: Quantitative analysis of blink reflexes in patients with hemiplegic disorders. Electroencephalogr Clin Neurophysiol 53:513–524, 1982.

20. Dengler, R, Wombacher, T, Schodel, M, and Struppler, A: Changes in the recruitment pattern of single motor units in the blink reflex of patients with Parkinsonism and hemiplegia. Electroencephalogr Clin Neurophysiol 61:16–22, 1985.

21. Desmedt, JE, and Godaux, E: Habituation of exteroceptive suppression and of exteroceptive reflexes in man as influenced by voluntary contraction. Brain Res 106:21–29, 1976.

22. Devi, S, Challenor, Y, Duarte, N, and Lovelace,

RE: Prognostic value of minimal excitability of facial nerve in Bell's palsy. J Neurol Neurosurg Psychiatry 41:649–652, 1978.

23. Eisen, A, and Danon, J: The orbicularis oculi reflex in acoustic neuromas: A clinical and electrodiagnostic evaluation. Neurology 24:306–311, 1974.

24. Esteban, A, and Gimenez-Roldan, S: Blink reflex in Huntington's Chorea and Parkinson's disease. Acta Neurol Scand 52:145–157, 1975.

25. Esteban, A, Mateo, D, and Gimenez-Roldan, S: Early detection of Huntington's disease: Blink reflex and levodopa load in presymptomatic and incipient subjects. J Neurol Neurosurg Psychiatry 44:43–48, 1981.

26. Ferguson, IT: Electrical study of jaw and orbicularis oculi reflexes after trigeminal nerve surgery. J Neurol Neurosurg Psychiatry 41:819–823, 1978.

27. Ferrari, E, and Messina, C: Blink reflexes during sleep and wakefulness in man. Electroencephalogr Clin Neurophysiol 32:55–62, 1972.

28. Fisher, MA, Shahani, BT, and Young, RR: Assessing segmental excitability after acute rostral lesions. II. The blink reflex. Neurology 29:45–50, 1979.

29. Gandiglio, G, and Fra, L: Further observations on facial reflexes. J Neurol Sci 5:273–285, 1967.

30. Gilliatt, RW, and Taylor, JC: Electrical changes following section of the facial nerve. Proc R Soc Med 52:1080–1083, 1959.

31. Goor, C, and Ongerboer de Visser, BW: Jaw and blink reflexes in trigeminal nerve lesions: An electrodiagnostic study. Neurology 26:95–97, 1976.

32. Hess, K, Kern, S, and Schiller, HH: Blink reflex in trigeminal sensory neuropathy. Electromyogr Clin Neurophysiol 24:185–190, 1984.

33. Hiraoka, M, and Shimamura, M: Neural mechanisms of the corneal blinking reflex in cats. Brain Res 125:265–275, 1977.

34. Hopf, HC, Hufschmidt, HJ, and Stroder, J: Development of the "trigeminofacial" reflex in infants and children. Ann Paediat (Paris) 204:52–64, 1965.

35. Kaplan, PE, and Kaplan, C: Blink reflex: Review of methodology and its application to patients with stroke syndromes. Arch Phys Med Rehabil 61:30–33, 1980.

36. Kayamori, R, Dickins, QS, Yamada, T, and Kimura, J: Brainstem auditory evoked potential and blink reflex in multiple sclerosis. 34:1318–1323, 1984.

37. Khoshbin, S, and Hallett, M: Multimodality evoked potentials and blink reflex in multiple sclerosis. Neurology 31:138–144, 1981.

38. Kimura, J: Alteration of the orbicularis oculi reflex by pontine lesions. Study in multiple sclerosis. Arch Neurol 22:156–161, 1970.

39. Kimura, J: An evaluation of the facial and trigeminal nerves in polyneuropathy: Electrodiagnostic study in Charcot-Marie-Tooth disease, Guillain-Barre syndrome, and diabetic neuropathy. Neurology 21:745–752, 1971.

40. Kimura, J: Electrodiagnostic study of brainstem strokes. Stroke 2:576–586, 1971.

41. Kimura, J: The blink reflex as a test for brainstem and higher central nervous system functions. In Desmedt, JE (ed): New Developments in Electromyography and Clinical Neurophysiology, Vol 3. Karger, Basel, 1973, pp 682–691.

42. Kimura, J: Disorder of interneurons in parkinsonism: The orbicularis oculi reflex to paired stimuli. Brain 96:87–96, 1973.

43. Kimura, J: Effect of hemispheral lesions on the contralateral blink reflex. Neurology 24:168–174, 1974.

44. Kimura, J: Electrically elicited blink reflex in diagnosis of multiple sclerosis: Review of 260 patients over a seven-year period. Brain 98:413–426, 1975.

45. Kimura, J: Conduction abnormalities of the facial and trigeminal nerves in polyneuropathy. Muscle Nerve 5:139–144, 1982.

46. Kimura, J: Clinical uses of the electrically elicited blink reflex. In Desmedt, JE (ed): Motor control mechanism in health and disease: New developments and clinical applications. Raven Press, New York, 1983, pp 773–786.

47. Kimura, J: Electromyography and nerve stimulation techniques: Clinical applications (Japanese). Egakushoin, Tokyo, 1989.

48. Kimura, J, Bodensteiner, J, and Yamada, T: Electrically elicited blink reflex in normal neonates. Arch Neurol 34:246–249, 1977.

49. Kimura, J, Giron, LT, Jr, and Young, SM: Electrophysiological study of Bell palsy: Electrically elicited blink reflex in assessment of prognosis. Arch Otolaryngol 102:140–143, 1976.

50. Kimura, J, and Harada, O: Excitability of the orbicularis oculi reflex in all night sleep: Its suppression in non-rapid eye movement and recovery in rapid eye movement sleep. Electroencephalogr Clin Neurophysiol 33:369–377, 1972.

51. Kimura, J, and Harada, O: Recovery curves of the blink reflex during wakefulness and sleep. J Neurol 213:189–198, 1976.

52. Kimura, J, and Lyon, LW: Orbicularis oculi reflex in the Wallenberg syndrome: Alteration of the late reflex by lesions of the spinal tract and nucleus of the trigeminal nerve. J Neurol Neurosurg Psychiatry 35:228–233, 1972.

53. Kimura, J, and Lyon, LW: Alteration of orbicularis oculi reflex by posterior fossa tumors. J Neurosurg 38:10–16, 1973.

54. Kimura, J, Powers, JM, and Van Allen, MW: Reflex response of orbicularis oculi muscle to supraorbital nerve stimulation: Study in normal subjects and in peripheral facial paresis. Arch Neurol 21:193–199, 1969.

55. Kimura, J, Rodnitzky, RL, and Okawara, S: Electrophysiologic analysis of aberrant regeneration after facial nerve paralysis. Neurology 25:989–993, 1975.

56. Kimura, J, Rodnitzky, RL, and Van Allen, MW: Electrodiagnostic study of trigeminal nerve. Orbicularis oculi reflex and masseter reflex in trigeminal neuralgia, paratrigeminal

syndrome, and other lesions of the trigeminal nerve. Neurology 20:574–583, 1970.

57. Kimura, J, Wilkinson, JT, Damasio, H, Adams, HR, Jr, Shivapour, E, and Yamada, T: Blink reflex in patients with hemispheric cerebrovascular accident (CVA). J Neurol Sci 67:15–28, 1985.

58. Kossev, A, Dengler, R, and Struppler, A: Quantitative assessment of the blink reflex in normals: Physiological side-to-side differences and frequency dependence. Electromyogr Clin Neurophysiol 23:501–511, 1983.

59. Kugelberg, E: Facial reflexes. Brain 75:385–396, 1952.

60. Lowitzsch, K, Kuhnt, U, Sakmann, CH, Maurer, K, Hopf, HC, Schott, D, and Thater, K: Visual pattern evoked responses and blink reflexes in assessment of multiple sclerosis diagnosis: A clinical study of 135 multiple sclerosis patients. J Neurol 213:17–32, 1976.

61. Lowitzsch, K, and Luder, G: Habituation of the blink reflex: Computer assisted quantitative analysis. Electroencephalogr Clin Neurophysiol 60:525–531, 1985.

62. Lyon, LW, Kimura, J, and McCormick, WF: Orbicularis oculi reflex in coma: Clinical, electrophysiological, and pathological correlations. J Neurol Neurosurg Psychiatry 35:582–588, 1972.

63. Lyon, LW, and Van Allen, MW: Alteration of the orbicularis oculi reflex by acoustic neuroma. Arch Otolaryngol 95:100–103, 1972.

64. Lyon, LW, and Van Allen, MW: Orbicularis oculi reflex. Studies in internuclear ophthalmoplegia and pseudointernuclear ophthalmoplegia. Arch Ophthalmol 87:148–154, 1972.

65. Magladery, JW, and Teasdall, RD: Corneal reflexes: An electromyographic study in man. Arch Neurol 5:269–274, 1961.

66. Martin, RC: Late results of facial nerve repair. Ann Otolaryngol 64:859–869, 1955.

67. McDonald, WI, and Sears, TA: The effects of experimental demyelination on conduction in the central nervous system. Brain 93:583–598, 1970.

68. Messina, C: L'abitudine dei riflessi trigeminofacciali in parkinsoniani sottoposti a trattamento con L-DOPA. Riv Neurol 40:327–336, 1970.

69. Messina, C, and Micalizzi, V: I riflessi trigemino-facciali nel corso del coma insulinico. Acta Neurologica 25:357–361, 1970.

70. Messina, C, and Quattrone, A: Comportamento dei riflessi trigemino-facciali in soggetti con lesioni emisferiche. Riv Neurol 43:379–386, 1973.

71. Moller, AR, and Jannetta, PJ: Blink reflex in patients with hemifacial spasm: Observations during microvascular decompression operations. J Neurol Sci 72:171–182, 1986.

72. Namerow, NS: Observations of the blink reflex in multiple sclerosis. In Desmedt, JE (ed): New Developments in Electromyography and Clinical Neurophysiology, Vol 3. Karger, Basel, 1973, pp 692–696.

73. Namerow, NS, and Etemadi, A: The orbicu-

laris oculi reflex in multiple sclerosis. Neurology 20:1200–1203, 1970.

74. Ongerboer de Visser, BW, and Goor, C: Electromyographic and reflex study in idiopathic and symptomatic trigeminal neuralgias: Latency of the jaw and blink reflexes. J Neurol Neurosurg Psychiatry 37:1225–1230, 1974.

75. Ongerboer de Visser, BW, and Kuypers, HGJM: Late blink reflex changes in lateral medullary lesions: An electrophysiological and neuro-anatomical study of Wallenberg's syndrome. Brain 101:285–294, 1978.

76. Ongerboer de Visser, BW, Melchelse, K, and Megens, PHA: Corneal reflex latency in trigeminal nerve lesions. Neurology 27:1164–1167, 1977.

77. Paty, DW, Blume, WT, Brown, WF, Jaatoul, N, Kertesz, A, and McInnis, W: Chronic progressive myelopathy: Investigation with CSF electrophoresis, evoked potentials, and CT scan. Ann Neurol 6:419–424, 1979.

78. Penders, C, and Boniver, R: Exploration electrophysiologique du reflexe de clignement dans la paralysie faciale a frigore. J Otolaryngol 34:17–26, 1972.

79. Penders, CA, and Delwaide, PJ: Interet de l'exploration du reflexe de clignement en cas de paralysie faciale. Electromyography 11:149–156, 1971.

80. Raffaele, R, Emery, P, Palmeri, A, and Perciavalle, V: Influences on blink reflex induced by IA afferents in human subjects. Eur Neurol 25:373–380, 1986.

81. Rodnitzky, RL, and Kimura, J: The effect of induced hyperthermia on the blink reflex in multiple sclerosis. Neurology 28:431–433, 1978.

82. Rossi, D, Buonaguidi, R, Muratorio, A, and Tusini, G: Blink reflexes in posterior fossa lesions. J Neurol Neurosurg Psychiatry 42:465–469, 1979.

83. Rushworth, G: Observations on blink reflexes. J Neurol Neurosurg Psychiatry 25:93–108, 1962.

84. Sanders, EACM, Ongerboer de Visser, BW, Barendswaard, EC, and Arts, RJHM: Jaw, blink and corneal reflex latencies in multiple sclerosis. J Neurol Neurosurg Psychiatry 48:1284–1289, 1985.

85. Sanders, EACM, Reulen, JPH, Hogenhuis, LAH, and Van der Velde, EA: Electrophysiological disorders in multiple sclerosis and optic neuritis. Can J Neurol Sci 12:308–313, 1985.

86. Sanes, JN, Foss, JA, and Ison, JR: Conditions that affect the thresholds of the components of the eyeblink reflex in humans. J Neurol Neurosurg Psychiatry 45:543–549, 1982.

87. Schenck, E, and Manz, F: The blink reflex in Bell's palsy. In Desmedt, JE (ed): New Developments in Electromyography and Clinical Neurophysiology, Vol 3. Karger, Basel, 1973, pp 678–681.

88. Sears, TA, Bostock, H, and Sheratt, M: The pathophysiology of demyelination and its implications for the symptomatic treatment of multiple sclerosis. Neurology 28:21–26, 1978.

89. Shahani, B: Effects of sleep on human reflexes with a double component. J Neurol Neurosurg Psychiatry 31:574–579, 1968.

90. Shahani, B: The human blink reflex. J Neurol Neurosurg Psychiatry 33:792–800, 1970.

91. Shahani, BT, and Parker, SW: Electrophysiological studies in patients with cerebellarpontine angle lesions (abstr). Neurology 29:582, 1979.

92. Shahani, BT, and Young, RR: Human orbicularis oculi reflexes. Neurology 22:149–154, 1972.

93. Shahani, BT, and Young, RR: The blink, H, and tendon vibration reflexes. In Goodgold, J, and Eberstein, A (eds): Electrodiagnosis of Neuromuscular Diseases, ed 2. Williams & Wilkins, Baltimore, 1977, pp 245–263.

94. Silverstein, LD, Graham, FK, and Calloway, JM: Preconditioning and excitability of the human orbicularis oculi reflex as a function of state. Electroencephalogr Clin Neurophysiol 48:406–417, 1980.

95. Stamboulis, E, Scarpalezos, S, Malliara-Loulakaki, S, Voudiklari, S, and Koutra, E: Blink reflex in patients submitted to chronic periodical hemodialysis. Electromyogr Clin Neurophysiol 27:19–23, 1987.

96. Taylor, N, Jebsen, RH, and Tenckoff, HA: Facial nerve conduction latency in chronic renal insufficiency. Arch Phys Med Rehabil 51:259–261, 1970.

97. Thatcher, DB, and Van Allen, MW: Corneal reflex latency. Neurology 21:735–737, 1971.

98. Trontelj, MA, and Trontelj, JV: Reflex arc of the first component of the human blink reflex: A single motoneurone study. J Neurol Neurosurg Psychiatry 41:538–547, 1978.

99. Walker, DD, and Kimura, J: A fast-recovery electrode amplifier for electrophysiology. Electroencephalogr Clin Neurophysiol 45:789–792, 1978.

100. Wartenberg, R: Associated movements in the oculomotor and facial muscles. Arch Neurol Psychiatry 55:439–488, 1946.

101. Waxman, SG, and Brill, MH: Conduction through demyelinated plaques in multiple sclerosis: Computer simulations of facilitation by short internodes. J Neurol Neurosurg Psychiatry 41:408–416, 1978.

102. Waylonis, GW, and Johnson, EW: Facial nerve conduction delay. Arch Phys Med Rehabil 45:539–541, 1964.

103. Wechsler, L, Stakes, J, Shahani, B, and Busis, N: Periodic leg movements of sleep (nocturnal myoclonus): An electrophysiological study. Ann Neurol 19:168–173, 1986.

104. Willer, JC, Boulu, P, and Bratzlavsky, M: Electrophysiological evidence for crossed oligosynaptic trigemino-facial connections in normal man. J Neurol Neurosurg Psychiatry 47:87–90, 1984.

Chapter 17

THE F WAVE

1 INTRODUCTION

Conventional nerve conduction studies seldom contribute to the investigation of more proximal lesions. To study the entire length of the sensory nerve, one may record somatosensory cerebral evoked potentials.[12] In contrast, measurement of the F wave helps in assessing motor conduction along the most proximal segment, because it results from the backfiring of antidromically activated anterior horn cells. Explored first in patients with Charcot-Marie-Tooth disease,[51] the method has since gained popularity in evaluation of a variety of neurologic disorders.[9,22,23,55,56,60,61,65,70,80,82,85–87,121]

The inherent variability of the latency and configuration makes the use of F wave less precise than that of the direct compound muscle action potential or M-response determination. Nonetheless, the technique usefully supplements the conventional nerve conduction studies, especially in characterizing demyelinating polyneuropathies, in which delay of the F wave often clearly exceeds the normal range. In addition to determination of F-wave latencies, calculation of conduction velocities and the F ratio permits comparison of conduction in the proximal versus the distal nerve segments.[55,56] The F wave also provides a measure of motor neuron excitability, which presumably dictates the probability of a recurrent response in individual axons. In this section the rapidly accumulating data on F-wave determination will be reviewed and its clinical value and limitations discussed.[58]

2 PHYSIOLOGY OF THE F WAVE

Recurrent Versus Reflexive Activation of the Motor Neuron

A supramaximal electric shock delivered to a nerve often elicits a late response in the innervated muscle. Since the original description by Magladery and McDougal,[74] who designated it as the F wave (presumably because they initially recorded it from intrinsic foot muscles), different authors have debated its neural source. The F wave occurs after the direct motor potential, or the M response. With more proximal stimulation, the latency of the M response increases, whereas that of the F wave decreases (Fig. 17–1). This indicates that the impulse destined to elicit the F wave first travels away from the recording electrodes toward the spinal cord before it returns to activate distal muscles.

This finding is consistent with either a reflex hypothesis[44,71,74] or a theory based on recurrent discharge of antidromically activated motor neurons,[11,77,78,110] or both.[39,41] The presence of the F wave in deafferentated limbs[35,77,78] and after transverse myelotomy[79] strongly suggests that it depends in part on backfiring of motor neurons. Studies using single-fiber electromyography[110,113] have also shown that the occurrence of the F wave requires prior activation of the motor axon. The evidence of its recurrent nature, however, does not necessarily preclude the presence of reflex components that may still contribute.

Block of Antidromic or Orthodromic Impulses

Motor neurons subject to recurrent activation fire only infrequently after a series of direct motor responses.[103] Thus, although antidromic activation and orthodromic activation of motor neurons usually follow the same physiologic principles,[6,16] additional mechanisms must prevent the motor neurons from generating the recurrent response with every stimulus.[16,93] Recurrent discharges develop in only a limited number of motor units, in part because the antidromic impulse fails to enter the somata in some of the motor neurons.[73] This type of block often takes place at the axon hillock, where membrane characteristics change; but it may also occur more distally, in the myelinated segment of the axons.

The impulse may also abate during orthodromic propagation of the spike potential generated in the soma-dendrite membrane (SD spike). The SD spike can travel orthodromically only after the axon hillock recovers from the refractory period induced by the passage of the antidromic impulse. The subliminally excited soma-dendrite membrane facilitates antidromic activation of the SD spike, resulting in increased probability of a recurrent response. Conversely, excessive depolarization of the soma-dendrite membrane may lead to early generation of the SD spike, which cannot propagate across the still depolarized axon hillock. The two opposing effects of subliminal depolarization of motor neurons have clinical implications: slight voluntary excitation of the motor neuron pool may enhance F-wave activation, but excessive effort may have the reverse effect.

Latency and Amplitude of the F Wave

The recurrent discharges probably encounter blockage at the initial segment more frequently in the smaller, lower

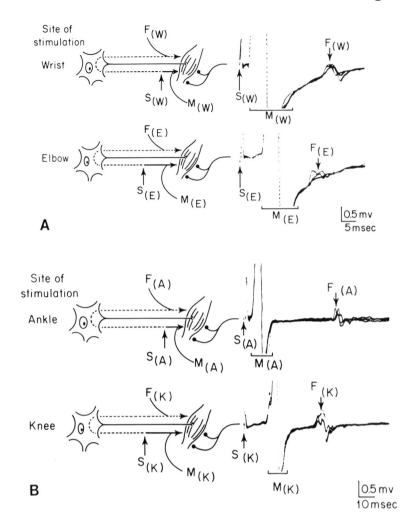

Figure 17–1. A. Normal M response (*horizontal brackets*) and F wave (*small arrows*) recorded from the thenar muscles after supramaximal stimulation of the median nerve at the wrist (*top*) or elbow (*bottom*). The shift of stimulus point proximally increased the latency of the M response and decreased that of F wave. The schematic diagrams illustrate the centrifugal (*solid arrows*) and centripetal impulses (*dotted arrows*). (Modified from Kimura,[51] with permission.) **B.** Normal M response (*horizontal brackets*) and F wave (*small arrows*) recorded from the abductor hallucis after supramaximal stimulation of the tibial nerve at the ankle (*top*) or knee (*bottom*). With a shift of stimulus site proximally, the latency of the M response increased, whereas that of the F wave decreased. (From Kimura et al.,[60] with permission.)

threshold motor neurons, which rapidly depolarize.[46,50] Preferential activation of the larger motor neurons may result if Renshaw cells inhibit the smaller motor neurons more effectively.[17,18,43] Hence the incidence of the F wave may, at least in theory, favor the larger motor neurons with faster conducting axons. This in turn provides a rationale for using the minimal latency of the F wave as a measure of the fastest conducting fibers. Because of a particular set of physiologic conditions required for generation and propagation of a recurrent discharge, the latency of successive F waves from a single motor axon varies only narrowly between 10 and 30 μs.[103] Parenthetically, the latency of consecutive H reflexes from a single motor axon may fluctuate by as

much as 2.5 ms, primarily because of variation in synaptic transmission (see Chapter 18.2).

Partial excitation of the nerve generates recurrent discharges in either larger anterior horn cells with lower threshold motor axons or smaller cells with higher thresholds.[27,47,50] Further, when the fast conducting axons are progressively blocked by use of a collision technique, the F wave continues to appear in proportion to the slow conducting motor axons that have escaped the collision.[64] Hence, recurrent discharges must occur not only in the larger motor neurons with fast conducting axons, but also in the smaller motor neurons with slower conducting fibers.

The F wave is elicited in approximately

1 to 5 percent of antidromically activated motor neurons regardless of their peripheral excitability or conduction characteristics. In normal subjects, F-wave frequency varies, with a mean of 79 percent, and most responses occur only once during a train of 200 stimuli.[88] In routine studies of the F wave, however, one occasionally encounters recurrence of the identical waveforms. For clinical studies, where both large and small axons are activated simultaneously, rather than selectively, therefore, anatomic or physiologic properties might predispose a given fraction of the more excitable motor neuron pool to backfiring.

A few-millisecond interval between the earliest and latest F wave, results, in part, from the difference between the fast and slow motor conduction.[91] The conduction time, however, is a function not only of the speed of the propagated impulse but also of the length of the fine terminal fibers innervating each muscle fiber.[92] The latter is in turn determined by the location of end plates within the muscle. The difference in the terminal length of the longest and shortest fibers is probably on the order of a few millimeters. Nonetheless, a slight change in the length of the terminal branch that becomes unmyelinated near the end plate results in a substantial latency difference. Another unknown variable is the distance between the recording electrodes and the motor end plate, where the muscle action potential originates. Because of these factors, the F wave representing slower conducting fibers may not necessarily be the ones showing a longer latency and vice versa.[64]

The amplitude and frequency of the F wave provide a measure of motor neuron excitability, but the relationship is physiologically complex.[21,25-27,32] No recurrent discharge is expected with an antidromic impulse producing subliminal depolarization in hypoexcitable cells. The F wave may also fail in hyperexcitable cells, which may discharge too rapidly during the refractory period of the initial axon segments. For example, despite spasticity, F-wave frequency is lower in motor neuron disease than in control subjects.[90] This finding, however, may in part reflect loss of lower motor neurons, because identical responses also occur more frequently than in controls.

In one study designed to analyze the constitution of the F wave, additional motor units contributed a greater potential when recruited at higher stimulus intensities.[120] In another study, however, no consistent correlation emerged between the latency and amplitude of the F wave.[29,30] Electrical stimulation of the dentate nucleus reduces the size of the F wave in man.[34] Intravenous or subcutaneous injections of thyrotropic-releasing hormone rapidly increase the amplitude of the F waves.[5]

3 AXON REFLEX AND OTHER LATE RESPONSES

Physiologic Characteristics

The axon reflex, as suggested by its original designation, the intermediate latency response, usually appears between the M response and F wave.[37] It presumably implies the presence of collateral sprouting in the proximal portion of the nerve. If a submaximal stimulus excites one branch of the axon but not the other, the antidromic impulse propagates up to the point of branching and turns around to proceed distally along a second branch. Shocks of higher intensity, activating both branches distally, eliminate the response, because two antidromic impulses collide as they turn around at the branching point (Fig. 17–2A and B). In contrast, the F wave varies in latency and waveform and appears with supramaximal stimuli, which normally abolish the axon reflex altogether. In some instances, however, the axon reflex may persist despite the use of very high-intensity stimuli. Here, fibrosis or other structural changes of the surrounding median may prevent the current from reaching the axons, thus, precluding a "supramaximal" activation.

The axon reflex most commonly appears with stimulation of the median or ulnar nerve at the wrist or the peroneal or tibial nerve at the ankle. Proximal stimu-

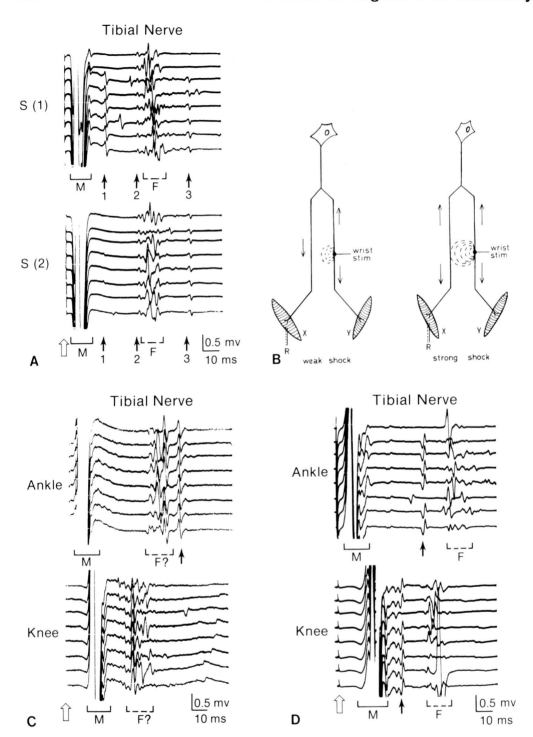

Figure 17–2. A. A 51-year-old man with low back pain. Stimulation of the right tibial nerve at the ankle elicited a number of axon reflexes. A series of eight tracings displayed with stepwise vertical shift of the baseline confirm the consistency. This type of display not only facilitates the selection of the F wave with minimal latency but also allows individual assessments of all the late responses. Of the three axon reflexes (*small arrows*, *1, 2,* and *3*) elicited by weak shocks, *S(1)*, stronger shocks, *S(2)*, eliminated only the earliest response. **B.** Collateral sprouting in the proximal part of the nerve. A strong shock, activating both branches, can eliminate the axon reflex generated by weak stimulation by collision. (From Fullerton and Gilliatt,[37] with

lation above the origin of the collateral sprout produces only an M response. Thus, series of stimuli applied along the course of the nerve may localize the site of bifurcation. Collateral sprouting, however, does not always develop at the level of the lesion, but frequently well below the actual site of involvement.[37] Distal and proximal stimuli may elicit the same axon reflex, allowing determination of conduction velocity for the short intersegment of that particular motor fiber.

The axon reflex has a constant latency and waveform, because it originates from the same portion of a single motor unit innervated by a collateral sprout. In the absence of synaptic connection along the pathway, the impulse can be generated successively by repetitive stimulation up to 40 per second. The point of axonal branching and the conduction velocity of the two branches of the axon involved determine the latency of the axon reflex. The unmyelinated regenerating collateral sprout may conduct the ascending or descending impulses of an axon reflex much slower than the intact axons in the vicinity that relay the F wave. Hence, occasional axon reflexes follow, rather than precede, the F wave (Fig. 17–2A and C).

A late motor response presumably mediated by an axon loop along the nerve may mimic the axon reflex.[99] Late muscle responses may be generated by various other pathophysiologic mechanisms such as reflection of an impulse and ephaptic transmission.[111] The late potentials elicited by repetitive discharges in the nerve trunk also resemble axon reflex. These components, however, fail at high repetition rates and tend to vary in latency and waveform even if they are elicited from a single axon. A late potential may also result from a scattered motor response with slow conduction in pathologic nerves. The latency of the axon reflex decreases, whereas that of a temporally dispersed M response increases, with proximal stimu-

lation (Figs. 17–2D and 17–3). Electric field of the muscle action potential could reexcite ephaptically an intramuscular axon, producing a muscle-nerve reverberating loop.[104] In this case, the original muscle potential and the repetitive discharge maintain the same interval regardless of the nerve stimulation point.

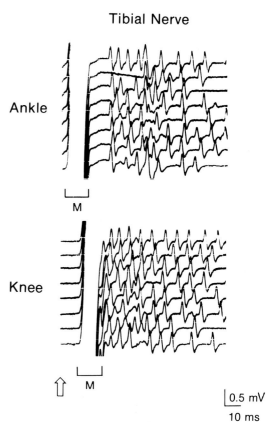

Figure 17–3. Incidental finding of unusual repetitive discharges resembling axon reflexes in a 38-year-old man with history of right pelvic fracture. Stimulation of the right tibial nerve at the ankle or knee elicited the repetitive discharge. Its onset latency shortened with proximal as opposed to distal stimulation, as expected in an axon reflex.

permission.) **C.** Axon reflexes after stimulation of the left tibial nerve at the ankle or knee in the same patient as in **A.** Proximal stimulation eliminated the axon reflex (*arrow*) that followed the F wave with distal stimulation. **D.** A 50-year-old man with recurrent backaches following laminectomy. Stimulation of the tibial nerve at the ankle or knee elicited the axon reflex (*arrow*). Like F-wave latency, the latency of the axon reflex decreased with proximal sites of stimulation. This indicates that the impulse first propagates in the centripetal direction.

Clinical Applications

The axon reflex occurs in patients with neurogenic atrophy and rarely, if at all, in healthy individuals. It has been observed in patients with tardy ulnar palsy, brachial plexus lesions, diabetic neuropathy, hereditary motor sensory neuropathy, facial neuropathy, amyotrophic lateral sclerosis, and cervical root lesions.[36,37,98,100,101] This sign of nerve regeneration abounds in chronic neuropathies and entrapment syndromes.[96,97] The diagnostic value of the axon reflex remains to be seen, although it generally implies a chronic neuropathic process.

4 DETERMINATION OF F-WAVE LATENCY

Recording Procedures

A supramaximal stimulus applied at practically any point along the course of a nerve elicits the F wave. Placing the anode distal to the cathode or off the nerve trunk avoids anodal block of the antidromic impulse. A surface electrode placed over the motor point of the tested muscle serves as the active lead (G_1) against the reference electrode (G_2) over the tendon. An optimal display of F waves requires an amplifier gain of 200 or 500 μV/cm and an oscilloscope sweep of 5 or 10 ms/cm, depending on the nerve length and stimulus point. These recording parameters truncate and compress the simultaneously recorded M response into the initial portion of the tracing. Thus, one must study the M response and F wave separately, using different gains and time bases.

F-wave latencies measured from the stimulus artifact to the beginning of the evoked potential vary by a few milliseconds from one stimulus to the next. Hence, for an adequate study, more than 10 F waves must be clearly identified among 16 to 20 trials displayed on a storage oscilloscope, with automatic shifting of successive sweeps vertically (Fig. 17–4). In addition, determination of the minimal and maximal latencies reveals the degree of scatter among consecutive

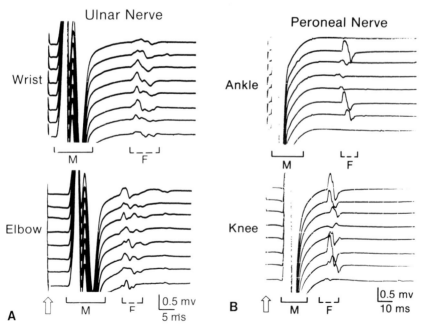

Figure 17–4. A. Eight consecutive tracings showing normal M responses and F waves recorded from the hypothenar muscles after stimulation of the ulnar nerve at the wrist and elbow. **B.** Eight consecutive tracings showing normal M responses and F waves recorded from the extensor digitorum brevis after stimulation of the peroneal nerve at the ankle and knee.

responses, providing a measure of temporal dispersion.[83] Electronic averaging of a large number of responses permits easy analysis of mean latency, although phase cancellation sometimes defeats its own purpose.[19,28,30,75]

Slight voluntary contraction enhances the F wave and is often employed in our laboratory to facilitate the response. During this maneuver, only a small number of axons carry a voluntary impulse at any given moment.[54] Despite the orthodromic activation in a few motor fibers, therefore, the antidromic impulse will reach the cell body in the remaining majority of axons and generate recurrent discharges. Therefore, the late response recorded during mild voluntary contraction may be taken as the F wave in evaluating motor conduction to and from the spinal cord. With greater effort to contract the muscle, voluntary impulses collide with antidromic activity in many axons, precluding the generation of the F wave. Instead, reflexively activated impulses will be propagated along the motor axons cleared of the antidromic impulse. Thus, the late response produced by this type of activation would contain a component analogous to the H reflex.[45,116]

Distal Versus Proximal Stimulation

With the use of F wave, distal stimulation at the wrist or ankle can be used in determining the motor conduction time along the entire length of the nerve. With diffuse or multisegmental lesions, the delay in nerve conduction increases in proportion to the length of the tested pathway. Thus, relatively mild slowing not identifiable by conventional motor nerve conduction studies may lead to delayed F waves. In the study of F waves, an increased latency detected by distal stimuli results from conduction delay anywhere along the course of the nerve. Localization of abnormality requires comparison of F-wave and M-response latencies using more proximal stimulation.

The F wave first travels in the centripetal direction toward the spinal cord before it turns around distally to activate the muscle. With more proximal stimulation, the F wave moves closer to the M response, because the latency of the M response increases, whereas that of the F wave decreases. The F wave occurs clearly after the M response with stimulation at the wrist, elbow, ankle, and knee. With axillary stimulation, however, the M response is superimposed on the F wave.[51,52] In this instance, simultaneous stimulation at the axilla and wrist helps to isolate the F wave. With this technique, the orthodromic impulse from the axilla and the antidromic impulse from the wrist collide, leaving the M response from the wrist and the F wave from the axilla intact. These two remaining evoked muscle potentials do not overlap, allowing detection of the F wave elicited by axillary stimulation.[51]

On the average, the decrease in latency of the F wave equals the increase in latency of the M response, when the stimulating point moves from the wrist to the elbow and then to the axilla. This observation allows the prediction of F-wave latency from the axilla without actual measurement, because it must equal the sum of the latencies of the F wave and M response elicited by distal stimulation minus the latency of the M response evoked by axillary stimulation.[61] Or $F(A) = F(W) + M(W) - M(A)$, where $F(A)$ and $F(W)$ represent the latencies of the F wave with stimulation at the axilla and wrist, and $M(A)$ and $M(W)$, latencies of the corresponding M response.[4]

For clinical studies, routine procedures include stimulation of the median and ulnar nerves at the wrist and elbow and of the tibial and peroneal nerves at the ankle and knee. When necessary, the equation described above provides the estimated latency of the F wave from any proximal sites. The F wave may also be elicited after stimulation of the facial nerve,[102] although superimposition of the M response usually makes its recognition difficult. Further, inadvertent stimulation of neighboring trigeminal afferent fibers may simultaneously activate reflex responses,[114] which may mimic the late response.

5 MOTOR CONDUCTION TO AND FROM THE SPINAL CORD

Central Latency

Central latency or conduction time from the stimulus point to and from the spinal cord equals $F - M$, where F and M are latencies of the F wave and the M response (Fig. 17–5). Subtracting an estimated delay of 1.0 ms for the turn-around time at the cell and dividing by two, $(F - M - 1)/2$ represents the conduction time along the proximal segment from the stimulus site to the spinal cord. Although the exact turnaround time at the anterior cells is not known in man,[113] animal data indicate a delay of nearly 1.0 ms.[40,73,93] The absolute refractory period of the fastest human motor fibers lasts

$$\text{F ratio} = \frac{(F-M-1)/2}{M} = \frac{F-M-1 \ (msec)}{M \times 2 \ (msec)}$$

$$\text{FWCV} = \frac{D}{(F-M-1)/2} = \frac{D \times 2 \ (mm)}{F-M-1 \ (msec)}$$

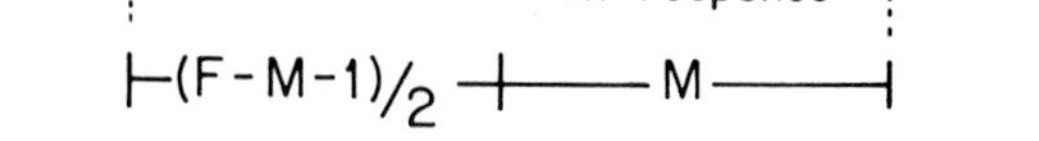

Figure 17–5. The latency difference between the F wave and the M response represents the passage of a motor impulse to and from the cord through the proximal segment. Considering an estimated minimal delay of 1.0 ms at the motor neuron pool, the proximal latency from the stimulus site to the cord equals $(F - M - 1)/2$, where F and M are latencies of the F wave and the M response. In the segment to and from the spinal cord, $FWCV = 2D/(F - M - 1)$, where D is the distance from the stimulus site to the cord and $(F - M - 1)/2$, the time required to cover the length D. Dividing the conduction time in the proximal segment to the cord by that of the remaining distal segment to the muscle, the F ratio $= (F - M - 1)/2M$, where $(F - M - 1)/2$ and M are proximal and distal latencies. (From Kimura,[56] with permission.)

about 1.0 ms or slightly less.[53,62] The recurrent discharge generated during the refractory period cannot propagate distally beyond the initial segment of the axon. In evaluating the minimal latency, therefore, it seems appropriate to assume a turnaround time of at least 1.0 ms and possibly greater.

A given F wave represents only a portion of the motor axons available for activation of the M response. The interval of a few milliseconds between the earliest and latest F wave probably results from the difference between the fast and slow conducting motor fibers. Selecting the shortest F-wave latency out of many trials usually, although perhaps not always, allows one to study the conduction properties of the fastest fibers.[9,22,51,60,66,86] In some diseased nerves, however, surviving fibers that contribute to the M response may not propagate antidromic impulses centripetally. In this instance, F waves are not generated at all, even though the same stimuli elicit an M response. In less extreme instances, if only the fast conducting fibers are blocked proximally, but not distally, the onset latency of the M response and minimal latency of the F wave, representing two separate groups of motor fibers, are not directly comparable for calculation of conduction velocity.

The error, however, must be small if the increase in latency of the M response elicited by a proximal stimulus equals the decrease in latency of the F wave.[51,60,77,86] One can test this relationship by comparing the sums of the F latency and M latency at distal and proximal stimulus sites. If the computed values are equal, the F wave of minimal latency travels in the centripetal direction with the same speed as the centrifugal impulse of the earliest portion of the M response. Thus, the same group of motor fibers, or at least those with the same conduction characteristics, contributes to the F wave and the fastest components of the M response. This, in turn, provides a rationale for directly comparing the latencies of these two muscle potentials in various assessments of proximal versus distal conduction characteristics.[55]

F-Wave Conduction Velocity

In the upper extremities, the surface distance is measured from the stimulus point to the C-7 spinous process via the axilla and midclavicular point.[51,56] In the lower extremities, surface measurement follows the nerve course from the stimulus site to the T-12 spinous process by way of the knee and greater trochanter of the femur.[60] The estimated nerve length then allows calculation of F-wave conduction velocity (FWCV) in the segment to and from the spinal cord:

$$FWCV = (2D)/(F - M - 1)$$

where D is the distance from the stimulus site to the cord and $(F - M - 1)/2$ is the time required to cover the length (see Fig. 17–5).

In clinical assessment of F-wave latency, it is necessary to determine surface distance to adjust for differing nerve lengths. Instead of measuring the extremity one may use the height of the patient if a nomogram is available.[112] The estimated length of a nerve segment by surface measurement correlates well with its F-wave latency. Observations in five cadavers showed good agreement between surface determinations and actual lengths of nerves in the upper[72] as well as lower extremities.[60] Absolute F-wave latencies may suffice in studying those extremities of average length.[10] With unilateral lesions affecting one nerve, comparison between the right and left sides in the same subject or one nerve with another in the same extremity provide the most sensitive measure of abnormality (Tables 17–1 and 17–2).[55,119]

The F Ratio

Comparative assessment between proximal and distal nerve segments is possible with the F/M ratio[22,23] where F represents the latency of the F wave and M, that of the M response. A similar ratio, $(F - M - 1)/2M$, directly compares the conduction time from the cord to the stimulus site, represented by $(F - M - 1)/2$, with that of the remaining distal nerve segment to the muscle, indicated by M. Designated the F ratio, it provides a simple means of evaluating conduction characteristics of the proximal versus distal segment (see Fig. 17–5). Studies of the F ratio circumvent the need for determining the nerve length. Clinical use of this ratio, however, assumes that the various limbs of different length have the same proportion for the proximal and distal segments.[56]

F ratios are close to unity with stimulation of the median nerve at the elbow, ulnar nerve, 3 cm above the medial epicondyle, the tibial nerve at the popliteal fossa, and the peroneal nerve immediately above the head of the fibula (see Table 17–1). With stimulation at these sites, therefore, one can grossly estimate the latency of the F wave as three times the latency of the M response plus one. This finding also indicates that the stimulus sites at the elbow or knee dissect the total length of the axon into two segments of equal conduction time. The conduction velocity is, therefore, faster along the longer proximal segment than along the shorter distal segment.[60] In fact, calculated F-wave conduction velocity (FWCV) suggests faster conduction proximally than distally.[9,51,60,61,65,80,86]

6   THE F WAVE IN HEALTH AND DISEASE

Clinical Value and Limitations

Clinical uses of the F wave suffer from inherent latency variability from one trial to the next. Determination of the shortest latency after a large number of trials can minimize this uncertainty. Recording as many as 100 F waves at each stimulus site proved useful in special studies[87] but is not practical in a routine clinical test. Determining the latency differences between two sides or between two nerves in the same limb serves as the most sensitive means of examining a patient with a unilateral disorder affecting a single nerve. Absolute latencies suffice for evaluating the entire course of the nerve in a diffuse process. Calculation of the central latency, the FWCV, and the F ratio provides additional information not other-

Table 17–1 F WAVES IN NORMAL SUBJECTS*

Number of Nerves Tested	Site of Stimulation	F-Wave Latency to Recording Site (ms)	Difference Between Right and Left (ms)	Central Latency† to and from the Spinal Cord (ms)	Difference Between Right and Left (ms)	Conduction Velocity‡ to and from the Spinal Cord (m/s)	F Ratio§ Between Proximal and Distal Segments
122 median nerves from 61 subjects	Wrist	26.6 ± 2.2 (31)**	0.95 ± 0.67 (2.3)**	23.0 ± 2.1 (27)**	0.93 ± 0.62 (2.2)**	65.3 ± 4.7 (56)††	
	Elbow	22.8 ± 1.9 (27)	0.76 ± 0.56 (1.9)	15.4 ± 1.4 (18)	0.71 ± 0.52 (1.8)	67.8 ± 5.8 (56)	0.98 ± 0.08 (0.82–1.14)**,††
	Axilla¶	20.4 ± 1.9 (24)	0.85 ± 0.61 (2.1)	10.6 ± 1.5 (14)	0.85 ± 0.58 (2.0)		
130 ulnar nerves from 65 subjects	Wrist	27.6 ± 2.2 (32)	1.0 ± 0.83 (2.7)	25.0 ± 2.1 (29)	0.84 ± 0.59 (2.0)	65.3 ± 4.8 (55)	
	Above elbow	23.1 ± 1.7 (27)	0.68 ± 0.48 (1.6)	16.0 ± 1.2 (18)	0.73 ± 0.52 (1.8)	65.7 ± 5.3 (55)	1.05 ± 0.09 (0.87–1.23)
	Axilla¶	20.3 ± 1.6 (24)	0.73 ± 0.54 (1.8)	10.4 ± 1.1 (13)	0.76 ± 0.52 (1.8)		
120 peroneal nerves from 60 subjects	Ankle	48.4 ± 4.0 (56)	1.42 ± 1.03 (3.5)	44.7 ± 3.8 (52)	1.28 ± 0.90 (3.1)	49.8 ± 3.6 (43)	
	Above knee	39.9 ± 3.2 (46)	1.28 ± 0.91 (3.1)	27.3 ± 2.4 (32)	1.18 ± 0.89 (3.0)	55.1 ± 4.6 (46)	1.05 ± 0.09 (0.87–1.23)
118 tibial nerves from 59 subjects	Ankle	47.7 ± 5.0 (58)	1.40 ± 1.04 (3.5)	43.8 ± 4.5 (53)	1.52 ± 1.02 (3.6)	52.6 ± 4.3 (44)	
	Knee	39.6 ± 4.4 (48)	1.25 ± 0.92 (3.1)	27.6 ± 3.2 (34)	1.23 ± 0.88 (3.0)	53.7 ± 4.8 (44)	1.11 ± 0.11 (0.89–1.33)

*Mean ± standard deviation (SD) in the same patients shown in Tables 6–1, 6–3, 6–10, and 6–12.
†Central latency = F − M where F and M are latencies of the F wave and M response, respectively.
‡Conduction velocity = 2D/(F − M − 1), where D is the distance from the stimulus point to C-7 or T-12 spinous process.
§F ratio = (F − M − 1)/2M with stimulation with the cathode on the volar crease at the elbow (median), 3 cm above the medial epicondyle (ulnar), just above the head of fibula (peroneal) and in the popliteal fossa (tibial).
¶F(A) = F(E) + M(E) − M(A) where F(A) and F(E) are latencies of the F wave with stimulation at the axilla and elbow, respectively, and M(A) and M(E) are latencies of the corresponding M response.
**Upper limits of normal calculated as mean + 2 SD.
††Lower limits of normal calculated as mean − 2 SD.

Table 17–2 COMPARISON BETWEEN TWO NERVES IN THE SAME LIMB*

Number of Nerves Tested	Site of Stimulation	F-Wave Latency to Recording Site			Central Latency† to and from the Spinal Cord		
		Median Nerve	Ulnar Nerve	Difference	Median Nerve	Ulnar Nerve	Difference
70 nerves from 35 patients	Wrist	26.6 ± 2.3 (31)‡	27.2 ± 2.5 (32)‡	1.00 ± 0.68 (2.4)‡	23.3 ± 2.2 (28)‡	24.5 ± 2.4 (29)‡	1.24 ± 0.75 (2.7)‡
	Elbow	22.9 ± 1.8 (26)	23.0 ± 1.7 (26)	0.84 ± 0.55 (1.9)	15.5 ± 1.4 (18)	16.0 ± 1.2 (18)	0.79 ± 0.65 (2.1)
		Peroneal Nerve	Tibial Nerve	Difference	Peroneal Nerve	Tibial Nerve	Difference
104 nerves from 52 patients	Ankle	47.7 ± 4.0 (55)	48.1 ± 4.2 (57)	1.68 ± 1.21 (4.1)	43.6 ± 4.0 (52)	44.1 ± 3.9 (52)	1.79 ± 1.20 (4.2)
	Knee	39.6 ± 3.7 (47)	40.1 ± 3.7 (48)	1.71 ± 1.19 (4.1)	27.1 ± 2.9 (33)	28.0 ± 2.7 (33)	1.75 ± 1.07 (3.9)

*Mean ± standard deviation (SD) in the same patients shown in Tables 6–2 and 6–11.
†Central latency = F − M, where F and M are latencies of the F wave and M response, respectively.
‡Upper limits of normal calculated as mean + 2 SD.

wise available, especially in the comparison of proximal and distal segments (see Tables 17–1 and 17–2).

The F wave is commonly abnormal in hereditary motor sensory neuropathy,[51,60,82] acute or chronic demyelinating neuropathy,[56,61,65] diabetic neuropathy,[9,63] uremic neuropathy,[1,84,85] alcoholic neuropathy,[70] and a variety of other neuropathies.[69] Other categories of disorders associated with F-wave changes include entrapment neuropathies,[22,119] amyotrophic lateral sclerosis,[2] and radiculopathies.[23,33]

Studies of the F wave help characterize polyneuropathies in general and those associated with prominent proximal disease in particular (Figs. 17–6, 17–7, and 17–8). In the early diagnosis of more localized nerve lesions such as radiculopathies, the remaining normal segment tends to dilute a conduction delay across the much shorter segment. Thus, relatively mild abnormalities over restricted segments may not alter F-wave latency beyond its inherent variability. In fact, in experimental allergic neuritis demyelination of the ventral root did not necessarily imply abnormalities in F-wave latency. Furthermore, only 14 percent of the guinea pigs and 7 percent of the rabbits showed an abnormal increase in F-wave latency in fibers with normal motor nerve conduction velocity.[115]

The F wave may provide a means of assessing motor neuron excitability.[76] In one study,[21] the amplitude of 32 F waves averaged 1 percent of the M response in normal subjects and was significantly greater in patients with spasticity, primarily because the F wave became more persistent. The largest F wave, 4.5 percent of the M response in normals, did not increase in the patient group with chronic paraparesis. The degree, duration, and type of spasticity may determine average as well as maximal amplitude of the F wave.[38] Unusually large F waves may appear in association with clinical spasticity and other upper motor neuron signs (Fig. 17–9). In these instances, reflex components may contribute to the late response, especially if the patient has prominent hyperreflexia. The amplitude of the F wave also increases in

disorders of the lower motor neuron, presumably because regenerated axons supply an increased number of muscle fibers.[105]

Normal Values

Tables 17–1 and 17–2 summarize the ranges of normal latency and other aspects of the F wave established in the same control subjects as described in the preceding section for conduction studies of the extremities. Placement of the cathode 3 cm more proximally in this study of the median and ulnar nerves has shorter than average F latency in this series, compared with previous studies.[51,60] In addition, an attempt to elicit three times as many F waves at each stimulus site and slight voluntary facilitation routinely employed also increased the chance of recording the fastest conducting fibers. The upper and lower limits of normal values are defined as 2 standard deviations above and below the mean.

In children, the minimal F-wave latency remains relatively constant during the first 3 years of life because rapid change in conduction velocity compensates for the increase in arm length. The F-wave latency then increases until about the 20th year of life, when it reaches 95 percent of its maximal value.[67]

Hereditary Motor Sensory Neuropathy

Patients with advanced illness have neither M response nor F wave in the lower extremities,[60,82] but relatively preserved responses in the upper extremities. These findings support the clinical impression that the disease affects the lower extremities more severely (Table 17–3). Mildly diseased nerves may show slow motor conduction in the distal segment and normal conduction in the proximal segment.[51] In advanced cases, conduction abnormalities affect both segments equally. A bimodal distribution of motor nerve conduction velocities (MNCVs)[108,109] supports the dichotomous

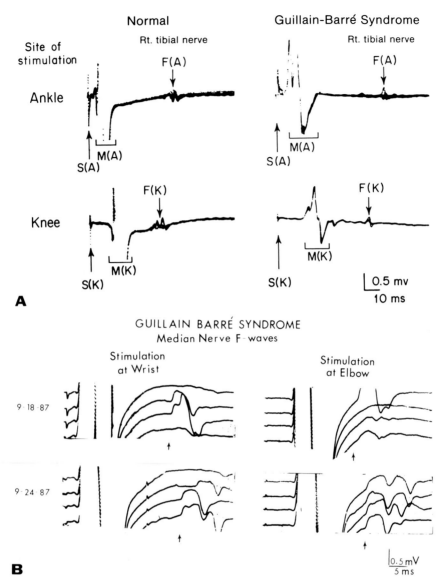

Figure 17–6. A. The M response (*brackets*) and the F wave (*small arrows*) recorded from the abductor hallucis in two subjects. The patient with the Guillain-Barré syndrome had increased F-wave latency. The M response was normal in latency, although reduced in amplitude. **B.** A 26-year-old man with progressive generalized weakness of 2 weeks' duration. He had difficulty rising from a chair and climbing stairs. Electrophysiologic studies on September 18 revealed normal nerve conduction studies, although the patient was unable to recruit motor unit potentials. On September 24, the minimal F-wave latency was increased by 4 ms from the previous measures with stimulation of the median nerve either at the wrist or at the elbow. Prolongation of minimal F latency to this degree, if reproducible, suggests a proximal conduction delay. This may be the only abnormality in some patients with the Guillain-Barré syndrome during an acute stage. (From Kimura,[59] with permission.)

separation into hypertrophic and neuronal types, or types I and II.[14,15] Intermediate F-wave latencies seen in the present series probably reflect extreme variability of conduction over a wide spectrum in each group (Fig. 17–10).

Guillain-Barré Syndrome

Conduction abnormalities may involve any segment of the peripheral nerve in this syndrome (Table 17–4). The disease commonly affects the most proximal, pos-

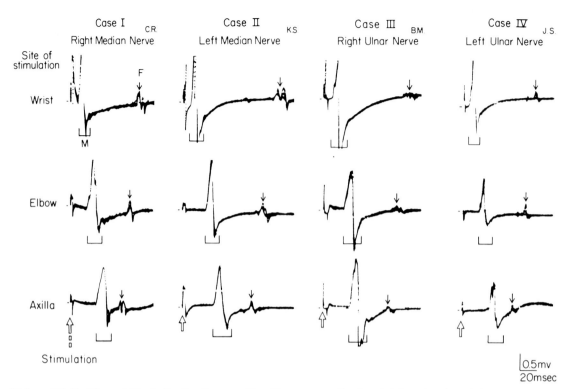

Figure 17–7. Charcot-Marie-Tooth disease. The M response (*horizontal brackets*) and F wave (*small arrows*) recorded from the thenar (cases 1 and 2) and hypothenar muscles (cases 3 and 4) in patients with hereditary motor sensory neuropathy type I. Three consecutive trials in each showed markedly increased latencies of the M response and F wave, requiring slower sweep speed of 20 ms/cm instead of the usual 5 ms/cm. Because of slowed conduction, the M response and F wave were separated even with stimulation at the axilla, rendering the collision technique unnecessary. (From Kimura,[51] with permission.)

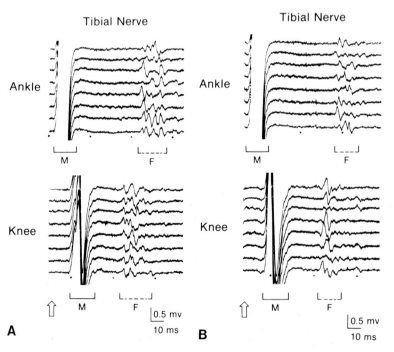

Figure 17–8. A 44-year-old man with adrenoleukodystrophy and diffuse weakness. Stimulation of the tibial nerve at the ankle or knee on the right (**A**) or left (**B**) elicited the F waves in the abductor hallucis. An increase in latency and duration and marked temporal dispersion of the F wave stand in sharp contrast to the normal M response.

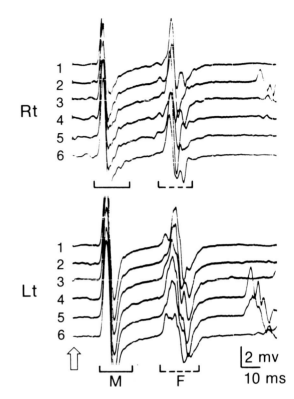

Figure 17–9. A 39-year-old man with chronic tetanus, diffuse hyperreflexia, and rigidity. Supramaximal stimulation of the peroneal nerve at the knee elicited large F waves in the extensor digitorum brevis. Six consecutive trials obtained on each side show consistency in the response. The average amplitude of the F wave was 57 percent of the corresponding M response on the *right* and 43 percent on the *left*. Reflex components may have contributed to the late response despite the use of supramaximal stimulation. (From Risk et al.,[94] with permission.)

Table 17–3 HEREDITARY MOTOR SENSORY NEUROPATHY (MEAN ± SD)

Number of Nerves Tested	Sites of Stimulation	M Latency (ms)	F Latency (ms)	MNCV Between Two Stimulus Sites (m/s)	FWCV from Cord to Stimulus Site (m/s)
36 median nerves	Wrist	6.4 ± 3.0	55.6 ± 26.1		33.7 ± 14.6
				30.4 ± 14.6	
	Elbow	15.6 ± 7.8	46.1 ± 21.4		36.4 ± 14.9
				38.9 ± 20.2	
	Axilla	22.2 ± 10.6	39.3 ± 17.8		38.4 ± 16.8
31 ulnar nerves	Wrist	5.2 ± 2.9	55.5 ± 35.1		39.2 ± 18.7
				38.0 ± 18.3	
	Below elbow	13.1 ± 7.9	48.2 ± 29.8		40.2 ± 19.0
				36.6 ± 19.3	
	Above elbow	18.0 ± 10.6	40.7 ± 27.2		42.3 ± 20.8
				42.5 ± 22.1	
	Axilla	21.3 ± 14.0	37.3 ± 23.6		43.7 ± 18.9
10 peroneal nerves	Ankle	5.6 ± 1.3	52.8 ± 10.6		47.2 ± 6.9
				40.7 ± 15.2	
	Knee	15.0 ± 4.8	50.8 ± 19.1		41.6 ± 6.8
12 tibial nerves	Ankle	5.4 ± 1.4	62.8 ± 21.3		42.9 ± 14.2
				40.3 ± 14.9	
	Knee	16.2 ± 6.3	52.5 ± 15.3		43.9 ± 12.3

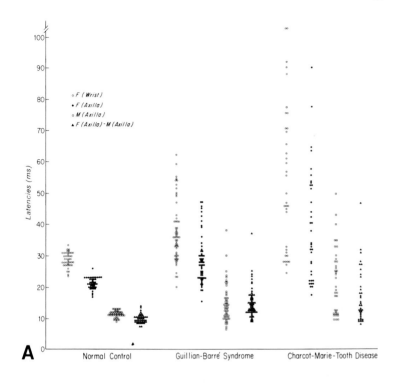

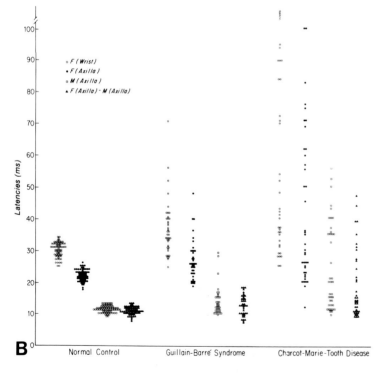

Figure 17–10. Latencies of the F wave and M response for median (**A**), ulnar (**B**), peroneal (**C**), and tibial nerves (**D**) in control, the Guillain-Barré syndrome, and Charcot-Marie-Tooth disease. The histogram includes only those nerves whose stimulation elicited both the M response and the F wave at sites of stimulation indicated in the key. The difference in latency between the F wave and the M response (*triangles*) equals the central latency required for passage of the impulses to and from the spinal cord. (From Kimura,[58] with permission.)

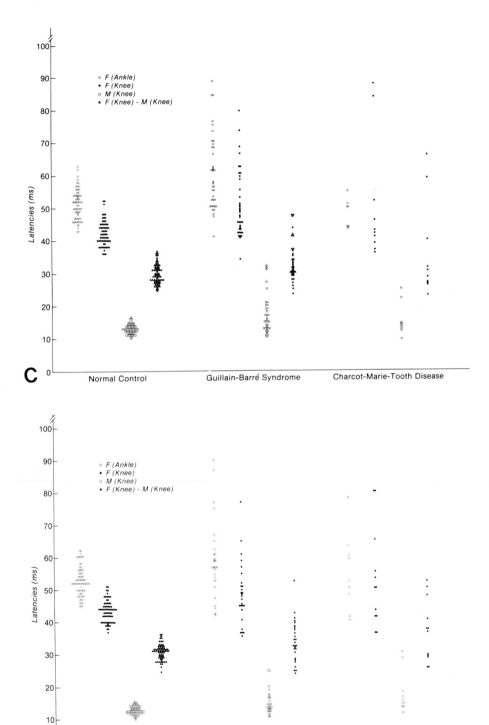

Figure 17–10. See legend on facing page.

Table 17–4 GUILLAIN-BARRÉ SYNDROME (MEAN ± SD)

Number of Nerves Tested	Sites of Stimulation	M Latency (ms)	F Latency (ms)	MNCV Between Two Stimulus Sites (m/s)	FWCV from Cord to Stimulus Site (m/s)
58 median nerves	Wrist	5.8 ± 3.1	38.1 ± 12.7		48.6 ± 11.1
				48.2 ± 12.1	
	Elbow	11.2 ± 4.8	32.6 ± 9.9		49.1 ± 11.4
				55.5 ± 14.1	
	Axilla	14.5 ± 5.7	29.4 ± 9.5		47.5 ± 14.5
40 ulnar nerves	Wrist	4.0 ± 2.0	36.8 ± 8.6		48.1 ± 9.7
				52.2 ± 10.7	
	Below elbow	8.3 ± 2.5	32.1 ± 7.1		47.4 ± 9.6
				47.7 ± 12.0	
	Above elbow	11.2 ± 3.5	29.7 ± 8.7		47.4 ± 10.7
				56.8 ± 14.9	
	Axilla	13.7 ± 4.8	27.2 ± 6.2		48.0 ± 12.3
39 peroneal nerves	Ankle	7.6 ± 4.8	59.9 ± 11.5		42.5 ± 8.7
				43.0 ± 8.2	
	Knee	16.9 ± 5.8	50.6 ± 10.3		43.9 ± 11.8
29 tibial nerves	Ankle	5.6 ± 2.3	56.4 ± 10.6		42.7 ± 8.8
				43.3 ± 9.0	
	Knee	14.6 ± 3.8	47.9 ± 9.4		43.8 ± 9.9

sibly radicular, portion of the nerve, and the most distal or terminal segment and relatively spares the main nerve trunk in early stages.[3,56,61,65] The MNCV may be normal in 15 to 20 percent of cases tested within the first few days of onset.[20,48,68] Some of these patients may have axonal neuropathies, but others probably have the lesion too proximal for detection with the use of ordinary techniques. In these cases, the F wave is typically absent initially during acute stages of illness. The return of the previously absent F wave indicates recovery of conduction across the proximal segment. The considerably increased F-wave latency, however, suggests demyelination of the involved segment (see Fig. 17–10).

Many patients have a normal F ratio, which indicates an equal slowing of conduction above and below the stimulus site at the elbow and knee. This does not necessarily mean uniform abnormalities along the entire length of the peripheral nerve. In our series, slowing of FWCV in the cord-to-axilla occurred more frequently than slowing in the elbow-to-wrist segment for both the median and ulnar nerves. In calculating the F ratio, a marked increase in terminal latency

compensated for the prominent proximal slowing.

Diabetic, Uremic, and Other Neuropathies

Clinical observations in diabetics have repeatedly shown that neuropathic symptoms usually appear in the distal extremities. These findings, however, do not necessarily indicate a distal pathologic process, because probability models can reproduce the same sensory deficit on the basis of randomly distributed axonal dysfunction.[117] In diabetic neuropathy, F-wave latency is increased and FWCV is decreased over both proximal and distal segments.[9,63] The average value and distribution of the F ratio indicate distally prominent conduction abnormalities despite slowing along the entire length of the nerve (Fig. 17–11). In contrast, patients with proximal amyotrophy may have an increased F ratio in the lower extremities.[7]

Patients undergoing hemodialysis for chronic renal failure have increased F-wave latency and a latency difference between the minimum and maximum

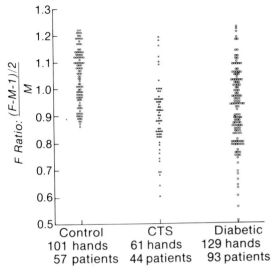

Figure 17–11. F ratio of the median nerve in the control group, carpal tunnel syndrome (*CTS*), and diabetic polyneuropathy. Statistical analysis showed significantly (*P*<0.01) reduced ratios in both disease groups, indicating disproportionate slowing of motor conduction distally. (From Kimura,[57] with permission.)

Plexopathy and Radiculopathy

A number of reports have suggested clinical value in assessing patients with root injuries.[22,23,31,33,49,81,112] The F wave usually remains normal in latency in mild cases of radiculopathy, especially if the lesion primarily affects the sensory fibers. The thoracic outlet syndrome with predominantly vascular symptoms rarely affects the F wave,[55,106,107] although the F-wave latency increases in the classical type with neuronal involvement.[13,42,118,119] F-wave latency may remain normal in clinically established cases of brachial or lumbosacral plexopathy. F-wave study has been found useful in some children with brachial plexus injury at birth.[66]

Thus, normal studies do not preclude the presence of radicular or plexus lesions. In general, the F-wave determination seems less helpful than might have been expected on theoretic grounds in early diagnosis of these conditions. An unequivocal delay of the F wave in conjunction with normal motor conduction distally, however, is a sign of a proximal lesion (Fig. 17–12). Right- versus left-sided comparison may provide a reliable means of assessing unilateral radicular or plexus lesions.[55] Further, the F wave is elicited less frequently on the affected side, compared with the normal side,

values.[75] In some of these cases an increased F ratio implies predominant affection of the proximal nerve segment[8]; but in others, slowing of nerve conduction involves both segments to the same extent.[24]

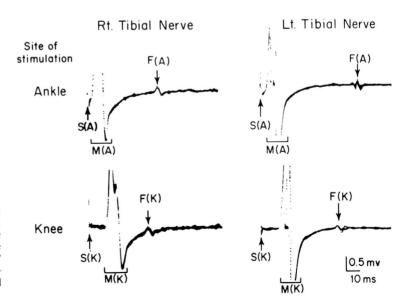

Figure 17–12. A patient with a sacral plexus lesion on the left. Stimulation of the tibial nerve at the ankle and knee elicited a normal M response (*brackets*) and a delayed F wave (*small arrows*) of the abductor hallucis on the affected side.

when the proximal lesion induces partial conduction block.

Other Disorders

A reduced F ratio of the median nerve in the carpal tunnel syndrome rivals that in diabetic neuropathy[63] (see Fig. 17–11). F waves also show abnormalities in compression neuropathies of the ulnar nerve. Differences between minimum and maximum F-wave latencies may provide a sensitive indicator for early detection of this syndrome.[106,107]

Some patients with cervical syringomyelia may have increased F-wave latencies of the median or ulnar nerve with normal peripheral conduction velocities.[89,95]

REFERENCES

1. Ackil, AA, Shahani, BR, and Young, RR: Sural nerve conduction studies and late responses in children undergoing hemodialysis. Arch Phys Med Rehabil 62:487–491, 1981.
2. Argyropoulos, CJ, Panayiotopoulos, CP, and Scarpalezos, S: F- and M-wave conduction velocity in amyotrophic lateral sclerosis. Muscle Nerve 1:479–485, 1978.
3. Asbury, AK, Arnason, BG, and Adams, RD: The inflammatory lesion in idiopathic polyneuritis: Its role in pathogenesis. Medicine 48:173–215, 1969.
4. Baba, M, Narita, S, and Matsunaga, M: F-wave conduction velocity from the spinal cord to the axilla without using collision technique: A simplified method. Electromyogr Clin Neurophysiol 20:19–25, 1980.
5. Beydoun, SR, and Engel, WK: F-wave amplitude is rapidly increased in patients receiving intravenous or subcutaneous thyrotropin releasing hormone (TRH) (abstr). Neurology 35 (Suppl 1):128, 1985.
6. Brock, LG, Coombs, JS, and Eccles, JC: Intracellular recording from antidromically activated motoneurones. J Physiol (Lond) 122:429–461, 1953.
7. Chokroverty, S: Proximal nerve dysfunction in diabetic proximal amyotrophy: Electrophysiology and electron microscopy. Arch Neurol 39:403–407, 1982.
8. Chokroverty, S: Proximal vs distal slowing of nerve conduction in chronic renal failure treated by long-term hemodialysis. Arch Neurol 39:53–54, 1982.
9. Conrad, B. Aschoff, JC, and Fischler, M: Der diagnostische Wert der F-Wellen-Latenz. J Neurol 210:151–159, 1975.
10. Daube, JR: F-wave and H-reflex measurements. American Academy of Neurology, Special Course #16, Clinical Electromyography, Chicago, 1979.
11. Dawson, GD, and Merton, PA: "Recurrent" discharges from motoneurones. XXth International Physiological Congress, Brussels, 1956.
12. Desmedt, JE, and Noel, P: Average cerebral evoked potentials in the evaluation of lesions of the sensory nerves and of the central somatosensory pathway. In Desmedt, JE (ed): New Developments in Electromyography and Clinical Neurophysiology, Vol 2. Karger, Basel, 1973, pp 352–371.
13. Dorfman, LJ: F-wave latency in the cervical-rib-and-band syndrome. Muscle Nerve 2:158–159, 1979.
14. Dyck, PJ: Inherited neuronal degeneration and atrophy affecting peripheral motor, sensory, and autonomic neurons. In Dyck, PJ, Thomas, PK, and Lambert, EH (eds): Peripheral Neuropathy, Vol 2. WB Saunders, Philadelphia, 1975.
15. Dyck, PJ, and Lambert, EH: Lower motor and primary sensory neuron diseases with peroneal muscular atrophy. II. Neurologic, genetic, and electrophysiologic findings in various neuronal degenerations. Arch Neurol 18:619–625, 1968.
16. Eccles, JC: The central action of antidromic impulses in motor nerve fibres. Pflugers Arch 260:385–415, 1955.
17. Eccles, JC: The inhibitory control of spinal reflex action. Electroencephalogr Clin Neurophysiol (Suppl 25):20–34, 1967.
18. Eccles, JC, Eccles, RM, Iggo, A, and Ito, M: Distribution of recurrent inhibition among motoneurones. J Physiol (Lond) 159:479–499, 1961.
19. Eisen, A, Hoirch, M, White, J, and Calne, D: Sensory group Ia proximal conduction velocity. Muscle Nerve 7:636–641, 1984.
20. Eisen, A, and Humphreys, P: The Guillain-Barré syndrome: A clinical and electrodiagnostic study of 25 cases. Arch Neurol 30:438–443, 1974.
21. Eisen, A, and Odusote, K: Amplitude of the F wave: A potential means of documenting spasticity. Neurology 29:1306–1309, 1979.
22. Eisen, A, Schomer, D, and Melmed, C: The application of F-wave measurements in the differentiation of proximal and distal upper limb entrapments. Neurology 27:662–668, 1977.
23. Eisen, A, Schomer, D, and Melmed, C: An electrophysiological method for examining lumbosacral root compression. Can J Sci Neurol 4:117–123, 1977.
24. Fierro, B, Modica, A, D'Arpa, A, Santangelo, R, and Raimondo, D: F-wave study in patients with chronic renal failure on regular haemodialysis. J Neurol Sci 74:271–277, 1986.
25. Fisher, MA: Electrophysiological appraisal of relative segmental motoneurone pool excitability in flexor and extensor muscles.

J Neurol Neurosurg Psychiatry 41:624–629, 1978.

26. Fisher, MA: F-waves: Comments on the central control of recurrent discharges (Corres). Muscle Nerve 2:406, 1979.

27. Fisher, MA: F-response latency-duration correlations: An argument for the orderly antidromic activation of motoneurons. Muscle Nerve 3:437–438, 1980.

28. Fisher, MA: F response latency determination. Muscle Nerve 5:730–734, 1982.

29. Fisher, MA: F response analysis of motor disorders of central origin. J Neurol Sci 62:13–22, 1983.

30. Fisher, MA: Cross correlation analysis of F response variability and its physiological significance. Electromyogr Clin Neurophysiol 23:329–339, 1983.

31. Fisher, MA, Kaur, D, and Houchins, J: Electrodiagnostic examination, back pain and entrapment of posterior rami. Electromyogr Clin Neurophysiol 25:183–189, 1985.

32. Fisher, MA, Shahani, BT, and Young, RR: Assessing segmental excitability after acute rostral lesions. I. The F response. Neurology 28:1265–1271, 1978.

33. Fisher, MA, Shivde, AJ, Teixera, C, and Grainer, LS: Clinical and electrophysiological appraisal of the significance of radicular injury in back pain. J Neurol Neurosurg Psychiatry 41:303–306, 1978.

34. Fox, JE, and Hitchcock, ER: Changes in F wave size during dentatomy. J Neurol Neurosurg Psychiatry 45:1165–1167, 1982.

35. Fox, JE, and Hitchcock, ER: F-wave size as a monitor of motor neuron excitability: Effect of deafferentation. J Neurol Neurosurg Psychiatry 50:453–459, 1987.

36. Fullerton, PM, and Gilliatt, RW: Intermediate latency responses to nerve stimulation (abstr). Electroencephalogr Clin Neurophysiol 17:94, 1964.

37. Fullerton, PM, and Gilliatt, RW: Axon reflexes in human motor nerve fibres. J Neurol Neurosurg Psychiatry 28:1–11, 1965.

38. Garcia-Mullin, R, and Mayer, RF: H reflexes in acute and chronic hemiplegia. Brain 95:559–572, 1972.

39. Gassel, MM: Monosynaptic reflexes (H-reflex) and motoneurone excitability in man. Dev Med Child Neurol 11:193–197, 1969.

40. Gassel, MM, Marchiafava, PL, and Pompeiano, O: Modulation of the recurrent discharge of alpha motoneurons in decerebrate and spinal cats. Arch Ital Biol 103:1–24, 1965.

41. Gassel, MM, and Wiesendanger, M: Recurrent and reflex discharges in plantar muscles of the cat. Acta Physiol Scand 65:138–142, 1965.

42. Gilliatt, RW, Willison, DM, Dietz, V, and Williams, IR: Peripheral nerve conduction in patients with a cervical rib and band. Ann Neurol 4:124–129, 1978.

43. Granit, R, Pascoe, JE, and Steg, G: The behavior of tonic (alpha) and (beta) motoneurones during stimulation of recurrent collaterals. J Physiol (Lond) 138:381–400, 1957.

44. Hagbarth, KE: Spinal withdrawal reflexes in the human lower limbs. J Neurol Neurosurg Psychiatry 23:222–227, 1960.

45. Hagbarth, KE: Post-tetanic potentiation of myotatic reflexes in man. J Neurol Neurosurg Psychiatry 25:1–10, 1962.

46. Henneman, E, Somjen, G, and Carpenter, DO: Excitability and inhibitibility of motoneurons of different sizes. J Neurophysiol 28:599–620, 1965.

47. Hopf, HC: Leitgeschwindigkeit motorischer Nerven bei der multiplen Sklerose und unter dem Eiflub hoher Cortisonmedikation. Dtsch Z Nervenheilk 187:522–526, 1978.

48. Humphrey, JG: Motor nerve conduction studies in the Landry-Guillain-Barré syndrome (acute ascending polyneuropathy). Electroencephalogr Clin Neurophysiol 17:96, 1964.

49. Kalyon, TA, Bilgic, F, and Ertem, O: The diagnostic value of late responses in radiculopathies due to disc herniation. Electromyogr Clin Neurophysiol 23:183–186, 1983.

50. Kernell, D: Input resistance, electrical excitability, and size of ventral horn cells in cat spinal cord. Science 152:1637–1640, 1966.

51. Kimura, J: F-wave velocity in the central segment of the median and ulnar nerves: A study in normal subjects and in patients with Charcot-Marie-Tooth disease. Neurology 24:539–546, 1974.

52. Kimura, J: Collision technique: Physiologic block of nerve impulses in studies of motor nerve conduction velocity. Neurology 26:680–682, 1976.

53. Kimura, J: A method for estimating the refractory period of motor fibers in the human peripheral nerve. J Neurol Sci 28:485–490, 1976.

54. Kimura, J: Electrical activity in voluntarily contracting muscle. Arch Neurol 34:85–88, 1977.

55. Kimura, J: Clinical value and limitations of F-wave determination: A comment (corres). Muscle Nerve 1:250–252, 1978.

56. Kimura, J: Proximal versus distal slowing of motor nerve conduction velocity in the Guillain-Barré syndrome. Ann Neurol 3:344–350, 1978.

57. Kimura, J: The carpal tunnel syndrome: Localization of conduction abnormalities within the distal segment of the median nerve. Brain 102:619–635, 1979.

58. Kimura, J: F-wave determination in nerve conduction studies. In Desmedt, JE (ed): Motor Control Mechanisms in Health and Disease. Raven Press, New York, 1983, pp 961–975.

59. Kimura, J: Electromyography and Nerve Stimulation Techniques: Clinical Applications. Egakushoin, Tokyo, 1989.

60. Kimura, J, Bosch, P, and Lindsay, GM: F-wave conduction velocity in the central segment of the peroneal and tibial nerves. Arch Phys Med Rehabil 56:492–497, 1975.

61. Kimura, J, and Butzer, JF: F-wave conduction velocity in Guillain-Barré syndrome: Assessment of nerve segment between axilla and spinal cord. Arch Neurol 32:524–529, 1975.

62. Kimura, J, Yamada, T, and Rodnitzky, RL: Refractory period of human motor nerve fibres. J Neurol Neurosurg Psychiatry 41:784–790, 1978.

63. Kimura, J, Yamada, T, and Stevland, NP: Distal slowing of motor nerve conduction velocity in diabetic polyneuropathy. J Neurol Sci 42:291–302, 1979.

64. Kimura, J, Yanagisawa, H, Yamada, T, Mitsudome, A, Sasaki, H, and Kimura, A: Is the F wave elicited in a select group of motoneurons? Muscle Nerve 7:392–399, 1984.

65. King, D, and Ashby, P: Conduction velocity in the proximal segments of a motor nerve in the Guillain-Barré syndrome. J Neurol Neurosurg Psychiatry 39:538–544, 1976.

66. Kwast, O: F wave study in children with birth brachial plexus paralysis. Electromyogr Clin Neurophysiol 24:457–467, 1984.

67. Kwast, O, Krajewska, G, and Kozlowski, K: Analysis of F wave parameters in median and ulnar nerves in healthy infants and children: Age related changes. Electromyogr Clin Neurophysiol 24:439–456, 1984.

68. Lambert, EH, and Mulder, DW: Nerve conduction in the Guillain-Barré syndrome. Electroencephalogr Clin Neurophysiol 17:86, 1964.

69. Lachman, T, Shahani, BT, Young, RR: Late responses as aids to diagnosis in peripheral neuropathy. J Neurol Neurosurg Psychiatry 43:156–162, 1980.

70. Lefebvre d'Amour, M, Shahani, BT, Young, RR, and Bird, KT: The importance of studying sural nerve conduction and late responses in the evaluation of alcoholic subjects. Neurology 29:1600–1604, 1979.

71. Liberson, WT, Gratzer, M, Zalis, A, and Grabinski, B: Comparison of conduction velocities of motor and sensory fibers determined by different methods. Arch Phys Med Rehabil 47:17–22, 1966.

72. Livingstone, EF, and DeLisa, JA: Electrodiagnostic values through the thoracic outlet using C8 root needle studies, F-waves, and cervical somatosensory evoked potentials. Arch Phys Med Rehabil 65:726–730, 1984.

73. Lloyd, DPC: The interaction of antidromic and orthodromic volleys in a segmental spinal motor nucleus. J Neurophysiol 6:143–151, 1943.

74. Magladery, JW, and McDougal, DB, Jr: Electrophysiological studies of nerve and reflex activity in normal man. 1. Identification of certain reflexes in the electromyogram and the conduction velocity of peripheral nerve fibres. Bull Johns Hopkins Hosp 86:265–290, 1950.

75. Marra, TR: F Wave measurements: A comparison of various recording techniques in health and peripheral nerve disease. Electromyogr Clin Neurophysiol 27:33–37, 1987.

76. Mastaglia, FL, and Carroll, WM: The effects of conditioning stimuli on the F-response. J Neurol Neurosurg Psychiatry 48:182–184, 1985.

77. Mayer, RF, and Feldman, RG: Observations on the nature of the F wave in man. Neurology 17:147–156, 1967.

78. McLeod, JG, and Wray, SH: An experimental study of the F wave in the baboon. J Neurol Neurosurg Psychiatry 29:196–200, 1966.

79. Miglietta, OE: The F response after transverse myelotomy. In Desmedt, JE (ed): New Developments in Electromyography and Clinical Neurophysiology, Vol 3. Karger, Basel, 1973, pp 323–327.

80. Muller, D: Die Bestimmung der F-Wellengeschwindigkeit am N. Ulnaris Gesunder. Psychiatr Neurol Med Psychol 27:619–623, 1975.

81. Ongerboer de Visser, BW, Van der Sande, JJ, and Kemp, B: Ulnar F-wave conduction velocity in epidural metastatic root lesions. Ann Neurol 11:142–146, 1982.

82. Panayiotopoulos, CP: F-wave conduction velocity in the deep peroneal nerve: Charcot-Marie-Tooth disease and dystrophia myotonica. Muscle Nerve 1:37–44, 1978.

83. Panayiotopoulos, CP: F chronodispersion: A new electrophysiologic method. Muscle Nerve 2:68–72, 1979.

84. Panayiotopoulos, CP, and Lagos, G: Tibial nerve H-reflex and F-wave studies in patients with uremic neuropathy. Muscle Nerve 3:423–426, 1980.

85. Panayiotopoulos, CP, and Scarpalezos, S: F-wave studies on the deep peroneal nerve. Part 2. 1. Chronic renal failure. 2. Limb-girdle muscular dystrophy. J Neurol Sci 31:331–341, 1977.

86. Panayiotopoulos, CP, Scarpalezos, S, and Nastas, PE: F-wave studies on the deep peroneal nerve. Part 1. Control subjects. J Neurol Sci 31:319–329, 1977.

87. Panayiotopoulos, CP, Scarpalezos, S, and Nastas, PE: Sensory (1a) and F-wave conduction velocity in the proximal segment of the tibial nerve. Muscle Nerve 1:181–189, 1978.

88. Peioglou-Harmoussi, S, Fawcett, PRW, Howel, D, and Barwick, DD: F-responses: A study of frequency, shape and amplitude characteristics in healthy control subjects. J Neurol Neurosurg Psychiatry 48:1159–1164, 1985.

89. Peioglou-Harmoussi, S, Fawcett, PRW, Howel, D, and Barwick, DD: F-responses in syringomyelia. J Neurol Sci 75:293–304, 1986.

90. Peioglou-Harmoussi, S, Fawcett, PRW, Howel, D, and Barwick, DD: F-response frequency in motor neuron disease and cervical spondylosis. J Neurol Neurosurg Psychiatry 50:593–599, 1987.

91. Peioglou-Harmoussi, S, Howel, D, Fawcett, PRW, and Barwick, DD: F-response behaviour in a control population. J Neurol Neurosurg Psychiatry 48:1152–1158, 1985.

92. Pinelli, P: Physical, anatomical and physiological factors in the latency measurement of the M response (abstr). Electroencephalogr Clin Neurophysiol 17:86, 1964.

93. Renshaw, B: Influence of discharge of motoneurons upon excitation of neighboring motorneurons. J Neurophysiol 4:167–183, 1941.

94. Risk, WS, Bosch, EP, Kimura, J, Cancillz, PA, Fischbeck, KH, and Layzer, RB: Chronic tetanus: Clinical report and histochemistry of muscle. Muscle Nerve 4:363–366, 1981.

95. Rossier, AB, Foo, D, Shillito, J, and Dyro, FM: Posttraumatic cervical syringomyelia incidence, clinical presentation, electrophysiological studies, syrinx protein and results of conservative and operative treatment. Brain 108:439–461, 1985.

96. Roth, G: Intravenous regeneration of lower motor neuron. 1. Study of 1153 motor axon reflexes. Electromyogr Clin Neurophysiol 18:225–288, 1978.

97. Roth, G: Intravenous regeneration: The study of motor axon reflexes. J Neurol Sci 41:139–148, 1979.

98. Roth, G: Reinnervation dans la paralysie plexulaire brachiale obstetricale. J Neurol Sci 58:103–115, 1983.

99. Roth, G, and Egloff-Baer, S: Motor axon loop: An electroneurographic response. Muscle Nerve 7:294–297, 1984.

100. Satoyoshi, E, Doi, Y, and Kinoshita, M: Pseudomyotonia in cervical root lesions with myelopathy. A sign of the misdirection of regenerating nerve. Arch Neurol 27:307–313, 1972.

101. Sawhney, BB, and Kayan, A: A study of axon reflexes in some neurogenic disorders. Electromyography 10:297–305, 1970.

102. Sawhney, BB, and Kayan, A: A study of the F wave from the facial muscles (abstr). Electroencephalogr Clin Neurophysiol 30:261, 1971.

103. Schiller, HH, and Stalberg, E: F responses studied with single fibre EMG in normal subjects and spastic patients. J Neurol Neurosurg Psychiatry 41:45–53, 1978.

104. Serra, G, Aiello, I, De Grandis, D, Tognoli, V, and Carreras, M: Muscle-nerve ephaptic excitation in some repetitive after-discharges. Electroencephalogr Clin Neurophysiol 57:416–422, 1984.

105. Shahani, BT, Potts, F, and Domingue, J: F response studies in peripheral neuropathies (abstr). Neurology 30:409–410, 1980.

106. Shahani, BT, Potts, F, Juguilon, A, and Young, RR: Maximal-minimal motor nerve conduction and F response studies in normal subjects and patients with ulnar compression neuropathies (abstr). Muscle Nerve 3:182, 1980.

107. Shahani, BT, Potts, F, Juguilon, A, and Young, RR: Electrophysiological studies in "thoracic outlet syndrome" (abstr). Muscle Nerve 3:182–183, 1980.

108. Thomas, PK, and Calne, DB: Motor nerve conduction velocity in peroneal muscular atrophy: Evidence for genetic heterogeneity. J Neurol Neurosurg Psychiatry 37:68–75, 1974.

109. Thomas, PK, Calne, DB, and Stewart, G: Hereditary motor and sensory polyneuropathy (peroneal muscular atrophy). Ann Hum Genet 38:111–153, 1974.

110. Thorne, J: Central responses to electrical activation of the peripheral nerves supplying the intrinsic hand muscles. J Neurol Neurosurg Psychiatry 28:482–495, 1965.

111. Tomasulo, RA: Aberrant conduction in human peripheral nerve: Ephaptic transmission? Neurology 32:712–719, 1982.

112. Tonzola, RF, Ackil, AA, Shahani, BT, and Young, RR: Usefulness of electrophysiological studies in the diagnosis of lumbosacral root disease. Ann Neurol 9:305–308, 1981.

113. Trontelj, JV: A study of the F response by single fibre electromyography. In Desmedt, JE (ed): New Developments in Electromyography and Clinical Neurophysiology, Vol 3. Karger, Basel, 1973, pp 318–322.

114. Trontelj, JV, and Trontelj, MJ: F-responses of human facial muscles: A single motoneurone study. J Neurol Sci 20:211–222, 1973.

115. Tuck, RR, Antony, JH, and McLeod, JG: F wave in experimental allergic neuritis. J Neurol Sci 56:173–184, 1982.

116. Upton, ARM, McComas, AJ, and Sica, REP: Potentiation of 'late' responses evoked in muscles during effort. J Neurol Neurosurg Psychiatry 34:699–711, 1971.

117. Waxman, SG, Brill, MH, Geschwind, N, Sabin, TD, and Lettvin, JY: Probability of conduction deficit as related to fiber length in random-distribution models of peripheral neuropathies. J Neurol Sci 29:39–53, 1976.

118. Weber, RJ, and Piero, DL: F-wave evaluation of thoracic outlet syndrome: A multiple regression derived F wave latency predicting technique. Arch Phys Med Rehabil 59:464–469, 1978.

119. Wulff, CH, and Gilliatt, RW: F waves in patients with hand wasting caused by a cervical rib and band. Muscle Nerve 2:452–457, 1979.

120. Yates, SK, and Brown, WF: Characteristics of the F response: A single motor unit study. J Neurol Neurosurg Psychiatry 42:161–170, 1979.

121. Young, RR, and Shahani, BT: Clinical value and limitations of F-wave determination (Corres). Muscle Nerve 1:248–250, 1978.

Chapter 18

H, T, MASSETER, AND OTHER REFLEXES

1 INTRODUCTION

Traditional nerve stimulation techniques used in an electromyography laboratory primarily assess the distal segments of the peripheral nerves. Methods of testing the proximal nerve segments or the central nervous system include, in addition to the blink reflex (see Chapter 16) and F wave (see Chapter 17), the H reflex, T reflex, tonic vibration reflex, and silent period. The reflex studies reveal conduction characteristics along the entire course of the sensory and motor axons as well as the excitability of the neuronal pool.

Extensive studies have proven the practical value of the H reflex in certain neurologic disorders. Clinical applications of the other techniques mentioned here await further clarification, even though they have contributed substantially as a means of quantitating physiologic studies of motor and sensory systems. This chapter will review the basic physiology and diagnostic usefulness of the newer techniques in evaluating the regions of the nervous system not accessible by the conventional methods.

2 H AND T REFLEXES

Neurologic examination exploits the muscle stretch reflex to measure motor neuron excitability in spasticity and other related conditions. Clinical observation, however, falls short in objectively evaluating the briskness, velocity, or symmetry of these responses. Electrophysiologic recordings offer these advantages by quantitating the response after a mechanical tap to the Achilles tendon or electrical stimulation of the tibial nerve. The electrically elicited spinal monosynaptic reflex, called the H reflex after Hoffmann, bypasses the muscle spindles, though otherwise identical in many respects to the stretch reflex induced by a mechanical tap to the tendon (T reflex).[77,106,107,131] Comparison of the H and T reflexes, therefore, provides an indirect measure of spindle sensitivity controlled by the gamma motor system.[11]

H Reflex Versus F Wave

Stimulation of most nerves in the limb, including the ulnar nerve, elicits an H reflex in newborn infants and during the first year of life.[75,161] In adults, however, the reflex can be evoked only in the calf muscles and flexor carpi radialis at rest.[28,86,129,147] Mild voluntary contraction may prime the motor neuron pool sufficiently to allow reflexive activation of other antigravity muscles and, to a lesser extent, physiologic flexors of both upper and lower extremities.[48,62,152] The limited distribution of the H reflex stands in contrast to an unrestricted elicitation of the F wave in practically any distal limb muscle.

The effect of increasing stimulus intensity also distinguishes the two (Fig. 18–1). H-reflex amplitude increases initially as the stimulus changes from subthreshold to submaximal. With a higher shock intensity, the H reflex diminishes progressively and is eventually replaced by the F wave when the stimulus elicits a maximal M response. An optimal elicitation of the H reflex requires maximal

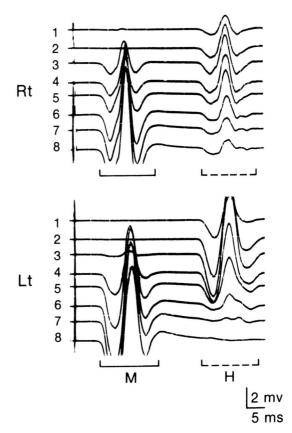

Figure 18–1. H reflex recorded from the soleus after stimulation of the tibial nerve at the knee. Shock intensity was gradually increased from the subthreshold level (1) to supramaximal stimulation (8). Note the initial increase and subsequent decrease in amplitude of the reflex potential with successive stimuli of progressively higher intensity. The H reflex normally disappears with shocks of supramaximal intensity, which elicit a maximal M response and F wave.

stimulation of the group IA afferent fibers without concomitant activation of motor fibers, although in practice few stimuli accomplish such selectivity. If the stimulus activates any motor axons eliciting an M response, the antidromic impulse in those axons can generate recurrent discharges. Thus, submaximal intensity does not guarantee the reflex origin of the late response.

Possible mechanisms for the extinction of the H reflex with increasing stimulus intensity include (1) collision of the reflex impulse with antidromic activity in the alpha motor axon[77,105]; (2) refractoriness of the axon hillock after the passage of the antidromic impulse[59]; and (3) Renshaw inhibition mediated by motor neuron axon collaterals via internuncial cells to the same and neighboring alpha motor neurons.[35,137,166,170] Gamma hydroxybutyrate, known to promote cataplexy, markedly suppresses the H reflex, presumably by presynaptic inhibition, but does not affect the F wave.[109]

Consecutive F waves characteristically vary in latency and waveform, because they represent recurrent discharges of different groups of motor neurons with different conduction characteristics. In contrast, H reflexes remain constant in response to repetitive stimuli, because each trial activates the same motor neuron pool (Fig. 18–2). When one is recording from single muscle fibers, however, the latency variability of consecutive H reflexes far exceeds that of the F waves. As mentioned earlier, this reflects a greater variability in synaptic transmission at a motor neuron, compared

with a relatively constant turnaround time for a recurrent discharge.[145,166] In one study, the latency of successive H reflexes recorded from single muscle fibers of the human triceps surae varied up to 2.5 ms.[166]

Recording Procedures

The H reflex recorded with the patient supine or prone suffices in clinical determination of reflex latencies (Fig. 18–3). For an accurate analysis of the amplitude or force of the reflex response, the subject sits upright in a modified dental chair. With this arrangement, a potentiometer monitors the movement of the feet and a force transducer measures the torque.[85] A soft cushion supports the knee, semi-

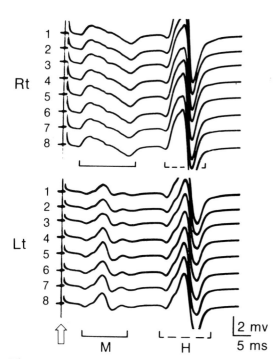

Figure 18–2. H reflex from the soleus after stimulation (*open arrow*) of the tibial nerve at the knee. Consecutive trials show consistency of the response on each side.

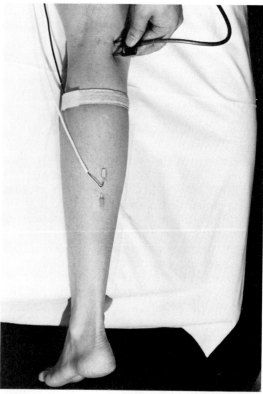

Figure 18–3. Recording of the H reflex from the soleus. A pair of surface electrodes is placed along the midline of the calf muscle with the active (G_1) electrode 2 cm distal to the insertion of the gastrocnemius and the inactive electrode (G_2) 3 cm further distally. The tibial nerve is stimulated at the popliteal fossa with the cathode placed 2 cm proximal to the anode.

flexed at about 120 degrees. Maintaining the angle of the ankle joint constant at about 110 degrees helps establish optimal relaxation of the calf muscle. The conventional recording uses the active electrode (G_1) placed 2 cm distal to the insertion of the gastrocnemius on the Achilles tendon and the reference electrode (G_2), 3 cm further distally. An alternative derivation consists of G_1 placed over the soleus just medial to the tibia, half the distance from the tibial tubercle to the medial malleolus, and G_2 over the Achilles tendon medial and proximal to the medial malleolus.[19] The H reflex appears as a triphasic potential with initial positivity with electrodes placed over the gastrocnemius and as a diphasic potential with initial negativity when recorded from the soleus. A second pair of electrodes, placed over the belly of the anterior tibialis muscle 3 cm apart, along the longitudinal axis and near midline, monitors the antagonistic muscle. A ground electrode is located between the stimulating and recording electrodes.

The effective modes of stimuli include (1) an electrical shock applied to the tibial nerve at the popliteal fossa (H reflex), (2) a tap of the Achilles tendon with a reflex hammer fitted to trigger the oscilloscope (T reflex), and (3) a mechanical stretch by quick displacement of the ankle. Standardization of stimulus conditions ensures reproducible results. Optimal intensities of mechanical or electrical stimuli are determined individually for obtaining the maximal responses. Maintaining the skin of the leg at more than 34°C guarantees the depth temperature of 35°C to 37°C along the nerve.

In studies under isometric conditions, one measures the force of induced muscle contraction (myogram) with a transducer placed against the foot plate. In isotonic conditions, one determines the degree and rate of foot displacement (motogram) using a potentiometer mounted on the axis of the foot plate. The common measurements of muscle action potentials recorded reflexively from the soleus include the onset latencies of the H and T reflexes determined to the initial deflection, either negative or positive, H_{max}/M_{max} and T_{max}/M_{max}, where H_{max}, M_{max}, and T_{max} repre-

sent the maximal amplitude of the H reflex, M response, and T reflex. In assessing these indices, the subject must control the degree of muscle contraction because the variability of baseline tension alters the H reflex magnitude.[173]

Excitability and the Recovery Curve

When elicited with an optimal mechanical or electrical stimulus, the amplitude of the H and T reflexes provides a measure of excitability in the soleus motor neuron.[6,120,130,172] This measurement helps in quantitatively evaluating supraspinal and segmental inputs on the alpha motor neurons[23,24,120] in studying the effect of spasticity,[103,136,159] postural changes,[25] or preparatory anticipation.[45] Caloric stimulation of the labyrinth facilitates the H reflex bilaterally,[22,26] whereas sleep in general and the rapid eye movement period in particular depress the reflex.[75] The background fusimotor activity plays little or no role in eliciting Achilles tendon jerk during complete relaxation.[14,15]

The paired-shock technique reveals the time course of alteration in motor neuron excitability by means of conditioning and test stimuli.[91,131,179] Shocks of suprathreshold intensity exert two opposing effects on the excitability of the motor neuron pool: Those motor neurons that have discharged in response to the conditioning stimulus become less responsive to a subsequent stimulus because of the refractory period, the Renshaw effect, and other inhibitory mechanisms. On the other hand, the remaining motor neurons, activated subliminally by the conditioning stimulus, become more excitable in response to the test stimulus as the result of partial depolarization. The presence of these two competing factors complicates the interpretation of the result.[72] The use of a subthreshold conditioning stimulus circumvents such ambiguity. The excitability curve plotted by this method consists of an early facilitation lasting 25 ms and a period of predominant depression for the next 500 ms before the excitability approaches the control level (Fig. 18–4). Superimposed on this long-lasting suppression, interceding potentiation begins from 50 to 200 ms or sometimes up to 300 ms, peaking at 150 ms. Initial facilitation coincides with excitatory postsynaptic potential in subliminally activated alpha motor neurons.[156,157] Subsequent depression presumably reflects presynaptic inhibition or transmitter depletion. The intervening relative facilitation, seen bilaterally with unilateral conditioning,[142] may result from interaction of segmental or long loop reflexes.[49,50]

Selective cutaneous stimulation of the peroneal or tibial nerve is another way of assessing supranuclear control of the H reflex.[134] In normal subjects, it results in marked amplitude reduction of the test response at an interstimulus interval of about 100 ms.[49,50,83,84] This physiologic inhibition may not occur in the presence of Parkinsonian rigidity.[112] Conditioning cutaneous stimulation may even facili-

tate the H reflex in patients with corticospinal lesions. The paired-shock technique also reveals the effects of reciprocal inhibition[5,92,121,123,178] and reflex interactions.[69,104]

Clinical Applications

H-reflex latency of tibial or median nerve provides a measure of nerve conduction along the entire length of the afferent and efferent pathways.[38,61] It increases in patients with alcoholic,[177] uremic,[67] and various other polyneuropathies.[147] In patients with diabetes, this test rivals the conventional nerve conduction studies in detecting early neuropathic abnormalities[176] and a clear-cut proximal-to-distal gradient of conduction slowing.[163,164] The test also helps establish maturational changes in the proximal versus distal segment of the tibial nerve.[171] The use of H latency and distal

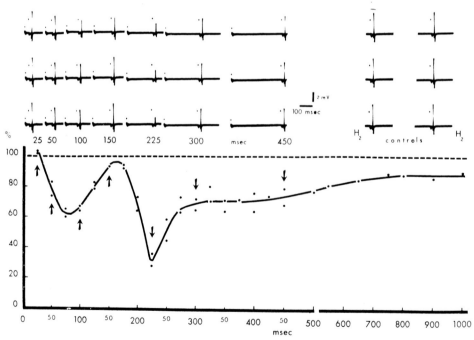

Figure 18–4. Conditioning of an H reflex by a subliminal H reflex stimulus. In the upper half are specimen records arranged in groups of three for each experimental situation. To the *right* are three H₂ control reflexes before and after the conditioning series. To the *left* are groups of three conditioned H₂ reflexes at testing intervals of 25, 50, 100, 150, 225, 300, and 450 ms, as indicated below. The S_1 stimulus was just below the threshold for evoking an H reflex, whereas the S_2 stimulus was just below the threshold for an M response. Below is plotting of the mean of the three H_2 reflexes at each testing interval (*abscissa*), the mean sizes being expressed as percentages of the mean H_2 reflex controls. (From Taborikova and Sax,[157] with permission.)

motor latency allows the calculation of a segmental conduction velocity along the reflex pathway.[163,164] In this computation, one divides the distance between the knee and T-11 by the latency difference between the H reflex and the M response. The result provides a mixed sensory and motor index or conduction velocity along the afferent and efferent fibers of the tibial nerve.[165]

Early studies revealed abnormalities of the T reflex in patients with lumbar and sacral root compression.[108] More recent work confirmed these results[30] and demonstrated clinical applications of the H reflex as a test for radiculopathy.[12,27,140,149] A delay or absence of the triceps surae reflex implicates the S-1 root, like a depression of the ankle stretch reflex in the neurologic examination.[3] In comparison, the H reflex recorded from the extensor digitorum longus after stimulation of the common peroneal nerve may show abnormalities in patients with L-5 radiculopathy.[27] In patients with cervical radiculopathy, abnormality of flexor carpi radialis reflex indicated lesions of the C-6 or C-7 root or both.[146]

Table 18–1 summarizes the normal values in our laboratory. In assessing a unilateral lesion, the latency difference between the two sides provides the most sensitive measure of the T or H reflex (Fig. 18–5).[12] Unilateral absence or a right-left latency difference greater than 1.5 ms supports the diagnosis of S-1 radiculopathy in the proper clinical context but does not by itself constitute sufficient evidence of a herniated disk or of a need for laminectomy.[56]

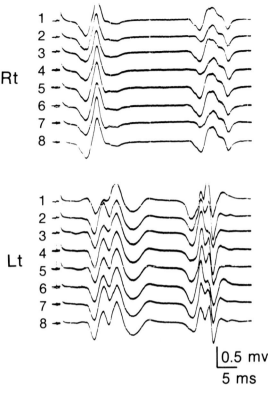

Figure 18–5. H reflex in a 77-year-old man with the cauda equina syndrome. The recording arrangement is the same as for Figure 18–2. The reflex was delayed by more than 2 ms on the right, compared with the left. The central latency as determined by the latency difference between the M response and the H reflex was also considerably greater on the right than on the left side.

3 THE MASSETER REFLEX

Sudden stretching of the muscle spindles from a sharp tap to the mandible activate the jaw reflex, or masseteric T re-

	Table 18–1 H REFLEX*		
Amplitude† (mV)	**Difference Between Right and Left (mV)**	**Latency‡ to Recording Site (ms)**	**Difference Between Right and Left (ms)**
2.4 ± 1.4	1.2 ± 1.2	29.5 ± 2.4 (35)§	0.6 ± 0.4 (1.4)§

*Mean ± standard deviation (SD) in the same 59 patients shown in Table 6–10.
†Amplitude of the evoked response measured from the baseline to the negative peak.
‡Latency measured to the onset of the evoked response.
§Upper limits of normal calculated as mean + 2 SD.

flex.[57,99] Electric stimulation of the masseter nerve elicits not only the direct motor responses[17] but also a masseteric H reflex.[18,53,55] This reflex, relayed via the mesencephalic nucleus of the trigeminal nerve, reflects conduction through the midbrain. The so-called motor root of the trigeminal nerve contains the sensory fibers of the muscle spindle that form the afferent arc of the masseter reflex and the motor axons to the extrafusal muscle fibers that form the efferent arc. The cell bodies of the proprioceptive spindle afferents lie in the mesencephalic trigeminal nucleus. The collateral branches from these cells make a monosynaptic connection with the motor neurons of the trigeminal nerve located in the pons. The physiology of the jaw reflex differs considerably from that of the spinal monosynaptic reflex. For example, muscle vibration that inhibits the soleus T and H reflexes potentiates the masseteric T and H reflexes.[54]

Methods and Normal Values

In eliciting the jaw reflex by a mechanical tap over the mandible, the closure of a microswitch attached to the percussion hammer triggers the oscilloscope sweep. The latency and amplitude vary with successive trials in the same subjects and among individuals. Thus, electrophysiologic evaluation depends on the side-to-side comparison of the reflex responses recorded simultaneously from the right and left masseter muscles, rather than the absolute values.

During repetitive testing, an increase in the weight supported by the mandible or Jendrassik's maneuver tends to facilitate the masseter reflex.[71] The amplitude ratio between simultaneously recorded right-sided and left-sided responses, however, remains relatively constant.[97] In one study,[127] using a needle recording electrode, the test was considered abnormal if the study showed unilateral absence of the reflex, a difference of more than 0.5 ms between the latencies of the two sides, or bilateral absence of the reflex up to the age of 70 years. Table 18–2 summarizes normal values in our laboratory.[97]

Clinical Applications

The jaw reflex poses technical problems as a diagnostic test in standardizing the mechanical stimulus and regulating the tonus of the masseter for optimal activation (Fig. 18–6). Nonetheless, an unequivocal unilateral delay or absence suggests a lesion of the trigeminal nerve or the brainstem.[97,144] Electromyographic study of the masseter muscle may document the presence of denervation, thus localizing the lesion within the motor pathway.[126] In one study, the use of the jaw reflex as a test of midbrain function revealed absence or increased latency in 12 of an unselected series of 32 patients with multiple sclerosis.[58,180]

Masseteric Silent Period

A jaw reflex elicited during voluntary clenching gives rise to a brief pause in the electromyographic activity of the masseter muscle (Fig. 18–7). This inactivity,

Table 18–2 LATENCY AND AMPLITUDE OF MASSETER REFLEX IN 20 NORMAL SUBJECTS

	Latency (ms)	Latency Difference (Large Value Minus Small Value)	Amplitude (mV)	Amplitude Ratio (Large Value Over Small Value)
Mean right	7.10		0.23	
Mean left	7.06		0.21	
Total	7.08	0.27	0.22	1.44
SD	0.62	0.15	0.24	0.42
Mean + 3 SD	9.0	0.8	Variable	2.7

From Kimura et al.,[97] with permission.

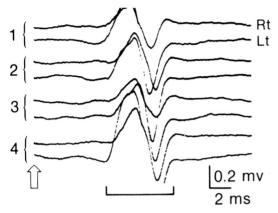

Rt
Lt

0.2 mv

2 ms

Figure 18–6. Jaw reflex recorded simultaneously from right (*top tracing of each frame*) and left (*bottom*) masseter after a mechanical tap on the chin (*open arrow*). Four trials were taken to show consistency in the response.

referred to as the masseteric silent period (SP), lasts about 30 ms in normal subjects.[154] The masseteric SP also occurs after acoustic or electric stimulation of the tongue, gums, oral mucosa, or belly of the muscle.[53,117,150] A unilateral stimulus causes SP on both sides, indicating the presence of crossed and uncrossed central pathways for this inhibition.[128] Anal-

ogous SP occurs in limb muscles after electrical stimulation of the nerve (see Chapter 18.5).

The force and direction of the tap and the magnitude of jaw clenching substantially influence the masseteric SP. In particular, a decrease in voluntary muscle contraction results in a major increase in its duration. Thus, stimulus and subject variables tend to limit its use as a clinical test of the masticatory system.[115] Some patients with tetanus lack the SP.[42,139,141,154] Conversely, the duration of the SP exceeds the normal range in patients with the temporomandibular joint syndrome.[7]

4 THE TONIC VIBRATION REFLEX

In contrast to the phasic activity of T and H reflexes, the tonic stretch reflex subserves postural and volitional movements. A vibratory stimulus applied to a tendon or a muscle excites the muscle spindles selectively and produces a sustained contraction of the muscle.[34,87]

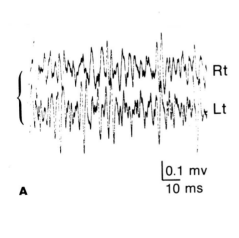

Rt

Lt

0.1 mv
10 ms

A

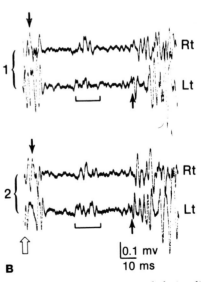

Rt
Lt

Rt

Lt

0.1 mv
10 ms

B

Figure 18–7. A. Voluntary contraction of the masseter. Electromyography was recorded simultaneously from right (*top tracing*) and left (*bottom*) sides with two pairs of surface electrodes placed on the belly of the muscle (G_1) and under the chin (G_2) on each side. **B.** Silent period (SP) of the masseter. The recording arrangement is the same as for **A**, but the mechanical tap was applied to the chin at the beginning of the sweep (*open arrow*). Electrical activity ceases immediately after the jaw reflex (*arrows from top*) elicited by the stimulus. Small voluntary potentials (*brackets*) break through in the midst of the SP before the return of full volitional activity (*arrows from bottom*) in approximately 80 ms after the tap.

This tonic vibration reflex (TVR) in many respects simulates a tonic stretch reflex,[54,66,122] although skin mechanoreceptors may also contribute.[1,40] Hence, the TVR provides a means of testing motor neuron reaction to tonic, rather than phasic, stimuli.[29,39,74,78,100,168] The TVR is elicited by a small vibrator that, attached over the tendon, oscillates at 150 Hz with an approximate amplitude of 0.5 to 1.5 mm. Intervals of at least 10 seconds should separate the stimuli to avoid cumulative depression of the reflex activities evoked segmentally. Surface electrodes placed over the belly (G_1) and tendon (G_2) of the muscle best register the TVR.

Normal and Abnormal Responses

The motor effects of tonic vibration include[8] (1) active and sustained muscle contraction,[2,64,100] (2) reciprocal inhibition of motor neurons innervating antagonistic muscles,[63] and (3) suppression of the T and H reflexes (Fig. 18–8).[20,21] The TVR involves more than a simple, spinal neural arc.[73] Studies in cat gastrocnemius muscle before and after lesions at preselected neural sites indicate that (1) generation of TVR requires an intact neural axis caudal to the midcolliculus, (2) facilitatory pathways ascend ipsilaterally in the ventral quadrant of the spinal cord, (3) the lateral vestibular nucleus and pontine reticular formation provide essential facilitation, and (4) the medullary reticular formation subserves inhibition.[4,13,52,135]

Abnormalities of the TVR occur in patients with a variety of motor disorders, including paretic muscles with involvement of the ventral quadrant of the spinal cord.[9,13,20,33,89,100,151] These abnormalities include (1) absence or diminution of the TVR, (2) loss of voluntary control over the TVR, (3) more abrupt development and termination of the TVR than in normals, (4) loss or diminution in TVR-induced suppression of the T and H reflexes, (5) asymmetries of the TVR of corresponding muscles in the two limbs, and (6) imbalances of the TVR in two antagonistic

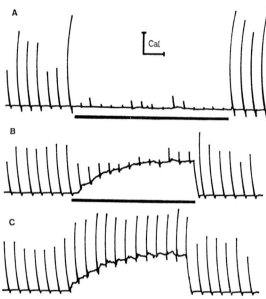

Figure 18–8. Effects of continuous muscle vibration in a normal subject showing suppression of phasic stretch reflexes with or without the generation of the tonic vibration reflex. **A.** Vibration of the quadriceps while knee reflexes are elicited every 5 seconds. Knee reflexes are depressed during the period of vibration (*bar*) even without the development of tonic contraction, probably because of the spread of the vibration wave to flexor muscles. **B.** Suppression of knee reflexes accompanying a tonic contraction induced by vibration. **C.** Voluntary contraction of quadriceps in the same subject as **B**, without suppression of knee reflexes. Calibration: vertical, 0.4 kg for **A**, 0.6 kg for **B** and **C**. Horizontal, 10 seconds. (From De Gail et al.,[20] with permission.)

muscles within the same limb. The TVR has also provided a means of assessing reciprocal inhibition, presynaptic inhibition, an inhibitory effect of acupuncture on the motor neurons,[80] and central control of voluntary movements.[43]

Clinical Applications

Clinical applications include early detection of incipient weakness, subclinical rigidity, spasticity, and involuntary movements such as tremors, clonus, and choreoathetosis.[64,100] The TVRs vary from patient to patient, depending upon the site of spinal cord lesions. Thus, the predictable pattern of abnormality, if clearly elucidated, would help localize the responsible lesion.[10]

A large number of papers have appeared describing the effect of tonic vibration on spasticity or rigidity.[64,79,100] In most reported series, vibration produced beneficial effects, e.g., (1) increased voluntary power of a weak muscle, (2) reduced resistance of the spastic antagonist, and (3) increased range of motion.[10] Unfortunately, these positive effects last only for the duration of vibration, which in practice cannot exceed a few minutes because of frictional generation of heat. Nonetheless, the technique holds therapeutic promise for patients with spinal cord injuries.

5 THE SILENT PERIOD AND THE LONG LATENCY (OR CORTICAL) RESPONSE

Despite continued effort, action potentials of a voluntarily contracting muscle undergo a transient suppression following electric stimulation of the nerve innervating that muscle.[76] This period of electrical inactivity, designated the silent period (SP), results from several physiologic mechanisms.[150] A number of investigators have studied the SP induced by electric stimulation[119,150] or by unloading the muscle spindles[153] in normal subjects[114] and in patients with neurologic disorders.[101]

Potentials That Break Through the Silent Period

The SP must represent a relative, rather than absolute, suppression because increasing voluntary muscle contraction can interrupt electrical inactivity. Two separate potentials, V_1 and V_2, appear.[167] At high levels of muscle contraction, where the antidromic activity collides with voluntary impulses in most axons, the first potential mainly comprises the H reflex.[113,152] At low levels of muscle contraction, based on few voluntary impulses, the first potential primarily represents the F wave, because sub-

stantial antidromic activity reaches the central motor neuron pool.[95] The second potential, V_2, which appears in the middle of the SP, is also designated the voluntary potential (VP), long latency reflex (LLR), long loop response or cortical (C) response.

Descending volitional inputs play an important role in the generation of V_2, normally seen only during tonic contraction of the muscle.[95] A similar LLR can also be elicited at rest in patients with posthypoxic intention myoclonus[181] and several other types of myoclonus,[93] presumably in response to segmental polysynaptic inputs to motor neurons.[167] Alternatively, some investigators equate V_2 with the transcortical reflex activity, or C response, elicited by brief stretching of arm muscles.[41,110] In contrast to total or partial absence of V_2 in hemiparetic patients and during cognitive tasks in normal subjects, repetitive trains of stimuli have a strong facilitatory effect.[16] In patients with Huntington's disease, V_1 is normal, but V_2 is elicited neither by displacements of the index finger nor by electrical stimulation of the median nerve.[124] In parkinsonism, the median latency component of V_2 in the stretched triceps surae muscle may be increased in latency.[148]

If V_1 occurs segmentally and V_2, cortically, the latency difference between them provides a measure of the central conduction along the spinal cord to and from the reflex center of V_2. The comparison between the arm and leg allows one to calculate mean spinal conduction time between the seventh cervical and fifth lumbar spinous process as[36,37]

$$(V_1 - V_2)_{leg}/2 - (V_1 - V_2)_{arm}$$

The practical value of this approach in assessing individual cases awaits further clarification, even though the measurements may be a reasonably accurate estimate of the mean conduction characteristics in a group of subjects.

Instead of electrical stimulation, sudden tilting of a platform around the axis of the human ankle joint also causes a regular pattern of short and medium latency discharges in the stretched triceps surae muscle and an LLR in its antagonist, the tibialis anterior muscle. Some authors have referred to these responses as "long loop" reflexes via transcortical pathways.[102,111,158] These discharges persist after spinal cord transection in cats and monkeys,[51,162] which suggests a segmental origin. Sudden stretching of the human wrist, giving rise to long loop stretch reflexes, accompanies a series of spindle discharges.[65] If repetitive segmental reflexes result from these group IA afferent bursts, LLR may not necessarily require transcortical pathways.

The same physiologic mechanisms may underlie LLR elicited by cutaneous stimulation and long loop reflex induced by stretching of the spindle. If so, both may represent activity at the segmental level modulated by descending impulses from the higher center, such as the cerebellum.[46] This point, however, requires further study, because some patients with multiple sclerosis have a delayed long loop reflex, a finding that implies the presence of a supraspinal pathway.[31,32]

Physiologic Mechanisms

Although recurrent Renshaw inhibition[138] follows the passage of an impulse along the motor axon to either direction,[133] antidromic activities produce more effective suppression.[143] The middle portion of the SP at least in part results from antidromic invasion of the Renshaw loop. Thus, the VP tends to occur with any maneuver that reduces the axonal volleys arriving at the central motor neuron pool.[96] For example, weaker stimuli, which activate fewer motor axons, favor the appearance of the VP.[150]

Even with supramaximal stimulation, not all antidromic impulses reach the central motor neuron pool, because during muscle contraction they collide with voluntary orthodromic impulses in some motor fibers. Greater effort increases the chance of collision, because more axons carry orthodromic impulses.[94,95] Stimulation of the nerve distally also enhances this probability, which increases in proportion to the length of the nerve segment between the stimulus site and the cell body (Fig. 18–9). Thus, the greater the voluntary effort and the weaker and more distal the nerve stimulation, the smaller the antidromic invasion and the weaker the recurrent inhibition of motor neurons responsible for the SP (Fig. 18–10).

In addition to the Renshaw effect, other mechanisms such as the unloading of the muscle spindle[88,119] and activation of the

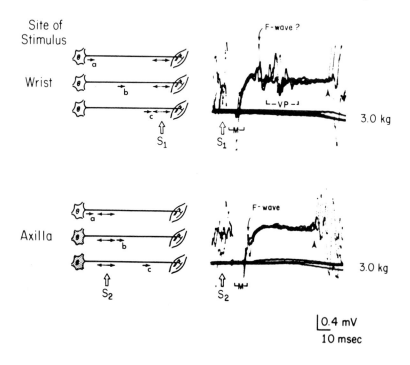

Figure 18–9. Simultaneous recording of muscle force of 3.0 kg (*straight line*) and the silent period (SP) from the voluntarily contracting first dorsal interosseous muscle (three trials superimposed). The SP was broken by the voluntary potential (*VP*) with a stimulus at the wrist but not with a stimulus at the axilla, indicating greater inhibition of motor neurons with proximal than with distal nerve stimulation. After distal stimulation, most antidromic activity is extinguished by collision with voluntary impulses (*a, b,* and *c*) before reaching the motor neuron pool. After proximal stimulation, antidromic activity, escaping collision, presumably invades recurrent axon collaterals and inhibits the motor neurons (*shaded*). (From Kimura,[96] with permission.)

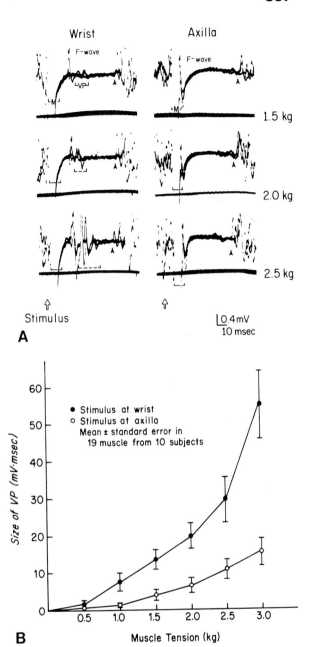

Figure 18–10. A. Stimulation and recording as in Figure 18–9 at muscle tension ranging from 1.5 to 2.5 kg. Stimulating at the wrist, the voluntary potential (VP) became progressively greater in size at increasingly higher muscle forces. With stimulation at the axilla, no VP was recorded at any level of muscle force, but the duration of the silent period (SP) was shortened as the muscle force was increased. **B.** Muscle tension and the size of the VP breaking through the SP. For muscle forces of 1.0 kg and above, the VP was significantly larger with stimulus at the wrist than at the axilla, indicating that motor neurons were more inhibited by proximal, as opposed to distal, nerve stimulation during voluntary muscle contraction. The difference in inhibitory effects of proximal versus distal stimulation became progressively larger as the muscle force was increased. (From Kimura,[96] with permission.)

Golgi tendon organ[60] contribute to the SP during muscle contraction.[150] Ascending cutaneous volleys may also have an inhibitory effect, although a well-defined SP on this basis results only from a high-intensity stimulus.[98] Sensory nerve stimulation could generate a reproducible SP, presumably through either group IB afferent fibers from tendon organs or through ascending reflex pathways.[81,82] In this case, proximal stimulation, which activates a greater number of afferent fibers, would inhibit the motor neurons more effectively.

6 OTHER REFLEXES

The flexor reflex elicited by stimulation of the peripheral nerve consists of two or more components usually demonstrating excitation-inhibition cycles.[44,118] Analogous to flexor reflexes in the limb muscles, stimulation of perianal skin elicits a two-component response in the external anal sphincter.[132,155] Stimulation of penis or clitoris also evokes reflex responses with a typical latency of 33 ms in the external anal and urethral sphincters.[175]

Stimulation of the pudendal nerve evokes reflexive contraction of the bulbocavernosus muscle, which may complement pudendal somatosensory evoked potentials in diagnostic evaluation of bowel, bladder, and sexual function (see Fig. 19–20A). With the active electrode (G_1) over the bulbocavernosus muscle beneath the scrotum and the reference electrode (G_2) over the iliac crest, stimuli applied at a rate of 1.5 per second elicits, after 30 to 50 averaging, an initially negative biphasic or triphasic response with onset latency of 35.9 ± 9.0 ms.[68] This technique may prove useful in the evaluation of spinal cord injury in general and the neurogenic bladder in particular.[90] The reflex latency is also increased in diabetic neuropathy[47] or impotence secondary to peripheral nerve involvement.[116] The bulbocavernosus reflex provides a more sensitive measure of the sacral nervous system than conventional or single-fiber electromyography of external urethral and anal sphincters.[174]

The abdominal stretch reflex is elicited in patients with upper motor neuron lesions bilaterally. Its latency ranges from 16.5 to 25 ms, with side-to-side variation not exceeding 3 ms.[160]

The auditory postauricular reflex generated in the posterior auricular muscle has two prominent components at latencies of 12 and 16 ms.[70] Voluntary contraction of the neck extensor or facial muscles enhances the response. A markedly enlarged reflex may help differentiate an upper motor neuron lesion in clinically equivocal cases.

The corneomandibular reflex, not seen in healthy subjects, may appear with lesions involving the precentrobulbar tract.[125] Electromyographic studies help differentiate this reflex from clinically similar corneomental reflex.

Stimulation of the dorsal genital nerve elicits reflex activation of the external anal sphincter with the latency of 38.5 ± 5.8 ms (mean $\pm$ standard deviation) in control subjects. Patients with fecal incontinence may have absence or delay of this pudendoanal reflex.[169]

REFERENCES

1. Abbruzzese, G, Hagbarth, KE, Homma, I, and Wallin, U: Excitation from skin receptors contributing to the tonic vibration reflex in man. Brain Res 150:194–197, 1978.
2. Agarwal, GC, and Gottlieb, GL: Effect of vibration of the ankle stretch reflex in man. Electroencephalogr Clin Neurophysiol 49:81–92, 1980.
3. Aiello, I, Rosati, G, Serra, G, and Manca, M: The diagnostic value of H-index in S1 root compression. J Neurol Neurosurg Psychiatry 44:171–172, 1981.
4. Andrews, C, Knowles, L, and Hancock, J: Control of the tonic vibration reflex by the brain stem reticular formation in the cat. J Neurol Sci 18:217–226, 1973.
5. Bathien, N, and Rondot, P: Reciprocal continuous inhibition in rigidity of parkinsonism. J Neurol Neurosurg Psychiatry 40:20–24, 1977.
6. Berardelli, A, Hallett, M, Kaufman, C, Fine, E, Berenberg, W, and Simon, SR: Stretch reflexes of triceps surae in normal man. J Neurol Neurosurg Psychiatry 45:513–525, 1982.
7. Bessette, R, Bishop, B, and Mohl, N: Duration of masseteric silent period in patients with TMJ syndrome. J Appl Physiol 30:864–869, 1971.
8. Bishop, B: Vibratory stimulation. Part I. Neurophysiology of motor responses evoked by vibratory stimulation. Phys Ther 54:1273–1282, 1974.
9. Bishop, B: Vibratory stimulation. Part II. Vibratory stimulation as an evaluation tool. Phys Ther 55:28–34, 1975.
10. Bishop, B: Vibratory stimulation. Part III. Possible applications of vibrations in treatment of motor dysfunctions. Phys Ther 55:139–143, 1975.
11. Bishop, B, Machover, S, Johnston, R, and Anderson, M: A quantitative assessment of gamma-motoneuron contribution to the Achilles tendon reflex in normal subjects. Arch Phys Med Rehabil 49:145–154, 1968.
12. Braddom, RI, and Johnson, EW: Standardization of H reflex and diagnostic use in S1 radiculopathy. Arch Phys Med Rehabil 55:161–166, 1974.

13. Burke, D, Knowles, L, Andrews, C, and Ashby, P: Spasticity, decerebrate rigidity and the clasp-knife phenomenon. An experimental study in the cat. Brain 95:31–48, 1972.

14. Burke, D, McKeon, B, and Skuse, NF: Irrelevance of fusimotor activity to the Achilles tendon jerk of relaxed humans. Ann Neurol 10:547–550, 1981.

15. Burke, D, McKeon, B, and Skuse, NF: Dependence of the achilles tendon reflex on the excitability of spinal reflex pathways. Ann Neurol 10:551–556, 1981.

16. Conrad, B and Aschoff, JC: Effects of voluntary isometric and isotonic activity on late transcortical reflex components in normal subjects and hemiparetic patients. Electroencephalogr Clin Neurophysiol 42:107–116, 1977.

17. Cruccu, G: Intracranial stimulation of the trigeminal nerve in man. I. Direct motor responses. J Neurol Neurosurg Psychiatry 49:411–418, 1986.

18. Cruccu, G, and Bowsher, D: Intracranial stimulation of the trigeminal nerve in man. II. Reflex responses. J Neurol Neurosurg Psychiatry 49:419–427, 1986.

19. Daube, JR: F-wave and H-reflex measurements. American Academy of Neurology, Course #16, Clinical Electromyography, 1979, pp 93–101.

20. De Gail, P, Lance, JW, and Neilson, PD: Differential effects on tonic and phasic reflex mechanisms produced by vibration of muscles in man. J Neurol Neurosurg Psychiatry 29:1–11, 1966.

21. Delwaide, PJ: Différences d'organisation fonctionnelle des arcs myotatiques du quadriceps et du court biceps chez l'homme. Revu Neurol (Paris) 128:39–46, 1973.

22. Delwaide, PJ: Excitability of lower limb myotatic reflex arcs under the influence of caloric labyrinthine stimulation: Analysis of the postural effects in man. J Neurol Neurosurg Psychiatry 40:970–974, 1977.

23. Delwaide, PJ: Electrophysiological analysis of the mode of action of muscle relaxants in spasticity. Ann Neurol 17:90–95, 1985.

24. Delwaide, PJ, and Crenna, P: Cutaneous nerve stimulation and motoneuronal excitability. II. Evidence for nonsegmental influences. J Neurol Neurosurg Psychiatry 47:190–196, 1984.

25. Delwaide, PJ, Figiel, C, and Richelle, C: Effects of postural changes of the upper limb on reflex transmission in the lower limb: Cervicolumbar reflex interactions in man. J Neurol Neurosurg Psychiatry 40:616–621, 1977.

26. Delwaide, PJ, and Juprelle, M: The effects of caloric stimulation of the labyrinth on the soleus motor pool in man. Acta Neurol Scand 55:310–322, 1977.

27. Deschuytere, J, and Rosselle, N: Diagnostic use of monosynaptic reflexes in L5 and S1 root compression. In Desmedt, JE (ed): New Developments in Electromyography and Clinical Neurophysiology, Vol 3. Karger, Basel, 1973, pp 360–366.

28. Deschuytere, J, Rosselle, N, and De Keyser, C: Monosynaptic reflexes in the superficial forearm flexors in man and their clinical significance. J Neurol Neurosurg Psychiatry 39:555–565, 1976.

29. Desmedt, JE, and Godaux, E: Vibration-induced discharge patterns of single motor units in the masseter muscle in man. J Physiol (Lond) 253:429–442, 1975.

30. De Weerd, AW, and Jonkman, EJ: Measurement of knee tendon reflex latencies in lumbar radicular syndromes. Eur Neurol 5:304–308, 1986.

31. Diener, HC, Dichgans, J, Bacher, M, and Gushchlbauer, B: Characteristic alterations of long-loop "reflexes" in patients with Friedreich's disease and late atrophy of the cerebellar anterior lobe. J Neurol Neurosurg Psychiatry 47:679–685, 1984.

32. Diener, HC, Dichgans, J, Hulser, PJ, Buettner, UW, Bacher, M and Guschlbauer, B: The significance of delayed long-loop responses to ankle displacement for the diagnosis of multiple sclerosis. Electroencephalogr Clin Neurophysiol 57:336–342, 1984.

33. Dimitrijevic, MR, Spencer, WA, Trontelj, JV, and Dimitrijevic, M: Reflex effects of vibration in patients with spinal cord lesions. Neurology 27:1078–1086, 1977.

34. Dindar, F, and Verrier, M: Studies on the receptor responsible for vibration induced inhibition of monosynaptic reflexes in man. J Neurol Neurosurg Psychiatry 38:155–160, 1975.

35. Eccles, JC: The inhibitory control of spinal reflex action. Electroencephalogr Clin Neurophysiol (Suppl 25): 20–34, 1967.

36. Eisen, A, Burton, K, Larsen, A, Hoirch, M, and Calne, D: A new indirect method for measuring spinal conduction velocity in man. Electroencephalogr Clin Neurophysiol 59:204–213, 1984.

37. Eisen, A, Hoirch, M, Fink, M, Goya, T, and Calne, D: Noninvasive measurement of central sensory and motor conduction. Neurology 35:503–509, 1985.

38. Eisen, A, Hoirch, M, White, J, and Calne, D: Sensory group Ia proximal conduction velocity. Muscle Nerve 7:636–641, 1984.

39. Eklund, G: Some physical properties of muscle vibrators used to elicit tonic proprioceptive reflexes in man. Acta Soc Med Upsal 76:271–280, 1971.

40. Eklund, G, Hagbarth, KE, and Torebjork, E: Exteroceptive vibration-induced finger flexion reflex in man. J Neurol Neurosurg Psychiatry 41:438–443, 1978.

41. Evarts, EV: Sensorimotor cortex activity associated with movements triggered by visual as compared to somesthetic inputs. In Schmitt, FO, and Worden, FG (eds): The Neurosciences, Third Study Program. MIT Press, Cambridge, Mass, 1974, pp 327–337.

42. Fernandez, JM, Ferrandiz, M, Larrea, L, Ramio, R, and Boada, M: Cephalic tetanus studied with single fibre EMG. J Neurol Neurosurg Psychiatry 46:862–866, 1983.

43. Fisher, MA, Shahani, BT, and Young, RR: Electrophysiologic analysis of motor system after stroke, suppressive effect of vibration. Arch Phys Med Rehabil 60:11–14, 1979.

44. Fisher, MA, Shahani, BT, and Young, RR: Electrophysiologic analysis of motor system after stroke, flexor reflex. Arch Phys Med Rehabil 60:7–11, 1979.

45. Frank, JS: Spinal motor preparation in humans. Electroencephalogr Clin Neurophysiol 63:361–370, 1986.

46. Friedemann, HH, Noth, J, Diener, HC, and Bacher, M: Long latency EMG responses in hand and leg muscles: Cerebellar disorders. J Neurol Neurosurg Psychiatry 50:71–77, 1987.

47. Gallai, V, and Mazaiotta, G: Electromyographical studies of the bulbo-cavernous reflex in diabetic men with sexual dysfunction. Electromyogr Clin Neurophysiol 26:521–527, 1986.

48. Garcia, HA, Fisher, MA, and Gilai, A: H reflex analysis of segmental reflex excitability in flexor and extensor muscles. Neurology 29:984–991, 1979.

49. Gassel, MM, and Ott, KH: Local sign and late effects on motoneuron excitability of cutaneous stimulation in man. Brain 93:95–106, 1970.

50. Gassel, MM, and Ott, KH: Patterns of reflex excitability change after widespread cutaneous stimulation in man. J Neurol Neurosurg Psychiatry 36:282–287, 1973.

51. Ghez, C, and Shinoda, Y: Spinal mechanisms of the functional stretch reflex. Brain Res 32:55–68, 1978.

52. Gillies, JD, Burke, DJ, and Lance, JW: Tonic vibration reflex in the cat. J Neurophysiol 34:252–262, 1971.

53. Godaux, E, and Desmedt, JE: Human masseter muscle: H- and tendon reflexes: Their paradoxical potentiation by muscle vibration. Arch Neurol 32:229–234, 1975.

54. Godaux, E, and Desmedt, JE: Evidence for a monosynaptic mechanism in the tonic vibration reflex of the human masseter muscle. J Neurol Neurosurg Psychiatry 38:161–168, 1975.

55. Godaux, E, and Desmedt, JE: Exteroceptive suppression and motor control of the masseter and temporalis muscles in normal man. Brain Res 85:447–458, 1975.

56. Goodgold, J: H reflex (corres). Arch Phys Med Rehabil 57:407, 1976.

57. Goodwill, CJ: The normal jaw reflex: Measurement of the action potential in the masseter muscles. Ann Phys Med 9:183–188, 1968.

58. Goodwill, CJ, and O'Tuama, L: Electromyographic recording of the jaw reflex in multiple sclerosis. J Neurol Neurosurg Psychiatry 32:6–10, 1969.

59. Gottlieb, GL, and Agarwal, GC: Extinction of the Hoffman reflex by antidromic conduction. Electroencephalogr Clin Neurophysiol 41:19–24, 1976.

60. Granit, R: Reflex self-regulation of muscle contraction and autogenetic inhibition. J Neurophysiol 13:351–372, 1950.

61. Guiheneuc, P, and Bathien, N: Two patterns of results in polyneuropathies investigated with the H reflex: Correlation between proximal and distal conduction velocities. J Neurol Sci 30:83–94, 1976.

62. Guiheneuc, P, and Ginet, J: Etude du reflexe de Hoffmann obtenu au niveau du muscle quadriceps de sujets humains normaux. Electroencephalogr Clin Neurophysiol 36:225–231, 1974.

63. Hagbarth, KE: EMG studies of stretch reflexes in man. Electroencephalogr Clin Neurophysiol Suppl 25:74–79, 1967.

64. Hagbarth, KE: The effect of muscle vibration in normal man and in patients with motor disorders. In Desmedt, JE (ed): New Developments in Electromyography and Clinical Neurophysiology, Vol 3. Karger, Basel, 1973, pp 428–443.

65. Hagbarth, KE, Hagglund, JV, Wallin, EW, and Young, RR: Grouped spindle and electromyographic response to abrupt wrist extension movements in man. J Physiol (Lond) 312:81–96, 1981.

66. Hagbarth, KE, Hellsing, G, and Lofstedt, L: TVR and vibration-induced timing of motor impulses in the human jaw elevator muscles. J Neurol Neurosurg Psychiatry 39:719–728, 1976.

67. Halar, EM, Brozovich, FV, Milutinovic, J, Inouye, VL, and Becker, VM: H-reflex latency in uremic neuropathy: Correlation with NCV and clinical findings. Arch Phys Med Rehabil 60:174–177, 1979.

68. Haldeman, S, Bradley, WE, Bhatia, NN, and Johnson, BK: Pudendal evoked responses. Arch Neurol 39:280–283, 1982.

69. Hamann, WC, and Morris, JGL: Effects of stretching the patella tendon on voluntary and reflex contractions of the calf muscles in man. Exp Neurol 55:405–413, 1977.

70. Hammond, EJ, and Wilder, BJ: Enhanced auditory postauricular evoked responses after corticobulbar lesions. Neurology 35:278–281, 1985.

71. Hannam, AG: Effects of voluntary contraction of the masseter and other muscles upon the masseteric reflex in man. J Neurol Neurosurg Psychiatry 35:66–71, 1972.

72. Hayes, KC, Robinson, KL, Wood, GAS, and Jennings, LS: Assessment of the H-reflex excitability curve using a cubic spline function. Electroencephalogr Clin Neurophysiol 46:114–117, 1979.

73. Hendrie, A, and Lee, RG: Selective effects of vibration on human spinal and long-loop reflexes. Brain Res 157:369–375, 1978.

74. Hirayama, K, Homma, S, Mizote, M, Nakajima, Y, and Watanabe, S: Separation of the contributions of voluntary and vibratory activation of motor units in man by cross-correlograms. Jpn J Physiol 24:293–304, 1974.

75. Hodes, R: Effects of age, consciousness, and other factors on human electrically induced reflexes (EIRs). Electroencephalogr Clin Neurophysiol Suppl 25:80–91, 1967.

76. Hoffmann, P: Demonstration eines Hem-

mungsreflexes im menschlichen Rucken-mark. Zeitschrift fur Biologie 70:515–524, 1919.

77. Hoffmann, P: Untersuchungen uber die Ei-genreflexe (Sehnenreflexe) menschlicher Mus-keln. Springer, Berlin, 1922.

78. Homma, S: A survey of Japanese research on muscle vibration. In Desmedt, JE (ed): New Developments in Electromyography and Clini-cal Neurophysiology, Vol 3. Karger, Basel, 1973, pp 463–468.

79. Homma, S, Ishikawa, K, and Stuart, DG: Mo-toneuron responses to linearly rising muscle stretch. Am J Phys Med 49:290–306, 1970.

80. Homma, S, Nakajima, Y, and Toma, S: Inhibi-tory effect of acupuncture on the vibration-in-duced finger flexion reflex in man. Electroen-cephalogr Clin Neurophysiol 61:150–156, 1985.

81. Hufschmidt, HJ: Wird die Silnet period nach direkter Muskelreizung durch die Golgi-Seh-nenorgane ausgelöst? Pflugers Arch 271:35–39, 1960.

82. Hufschmidt, HJ, and Linke, D: A damping fac-tor in human voluntary contraction. J Neurol Neurosurg Psychiatry 39:536–537, 1976.

83. Hugon, M: Methodology of the Hoffmann reflex in man. In Desmedt, JE (ed): New Develop-ments in Electromyography and Clinical Neu-rophysiology, Vol 3. Karger, Basel, 1973, pp 277–293.

84. Hugon, M, and Bathien, N: Influence de la stimulation du nerf sural sur divers réflexes monosynaptiques de l'homme (abstr). J Phy-siol (Paris) 59:244, 1967.

85. Hugon, M, Delwaide, P, Pierrot-Deseilligny, E, and Desmedt, JE: A discussion of the method-ology of the triceps surae T- and H-reflexes. In Desmedt, JE (ed): New Developments in Elec-tromyography and Clinical Neurophysiology, Vol 3. Karger, Basel, 1973, pp 773–780.

86. Jabre, JF: Surface recording of the H-reflex of the flexor carpi radialis. Muscle Nerve 4:435–438, 1981.

87. Jack, JJB, and Roberts, RC: The role of mus-cle spindle afferents in stretch and vibration reflexes of the soleus muscle of the decere-brate cat. Brain Res 146:366–372, 1978.

88. Jacobs, MB, Andrews, LT, Iannone, A, and Greninger, L: Antagonist EMG temporal pat-terns during rapid voluntary movement. Neu-rology 30:36–41, 1980.

89. Kanda, K, Homma, S, and Watanabe, S: Vi-bration reflex in spastic patients. In Desmedt, JE (ed): New Developments in Electromyogra-phy and Clinical Neurophysiology, Vol 3. Karger, Basel, 1973, pp 469–474.

90. Kaplan, PE: Somatosensory evoked response obtained after stimulation of the pelvic and pudendal nerve. Electromyogr Clin Neurophy-siol 23:99–102, 1983.

91. Katz, R, Morin, C, Pierrot-Deseilligny, E, and Hibino, R: Conditioning of H reflex by a pre-ceding subthreshold tendon reflex stimulus. J Neurol Neurosurg Psychiatry 40:575–580, 1977.

92. Katz, R, and Pierrot-Deseilligny E: Recurrent

inhibition of a-motoneurons in patients with upper motor neuron lesions. Brain 105:103–124, 1982.

93. Kelley, JJ, Sharbrough, FW, and Daube, JR: A clinical and electrophysiological evaluation of myoclonus. Neurology 31:581–589, 1981.

94. Kimura, J: A method for estimating the refrac-tory period of motor fibers in the human pe-ripheral nerve. J Neurol Sci 28:485–490, 1976.

95. Kimura, J: Electrical activity in voluntarily contracting muscle. Arch Neurol 34:85–88, 1977.

96. Kimura, J: Recurrent inhibition of moto-neurons during the silent period in man. In Desmedt, JE (ed): Motor Control Mechanisms in Health and Disease. Raven Press, New York, 1983, pp 459–465.

97. Kimura, J, Rodnitzky, RL, and Van Allen, MW: Electrodiagnostic study of trigeminal nerve: Orbicularis oculi reflex and masseter reflex in trigeminal neuralgia, paratrigeminal syndrome, and other lesions of the trigeminal nerve. Neurology 20:574–583, 1970.

98. Kranz, H, Adorjani, C, and Baumgartner, G: The effect of nociceptive cutaneous stimuli on human motorneurons. Brain 96:571–590, 1973.

99. Kugelberg, E: Facial reflexes. Brain 75:385–396, 1952.

100. Lance, JW, Burke, D, and Andrews, CJ: The reflex effects of muscle vibration: Studies of tendon jerk irradiation, phasic reflex inhibi-tion and the tonic vibration reflex. In Des-medt, JE (ed): New Developments in Electro-myography and Clinical Neurophysiology, Vol 3. Karger, Basel, 1973, pp 444–462.

101. Laxer, K, and Eisen, A: Silent period measure-ment in the differentiation of central demye-lination and axonal degeneration. Neurology 25:740–744, 1975.

102. Lee, RG, and Tatton, WG: Motor responses to sudden limb displacements in primates with specific CNS lesions and in human patients with motor system disorders. Can J Neurol 2:285–293, 1975.

103. Little, JW, and Halar, EM: H-reflex changes following spinal cord injury. Arch Phys Med Rehabil 66:19–22, 1985.

104. Lundberg, A, Malmgren, K, and Schomburg, ED: Group II excitation in motoneurones and double sensory innervation of extensor digi-torum brevis. Acta Physiol Scand 94:398–400, 1975.

105. Magladery, JW, and McDougal, DB, Jr: Elec-trophysiological studies of nerve and reflex ac-tivity in normal man. I. Identification of cer-tain reflexes in the electromyogram and the conduction velocity of peripheral nerve fibres. Bull Johns Hopkins Hosp 86:265–290, 1950.

106. Magladery, JW, Porter, WE, Park, AM, and Teasdall, RD: Electrophysiological studies of nerve and reflex activity in normal man. IV. The two-neurone reflex and identification of certain action potentials from spinal roots and cord. Bull Johns Hopkins Hosp 88:499–519, 1951.

107. Magladery, JW, Teasdall, RD, Park, AM, and Languth, HW: Electrophysiological studies of reflex activity in patients with lesions of the nervous system. 1. A comparison of spinal motoneurone excitability following afferent nerve volleys in normal persons and patients with upper motor neurone lesions. Bull Johns Hopkins Hosp 91:219–244, 1952.

108. Malcolm, DS: A method of measuring reflex times applied in sciatica and other conditions due to nerve-root compression. J Neurol Neurosurg Psychiatry 14:15–24, 1951.

109. Mamelak, M, and Sowden, K: The effect of gammahydroxybutyrate on the H-reflex: Pilot study. Neurology 33:1497–1500, 1983.

110. Marsden, CD, Merton, PA, and Morton, HB: Is the human stretch reflex cortical rather than spinal? Lancet 1:759–761, 1973.

111. Marsden, CD, Merton, PA, and Morton, HB: Stetch reflex and servo-action in a variety of human muscles. J Physiol (Lond) 259:531–560, 1976.

112. Martinelli, P, and Montagna, P: Conditioning of the H reflex by stimulation of the posterior tibial nerve in Parkinson's disease. J Neurol Neurosurg Psychiatry 42:701–704, 1979.

113. McComas, AJ, Sica, REP, and Upton, ARM: Excitability of human motoneurones during effort. J Physiol (Lond) 210:145–146, 1970.

114. McLellan, DL: The electromyographic silent period produced by supramaximal electrical stimulation in normal man. J Neurol Neurosurg Psychiatry 36:334–341, 1973.

115. McNamara, DC, Crane, PF, McCall, WD, Jr, and Ash, M, Jr: Duration of the electromyographic silent period following the jaw-jerk reflex in human subjects. J Dent Res 56:660–664, 1977.

116. Mehta, A, Viosca, S, Korenman, S, and Davis, S: Peripheral nerve conduction studies and bulbocavernosus reflex in the investigation of impotence. Arch Phys Med Rehabil 67:332–335, 1986.

117. Meier-Ewert, K, Gleitsmann, K, and Reiter, F: Acoustic jaw reflex in man: Its relationship to other brain-stem and microreflexes. Electroencephalogr Clin Neurophysiol 36:629–637, 1974.

118. Meinck, H-M, Kuster, S, Benecke, R, and Conrad, B: The flexor reflex: Influence of stimulus parameters on the reflex response. Electroencephalogr Clin Neurophysiol 61:287–298, 1985.

119. Merton, PA: The silent period in a muscle of the human hand. J Physiol (Lond) 114:183–198, 1951.

120. Milner-Brown, SH, Girvin, JP, and Brown, WF: The effects of motor cortical stimulation on the excitability of spinal motoneurons in man. Can J Neurol Sci 2:245–253, 1975.

121. Mizuno, Y, Tanaka, R, and Yanagisawa, N: Reciprocal group I inhibition on triceps surae motoneurons in man. J Neurophysiol 34:1010–1017, 1971.

122. Moddel, G, Best, B, and Ashby, P: Effect of differential nerve block on inhibition of the monosynaptic reflex by vibration in man. J Neurol Neurosurg Psychiatry 40:1066–1071, 1977.

123. Morin, C, and Pierrot-Deseilligny, E: Role of Ia afferents in the soleus motoneurones inhibition during a tibialis anterior voluntary contraction in man. Exp Brain Res 27:509–522, 1977.

124. Noth, J, Podoll, K, and Friedemann, HH: Long-loop reflexes in small hand muscles studied in normal subjects and in patients with Huntington's disease. Brain 108:65–80, 1985.

125. Ongerboer de Visser, BW: The recorded corneo-mandibular reflex. Electroencephalogr Clin Neurophysiol 634:25–31, 1986.

126. Ongerboer de Visser, BW, and Goor, C: Electromyographic and reflex study in idiopathic and symptomatic trigeminal neuralgias: Latency of the jaw and blink reflexes. J Neurol Neurosurg Psychiatry 37:1225–1230, 1974.

127. Ongerboer de Visser, BW, and Goor, C: Jaw reflexes and masseter electromyograms in mesencephalic and pontine lesions: An electrodiagnostic study. J Neurol Neurosurg Psychiatry 39:90–92, 1976.

128. Ongerboer de Visser, BW, and Goor, C: Cutaneous silent period in masseter muscles: A clinical and electrodiagnostic evaluation. J Neurol Neurosurg Psychiatry 39:674–679, 1976.

129. Ongerboer de Visser BW, Schimsheimer, RJ, and Hart, AAM: The H-reflex of the flexor carpi radialis muscle: A study in controls and radiation-induced brachial plexus lesions. J Neurol Neurosurg Psychiatry 47:1098–1101, 1984.

130. Owens, LA, Peterson, CR, and Burdick, AB: Familial spastic paraplegia: A clinical and electrodiagnostic evaluation. Arch Phys Med Rehabil 63:357–361, 1982.

131. Paillard, J: Reflexes et Regulations d'Origine Proprioceptive chez l'Homme. Etude Neurophysiologique et Psychophysiologique. Arnette, Paris, 1955.

132. Pedersen, E, Klemar, B, Schroder, HD, and Torring, J: Anal sphincter responses after perianal electrical stimulation. J Neurol Neurosurg Psychiatry 45:770–773, 1982.

133. Pierrot-Deseilligny, E, and Bussel, B: Evidence for recurrent inhibition by motoneurons in human subjects. Brain Res 88:105–108, 1975.

134. Pierrot-Deseilligny, E, Bussel, B, and Morin, C: Supraspinal control of the changes induced in H-reflex by cutaneous stimulation, as studied in normal and spastic man. In Desmedt, JE (ed): New Developments in Electromyography and Clinical Neurophysiology, Vol 3. Karger, Basel, 1973, pp 550–555.

135. Pompeiano, O, and Barnes, CD: Response of brain stem reticular neurons to muscle vibration in the decerebrate cat. J Neurophysiol 34:709–724, 1971.

136. Rack, PMH, Ross, HF, and Thilmann, AF: The ankle stretch reflexes in normal and spastic subjects: The response to sinusoidal movement. Brain 107:637–654, 1984.

137. Renshaw, B: Influence of the discharge of

motoneurons upon excitation of neighboring motoneurons. J Neurophysiol 4:167–183, 1941.

138. Renshaw, B: Central effects of centripetal impulses in axons of spinal ventral roots. J Neurophysiol 9:191–204, 1946.

139. Ricker, K, Eyrich, K, and Zwirner, R: Seltenere Formen von Tetanuserkrankung: Klinische und electromyographische Untersuchung. Arch Psychiatr Nervenkr 215:75–91, 1971.

140. Rico, RE, and Jonkman, EJ: Measurement of the achilles tendon reflex for the diagnosis of lumbosacral root compression syndromes. J Neurol Neurosurg Psychiatry 45:791–795, 1982.

141. Risk, W, Bosch, EP, Kimura, J, Cancilla, PA, Fischbeck, K, and Layzer, RB: Chronic tetanus: Clinical report and histochemistry of muscle. Muscle Nerve 4:363, 1981.

142. Robinson, KL, McIlwain, JS, and Hayes, KC: Effects of H-reflex conditioning upon the contralateral alpha motoneuron pool. Electroencephalogr Clin Neurophysiol 46:65–71, 1979.

143. Ryall, RW, Piercey, MF, Polosa, C, and Goldfarb, J: Excitation of Renshaw cells in relation to orthodromic and antidromic excitation of motoneurons. J Neurophysiol 35:137–148, 1972.

144. Schenk, E, and Beck, U: Somatic brain stem reflexes in clinical neurophysiology. Electromyogr Clin Neurophysiol 15:107–116, 1975.

145. Schiller, HH, and Stalberg, E: F responses studied with single fibre EMG in normal subjects and spastic patients. J Neurol Neurosurg Psychiatry 41:45–53, 1978.

146. Schmisheimer, RJ, Ongerboer de Visser, BW, and Kemp, B: The flexor carpi radialis H-reflex in lesions of the sixth and seventh cervical nerve roots. J Neurol Neurosurg Psychiatry 48:445–449, 1985.

147. Schimsheimer, RJ, Ongerboer de Visser, BW, Kemp, B, and Bour, LJ: Flexor carpi radialis H-reflex in polyneuropathy: Relations to conduction velocities of the median nerve and the soleus H-reflex latency. J Neurol Neurosurg Psychiatry 50:447–452, 1987.

148. Scholz, E, Diener, HC, Noth, J, Friedemann, H, Dichgans, J, and Bacher, M: Medium and long latency EMG responses in leg muscles: Parkinson's disease. J Neurol Neurosurg Psychiatry 50:66–70, 1987.

149. Schuchmann, JA: H reflex latency in radiculopathy. Arch Phys Med Rehabil 59:185–187, 1978.

150. Shahani, BT, and Young, RR: Studies of the normal human silent period. In Desmedt, JE (ed): New Developments in Electromyography and Clinical Neurophysiology, Vol 3. Karger, Basel, 1973, pp 589–602.

151. Somerville, J, and Ashby, P: Hemiplegic spasticity: Neurophysiologic studies. Arch Phys Med Rehabil 59:592–596, 1978.

152. Stanley, EF: Reflexes evoked in human thenar muscles during voluntary activity and their conduction pathways. J Neurol Neurosurg Psychiatry 41:1016–1023, 1978.

153. Struppler, A: Silent period. Electromyogr Clin Neurophysiol 15:163–168, 1975.

154. Struppler, A, Struppler, E, and Adams, RD: Local tetanus in man: Its clinical and neurophysiological characteristics. Arch Neurol 8:162–178, 1963.

155. Swash, M: Early and late components in the human anal reflex. J Neurol Neurosurg Psychiatry 45:767–769, 1982.

156. Taborikova, H: Fraction of the motoneurone pool activated in the monosynaptic H-reflexes in man. Nature 209:206–207, 1966.

157. Taborikova, H, and Sax, DS: Conditioning of H-reflexes by a preceding subthreshold H-reflex stimulus. Brain 92:203–212, 1969.

158. Tatton, WG, Forner, SD, Gerstein, GL, Chambers, WW, and Liu, CN: The effect of post-central cortical lesions on motor responses to sudden upper limb displacements in monkeys. Brain Res 96:108–113, 1975.

159. Taylor, S, Ashby, P, and Verrier, M: Neurophysiological changes following traumatic spinal lesions in man. J Neurol Neurosurg Psychiatry 47:1102–1108, 1984.

160. Teasdall, RD, and Van Den Ende, H: A note on the deep abdominal reflex. J Neurol Neurosurg Psychiatry 45:382–383, 1982.

161. Thomas, JE, and Lambert, EH: Ulnar nerve conduction velocity and H-reflex in infants and children. J Appl Physiol 15:1–9, 1960.

162. Tracey, DJ, Walmsey, B, and Brinkman, J: "Long loop" reflexes can be obtained in spinal monkeys. Neurosci Lett 18:59–65, 1978.

163. Troni, W: Analysis of conduction velocity in the H pathway. Part 1. Methodology and results in normal subjects. J Neurol Sci 51:223–233, 1981.

164. Troni, W: Analysis of conduction velocity in the H pathway. Part 2. An electrophysiological study in diabetic polyneuropathy. J Neurol Sci 51:235–246, 1981.

165. Troni, W, Cantello, R, and Rainero, E: The use of the H reflex in serial evaluation of nerve conduction velocity. Electroencephalogr Clin Neurophysiol 55:82–90, 1983.

166. Trontelj, JV: A study of the H-reflex by single fibre EMG. J Neurol Neurosurg Psychiatry 36:951–959, 1973.

167. Upton, ARM, McComas, AJ, and Sica, REP: Potentiation of 'late' responses evoked in muscles during effort. J Neurol Neurosurg Psychiatry 34:699–711, 1971.

168. Van Boxtel, A: Selective effects of vibration on monosynaptic and late EMG responses in human soleus muscle after stimulation of the posterior tibial nerve or a tendon tap. J Neurol Neurosurg Psychiatry 42:995–1004, 1979.

169. Varma, JS, Smith, AN, and McInnes, A: Electrophysiological observation of the human pudendo-anal reflex. J Neurol Neurosurg Psychiatry 49:1411–1416, 1986.

170. Veale, JL, and Rees, S: Renshaw cell activity in man. J Neurol Neurosurg Psychiatry 36:674–683, 1973.

171. Vecchierini-Blineau, MF, and Guiheneuc, P: Electrophysiological study of the peripheral nervous system in children: Changes in proxi-

mal and distal conduction velocities from birth to age 5 years. J Neurol Neurosurg Psychiatry 42:753–759, 1979.

172. Vecchierini-Blineau, MF, and Guiheneuc, P: Excitability of the monosynpatic reflex pathway in the child from birth to four years of age. J Neurol Neurosurg Psychiatry 44:309–314, 1981.

173. Verrier, MC: Alterations in H-reflex magnitude by variations in baseline EMG excitability. Electroencephalogr Clin Neurophysiol 60:492–499, 1985.

174. Vodusek, DB, Janko, M, and Lokar, J: EMG, single fibre EMG and sacral reflexes in assessment of sacral nervous system lesions. J Neurol Neurosurg Psychiatry 45:1064–1066, 1982.

175. Vodusek, DB, Janko, M, and Lokar, J: Direct and reflex responses in peroneal muscles on electrical stimulation. J Neurol Neurosurg Psychiatry 46:67–71, 1983.

176. Wager, EW, Jr, and Buerger, AA: A linear re-

lationship between H-reflex latency and sensory conduction velocity in diabetic neuropathy. Neurology 24:711–714, 1974.

177. Willer, JC, and Dehen, H: Le reflexe h du muscle pedieux: Etude au cours des neuropathies alcooliques latentes. Electroencephalogr Clin Neurophysiol 42:205–212, 1977.

178. Yanagisawa, N, Tanaka, R, and Ito, Z: Reciprocal Ia inhibition in spastic hemiplegia of man. Brain 99:555–574, 1976.

179. Yap, CB: Spinal segmental and long-loop reflexes on spinal motoneurone excitability in spasticity and rigidity. Brain 90:887–896, 1967.

180. Yates, SK, and Brown, WF: The human jaw jerk: electrophysiologic methods to measure the latency, normal values, and changes in multiple sclerosis. Neurology 31:632–634, 1981.

181. Young, RR, and Shahani, BT: Clinical neurophysiological aspects of post-hypoxic intention myoclonus. Adv Neurol 26:85–105, 1979.

Chapter 19

SOMATOSENSORY AND MOTOR EVOKED POTENTIALS

1 INTRODUCTION

Conventional sensory nerve conduction techniques are primarily used in evaluating the more distal portions of the peripheral nerve and seldom contribute to the study of the less accessible proximal segments. In contrast, studies of somatosensory evoked potentials (SEPs) assess the entire length of the afferent pathways. Early SEP works emphasized changes in amplitude and waveform of the recorded potentials in diseases affecting the cerebrum or spinal cord.[137,238,316,327] Other studies have focused on the evaluation of central neural conduction determined by latencies of the SEPs recorded over the spine or scalp.[14,63,74,75,81,85,107,165,240,247]

This chapter will review recording techniques and neural sources of spinal and scalp-recorded SEPs and discuss their practical value and limitations in the diagnosis of certain disorders of the nervous system. Published studies have dealt mainly with the median or tibial nerves, less frequently with ulnar or peroneal nerve, and only occasionally with nonlimb nerves such as the trigeminal and pudendal nerves.

2 TECHNIQUES AND GENERAL PRINCIPLES

Stimulation

Electrical or mechanical stimuli applied at any level can elicit SEPs.[98,235,248,249,269,295] The common sites of stimulation include the median or ulnar nerve at the wrist, the tibial nerve at the ankle, and the peroneal nerve at the knee. A shock with intensity adjusted to cause a small twitch of the innervated muscle suffices to activate all the large myelinated, more easily excitable sensory fibers. Thus, it is convenient to specify the stimulus in milliamperes above the muscle twitch threshold.[87] When one is using square wave pulses of 0.1- to 0.2-ms duration, the usual intensity ranges from 10 to 30 mA, or for a skin resistance of 5kΩ, from 50 to 150 V. Subcutaneous shocks, with the use of a needle electrode inserted close to the nerve, require considerably less current.[189]

The optimal frequency and number of stimuli vary a great deal, depending on the components under study.[216] Small spinal potentials and scalp-recorded short-latency SEPs need a greater number of trials than the later components to achieve the same resolution. For short-latency potentials during the first 20 ms, one trial requires up to 4000 stimuli. The rate of stimulation should not exceed 4 per second, because most subjects tolerate higher frequencies poorly.[189] Medium and long-latency components in the range of 20- to 200-ms intervals require only 200 to 400 stimuli delivered randomly at lower rates of 1 to 2 per second. With shocks given every 30 seconds or less, the initial stimuli of a train may give rise to a disproportionately larger response because the contribution of later stimuli diminishes, presumably because of habituation.

Unilateral stimulation elicits short-latency components symmetrically over both hemispheres. Long-latency responses show obvious asymmetry with major contralateral components that can vary considerably from one trial to the next.[137,238,240,308,316,327,335] In contrast, simultaneous bilateral stimulation gives rise to symmetric responses of all the SEP components, allowing instantaneous comparison between the two hemispheres. A routine evaluation in our laboratory consists of right and left unilateral stimulation to delineate abnormalities of short-latency peaks and bilateral stimulation for assessment of any asymmetry of medium- and long-latency components.[336]

Recording

The analysis of the SEP topography requires simultaneous recording from 16 to 32 scalp areas. For clinical testing, however, two to four well-selected channels covering optimal recording sites suffice. The international 10–20 system (Fig. 19–1A) designates the scalp positions ac-

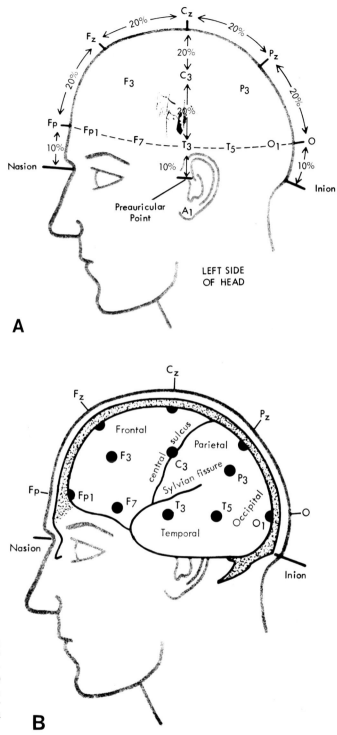

Figure 19–1. A. The 10–20 system based on electrode placement at either 10 or 20 percent of the total distance between skull landmarks. **B.** Relationship between central sulcus, sylvian fissure, lobes of the brain, and electrode positions. (From Harner and Sannit,[143] with permission.)

cording to their specific anatomic locations. It derives its name from spacing the electrodes 10 to 20 percent of the total distance between the nasion and inion in the sagittal plane and between right and left preauricular points in the coronal plane.[143] The use of percentages, rather than absolute distances, provides flexibility for normal variations in head size and shape. On the basis of the anatomic relationship between electrodes placed according to the 10–20 system and cortical landmarks, the C_3 electrode, for instance, lies within 1 cm of the central sulcus (Fig. 19–1B).

Optimal scalp electrodes (G_1) include P_3, P_4, C_3, or C_4 contralateral to the side of stimulus for median or ulnar SEPs and C_1, C_2, or C_z for peroneal or tibial SEPs. A common reference electrode (G_2) usually lies at F_z, the chin, or connecting the ears, A_1 and A_2. Far-field potentials (FFPs) typically affect all scalp points nearly equally. Thus, they tend to cancel, if recorded between two cephalic leads. In contrast, a knee or other noncephalic references provide good resolution of FFPs in median[189] or tibial SEPs.[337] Although the greater separation between G_1 and G_2 generally results in larger amplitude of the recorded response, background noises also amplify, thereby obscuring the signal. The size and shape of evoked potentials depend not only on the potential at G_1 but also on the activity of G_2. Superfluous peaks generated by an "active" G_2 may confuse SEP analyses, especially in assessing short-latency peaks. The activity of G_2, if opposite in polarity to G_1, helps enhance the signal under study. For example, short-latency median SEPs amplify substantially if registered with G_1 placed on the neck and G_2 connecting the ears. Here the recorded reponse represents summation of the neative field registered by G_1 and the concomitant positive potentials from the same source detected at G_2.

The majority of healthy subjects have recordable cervical potentials and short-latency scalp peaks after stimulation of median or ulnar nerve at the wrist. Stimulation of lower limbs gives rise to less consistent cervical and short-latency scalp SEPs. Electrodes placed on the lumbosacral spinous process regularly register spinal potentials after unilateral or bilateral stimulation of the peroneal or tibial nerve.[61,85,165,247,328] The amplitude of evoked spinal potentials increases substantially if recorded with G_1 inserted in the subdural[110] or epidural space,[285] although surface recordings provide a more practical means for clinical use.

Averaging Procedure

A commonly used instrument averages cerebral potentials after amplification by a factor of 10^4 and 10^5 with a frequency cutoff of from 5 to 10 Hz to 3 to 10 kHz. In special studies, a high-pass (low-frequency) restriction of 200 to 300 Hz may aid in selectively eliminating slowly changing events such as synaptic discharges.[208] A computer or averager will then process the amplified potential, converting an analog signal into a digital one, for on-line measurement or for later off-line analysis of the stored data. In general, an adequate analysis will require analog-to-digital (A/D) conversion with 10 to 12-bit resolution (2^{10} to 2^{12} voltage levels or 1024 to 4096 separate voltage steps) and an intersample interval (dwell time per address) of 100 to 500 μs (measurement taken every 100 to 500 μs or 10 to 20 separate points per millisecond). For the study of medium- and long-latency components, a sampling rate as slow as 1 to 2 ms per address (1 kHz to 500 Hz) may suffice.

The memory core available, the number of channels employed, and the duration of total sweep time for each channel determine the sampling rate. Accurate resolution of a waveform requires a minimal sampling at twice the frequency of the signals under study. This allows for the definition of the peak and trough of each complete cycle of a sine wave. For example, an analysis of 5-kHz components calls for sampling rate of 10,000 times per second (10 kHz). Thus, the size of memory core divided by twice the analysis time will dictate the limit of high-frequency analysis. Sampling for a duration of 1 second at a 5-kHz cutoff, therefore, requires 10K memory, whereas only 5K memory would suffice for a shorter analysis time of 500 ms.

To exclude electrocardiogram (ECG), muscle potentials, and other artifacts from averaging, the operator must either study each tracing separately and select only acceptable trials or use a computer program for automatic editing. In our laboratory, the edit program at a sampling rate of 100 to 500 μs/address rejects any trials with five successive equipotential points, which usually indicate an overloaded response or mistrigger. A more commonly used program, based on amplitude criteria, eliminates any unrealistically large potentials exceeding a predetermined level. Artifacts increase nearly in proportion to the distance separating G_1 and G_2, with a higher rate of rejection when referenced to the knee than the ear. A computer program can provide random triggering of the stimulus while automatically avoiding the ECG artifact. Here, QRS complexes, defined as overloaded artifacts exceeding the duration of 100 ms, trigger the stimulus with a varying time delay of 0 to 200 ms following the overloaded period. A high-pass (low-frequency) filter setting of 30 to 100 Hz largely eliminates ECG T waves.

The amplitude resolution of the computer system dictates the degree of accuracy in analyzing small neural responses. Dividing the sum of the responses by an artificial number lower than the actual trial count used in averaging tends to amplify small peaks that would otherwise barely exceed the baseline. The use of a small divisor, however, would excessively amplify the remaining larger component responses, truncating the peaks, which would fall outside the range of the oscilloscope display. A computer program can circumvent this difficulty by determining the smallest divisor that will retain the largest point within the display range. The oscilloscope displays the sum divided by this "adjusted trial count," with a correction factor applied to the computer measurement. Typically, the divisors range from 1/15 to 1/30 of the actual trial count. Other computer manipulations for optimal recording of unstable or rapidly changing evoked potentials include real-time reconstruction using a two-dimensional filter method that stacks successive responses for easy tracking.[272]

3 FIELD THEORY

Near-Field Potential Versus Far-Field Potential

The near-field potentials (NFPs) and FFPs distinguish two different manifestations of the volume-conducted field.[160] The NFP represents the propagating action potentials as detected when the impulse passes under the pickup electrodes, whereas the FFP relates to a stationary potential generated by the signal away from the recording site. A bipolar derivation, used in conventional nerve conduction studies, registers primarily, though not exclusively, the NFP from the axonal volley along the course of the nerve. In contrast, a referential montage preferentially detects the FFP, although it may also register the NFP if the impulse passes near the active (G_1) or indifferent (G_2) electrode. Far-field recording has gained popularity in the study of evoked potentials for detection of a voltage source generated at a distance.[41,44,62,65,77,78,79,95,108,109,204,207,208]

Earlier studies on short-latency auditory evoked potentials suggested that neural discharges from the brainstem might account for FFPs.[159,160,293] This assumption led to the common belief that stationary peaks of cerebral evoked potentials generally originated from fixed neural generators, such as those that occur at relay nuclei. Subsequent animal experiments,[319] however, emphasized the role of a synchronized volley of action potentials within afferent fiber tracts as the source of FFPs. Furthermore, the initial positive peak of the scalp-recorded median (P_9) and tibial (P_{17}) SEP is generated before the propagating sensory nerve action potentials reach the second order neurons in the dorsal column.[44,62,65,76,77,79,95,161,172,184,186,203,204,207,221,241,325,334,337]

These peaks, therefore, must result from axonal volleys of the first-order afferents.[184,186]

Hence, two types of standing peaks can

develop in far-field recording: a volume-conducted potential representing a fixed neural discharge and a stationary peak from an advancing front of axonal depolarization. Why, then, does the far-field activity from a moving source appear as a nonpropagating potential at certain fixed points in time? The traveling volley along the short sequential segments of the brainstem pathways may summate in far-field recording, with the result that the recorded potentials appear as discrete peaks.[319] This mechanism alone does not seem to account for the stationary potentials derived from propagating volleys along the much greater length of the afferent pathways, such as the median or tibial nerve.[103,177,182,183,186,199]

Animal and Human Studies of Peripheral Nerve Volleys

A series of important animal experiments[235] revealed interesting observations on the bullfrog's action potentials, recorded by fluid electrodes, that is, Ringer's solution containing a nerve immersed through a slot of the partition. Stimulation of the nerve at the initial chambers gave rise to a biphasic action potential recorded by adjacent fluid electrodes in the subsequent chambers. With wider separation of the two recording electrodes, the number of action potentials increased to equal the number of the partition between the electrodes. A subsequent experiment[233] demonstrated that the biphasic action potential recorded between the adjacent fluid electrodes became monophasic after sectioning of the nerve at the point of exit from the slot to the next compartment. Cutting the nerve at the point of its entrance into the slot totally abolished the evoked potential.

Studies of the peripheral sensory potentials in man, as simple models of far-field recording, elucidated the possible physiologic mechanisms for the generation of stationary peaks from a moving source.[180,182,183,338] In referential recording of the antidromic median sensory potentials along the third digit, for example, a stationary positive peak developed coincident with the entry of the propagat-

ing sensory potential into the palm-digit junction.[182] In referential recording of antidromic radial sensory potentials (Fig. 19–2A), the digital electrodes detected two stationary FFPs, PI-NI and PII-NII.[181,183] When compared with bipolar recording of the traveling source, P_I occurred with the passage of the propagating sensory impulse at the wrist, and PII, at the base of the digit (Fig. 19–2B). Systematic alteration of stimulus intensity has revealed that FFP occurs in proportion to the propagating volley detected at the boundary of the volume conductor (Fig. 19–3).[179]

One traditionally regards the FFP as a monophasic positivity reflecting the approaching wavefront of depolarization.[160,200,329] Our findings indicate, however, that stationary activity from a moving source often contains a major negative component, that sometimes far exceeds the preceding positivity in amplitude and duration.[180,183] On theoretic grounds, the direction of the traveling impulse in relation to the size of the volume being left and entered may determine the polarity of the FFP. A computer model[296] predicts that the volume entered becomes initially positive or negative compared to the volume departed depending on the relative size of the adjoining conductors. In the case of a boundary constriction, a consensus has emerged that points on the far side begin to go positive when the generator approaches the boundary (Fig. 19–4).[67] Other major determining factors include the direction of axonal volleys as documented in an analysis of P_9 of the median SEP.[79]

Concept of Junctional Potential

Why does a potential difference develop at the boundary with the arrival of the propagating volley? The external field induced by the traveling impulse probably undergoes an abrupt change in current density based on the geometric contour of the volume conductor entered.[182,183] The FFP recorded by surface electrodes in our model bears a great resemblance to the NFP registered by fluid electrodes in an

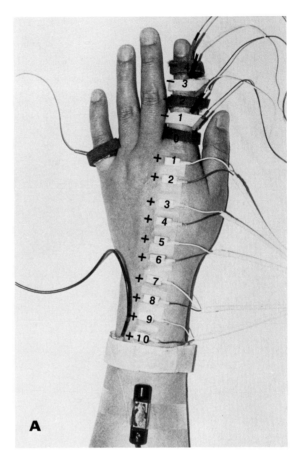

Figure 19–2. A. Stimulation of the radial nerve 10 cm proximal to the styloid process of the radius and serial recording of antridromic sensory potentials in 1.5-cm increments along the length of the radial nerve. The *0* level at the base of the second digit indicates the site where the volume conductor changes abruptly. In most hands, +6 lies near the distal crease of the wrist, where another geometric transition takes place. The ring electrode around the fifth digit served as an indifferent lead for referential recordings. **B.** Sensory nerve potentials across the hand and along the second digit in a normal subject recorded antidromically after stimulation of the superficial sensory branch of the radial nerve 10 cm proximal to the styloid process of the radius (compare **A**). In a bipolar recording (*left*), the initial negative peaks, N_1 (*arrow pointing up*), showed a progressive increase in latency and reduction in amplitude distally and no response beyond −1. In a referential recording (*right*), biphasic peaks, P_W-N_W and P_D-N_D (*arrows pointing down*) showed greater amplitude distally, with a stationary latency irrespective of the recording sites along the digit. The onset of P_W extended proximally to the recording electrodes near the wrist (*small arrows pointing down*), whereas P_D first appeared at the base of the digit. (From Kimura et al.,[183] with permission.)

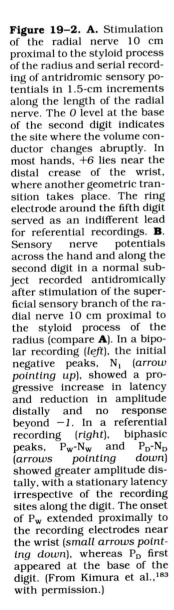

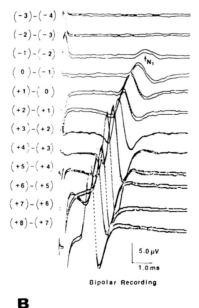

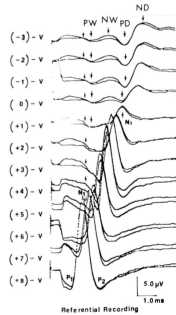

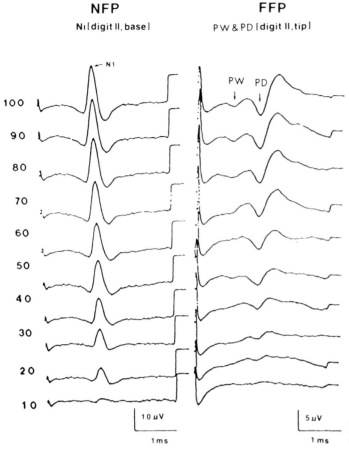

Figure 19–3. The FFP recorded referentially, with G_1 at the tip of the second digit and G_2 at the fifth digit, and NFP registered bipolarly, with G_2 at the base of the digit and G_1 1 cm proximally, after stimulation of the radial nerve. With reduction of stimulus from a maximal (*top*) to a threshold intensity (*bottom*) in 10 steps, the amplitude of FFP (P_W and P_D) declined in proportion to that of NFP (N_1). (From Kimura et al.,[180] with permission.)

in-vitro experiment.[232,233,236] What constitutes an effective partition for this phenomenon, however, remains to be elucidated. The voltage step, once developed at the partition, appears instantaneously as a steady potential between the two compartments. To draw an analogy, an oncoming train (axonal volley) becomes simultaneously visible (FFP) to all bystanders at a distance (series of recording electrodes) as it emerges from a tunnel (partition of the volume conductor), whereas the same bystanders see the train pass by at different times (NFP), depending on their position along the railroad.[180] The designation *junctional* or *intercompartmental* potential specifies the source of the voltage step by location and differentiates this type of FFP from fixed neural generators. A pair of electrodes only a short distance apart best

detects such a stationary potential, so long as they are placed across the partition in question. This observation calls for reassessment of the commonly used dichotomy, equating a referential recording with the FFP and a bipolar recording with the NFP.

Clinical Implications

The complex waveform of the FFP with both positive and negative phases reflects a number of diverse physiologic mechanisms, including those dependent upon the physical relationships between the nerve and the surrounding conducting medium. The animal and human data provide strong, albeit indirect, support for the contention that some of the stationary peaks of scalp-recorded SEPs may re-

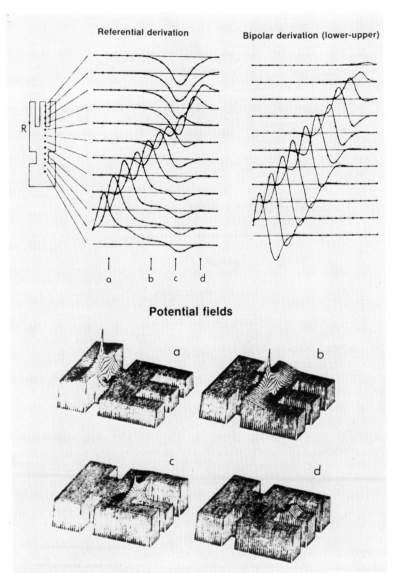

Figure 19–4. Representation of antidromic sensory action potentials propagating through a 3-"fingered" "hand" with independent attenuation of sources and sinks on propagation through the hand, such that the initial source reaches zero amplitude first. The examples include calculated potential fields (*bottom*) and referential (*left*) and bipolar (*right*) waveforms of potential at 12 recording sites against generator position. By field *a* the whole hand has acquired a potential of the same polarity as the initial generator source. In field *b* a stationary potential begins to appear throughout the length of the middle digit, reaching a peak at field *c*. In field *d* the final potential present at the tip of the middle digit is of negative polarity relative to the reference on the lateral digit, as in the actual recordings (compare Fig. 19–2B). (From Cunningham et al.,[67] with permission.)

sult from an abrupt alteration in current flow at various boundaries of the volume conductor. For example, the initial positive peaks of the median (P_9) and tibial SEPs (P_{17}) may arise when the propagating volleys enter the shoulder and pelvic girdles.[182,183] Similarly, the second positive peaks of the median (P_{11}) and tibial SEPs (P_{24}) may, in part, reflect changes in geometry as the impulses reach the cervical cord and conus medullaris. The latencies of these early components support this view. Indeed, a few studies[103,119,120] have documented that P_9 of the median

SEP represents the FFP generated at the shoulder, and change in the position of shoulder girdle slightly but significantly alters the latency of P_9.[79]

Hence, some FFPs used in clinical analysis of the afferent system may not exclusively relate to a specific neural generator. As an inference, certain abnormalities of somatosensory and other evoked potentials could result from changes affecting the surrounding tissue and not necessarily from the sensory pathways per se. Clinical studies of cerebral evoked potential exploit far-field recording in the evaluation of subcortical pathways not otherwise accessible. In these instances, junctional potentials may render clinically useful information disclosing the arrival of the impulse at a given anatomic landmark forming a partition of the volume conductor. This type of recording, however, fails to provide a direct measurement of neural activities responsible for sensory transmission.

4 NEURAL SOURCES OF VARIOUS PEAKS

Nomenclature

Considerable confusion exists in the analysis of SEPs because various authors use different nomenclature for the same waveforms. Some describe the components by location and sequence, i.e., CP for cervical potential and IP, NI, PI and NII for initial positive and subsequent negative and positive scalp-recorded potentials. Others specify the average peak latency to the nearest millisecond, i.e., cervical N_{13} or scalp-recorded P_{14}, N_{17}, P_{20}, and N_{29}.[87] Unfortunately, the latency of the same component varies individually, reflecting the different lengths of the somatosensory pathways, most peaks showing a good correlation with height.[49,50] Ideally, the name of the various components should indicate the respective neural sources, but the exact generator sites of most peaks still remain to be determined.

Median and Ulnar Nerves

Several studies have confirmed the presence of short-latency SEPs in man.[10,44,62,65,76,109,161,189,192,196,207,310] The SEPs recorded simultaneously from the scalp and cervical electrodes help delineate the field distribution of such short latency components (Figs. 19–5 through 19–7, Table 19–1). Most of these studies have dealt with median SEPs, but studies of the ulnar nerve have revealed comparable results.[98,122,123,158]

Stimulation of the median nerve at the

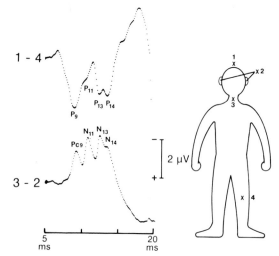

Figure 19–5. Simultaneous recording from C_z (1) referenced to the knee (4) and low cervical electrode (3) referenced to the ear (2) after stimulation of the median nerve at the wrist in a normal subject. Four positive peaks, P_9, P_{11}, P_{13}, and P_{14}, recorded at C_z, were nearly identical in latency to four negative peaks, N_9 (P_{C9}), N_{11}, N_{13}, and N_{14}, recorded at the low cervical electrode. (Compare Fig. 19–6 and 19–7.) (From Yamada et al.,[334] with permission.)

wrist elicits cervical potential (CP) consisting of four negative peaks N_9, N_{11}, N_{13}, and N_{14} when referenced to tied ears (see Fig. 19–5).[334] The earliest component, however, shows relative positivity if recorded with a noncephalic reference (see Fig. 19–6). With the use of the knee reference, the initial positive potential (IP) of scalp-recorded SEPs contain four positive peaks: P_9, P_{11}, P_{13}, and P_{14} (see Fig. 19–7). When referenced to the ear (see Fig. 19–7), the IP consists only of P_{13} and P_{14} because the first two peaks show no potential difference between the two recording sites. These four peaks of the cervical and scalp-recorded SEPs normally occur within the first 15 ms, followed by a small but distinct negative peak, N_{18}, recorded bilaterally in the frontal region. The medium and long-latency components of scalp-recorded SEPs from the central areas during the next 100 ms consist of N_{19}, P_{22}, N_{30}, P_{40}, and N_{60}, or, according to another nomenclature, NI, PI, NII, PII, and NIII.

The earliest scalp potential, P_9,[12,65,189] originates from a distal portion of the brachial plexus and corresponds to N_9 of

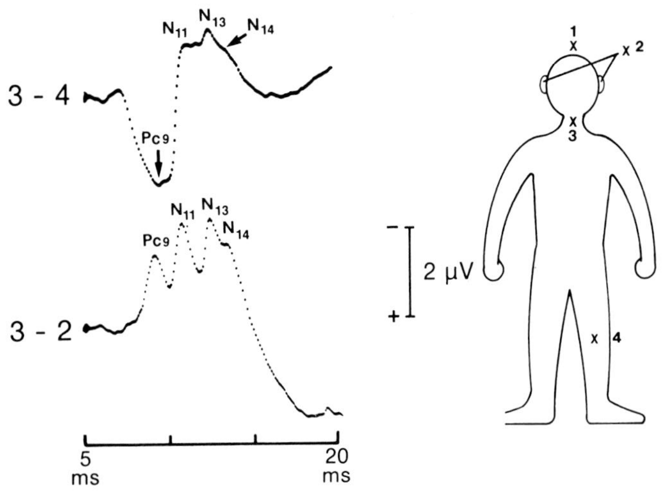

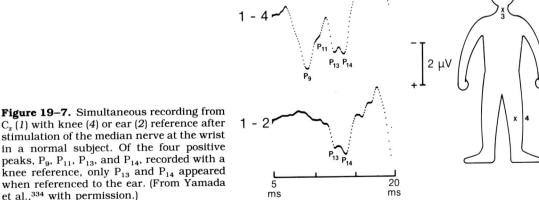

Figure 19–6. Simultaneous recording from a low cervical electrode (3) with knee (4) or ear (2) reference after stimulation of the median nerve at the wrist in a normal subject. The recording with a knee reference showed the initial positive peak, P_{c9}, followed by three negative peaks, N_{11}, N_{13}, and N_{14}. The use of an ear reference reversed the polarity of the first peak and enhanced the subsequent negative peaks. (From Yamada et al.,[334] with permission.)

the cervical potential, by means of a scalp reference (see Fig. 19–5; Fig. 19–8).[161] As mentioned above, the field distribution of the first component shows a diagonal orientation with negativity at the shoulder and axilla and positivity over the entire scalp and neck (see Fig. 19–6; Fig. 19–8). The propagating impulse gives rise to a junctional potential at this level, presumably because of the abrupt geometric change of the volume conductor, although anatomic orientation of the impulse and branching of the nerve may also contribute.[79,182,183,184,237,334]

According to an estimation based on nerve conduction studies, sensory impulses reach the spinal cord in 10 to 11 ms after stimulation of the median nerve at the wrist.[100,185] Thus, N_{11} starts upon arrival of the peripheral nerve volley at the spinal cord level.[76] It closely relates to the activity recorded from the side of the neck ipsilateral to the stimulation (Fig. 19–9). The characteristics of the refractory period indicate the presynaptic nature of this component.[112] The neural source of N_{11}, therefore, must lie near the entry zone, with scalp-recorded P_{11} reflecting the positive end of the same field.[334] Some investigators, however, have observed a delayed N_{11} in patients with cervical cord and medullary lesions. This finding would imply a more rostral origin.[9]

Figure 19–7. Simultaneous recording from C_z (1) with knee (4) or ear (2) reference after stimulation of the median nerve at the wrist in a normal subject. Of the four positive peaks, P_9, P_{11}, P_{13}, and P_{14}, recorded with a knee reference, only P_{13} and P_{14} appeared when referenced to the ear. (From Yamada et al.,[334] with permission.)

**Table 19–1 LATENCY OF ERB'S POTENTIAL AND
SHORT-LATENCY MEDIAN SEP IN 34 NORMAL SUBJECTS**

Components	Latency (Left and Right Combined)			Latency Difference (Between Left and Right)		
	Number Identified	Mean ± SD (ms)	Mean + 3 SD	Number Identified	Mean ± SD (ms)	Mean + 3 SD
Erb's potential	68	9.8 ± 0.8	12.2	34	0.4 ± 0.2	1.0
P_9*	68	9.1 ± 0.6	10.9	34	0.4 ± 0.2	1.0
N_{11}	43	11.2 ± 0.6	13.0	19	0.4 ± 0.3	1.3
N_{13}*	68	13.2 ± 0.9	15.9	34	0.5 ± 0.4	1.7
P_{14}	55	14.1 ± 0.9	16.8	25	0.5 ± 0.4	1.7
N_{18}	68	18.3 ± 1.5	22.8	34	0.5 ± 0.5	2.0
Interwave peaks						
P_9-P_{11}	43	2.2 ± 0.3	3.1	19	0.2 ± 0.2	0.8
N_{11}-N_{13}	43	1.9 ± 0.4	3.1	19	0.2 ± 0.2	0.8
N_{13}-P_{14}	55	1.0 ± 0.4	2.2	25	0.3 ± 0.2	0.9
P_{14}-N_{18}	55	4.2 ± 0.9	6.9	25	0.7 ± 0.5	2.2
P_9-N_{13}*	68	4.0 ± 0.4	5.2	34	0.3 ± 0.3	1.2
N_{13}-N_{18}*	68	5.1 ± 0.9	7.8	34	0.6 ± 0.5	2.1

*Consistently measurable components and interwave peaks.
From Yamada et al.,[340] with permission.

Despite considerable clarification during recent years, the origin and identity of N_{13}/P_{13} components still rank as one of the most controversial SEP topics. The negativity reaches a maximum at the cervical level with decreasing amplitude rostrally and caudally.[29,334] A slight delay of N_{13} at higher cervical electrodes suggests the presence of a traveling wave.[63,185] Recordings of N_{13} from esophageal electrodes or from anterior neck electrodes clearly establish the existence of an

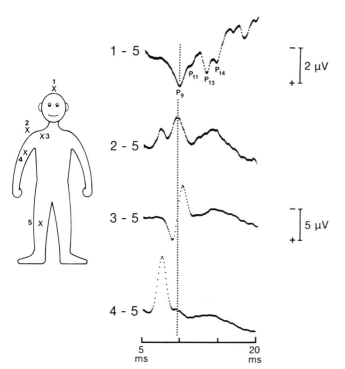

Figure 19–8. Relationship between scalp-recorded P_9 (1) and potentials recorded at shoulder (2), Erb's point (3), and 5 cm distal to the axilla (4). The positive peak, P_9, at the scalp corresponded in latency to the negative peak recorded from the shoulder. (From Yamada et al.,[334] with permission.)

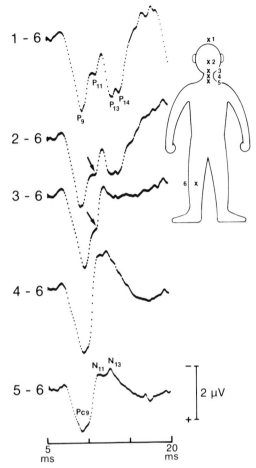

Figure 19–9. Responses recorded from a series of electrodes placed longitudinally at C_z (*1*), O_z (*2*), and high (*3*), mid (*4*), and low (*5*) cervical regions, with a reference lead at the knee (*6*). The amplitude of P_{11} decreased progressively from C_z to high cervical electrodes (*arrows*) with phase reversal to N_{11} at mid and low cervical electrodes. In contrast, the positive field of the first component (P_9) extended from C_z to low cervical electrodes (P_{c9}). (From Yamada et al.,[334] with permission.)

anteroposterior field with positivity anteriorly and maximum amplitude below the foramen magnum.[77,167] These findings suggest that the near-field N_{13} recorded over the cervical spine probably originates in the dorsal horns, although ascending volleys in the dorsal column may also contribute. Lesions at the cervicomedullary junction spare N_{13}, while abolishing subsequent components.[215,218] Some investigators have recorded two subcomponents with different orientations, N_{13a}/P_{13a} and N_{13b}/P_{13b}, possibly

corresponding to generators in the dorsal horns and the cuneate nucleus.[2,167,168]

The third positive scalp potential consists of two different generator sources, P_{13} and P_{14}, as evidenced by its bilobed appearance.[65,189] Debates continue on whether scalp-recorded P_{13} represents the phase reversal of N_{13} from the dorsal horn[77] or corresponds to the ascending volley of the posterior column.[9,161,229,334] A small and at times equivocal P_{13}, recorded over the scalp, stands in contrast to N_{13}, which represents the largest cervical potential. Although some believe that P_{13} originates below the foramen magnum,[75,76,216,218] others,[144,173] on the basis of intracranial recordings in man, propose that P_{13}, like P_{14}, arises from volleys ascending in the medial lemniscus at the brainstem level. Although the origin of P_{13} remains uncertain, it probably corresponds to N_{13} arising from the cervical cord, which in turn consists of at least two subcomponents, as mentioned above.

Earlier studies suggested that P_{14} might arise in the thalamus.[65,161,189] In fact, SEPs recorded in man from the nucleus ventralis caudalis consist of monophasic or diphasic potentials with mean onset latency of 13.8 ms.[40] The preservation of P_{14} in patients with cerebral,[9] thalamic,[152,234] or mesencephalic lesions[44] suggests a more caudal location of the neural source. Further, this component shows no phase reversal between scalp and nasopharyngeal recordings (Fig. 19–10),[334] as might be expected if it were generated in the thalamus. Unlike N_{11} and N_{13}, however, a cervical electrode barely detects N_{14} in most subjects, which indicates that its neural source is rostral to the cervical spine. All of these observations together suggest that P_{14} originates in the medial lemniscus.[76,108,144]

The polarity characteristics of the short-latency SEPs suggest that a negative field near the generator site gives rise to the cervical potentials and that scalp-recorded peaks reflect primarily, although not exclusively, FFPs from the same source. Based on the polarity and mean latency, the presumed generator sites include (1) the entry to the brachial

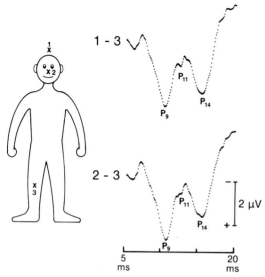

Figure 19–10. Responses recorded from C_z (*1*) and nasopharyngeal electrode (*2*) with a knee reference. The identical waveform of P_{14} in both tracings indicates its generation source caudal to both recording sites, i.e., at or below the base of the skull. (From Yamada et al.,[334] with permission.)

cephalic reference best delineates short-latency SEPs (see Fig. 19–7).[334]

Negative-positive peaks, NI, PI, and NII, subsequent to P_{14} (Fig. 19–11; Table 19–2) show the shortest latency at the frontal electrodes (N_{18}, P_{20}, and N_{29}), with a progressive delay toward the central (N_{19}, P_{22}, and N_{32}) and parietal areas (N_{20}, P_{26}, and N_{34}). In contrast to a small N_{18} recorded bilaterally in the frontal region, the first major negative peaks, N_{19} and N_{20}, skew to the hemisphere contralateral to the side of stimulation. The vertex and ipsilateral, and occasionally contralateral, central electrodes may also register the first negative peak, N_{18}. In this case, N_{18} precedes N_{19} as an additional separate peak, suggesting the presence of two distinct components of separate neural origin. Despite a similarity in latency, close scrutiny reveals that P_{20} has a slightly later onset than N_{20}.[76,184] Topographic analysis otherwise supports the hypothesis of a dipole relationship between these two components of opposite polarity.[3,72,80,315]

If P_{14} arises in the medial lemniscus, then N_{18} originates in a subthalamic structure. In fact, P_{14}-N_{18} may represent a diphasic waveform generated near the foramen magnum.[179] This may explain why N_{18} appears on the scalp with a latency shorter than that of the negativity in the thalamus.[1] An extensive thalamic lesion may spare N_{18}, abolishing N_{20} and subsequent components.[217] The first cortical potential, N1 (N_{19} and N_{20}), and the subsequent negative peak, NII (N_{32} and N_{34}), may have separate cortical generators because of a different potential distribution.[289] The last negative peak, NIII

plexus at the shoulder (N_9 and P_9), (2) the entry to the cervical cord at the neck (N_{11} and P_{11}), (3) dorsal column volley with possible contribution from dorsal horn interneurons and the cuneate nucleus (N_{13} and P_{13}), and (4) entry to the medial lemniscus at the foramen magnum (N_{14} and P_{14}). Of these, P_9, P_{11}, and perhaps P_{14} may in part represent a junctional potential generated by propagating volleys crossing the geometric partition at the shoulder, neck, and foramen magnum. For clinical application, a combined recording from the scalp with a noncephalic reference or from the neck with a

Table 19–2 LATENCY OF MEDIUM- AND LONG-LATENCY MEDIAN SEP IN 34 NORMAL SUBJECTS

Components	Latency (Left and Right Combined)			Latency Difference (Between C3 and C4)		
	Number Identified	*Mean ± SD (ms)*	*Mean + 3 SD*	*Number Identified*	*Mean ± SD (ms)*	*Mean + 3 SD*
N_{18} (NI)	68	18.1 ± 1.6	22.9	34	0.4 ± 0.4	1.6
P_{22} (PI)	68	22.8 ± 2.3	29.7	34	0.6 ± 0.4	1.8
N_{30} (NII)	68	31.6 ± 2.6	39.4	34	0.5 ± 0.4	1.7
P_{40} (PII)	68	43.6 ± 3.6	54.4	34	0.6 ± 0.5	2.1
P_{60} (NIII)	64	62.8 ± 9.3	90.7	32	1.5 ± 1.1	4.8

From Yamada et al.,[340] with permission.

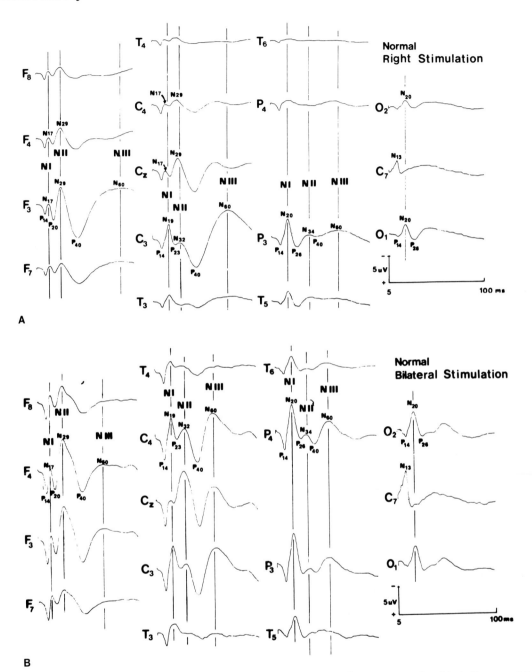

Figure 19–11. A. Median SEP after unilateral stimulation on the right in a normal subject. Topographic analysis indicates N_{17}-P_{20}-N_{29} peaks distributed bifrontally and at the ipsilateral central (C_4) and C_z electrodes, N_{19}-P_{23}-N_{32} at the contralateral central electrode (C_3), and N_{20}-P_{26}-N_{34} at the parietal and occipital electrodes. In this and the subsequent figure, C_7 indicates a cervical electrode just above the C-7 spinal process. (From Yamada et al.,[332] with permission.) **B.** Median SEP after bilateral stimulation in the same subject as in **A.** Potentials recorded over homologous electrodes in the two hemispheres show symmetric patterns that resemble contralateral responses elicited by unilateral stimulation. (From Yamada et al.,[332] with permission.)

Figure continued on next page.

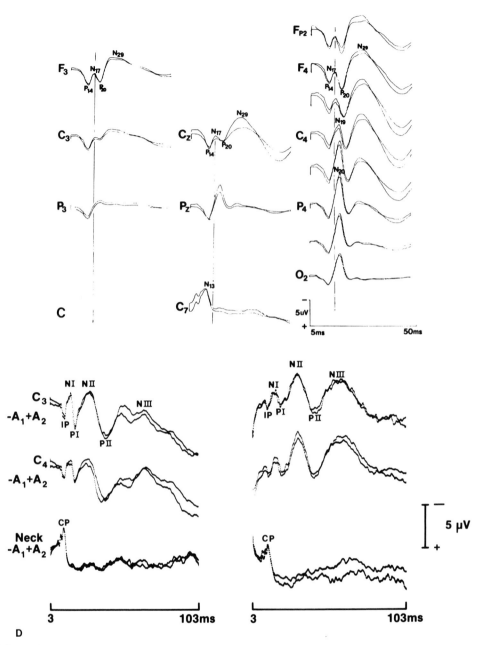

Figure 19–11 (Cont.). C. Topographic display of scalp (10–20 international system) and cervical potentials (C_7) to stimulation of the left median nerve. Frontal N_{17} (F_{Pa}, F_4) preceded central N_{19} (C_4) and parietal N_{20} (P_4) contralateral to the stimulus. N_{17} also appeared at the vertex (C_z) and frontal and central areas ipsilaterally (F_3, C_3). (From Kimura and Yamada,[184] with permission.) **D.** Cervical and scalp-recorded SEPs in two normal subjects after simultaneous bilateral stimulation. Tracings were recorded from the left (C_3) and right (C_4) central regions of the scalp and the mid neck, all referenced to the connected ears. The initial positive potential (*IP*) consists of P_{13} and P_{14}, and cervical potential (*CP*), N_9, N_{11}, N_{13}, and N_{14}. The subsequent negative and positive peaks, NI, PI, NII, PII, and NIV, correspond to N_{19}, P_{22}, N_{32}, P_{40}, and N_{60}. (From Yamada et al.,[337] with permission.)

(N_{60}), shows a wider distribution over the cortex with greater temporal variability than the earlier peaks. In contrast to the medium-latency responses relayed by specific oliogosynaptic routes, a nonspecific polysynaptic pathway probably mediates the long-latency component.

Tibial and Peroneal Nerves

The scalp-recorded potentials usually begin with P_{35} after stimulation of the peroneal nerve at the knee and P_{40} after stimulation of the tibial nerve at the ankle (Fig. 19–12).[88,185] The peroneal or tibial SEP also contains earlier peaks that correspond to the short-latency components of the median SEP (Fig. 19–13).[172,203,228,258,271,291,318] Recording these small potentials requires adequate care for technical details and a greater number of trials for averaging. These components, distributed diffusely over both hemispheres, consist of three regularly elicited components, P_{17}, P_{24}, and P_{31}, and three less consistent peaks, P_{11}, P_{21}, and P_{27}. Of these, only P_{31} is recorded

with the use of the ear or shoulder as a reference; and only P_{24} and P_{31}, with the iliac crest (Table 19–3).

Simultaneous recordings from multiple levels along the somatosensory pathway suggest that P_{17} originates in the peripheral nerve, P_{24}, spinal cord, and P_{31}, brainstem (Fig. 19–14). The initial major component reverses its polarity near the pelvis, rendering the trunk and scalp more positive (P_{17}) than the leg, concomitant with the arrival of the propagating nerve potential at the gluteus (N_{16}). The second component shows the largest negativity over the T-11 to T-12 spinous processes (N_{23}) associated with stationary positive peaks rostrally (P_{24}), with a latency slightly longer than the estimated nerve conduction time from the ankle to the spinal cord. The last component, best recorded as a positive peak at the scalp (P_{31}), coincides with the negative source located in the brainstem (N_{30}). The less consistent peaks, P_{11} and P_{21}, are generated with the arrival of the peripheral nerve potential, N_{11}, at the popliteal fossa, and the spinal potential, N_{21}, at the L-4 spinous process. The other peak, P_{27},

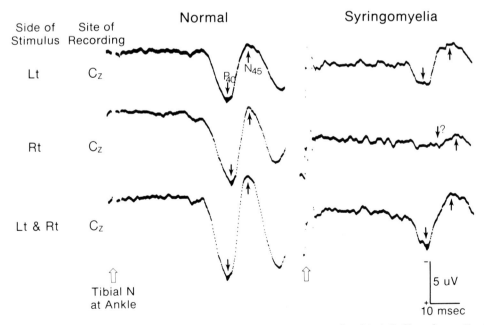

Figure 19–12. Tibial SEPs after stimulation at the ankle in a normal subject (*left*) and a patient with syringomyelia and loss of vibration sense in the right leg (*right*). Note markedly reduced P_{40} and N_{45} to right-sided stimulation in the patient (*middle tracing*). The use of ear reference precluded the recording of short latency positive peaks, P_{17} and P_{24} and minimized P_{31} and the subsequent negative peak, N_{37}, preceding P_{40} (compare Fig. 19–15).

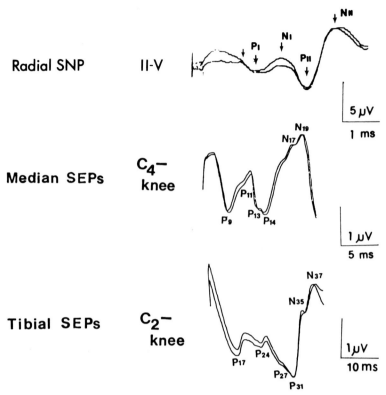

Figure 19–13. Scalp-recorded SEPs using a noncephalic reference after stimulation of the median nerve at the wrist (*middle*) and tibial nerve at the ankle (*bottom*). The waveforms consist of four positive peaks initially and two negative peaks thereafter, all within the first 20 ms in median and 40 ms in tibial SEPs, following the stimulus. For comparison, the top tracing shows FFP, PI-NI, and PII-NII, recorded from digit II referenced to digit V, after stimulation of the radial sensory fibers at the forearm. (From Kimura,[187] with permission.)

**Table 19–3 LATENCY OF SHORT-LATENCY TIBIAL SEP (A)
AND NEGATIVE PEAKS ALONG THE SOMATOSENSORY
PATHWAY (B) IN 21 HEALTHY SUBJECTS**

Recording Site	Scalp					
(A) Components	P_{11}	P_{17}	P_{21}	P_{24}	P_{27}	P_{31}
Mean ± SD (ms)	11.4 ± 2.7	17.3 ± 1.9	20.8 ± 1.9	23.8 ± 2.0	27.4 ± 2.1	31.2 ± 2.1
Number recorded	22	40	21	39	30	40
Number tested	40	40	40	40	40	40

Recording Site	Gluteus	L-4	T-12	C-7	C-2
(B) Components	N_{16}	N_{21}	N_{23}	N_{28}	N_{30}
Mean ± SD (ms)	16.4 ± 3.2	20.9 ± 2.2	23.2 ± 2.1	27.6 ± 1.8	30.2 ± 1.9
Number recorded	20	40	40	18	25
Number tested	22	40	40	22	26

From Yamada et al.,[337] with permission.

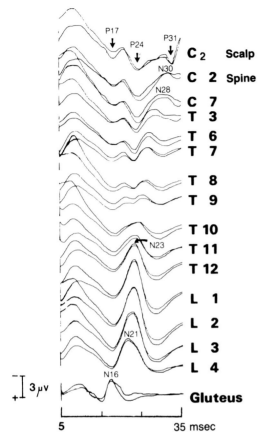

Figure 19–14. Tibial SEPs recorded from a scalp lead and longitudinally placed electrodes over the spine. The first two positive peaks, P_{17} and P_{24}, appeared diffusely not only over the scalp but also along the entire spine. The gluteal lead registered a negative peak, N_{16}, which preceded P_{17} slightly. The second component, P_{24}, extended caudally to the T-11 spine, corresponding to negativity, N_{23}, best recorded at the T-12 spine. A negative peak, N_{30}, recorded at the C-2 spine, slightly preceded P_{31}. (From Yamada et al.,[337] with permission.)

coincides with the arrival of the negative cervical potential, N_{28}, at the C-7 spinous process.

In contrast to the diffuse distribution of the early positive components, the first negative peak shows interhemispheric asymmetry with the ipsilateral response, N_{35}, appearing before the contralateral response, N_{37} (Fig. 19–15). These two peaks probably represent the subthalamic or subcortical responses generated by two independent sources in each hemisphere. In clinical studies, the sub-

sequent positivity, P_{40}, is used in measuring the conduction time to the cortex because of its consistency. The cortical potentials often, though not always, show a paradoxical lateralization with higher amplitude ipsilaterally. This finding may reflect transverse, rather than perpendicular, orientation of the generators located in the mesial surface of the postcentral sulcus.[66] Intrathecal stimulation of the lumbosacral cord elicits similar cortical potentials, although 10 to 15 ms shorter in latency.[111]

Tibial or peroneal SEPs recorded over sacral, lumbar, or low thoracic levels correspond to median or ulnar SEPs at the cervical level.[14,15,73,85,107,165,190,204,247] With the reference electrode (G_2) placed at the T-6 spinous process or the iliac crest, the lumbosacral potential usually attains the maximal amplitude with the active electrode (G_1) over the upper lumbar-lower thoracic vertebrae. Stimulation of the sciatic nerve in the monkey also elicits predominantly negative triphasic spinal potentials along the cauda equina and caudal spinal cord.[113]

The lumbosacral potentials recorded from the surface after stimulation of peroneal or tibial nerve consist of two negative components (Fig. 19–16). The latency of the early peak increases from sacral to upper lumbar levels, but that of the second peak remains constant (Fig. 19–17). Thus, the latency separation between the peaks maximizes in recordings from the lower lumbar or upper sacral sites.[247] The first peak probably represents a traveling wave ascending through the nerve roots of the cauda equina (cauda peak, or R wave); and the second peak, a standing potential generated in the conus medullaris, located at the level of the T-12 spinous process (cord peak, or S wave). The stability of the cord peak in response to short-interval paired stimuli suggests a presynaptic origin.[247]

The cord peak recorded at the level of the T-12 spinous process probably consists of several components, including volleys in the dorsal root and dorsal column,[78] orthodromic and antidromic discharges in the ventral roots,[251] and activities generated locally in interneurons.[24]

Reference:Knee   **Reference:Ear**

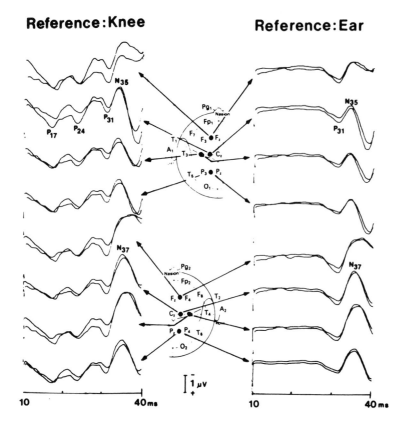

Figure 19–15. Tibial SEPs after unilateral stimulation on the left. The major components consist of symmetrically distributed P_{17}, P_{24}, and P_{31}, recorded with the use of a knee reference (*left column*), and the subsequent asymmetric negative peaks, ipsilateral N_{35} and contralateral N_{37}. The use of ear reference precluded the recording of the short-latency positive peaks, P_{17} and P_{24} (*right column*). (From Yamada,[333] with permission.)

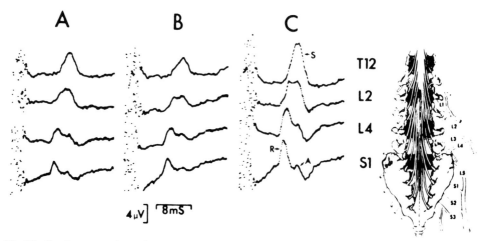

Figure 19–16. Surface recording of lumbosacral evoked potentials after stimulation of the right (**A**), left (**B**), and bilateral (**C**) tibial nerves at the popliteal fossa using a common reference electrode placed over the T-6 spine. The responses consist of spinal (S) and double-peaked (R and A) cauda equina potentials as labeled in **C**. In the diagram of the lower spine and pelvis on the *right*, the shaded areas indicate the location of recording electrodes at T-12, L-2, L-4, and S-1 vertebral levels. All traces represent averages of 64 responses. (From Dimitrijevic, et al.,[85] with permission.)

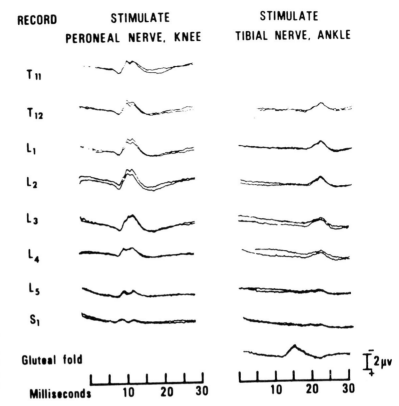

RECORD   STIMULATE   STIMULATE
PERONEAL NERVE, KNEE   TIBIAL NERVE, ANKLE

T 11

T 12

L 1

L 2

L 3

L 4

L 5

S 1

Gluteal fold

Milliseconds   10   20   30   10   20   30

Figure 19–17. Cauda (first) and cord (second) peaks in spinal evoked potentials recorded from multiple spinal levels with a reference electrode over the iliac crest contralateral to the side of stimulation. Each of the superimposed traces represents the average of 128 responses. For comparison, the last tracing of the right column shows a potential recorded from an electrode over the sciatic nerve at the gluteal fold in response to tibial nerve stimulation. (From Phillips and Daube,[247] with permission.)

When recorded caudal to the T-12 spinous process, the potential that occurs synchronously with the cord peak and labeled as the A wave by some may represent efferent motor activity in the anterior root.[73,85,251] It may also be a reflection of a junctional potential recorded at the reference electrode, which becomes positive (P_{24}) when traveling volleys arrive at the conus medullaris (N_{23}) (Fig. 19–18). Such a positive field extends over the entire trunk, head, and arm, affecting any reference electrode placed rostral to the T-12 spinous process.[337]

Trigeminal Nerve

In eliciting SEPs from the trigeminal nerve, the sites of stimulation include the peripheral nerve bundle,[264,288] the upper or lower lip,[114,300] the gums,[26,27,28,42] and other parts of the face.[13,38,91] Each of these methods gives rise to a major triphasic waveform, although the evoked responses vary considerably, depending on the technique used. In one study, scalp SEPs elicited by stimulation of the second division (upper lip) consisted of N_8, P_{14}, and N_{18}, whereas the polarity was reversed to P_8, N_{13}, and P_{19} with stimulation of the third division (lower lip).[114] A bipolar recording between C_3 (G_1) and F_3 (G_2) also revealed an inverted sequence, N_{13}, P_{19}, and N_{26} following simultaneous stimulation of both the upper and lower lips unilaterally (Fig. 19–19, Table 19–4).[300] Using an ear reference, stimulation of the gum above the first maxillary bicuspid elicited scalp responses, N_{20}, P_{34}, and N_{51}.[26,27] Stimulation of the infraorbital nerve elicited three peaks over the scalp, W_1, W_2, and W_3, corresponding to the activity at the entry into the gasserian ganglion, into the pons, and into the trigeminal spinal tract.[193] Awake subjects also had additional components P_4, N_5, P_6, and N_7 when recorded with the use of a noncephalic reference.[143]

The dependence of the waveform on the

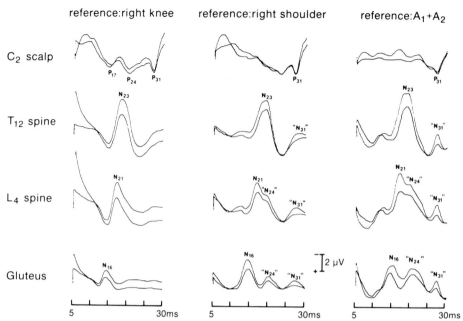

Figure 19–18. FFP recorded over the scalp and NFP along the thoracic and lumbar spine and gluteal fold after stimulation of the tibial nerve at the ankle. The scalp peaks consist of P_{17}, P_{24}, and P_{31} (*left, top*), which coincide with the arrival of NFP at the hip (N_{16}), conus medullaris (N_{23}), and brainstem, respectively. The first lumbar potential (N_{21}), which probably originates from the cauda equina, gives rise to an inconsistent FFP over the scalp. The second (“N_{24}”) and the third (“N_{31}”) peaks in the *right* and *middle columns* represent FFPs registered by an “active” reference electrode. In the *left column*, the use of the “inactive” knee reference eliminated the superfluous peaks.

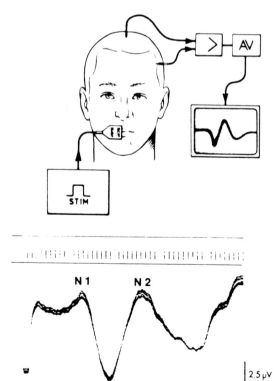

Figure 19–19. A cortical SEP of the trigeminal nerve elicited in a healthy subject following stimulation of the lips. (From Stohr et al.,[301] with permission.)

Table 19-4 PI LATENCY AND NI/PI AMPLITUDE OF TRIGEMINAL SEP IN 82 HEALTHY SUBJECTS

Latency (ms) (Mean ± SD)	Upper Limit (ms) (Mean + 2 SD)	Side to Side Latency Difference (ms) (Mean ± SD)	Upper Limit (ms) (Mean + 2 SD)	Amplitude (μV) (Mean)	Side to Side Amplitude Difference (μV) (Mean)
18.5 ± 1.51	22.3	0.55 ± 0.55	1.93	2.6	0.51

Modified from Stohr and Petruch.[300]

mode of stimulation and recording montage makes it imperative to standardize the test for clinical use in each laboratory. Each published method has advantages and disadvantages. Surface stimulation of the nerve bundle or the lip tends to activate facial muscles, causing major interference with the signal. Needle stimulation of the peripheral division, though invasive, accomplishes more selective activation of the sensory fibers. Stimulation of the gum requires a special supporter to maintain optimal contact between the electrodes and the surface. Regardless of the method selected, stimulus current readily spreads to the pickup electrodes because of their proximity. This results in a large stimulus artifact that tends to preclude accurate analysis of short-latency components. Technical problems thereby limit the clinical usefulness of trigeminal SEPs, despite its theoretic applicability to a number of entities, such as trigeminal neuralgia.[27]

Pudendal Nerve

Stimulation applied either to the base of the penis through a pair of ring electrodes or to the clitoral branch of the pudendal nerve elicits SEPs over the sensory cortex and spinal cord.[135] The concurrent measurement of the cortical and spinal potentials and bulbocavernosus reflexes (see Chapter 18.6) permits the evaluation of the peripheral and central sensory and motor pathways. Stimulation of the vesicourethral junction also elicits cerebral evoked responses with a late prominent negativity.[266] In contrast to distal urethral or pudendal nerve stimulation that activates the somatic afferents, this technique probably excites the visceral afferents arising from the vesicourethral junction.[267]

The pudendal SEPs recorded with G_1 2 cm behind C_z and G_2 over the forehead resemble that of the tibial SEPs (Fig. 19-20A), with an initial positive deflection and subsequent negative and positive sequence.[135] Table 19-5 summarizes the mean latencies and standard deviations of these waves in each of the populations studied. The peak-to-peak amplitude of the maximal response recorded over the midline ranges from 0.5 to 2 μV in men and 0.2 to 1 μV in women, as compared with 1 to 5 μV in the tibial SEP. After stimulation of the pudendal nerve, the spinal potential recorded by G_1 over the L-1 and G_2 over the L-5 spinous process consists of a dominant negative peak with the onset latency 9.9 ± 3.4 ms (Fig. 19-20B).[135] The amplitude ranges from 0.1 to 0.5 μV, but the response is inconsistent in overweight subjects. In comparison, stimulation of the tibial nerve at the ankle elicits spinal response with an onset latency of 20.8 ± 1.8 ms and amplitude of 0.25 to 1 μV.

Based on the latency of spinal potentials, the impulses arrive at the L-1 level about 10 ms earlier after stimulation of the dorsal nerve of the penis than after stimulation of the tibial nerve at the ankle. Pudendal and tibial SEPs over the scalp, however, show similar latencies, presumably because the muscle afferents of the posterior tibial nerve conduct much faster than the cutaneous afferents of the pudendal nerve.

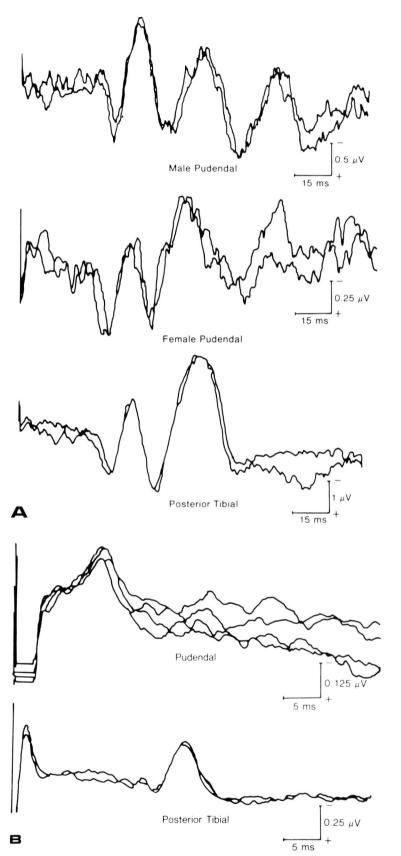

Male Pudendal

0.5 μV

15 ms

Female Pudendal

0.25 μV

15 ms

Posterior Tibial

1 μV

15 ms

A

Pudendal

0.125 μV

5 ms

Posterior Tibial

0.25 μV

5 ms

B

Figure 19–20. A. Cortical SEPs recorded 2 cm behind C$_z$ on stimulation of pudendal and posterior tibial nerves. **B.** Spinal SEPs recorded at the L-1 vertebral spinous process on stimulation of pudendal (*top*) and posterior tibial nerves (*bottom*). (From Haldeman et al.,[135] with permission.)

Table 19–5 LATENCY COMPARISON BETWEEN TIBIAL AND PUDENDAL EVOKED POTENTIAL IN HEALTHY SUBJECTS (MEAN ± SD)

	Onset (ms)	P_1 (ms)	N_1 (ms)	P_2 (ms)	N_2 (ms)	P_3 (ms)	N_3 (ms)
Men (13)							
Tibial	34.0 ± 2.8	41.2 ± 2.9	50.5 ± 3.0	62.7 ± 3.3	78.5 ± 4.4	99.5 ± 6.0	117.9 ± 9.0
Pudendal	35.2 ± 3.0	42.3 ± 1.9	52.6 ± 2.6	64.9 ± 3.4	79.3 ± 4.0	96.6 ± 4.7	116.0 ± 7.2
Women (7)							
Tibial	32.7 ± 1.7	39.3 ± 1.4	49.4 ± 2.1	60.0 ± 2.0	76.1 ± 4.2	96.1 ± 5.8	119.2 ± 7.9
Pudendal	32.9 ± 2.9	39.8 ± 1.3	49.1 ± 2.3	59.4 ± 2.8	73.4 ± 4.6	90.1 ± 5.8	110.0 ± 10.2

From Haldeman et al.,[135] with permission.

5 PATHWAYS FOR SOMATOSENSORY POTENTIALS

Peripheral Inputs and Their Interaction

Early clinical studies revealed abnormal SEPs only in patients with impaired vibration or position sense, whether the lesions involved the spinal cord,[137] cerebral hemisphere,[327] or brainstem.[240] These findings suggested the dependency of all SEP components upon the integrity of the dorsal column-medial lemniscal system in man. Activity carried in the anterolateral column, however, also reaches the cortex in monkeys as well as in man.[7,31,344] Indeed, stimulation with an intensity great enough to activate both large and small diameter fibers in the peroneal nerve produces SEPs even after transection of the dorsal column and spinocervical tract in cats.[175]

Clinical observations also support the experimental evidence in favor of separate sensory pathways mediating various SEP peaks. For example, occasional patients with selective impairment of pain-temperature sensation without loss of position-vibration sense have a depressed or absent NII despite relative preservation of NI (Figs. 19–21 and 19–22). Conversely, lesions of the brainstem, cervical cord, or brachial plexus[341] may affect NI and earlier peaks selectively, sparing NII and subsequent components (Figs. 19–23, 19–24, and 19–25). Such dissociated abnormalities of early or late com-

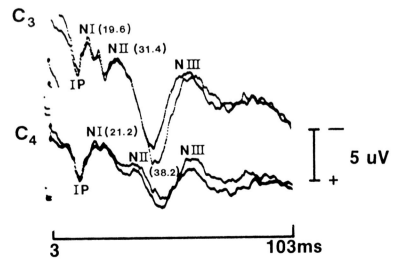

Figure 19–21. Scalp-recorded potential in response to bilateral stimulation of the median nerve in a 33-year-old man with traumatic avulsion of the C-8, T-1, and probably T-2 roots on the left. Myelography demonstrated a large meningocele at C-7. Interhemispheric comparison revealed no asymmetry for the initial positive peak, IP, and the subsequent negative peaks, NI, and NIII despite an obvious delay of NII on the right (C_4).

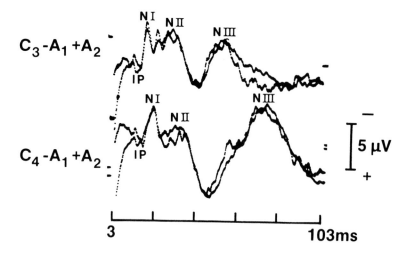

Figure 19–22. Scalp-recorded potential in response to bilateral stimulation of the median nerve in a 46-year-old woman with multiple sclerosis. Interhemispheric comparison showed a slight delay of IP and NI and far greater delay of NII and NIII on the right (C_4). (From Yamada et al.,[337] with permission.)

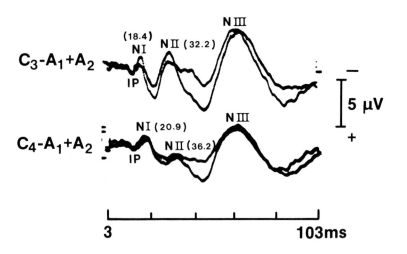

Figure 19–23. Scalp-recorded potential to bilateral stimulation of the median nerve in a 59-year-old woman with multiple sclerosis. Despite a delay of NI and NII on the right (C_4) following a normal IP, NIII showed no difference between the two hemispheres. (From Yamada et al.,[337] with permission.)

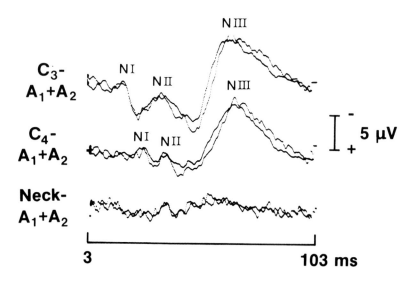

Figure 19–24. Scalp-recorded potential to bilateral stimulation of the median nerve in a 55-year-old woman with multiple sclerosis. The tracings consisted of bilaterally absent IP, substantially delayed NI, borderline NII, and normal NIII on the right.

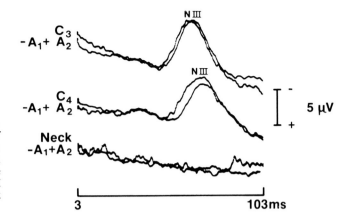

$$-A_1+ \overset{C_3}{A_2}$$
$$-A_1+ \overset{C_4}{A_2}$$
Neck
$$-A_1+A_2$$

N III
N III

5 µV

3   103ms

Figure 19–25. Scalp-recorded and cervical potentials to bilateral stimulation of the median nerve in a 47-year-old man with multiple sclerosis. A well-preserved NIII occurred as the initial potential in the absence of preceding peaks, IP, NI, and NII (Compare Fig. 19–11D.) (From Yamada et al.,[337] with permission.)

poneats suggest the presence of at least partially independent central pathways, mediating NI, NII, and NIII. These findings also tend to refute the traditional view that successive peaks of the SEP represent the sequential activation of a unitary somatosensory pathway.

The fast conducting, large myelinated sensory fibers such as muscle afferents primarily, though perhaps not exclusively, mediate SEP components.[121] A pinprick, but neither touch nor tactile tap, elicits SEP in patients with loss of vibration and touch sensations.[284] Mechanical stimuli also evoke lower amplitude responses with fewer components than electrical stimulation, which activates more fibers synchronously.[52,53,249] Passive plantar flexion of the ankle can also elicit cerebral potentials in man, presumably via the afferent fibers that originate from muscle mechanoreceptors.[295] These findings all support the contention that first-order afferents outside the posterior column also contribute to some of the SEP peaks.[316]

During gating experiments, which test input interactions, different kinds of movement primarily affect the late components of SEP, with only minimal changes in the early cortical responses.[8,263] In one study, movement of the first digit, but not the fifth digit, attenuated P_{27} cerebral potential elicited by stimulation of the first and second digit, or the median nerve. Conversely, the other component, evoked by stimulation of the fifth digit, or the ulnar nerve, was attenuated during movement of the fifth

digit but not of the first digit. Thus, gating of SEPs occurs selectively with movement that involves the areas of stimulation.[309] Vibration attenuates spinal and cerebral potentials evoked by stimulation of the mixed nerve or muscle spindle but has no effect on cutaneous input.[51] The final waveform of the recorded potential depends on complex interaction of varied sensory inputs from different sources, some facilitatory and others inhibitory.[169,170,171]

Central Mechanisms for Integration

The brief effect of an inflated cuff on the nerve, caused by ischemia, rather than mechanical compression, involves the largest myelinated fibers first.[116,118] Such tourniquet-induced ischemia diminishes the short-latency SEP peaks, P_9, P_{14}, and the first cortical response, NI, along with the nearly parallel loss of the potential recorded at Erb's point.[339] Thus, the large myelinated fibers responsible for nerve action potential must subserve the early SEP components. Relative sparing of the later components, PII, NII, PIII, and NIII, implies the presence of independent routes, possibly involving different peripheral axons, for example, smaller myelinated fibers. Interestingly, ischemia prolongs the latencies of PII and later peaks more than those of the earlier peaks (Fig. 19–26), again indicating the heterogeneity of the afferent fibers contributing to the SEPs.

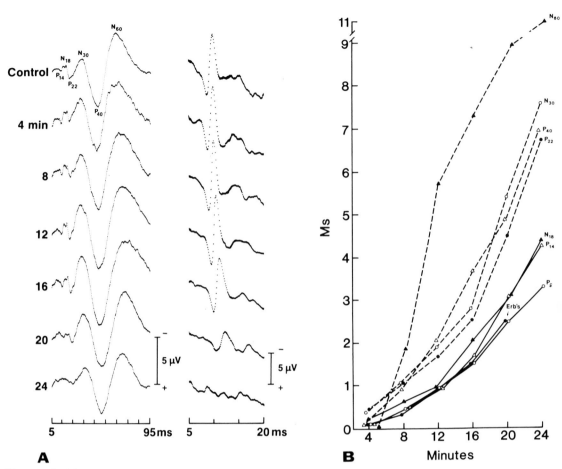

Figure 19–26. A. Sequential changes of scalp-recorded SEPs (*left*) and Erb's potential (*right*) during mechanical application of pressure cuff around the upper arm in a normal subject. Ischemia affected the initial positive and negative components, P_{14} and N_{19}, along with Erb's potentials earlier than the subsequent components, P_{22}, N_{32}, P_{40}, and N_{60}. A 24-minute compression abolished the "ischemia sensitive" peaks while preserving the "ischemia resistant" peaks relatively intact. **B.** Effect of ischemia on the SEPs and Erb's potential. The change in latency in milliseconds (*ordinate*) plotted against duration of ischemia in minutes (*abscissa*) showed a clear dissociation in the time course of latency change between the "ischemia-sensitive" and "ischemia-resistant" components. (From Yamada et al.,[339] with permission.)

Such discrepancy between early and late SEP components may also reflect central amplification that compensates for peripheral conduction block.[104] In one study, early components of SEPs attained a maximum amplitude before the responsible muscle afferent volley reached 50 percent of its maximum.[121] Therefore, a few large afferent fibers which survive the effects of ischemia may suffice to evoke late SEPs. In addition, apparent dissociation between central and peripheral sensory response may result from a differential effect of desynchronization on peripheral axons and central synaptic relays. The nerve action potential undergoes substantial diminution based solely on phase cancellation between unit discharges of fast and slow conducting fibers.[181] Similarly, the diminution of early SEPs may initially result from temporal dispersion of axonal volleys, rather than conduction block. If so, the cortex, operating as an itegrator, may generate a sizable evoked response after several synaptic relays, which tend to resynchronize the incoming inputs.[74]

On the other hand, SEPs may favor the inputs from the fast conducting fibers that reach the synapse first, inhibiting those from the slow conducting fibers by prior activation of the common pathway shared by the afferent fibers. This phenomenon would also explain the commonly encountered discrepancy between a very abnormal sensory nerve action potential and a relatively normal SEP recorded in severe neuropathies.

Regardless of the underlying physiologic mechanisms, these observations have practical implication in the clinical assessment of SEP abnormalities. Patients with severe sensory neuropathy may have absent peripheral nerve potentials with preserved, albeit delayed, SEP peaks. These disorders may affect the amplitude of the initial SEP peaks selectively without concomitant diminution of the later components. More important, conduction abnormalities of the peripheral nerve can lead to increased interpeak latencies of scalp responses as the result of disproportionate delay of the late components. Thus, a latency dissociation between early and late SEP peaks does not necessarily imply a central lesion. This possibility underscores the importance of demonstrating the integrity of the peripheral nervous system by appropriate conduction studies as part of SEP evaluation.

Measurement of Conduction Time and Various Factors

In the clinical assessment of SEPs, two separate trials with the same stimulus setting serve to confirm the consistency of the recorded response. Repeat studies on successive occasions show better stability for SEPs elicited by stimulation of the upper limbs than of the lower limbs.[274] The usual measurements include onset and peak latencies and peak-to-peak amplitudes (see Tables 19-1 to 19-3). Available data suggest a linear relationship between the subject's height and the latency of a given peak elicited by stimulation of lower limb.[174] SEP latencies also change as a function of body temperature, affecting central, more than peripheral, conduction time, which includes synaptic delay.[154]

Group means of the median SEPs indicate minor differences in waveform and latency between the sexes.[155] During normal postnatal development up to 8 years of age, scalp-recorded tibial SEPs show latency changes that reflect complex maturation of central pathways. In contrast, the latencies of the peripheral and lumbar potentials correlate positively with age and height, yielding a predictable homogram.[128] Short-latency SEPs in infants and children resemble those of adults but show great maturational changes until adolescence. The peripheral part of the sensory pathway reaches the adult range at 3 to 4 years of age; and the central part, at school age.[313] The central conduction time (mean $\pm$ SD), measured from the cervical area (N_{13}) to the primary cortical response (NI), remains relatively constant (5.66 ± 0.44 ms) between 10 and 49 years of age. It increases by approximately 0.3 ms between the fifth and sixth decades, with no further change thereafter.[153]

The conduction time over the somatosensory pathways can be divided into two parts, peripheral and central latencies. The peripheral latency corresponds to peripheral sensory nerve conduction of the first neuron, that is, from the stimulus site to the spinal cord entry. The central latency consists of spinal conduction along the remaining segment of the first-order afferent up to the dorsal column nuclei and subsequent relay through the lemniscal system and thalamocortical fibers over at least three synapses. The central latency for the median nerve is a measure of the sensory pathways from the cervical enlargement (C-7 spinous process); and that of the tibial nerve, from the conus medullaris (T-12 spinous process). The difference between the two provides the spinal cord conduction from the conus medullaris to the cervical enlargement.[88,89,96] The same spinal cord segment can be assessed on the basis of measurement of cortical potentials elicited by epidural stimulation of the cervical and thoracic spinal cord.[30]

The spinal potentials may be recorded over the C-7 spinous process for assessment of the peripheral conduction time to the spinal cord in the upper extremities and over the T-12 spinous process, in the lower limb. Despite the technical difficulty, such direct measurements provide more reproducible results than an indirect estimate of peripheral latency that causes considerable variability because of a cumulative error.[88,101,185] Thus, the currently available methods provide only a gross approximation of spinal cord conduction. In addition, the technique applies only to SEP components mediated by large myelinated, fast conducting fibers. Changes in conduction characteristics of slower conducting fibers, not assessed by conventional nerve conduction studies, could alter the latency and waveform of SEPs.

6　CLINICAL APPLICATION

Studies of SEPs have made a steady progress since the original description by Dawson[69] four decades ago. The advent of microcomputers and digital processors has freed the student of clinical neurophysiology from the limitations of analog analysis. This in turn has led to a rapid escalation in the use of SEP and other evoked potential studies in the clinical domain, and a great number of patients currently undergo such a test as a routine procedure. Important questions remain, however, to clearly delineate the practical scope of SEP and its proper usage. These include standardization of the technique and nomenclature, precise localization of neural generators, and elucidation of various factors that affect the measurements.[139]

Lesions of the Peripheral Nerve

The studies of SEPs supplement conventional sensory nerve conduction tests in general and assessment of the proximal sensory fibers in particular.[81,140] Selective stimulation of the afferent fibers elicits only a small peripheral sensory response, especially in diseased nerves.[253] Mixed-nerve potentials, though relatively large, contain not only the sensory volleys from skin, joint, and muscle afferent fibers but also antidromic motor im-

pulses. In contrast, spinal or scalp-recorded responses are solely derived from sensory potentials, primarily mediated by the large afferent fibers, even after stimulation of a mixed nerve. Selective stimulation of digital nerves is used to elicit SEPs, with the first, third, and fifth digits corresponding to the C-6, C-7, and C-8 roots, in the differentiation of radicular lesions.[304]

Disorders commonly tested by this means include lesions involving the root, plexus or thoracic outlet (see Fig. 19–21).[98,99,245,342,343] Surgical decompression of lumbar spinal stenosis may shorten the latency of tibial, peroneal, or sural SEPs.[131] Pudendal SEPs, together with the bulbocavernosus reflex, help in evaluating sacral nerve root or plexus injuries and bowel, bladder, and sexual dysfunction.[135,231]

SEP measurement also provides sensory studies of the median, ulnar, radial, musculocutaneous, sural, superficial peroneal, and saphenous nerves.[98,305] These studies help characterize peripheral sensory conduction, especially in the absence of sensory nerve action potentials, in peripheral neuropathies.[257] In one series of eight patients with chronic acquired demyelinating neuropathy, however, the results often provided misleadingly normal data, presumably because of central amplification of an attenuated response arising from a few axons that conduct normally.[244] In 27 patients with the Guillain-Barré syndrome, 7 had normal SEPs despite an abnormal F wave from the same nerve, and none had normal late responses and abnormal SEPs.[254]

Unfortunately, the test improves the accuracy of diagnosis less than one might have expected on theoretic grounds in many instances. For example, in a combined study of SEPs and peripheral sensory nerve action potentials, preoperative findings correlated well with the discovered locus of brachial plexus lesions in only 8 of 16 patients.[166] In the remaining eight, five patients had only minor discrepancy between electrophysiologic and operative data, but the other three patients had unexpected root avulsions at surgery despite a prediction of a purely postganglionic lesion. The use of SEP alone would have been less helpful because abnormalities of peripheral sensory nerve action potentials contributed substantially to the accurate localization of the pathologic process. A major limitation of this technique stems from its inability to test preganglionic involvement in the presence of a postganglionic lesion, which precludes the evaluation of a more proximal segment.

Lesions of the Spinal Cord and Brainstem

Simultaneous recording of a sensory nerve action potential and SEPs allows comparison of peripheral and central sensory conduction times. SEP abnormalities indicative of central involvement are seen in various neuropathies[261,279] and many systemic disorders, such as myotonic dystrophy.[22] In Freidreich's ataxia, studies of the sural nerve show normal conduction velocity despite reduced amplitude.[92,219] Similarly, the median sensory potentials recorded at the clavicular fossa show a marked attenuation but little evidence of delay. Studies of SEPs reveal a dispersed and delayed cortical response, suggesting slowed conduction in central pathways.[81,162] Spinal SEPs also show frequent defects in spinal afferent transmission in diabetes[60] and Charcot-Marie-Tooth disease.[163]

SEPs provide a unique means of assessing spinal cord injury,[344] subacute combined degeneration,[115] spondylotic myelopathy,[342,343] hereditary spastic paraplegia,[311] and metachromatic leukodystrophy.[330] In syringomyelia (see Fig. 19–12) abnormalities of SEPs accompany clinical sensory loss despite normal sensory nerve conduction studies.[6] Patients with subacute myelopticoneuropathy (SMON) also have marked attenuation of the cortical component and delayed central conduction of tibial SEPs with normal peripheral conduction.[277]

A focal compression of the spinal cord generally results in slowing of spine-to-spine and spine-to-scalp propagation velocities. In contrast, diffuse or multifocal lesions of the spinal cord often lead to the absence of scalp response.[268] Cervical cord lesions attenuate or abolish cervical responses if evoked by stimulation of an appropriate nerve, for example, the musculocutaneous nerve for the C-5 or C-6 segment, the median nerve for the C-6 or C-7 segment, and the ulnar nerve for the C-8 or T-1 segment.[201] High cervical lesions that spare early cervical potentials may abolish or delay the later components.[270] Finally, the electrophysiological characteristics of lumbosacral evoked potentials suggest a degree of spinal cord dysfunction caudal to the area of injury in a substantial number of patients with established spinal cord injury.[195]

SEPs also provide useful data in patients with vascular lesions of the brainstem[81] and infratentorial space-occupying lesions.[324] Motor neuron disease may also exhibit various SEP abnormalities, despite sparing of the sensory system clinically.[33,56,250]

Lesions of the Diencephalon and Cerebrum

In patients with localized cerebral lesions, SEP abnormalities vary considerably, depending on the site of involvement. The pattern of SEP changes, therefore, help localize lesions within the cerebral hemispheres.[299] Thalamic lesions may spare N_{18}[217] or totally eliminate NI, NII, and NIII components with preservation of only P_{14}.[47,142] Capsular lesions tend to spare P_{14} and frontal N_{18} but alter all the subsequent SEP components or involve NII or NIII selectively. In contrast, a sizable lesion in the frontal or parietal lobe may affect only NII or NIII.[335] In some cases, the corresponding peaks at the central and parietal electrodes may show independent abnormalities after stroke[332] or resuscitation from cardiac arrest.[37] Similarly, frontoparietal tumors may result in complete absence of N_{70} of the tibial SEP, whereas a frontal meningioma leads only to a slight alteration.[93] In patients with occlusive cerebral vascular disease, SEP studies may reveal increased amplitude on the affected side, perhaps reflecting a disturbance of the suppressor cortex.[297]

A variety of SEP abnormalities observed in restricted nonhemorrhagic thalamic lesions reflect the presumed vascular territories.[331] Involvement of the primary sensory nuclei, causing the thalamic syndrome or the loss of all modalities of sensation, characteristically eliminates all SEP components after P_{14} and N_{18}. Anterior thalamic lesions not involving primary sensory nuclei often delay NI, whereas medial thalamic lesions tend to affect central NIII. Posterior capsular or lateral thalamic lesions may involve both NII and NIII or NIII alone. The complex relationship between the type of SEP abnormalities and the location of thalamic lesions suggests the presence of multiple, at least partially independent, thalamocortical projections mediating regionally specific somatosensory inputs.[217,334] Change in median SEP noted while monitoring carotid endarterectomy may signal cerebral ischemia and the need for a shunt during the surgery.[210]

Patients with Huntington's disease show a drastic diminution in amplitude of early cortical components, especially the N_{20}-P_{25} component of median SEPs and the N_{33}-P_{40} component of tibial SEPs.[32,94,243] In contrast, those with cortical myoclonus characteristically have grossly enlarged responses that persist after administration of clonazepam or lisuride, which is known to reduce myoclonus.[262] Also, some patients with adult ceroid lipofuscinosis have nearly monophasic, very high amplitude SEPs totally unlike those found in normal control subjects.[322] In Wilson's disease, the majority of patients with neurologic manifestations have some abnormalities of median or tibial SEPs, as expected from widespread degeneration of the brain.[48]

Multiple Sclerosis

Symptoms and signs of multiple sclerosis result from abnormal conduction of central nerve fibers across areas of demyelination. Delayed median SEPs in patients with impairment of position or vibration sense indicate conduction abnormalities of the posterior column (see Figs. 19–22, 19–23, 19–24, and 19–25).[55,238] Studies of SEPs can also uncover clinically silent lesions and document dissemination of disease in patients with clinical signs confined to a single site.[43,54,81,340] Such studies also help quantitate any known abnormalities and localize the level of the sensory disturbance in patients with paraparesis.[89]

Scalp-recorded SEPs show an overall incidence of abnormality ranging from 50 to 86 percent in patients with an established diagnosis.[12,23,43,81,214,314] Subclinical abnormalities are found in 20 to 40 percent of suspected or possible cases,[43,54,340] with greater sensitivity after stimulation of the lower limb.[4,337] A substantial number of patients have major asymmetries in the medium- and long-latency components (after NI) elicited by bilateral stimulation of the median nerve, despite normal short latency components (up to NI).[340] Rising body temperature causes conduction block in demyelinated axons in the sensory pathway, thereby distorting the cervical[213] and scalp SEPs.[246]

Recording a short-latency median SEP (N_{13}) from the neck revealed abnormalities in 69 to 94 percent of those with a definite diagnosis and in 44 to 58 percent of patients with a possible diagnosis.[106,212,290] The latency difference between cervical and scalp-recorded negative peaks showed an 83 percent incidence in the definitive group and 68 percent overall abnormality.[101,102] Stimulation of the tibial nerve commonly fails to elicit cervical potentials in definite multiple sclerosis even with minimal clinical signs.[291]

The incidence of evoked potential abnormalities generally increases in proportion to the duration of clinical illness.[34] Unfortunately, intertrial variability sometimes exceeds the expected changes brought about by disease processes, leading to a tenuous temporal correlation between clinical and electrical changes.[5,54,57,226] Indeed, evoked potential studies may not provide information for monitoring progression of disease, and there may be frequent disparity between the clinical and electrophysiologic courses.[5,68,83] Furthermore, some SEP abnormalities are not directly correlated with the presence or degree of clinical sensory impairment.[11]

As a dignostic study of multiple sclerosis, SEPs and visual evoked potentials (VEPs) contribute more than brainstem auditory evoked potentials (BAEPs) or electrically elicited blink reflexes. Waveform analyses may yield a higher incidence of abnormality than latency measurement alone.[105] Serial studies of multimodality evoked potentials, if properly selected on the basis of clinical findings, can establish temporal or anatomic dissemination but not necessarily the specific diagnosis of multiple sclerosis. Combined evoked potential testing yields a higher sensitivity than magnetic resonance imaging (MRI),[127] although MRI is positive more often than is any single evoked potential study.[298] Although the incidence of SEP abnormalities is high in multiple sclerosis, patients with acute inflammatory transverse myelopathy tend to have entirely normal responses.[255]

Spinal Cord Monitoring

Another clinical application of SEPs relates to its use as an intraoperative spinal cord monitor.[133,294] During scoliosis surgery or removal of a spinal cord tumor, general or local anesthesia precludes clinical examination of spinal cord function. Tibial or peroneal SEPs, however, can be recorded under halogenated inhalational anesthesia, but average amplitude is slightly reduced.[265] In patients with preoperative evidence of cervical cord damage, the cortical response tends to fluctuate. In fact, it could be lost without a major change in the concentration of the anesthetic agent or surgical manipulation.[321] Even though a normal SEP offers no guarantee for the integrity of the entire pathways of the spinal cord, a markedly distorted or delayed response signals a warning and an impending risk.

Most initial studies dealt with cortical potentials evoked by peripheral nerve stimulation.[36] The major disadvantage of this type of recording centers around its dependency on fluctuating levels of consciousness during anesthesia. Spinal cord potentials show less variability when recorded either from Kirshner wire electrodes inserted in the spinous processes[242] or from needles in the interspinous ligament.[205] It is also possible to record FFPs with the use of a pair of surface electrodes placed over the neck and scalp.[205] In some patients, however, both the spinal cord potentials and FFPs have an amplitude too low to permit a satisfactory assessment. In these cases, cauda equina stimulation produces significantly higher evoked potentials, permitting reliable monitoring of spinal cord function.[86]

Stainless steel wire electrodes inserted into the epidural space register two to three negative potentials after stimulation of the peripheral nerve in man (Fig. 19–27).[164,286] Estimated conduction velocity ranges between 65 and 80 m/s for the fastest activity and 30 and 50 m/s for the slower waves.[164,209] In animals, spinal evoked potentials also consist of two negative peaks after direct cord stimulation.[306,317] Transection of the lateral column attenuates the first peak; and that of the posterior column, the second peak. The subsequent polyphasic waves probably result from slower conducting sensory ascending pathways.

Epidurally applied shock to the spinal cord yields better spinal or scalp potentials than surface stimulation of the peripheral nerve. Spinal potentials elicited by this means consist of two major negative peaks, NI and NII, and subsequent multiple smaller components (Fig. 19–28).[209] The same spinal stimulation also elicits a compound muscle action potential in the lower limb as a measure of motor function during surgery. Individual variability in the waveform and amplitude of the spinal potential reflect inconsistency in the placement of the stimulating or recording electrodes. Precise positioning of electrodes at optimal locations would minimize this difficulty by selective stimulation of or recording from the spinal pathway in question. The facilitatory or inhibitory effect on the spinal motor neurons, however, may spread many segments below the level of the cathode.[134] The various recording techniques described here complement each

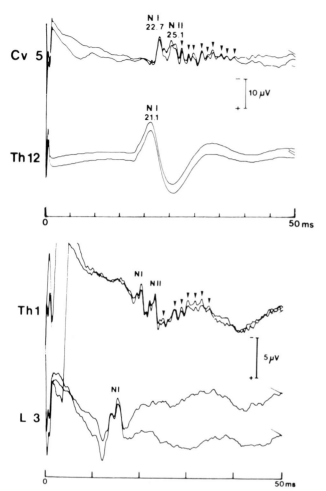

Figure 19–27. SEPs recorded from an epidural electrode placed at the rostral and caudal spine after stimulation of the tibial nerve in two subjects with scoliosis. The response recorded at T-12 spine level consisted of a single diphasic potential with the initial negativity. The waveform varied considerably when recorded further caudally. Polyphasic waves followed the major negative peaks, NI and NII, at the rostral spine. (From Machida et al.,[209] with permission.)

other in the assessment of spinal cord function in the operating room. Postoperative neurologic deficits, however, may ensue despite unchanged intraoperative SEPs.[197]

Clinical Value and Limitations

Although a number of neurologic conditions not discussed above accompany abnormal SEPs, the value of the technique as a clinical test in these entities awaits further analysis.[46,178] This category includes head trauma[39,273] coma,[124,323] brain death,[25,125,130] cord injury,[90] cervical spondylosis,[287] degenerative diseases in children,[59] adrenoleukodystrophy,[126] trigeminal neuralgia,[301] neurogenic bladder,[135] cerebral aneurysm,[117] cerebrovascular ischemic disease,[252] olivopontocerebellar atrophy,[141] achondroplasia,[239] myotonic dystrophy,[21,132] maturational changes,[20] spasticity,[73] hypoglycemia,[82] and progressive muscular dystrophy.[302]

Studies of SEPs have helped in delineating the pathophysiology in a variety of disorders affecting the peripheral or central nervous system. Clinical correlation, however, does not necessarily lead to practical application. A statistical difference between control and patient groups may add little in evaluating individual cases. In the clinical domain, the test must unveil relevant information pertinent to the diagnosis or management of the patient in question. Even unequivocal SEP abnormalities often fail to clearly lo-

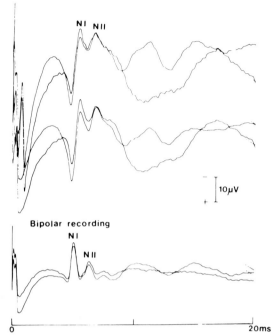

Figure 19–28. Comparison of monopolar and bipolar recording of spinal evoked potentials. The *two top tracings* show monopolar recordings from a pair of epidural electrodes placed 1 cm apart at the level of the T-5 spine, referenced to a surface electrode at the paraspinal muscle. The *bottom tracing* represents a bipolar derivation between the two epidural leads used for the top montage. Bipolar recording yielded a better defined, more stable potential with fewer technical problems, such as muscle artifacts or stimulus-related baseline shift. (From Machida et al.,[209] with permission.)

calize the lesion, because the neuroanatomic origin of each peak still awaits elucidation. Abuse and misuse are common with any new diagnostic procedure. This is a particular problem, however, in SEP studies, which have become routine before their time, while the technique still continues to evolve rapidly.[97,177] Despite widely publicized clinical applications, these investigative procedures at present can provide relatively limited information relevant to the individual patient in many instances.

Conservative and selective use of the test in proper clinical contexts would maximize its impact in electrodiagnostic medicine. Only with such precaution will SEP studies play a meaningful role as a diagnostic procedure in neurology.

With SEPs one can directly assess the transmission of the impulses that underlie the fundamental function of the nervous system. Thus, the technique has a wide range of applicability in physiologic studies of the peripheral and central nervous system in man. Such clinical and experimental data will help define precisely its diagnostic value and limitations.[16,45,57,64,75,138,202,230] With better understanding of the anatomy and physiology of the sensory pathways and standardization of the technique, SEP will secure its unique position as an important electrophysiologic measure for a number of neural dysfunctions.

7 MOTOR EVOKED POTENTIALS

Electrical Stimulation of the Brain and Spinal Cord

Single electrical stimuli delivered to the scalp can excite the motor pathway, inducing muscle action potentials from hand and leg.[220] In stimulating the motor cortex from the surface, anodal current, which hyperpolarizes dendrites, depolarizes the axon and cell body more effectively than cathodal current.[146] This noninvasive method allows evaluation of central conduction with efferent systems in man. Initial studies used bipolar stimulation with a specially made stimulator capable of delivering a high-voltage (2000 V) pulse of short duration (10 μs). With this technique, a single stimulus to the scalp elicited a submaximal muscle action potential of 1 mV or more. With moderate voluntary contraction of the muscle under study, a single scalp stimulus not much above threshold yielded a muscle action potential of near maximal amplitude.

Surface stimulation of the cervical spinal cord near the C-6 spinous process elicits muscle action potentials in the upper limbs. Similarly, shocks applied to the lumbosacral enlargement near the T-12 spinous process evoke muscle action potentials in the lower extremity.[210] Voluntary contraction does not appreciably facilitate the effect of spinal, as opposed to cortical, stimulation. Stimuli delivered over the lumbosacral spine overlying the cauda equina elicits the response less effectively than those delivered over the conus medullaris. These two observations together suggest that depolarization originates distal to the anterior horn cell, probably in the axon hillock known to have the lowest threshold for excitation.

A modified technique using commercially available stimulators allows electrical stimulation of the motor cortex with a fraction of shock intensities emloyed previously[145,260] (Fig. 19–29). This method utilizes unifocal stimulation with a flat anode (4.8 cm²) placed on the scalp over the motor cortex and a flexible stainless steel belt cathode wrapped around the head 2 to 3 cm above the nasion-inion plane. A unifocal stimulation tends to concentrate the electrical field from anode to cathode on the motor cortex, whereas a bifocal method orients the field tangentially along the surface.

Evaluating motor systems complements SEP studies in assessing a lesion of the spinal cord or monitoring an operative procedure.[259] In one study,[211] latency comparison between cortical and spinal stimulation yielded a conduction velocity of 48 m/s from cortex to cervical spinal cord and 47 m/s from cortex to lumbosa-

cral enlargement. The formula used in calculating conduction velocity is valid only if cortical and spinal stimulation activates the same group of motor fibers. The cortex-to-hand latency of 22.5 ms obtained by this method slightly exceeded that of 18 to 21 ms after stimulation of exposed human cortex during neurosurgical procedures.[227]

Patients undergoing surgery may show an abnormal motor potential recorded directly from the spinal cord despite normal SEPs.[35,198] A few studies have revealed slowing or block of corticospinal conduction in radiation myelopathy,[292] cervical cord trauma,[312] multiple sclerosis,[224,256,292] and motor neuron disease.[157] Transcutaneous stimulation of the cauda equina at the L-1 spine elicits a compound muscle action potential in the external anal sphincter.[303] The pudendal nerve latency (mean ± standard deviation [SD]) was greater in patients with idiopathic neurogenic fecal incontinence (7.3 ± 0.7 ms), compared with normal subjects (5.6 ± 0.6 ms). The proximal conduction between the L-1 and L-4 vertebral levels, however, showed no difference between the two groups.[176,303] These observations demonstrate the clinical utility of evaluating not only the afferent but also the efferent system in man.

High-intensity stimulation of the scalp

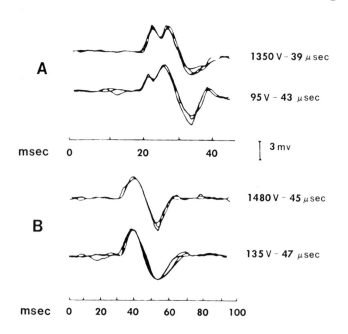

A

1350 V – 39 μsec

95 V – 43 μsec

msec 0   20   40   ⎸ 3 mv

B

1480 V – 45 μsec

135 V – 47 μsec

msec 0 20 40 60 80 100

Figure 19–29. Compound muscle action potentials (MAPs) evoked by electrically stimulating hand (**A**) and leg (**B**) motor areas over the scalp in the same subject. Comparison between bipolar (first and third tracings) and unipolar (second and fourth) stimulation show a substantial difference in stimulus voltage and duration required to elicit similar MAP waveform, amplitude, and latency in the two conditions. (From Rossini,[256] with permission.)

causes discomfort with contraction of scalp and facial muscles. It also poses considerable concern regarding electrical hazards. The unifocal method requires relatively low-voltage stimuli that can be delivered from an ordinary stimulator built according to established safety standards. Electric shocks used to produce convulsions as a therapeutic regimen far exceed those required to evoke motor potentials. Seizures as a result of kindling typically develop after trains of long-duration stimuli of about 1.0 ms. The delivery of single stimuli of very short duration (50 μs) on several occasions will probably produce very limited side effects. Nevertheless, one must seek unequivocal evidence from an animal model that no permanent adverse changes result from cortical stimulation.

Magnetic Coil Stimulation of the Brain

Magnetic coil stimuli applied to the human brain through the intact scalp and skull in conscious, alert subjects can also evoke compound muscle action potentials.[17,70,148,150,151] A magnetic field virtually unattenuated by the scalp and skull induces intracranial currents that, in turn, excite neural elements of the brain. The muscle responses elicited on the contralateral side of the body have a latency consistent with conduction in fast central pathways (Fig. 19–30). The factors that dictate the size of a muscle response include the intensity of stimuli, location and orientation of the stimulating coil, and intrinsic excitability of neural elements.[150] For reasons not completely understood, the stimulating coil centered near the vertex, rather than laterally over the motor cortex, gives rise to a maximal response. Voluntary contraction of a muscle, by increasing the neural excitability, greatly enhances its response to brain stimulation but not to cord stimulation.[147,221] A slight contraction of the muscle also shortens the onset latency of the compound muscle action potential by about 3 ms without further change when the background contraction increases (Fig. 19–31). In slightly

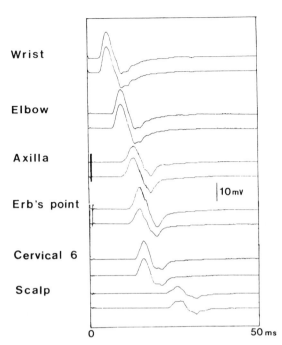

Figure 19–30. Compound muscle action potentials recorded from abductor pollicis brevis after magnetic coil stimulation at various points along the motor pathways. Scalp stimulation characteristically evokes less than a maximal response, despite the use of an optimal stimulus.

contracted muscle, neither a wide range of stimulus intensity nor the position of the stimulating coil within an area of 6 cm² over the vertex alters the onset latency substantially.[150]

Magnetic brain stimulation causes multiple firing of some spinal motor neurons in the muscle, exerting a slight voluntary background contraction.[150] Hence, a maximal antidromic volley set up by stimulation at the wrist fails to eliminate entirely the orthodromic volley of the peripheral nerve induced by magnetic stimulation of the brain. Here, the remaining response corresponds to the spinal motor neurons that fired more than once. These findings suggest that the enhancement of responses by voluntary background contraction depends not only on the additional recruitment of higher threshold motor units in the motor neuron pool but also on multiple firing of the same motor units.

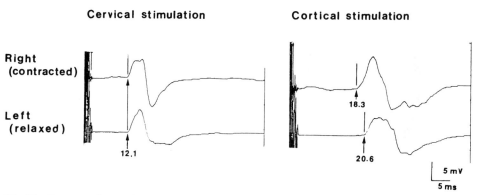

Figure 19–31. Compound muscle action potentials recorded from abductor digiti minimi after magnetic stimulation over the neck and scalp (C$_z$). Responses in each column represent simultaneous recording from the minimally contracted muscle on the right and relaxed muscle on the left. Note the effect of voluntary facilitation with cortical but not with cervical stimulation.

Table 19–6 summarizes the onset latency of the compound muscle action potentials elicited by magnetic stimulation. The total conduction time comprises activation of cortical structures, conduction down the cortical spinal pathway, activation of spinal motor neurons, and conduction along the peripheral nerve to the muscle. Stimulation over the cervical area with the cathode between the C-7 and T-1 spinous processes excites the motor roots at the foramina where they leave the spinal canal.[225] The conduction time, calculated as the difference in latency between scalp-evoked compound muscle action potentials and root-evoked compound muscle action potentials, therefore, contains a small peripheral component. Thus, it is estimated that the total motor conduction time of about 20 ms from the scalp to the intrinsic hand muscle involves peripheral latency of 13 ms, synaptic and root delay of 1.5 ms, and central motor conduction time (CMCT) of 5.5 ms. The use of F waves[260]

and magnetic coil stimulation over Erb's point[17] has yielded a similar peripheral latency and calculated value for CMCT.

Threshold brain stimuli can elicit single motor unit discharges in the intrinsic hand muscles at a constant latency. The size principle of Henneman applies to brain stimulation, which, even from different coil position up to 7 cm apart, initially activates those motor units with the lowest threshold for voluntary activation. Stronger stimuli cause the same motor units to discharge with less latency and recruit other motor units. The onset latency shortens and the amplitude increases with voluntary contraction of the muscle itself or of the ipsilateral intrinsic hand muscles, but not with activation of the contralateral intrinsic hand muscles or ipsilateral leg muscles.

Magnetic stimulation capable of painless excitation of the motor system has an obvious advantage over electrical stimulation in the clinical domain.[58,149,224] No investigator thus far has reported side ef-

Table 19–6 NORMATIVE DATA (n = 36 SIDES)

Measurement	Mean	SD	Range	Mean + 2.5 SD
Conduction time C-7/T-1 to ADM (ms)	13.60	1.35	10.9–16.9	16.3
Conduction time C-7/T-1 to wrist (ms)	11.18	1.19	8.7–13.8	13.56
Conduction time scalp to ADM (ms)	19.73	1.25	17.5–23.1	22.23
Central conduction time (ms)	6.13	0.89	4.5–7.7	8.35
R/L difference in onset latency (ms) (n = 12)	0.69	0.58	0–1.8	2.14
Amplitude as % of amplitude from wrist	—	—	18.6–96.6	—

From Mills,[223] with permission.

fects. The possibility of adverse effects, nonetheless, must be borne in mind with the introduction of new techniques. For now, it is advisable to exclude patients with a history of epilepsy, those with a cardiac pacemaker, and those who have undergone neurosurgery. Intracranial metallic objects such as aneurysm clips and shunts could conceivably be dislodged during magnetic stimulation. A train of high-frequency stimuli at a rate of three or more per second could kindle the motor cortex to induce epileptic foci,[129] but magnetic stimulation cannot be given at such high frequencies because of the time required to charge the capacitors.

In normal subjects maintaining a small voluntary contraction, magnetic stimulation with intensity 20 percent above threshold for relaxed muscles evokes compound muscle action potential of at least 18 percent of the maximal response elicited by electrical stimulation of the nerve (Fig. 19–30). Therefore, any response reduced to a level below 15 percent of the maximum compound muscle action potential suggests conduction block along the central or peripheral pathways.[223] In one series dealing with 83 sides from 44 patients with definite or probable multiple sclerosis, central motor conduction was prolonged on 47 sides and amplitude was below normal on 29. On five sides, central motor conduction time was normal, but amplitude of the compound muscle action potential was reduced. One patient had no response at all. On 31 sides, both central motor conduction time and compound muscle action potential amplitude were normal.

A number of other motor system diseases, such as motor neuron disease[156] and radiation myelopathy,[292] show similar conduction abnormalities along the central motor system. Therefore, these findings are by no means specific to multiple sclerosis, although other conditions do not cause the extreme prolongation of central motor conduction time characteristic of demyelination.[223] Experience with the patient population is still limited but includes, in addition to those results discussed above, studies of cord compres-

sion, neurosarcoidosis, cerebral infarct, and degenerative disorders.[17,147]

Jerk-Locked Averaging

Jerk-locked backward averaging of the scalp electroencephalogram (EEG) helps identify cerebral events that are time-locked to a voluntary or involuntary muscle contraction. With this technique, rectified electromyographic (EMG) signals serve as the trigger for averaging the cerebral activity preceding the movement by means of a delay line (Fig. 19–32). Since the initial description,[188,320] a number of investigators have

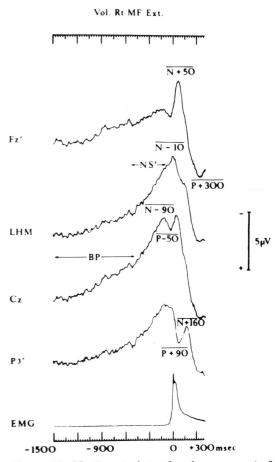

Figure 19–32. Terminology of each component of cortical potentials associated with voluntary, self-paced middle finger extension. The record shows a grand average in 14 healthy subjects, with 200 trials for each subject. (From Shibasaki et al.,[276] with permission.)

used the method in assessing movement-related cortical potential,[18,19,194,275] the mechanisms of synkinesis,[278] the pathophysiology of myoclonus,[136,293] parkinsonism,[84] and other involuntary movements.[280]

Movement-related cortical potential consists of at least eight separate components.[276,307] Those preceding the EMG onset include a symmetric early negative shift called Bereitshaftspotential (BP); intermediate shift (IS); a negative slow wave maximal over the contralateral precentral region (NS); P−50, or premotion positivity (PMP); and N−10, or motor potential (MP). The components occurring after the EMG onset include N+50, or a sharp negative wave over the contralateral frontal region; P+90; N+160; and P+300, or a widely distributed large positivity maximal over the contralateral precentral region (see Fig. 19–32). In the clinical domain, patients with Parkinson's disease show abnormal topography of premotion slow negativity, or BP/NS' complex with reduced amplitude on the side of the affected basal ganglia.[71,281] This component also undergoes predictable reduction of amplitude with cerebellar ataxia in general and Ramsay Hunt syndrome in particular, presumably reflecting the dysfunction of the cerebellofugal or dentatothalamic pathway.[281]

Averaging the EEG time-locked to a myoclonic discharge helps in identifying the responsible cortical spike, which otherwise might escape detection, or in determining cortical excitability after myoclonus.[283] The EEG correlates of myoclonus established by this means resemble the giant early cortical component of the somatosensory evoked potentials in waveform, topography, and time relationship to spontaneous myoclonus.[282] In such cortical reflex myoclonus, cortical spikes precede the movement of the upper extremity by 6 to 22 ms. In contrast, periodic synchronous discharges start 50 to 85 ms before the myoclonus in patients with Creutzfeldt-Jakob disease. Patients with Alzheimer's disease, including the cases of trisomy 21, demonstrate a focal, negative cerebral potential over the contralateral central region antecedent to the myoclonic

jerks.[326] This EEG event differs from that associated with the myoclonus of spongiform encephalopathy.

Cortical slow negativity similar to BP/NS' precedes choreic movement in patients with chorea-acanthocytosis but not in those with Huntington's disease.[280] In patients with Gilles de la Tourette's syndrome, spontaneous tics do not accompany any slow negativity, although a premotion negativity precedes voluntary jerks, mimicking their tics. Patients with mirror movement may show abnormal topography of NS' which appears bilaterally, indicating unintended participation of the opposite motor cortex.[278]

REFERENCES

1. Albe-Fessard, D, Tasker, R, Yamashiro, K, Chodakiewitz, J, and Dostrovsky, J: Comparison in man of short latency averaged evoked potentials recorded in thalamic and scalp hand zones of representation. Electroencephalogr Clin Neurophysiol 65:405–415, 1986.
2. Allison, T: Scalp and cortical recordings of initial somatosensory cortex activity to median nerve stimulation in man. Ann NY Acad Sci 388:671–678, 1982.
3. Allison, T, Goff, WR, Williamson, PD, and VanGilder, JC: On the neural origin of early components of the human somatosensory evoked potential. In Desmedt, JE (ed): Progress in Clinical Neurophysiology, Vol 7. Karger, Basel, 1980, pp 51–68.
4. Aminoff, MJ: AAEE Minimonograph #22: The clinical role of somatosensory evoked potential studies: A critical appraisal. Muscle Nerve 7:345–354, 1984.
5. Aminoff, MJ, Davis, SL, and Panitch, HS: Serial evoked potential studies in patients with definite multiple sclerosis. Arch Neurol 41:1197–1202, 1984.
6. Anderson, NE, Frith, RW, and Synek, VM: Somatosensory evoked potentials in syringomyelia. J Neurol Neurosurg Psychiatry 49:1407–1410, 1986.
7. Andersson, SA, Norrsell, K, and Norrsell, U: Spinal pathways projecting to the cerebral first somatosensory area in the monkey. J Physiol (Lond) 225:589–597, 1972.
8. Angel, R, Weinrich, M, and Rodnitzky, R: Recovery of somatosensory evoked potentials amplitude after movement. Ann Neurol 19:344–348, 1986.
9. Anziska, BJ, and Cracco, RQ: Short latency SEPs to median nerve stimulation: Studies in patients with focal neurologic disease. Electroencephalogr Clin Neurophysiol 49:227–239, 1980.
10. Anziska, BJ, and Cracco, RQ: Short latency

SEPs to median nerve stimulation: Comparison of recording methods and origin of components. Electroencephalogr Clin Neurophysiol 52:531–539, 1981.

11. Anziska, BJ, and Cracco, RQ: Short-latency somatosensory evoked potentials to median nerve stimulation in patients with diffuse neurologic disease. Neurology 33:989–993, 1983.

12. Anziska, B, Cracco, RQ, Cook, AW, and Field SW: Somatosensory far-field potentials: Studies in normal subjects. Electroencephalogr Clin Neurophysiol 45:602–610, 1978.

13. Badr, GG, Hanner Pand, S, and Edstrom, L: Cortical evoked potentials to trigeminus nerve stimulation in humans. Electroencephalogr Clin Neurophysiol 14:61–66, 1983.

14. Baran, EM: Evoked spinal responses in man (abstr). Electroencephalogr Clin Neurophysiol 45:16P, 1978.

15. Baran, EM: Spinal cord responses to peripheral nerve stimulation in man. Arch Phys Med Rehabil 61:10–17, 1980.

16. Barber, C, and Blum, T: Evoked Potentials, III. Butterworths, Boston, 1987.

17. Barker, AT, Freestone, IL, Jalinous, T, Merton, PA, and Morton, HB: Magnetic stimulation of the human brain (abstr). J Physiol 369:3P, 1985.

18. Barrett, G, Shibasaki, H, and Neshige, R: A computer-assisted method for averaging movement-related cortical potentials with respect to EMG onset. Electroencephalogr Clin Neurophysiol 60:276–281, 1985.

19. Barrett, G, Shibasaki, H, and Neshige, R: Cortical potentials preceding voluntary movement: Evidence for three periods of preparation in man. Electroencephalogr Clin Neurophysiol 62:327–339, 1986.

20. Bartel, PR, Conradie, J, Robinson, E, Prinsloo, J, and Becker, P: The relationship between median nerve somatosensory evoked potential latencies and age and growth parameters in young children. Electroencephalogr Clin Neurophysiol 68:180–186, 1987.

21. Bartel, PR, Lotz, BP, Robinson, E, and Van der Meyden, C: Posterior tibial and sural nerve somatosensory evoked potentials in dystrophia myotonica. J Neurol Sci 70:55–65, 1985.

22. Bartel, PR, Lotz, BP, and Van der Meyden, CH: Short-latency somatosensory evoked potentials in dystrophia myotonica. J Neurol Neurosurg Psychiatry 47:524–529, 1984.

23. Bartel, PR, Markand, ON, and Kolar, OJ: The diagnosis and classification of multiple sclerosis: Evoked responses and spinal fluid electrophoresis. Neurology 33:611–617, 1983.

24. Beal, JE, Applebaum, AE, Foreman, RD, and Willis, WD: Spinal cord potentials evoked by cutaneous afferents in the monkey. J Neurophysiol 40:199–211, 1977.

25. Belsh, JM, and Chokroverty, S: Short-latency somatosensory evoked potentials in brain-dead patients. Electroencephalogr Clin Neurophysiol 68:75–78, 1987.

26. Bennett, MH, and Janetta, PJ: Trigeminal evoked potentials in humans. Electroen-

cephalogr Clin Neurophysiol 48:517–526, 1980.

27. Bennett, MH, and Janetta, PJ: Evoked potentials in trigeminal neuralgia. Neurosurgery 13:242–247, 1983.

28. Bennett, MH, and Lunsford, DL: Percutaneous retrogasserian glycerol rhizotomy for tic douloureux: Part 2. Results and implications of trigeminal evoked potentials studies. Neurosurgery 14:431–435, 1984.

29. Beric, A, Dimitrijevic, MR, Prevec, TS, and Sherwood, AM: Epidurally recorded cervical somatosensory evoked potential in humans. Electroencephalogr Clin Neurophysiol 65:94–101, 1986.

30. Beric, A, Dimitrijevic, MR, Sharkey, PC, and Sherwood, AM: Cortical potentials evoked by epidural stimulation of the cervical and thoracic spinal cord in man. Electroencephalogr Clin Neurophysiol 65:102–110, 1986.

31. Blackburn, JG, Perot, PL, and Katz, S: Effects of low spinal transection on somatosensory evoked potentials from forelimb stimulation (abstr). Fed Proc 33:400, 1974.

32. Bollen, EL, Arts, RJ, Roos, RA, Van der Velde, EA, and Burums, OJ: Somatosensory evoked potentials in huntington's chorea. Electroencephalogr Clin Neurophysiol 62:235–240, 1985.

33. Bosch, EP, Yamada, T, and Kimura, J: Somatosensory evoked potentials in motor neuron disease. Muscle Nerve 8:556–562, 1985.

34. Bottcher, J, and Trojaborg, W: Follow-up of patients with suspected multiple sclerosis: A clinical and electrophysiological study. J Neurol Neurosurg Psychiatry 45:809–814, 1982.

35. Boyd, SG, Rothwell, JC, Cowan, JMA, Webb, PJ, Morley, T, Asselman, P, and Marsden, CD: A method of monitoring function in corticospinal pathways during scoliosis surgery with a note on motor conduction velocities. J Neurol Neurosurg Psychiatry 49:251–257, 1986.

36. Brown, RH, and Nash, CL, Jr: Current status of spinal cord monitoring. Spine 4:466–470, 1979.

37. Brunko, E, and Zegers de Beyl, D: Prognostic value of early cortical somatosensory evoked potentials after resuscitation from cardiac arrest. Electroencephalogr Clin Neurophysiol 66:15–24, 1987.

38. Buettner, UN, Petruch, F, Scheglemann, K, and Stohr, M: Diagnostic significance of cortical somatosensory evoked potentials following trigeminal nerve stimulation. In Courjon, J, Manuguirere, F, and Revol, M (eds): Clinical Applications of Evoked Potentials in Neurology. Raven Press, New York, 1982, pp 339–345.

39. Cant, B, Hume, A. Judson, J, and Shaw, N: The assessment of severe head injury by short-latency somatosensory and brain-stem auditory evoked potentials. Electroencephalogr Clin Neurophysiol 65:188–195, 1986.

40. Celesia, GG: Somatosensory evoked potentials recorded directly from human thalamus and

SM I cortical area (all). Arch Neurol 36:399–405, 1979.

41. Celesia, GG: Somatosensory evoked potentials: A quest for relevance. J Clin Neurophysiol 2(1):77–82, 1985.

42. Chapman, CR, Gerlach, R, Jacobson, R, Buffington, V, and Kaufman, E: Comparison of short-latency trigeminal evoked potentials elicited by painful dental and gingival stimulation. Electroencephalogr Clin Neurophysiol 65:20–26, 1986.

43. Chiappa, KH: Pattern shift visual, brainstem auditory, and short-latency somatosensory evoked potentials in multiple sclerosis. Neurology 30:110–123, 1980.

44. Chiappa, KH, Choi, SK, and Young, RR: Short-latency somatosensory evoked potentials following median nerve stimulation in patients with neurological lesions. In Desmedt, JE (ed): Clinical Uses of Cerebral, Brainstem and Spinal Somatosensory Evoked Potentials, Progress in Clinical Neurophysiology, Vol 7. Karger, Basel, 1980, pp 264–281.

45. Chiappa, KH, and Yiannikas, C: Evoked Potentials in Clinical Medicine. Raven Press, New York, 1983.

46. Chiappa, KH, and Young RR: Evoked responses: Overused, underused or misused? Arch Neurol 42:76–77, 1985.

47. Chu, NS: Median and tibial somatosensory evoked potentials: Changes in short- and long-latency components in patients with lesions of the thalamus and thalamo-cortical radiations. J Neurol Sci 76:199–219, 1986.

48. Chu, NS: Sensory evoked potentials in Wilson's disease. Brain 109:491–507, 1986.

49. Chu, NS: Somatosensory evoked potentials: Correlations with height. Electroencephalogr Clin Neurophysiol 65:169–176, 1986.

50. Chu, NS, and Hong, CI: Erb's and cervical somatosensory evoked potentials: Correlations with body size. Electroencephalogr Clin Neurophysiol 62:319–322, 1985.

51. Cohen, LG, and Starr, A: Vibration and muscle contraction affect somatosensory evoked potentials. Neurology 35:691–698, 1985.

52. Cohen, LG, and Starr, A: About the origin of cerebral somatosensory potentials evoked by Achilles tendon taps in humans. Electroencephalogr Clin Neurophysiol 62:108–116, 1985.

53. Cohen, LG, Starr, A, and Pratt, H: Cerebral somatosensory potentials evoked by muscle stretch, cutaneous taps and electrical stimulation of peripheral nerves in the lower limbs in man. Brain 108:103–121, 1985.

54. Cohen, SN, Syndulko, K, Hansch, E, et al: Variability on serial testing of visual evoked potentials in patients with multiple sclerosis. In Courjon, J, Mauguiere, F, and Revol, M (eds): Clinical Applications of Evoked Potentials in Neurology. Raven Press, New York, 1982, pp 559–565.

55. Conrad, B, and Bechinger, D: Sensorische und motorische Nervenleitgeschwindigkeit und distale Latenz bei Multipler Sklerose. Arch Psychiatr Nervenkr 212:140–149, 1969.

56. Cosi, V, Poloni, M, Mazzini, and Callieco, R: Somatosensory evoked potentials in amyotrophic lateral sclerosis. J Neurol Neurosurg Psychiatry 47:857–861, 1984.

57. Courjon, J, Mauguiere, F, and Revol, M: Clinical applications of evoked potentials in neurology. Raven Press, New York, 1982.

58. Cowan, JMA, Dick, JPR, Day, BL, Rothwell, JC, Thompson, PD, and Marsden, CD: Abnormalities in central motor pathway conduction in multiple sclerosis. Lancet 2:304–307, 1984.

59. Cracco, JB, Bosch, VV, and Cracco, RQ: Cerebral and spinal somatosensory evoked potentials in children with CNS degenerative diseases. Electroencephalogr Clin Neurophysiol 49:437–445, 1980.

60. Cracco, JB, Castells, S, and Mark, E: Spinal somatosensory evoked potentials in juvenile diabetes. Ann Neurol 15:55–58, 1984.

61. Cracco, JB, Cracco, RQ, and Stolove, R: Spinal evoked potential in man: A maturational study. Electroencephalogr Clin Neurophysiol 45:58–64, 1979.

62. Cracco, RQ: The initial positive potential of human scalp-recorded somatosensory evoked response. Electroencephalogr Clin Neurophysiol 32:623–629, 1972.

63. Cracco, RQ: Spinal evoked response: Peripheral nerve stimulation in man. Electroencephalogr Clin Neurophysiol 35:379–386, 1973.

64. Cracco, RQ, and Bodis-Wollner, I: Evoked potentials. In Frontiers of Clinical Neuroscience, Vol 3. Alan R. Liss, New York, 1986.

65. Cracco, RQ, and Cracco, JB: Somatosensory evoked potential in man: far field potentials. Electroencephalogr Clin Neurophysiol 41:460–466, 1976.

66. Cruse, R, Klem, G, Lesser, RP, and Lueders, H: Paradoxical lateralization of cortical potentials evoked by stimulation of posterior tibial nerve. Arch Neurol 39:222–225, 1982.

67. Cunningham, K, Halliday, AM, and Jones, SJ: Stimulation of "stationary" SAP and SEP phenomena by 2-dimensional potential field modelling. Electroencephalogr Clin Neurophysiol 65:416–428, 1986.

68. Davis, SL, Aminoff, MJ, and Panitch, HS: Clinical correlations of serial somatosensory evoked potentials in multiple sclerosis. Neurology 35:359–365, 1985.

69. Dawson, GD: Cerebral responses to electrical stimulation of peripheral nerve in man. Electroencephalogr Clin Neurophysiol 10:134–140, 1947.

70. Day, BL, Dick, JPR, Marsden, CD, and Thompson, PD: Differences between electrical and magnetic stimulation of the human brain (abstr). J Physiol 378:36P, 1986.

71. Deecke, L, Englitz, HG, Kornhuber, HH, and Schmitt, G: Cerebral potentials preceding voluntary movement in patients with bilateral or unilateral Parkinson akinesia. In Desmedt, JE (ed): Attention, Voluntary Contraction and

Event-Related Cerebral Potentials. Progress in Clinical Neurophysiology, Vol 1. Karger, Basel, 1977, pp 151–163.

72. Deiber, MP, Giard, MH, and Mauguiere, F: Separate generators with distinct orientations for N20 and P22 somatosensory evoked potentials to finger stimulation? Electroencephalogr Clin Neurophysiol 65:321–334, 1986.

73. Delwaide, PJ, Schoenen, J, and De Pasqua, V: Lumbosacral spinal evoked potentials in patients with multiple sclerosis. Neurology 35:174–179, 1985.

74. Desmedt, JE: Somatosensory cerebral evoked potentials in man. In Rémond, A (ed): Handbook of Electroencephalography and Clinical Neurophysiology, Vol 9. Elsevier, Amsterdam, 1971, pp 55–82.

75. Desmedt, JE: Clinical Uses of Cerebral, Brain Stem and Spinal Somatosensory Evoked Potentials. Progress in Clinical Neurophysiology. Karger, Basel, 1980.

76. Desmedt, JE, and Cheron, G: Central somatosensory conduction in man: Neural generators and interpeak latencies of the far-field components recorded from neck and right or left scalp and earlobes. Electroencephalogr Clin Neurophysiol 50:382–403, 1980.

77. Desmedt, JE, and Cheron, G: Non-cephalic reference recording of early somatosensory potentials to finger stimulation in adult or aging normal man: Differentiation of widespread N_{18} and contralateral N_{20} from prerolandic P_{22} and N_{30} components. Electroencephalogr Clin Neurophysiol 52:553–570, 1981.

78. Desmedt, JE, and Cheron, G: Spinal and far-field components of human somatosensory evoked potential to posterior tibial nerve stimulation analysed with oesophageal deviations and non-cephalic reference recording. Electroencephalogr Clin Neurophysiol 56:635–651, 1983.

79. Desmedt, JE, Huy, NT, and Carmeliet, J: Unexpected latency shifts of the stationary P9 somatosensory evoked potential far field with changes in shoulder position. Electroencephalogr Clin Neurophysiol 56:623–627, 1983.

80. Desmedt, JE, Nguyen, TH, and Bourguet, M: Bit-mapped color imaging of human evoked potentials with reference to the N_{20}, P_{22}, P_{27} and N_{30} somatosensory responses. Electroencephalogr Clin Neurophysiol 68:1–19, 1987.

81. Desmedt, JE, and Noel, P: Average cerebral evoked potentials in the evaluation of lesions of the sensory nerves and of the central somatosensory pathway. In Desmedt, JE (ed): New Developments in Electromyography and Clinical Neurophysiology, Vol 2, Karger, Basel, 1973, pp 352–371.

82. Deutsch, E, Freeman, S, Sohmer, H, and Gafni, M: The persistence of somatosensory and auditory pathway evoked potentials in severe hypoglycemia in the cat. Electroencephalogr Clin Neurophysiol 61:161–164, 1985.

83. De Weerd, AW: Variability of central conduction in the course of multiple sclerosis: Serial recordings of evoked potentials in the evaluation of therapy. Clin Neurol Neurosurg 89:9–15, 1987.

84. Dick, JPR, Cantello, R, Buruma, O, Gioux, M, Benecke, R, Day, BL, Rothwell, JC, Thompson, PD, and Marsden, CD: The bereitschaftspotential, L-DOPA and parkinson's disease. Electroencephalogr Clin Neurophysiol 66: 263–274, 1987.

85. Dimitrijevic, MR, Larsson, LE, Lehmkuhl, D, and Sherwood, A: Evoked spinal cord and nerve root potentials in humans using a non-invasive recording technique (part). Electroencephalogr Clin Neurophysiol 45:331–340, 1978.

86. Dinner, DS, Luders, H, Lesser, RP, and Morris, HH: Invasive methods of somatosensory evoked potential monitoring. J Clin Neurophysiol 3:113–130, 1986.

87. Donchin, E, Callaway, E, Cooper, R, Desmedt, JE, Goff, WR, Hillyard, SA, and Sutton, S: Publication criteria for studies of evoked potentials (EP) in man. In Desmedt, JE (ed): Progress in Clinical Neurophysiology, Vol 1. Karger, Basel, 1977, pp 1–11.

88. Dorfman, LJ: Indirect estimation of spinal cord conduction velocity in man. Electroencephalogr Clin Neurophysiol 42:26–34, 1977.

89. Dorfman, LJ, Bosley, TM, and Cummins, KJ: Electrophysiological localization of central somatosensory lesions in patients with multiple sclerosis. Electroencephalogr Clin Neurophysiol 44:742–753, 1978.

90. Dorfman, LJ, Donaldson, SS, Gupta, PR, et al: Electrophysiologic evidence of subclinical injury to the posterior columns of the human spinal cord after therapeutic radiation. Cancer 50:2815–2819, 1982.

91. Drechscler, F: Short and long latency cortical potentials following trigeminal nerve stimulation in man. In Barber, C (ed): Evoked Potentials. University Park Press, Baltimore, 1980, pp 415–422.

92. Dyck, PJ, Lambert, EH, and Nichols, PC: Quantitative measurement of sensation related to compound action potential and number and size of myelinated fibres of sural nerve. In Remond, A (ed): Handbook of Electroencephalography and Clinical Neurophysiology, Vol 9. Elsevier, Amsterdam, 1971, pp 83–118.

93. Ebner, A, Eisiedel-Lechtape, H, and Tucking, CH: Somatosensory tibial nerve evoked potentials with parasagittal tumors: A contribution to the problem of generators. Electroencephalogr Clin Neurophysiol 54:508–515, 1982.

94. Ehle, AL, Stewart, RM, Lellelid, NA, and Leventhal, NA: Evoked potential in Huntington's disease. Arch Neurol 41:379–382, 1984.

95. Eisen, A: The somatosensory evoked potential. Can J Neurol Sci 9:65–77, 1982.

96. Eisen, A: AAEE Minimonograph #24: Noninvasive measurement of spinal cord conduction, review of presently available methods. Muscle Nerve 2:95–103, 1986.

97. Eisen, A, and Cracco, RQ: Overuse of evoked potentials. Caution. Neurology 33:618–621, 1983.

98. Eisen, A, and Elleker, C: Sensory nerve stimulation and evoked cerebral potentials. Neurology 30:1097–1105, 1980.

99. Eisen, A, Hoirch, M, and Moll, A: Evaluation of radiculopathies by segmental stimulation and somatosensory evoked potentials. Can J Neurol Sci 10:178–182, 1983.

100. Eisen, A, and Nudleman, K: F-wave and cervical somatosensory response conduction from the seventh cervical spinous process to cortex in multiple sclerosis. Can J Neurol Sci 5:289–295, 1978.

101. Eisen, A, and Nudleman, K: Cord to cortex conduction in multiple sclerosis. Neurology 29:189–193, 1979.

102. Eisen, A, and Odusote, K: Central and peripheral conduction times in multiple sclerosis. Electroencephalogr Clin Neurophysiol 48:253–265, 1980.

103. Eisen, A, Odusote, K, Bozek, C, and Hoirch, M: Far-field potentials from peripheral nerve: Generated at sites of muscle mass change. Neurology 36:815–818, 1986.

104. Eisen, A, Purves, S, and Hoirch, M: Central nervous system amplification: its potential in the diagnosis of early multiple sclerosis. Neurology 32:359–364, 1982.

105. Eisen, A, Roberts, K, and Lawrence, P: Morphological measurement of the SEP using a dynamic time warping algorithm. Electroencephalogr Clin Neurophysiol 65:136–141, 1986.

106. Eisen, A, Stewart, J, Nudleman, K, and Cosgrove, JBR: Short-latency somatosensory responses in multiple sclerosis. Neurology 29:827–834, 1979.

107. El-Negamy, E, and Sedgwick, EM: Properties of spinal somatosensory evoked potential recorded in man. J Neurol Neurosurg Psychiatry 41:762–768, 1978.

108. Emerson, RG, and Pedley, TA: Generator sources of median somatosensory evoked potentials. J Clin Neurophysiol 1:203–218, 1984.

109. Emerson, RG, Seyal, M, and Pedley, TA: Somatosensory evoked potentials following median nerve stimulation. I. The cervical components. Brain 107:169–182, 1984.

110. Ertekin, C: Evoked electrospinogram in spinal cord and peripheral nerve disorders. Acta Neurol Scand 57:329–344, 1978.

111. Ertekin, C, Sarica, Y, and Uckardesler, L: Somatosensory cerebral potentials evoked by stimulation of the lumbo-sacral spinal cord in normal subjects and in patients with conus medullaris and cauda equina lesions. Electroencephalogr Clin Neurophysiol 59:57–60, 1984.

112. Favale, E, Ratto, S, Leandri, M, and Abbruzzese, M: Investigations on the nervous mechanisms underlying the somatosensory cervical response in man. J Neurol Neurosurg Psychiatry 45:796–801, 1982.

113. Feldman, MH, Cracco, RQ, Farmer, P, and Mount, F: Spinal evoked potential in the monkey. Ann Neurol 7:238–244, 1980.

114. Findler, G, and Fiensod, M: Sensory evoked response to electrical stimulation of the trigeminal nerve in humans. J Neurosurg 56:545–549, 1982.

115. Fine, EJ, and Hallet, M: Neurophysiological study of subacute combined degeneration. J Neurol Sci 45:331–336, 1980.

116. Fowler, TJ, and Ochoa, J: Unmyelinated fibres in normal and compressed peripheral nerves of the baboon: A quantitative electron microscopic study. Neuropathol Appl Neurobiol 1:247–265, 1975.

117. Fox, JE, and Williams, B: Central conduction time following surgery for cerebral aneurysm. J Neurol Neurosurg Psychiatry 47:873–875, 1984.

118. Fox, JL, and Kenmore, PI: The effect of ischemia on nerve conduction. Exp Neurol 17:403–419, 1967.

119. Firth, RW, Benstead, TB, and Daube, JR: The SEP standing waveform at the shoulder due to a change in volume conductor (abstr). Electroencephalogr Clin Neurophysiol 61:S272, 1985.

120. Frith, RW, Benstead, TJ, and Daube, JR: Stationary waves recorded at the shoulder after median nerve stimulation. Neurology 36:1458–1464, 1986.

121. Gandevia, SC, Burke, D, and McKeon, B: The projection of muscle afferents from the hand to cerebral cortex in man. Brain 107:1–13, 1984.

122. Ganes, T: A study of peripheral, cervical and cortical evoked potentials and afferent conduction times in the somatosensory pathway. Electroencephalogr Clin Neurophysiol 49:446–451, 1980.

123. Ganes, T: Somatosensory conduction times and peripheral, cervical and cortical evoked potentials in patients with cervical spondylosis. J Neurol Neurosurg Psychiatry 43:683–689, 1980.

124. Ganes, T, and Lundar, T: The effect of thiopentone on somatosensory evoked responses and EEGs in comatose patients. J Neurol Neurosurg Psychiatry 46:509–514, 1983.

125. Ganes, T, and Nakstad, P: Subcomponents of the cervical evoked response in patients with intracerebral circulatory arrest. J Neurol Neurosurg Psychiatry 47:292–297, 1984.

126. Garg, BP, Markand, ON, Demyer, WE, and Warren, C, Jr: Evoked response studies in patients with adrenoleukodystrophy and heterozygous relatives. Arch Neurol 40:356–359, 1983.

127. Giesser, BS, Kurtzberg, D, Vaughan, HG, Jr, Arezzo, JC, Aisen, ML, Smith, CR, Larocca, NG, and Scheinberg, LC: Trimodal evoked potentials compared with magnetic resonance imaging in the diagnosis of multiple sclerosis. Arch Neurol 44:281–284, 1987.

128. Gilmore, RL, Bass, NH, Wright, EA, Greathouse, D, Stanback, K, and Norvell, E: Develop-

mental assessment of spinal cord and cortical evoked potentials after tibial nerve stimulation: Effects of age and stature on normative data during childhood. Electroencephalogr Clin Neurophysiol 62:241–251, 1985.

129. Goddard, GV, McIntyre, DC, and Leech, CK: A permanent change in brain function from daily electrical stimulation. Exp Neurol 25:295–330, 1969.

130. Goldie, WD, Chiappa, KH, Young, RR, and Brooks, EB: Brainstem auditory and short-latency somatosensory evoked responses in brain death. Neurology 31:248–256, 1981.

131. Gonzalez, EG, Hajdu, M, Bruno, R, Keim, H, and Brand, L: Lumbar spinal stenosis: Analysis of pre- and postoperative somatosensory evoked potentials. Arch Phys Med Rehabil 66:11–15, 1985.

132. Gott, PS, and Karnaze, DS: Short-latency somatosensory evoked potentials in myotonic dystrophy: Evidence for a conduction disturbance. Electroencephalogr Clin Neurophysiol 62:455–458, 1985.

133. Grundy, BL, Nash, CL, and Brown, RH: Arterial pressure manipulation alters spinal cord function during correction for scoliosis. Anesthesiology 54:249–253, 1981.

134. Guru, K, Mailis, A, Ashby, P, and Vanderlinden, G: Postsynaptic potentials in motoneurons caused by spinal cord stimulation in humans. Electroencephalogr Clin Neurophysiol 66:275–280, 1987.

135. Haldeman, S, Bradley, WE, Bhatia, NN, and Johnson, BK: Pudendal evoked responses. Arch Neurol 39:280–283, 1982.

136. Hallett, M, Chadwick, D, and Marsden, CD: Cortical reflex myoclonus. Neurology 29:1107–1125, 1979.

137. Halliday, AM: Changes in the form of cerebral evoked responses in man associated with various lesions of the nervous system. Electroencephalogr Clin Neurophysiol 25:178–192, 1967.

138. Halliday, AM: Somatosensory evoked responses. In Remond, A (ed): Handbook of ECN, Volume 8A. Elsevier, Amsterdam, 1975, pp 60–67.

139. Halliday, AM: Current status of the SEP. International Symposium on Somatosensory Evoked Potentials, Kansas City, Missouri, September 22–23, 1984. American EEG Society and American Association of EMG and Electrodiagnosis, Rochester, Minn, 1984, pp 33–37.

140. Halliday, AM, and Wakefield, GS: Cerebral evoked potentials in patients with dissociated sensory loss. J Neurol Neurosurg Psychiatry 26:211–219, 1963.

141. Hammond, EJ, and Wilder, BJ: Evoked potentials in olivopontocerebellar atrophy. Arch Neurol 40:366–369, 1983.

142. Hammond, EJ, Wilder, BJ, and Ballinger, WE, Jr: Electrophysiologic recordings in a patient with a discrete unilateral thalamic infarction. J Neurol Neurosurg Psychiatry 45:640–643, 1982.

143. Harner, PF, and Sannit, T: A review of the international ten-twenty system of electrode placement. Grass Instrument Co, Quincy, Mass, 1974.

144. Hashimoto, I: Somatosensory evoked potentials from the human brain-stem: Origins of short latency potentials. Electroencephalogr Clin Neurophysiol 57:221–227, 1984.

145. Hassan, NF, Rossini, PM, Cracco, RQ, and Cracco, JB: Unexposed motor cortex activation by low voltage stimuli. In Morrocutti, C, and Rizzo, PA (eds): Evoked Potentials: Neurophysiological and Clinical Aspects. Elsevier, Amsterdam, 1985, pp 3–13.

146. Hern, JEC, Landgren, S, Phillips, CG, and Porter, R: Selective excitation of corticofugal neurones by surface anodal stimulation of the baboon's motor cortex. J Physiol 161:73–90, 1962.

147. Hess, CW, Mills, KR, and Murray, WF: Magnetic stimulation of the human brain: The effects of voluntary muscle activity (abstr). J Physiol 378:37P, 1986.

148. Hess, CW, Mills, KR, and Murray, WF: Percutaneous stimulation of the human brain: A comparison of electrical and magnetic stimuli (abstr). J Physiol 378:35P, 1986.

149. Hess, CW, Mills, KR, and Murray, WF: Measurement of central motor conduction in multiple sclerosis using magnetic brain stimulation. Lancet 2:355–358, 1986.

150. Hess, CW, Mills, KR, and Murray, WF: Methodological considerations for magnetic brain stimulation. In Barber, C, and Blum, T (eds): Evoked Potentials. III. The Third International Evoked Potential Symposium. Butterworths, London, 1987, pp 456–461.

151. Hess, CW, Mills, KR, and Murray, MF: Responses in small hand muscles from magnetic stimulation of the human brain. J Physiol 338:397–419, 1987.

152. Hume, AL, and Cant, BR: Conduction time in central somatosensory pathways in man. Electroencephalogr Clin Neurophysiol 45:361–375, 1978.

153. Hume, AL, Cant, BR, Shaw, NA, and Cowan, JC: Central somatosensory conduction time from 10 to 79 years. Electroencephalogr Clin Neurophysiol 54:49–54, 1982.

154. Hume, AL, and Durkin, MA: Central and spinal somatosensory conduction times during hypothermic cardiopulmonary bypass and some observations on the effects of fentanyl and isoflurane anesthesia. Electroencephalogr Clin Neurophysiol 65:46–58, 1986.

155. Ikuta, T, and Furuta, N: Sex differences in the human group mean SEP1. Electroencephalogr Clin Neurophysiol 54:449–457, 1982.

156. Ingram, DA, and Swash, M: Central motor conduction is abnormal in motor neuron disease (abstr). Muscle Nerve 9(5S):101, 1986.

157. Ingram, DA, and Swash, M: Central motor conduction is abnormal in motor neuron disease. J Neurol Neurosurg Psychiatry 50:159–166, 1987.

158. Jerrett, SA, Cuzzone, LJ, and Pasternak, BM:

Thoracic outlet syndrome: Electrophysiologic reappraisal. Arch Neurol 41:960–963, 1984.

159. Jewett, DL: Volume-conducted potentials in response to auditory stimuli as detected by averaging in the cat. Electroencephalogr Clin Neurophysiol 28:609–618, 1970.

160. Jewett, DL, and Williston, JS: Auditory-evoked far fields averaged from the scalp of humans. Brain 94:681–696, 1971.

161. Jones, SJ: Short latency potentials recorded from the neck and scalp following median stimulation in man. Electroencephalogr Clin Neurophysiol 43:853–863, 1977.

162. Jones, SJ, Baraitser, M, and Halliday, AM: Peripheral and central somatosensory nerve conduction defects in Freidreich's ataxia. J Neurol Neurosurg Psychiatry 43:495–503, 1980.

163. Jones, SJ, Carroll, WM, and Halliday, AM: Peripheral and central sensory nerve conduction in Charcot-Marie-Tooth disease and comparison with Friedreich's ataxia. J Neurol Sci 61:135–148, 1983.

164. Jones, SJ, Edgar, MA, Ransford, AI, and Thomas, NP: A system for the electrophysiological monitoring of the spinal cord during operations for scoliosis. J Bone Joint Surg 65-B:134–139, 1983.

165. Jones, SJ, and Small, DG: Spinal and sub-cortical evoked potentials following stimulation of the posterior tibial nerve in man. Electroencephalogr Clin Neurophysiol 44:299–306, 1978.

166. Jones, SJ, Wynn Parry, CB, and Landi, A: Diagnosis of brachial plexus traction lesions by sensory nerve action potentials and somatosensory evoked potentials. Injury 12:376–382, 1981.

167. Kaji, R, Kawaguchi, S, Tanaka, R, Kojima, J, McCormick, F, and Kameyama, M: Short latency vector somatosensory evoked potentials to median nerve stimulation: Two generators of N_{13} potential. Reviewed paper.

168. Kaji, R, and Sumner, AJ: Bipolar recording of short-latency somatosensory evoked potentials after median nerve stimulation. Neurology 37:410–418, 1987.

169. Kakigi, R: Ipsilateral and contralateral SEP components following median nerve stimulation: Effects of interfering stimuli applied to the contralateral hand. Electroencephalogr Clin Neurophysiol 64:246–259, 1986.

170. Kakigi, R, and Jones, SJ: Effects on median nerve SEPs of tactile stimulation applied to adjacent and remote areas of the body surface. Electroencephalogr Clin Neurophysiol 62:252–265, 1985.

171. Kakigi, R, and Jones, SJ: Influence of concurrent tactile stimulation on somatosensory evoked potentials following posterior tibial nerve stimulation in man. Electroencephalogr Clin Neurophysiol 65:118–129, 1986.

172. Kakigi, R, Shibasaki, H, Hashizume, A, and Kuroiwa, Y: Short latency somatosensory evoked spinal and scalp-recorded potentials following posterior tibial nerve stimulation in

man. Electroencephalogr Clin Neurophysiol 53:602–611, 1982.

173. Katayama, Y, and Tsubokawa, T: Somatosensory evoked potentials from the thalamic sensory relay nucleus (VPL) in humans: Correlations with short latency somatosensory evoked potentials recorded at the scalp. Electroencephalogr Clin Neurophysiol 68:187–201, 1987.

174. Katifi, H, and Sedgwick, E: Somatosensory evoked potentials from posterior tibial nerve and lumbosacral dermatomes. Electroencephalogr Clin Neurophysiol 65:249–259, 1986.

175. Katz, S, Martin, HF, and Blackburn, JG: The effect of interaction between large and small diameter fiber systems on the somatosensory evoked potential. Electroencephalogr Clin Neurophysiol 45:45–52, 1978.

176. Kiff, ES, and Swash, M: Normal proximal and delayed distal conduction in the pudendal nerves of patients with idiopathic (neurogenic) faecal incontinence. J Neurol Neurosurg Psychiatry 47:820–823, 1984.

177. Kimura, J: Field theory: The origin of stationary peaks from a moving source. International Symposium on Somatosensory Evoked Potentials, Kansas City, Missouri, September 22–23, 1984. American EEG Society and American Association of EMG and Electrodiagnosis, Rochester, Minn, 1984, pp 39–50.

178. Kimura, J: Abuse and misuse of evoked potentials as a diagnostic test. Arch Neurol 42:78–80, 1985.

179. Kimura, J, Ishida, T, Suzuki, S, Kudo, Y, Matsuoka, H, and Yamada, T: Far-field recording of the junctional potential generated by median nerve volleys at the wrist. Neurology, 36:1451–1457, 1986.

180. Kimura, J, Kimura A, Ishida, T, Kudo, Y, Suzuki, S, Machida, M, and Yamada, T: What determines the latency and the amplitude of stationary peaks in far-field recordings? Ann Neurol, 19:479–486, 1986.

181. Kimura, J, Machida, M, Ishida, T, Yamada, T, Rodnitzky, RL, Kudo, Y, and Suzuki, S: Relation between size of compound sensory or muscle action potentials, and length of nerve segment. Neurology 36:647–652, 1986.

182. Kimura, J, Mitsudome, A, Beck, DO, Yamada, T, and Dickins, QS: Field distribution of antidromically activated digital nerve potentials: Model for far-field recording. Neurology 33:1164–1169, 1983.

183. Kimura, J, Mitsudome, A, Yamada, T, and Dickins, QS: Stationary peaks from a moving source in far-field recording. Electroencephalogr Clin Neurophysiol 58:351–361, 1984.

184. Kimura, J, and Yamada, T: Short-latency somatosensory evoked potentials following median nerve stimulation. Ann NY Acad Sci 388:689–694, 1982.

185. Kimura, J, Yamada, T, and Kawamura, H: Central latencies of somatosensory evoked potentials. Arch Neurol 35:683–688, 1978.

186. Kimura, J, Yamada, T, Shivapour, E, and

Dickins, QS: Neural pathways of somatosensory evoked potentials: Clinical implication. Electroencephalogr Clin Neurophysiol (Suppl) 36:336–348, 1982.

187. Kimura, J, Yamada, T, and Walker, DD: Theory of near- and far-field evoked potentials. In Lüders H (ed): Advanced Evoked Potentials. Martinus Nijhoff, Boston, 1988 (in press).

188. Kornhüber, HH, and Deecke, L: Hirnpotentialanderungen bei Willkurbewegungen und passiven Bewegungen deds Menschen: Bereitschaftspotential und reafferent Potentiale. Pflugers Arch Ges Physiol 284:1–17, 1965.

189. Kritchevsky, M, and Wiederholt, WC: Short-latency somatosensory evoked potentials. Arch Neurol 35:706–711, 1978.

190. Lastimosa, ACB, Bass, NH, Stanback, K, and Norvell, EE: Lumbar spinal cord and early cortical evoked potentials after tibial nerve stimulation: Effects of stature on normative data. Electroencephalogr Clin Neurophysiol 54:499–507, 1982.

191. Leandri, M, and Campbell, JA: Origin of early waves evoked by infraorbital nerve stimulation in man. Electroencephalogr Clin Neurophysiol 65:13–19, 1986.

192. Leandri, M, Favale, E, Ratto, S, and Abbruzzese, M: Conducted and segmental components of the somatosensory cervical response. J Neurol Neurosurg Psychiatry 44:718–722, 1981.

193. Leandri, M, Parodi, CI, Zattoni, J, and Favale, E: Subcortical and cortical responses following infraorbital nerve stimulation in man. Electroencephalogr Clin Neurophysiol 66:253–262, 1987.

194. Lee, B, Luders, H, Lesser, R, Dinner, D, and Morris, H, III: Cortical potentials related to voluntary and passive finger movements recorded from subdural electrodes in humans. Ann Neurol 20:32–37, 1986.

195. Lehmkuhl, D, Dimitrijevic, MR, and Renouf, F: Electrophysiological characteristics of lumbosacral evoked potentials in patients with established spinal cord injury. Electroencephalogr Clin Neurophysiol 59:142–155, 1984.

196. Lesser, RP, Luders, H, Hahn, J, and Klem, G: Early somatosensory potentials evoked by median nerve stimulation: Intraoperative monitoring. Neurology 31:1519–1523, 1981.

197. Lesser, RP, Raudzens, P, Luders, H, Nuwer, MR, Goldie, WD, Morris, HH, III, Dinner, DS, Klem, G, Hahn, JF, Shetter, AG, Ginsburg, HH, and Gurd, AR: Postoperative neurological deficits may occur despite unchanged intraoperative somatosensory evoked potentials. Ann Neurol 19:22–25, 1986.

198. Levy, WJ, York, DH, McCaffrey, M, and Tanzer, F: Motor evoked potentials from transcranial stimulation of the motor cortex in humans. Neurosurgery 15:287–302, 1984.

199. Lin, JT, Phillips, LH, II, and Daube, JR: Far-field potentials recorded from peripheral nerves (abstr). Electroencephalogr Clin Neurophysiol 50:174P, 1980.

200. Lorente de NO, R: A study of nerve physiology. Studies from the Rockefeller Institute, Vol 132, 1947, Chapter 16.

201. Louis, AA, Gupta, P, and Perkash, I: Localization of sensory levels in traumatic quadriplegia by segmental somatosensory evoked potentials. Electroencephalogr Clin Neurophysiol 62:313–316, 1985.

202. Luders, H: Advanced evoked potentials. Martinus Nijhoff, Boston, 1988.

203. Luders, H, Andrish, J, Gurd, A, Weiker, G, and Klem, G: Origin of far-field subcortical potentials evoked by stimulation of the posterior tibial nerve. Electroencephalogr Clin Neurophysiol 52:336–344, 1981.

204. Luders, H, Dinner, SD, Lesser, RP, and Klem, G: Origin of far field subcortical evoked potentials to posterior tibial and median nerve stimulation. Arch Neurol 40:93–97, 1983.

205. Luders, H, Gurd, A, Hahn, J, et al: A new technique for intraoperative monitoring of spinal cord function evoked potentials. Spine 7:110–115, 1982.

206. Luders, H, Lesser, RP, Dinner, DS, Hahn, JF, Salanga, V, and Morris, HH: The second sensory area in humans: Evoked potential and electrical stimulation studies. Ann Neurol 17:177–184, 1985.

207. Luders, H, Lesser, R, Hahn, J, Little, J, and Klem, G: Subcortical somatosensory evoked potentials to median nerve stimulation. Brain 106, 341–372, 1983.

208. Maccabee, PJ, Pinkhasov, EI, and Cracco, RQ: Short latency evoked potentials to median nerve stimulation: Effect of low frequency filter. Electroencephalogr Clin Neurophysiol 55:34–44, 1983.

209. Machida, M, Weinstein, SL, Yamada, T, and Kimura, J: Spinal cord monitoring—electrophysiological measures of sensory and motor function during spinal surgery. Spine 10:407–413, 1985.

210. Markan, ON, Dilley, RS, Moorthy, SS, and Warren, C: Monitoring of somatosensory evoked responses during carotid endartectomy. Arch Neurol 41:375–378, 1984.

211. Marsden, CD, Merton, PA, and Morton, HB: Percutaneous stimulation of spinal cord and brain: Pyramidal tract conduction velocities in man (abstr). J Physiol 238:6P, 1982.

212. Mastaglia, FI, Black, JL, and Collins, DWK: Visual and spinal evoked potentials in diagnosis of multiple sclerosis (abstr). Br Med J 2:732, 1976.

213. Matthews, WB, Read, DJ, and Pountney, E: Effect of raising body temperature on visual and somatosensory evoked potentials in patients with multiple sclerosis. J Neurol Neurosurg Psychiatry 42:250–255, 1979.

214. Matthews, WB, Wattam-Bell, JRB, and Pountney, E: Evoked potentials in the diagnosis of multiple sclerosis. A follow-up study. J Neurol Neurosurg Psychiatry 45:303–307, 1982.

215. Mauguiere, F, and Courjon, J: The origins of short-latency somatosensory evoked potentials in humans. Ann Neurol 9:607–611, 1981.

216. Mauguiere, F, Courjon, J, and Schott, B: Dissociation of early SEP components in unilateral traumatic section of the lower medulla. Ann Neurol 13:309–313, 1983.

217. Mauguiere, F, Desmedt, JE, and Courjon, J: Astereggnosis and dissociated loss of frontal or parietal components of somatosensory evoked potentials in hemispheric lesions. Brain 106:271–311, 1983.

218. Mauguiere, F, and Ibanez, V: The dissociation of early SEP components in lesions of the cervico-medullary junction: A cue for routine interpretation of abnormal cervical responses to median nerve stimulation. Electroencephalogr Clin Neurophysiol 62:406–420, 1985.

219. McLeod, JG: An electrophysiological and pathological study of peripheral nerves in Friedreich's ataxia. J Neurol Sci 12:333–349, 1971.

220. Merton, PA, and Morton, HB: Stimulation of the cerebral cortex in the intact human subject. Nature 285:227, 1980.

221. Merton, PA, Morton, HB, Hills, DK, and Marsden, CD: Scope of a technique for electrical stimulation of human brain, spinal cord and muscle. Lancet 11:579–600, 1982.

222. Meyer-Hardting, E, Wiederholt, WC, and Budnick, B: Recovery function of short-latency components of the human somatosensory evoked potential. Arch Neurol 40:290–293, 1983.

223. Mills, KR: Handout, Clinical Electromyography, American Academy of Neurology, 1987.

224. Mills, KR, and Murray, NMF: Corticospinal tract conduction time in multiple sclerosis. Ann Neurol 18:601–605, 1985.

225. Mills, KR, and Murray, NMF: Electrical stimulation over the human vertebral column: Which neural elements are excited? Electroencephalogr Clin Neurophysiol 63:582–589, 1986.

226. Mills, KR, and Murray, NMF: Neurophysiological evaluation of associated demyelinating peripheral neuropathy and multiple sclerosis: A case report. J Neurol Neurosurg Pyschiatry 49:320–323, 1986.

227. Milner-Brown, SH, Girvin, JP, and Brown, WF: The effects of motor cortical stimulation on the excitability of spinal motoneurons in man. Can J Neurol Sci August: 245–253, 1975.

228. Molaie, M: Scalp-recorded short and middle latency peroneal somatosensory evoked potentials in normals: Comparison with peroneal and median nerve SEPs in patients with unilateral hemispheric lesions. Electroencephalogr Clin Neurophysiol 68:107–118, 1987.

229. Moller, A, Jannetta, F, and Burgess, J: Neural generators of the somatosensory evoked potentials: Recording from the cuneate nucleus in man and monkeys. Electroencephalogr Clin Neurophysiol 65:241–248, 1986.

230. Moor, EJ: Bases of Auditory Brain-stem Evoked Responses. New York, Grune & Stratton, 1983.

231. Nainzadeh, N, and Lane, ME: Somatosensory evoked potentials following pudendal nerve stimulation as indicators of low sacral root involvement in a postlaminectomy patient. Arch Phys Med Rehabil 68:170–172, 1987.

232. Nakanishi, T: Action potentials recorded by fluid electrodes. Electroencephalogr Clin Neurophysiol 53:343–345, 1982.

233. Nakanishi, T: Origin of action potential recorded by fluid electrodes. Electroencephalogr Clin Neurophysiol 55:114–115, 1983.

234. Nakanishi, T, Shimada, Y, Sakuta, M, and Toyokura, Y: The initial positive component of the scalp-recorded somatosensory evoked potential in normal subjects and in patients with neurological disorders. Electroencephalogr Clin Neurophysiol 45:26–34, 1978.

235. Nakanishi, T, Takita, K, and Toyokura, Y: Somatosensory evoked responses to the tactile tap in man. Electroencephalogr Clin Neurophysiol 34:1–6, 1973.

236. Nakanishi, T, Tamaki, M, Arasaki, K, and Kudo, N: Origins of the scalp-recorded somatosensory far-field potentials in man and cat. Electroencephalogr Clin Neurophysiol Suppl 36:336–348, 1982.

237. Nakanishi, T, Tamaki, M, and Kudo, K: Possible mechanism of generation of SEP-far field component in the brachial plexus in the cat. Electroencephalogr Clin Neurophysiol 63:68–74, 1986.

238. Namerow, NS: Somatosensory evoked responses in multiple sclerosis patients with varying sensory loss. Neurology 18:1197–1204, 1968.

239. Nelson, FW, Goldie, WD, Hecht, JT, Butler, IJ, and Scott, CI: Short-latency somatosensory evoked potentials in the management of patients with achondroplasia. Neurology 34:1053–1058, 1984.

240. Noel, P, and Desmedt, JE: Somatosensory cerebral evoked potentials after vascular lesions of the brainstem and diencephalon. Brain 98:113–128, 1975.

241. Noel, P, and Desmedt, JE: Cerebral and far-field somatosensory evoked potentials in neurological disorders. In Desmedt, JE (ed): Clinical Use of Cerebral, Brainstem and Spinal Somatosensory Evoked Potentials. Progress in Clinical Neurophysiology, Vol 7. Karger, Basel, pp 205–230, 1980.

242. Nordwall, A, Axellgaard, J, Harada, Y, et al: Spinal cord monitoring using evoked potentials recorded from vertebral bone in cat. Spine 4:486–494, 1979.

243. Noth, J, Engel, L, Friedmann, HH, and Lange, HW: Evoked potentials in patients with Huntington's disease and their offspring. I. Somatosensory evoked potentials. Electroencephalogr Clin Neurophysiol 59:134–141, 1984.

244. Parry, GJ, and Aminoff, MJ: Somatosensory evoked potentials in chronic acquired demyelinating peripheral neuropathy. Neurology 37:313–316, 1987.

245. Perlik, S, Fisher, M, Patel, D, and Slack, C: On the usefulness of somatosensory evoked responses for the evaluation of lower back pain. Arch Neurol 43:907–913, 1986.

246. Phillips, KR, Potvin, AR, Syndulko, K, Cohen, SN, Tourtellotte, WW, and Potvin, JH: Multimodality evoked potentials and neurophysiological tests in multiple sclerosis: Effects of hyperthermia on test results. Arch Neurol 40:159–164, 1983.

247. Phillips, LH, II, and Daube, JR: Lumbosacral spinal evoked potentials in humans. Neurology 30:1175–1183, 1980.

248. Pratt, H, and Starr, A: Mechanically and electrically evoked somatosensory potentials in humans: Scalp and neck distributions of short latency components. Electroencephalogr Clin Neurophysiol 51:138–147, 1981.

249. Pratt, H, Starr, A, Amile, RN, and Politoske, D: Mechanically and electrically evoked somatosensory potentials in normal humans. Neurology 29:1236–1244, 1979.

250. Radtke, R, Erwin, A, and Erwin, C: Abnormal sensory evoked potentials in amyotrophic lateral sclerosis. Neurology 36:796–801, 1986.

251. Ratto, S, Abbruzzese, M, Abbruzzese, G, and Favale, E: Surface recording of the spinal ventral root discharge in man. Brain 106:897–909, 1983.

252. Reisecker, F, Witzmann, A, and Deisenhammer, E: Somatosensory evoked potentials (SSEPs) in various groups of cerebro-vascular ischaemic disease. Electroencephalogr Clin Neurophysiol 65:260–268, 1986.

253. Robertson, WC, Jr, and Lambert, EH: Sensory nerve conduction velocity in children using evoked potentials. Arch Phys Med Rehabil 59:1–4, 1978.

254. Ropper, AH, and Chiappa, KH: Evoked potentials in Guillain-Barré syndrome. Neurology 36:587–590, 1986.

255. Ropper, AH, Miett, T, and Chiappa, KH: Absence of evoked potential abnormalities in acute transverse myelopathy. Neurology 32:80–82, 1982.

256. Rossini, PM: Evaluation of sensory-motor "central" conduction in normals and in patients with demyelinating diseases. In Morrocutti, C and Rizzo, PA (ed): Evoked Potentials: Neurophysiological and Clinical Aspects. Elsevier, Amsterdam, 1986.

257. Rossini, PM, and Cracco, JB: Somatosensory and brainstem auditory evoked potentials in neurodegenerative system disorders. Eur Neurol 26:176–188, 1987.

258. Rossini, PM, Cracco, RQ, Cracco, JB, and House, WJ: Short latency somatosensory evoked potentials to peroneal nerve stimulation: Scalp topography and the effect of different frequency filters. Electroencephalogr Clin Neurophysiol 52:540–552, 1981.

259. Rossini, PM, Di Stefano, E, and Stanzione, P: Nerve impulse propagation along central and peripheral fast conduction motor and sensory pathways in man. Electroencephalogr Clin Neurophysiol 60:320–334, 1985.

260. Rossini, PM, Marciani, MG, Caramia, M, Roma, V, and Zarola, F: Nervous propagation along "central" motor pathways in intact man: Characteristics of motor responses to "bifocal" and "unifocal" spine and scalp non-invasive stimulation. Electroencephalogr Clin Neurophysiol 61:272–286, 1985.

261. Rossini, PM, Treviso, M, Di Stefano, E, and Di Paolo, B: Nervous impulse propagation along peripheral and central fibres in patients with chronic renal failure. Electroencephalogr Clin Neurophysiol 56:293–303, 1983.

262. Rothwell, JC, Obeso, JA, and Marsden, CD: On the significance of giant somatosensory evoked potentials in cortical myoclonus. J Neurol Neurosurg Psychiatry 47:33–42, 1984.

263. Rushton, DN, Rothwell, JC, and Craggs, MD: Gating of somatosensory evoked potentials during different kinds of movement in man. Brain 104:465–491, 1981.

264. Salar, G, Iob, I, and Mingrino, S: Somatosensory evoked potentials before and after percutaneous thermocoagulation of the gasserian ganglion for trigeminal neuralgia. In Courjon, J, Mauguire, F, and Revol, M (eds): Clinical Applications of Evoked Potentials in Neurology, Raven Press, New York, 1982, pp 359–365.

265. Salzman, SK, Beckman, AL, Marks, HG, Naidu, R, Bunnell, WP, and Macewen, GD: Effects of halothane on intraoperative scalp-recorded somatosensory evoked potentials to posterior tibial nerve stimulation in man. Electroencephalogr Clin Neurophysiol 65:36–45, 1986.

266. Sarica, Y, and Karacan, I: Cerebral responses evoked by stimulation of the vesico-urethral junction in normal subjects. Electroencephalogr Clin Neurophysiol 65:440–446, 1986.

267. Sarica, Y, Karacan, I, Thornby, JI, and Hirshkowitz, M: Cerebral responses evoked by stimulation of vesico-urethral junction in man: Methodological evaluation of monopolar stimulation. Electroencephalogr Clin Neurophysiol 65:130–135, 1986.

268. Schiff, JA, Cracco, RQ, Rossini, PM, and Cracco, JB: Spine and scalp somatosensory evoked potentials in normal subjects and patients with spinal cord disease: Evaluation of afferent transmission. Electroencephalogr Clin Neurophysiol 59:374–387, 1984.

269. Schramm, J: Clinical experience with the objective localization of the lesion in cervical myelopathy. In Grote, W, Brock, M, Clar, HE, Klinger, M, and Nau, HE (eds): Advances in Neurosurgery, Vol 8, Surgery of Cervical Myelopathy Infantile Hydrocephalus: Long Term Results, Springer-Verlag, Berlin, 1980, pp 26–32.

270. Sedgwick, EM, El-Negamy, E, and Frankel, H: Spinal cord potentials in traumatic paraplegia and quadriplegia. J Neurol Neurosurg Psychiatry 43:823–830, 1980.

271. Seyal, M, Emerson, RG, and Pedley, TA: Spinal and early scalp-recorded components in the somatosensory evoked potential following stimulation of the posterior tibial nerve. Electroencephalogr Clin Neurophysiol 55:320–330, 1983.

272. Sgro, JA, Emerson, RG, and Pedley, TA: Real-time reconstruction of evoked potentials using a new two-dimensional filter method. Elec-

troencephalogr Clin Neurophysiol 62:372–380, 1985.

273. Shaw, N: Somatosensory evoked potentials after head injury: The significance of the double peak. J Neurol Sci 73:145–153, 1986.

274. Shaw, NA, and Synek, VM: Intersession stability of somatosensory evoked potentials. Electroencephalogr Clin Neurophysiol 66:281–285, 1987.

275. Shibasaki, H, Barrett, G, Halliday, E, and Halliday, AM: Components of the movement-related cortical potential and their scalp topography. Electroencephalogr Clin Neurophysiol 49:213–226, 1980.

276. Shibasaki, H, Barrett, G, Halliday, E, and Halliday, AM: Cortical potentials following voluntary and passive finger movements. Electroencephalogr Clin Neurophysiol 50:201–213, 1980.

277. Shibasaki, H, Kakigi, R, Ohnishi, A, and Kuroiwa, Y: Peripheral and central nerve conduction in subacute myelo-optico-neuropathy. Neurology 32:1186–1189, 1982.

278. Shibasaki, H, and Nagae, K: Application of movement-related cortical potentials. Ann Neurol 15:299–302, 1984.

279. Shibasaki, H, Ohnishi, A, and Kuroiwa, Y: Use of SEPs to localize degeneration in a rare polyneuropathy: Studies on polyneuropathy associated with pigmentation, hypertrichosis, edema, and plasma cell dyscrasia. Ann Neurol 12:355–360, 1982.

280. Shibasaki, H, Sakai, T, Nishimura, H, Sato, Y, Goto, I, and Kuroiwa Y: Involuntary movements in chorea-acanthocytosis: A comparison with Huntington's chorea. Ann Neurol 12:311–314, 1982.

281. Shibasaki, H, Shima, F, and Kuroiwa, Y: Clinical studies of the movement-related cortical potential (MP) and the relationship between the dentatorubrothalamic pathway and readiness potential. J Neurol 219:15–25, 1978.

282. Shibasaki, H, Yamashita, Y, and Kuroiwa, Y: Electroencephalographic studies of myoclonus: Myoclonus-related cortical spikes and high amplitude somatosensory evoked potentials. Brain 101:447–460, 1978.

283. Shibasaki, H, Yamashita, Y, Neshige, R, Tobimatsu, S, and Fukui, R: Pathogenesis of giant somatosensory evoked potentials in progressive myoclonic epilepsy. Brain 108:225–240, 1985.

284. Shimada, Y, and Nakanishi, T: Somatosensory evoked responses to mechanical stimulation in man. Adv Neurol Sci 23:282–293, 1979.

285. Shimizu, H, Shimoji, K, Maruyama, Y, Sato, Y, Harayama, H, and Tsubaki, T: Slow cord dorsum potentials elicited by descending volleys in man. J Neurol Neurosurg Psychiatry 42:242–246, 1979.

286. Shimoji, K, Higashi, H, and Kano, T: Epidural recording of spinal electrogram in man. Electroencephalogr Clin Neurophysiol 30:236–239, 1971.

287. Siivola, J, Sulg, I, and Heiskari, M: Somatosensory evoked potentials in diagnostics of cervical spondylosis and herniated disc. Electroencephalogr Clin Neurophysiol 52:276–282, 1981.

288. Singh, N, Sachdev, KK, and Brisman, R: Trigeminal nerve stimulation: Short latency somatosensory evoked potentials. Neurology 32:97–101, 1982.

289. Slimp, JC, Tamas, LB, Stolov, WC, and Wyler, AR: Somatosensory evoked potentials after removal of somatosensory cortex in man. Electroencephalogr Clin Neurophysiol 65:111–117, 1986.

290. Small, DG, Matthews, WB, and Small, M: The cervical somatosensory evoked potential (SEP) in the diagnosis of multiple sclerosis. J Neurol Sci 35:211–224, 1978.

291. Small, M, and Matthews, WB: A method of calculating spinal cord transit time from potentials evoked by tibial nerve stimulation in normal subjects and in patients with spinal cord disease. Electroencephalogr Clin Neurophysiol 59:156–164, 1984.

292. Snooks, SJ, and Swash, M: Motor conduction velocity in the human spinal cord: Slowed conduction in multiple sclerosis and radiation myelopathy. J Neurol Neurosurg Psychiatry 48:1135–1139, 1985.

293. Sohmer, H, and Feinmesser, M: Cochlear and cortical audiometry conveniently recorded in the same subject. Israel J Med Sci 6:219–223, 1970.

294. Spielholz, NI, Benjamin, MV, Engler, GL, and Ransohoff, J: Somatosensory evoked potentials during decompression and stabilization of the spine: Methods and findings. Spine 4:500–505, 1979.

295. Starr, A, McKeon, B, Skuse, N, and Burke, D: Cerebral potentials evoked by muscle stretch in man. Brain 104:149–166, 1981.

296. Stegeman, D, Van Oosteron, A, and Colon, E: Far field evoked potential components induced by a propagating generator: Computational evidence. Electroencephalogr Clin Neurophysiol 67:176–187, 1987.

297. Stejskal, L, and Sobota, J: Somatosensory evoked potentials in patients with occlusions of cerebral arteries. Electroencephalogr Clin Neurophysiol 61:482–490, 1985.

298. Stewart, JM, Houser, OW, Baker, HL, Jr, O'Brien, PC, and Rodriguez, M: Magnetic resonance imaging and clinical relationships in multiple sclerosis. Mayo Clin Proc 62:174–184, 1987.

299. Stöhr, M, Dichgans, J, Voigt, K, and Buettner, UW: The significance of SEPs for localization of unilateral lesions within the cerebral hemispheres. J Neurol Sci 61:49–63, 1983.

300. Stöhr, M, and Petruch, F: Somatosensory evoked potentials following stimulation of the trigeminal nerve in man. Neurology 220:95–98, 1979.

301. Stöhr, M, Petruch, F, and Scheglmann, K: Somatosensory evoked potentials following trigeminal nerve stimulation in trigeminal neuralgia. Ann Neurol 9:63–66, 1981.

302. Sugimoto, SU, Tsuruta, K, Kurihara, T, Ono, S, Morotomi, Y, Inoue, K, and Matsukura, S:

Posterior tibial somatosensory evoked potentials in duchenne-type progressive muscular dystrophy. Electroencephalogr Clin Neurophysiol 64:525–527, 1986.

303. Swash, M, and Snooks, S: Slowed motor conduction in lumbosacral nerve roots in cauda equina lesions: A new diagnostic technique. J Neurol Neurosurg Psychiatry 49:808–816, 1986.

304. Synek, VM: Somatosensory evoked potentials after stimulation of digital nerves in upper limbs: Normative data. Electroencephalogr Clin Neurophysiol 65:460–463, 1986.

305. Synek, VM, and Cowan, JC: Saphenous nerve evoked potentials and the assessment of intraabdominal lesions of the femoral nerve. Muscle Nerve 6:453–456, 1983.

306. Tamaki, T, Tsuji, H, Inoue, S, and Kobayashi, H: The prevention of iatrogenic spinal cord injury utilizing the evoked spinal cord potential. Int Orthop 4:313–317, 1981.

307. Tamas, LB, and Shibasaki, H: Cortical potentials associated with movement: A review. J Clin Neurophysiol 2:157–171, 1985.

308. Tamura, K: Ipsilateral somatosensory evoked responses in man. Folia Psychiatr Neurol Jpn 26:83–94, 1972.

309. Tapia, MC, Cohen, LG, and Starr, A: Selectivity of attenuation (i.e., gating) of somatosensory potentials during voluntary movement in humans. Electroencephalogr Clin Neurophysiol 68:226–230, 1987.

310. Taylor, MJ, and Black, SE: Lateral asymmetries and thalamic components in far-field somatosensory evoked potentials. Can J Neurol Sci 11:252–256, 1984.

311. Thomas, PK, Jefferys, JGR, Smith, IS, and Loulakakis, D: Spinal somatosensory evoked potentials in hereditary spastic paraplegia. J Neurol Neurosurg Psychiatry 44:243–246, 1981.

312. Thompson, PD, Dick, JPR, Asselman, P, Griffin, GB, Day, BL, Rothwell, JC, Sheehy, MP, and Marsden, CD: Examination of motor function in lesions of the spinal cord by stimulation of the motor cortex. Ann Neurol 21:389–396, 1987.

313. Tomita, Y, Nishimura, S, and Tanaka, T: Short latency SEPs in infants and children: Developmental changes and maturational index of SEPs. Electroencephalogr Clin Neurophysiol 65:335–343, 1986.

314. Trojaborg, W, and Petersen, E: Visual and somatosensory evoked cortical potentials in multiple sclerosis. J Neurol Neurosurg Psychiatry 42:323–330, 1979.

315. Tsuji, S, and Murai, Y: Scalp topography and distribution of cortical somatosensory evoked potentials to median nerve stimulation. Electroencephalogr Clin Neurophysiol 65:429–439, 1986.

316. Tsumoto, T, Hirose, N, Nonaka, S, and Takahashi, M: Cerebrovascular disease: Changes in somatosensory evoked potentials associated with unilateral lesions. Electroencephalogr Clin Neurophysiol 35:463–473, 1973.

317. Tsuyama, N, Tsuzuki, N, Kurokawa, T, and Imai, T: Clinical application of spinal cord action potential measurement. Int Orthop 2:39–46, 1978.

318. Vas, GA, Cracco, JB, and Cracco, RQ: Scalp-recorded short latency cortical and subcortical somatosensory evoked potentials to peroneal nerve stimulation. Electroencephalogr Clin Neurophysiol 52:1–8, 1981.

319. Vaughan, HG, Jr: The neural origins of human event-related potentials. Ann NY Acad Sci 388:125–138, 1982.

320. Vaughan, HG, Jr, Costa, LD, Gilden, L, and Schimmel, H: Identification of sensory and motor components of cerebral activity in simple reaction-time tasks. Proceedings of the 73rd Convention of the American Psychological Association, Vol 1, 1965, pp 179–180.

321. Veilleux, M, Daube, JR, and Cucchiara, RF: Monitoring of cortical evoked potentials during surgical procedures on the cervical spine. Mayo Clin Proc 62:256–264, 1987.

322. Vercruyssen, A, Martin, JJ, Ceuterick, C, Jacobs, K, and Swerts, L: Adult ceroid-lipofuscinosis: Diagnostic value of biopsies and of neurophysiological investigations. J Neurol Neurosurg Psychiatry 45:1056–1059, 1982.

323. Walser, H, Mattle, H, Keller, HM, and Janzer, R: Early cortical median nerve somatosensory evoked potentials. Prognostic value in anoxic coma. Arch Neurol 42:32–38, 1985.

324. Wang, AD, Symon, L, and Gentili, F: Conduction of sensory action potentials across the posterior fossa in infratentorial space-occupying lesions in man. J Neurol Neurosurg Psychiatry 45:440–445, 1982.

325. Wiederholt, WC, and Iragui-Madoz, VJ: Far field somatosensory potentials in the rat. Electroencephalogr Clin Neurophysiol 42:456–465, 1977.

326. Wilkins, DE, Hallett, M, Berardelli, A, Walshe, T, and Alvarez, N: Physiologic analysis of the myoclonus of Alzheimer's disease. Neurology 34:898–903, 1984.

327. Williamson, PD, Goff, WR, and Allison, T: Somatosensory evoked responses in patients with unilateral cerebral lesions. Electroencephalogr Clin Neurophysiol 28:566–575, 1970.

328. Wong, PKH, Lombroso, CT, and Matsumiya, Y: Somatosensory evoked potentials: Variability analysis in unilateral hemispheric disease. Electroencephalogr Clin Neurophysiol 54:266–274, 1982.

329. Wood, CC, and Allison, T: Interpretation of evoked potentials: a neurophysiological perspective. Can J Pyschol/Rev 35(2):113–135, 1981.

330. Wulff, CH, and Trojaborg, W: Adult metachromatic leukodystrophy: Neurophysiologic findings. Neurology 35:1776–1778, 1985.

331. Yamada, T, Graff-Radford, NR, Kimura, J, Dickins, QS, and Adams, HP, Jr: Topographic analysis of somatosensory evoked potentials in patients with well-localized thalamic infarctions. J Neurol Sci 68:31–46, 1985.

332. Yamada, T, Kayamori, R, Kimura, J, and

Beck, DO: Topography of somatosensory evoked potential after stimulation of the median nerve. Electroencephalogr Clin Neurophysiol 59:29–43, 1984.

333. Yamada, T, Kimura, J, and Machida, M; Scalp-recorded far-field potentials and spinal potentials after stimulation of the tibial nerve. In Nodar, RH, and Barber, C (eds): Evoked Potentials II. Butterworth Publishers, Boston, 1983, pp 353–362.

334. Yamada, T, Kimura, J, and Nitz, DM: Short latency somatosensory evoked potentials following median nerve stimulation in man. Electroencephalogr Clin Neurophysiol 48: 367–376, 1980.

335. Yamada, T, Kimura, J, Wilkinson, T, and Kayamori, R: Short- and long-latency median somatosensory evoked potentials: Findings in patients with localized neurological lesions. Arch Neurol 40:215–220, 1983.

336. Yamada, T, Kimura, J, Young, S, and Powers, M: Somatosensory evoked potentials elicited by bilateral stimulation of the median nerve and its clinical application. Neurology 28:218–223, 1978.

337. Yamada, T, Machida, M, and Kimura, J: Farfield somatosensory evoked potentials after stimulation of the tibial nerve in man. Neurology 32:1151–1158, 1982.

338. Yamada, T, Machida, M, Oishi, M, Kimura, A, Kimura, J, and Rodnitzky, RL: Stationary negative potentials near the source vs positive far-field potentials at a distance. Electroencephalogr Clin Neurophysiol 60:509–524, 1985.

339. Yamada, T, Muroga, T, and Kimura, J: The effect of tourniquet induced ischemia on somatosensory evoked potentials. Neurology 31:1524–1529, 1981.

340. Yamada, T, Shivapour, E, Wilkinson, JT, and Kimura, J: Short- and long-latency somatosensory evoked potentials in multiple sclerosis. Arch Neurol 38:88–94, 1982.

341. Yamada, T, Wilkinson, JT, and Kimura, J: Are there multiple pathways for the short and long latency SEPs. Electroencephalogr Clin Neurophysiol 51:43P–44P, 1981.

342. Yiannikas, C, Shahani, BT, and Young, RR: Short-latency somatosensory-evoked potentials from radial, median, ulnar and peroneal nerve stimulation in the assessment of cervical spondylosis: Comparison with conventional electromyography. Arch Neurol 43: 1264–1271, 1986.

343. Yu, YL, and Jones, SJ: Somatosensory evoked potentials in cervical spondylosis correlation of median, ulnar and posterior tibial nerve responses with clinical and radiological findings. Brain 108:273–300, 1985.

344. Ziganow, S: Neurometric evaluation of the cortical somatosensory evoked potential in acute incomplete spinal cord. Electroencephalogr Clin Neurophysiol 65:86–93, 1986.

Part VI

DISORDERS OF THE SPINAL CORD AND PERIPHERAL NERVOUS SYSTEM

Chapter 20

DISEASES OF THE MOTOR NEURON

1 INTRODUCTION

Many disorders affect the spinal cord, but the most commonly encountered in an electromyographic laboratory are degenerative diseases of the anterior horn cells. Of various classifications proposed, those based on clinical as well as genetic features provide the most satisfactory categorization, pending the elucidation of the basic biochemical defects.[77] Classical motor neuron disease characteristically shows combined involvement of the upper and lower motor neurons. This group includes progressive bulbar palsy, progressive muscular atrophy, amyotrophic lateral sclerosis, and its variant, primary lateral sclerosis. In contrast, the patients with spinal muscular atrophies have genetically determined degeneration of the anterior horn cells without corticospinal tract involvement.

A number of other conditions, infectious and toxic in nature, affect the motor neurons in the spinal cord selectively or in conjunction with the corticospinal tract. Despite the advent of a vaccine in the 1950s, poliomyelitis still prevails in the tropics and, less frequently, in the

United States. With diminishing public awareness of the need for vaccination, new epidemics may develop. Sequelae of poliomyelitis, although relatively common, may escape detection unless clinically suspected. Experimental studies have shown degeneration of anterior horn cells after the administration of aluminum, vincristine, and acetyl ethyl tetramethyl tetralin. Syringomyelia, another classic neurologic disorder, also affects the spinal cord. The disease often mimics motor neuron disease, because cutaneous touch sensation may remain completely normal. Careful sensory examination will, however, reveal a selective loss of pain perception in the involved cervical or lumbosacral dermatomes.

Amyotrophy may occur as a feature of familial multisystem atrophies such as familial motor neuron disease and familial spastic paraplegia. Some subtypes of spinocerebellar degeneration also show clinical and electromyographic evidence of lower motor neuron disease as a major finding. They include autosomal dominant olivopontocerebellar atrophy, autosomal recessive glutamate dehydrogenase deficiency, and a Portuguese variant, Joseph disease. Other systemic disorders may present with amyotrophy and denervation, notably, Parkinson's disease, Huntington's chorea, Pick's disease, and xeroderma pigmentosum. Juvenile spinal muscular atrophy with hexosaminidase deficiency resembles the Kugelberg-Welander phenotype.[72]

This section will discuss certain diseases of the anterior horn cells as they pertain to electromyography and nerve conduction studies. Readers interested in more comprehensive clinical reviews should consult existing texts.[14,129]

2 MOTOR NEURON DISEASE

Motor neuron disease, together with Parkinsonian syndrome and Alzheimer's disease, constitutes a triad of degenerative disorders of the aging nervous system.[114] In these disorders, selective vulnerability of a special set of cells leads primarily to degeneration of the upper and lower motor neurons. Despite certain clinical resemblance to transmissible Jakob-Creutzfeldt disease, attempts to isolate a virus have consistently failed.[117] Unidentified virus might have caused a motor neuron disease with the clinical and pathologic appearance of amyotrophic lateral sclerosis in a woman severely bitten by a cat.[66] Another clue for the pathogenesis may lie in the lack of hexosaminidase in some patients with recessively inherited motor neuron disease.[91,112] Other members of afflicted families have had typical Tay-Sachs disease. Interestingly, the enzyme may fall to a very low level in those with motor neuron disease, as in the relatives with Tay-Sachs disease.[72]

A monoclonal protein, which usually produces a sensory motor peripheral neuropathy, sometimes induces a motor system disorder resembling motor neuron disease. This observation suggests that the antibodies may impair the function of the cell body itself or the axons that extend from the cell body.[113,115] A search for an immunologic abnormality has led to the demonstration of serum antibodies against a growth factor in some patients.[52] This finding, if confirmed, implies that motor neuron disease results from deficiency of nerve growth factor.[5] In still another study, uncontrolled trials of thyrotropin-releasing hormone (TRH) have improved motor function in some patients.[38] In-vitro application of TRH to rat muscles increases the frequency of fibrillation potentials and miniature end-plate potentials. Such influence on nerve terminals may account for its effect on muscle strength.[126]

The various syndromes, though described as separate nosologic entities, may represent a disease spectrum according to the sites of maximal neuronal involvement. This entity consists of common sporadic disorders and 5 to 10 percent of familial cases with an autosomal dominant pattern of inheritance. Patients with progressive muscular atrophy present with only lower motor neuron impairment, whereas those with amyotrophic lateral sclerosis have features of upper motor neuron lesions as well. On the other hand, prominent corticospinal tract signs without lower motor neuron involvement characterize primary lateral sclerosis. Progressive bulbar palsy shows a combination of brainstem dysfunction and spasticity of the extremities.

Electromyography and nerve stimulation techniques help establish the differential diagnosis of these disease entities. Reduced recruitment suggests loss of motor neurons during early stages. Fibrillation potentials appear at least 2 to 3 weeks after the onset of illness. Motor unit potentials of large amplitude and long duration develop later as the consequence of reinnervation. The clinical severity of the disease correlates approximately with the magnitude and distribution of fibrillation potentials and the degree of reduction in amplitude of the compound muscle action potentials. Despite reduced amplitude of muscle potential and slowed motor conduction, sensory action potentials remain normal in most cases.

Amyotrophic Lateral Sclerosis

The adjective amyotrophic implies muscle wasting as a result of an anterior horn cell disorder. The term in other contexts may refer to any neurogenic atrophy, including those resulting from radicular lesions or localized injuries of the peripheral nerve. The disease has a prevalence of 2 to 7 and an incidence of 1.4 per population of 100,000.[9,84] Etiologic possibilities include genetic, toxic, and viral causes, although none has been proven.[79,85,90,107]

The essential pathologic change consists of relatively selective degeneration of the motor cells in the spinal cord, brainstem, and, to a much lesser extent, the cortex. The most extensive cellular damage occurs at the cervical and lumbar levels, primarily affecting the large motor cells. Studies of the ventral spinal root reveal axonal degeneration of the large myelinated fibers.[55] In the brainstem, histologic changes predominate in the motor nuclei of the 10th, 11th, and 12th cranial nerves and, less frequently, those of the 5th and 7th nerves; rarely, the nuclei of the 3rd, 4th, and 6th cranial nerves may also degenerate.[59] The cellular damage consistently leads to secondary changes of the corticospinal tracts in the lateral and ventral funiculi of the spinal cord. Indeed, autopsy studies reveal these pathologic alterations even when the patient had no clinical signs of upper motor neuron lesions in life. Although the anterior horn cells and the corticospinal tracts undergo most severe degeneration, a wide spectrum of changes affects the entire spinal cord.

Symptoms usually begin in the fifth to seventh decades, affecting men two to four times as frequently as women.[94] Commonly, distal weakness develops as an early symptom. Despite asymmetric initial manifestations, at times limited to only one extremity, the disease progresses rapidly to involve muscles of the trunk and those innervated by the cranial nerves. Bulbar signs tend to appear late in the course of the disease, but up to one third of the patients may have dysarthria and dysphagia as the initial symptoms.[12] Although patients frequently complain of aching and other vague sensory problems, they usually have no clear objective loss of sensation. In one series, however, thermal threshold tests were abnormal in 80 percent of patients with motor neuron disease.[69] Pathologic examination of the peripheral nerves also shows some involvement of sensory axons, but not as an essential part of the disease.[32] Spasms and cramps of the leg muscles occur early, often appearing at night or after exercise. Urinary symptoms are rare at the onset of the disease, but a neurogenic bladder may develop terminally. Pathologic laughter and crying signal pseudobulbar manifestations at some stage of the illness.

Clinical signs include widespread atrophy affecting the muscles of the extremities and face, usually in proportion to the degree of weakness. The sparing of the extraocular muscles stands in contrast to the frequent involvement of the tongue. Most patients have hyperreflexia with ankle clonus and extensor plantar responses in some cases. Often the patient is unaware of spontaneous muscle twitching. A paucity of fasciculations may suggest slow progression of the illness,[98] but their abundance does not necessarily imply a worse prognosis. Benign fasciculations, not infrequently seen in healthy subjects, usually involve the eyelid, calf, or intrinsic hand muscles, especially after strong contraction. In contrast to motor neuron disease, neither muscle weakness nor atrophy develops, and electromyography shows no evidence of denervation.

The signs and symptoms may wax and wane, with an apparent improvement presumably after reinnervation and collateral sprouting. In one study 32 of 74 patients showed a fluctuating course.[125] Despite this pattern, the disease usually progresses without remission, leading to death in 2 to 3 years, most often as the result of respiratory difficulties. Perhaps as many as 20 percent of all patients, however, have a more favorable course, with survival in excess of 5 years.[14] The "benign" form lacks bulbar signs in the early stages but otherwise shares the same clinical features with the classic variety. It is distinguished from the slowly progressive Kugelberg-Welander form of spinal muscular atrophy by the absence of proximal atrophy.

Clinical diagnosis depends on the combined features of widespread muscular atrophy, weakness, fasciculations, and evidence of damage to corticospinal and bulbar tracts. Differential diagnoses include any condition associated with diffuse muscle atrophy. A syndrome clinically resembling amyotrophic lateral sclerosis has been reported in association with lead intoxication,[11] chronic mercurialism,[74] and proximal motor neuropathy.[20] Cervical spondylosis and developmental anomalies in the region of the foramen magnum sometimes simulate the disease closely, with presenting symptoms of muscular weakness in the upper extremities and evidence of spasticity in the lower extremities. When motor neuron disease and cervical or lumbar spondylosis coexist, sensory symptoms of radiculopathy alter the picture of pure motor dysfunction. A myelogram helps distinguish these diagnostic possibilities. Elevated muscle enzyme levels do not exclude the diagnosis be-

cause the serum level reaches two or three times normal value in about half of the patients with motor neuron disease.[130]

Degeneration of the anterior horn cells results in denervation of muscle fibers. Collateral sprouts from surviving motor neurons then reinnervate the affected motor units. Histochemical studies of fresh frozen specimens, thus, show characteristic denervation atrophy with fiber grouping that represents a compensatory mechanism.[30] Myopathic changes also appear, presumably as part of the denervation process,[1] although most biopsies show a relatively intact intermyofibrillar network and cellular architecture of the fibers.[17] Type I grouping correlates with the best prognosis, whereas a high density of atrophic fibers implies a rapid progression.[100] According to a quantitative study of the terminal innervation ratio and fiber type grouping, collateral reinnervation occurs less in amyotrophic lateral sclerosis than in the more slowly progressing Charcot-Marie-Tooth disease.[124]

Electromyographic abnormalities found during various stages of the illness reflect the sequence of pathologic changes in the muscle.[24] Diffuse denervation gives rise to widespread fibrillation potentials and positive sharp waves (see Fig. 13–7B). Fasciculation potentials are less specific, although they imply motor neuron irritability in the appropriate clinical context.[26,86] The distribution of findings is typically asymmetric, particularly early. The presence of both large and small fibrillation potentials suggests both recent and chronic denervation. Many motor unit potentials have large amplitude and polyphasic waveforms, some with late components.[18] The motor unit potentials, reduced in number, recruit poorly and discharge rapidly, producing a less-than-full interference pattern (see Figs. 12–8B and 12–9).

The common complaint of easy fatigability suggests impairment of neuromuscular transmission that may result from decreased trophic function of the neuron.[120] In these cases, needle examination reveals small unstable motor unit potentials with temporal amplitude variability (see Fig. 13–12). Discharging units usually show more stability in the relatively chronic forms. Many patients with a rapidly progressive form of the disease show abnormalities of the compound muscle action potentials elicited by slow repetitive nerve stimulation.[8] In one series,[25] 67 percent of 55 patients showed a decremental response, especially in the muscles showing atrophy or frequent fasciculations. As in myasthenia gravis, local cooling or administration of edrophonium (Tensilon) normalizes the findings, and exercise induces posttetanic exhaustion.

Additional physiologic findings include increased fiber density and jitter values determined by single-fiber electromyography (SFEMG).[122] These subtle signs of reinnervation and immature motor nerve terminals may occur in muscle showing no abnormality either clinically or by conventional needle examination. Despite active reinnervation, progressive denervation produces a deteriorating clinical course. A computer-assisted quantitative measure of motor unit function showed that reinnervation compensated only for up to a 50 percent loss of the motor neuron pool.[54]

In most cases, sensory action potentials remain normal in amplitude and latency. Thus, any abnormalities in sensory conduction studies suggest another disorder. Studies of the motor nerve may reveal slight slowing in association with the reduction in amplitude of the muscle action potential, as shown in the proximal and distal segments of the sciatic nerve.[80] The values rarely fall below 70 to 80 percent of the normal lower limits, and some studies found little or no change in maximal conduction velocity.[54] These findings suggest at least partial preservation of the fastest fibers for a long time with no evidence indicating their preferential loss. Pathologic slowing of normally slow fibers may increase the scatter of velocities.

Increased excitability of the spinal motor neuron pool results in a higher incidence of the H reflex in the soleus muscle after stimulation of the tibial nerve. As is not the case in normal subjects, the H reflex can also be elicited in the intrinsic hand muscles after stimulation of the ulnar or median nerve, and in the extensor digitorum or tibialis anterior muscle after stimulation of the peroneal nerve.[98]

A variety of focal or diffuse neuropathic disorders may mimic amyotrophic lateral sclerosis. So that one can avoid falsely diagnosing this universally fatal disease, a set of electrophysiologic criteria has been proposed[86]: (1) fibrillation and fasciculation in muscles of the lower and the upper extremities or in the extremities and the head; (2) a reduction in number and an increase in amplitude and duration of motor unit action potentials; (3) normal electrical excitability of the surviving motor nerve fibers; (4) motor fiber conduction velocity within the normal range in nerves of relatively unaffected muscles and not less than 70 percent of the average normal value according to age in nerves of more severely affected muscles; and (5) normal excitability and conduction velocity of afferent nerve fibers, even in severely affected extremities.

Typical cases show asymmetric and multifocal abnormalities involving more than two muscles innervated by different nerves and spinal roots in at least three limbs, counting the head as one extremity. The involvement of the upper and lower limbs serves to differentiate this entity from a syrinx or spondylosis with segmental abnormalities. An optimal selection of the muscles for examination can minimize the ambiguity regarding the possible effect of compressive neuropathies, such as the carpal tunnel syndrome or tardy ulnar palsy. Thus, for example, one should examine the flexor pollicis longus, rather than the thenar or hypothenar muscles. Similarly, denervation of the extensor digitorum brevis may result from nerve entrapment by a tight shoe. Studies should include, in addition to electromyography, sensory as well as motor conduction measurements and, when appropriate, tests of neuromuscular transmission. Sparing of sensory nerves provides an important clue, especially if demonstrated in one of the weaker extremities. Evidence of defective neuromuscular transmission with either repetitive stimulation or SFEMG suggests active disease with recent reinnervation and immature end plates, and therefore a poor prognosis.

Progressive Muscular Atrophy

In this rare syndrome of Aran-Duchenne, clinical signs and symptoms suggest a selective disorder of the anterior horn cells, although pathologic studies may show some changes in the corticospinal tract as well. Most cases occur sporadically. Familial forms, reported in a small percentage, have a more benign course. Atrophy and weakness of the extremities develop without accompanying features of spasticity or other evidence of upper motor neuron involvement. The patients initially have asymmetric wasting and weakness of the intrinsic hand muscles. Atrophy of the muscles of the shoulder girdle and lower extremities and bulbar paralysis then develop. Less commonly, the clinical signs may resemble Charcot-Marie-Tooth disease or peroneal nerve palsy, with preferential involvement of the anterior leg compartment in early stages. Diaphragmatic paralysis, although rare, may cause respiratory insufficiency as a prominent presenting symptom.[99] Despite generalized wasting and weakness, the stretch reflexes usually remain normal or only slightly decreased. The disease runs a slower course than classic amyotrophic lateral sclerosis. Nonetheless, the symptoms and signs steadily progress without remission, leading to eventual demise, often from aspiration pneumonia.

Progressive Bulbar Palsy

The signs and symptoms predominantly involving the bulbar muscles justify the name progressive bulbar palsy.[3] The presence of disease in siblings suggests an autosomal recessive form of inheritance.[6] The disease usually begins in the fifth or sixth decade with initial symptoms of progressive dysarthria and dysphagia. The tongue becomes atrophic, with visible fasciculations. Troublesome signs include pooling of saliva, nasal regurgitation of fluids, and inability to chew or swallow. Most of these patients will eventually show signs of pseudobulbar palsy from lesions affecting the brainstem at higher levels or the cerebral cortex. Despite the often localized initial symptoms, widespread involvement of motor neurons ensues in the terminal stage.

Primary Lateral Sclerosis

As the name indicates, pathologic studies in typical cases show selective loss of the corticospinal and corticobulbar tracts, with sparing of the anterior horn cells, in primary lateral sclerosis. The clinical signs include spasticity, diffuse hyperreflexia, Babinski signs, and pseudobulbar palsy. In the conspicuous absence of atrophy and weakness of distal musculature, the disease may simulate cord compression with a spastic paraparesis. Neither electromyography nor motor and sensory nerve conduction studies disclose abnormalities.[116] These negative findings distinguish this disorder from other motor neuron diseases as a distinct entity.

Disorders With Geographic Predilection

Geographic foci of motor neuron disease described in the literature include the island of Guam[13]; the Kii peninsula of Japan[119]; and the Ryukyu Islands, south of Japan.[78] The Guamanian motor neuron disease in the Chamorro population[49] shows a high familial incidence. Nearly 10 percent of the adult population on the island die from the disease. The Parkinson dementia complex affects the same population, but the two entities have no etiologic relationship. Some patients with motor neuron disease in Japan also suffer from presenile dementia.[92]

3 SPINAL MUSCULAR ATROPHY

A modification of the 1968 proposal by the World Commission on Neuromuscular Diseases classifies spinal muslar atrophy in eight subgroups: infantile spinal muscular atrophy (Werd-nig-Hoffmann disease), juvenile spinal muscular atrophy (Kugelberg-Welander disease), juvenile progressive bulbar palsy (Fazio-Londe disease), scapuloperoneal spinal muscular atrophy, facioscapulohumeral spinal muscular atrophy, arthrogryposis multiplex with anterior horn cell disease, and distal spinal muscular atrophy.[33,48] Although the term spinal has now gained wide acceptance, the unfortunate choice of the word may cause clinical confusion, in view of prevalent bulbar involvement in many of these entities. It is used here not to exclude such cases but to distinguish diseases of anterior horn cells from those of the nerve root or peripheral nerves. Indeed, in few of these conditions does the disease limit itself to the spinal cord.[87] Spinal muscular atrophy causes one of the most devastating outcomes of all the genetically determined neurologic disorders in childhood. In a series of 108 patients seen at the Mayo Clinic between 1955 and 1975, the mortality rate reached 31 percent, with a mean age of 65 months at the time of death.[83] Furthermore, only 35 percent of these patients could ambulate unassisted.

No consensus has emerged on whether various subdivisions represent independent entities or spectrum of the same disorder. At least two categories have clinically and genetically distinct features: the rapidly progressive infantile form, or Werdnig-Hoffmann disease, with death before 3 years of age[103] and the late childhood or juvenile form, called the Kugelberg-Welander syndrome.[16,37,47,101] An intermediate type has an onset between 3 and 18 months of age.[28,96] Despite the overlap in onset, the infantile, juvenile, and intermediate forms have a different time course with regard to the disease and age at death. Table 20–1 summarizes these and other features useful in separating the three types of spinal muscular atrophy.[77]

Another form with adult onset, once reported as a variant of the late juvenile type,[16,37,96] may be a separate entity, according to a survey over a 10-year period in northeast England.[105] The distribution of affected muscles distinguishes amyotrophic lateral sclerosis with distal weakness from the adult form of spinal muscular atrophy with more proximal involvement. A variety called juvenile amyotrophic lateral sclerosis affects older children. Its characteristic features include late onset, rapid progression, and the presence of both upper and lower motor neuron signs.[97]

Various types of spinal muscular atrophy share the same or similar electromyographic findings, consisting of fibrillation potentials, positive sharp waves, fasciculation potentials, large motor unit potentials, and a reduced interference pattern.[15,47,88,89] In a rapidly progressing infantile spinal muscular atrophy, electromyography suggests a mixture of denervation and regeneration with small motor unit potentials that vary temporally in configuration.

Table 20–1 DISTINGUISHING FEATURES OF THE VARIOUS FORMS OF PROXIMAL SPINAL MUSCULAR ATROPHY

Type	Age (Usual)		Ability to Sit Without Support*	Fasciculations of Skeletal Muscles	Serum Creatine Kinase
	Onset	Survival			
Infantile	<9 months	<4 years	Never	+/−	Normal
Intermediate	3–18 months	>4 years	Usually	+/−	Usually normal
Juvenile	>2 years	adulthood	Always	++	Often raised
Adult	>30 years	50 years +	Always	++	Often raised

*At some time during the course of the illness.
From Kloepfer and Emery,[77] with permission.

Infantile Spinal Muscular Atrophy

This disease, first described by Werdnig[131] and Hoffmann,[64] has an autosomal recessive trait. Parents of affected children have a significantly higher rate of consanguinity, compared with controls. The estimated incidence ranges from 1 in 15,000 to 1 in 25,000 live births in Britain.[103] One third of the affected children have the disease already manifested at birth by decreased fetal movements or congenital arthrogryposis.[106] The remainder usually show the onset of illness by 3 months, and certainly before 6 months, after birth, with delayed developmental milestones. The infant dies of pneumonia often before the first birthday and usually by the age of 3 years, although not all cases of neurogenic muscular atrophy in infancy have this malignant course.[28] In chronic spinal muscular atrophy of childhood, clinical signs first appear at about 6 months but occasionally as late as 8 years of age, with the median age of death later than 10 years.[16,34,35,47,104]

The clinical features comprise progressive muscle weakness, atrophy of the trunk and extremities, hypotonia, and feeding difficulties. The infants characteristically lie motionless with limbs abducted in the frog-leg position. They are unable to hold their head up or sit, and have difficulty with any type of locomotion with the loss of previously developed motor skills. About half of the patients have fasciculations in the tongue and, much less frequently, in the atrophic muscles of the limbs. Although skeletal deformities are uncommon at birth,[106] children with the chronic form of spinal muscular atrophy may develop kyphoscoliosis, contractures of the joints, and dislocation of the hip as the disease progresses. Bulbar signs appear later in the course of the rapidly progressive illness. The facial muscles are affected mildly, if at all, giving the infant an alert expression, despite severe generalized hypotonia with reduced or absent stretch reflexes. The patients have normal sphincter functions and an intact sensory system even in the terminal stages of illness.

Muscle biopsy reveals sheets of round atrophic fibers intermixed with clumps of hypertrophic type I fibers. The chronic form shows fiber type grouping with large type II fibers and elevated levels of serum creatine kinase (CK).

The incidence of fibrillation potentials and sharp positive waves depends on the progression and severity of the disease. It reached 100 percent in one study[61] but considerably less in another.[83] Fasciculation potentials may be recorded, but only infrequently. In one series,[15] electromyography revealed unique potentials regularly discharging at a rate of 5 to 15 impulses per second in 75 percent of 30 patients. This finding had no correlation with the patient's age, duration and severity of the disease, or biopsy findings. Subsequent studies have failed to confirm the specificity of this activity, as once claimed.

The motor unit potentials recruit poorly, reflecting the loss of anterior horn cells. The maximal effort produces an incomplete interference pattern with a limited number of potentials discharging at a rapid rate. In extreme instances, only one or two motor units fire at 40 to 50 impulses per second. As expected from collateral sprouting and a high fiber density, quantitative survey shows motor unit potentials of high amplitude and long duration. Regenerating axons, however, may give rise to potentials of low amplitude and short duration. The temporal variability of the waveform suggests instability of neuromuscular transmission. In advanced stages, the motor unit potentials are either abnormally large or small, with no normal units between the two extremes.[61]

Nerve conduction studies show normal or nearly normal velocities but reduced compound muscle action potentials. In one study,[83] 94 percent of the patients showed a reduction of amplitude to less than 50 percent of the normal mean. Mild slowing of conduction velocity results from the loss of fast conducting axons.[93] Repetitive stimulation of the nerve at either slow or fast rates shows a decrementing muscle response during ongoing reinnervation, suggesting defective neuromuscular transmission. In contrast to the motor responses, sensory nerve studies usually reveal normal amplitudes and velocities. Some patients, however, have minor abnormalities of the sensory nerves electrophysiologically[109] and histologically.[19] Rare cases of infantile neuronal degeneration clinically resemble infantile spinal muscular atrophy. In this disorder, characterized by markedly decreased conduction velocity, however, a demyelinative neuropathy constitutes part of the widespread extensive neuronal degeneration.[123]

Juvenile Spinal Muscular Atrophy

The juvenile form of spinal muscular atrophy[82] inherited in an autosomal dominant or recessive fashion, begins with proximal muscle weakness and atrophy in the lower extremities.[33,36,47,62,96] Two thirds of the patients have a family history. The disease progresses more slowly, with less predilection for proximal muscles in the dominant variety, compared with the recessive type.[16,102,134] Compared with the infantile form, it has a later onset throughout childhood or adolescence, but most commonly between the ages of 5 and 15 years. The symptoms initially involve the extensor muscles of the hip and knees, and later, the shoulder girdle muscles.

The patient has a characteristic lordotic posture with protuberant abdomen, hyperextended knees, and hypertrophic calves. Involvement of the cranial musculature is rare, although ptosis has been described. Half of the patients have fasciculations in the proximal muscles. This abnormality affects the legs more than the arms, and spares the distal muscles and the tongue except in the advanced stages. Examination usually reveals hyporeflexia with atrophy, although some patients have hyperreflexia and Babinski signs. The disease follows a relatively benign course, with frequent survival into adulthood; but most patients are confined to a wheelchair by their mid-30s. In some patients chronic neurogenic quadriceps amyotrophy develops as a forme fruste of Kugelberg-Welander disease.[10,45] The differential diagnoses otherwise include polymyositis and muscular dystrophy.[129]

Levels of serum enzymes such as CK, though elevated modestly, remain nearly constant as the disease progresses. In Duchenne's muscular dystrophy, an initially very high level of CK gradually declines later. Muscle biopsies show fascicular atrophy and fiber type grouping characteristic of a neurogenic disorder with occasional mixture of myopathic features.

An overall incidence of fibrillation potentials ranged from 20 to 40 percent in one series[62] to 64 percent in another.[83] The percentage is even higher in a group of more severely affected patients,[96] although it does not match the level seen in the Werdnig-Hoffmann disease. Fasciculation potentials were common in one series,[47] but not in another.[83] Complex repetitive discharges may be present at a late stage. Spontaneous activities involve the lower limbs more than the upper limbs and proximal muscles more than distal muscles.[62]

Voluntary contraction gives rise to motor unit potentials of high amplitude and long duration, which recruit poorly even at maximal effort.[15] Late components indicate the presence of slow conducting regenerating axons. The percentage of large motor unit potentials increases with the duration of the disease.[62] In advanced cases, small polyphasic potentials also appear, suggesting secondary myopathic changes of atrophic muscles. These potentials show constant configuration, unlike the varying waveforms seen in the more rapidly progressive infantile cases.[83]

Motor and sensory nerve conduction studies, though usually normal,[93,118] may reveal a moderate reduction in amplitude of the compound muscle action potential. As in Werdnig-Hoffmann disease, this abnormality shows a strong correlation to the patient's functional capacity. In one series,[83] 54 percent were bedridden if the amplitude fell below half of normal, compared with only 7 percent in the remainder.

Juvenile Progressive Bulbar Palsy

Slowly progressive bulbar palsy characterizes this very rare disorder of Fazio-Londe inherited as an autosomal recessive trait.[3,27] The clinical features consist of ophthalmoplegia, facial diplegia, laryngeal palsy, and other cranial nerve paralyses with onset in early childhood. Facial diplegia, if present at birth, suggests other entities such as infantile myotonic dystrophy, infantile facioscapulohumeral dystrophy, and the Mobius syndrome.[14] Progressive ophthalmoplegia and dysphagia may also develop in some cases of juvenile spinal muscular atrophy as late manifestations but not as the presenting features. Electromyographic abnormalities, prominent in bulbar and pontine musculature,. consist of fibrillation potentials, positive sharp waves, and impaired recruitment of motor unit potentials.

Scapuloperoneal Spinal Muscular Atrophy

As indicated by the name, a unique pattern of muscular weakness distinguishes this type of spinal muscular atrophy from the others.[39,73] A form of muscular dystrophy also exhibits the same distribution of weakness with features often indistinguishable from muscular atrophy. Because of this, some prefer the term scapuloperoneal syndrome to include both neurogenic and myogenic forms. In addition, hereditary motor and sensory neuropathy (HMSN) type I may present as scapuloperoneal atrophy associated with distal sensory loss.[110]

This variety of muscular atrophy slowly progresses after its usual onset in early adulthood. It is inherited as an autosomal dominant trait, although sporadic cases have also been described. Atrophy and weakness initially affect the anterior tibial and peroneal muscles and later the musculature of the pectoral girdle, producing winging of the scapulae. Muscle biopsies show a mixture of neuropathic and myopathic patterns in most cases. Electromyographic studies demonstrate low-amplitude, short-duration motor unit potentials, fibrillation potentials, and positive sharp waves. Nerve conduction studies reveal normal motor and sensory responses.

Facioscapulohumeral Spinal Muscular Atrophy

Like the scapuloperoneal atrophy, this condition has a unique distribution of weakness and a similar counterpart among the muscular dystrophies.[40] When inherited, it follows an autosomal dominant pattern. Atrophy primarily affects the muscles of the face and pectoral girdle musculature. The weakness begins in early adult life and takes a slowly progressive course. Clinical features resemble those of facioscapulohumeral muscular dystrophy. A descriptive term, facioscapulohumeral syndrome, used in some cases, suggests inability to distinguish between neurogenic and myogenic forms.

Arthrogryposis Multiplex Congenita

This is a syndrome defined as congenital contractures of at least two different joints and major muscle wasting not associated with a progressive neurologic disorder.[43] The condition may result from a number of different neuromuscular and bony disorders, causing immobilization of the limbs at the time of the embryonic formation of joints. One study describes a dominantly inherited lower motor neuron disorder as the cause of arthrogryposis present at birth.[44] Disorders of the motor neuron probably predominate, although different investigators postulate myogenic or neurogenic origins.[21,81] Electromyography may show spontaneous discharges such as fibrillation potentials or complex repetitive discharges. Motor unit potentials show reduction in number and poor recruitment. The nerve conduction studies reported in a few cases have been normal.

Focal Amyotrophy

In some patients distal amyotrophy of the upper limbs develops.[127] Those reported from Japan have distal and segmental muscular atrophy of juvenile onset.[121] The clinical features include male preponderance, localized atrophy uniquely affecting the hand and the forearm, sparing of the lower extremities and cranial nerves, and rapid progression at first followed by a slower change. The age of onset, distribution of atrophy, and benign course distinguish it from motor neuron disease.[60] Electromyography shows motor unit potentials of large amplitude and long duration, with impaired recruitment. Nerve conduction studies reveal reduction in amplitude of compound muscle potentials but normal velocities.

When atrophy involves part of the body, one must entertain the diagnosis of focal motor neuron disease with extreme caution and only after excluding alternative possibilities such as spinal cord tumors, radiculopathy, plexopathy, and mononeuropathy. Sensory abnormalities, if present, help distinguish these conditions from focal motor neuron disease.

Other Disorders

Distal spinal muscular atrophy resembles HMSN types I and II except for preservation of stretch reflexes, relative sparing of the upper limb, and a normal sensory examination. In one study of 34 patients,[57] motor and sensory conduction studies revealed no abnormality.

Patients with X-linked recessive bulbospinal atrophy or neuropathy have mild facial weakness, severe atrophy of the tongue without prominent bulbar symptoms, and a high serum CK level. Electromyography typically shows fibrillation potentials, complex repetitive discharges, and large motor unit potentials. Despite the clinical resemblance, this entity carries a much better prognosis than motor neuron disease.[58]

One study reports three patients from a large family, who had an autosomal dominant scapulohumeral form of spinal muscular atrophy.[70] The disease progressed rapidly, without evidence of corticospinal tract dysfunction; and the patients died from respiratory failure within 3 years.

Chronic asymmetric spinal muscular atrophy typically shows asymmetric neurogenic atrophy involving one or more limbs without evidence of pyramidal tract dysfunction or bulbar signs.[56] Patients with this disease have no evidence of generalized neuropathy, although the motor nerve conduction velocities may be slightly reduced because of muscle wasting.

4 JAKOB-CREUTZFELDT DISEASE

Despite the very early recognition of this entity,[22,68] only recent studies have proven its transmissibility in man and the chimpanzee.[51] Accidental inoculation occurred after a corneal transplant in one patient[31] and following a surgical procedure using contaminated stereotactic electrodes in two others.[7] Although the organism has not been isolated, brain tissue from dying patients causes scrapielike encephalopathy in goats.[53] The pathologic features resemble those of kuru, a transmissible disease seen in New Guinea,[50] and consist of widespread spongiform degeneration with loss of nerve cells in the cortex, basal ganglia, and spinal cord.

The disease occurs sporadically or familially. It affects both sexes equally, with onset in middle age or later. Following vague prodromal symptoms, mental deterioration, anxiety, depression, memory loss, and confusion develop. A variety of neurologic disturbances indicate cortical degeneration and upper and lower motor neuron involvement. The most commonly encountered symptoms include weakness, rigidity, spasticity with hyperreflexia, muscular atrophy, incoordination, tremor, and visual loss. Wasting of the muscles with fasciculations during late stages of illness mimics the typical appearance of motor neuron disease. The patient usually has spontaneous myoclonus, which may become less prominent in the advanced stages. The disease follows a rapidly progressive course, leading to severe dementia, blindness, lethargy, and eventually coma and death within a year after onset.

A characteristic electroencephalographic abnormality seen in 90 percent of cases consists of localized or diffuse bursts of high-voltage sharp or slow waves. The evidence of denervation in electromyography indicates muscular atrophy with involvement of motor cells in the medulla or spinal cord. Motor and sensory nerve conduction studies reveal no abnormality unless the patient has a compressive or diffuse nutritional neuropathy in chronic stages.

Electromyographers have increasing concern about the risks involved in examining patients with Jakob-Creutzfeldt disease. With this disease, in contrast to the acquired immunodeficiency syndrome (AIDS), exposure to saliva, nasopharyngeal secretions, urine, or feces should cause no special alarm.[46] After such contact, however, one must thoroughly wash one's hands and other exposed body parts with hospital detergent or ordinary soap. One must also discard needle electrodes used for electromyography after incineration (see Chapter 3.2).

5 POLIOMYELITIS

Poliomyelitis no longer prevails as summer epidemics in this country, but sporadic cases still occur throughout the year. Most clinical illness develops after infection by type I virus, but at times also by type II or III. The intestinal and respiratory tracts initially invaded by the virus transmit the agent to the nervous system via the bloodstream. Affected anterior horn cells in the spinal cord and brainstem undergo degenerative changes, causing an inflammatory reaction in the meninges. Isolation of the poliomyelitis virus confirms the diagnosis in about 90 percent of patients with paralytic illness.

The clinical features of systemic infection are flulike symptoms such as fever, general malaise, diarrhea, and loss of appetite. Only a small percentage of patients in whom meningeal irritation develops complain of headaches, a stiff neck, and vomiting. In some cases, paralytic illness follows the predromal symptoms. It progresses over a period of several days to a week, affecting one or more extremities or, in a small number of children, bulbar musculature. Respiration weakens with the involvement of the diaphragm, intercostal muscles, and abdominal muscles, requiring assisted ventilation in advanced cases. Neurologic examination shows widespread atrophy, diminished or absent stretch reflexes in the affected limbs, and a normal sensory system. The spinal fluid examination reveals mild pleocytosis.

Considerable recovery takes place even if severe generalized weakness develops. Late deterioration of function occurs in some survivors suggesting the possibility of latent virus infection.[4,95] A few studies have shown a statistically significant association between poliomyelitis and motor neuron disease.[75,108] Histopathologic and virologic studies in one patient with amyotrophic lateral sclerosis and antecedent poliomyelitis, however, provided no evidence of the continuing presence of poliovirus.[111] If poliomyelitis has already depleted motor neurons, minor additional damage to the surviving anterior horn cells during advancing age might result in exaggerated clinical signs. In addition, the diseased neurons may have certain predisposition to senile degeneration, or some surviving motor neurons may have incorporated too many muscle fibers from the denervated units beyond the metabolic capability. A long-term follow-up study of poliomyelitis patients with apparent late progression has shown a relatively benign course, with the development of fasciculation but few upper motor neuron signs.[41,95]

Electromyography initially shows only a reduced recruitment pattern during the acute phase of poliomyelitis. Fibrillation potentials develop as the motor axons degenerate. Reinnervation results in diminution of spontaneous discharges and the appearance of motor unit potentials of large amplitude and long duration. Weak muscles may have only a few extremely large motor unit potentials. A prospective study of 24 patients with a history of paralytic poliomyelitis revealed evidence of widespread chronic partial denervation despite restricted clinical weakness.[63] Electromyography showed a substantially increased mean interference amplitude not only in weak muscles but also in apparently unaffected muscles contralateral to the clinically involved spinal segments. Nerve conduction studies revealed normal velocities but reduced amplitude of the compound muscle action potentials, approximately in proportion to the degree of muscle atrophy.[71]

In the absence of adequate reinnervation, fibrillation potentials may persist many years after the acute episode. In these cases, the spontaneous discharges of very low amplitude indicate small atrophic muscle fibers. Even after reinnervation, diseased anterior horn cells may degenerate prematurely and cause the reappearance of spontaneous discharges. Alternatively, muscle fibers may drop out from the motor neuron that can no longer meet the increased metabolic demand of an enlarged motor unit. Single-fiber electromyography in survivors of poliomyelitis has shown a significant increase in jitter and fiber density without neurogenic blocking.[133] These findings of defective neuromuscular transmission may represent disintegration with aging of the reinnervated motor units.

A poliomyelitis-like syndrome may develop in association with asthma.[23,65,67,132] The disease predominantly affects boys 10 years old or younger. The patient develops acute flaccid monoplegia involving a single upper or lower limb without sensory deficits. Marked atrophy in the involved extremity signals a poor prognosis for recovery. Cerebrospinal fluid examination reveals a pleocytosis and slight protein elevation, but no rise in poliovirus antibody titers. The lesion may lie in the brachial plexus, but the absence of sensory abnormalities favors the motor roots[23,29] or anterior horns[132] as the locus of the disease. Despite clinical similarities to poliomyelitis, the disease can affect previously vaccinated children. Electromyographic features also resemble those seen in poliomyelitis. In one patient,[132] C-5 root synkinesis developed between biceps and inspiratory muscles from aberrant regeneration.[76]

The patients with acute hemorrhagic conjunctivitis caused by enterovirus 70 may have poliolike paralysis of the limb and cranial muscles.[128] Early complaints include root pain and weakness. Electromyography of affected and some unaffected muscles show fibrillation potentials early and large polyphasic motor unit potentials later. Nerve conduction studies reveal no specific abnormalities.

6 SYRINGOMYELIA

Signs and symptoms of syringomyelia result from cavitation and gliosis of unknown pathogenesis affecting the spinal cord and medulla. The disease may begin at any age, but most often in the third or fourth decade. It may occur sporadically or familially, affecting both sexes equally. The patient frequently has other congenital defects, such as spina bifida or Arnold-Chiari malformation. Other associated features consist of scoliosis, trophic changes, and intramedullary tumors found in conjunction with a syrinx. Secondary cavitation may develop after traumatic, vascular, or infectious lesions of the spinal cord. A slowly progressive course extends over a period of many years, although damage to medullary nuclei may lead to a rapid demise. The differential diagnoses include motor neuron disease, multiple sclerosis, spinal cord tumor, anomalies of the cervical spine, and posterior fossa lesions.[2]

The cavities vary in location and in longitudinal extent, but most frequently affect the cervical cord, which may be distended with the fluid in the cavity or, conversely, flattened. Irregularly shaped gliosis and cavities, though ordinarily located near the central canal, may involve the entire white and gray matter, affecting motor and sensory cells and various fiber tracts in any combination. Damage to the anterior commissure of the spinal cord causes the characteristic dissociation of sensory abnormalities. Other common sites of involvement include the posterior and lateral funiculi, with damage to the corticospinal tract.

Clinical symptoms and signs depend on the location and extent of the pathologic changes. A syrinx in the cervical region causes atrophy and weakness of intrinsic hand muscles and dissociated loss of pain sensation with preservation of light touch in the lower cervical or upper thoracic dermatomes. A syrinx at the root entry zone gives rise to a segmental loss in all modalities of cutaneous sensation, whereas lesions of the posterior column selectively affect the vibratory sense. Other signs include spasticity, hyperreflexia, Babinski signs, ataxia of the lower extremities, and a neurogenic bladder. A syrinx may affect the lumbosacral region alone or in association with lesions at the cervical level. The clinical features, then, include muscular atrophy and dissociated sensory loss of the lower extremities and paralysis of the bladder. The loss of stretch reflexes suggests lesions at the root entry zone or the anterior horn cells in the lumbar region.

Syringobulbia denotes a syrinx formed in the medulla that commonly involves the descending nucleus of the fifth nerve and nuclei of the lower medulla either unilaterally or bilaterally. Common features include atrophy of the tongue, loss of pain and temperature sensation in the face, abnormalities of extraocular muscles, and respiratory difficulties. A lesion of the spinal accessory nuclei causes atrophy of the trapezius and sternocleidomastoid muscles. Spastic paraparesis results from interruption of the upper motor neuron tracts.

Electromyography reveals fibrillation potentials and positive sharp waves in the atrophic muscle. Sparing of the lower limbs serves to distinguish syrinx from motor neuron disease. Motor nerve conduction studies show normal velocities but reduced amplitude of the compound muscle action potentials in the affected limb. The finding of normal sensory nerve potentials despite clinical sensory loss confirms a preganglionic involvement of the sensory pathway.[42] In these instances, somatosensory evoked potentials may reveal central conduction block (see Fig. 19–12). A lesion of the spinal tract or nucleus of the trigeminal nerve causes an afferent abnormality of the blink reflex with the absence of R_2 bilaterally after stimulation on the affected side of the face (see Fig. 16–14).

REFERENCES

1. Achari, AN, and Anderson, MS: Myopathic changes in amyotrophic lateral sclerosis: Pathologic analysis of muscle biopsy changes in 111 cases. Neurology 24:477–481, 1974.
2. Alani, SM: Denervation in wasted hand muscles in a case of primary cerebellar ectopia without syringomyelia. J Neurol Neurosurg Psychiatry 48:84–85, 1985.
3. Alexander, MP, Emery, ES, III, and Koerner, FC: Progressive bulbar paresis in childhood. Arch Neurol 33:66–68, 1976.
4. Anderson, AD, Levine, SA, and Gellert, H: Loss of ambulatory ability in patients with old anterior poliomyelitis. Lancet 2:1061–1063, 1972.
5. Appel, SH: A unifying hypothesis for the cause of amyotrophic lateral sclerosis, parkinsonism and Alzheimer's disease. Ann Neurol 10:499–505, 1981.
6. Benjamins, D: Progressive bulbar palsy of childhood in siblings (corres). Ann Neurol 8:203, 1980.
7. Bernoulli, C, Siegfried, J, Baumgartner, G, Regli, F, Rabinowicz, T, Gajdusek, DC, and Gibbs, CJ, Jr: Danger of accidental person-to-person transmission of Creutzfeldt-Jacob disease by surgery. Lancet 1:478–479, 1977.
8. Bernstein, LP, and Antel, JP: Motor neuron disease: Decremental responses to repetitive nerve stimulation. Neurology 31:202–204, 1981.
9. Bobowick, AR, and Brody, JA: Epidemiology of motor-neuron diseases. N Engl J Med 288:1047–1055, 1973.
10. Boddie, HG, and Stewart-Wynne, EG: Quadriceps myopathy: Entity or syndrome? Arch Neurol 31:60–62, 1974.
11. Boothby, JA, DeJesus, PV, and Rowland, LP: Reversible forms of motor neuron disease: Lead "neuritis." Arch Neurol 31:18–23, 1974.
12. Brain, WR, Croft, P, and Wilkinson, M: The course and outcome of motor neuron disease. In Norris, FH, Jr, and Kurland, LT (eds): Motor Neuron Diseases: Research on Amyotrophic Lateral Sclerosis and Related Disorders. Grune & Stratton, New York, 1969, pp 20–27.
13. Brody, JA, and Chen, KM: Changing epidemiologic patterns of amyotrophic lateral sclerosis and parkinsonism-dementia on Guam. In Norris, FH, Jr, and Kurland, LT (eds): Motor Neuron Diseases: Research on Amyotrophic Lateral Sclerosis and Related Disorders. Grune & Stratton, New York, 1969, pp 61–79.
14. Brooke, MH: A Clinician's View of Neuromuscular Diseases. Williams & Wilkins, Baltimore, 1977.
15. Buchthal, F, and Olsen, PZ: Electromyography and muscle biopsy in infantile spinal muscular atrophy. Brain 93:15–30, 1970.
16. Bundey, S, and Lovelace, RE: A clinical and genetic study of chronic proximal spinal muscular atrophy. Brain 98:455–472, 1975.
17. Butler, RC, Gawel, M, Rose, FC, and Sloper, JC: Muscle biopsy in motor neurone disease. In Rose, FC (ed): Motor Neurone Disease.

Grune & Stratton, New York, 1977, pp 79–93.

18. Carleton, SA, and Brown, WF: Changes in motor unit populations in motor neurone disease. J Neurol Neurosurg Psychiatry 42:42–51, 1979.

19. Carpenter, S, Karpati, G, Rothman, S, Watters, G, and Andermann, F: Pathological involvement of primary sensory neurons in Werdnig-Hoffmann disease. Acta Neuropathol 42:91–97, 1978.

20. Chad, D, Hammer, K, and Sargent, J: Slow resolution of multifocal weakness and fasciculation: A reversible motor neuron syndrome. Neurology 36:1260–1263, 1986.

21. Clarren, SK, and Hall, JG: Neuropathologic findings in the spinal cords of 10 infants with arthrogryposis. J Neurol Sci 58:89–102, 1983.

22. Creutzfeldt, HG: Uber eine eigenartige herdformige Erkrankung des Zentralnervensystems. In Nissl, F, and Alzheimer, A (eds): Histologische und Histopathologische. G Fisher, Jena, 1921, pp 1–48.

23. Danta, G: Electrophysiological study of amyotrophy associated with acute asthma (asthmatic amyotrophy). J Neurol Neurosurg Psychiatry 38:1016–1021, 1975.

24. Daube, J: AAEE Minimonograph #18: EMG in motor neuron disease. American Association of Electromyography and Electrodiagnosis, Rochester, Minn, 1982.

25. Denys, EH, and Norris, FH, Jr: Amyotrophic lateral sclerosis: Impairment of neuromuscular transmission. Arch Neurol 36:202–205, 1979.

26. Desmedt, JE, and Borenstein, S: Interpretation of electromyographical data in spinal muscular atrophy. In Rose, FC (ed): Motor Neurone Disease. Grune & Stratton, New York, 1977, pp 112–120.

27. Dobkin, BH, and Verity, MA: Familial progressive bulbar and spinal muscular atrophy: Juvenile onset and late morbidity with ragged-red fibers. Neurology 26:754–763, 1976.

28. Dubowitz, V: Infantile muscular atrophy: A prospective study with particular reference to a slowly progressive variety. Brain 87:707–718, 1964.

29. Dubowitz, V: Muscle Disorders in Childhood. In Schaffer, AJ, and Markowitz, M (eds): Major Problems in Clinical Pediatrics, Vol 16. WB Saunders, Philadelphia, 1978.

30. Dubowitz, V, and Brooke, MH: Muscle Biopsy: A Modern Approach. WB Saunders, Philadelphia, 1973.

31. Duffy, P, Wolf, J, Collins, G, Devoe, AV, Streeten, B, and Cowen, D: Possible person-to-person transmission of Creutzfeldt-Jakob disease. N Engl J Med 290:692–693, 1974.

32. Dyck, PJ, Stevens, JC, Mulder, DW, and Espinosa, RE: Frequency of nerve fiber degeneration of peripheral motor and sensory neurons in amyotrophic lateral sclerosis: Morphometry of deep and superficial peroneal nerves. Neurology 25:781–785, 1975.

33. Emery, AEH: The nosology of the spinal muscular atrophies. J Med Genet 8:481–495, 1971.

34. Emery, AEH, Davie, AM, Holloway, S and Skinner, R: International collaborative study of the spinal muscular atrophies. Part 1. Analysis of clinical and laboratory data. J Neurol Sci 29:83–94, 1976.

35. Emery, AEH, Davie, AM, Holloway, S, and Skinner, R: International collaborative study of the spinal muscular atrophies. Part 2. Analysis of genetic data. J Neurol Sci 30:375–384, 1976.

36. Emery, AEH, Davie, AM, and Smith, C: Spinal muscular atrophy—resolution of heterogeneity. In Bradley, WG, Gardner-Medwin, D, and Walton, JN (eds): Recent Advances in Myology. Excerpta Medica, Amsterdam, 1975, pp 557–565.

37. Emery, AEH, Hausmanowa-Petrusewicz, I, Davie, AM, Holloway, S, Skinner, R, and Borkowska, J: International collaborative study of the spinal muscular atrophies. Part 1. Analysis of clinical and laboratory data. J Neurol Sci 29:83–94, 1976.

38. Engel, WK, Siddique, T, and Nicoloff, JT: Effect on weakness and spasticity in amyotrophic lateral sclerosis of thyrotopin-releasing hormone. Lancet 2:73–75, 1983.

39. Feigenbaum, JA, and Munsat, TL: A neuromuscular syndrome of scapuloperoneal distribution. Bull LA Neurol Soc 35:47–57, 1970.

40. Fenichel, GM, Emery, ES, and Hunt, P: Neurogenic atrophy simulating facioscapulohumeral dystrophy: A dominant form. Arch Neurol 17:257–260, 1967.

41. Fetell, MR, Smallberg, G, Lewis, LD, Lovelace, RE, Hays, AP, and Rowland, LP: A benign motor neuron disorder: Delayed cramps and fasciculation after poliomyelitis or myelitis. Ann Neurol 11:423–427, 1982.

42. Fincham, RW, and Cape, CA: Sensory nerve conduction in syringomyelia. Neurology 18:200–201, 1968.

43. Fisher, RL, Johnstone, WT, Fisher, WH, Jr, and Goldkamp, OG: Arthrogryposis multiplex congenita: A clinical investigation. J Pediatr 76:255–261, 1970.

44. Fleury, P, and Hageman, G: A dominantly inherited lower motor neuron disorder presenting at birth with associated arthrogryposis. J Neurol Neurosurg Psychiatry 48:1037–1048, 1985.

45. Furukawa, T, Akagami, N, and Maruyama, S: Chronic neurogenic quadriceps amyotrophy. Ann Neurol 2:528–530, 1977.

46. Gajdusek, DC, Gibbs, CJ, Jr, Asher, DM, Brown, P, Diwan, A, Hoffman, P, Nemo, G, Rohwer, R, and White, L: Precautions in medical care of, and in handling materials from, patients with transmissible virus dementia (Creutzfeldt-Jakob Disease). N Engl J Med 297:1253–1258, 1977.

47. Gardner-Medwin, D, Hudgson, P, and Walton, JN: Benign spinal muscular atrophy arising in childhood and adolescence. J Neurol Sci 5:121–158, 1967.

48. Gardner-Medwin, D, and Walton, JN: The clinical examination of the voluntary muscles. In Walton, JN (ed): Disorders of Voluntary

Muscle, ed 3. Churchill Livingstone, Edinburgh, 1974, pp 517–560.

49. Garruto, RM, Gajdusek, C, and Chen, KM: Amyotrophic lateral sclerosis among Chamorro migrants from Guam. Ann Neurol 8:612–619, 1980.

50. Gibbs, CJ, Jr, and Gajdusek, DC: Kuru—A prototype subacute infectious disease of the nervous system as a model for the study of amyotrophic lateral sclerosis. In Norris, FH, Jr, and Kurland, LT (eds): Motor Neuron Diseases: Research on Amyotrophic Lateral Sclerosis and Related Disorders. Grune & Stratton, New York, 1969, pp 269–279.

51. Gibbs, CJ, Jr, Gajdusek, DC, Asher, DM, Alpers, MP, Beck, E, Daniel, PM, and Matthews, WB: Creutzfeldt-Jakob disease (spongiform encephalopathy): Transmission to the chimpanzee. Science 161:388–399, 1968.

52. Gurney, ME: Suppression of sprouting at the neuromuscular junction by immunoassay sera. Nature 307:546–548, 1984.

53. Hadlow, WJ, Prusiner, SB, Kennedy, RC, and Race, RE: Brain tissue from persons dying of Creutzfeldt-Jakob disease causes scrapie-like encephalopathy in goats. Ann Neurol 8:628–631, 1980.

54. Hansen, S, and Ballantyne, JP: A quantitative electrophysiological study of motor neurone disease. J Neurol Neurosurg Psychiatry 41:773–783, 1978.

55. Hanyu, N, Oguchi, K, Yanagisawa, N, and Tsukagoshi, H: Degeneration and regeneration of ventral root motor fibres in amyotrophic lateral sclerosis: Morphometric studies of cervical ventral roots. J Neurol Sci 55:99–115, 1982.

56. Harding, AE, Bradbury, PG, and Murray, NMF: Chronic asymmetrical spinal muscular atrophy. J Neurol Sci 59:69–83, 1983.

57. Harding, AE, and Thomas, PK: Hereditary distal spinal muscular atrophy. J Neurol Sci 45:337–348, 1980.

58. Harding, AE, Thomas, PK, Baraitser, M, Bradburg, PG, Morgan-Hughes, JA, and Ponsford, JR: X-linked recessive bulbospinal neuropathy: A report of ten cases. J Neurol Neurosurg Psychiatry 45:1012–1019, 1982.

59. Harvey, DG, Torack, RM, and Rosenbaum, HE: Amytrophic lateral sclerosis with ophthalmoplegia: A clinicopathologic study. Arch Neurol 36:615–617, 1979.

60. Hashimoto, O, Asada, M, Ohta, M, and Kuroiwa, Y: Clinical observations of juvenile nonprogressive muscular atrophy localized in hand and forearm. J Neurol 211:105–110, 1976.

61. Hausmanowa-Petrusewicz, I: Spinal Muscular Atrophy. Infantile and Juvenile Type. US Department of Commerce, National Technical Information Service, Springfield, Va, 1978.

62. Hausmanowa-Petrusewicz, I, Fidzianska, A, Niebroj-Dobosz, I, and Strugalska, MH: Is Kugelberg-Welander spinal muscular atrophy a fetal defect? Muscle Nerve 3:389–402, 1980.

63. Hayward, M, and Seaton, D: Late sequelae of paralytic poliomyelitis: A clinical and electromyographic study. J Neurol Neurosurg Psychiatry 42:117–122, 1979.

64. Hoffmann, J: Ueber chronische spinale Muskelatrophie im Kindesalter, auf familiärer Basis. Deutsche Zeitschrift für Nervenheilkunde 3:427–470, 1893.

65. Hopkins, IJ: A new syndrome: Poliomyelitis-like illness associated with acute asthma in childhood. Aust Paediatr J 10:273–276, 1974.

66. Hudson, A, Vinters, H, Povey, R, Hatch, L, Percy, D, Noseworthy, J, and Kaufmann, J: An unusual form of motor neuron disease following a cat bite. Can J Neurol Sci 13:111–116, 1986.

67. Ilett, SJ, Pugh, RJ, and Smithells, RW: Poliomyelitis-like illness after acute asthma. Arch Dis Child 52:738–740, 1977.

68. Jakob, A: Über eigenartige Erkrankungen des Zentralnervensystems mit bemerkenswertem anatomischen Befunde. Z Neurol Psychiatr 64:147–228, 1921.

69. Jamal, GA, Weir, AI, Hansen, S, and Ballantyne, JP: Sensory involvement in motor neuron disease: Further evidence from automated thermal threshold determination. J Neurol Neurosurg Psychiatry 48:906–910, 1985.

70. Jansen, PHP, Joosten, EMG, Jaspar, HHJ, and Vingerhoets, HM: A rapidly progressive autosomal dominant scapulohumeral form of spinal muscular atrophy. Ann Neurol 20:538–540, 1986.

71. Johnson, EW, Guyton, JD, and Olsen, KJ: Motor nerve conduction velocity studies in poliomyelitis. Arch Phys Med Rehabil 41:185–190, 1960.

72. Johnson, WG, Wigger, HJ, Karp, HR, Glaubiger, LM, and Rowland, LP: Juvenile spinal muscular atrophy: A new hexosaminidase deficiency phenotype. Ann Neurol 11:11–16, 1982.

73. Kaeser, HE: Scapuloperoneal muscular atrophy. Brain 88:407–418, 1965.

74. Kantarjian, AD: A syndrome clinically resembling amyotrophic lateral sclerosis following chronic mercurialism. Neurology 11:639–644, 1961.

75. Kayser-Gatchalian, MC: Late muscular atrophy after poliomyelitis. Eur Neurol 10:371–380, 1973.

76. Kerr, FWL: Structural and functional evidence of plasticity in the central nervous system. Exp Neurol 48 (3):16–31, 1975.

77. Kloepfer, HW, and Emery, AEH: Genetic aspects of neuromuscular disease. In Walton, JN (ed): Disorders of Voluntary Muscle, ed 3. Churchill Livingstone, Edinburgh, 1974, pp 852–885.

78. Kondo, K, Tsubaki, T, and Sakamoto, F: The Ryukyuan muscular atrophy: An obscure heritable neuromuscular disease found in the islands of southern Japan. J Neurol Sci 11:359–382, 1970.

79. Kott, E, Livni, E, Zamir, R, and Kuritzky, A: Cell-mediated immunity to polio and HLA antigens in amyotrophic lateral sclerosis. Neurology 29:1040–1044, 1979.

80. Koutlidis, RM, deRecondo, J, and Bathien, N: Conduction of the sciatic nerve in its proximal and distal segment in patients with ALS (amyotrophic lateral sclerosis). J Neurol Sci 64:183–191, 1984.

81. Krugliak, L, Gadoth, N, and Behar, AJ: Neuropathic form of arthrogryposis multiplex congenita: Report of 3 cases with complete necropsy, including the first reported case of agenesis of muscle spindles. J Neurol Sci 37:179–185, 1978.

82. Kugelberg, E, and Welander, L: Heredofamilial juvenile muscular atrophy simulating muscular dystrophy. Arch Neurol Psychiatry 75:500–509, 1956.

83. Kuntz, NL, Gomez, MR, and Daube, JR: Prognosis in childhood proximal spinal muscular atrophy (abstr). Neurology 30:378, 1980.

84. Kurland, LT, Choi, NW, and Sayre, GP: Implications of incidence and geographic patterns on the classification of amyotrophic lateral sclerosis. In Norris, FH, Jr, and Kurland, LT (eds): Motor Neuron Diseases: Research on Amyotrophic Lateral Sclerosis and Related Disorders. Grune & Stratton, New York, 1969, pp 28–50.

85. Kurlander, HM, and Patten, BM: Metals in spinal cord tissue of patients dying of motor neuron disease. Ann Neurol 6:21–24, 1979.

86. Lambert, EH: Electromyography in amyotrophic lateral sclerosis. In Norris, FH, Jr, and Kurland, LT (eds): Motor Neuron Diseases: Research on Amyotrophic Lateral Sclerosis and Related Disorders. Grune & Stratton, New York, 1969, pp 135–153.

87. Liversedge, LA, and Campbell, MJ: Motor neurone diseases. In Walton, JN (ed): Disorders of Voluntary Muscle, ed 3. Churchill Livingstone, Edinburgh, 1974, pp 775–803.

88. Meadows, JC, Marsden, CD, and Harriman, DGF: Chronic spinal muscular atrophy in adults. Part 1. The Kugelberg-Welander syndrome. J Neurol Sci 9:527–550, 1969.

89. Meadows, JC, Marsden, CD, and Harriman, DGF: Chronic spinal muscular atrophy in adults. Part 2. Other forms. J Neurol Sci 9:551–566, 1969.

90. Miller, JR, Guntaka, RV, and Myers, JC: Amyotrophic lateral sclerosis: Search for polio-virus by nucleic acid hybridization. Neurology 30:884–886, 1980.

91. Mitsumoto, H, Sliman, RJ, Schafer, IA, Sternick, CS, Kaufman, B, Wilbourn, A, and Horwitz, SJ: Motor neuron disease and adult hexosaminidase-A deficiency in two families: Evidence for multisystem degeneration. Ann Neurol 17:378–385, 1985.

92. Mitsuyama, Y: Presenile dementia with motor neuron disease in Japan: Clinico-pathological review of 26 cases. J Neurol Neurosurg Psychiatry 47:953–959, 1984.

93. Moosa, A, and Dubowitz, V: Motor nerve conduction velocity in spinal muscular atrophy of childhood. Arch Dis Child 51:974–977, 1976.

94. Mulder, DW, and Espinosa, RE: Amyotrophic lateral sclerosis: Comparison of the clinical syndrome in Guam and the United States. In Norris, FH, Jr, and Kurland, LT (eds): Motor Neuron Diseases: Research on Amyotrophic Lateral Sclerosis and Related Disorders. Grune & Stratton, New York, 1969, pp 12–19.

95. Mulder, DW, Rosenbaum, RA, and Layton, DD, Jr: Late progression of poliomyelitis or forme fruste amyotrophic lateral sclerosis. Mayo Clin Proc 47:756–761, 1972.

96. Namba, T, Aberfeld, DC, and Grob, D: Chronic proximal spinal muscular atrophy. J Neurol Sci 11:401–423, 1970.

97. Nelson, JS, and Prensky, AL: Sporadic juvenile amyotrophic lateral sclerosis: A clinicopathological study of a case with neuronal cytoplasmic inclusions containing RNA. Arch Neurol 27:300–306, 1972.

98. Norris, FH, Jr: Adult spinal motor neuron disease. Progressive muscular atrophy (Aran's disease) in relation to amyotrophic lateral sclerosis. In Vinken, PJ, and Bruyn, GW (eds): Handbook of Clinical Neurology, Vol 22. System Disorders and Atrophies. North-Holland, Amsterdam, 1975, pp 1–56.

99. Parhad, IM, Clark, WA, Barron, KD, and Staunton, SB: Diaphragmatic paralysis in motor neuron disease: Report of 2 cases and a review of the literature. Neurology 28:18–22, 1978.

100. Patten, BM, Zito, G, and Harati, Y: Histologic findings in motor neuron disease: Relation to clinically determined activity, duration and severity of disease. Arch Neurol 36:560–564, 1979.

101. Pearn, JH: The spinal muscular atrophies of childhood: A genetic and clinical study. PhD thesis, University of London, 1974.

102. Pearn, JH: Autosomal dominant spinal muscular atrophy: A clinical and genetic study. J Neurol Sci 38:263–275, 1978.

103. Pearn, JH, Carter, CO, and Wilson, J: The genetic identity of acute infantile spinal muscular atrophy. Brain 96:463–470, 1973.

104. Pearn, JH, Gardner-Medwin, D, and Wilson, J: A clinical study of chronic childhood spinal muscular atrophy: A review of 141 cases. J Neurol Sci 38:23–37, 1978.

105. Pearn, JH, Hudgson, P, and Walton, JN: A clinical and genetic study of spinal muscular atrophy of adult onset. The autosomal recessive form as a discrete disease entity. Brain 101:591–606, 1978.

106. Pearn, JH, and Wilson, J: Acute Werdnig-Hoffmann disease: Acute infantile spinal muscular atrophy. Arch Dis Child 48:425–430, 1973.

107. Pierce-Ruhland, R, and Patten, BM: Muscle metals in motor neuron disease. Ann Neurol 8:193–195, 1980.

108. Poskanzer, DC, Cantor, HM, and Kaplan, GS: The frequency of preceding poliomyelitis in amyotrophic lateral sclerosis. In Norris, FH, Jr, and Kurland, LT (eds): Motor Neuron Diseases: Research on Amyotrophic Lateral Sclerosis and Related Disorders. Grune & Stratton, New York, 1969, pp 286–290.

109. Raimbault, J, and Laget, P: Electromyography in the diagnosis of infantile spinal amyo-

trophy of Werdnig-Hoffmann type. Pathol Biol (Paris) 20:287–296, 1972.

110. Ronen, G, Lowry, N, Wedge, J, Sarnat, H, and Hill, A: Hereditary motor sensory neuropathy type I presenting as scapuloperoneal atrophy (Davidenkow syndrome) electrophysiological and pathological studies. Can J Neurol Sci 13:264–266, 1986.

111. Roos, RP, Viola, MV, Wollmann, R, Hatch, MH, and Antel, JP: Amyotrophic lateral sclerosis with antecedent poliomyelitis. Arch Neurol 37:312–313, 1980.

112. Rowland, LP: Hexosaminidase deficiency: A cause of recessive inherited motor neuron disease. In Rowland, LP (ed): Human Motor Neuron Diseases. Raven Press, New York, 1982, pp 159–164.

113. Rowland, LP: Peripheral neuropathy, motor neuron disease, or neuropathy? In Battistin, L, Hashim, GA, and Lajtha, A (eds): Clinical and Biological Aspects of Peripheral Nerve Diseases. Alan R Liss, New York, 1983, pp 27–41.

114. Rowland, LP: Looking for the cause of amyotrophic lateral sclerosis (editorial). N Engl J Med 311:979–981, 1985.

115. Rowland, LP, Defendini, R, Sherman, W, Hirano, A, Olarte, MR, Laton, N, Lovelace, RE, Inone, K, and Osserman, EF: Macroglobulinemia with peripheral neuropathy simulating motor neuron disease. Ann Neurol 11:532–536, 1982.

116. Russo, LS, Jr: Clinical and electrophysiologic studies in primary lateral sclerosis. Arch Neurol 39:662–664, 1982.

117. Salazar, AM, Masters, CL, Gajdusek, DC, and Gibbs, CJ, Jr: Syndromes of amyotrophic lateral sclerosis and dementia: Relation to transmissible Creutzfeldt-Jakob disease. Ann Neurol 14:17–26, 1983.

118. Schwartz, MS, and Moosa, A: Sensory nerve conduction in the spinal muscular atrophies. Dev Med Child Neurol 19:50–53, 1977.

119. Shiraki, H: The neuropathology of amyotrophic lateral sclerosis (ALS) in the Kii Peninsula and other areas of Japan. In Norris, FH, Jr, and Kurland, LT (eds): Motor Neuron Diseases: Research on Amyotrophic Lateral Sclerosis and Related Disorders, Vol 2. Grune & Stratton, New York, 1969, pp 80–84.

120. Simpson, JA: Disorders of neuromuscular transmission. Proc R Soc Med 59:993–998, 1966.

121. Sobue, I, Saito, N, Iida, M, and Ando, K: Juvenile type of distal and segmental muscular atrophy of upper extremities. Ann Neurol 3:429–432, 1978.

122. Stalberg, E, Schwartz, MS, and Trontelj, JV: Single fibre electromyography in various processes affecting the anterior horn cells. J Neurol Sci 24:403–415, 1975.

123. Steiman, GS, Rorke, LB, and Brown, MJ: Infantile neuronal degeneration masquerading as Werdnig-Hoffmann disease. Ann Neurol 8:317–324, 1980.

124. Telerman-Toppet, N, and Coers, C: Motor innervation and fiber type pattern in amyotrophic lateral sclerosis and in Charcot-Marie-Tooth disease. Muscle Nerve 1:133–139, 1978.

125. Tyler, HR: Double-blind study of modified neurotoxin in motor neuron disease. Neurology 29:77–81, 1979.

126. Uchida, H, Nemoto, H, and Kinoshita, M: Action of thyrotropin-releasing hormone (TRH) on the occurrence of fibrillation potentials and miniature end-plate potentials (MEPPs): An experimental study. J Neurol Sci 76:125–130, 1986.

127. Van Gent, EM, Hoogland, RA, and Jennekens, FGI: Distal amyotrophy of predominantly the upper limbs with pyramidal features in a large kinship. J Neurol Neurosurg Psychiatry 48:266–269, 1985.

128. Wadia, NH, Wadia, PN, Katrak, SM, and Misra, VP: A study of the neurological disorder associated with acute haemorrhagic conjunctivitis due to enterovirus 70. J Neurol Neurosurg Psychiatry 46:599–610, 1983.

129. Walton, JN (ed): Disorders of Voluntary Muscle, ed 3. Churchill Livingstone, Edinburgh and London, 1974.

130. Welch, KMA, and Goldberg, DM: Serum creatine phosphokinase in motor neuron disease. Neurology 22:697–701, 1972.

131. Werdnig, G: Zwei frühinfantile hereditäre Fälle von progressiver Muskelatrophie unter dem Bilde der Dystrophie, aber auf neurotischer Grundlage. Archiv für Psychiatrie und Nervenkrankheitnen 22:437–480, 1891.

132. Wheeler, SD, and Ochoa, J: Poliomyelitis-like syndrome associated with asthma: A case report and review of the literature. Arch Neurol 37:52–53, 1980.

133. Wiechers, DO, and Hubbell, SL: Late changes in the motor unit after acute poliomyelitis. Muscle Nerve 4:524–528, 1981.

134. Zellweger, H, Simpson, J, McCormick, WF, and Ionasescu, V: Spinal muscular atrophy with autosomal dominant inheritance. Report of a new kindred. Neurology 22:957–963, 1972.

Chapter 21

DISEASES OF THE ROOT AND PLEXUS

1 INTRODUCTION

Proximal lesions at the level of the root or plexus affect either the motor or sensory fibers, or both. The features of motor involvement include weakness and atrophy of the muscle, hyporeflexia, fatigue, cramps, and fasciculations. Sensory abnormalities, which usually accompany motor deficits, sometimes dominate the picture. Such symptoms range from mild distal paresthesias to complete loss of sensation, dysesthesias, and severe pain. Peripheral lesions such as the carpal tunnel syndrome can mimic proximal abnormalities of the root or plexus. Selec-

tive damage to these anatomic regions occurs as a result of trauma, mechanical compression, and, less frequently, neoplastic and inflammatory processes. Thus, differential diagnosis is limited in the proximal lesions, compared with a much wider range of possibilities encountered in neuropathies and other distal involvement (see Chapter 22).

In the evaluation of radicular or plexus injuries, electrophysiologic studies help delineate the distribution of the affected muscles, localize the level, and elucidate the extent and chronicity. Needle examination initially reveals poor recruitment of motor unit potentials, indicating structural or functional loss of axons. Subsequent appearance of fibrillation potentials and positive sharp waves in 2 to 3 weeks suggests axonal degeneration. Low-amplitude, polyphasic motor unit potentials have temporal instability during active regeneration of motor axons. In contrast, high-amplitude, long-duration motor unit potentials with stable configuration appear later, after completion of reinnervation. Nerve conduction studies reveal reduced amplitude of the muscle or sensory action potentials in appropriate distribution, depending on the site of involvement.

2 CERVICAL AND THORACIC ROOTS

There are eight cervical roots and only seven cervical vertebrae. The C-1 through C-7 roots emerge above their respective vertebrae, but the C-8 root exits between the C-7 and T-1 vertebrae. Cervical radiculopathy results from spondylosis, a herniated disk, or traumatic avulsion.[57]

In compression of the C-5 root, pain in the interscapular region radiates along the lateral aspect of the arm to the elbow. With involvement of the C-6 root, pain extends over the shoulder to the lateral aspect of the arm, forearm, and thumb. Pain induced by irritation of the C-7 root typically involves the entire arm and forearm, with radiation into the third digit and, to a lesser extent, the second and fourth digits. Less commonly encountered are C-8 root pain radiating to the fourth and fifth digits and T-1 root pain localized deep in the shoulder, axilla, and medial aspect of the arm. Although sensory symptoms help in evaluating radiculopathy, they often fail to provide the exact localization because of the dermatomal overlap and variability.

The distribution of motor deficits and changes in the stretch reflexes provide more reliable localization. Clinical assessment of radiculopathy depends on testing movements of the arm that rely upon almost exclusive control by single roots. Recommended maneuvers include shoulder abduction to 180 degrees (C-5);

elbow flexion in full and half supination (C-6); and adduction of the shoulder, extension of the elbow, and extension and flexion of the wrist (C-7).[89] A C-8 root lesion affects the long extensors and flexors of the fingers and, to a lesser degree, the intrinsic hand muscles, which receive substantial supply from the T-1 root. An ulnar nerve lesion spares the median-innervated thenar muscles, whereas a T-1 root lesion affects all the small hand muscles. The abnormalities of certain muscle stretch reflexes assist in determining the level of root lesions, for example, biceps brachii (C-5 or C-6), supinator (C-6), triceps (C-7), and finger flexors (C-8).

Electromyographic studies provide an objective means of corroborating clinical diagnosis of a radicular lesion (see Tables 1–1, 1–2, and 1–3). Affected muscles show reduced recruitment and incomplete interference patterns at the beginning, and fibrillation potentials, positive sharp waves, and high-amplitude, long-duration motor unit potentials later in the course of the disease. Nerve conduction studies often fail to reveal substantial abnormalities, even with the addition of late response and somatosensory evoked potentials. The paucity of conduction defects probably reflects the nature of lesions that affect a very restricted segment of the roots. Preganglionic involvement spares the sensory nerve action potentials, although degeneration of motor axons leads to muscle atrophy and reduction in amplitude of compound muscle potentials. Cervical root stimulation may reveal conduction abnormalities in patients with symptoms of radiculopathy.[11]

Cervical Spondylosis

This condition results from bony overgrowth of the vertebrae following degeneration of the intervertebral disk. A spondylotic bar, protruding posteriorly, most commonly impinges on the C-5 and C-6 roots and, less frequently, on the C-7 root. Other cervical and thoracic roots are rarely affected. In most typical cases, neck movement triggers pain in the appropriate dermatome. Some patients have asymptomatic bars, and others suffer from constant pain not alleviated by postural maneuvers. A C-5 or C-6 root lesion suppresses the biceps and supinator stretch reflexes, whereas a C-7 radiculopathy diminishes the triceps reflex. Compressive cervical myelopathy just rostral to the origin of the C-7 root may enhance the triceps response and suppress the biceps and supinator reflexes.

Herniated Cervical Disk

Herniated disks are less frequent in the cervical than in the lumbar region. Cervical disks lesions usually occur unilaterally in patients with a history of neck trauma. Injury to a spine with preexisting cervical spondylosis may cause bilateral symptoms, multiple root involvement, or myelopathy secondary to compression of the spinal cord. The most common herniation between C-5 and C-6 vertebrae compresses the C-6 root; and those between C-6 and C-7 vertebrae, the C-7 root. Movement of the neck or the arm aggravates the initial symptom of pain over a typical root distribution. Compression of the ventral root causes weakness in the muscles innervated by the affected root.

Root Avulsion

Erb-Duchenne palsy results from avulsion of the C-5 and C-6 roots.[2] This type of injury occurs with downward traction on the plexus, which increases the angle between the head and shoulder, for example, following a forceps delivery with the shoulder fixed in position. The palsy produces a characteristic posture, with adduction and internal rotation of the arm and extension and pronation of the forearm. Despite the preservation of the intrinsic hand muscles, the patient cannot abduct the arm or supinate the forearm to bring the hand into a useful position. The muscles innervated by the C-5 and C-6 roots atrophy; but the sensory examination, though often limited in infants, usually reveals only mild changes.

Klumpke's palsy, with avulsion of the C-8 and T-1 roots, occurs much less frequently, from forced upward traction on the plexus. An attempt to grasp an overhead support during a fall increases the angle between the arm and thorax beyond the ordinary limit. This type of injury degenerates the ulnar nerve, the inner head of the median nerve, and a portion of the radial nerve. The intrinsic hand muscles and long flexors and extensors of the fingers atrophy, producing a partial clawhand. The patient also has anesthesia along the inner aspect of the hand, forearm, and arm. Horner's syndrome indicates damage of the cervical sympathetic fibers.

Myelography usually delineates the extent of root injury.[64] It must be interpreted with caution, however, because pseudomeningoceles may accompany intact roots, on the one hand, and root avulsion may fail to produce detectable meningoceles, on the other.[58,64] Preganglionic separation of the cell body with lesions at the root level preserves the anatomic and physiologic integrity of the peripheral axon. Thus, intradermal histamine injection induces a physiologic reflex skin reaction[13]; and despite sensory loss, nerve stimulation elicits a normal sensory action potential.[24,117] These findings stand in sharp contrast to the loss of chemical or electrical reactivity along the distal nerve segments in patients with plexus lesions.

The deep cervical muscles receive innervation from the posterior, as opposed to anterior, rami of the spinal nerves. The evidence of denervation here, therefore, indicates an intraforaminal lesion affecting the root or spinal nerve prior to the division into the two rami. Other muscles innervated at a position proximal to the brachial plexus include the rhomboids, supplied by the dorsal scapular nerve, and the serratus anterior, subserved by the long thoracic nerves. Spontaneous activity in these muscles also serves to distinguish between root and plexus lesions.

Thoracic Radiculopathy

Isolated involvement of lower thoracic or upper lumbar roots, though rare, may result from collapsed vertebral bodies.[79]

With lesions at this level, proximal weakness of the legs may lead to an erroneous diagnosis of myopathy. Electromyography shows spontaneous discharges localized to the affected myotomes in the limb and paraspinal muscles.

3 BRACHIAL PLEXUS

In peacetime, brachial plexus lesions are an infrequent cause of arm weakness.[30] Penetrating injuries from bullet wounds often involve the upper and lower trunk and posterior cord. A difficult birth or sudden traction applied to the arm or neck can also damage the plexus. In addition to direct injuries, indirect trauma results from fractures of the humerus or dislocation of the shoulder.[33] Plexopathy may develop after prolonged anesthesia with the patient in an unusual posture. Hemiplegics may sustain an injury from repeated pressure under the arms for lifting. Other possible causes include complications during axillary arteriography,[81] median sternotomy,[52] and jugular vein cannulation for coronary artery bypass graft surgery.[75] Appropriate radiologic and electrophysiologic studies help in determining the indication of surgical intervention, which can be of value only for well-selected patients.[68]

Idiopathic plexopathy ranks first in incidence among nontraumatic conditions affecting the brachial plexus.[9] Differential diagnosis includes Hodgkin's disease,[90] desensitizing injections,[123] the Ehlers-Danlos syndrome,[62] systemic lupus erythematosus,[12] and familial pressure-sensitive neuropathy.[15] Chronic compressive lesions of the brachial plexus range from primary nerve tumors to metastatic breast cancer and lymphoma. The patient with neoplastic invasion tends to have pain and Horner's syndrome.[76] Radiation therapy of the axillary region also causes plexopathy, mimicking tumor recurrence.[108] Intermittent compression, seen in some cases of the thoracic outlet syndrome, produces less well defined neurologic symptoms with little or no objective abnormality.[29]

The clinical features depend on the area of primary disease or injury. The upper trunk bears the brunt of damage from injury by firearm recoil, which forcefully retracts the clavicle against the underlying scalene muscles[115]; heavy backpack[27]; or a common football injury called "stinger."[33] The damage here causes a distribution of weakness similar to that seen in Erb-Duchenne palsy, with involvement of the shoulder and upper arm, and sparing of the hand function. The patient cannot abduct the arm, internally or externally rotate the shoulder, flex the elbow, or extend the wrist radially. Other clinical features include sparing of the rhomboid and serratus anterior innervated by more proximal branches; sensory changes over the lateral aspect of the arm, forearm, and hand; and reduced or absent biceps and supinator stretch reflexes.

Rare isolated injury to the middle trunk produces weakness in the general distribution of the radial nerve, involving the triceps only partially and sparing the brachioradialis entirely. There can be metastasis to any portion of the plexus, but predominantly to the lower trunk and medial cord, as expected from the location of lymph nodes. Selective damage to the lower trunk also results from local trauma or direct invasion from Pancoast's tumor in the apex of the lung. Lesions affecting the C-8 and T-1 roots impair hand function and cause Horner's syndrome. These clinical features bear resemblance to those seen in Klumpke's palsy. In addition to the intrinsic hand muscles, finger flexors and extensors are weak. Sensory changes involve the medial aspect of the arm; the forearm; and the hand, including the fourth and fifth digits.

Injury to the posterior cord seen in shoulder dislocation gives rise to the clinical picture of combined axillary and radial nerve palsies. The patient cannot extend the elbow, wrist, or fingers. The weak deltoid causes limited arm abduction after the first 30 degrees, the range subserved by the supraspinatus. Sensory changes involve the lateral aspect of the shoulder and arm; the posterior portion of the forearm; and the dorsal aspects of the lateral half of the hand, including the first two digits. Compressive lesions in the thoracic outlet tend to affect the medial cord. Motor and sensory deficits develop in the median- and ulnar-innervated region that receives supplies from the C-8 root. Though rare, local trauma can selectively damage the lateral cord, causing weakness in musculocutaneous and median-innervated muscles that receive axons from the C-6 and C-7 roots.

The nerve conduction abnormalities commonly seen in brachial plexus lesions include (1) severe amplitude attenuation of muscle and antidromic sensory nerve action potentials evoked with stimulation proximal to the site of nerve injury, compared with those evoked at a more distal site; and (2) slowing of conduction across the site of injury.[109] These findings suggest that the palsies result from a local demyelinating block with or without axonal loss.[109] The pattern of sensory potential abnormality from each digit may help localize the site of injury.[25,26,85] Electrophysiologic studies in radiation plexopathy often reveal abnormal sensory conduction, normal motor conduction, and myokymic discharges.[76] In traumatic plexopathies electromyography renders more information than do nerve conduction studies in delineating the degree, distribution, and time course of the disease.[102]

Aberrant regeneration of phrenic motor neurons may induce arm-diaphragm synkinesis after injury to the proximal portion of the brachial plexus or cervical nerve roots.[104] Synkinetic movements and axon reflexes may involve different, sometimes antagonistic, muscles in patients with brachial plexus injury at birth.[31] Simultaneous needle studies from multiple muscles help document such misdirected reinnervation.

Idiopathic Brachial Neuropathy

Idiopathic brachial neuritis,[100] also known as neuralgic amyotrophy[88,125] or brachial neuralgia, probably originates in the roots, although the exact site of the lesion remains unknown.[16,101,110,111] Most cases occur sporadically after the third decade, affecting men more than twice as frequently as women. The symptoms may develop as a complication of vaccination procedures, especially with injection into the deltoid. Trauma, infection, or serum sickness may precede acute onset of pain and other symptoms of neuralgia. Most patients have unilateral symptoms, but the condition may occasionally occur bilaterally and, in rare incidences, recurrently. Prognosis is generally good in the majority of cases.[111] Maximal recovery may, however, take a few years if the patient shows no improvement during the first few months after onset.

The disease typically begins with pain localized in the distribution of C-5 and C-6 dermatomes.[16,78,88,100,101,110,112] The clinical picture varies considerably, some patients having a chronic and painless form.[96] An intense aching sensation may radiate along the arm. Two thirds of patients experience relatively mild sensory impairment. Within a few days, the shoulder girdle musculature becomes weak and atrophic. The disease most severely affects C-5 and C-6 myotomes, and to a lesser extent, the muscles innervated by the spinal accessory nerve and C-7 root. Pain usually subsides with the onset of weakness but may last much longer. The characteristic posture with the arm flexed at the elbow and adducted at the shoulder[118] sometimes leads to a frozen-shoulder syndrome.[63]

The disease may cause selective paralysis in the distribution of a single root, trunk, cord, or peripheral nerve.[29] Such mononeuropathies tend to involve the radial, long thoracic, phrenic, suprascapular, or accessory nerve.[6,22,70,86,114] Occasionally the initial presenting symptoms mimic an anterior interosseous nerve palsy.[66,112] Concurrent involvement of the shoulder muscles in neuralgic amyotrophy suggests one of two possibilities[93]: (1) spatial scatter of the underlying abnormality to the forearm or (2) selective damage of the brachial plexus nerve bundle with topographic grouping at the level of the cord.[103]

Electromyography usually shows evidence of denervation, on the affected side and may also reveal subtle changes on the clinically asymptomatic side.[111] Typical findings seen in the involved muscles include fibrillation potentials, positive sharp waves, high-amplitude polyphasic motor unit potentials, and reduced interference pattern.[17] This, together with the course of clinical recovery, suggests axonal interruption and wallerian degeneration.

Conduction studies reveal slightly to moderately increased latencies from Erb's point to severely affected muscles. This change is usually attributable to the loss of fast conducting fibers that accompanies any condition characterized by reduced amplitude of the compound muscle potentials. Mild injury may lead to pure demyelination, from which the patient recovers rapidly without loss of axons.[93] A selective latency increase from Erb's point to individual muscles of the shoulder girdle suggests multiple mononeuropathies.[83] Conduction abnormalities may become more conspicuous after reinnervation has begun. The nerves in clinically unaffected extremities sometimes show widespread changes.[119] F-wave studies may show increased latency and slow conduction velocity in the segment above the axilla, but not as a consistent finding, especially in the early stages of illness.[67]

The diagnosis often depends on the combination of amplitude abnormalities of median or ulnar sensory studies, slowed conduction of musculocutaneous motor fibers, and lack of paraspinal fibrillation potentials on needle examination.[44] The loss or diminution of the sensory action potentials localizes the lesion distal to the dorsal root ganglion.[14,24] Normal paraspinal examination favors plexopathy but does not rule out radiculopathy.[69]

Familial Brachial Plexopathy

Nontraumatic brachial plexus neuropathy may develop on a familial basis in association with lesions outside the plexus.[15,17,36,116,120] Acute episodes have features indistinguishable from sporadic idiopathic neuralgic amyotrophy, but patients with familial variety have less pain.[46,98] Inherited as an autosomal dominant trait, the disease tends to affect a younger age group with no preference for either sex, although pregnancy may herald the onset.[55] The symptoms recur more frequently in the familial than in the sporadic variety. The lesions outside the plexus cause additional signs, such as Horner's syndrome and dysphonia.[47] The disease can also involve the lumbosacral plexus, cranial nerves, individual peripheral nerves such as the long thoracic nerve,[91] and the autonomic nervous system.[5,46,55,106]

Nerve conduction studies show normal or reduced amplitude of the recorded response. Electromyography reveals fibrillation potentials, positive sharp waves, and reduced recruitment, suggesting axonal damage.[36]

Some patients with familial pressure-sensitive neuropathy may also present with acute attacks of brachial plexopathy (see Chapter 22.5). This condition affects the peripheral nerves diffusely,[10,15,37] although conduction abnormalities have a predilection for the common sites of compression.[17] Sural nerve biopsies reveal bizarre focal swelling of the nerve fiber, mild reduction in the total myelinated fiber count, and an abnormal fiber diameter spectrum with loss of the normal bimodal distribution. The term tomaculous neuropathy, used to describe this pathologic condition, implies the sausage-shaped thickenings of the myelin sheaths.[17]

Plexopathy Secondary to Radiation

Plexopathy may develop months to years after radiation treatment and take a progressive course.[108] In patients with cancer and brachial plexus signs, radiation injury must be differentiated from tumor infiltration. According to a study of 100 cases, painless upper trunk lesions with lymphedema suggest radiation injury, whereas painful lower trunk lesions with the Horner syndrome tend to imply tumor infiltration.[71]

Electrophysiologic studies reveal mildly increased latency in proportion to reduced amplitude of the evoked potentials.

This stands in contrast to neoplastic infiltration that may cause considerable slowing of conduction across the plexus. Electromyography shows fibrillation potentials, positive sharp waves, and large, polyphasic motor unit potentials. The presence of myokymic discharges favors the diagnosis of radiation plexopathies, as opposed to tumor infiltration.[3]

Cervical Rib and Thoracic Outlet Syndrome

A variety of anomalous structures in the neck may affect the roots or trunks of the brachial plexus. A rudimentary cervical rib tends to compress the lower trunk of the brachial plexus as well as the subclavian or axillary artery. A compression syndrome may also result from a fibrous band or the first thoracic rib pressed upward by distortion of the thorax. The once widely publicized compression by the scalenus anticus muscle fell into disrepute. The cervical rib, though rare, does give rise to compression of the neurovascular structures, especially in women with a sagging shoulder girdle. Patients with the thoracic outlet syndrome often have low-set "droopy" shoulders and a long swan neck.[105] They usually complain of unilateral symptoms, even in the presence of bilateral cervical ribs.

Vascular features result from upward displacement of the axillary or subclavian artery by the cervical rib. Stenosis of the compressed artery may cause intermittent embolic phenomena of the brachial artery, with ischemic changes in the fingers. The hand turns cold and blue, with diminished or absent pulsations in the radial and ulnar arteries. Although scalenotomies for the disputed scalenus anticus syndrome declined in number, removal of the first rib abounds in surgical practice.[72,113] The procedure has limited indication for most patients with vascular symptoms.[48] If such an intervention offers a beneficial effect in the management of arm pain, the initially normal electrophysiologic studies usually fail to substantiate the subjective change.[74] Some have advocated that in

this condition ulnar nerve studies reveal consistent abnormalities with stimulation at Erb's point,[32,80,94,113] but without subsequent confirmation.[25,72,121,122]

Apart from the poorly defined condition described above, there is a rare, but more clearly recognizable neurologic entity, sometimes called classic thoracic outlet syndrome. It usually affects women with a rudimentary cervical rib.[49,50] The neural symptoms include local and referred pain secondary to pressure, paresthesias in the hand and forearm along the medial aspect, and weakness of the intrinsic hand muscles. Prominent atrophy of the abductor pollicis brevis may lead to erroneous diagnosis of the carpal tunnel syndrome. The thoracic outlet syndrome, however, gives rise to pain and sensory changes in the ulnar-innervated fingers.

Nerve conduction abnormalities found in patients with clear neurologic deficit consist of reduced or absent sensory action potentials of the ulnar nerve and an increase in the latency of the F wave on the affected side, when compared with the normal side.[35,124] Reduced amplitude of the ulnar sensory action potential indicates that the lesion is distal to the dorsal root ganglia.[50] Normal conduction velocities of the median and ulnar nerve help exclude the possibility of distal entrapment. Electromyography shows evidence of denervation in the intrinsic hand muscles, especially the abductor pollicis brevis. Patients free of neurologic deficits have none of these abnormalities even when vascular symptoms appear with postural maneuvers.[28,67] Some advocate the use of median versus ulnar somatosensory evoked potential studies, but their clinical usefulness awaits further clarification.[59]

4 LUMBOSACRAL ROOTS

Injury at this level usually takes place as the root exits through its foramin. Preganglionic damage, however, can occur anywhere along the long subarachnoid pathway of the cauda equina within the spinal canal. This anatomic peculiarity makes clinical and electrophysiologic localization of radicular lesions more difficult in the lower than in the upper extremities. Unlike the cervical roots, the lumbar roots emerge from the intervertebral spaces below their respective vertebrae. In the upper extremities, motor deficits serve as a more reliable localizing sign than sensory impairments. The reverse seems to hold in the lower extremities.

Radiculopathies rarely involve the first three lumbar roots that supply the skin of the anterior thigh. With compression of the L-4 root, pain radiates from the knee to the medial malleolus along the medial aspect. With L-5 root irritation, pain originates in the buttock and radiates along the posterior lateral aspect of the thigh, lateral aspect of the leg, dorsum of the foot, and first four toes. A lesion of the S-1 root causes pain to radiate down the back of the thigh, leg, and lateral aspect of the foot. Irritation of the S-2 through S-5 roots results in pain along the posteromedial aspect of the thigh, over the perianal area of the buttock, and in the genital region.

In the lower extremities, involvement of a single root does not necessarily cause prominent weakness or wasting, reflecting multiplicity of root supply. In most leg muscles, however, a single root primarily controls certain movements. These include hip flexion by L-2, knee extension and thigh adduction by L-3 inversion of the foot by L-4, toe extension by L-5, and eversion of the foot by S-1.[89] Lesions of a single root affect dorsiflexion of the foot to a lesser extent because of dual control by the L-4 and L-5 roots. Similarly, plantar flexion is subserved by the S-1 and S-2 roots. A lesion of the L-4 root depresses the knee stretch reflex, whereas an S-1 root lesion affects the ankle jerk and its electrical counterpart, the H reflex.

Conus Lesion

Tumors known to involve the conus medullaris include ependymoma, dermoid cyst, lipoma, and arteriovenous malformation.[77] They typically invade the sacral roots from below, beginning

with the S-5 root. Thus, the usual presenting features consist of a dull backache and sensory disturbances in the genital and perianal region that may escape detection even by careful examination. Impotence and impaired sphincter control soon develop. The bilateral diminution of the ankle jerk indicates upward extension of the tumor to the origin of the S-1 root. The lesion typically spares the knee reflex. Initially unilateral weakness soon spreads to the other extremity, leading to relatively symmetric involvement.

Electromyographic abnormalities often indicate a bilateral involvement of multiple roots despite asymmetric clinical signs. The anal sphincter also shows evidence of denervation and loss of tonus. Nerve conduction studies may reveal reduced muscle action potentials but normal sensory nerve potential, as predicted from preganglionic site of involvement. Some ascending spinal fibers undergo degeneration, as evidenced by abnormal somatosensory cerebral evoked potentials elicited by intrathecal stimulation of the lumbosacral cord.[41] Electrophysiologic studies should reveal no abnormalities in the upper extremities.

Cauda Equina Lesion

The lesions responsible for the lateral cauda equina syndrome include meningioma, neurofibroma, and a herniated disk. Such a mass in the spinal canal below the T-12 vertebra can affect any one of the lumbar or sacral roots singly or in combination. With a laterally located lesion at the level of the L-1, L-2, and L-3 vertebrae, pain typically radiates over the anterior thigh. Involvement of the L-4 root results in atrophy and weakness of the quadricep muscle and foot inverters, with a diminished knee reflex. A high, laterally located lesion may simultaneously compress the cord, giving rise to a hyperactive ankle reflex and other upper motor neuron signs. One must distinguish this rare, confusing presentation from amyotrophic lateral sclerosis.

Midline or diffuse involvement of the cauda equina suggests metastasis from prostate cancer, direct spread of tumors in the pelvic floor, or chondromas of the sacral bone. Similar clinical features may result from leukemic or lymphomatous infiltration or seeding with medulloblastoma, pinealoma, or other malignant tumors of the nervous system. Lower motor neuron syndromes may also follow radiation therapy,[54] redundant nerve root syndrome,[92] spinal arachnoiditis,[87] or ankylosing spondylitis.[8]

Except for asymmetric distribution and severe pain, signs and symptoms of a cauda equina lesion resemble those of a conus medullaris lesion. This lesion often causes bilateral involvement of the dermatomes ordinarily unaffected by a herniated lumbar disk. Unlike the compression at the intervertebral space, changing positions of the lower extremities fails to alleviate the discomfort. Reduced muscle stretch reflexes at both the knee and the ankle also tend to localize the lesion at the cauda equina, rather than conus medullaris.

Electromyography shows fibrillation potentials and large motor unit potentials in the distribution of several lumbosacral roots, including paraspinal muscles[8] and urethral sphincter.[45] Again the findings mimic those of an intrinsic cord involvement except for an asymmetric distribution of the abnormalities with spread above the sacral myotomes. Thus, a substantial side-to-side difference in amplitude of the compound muscle action potentials favors the diagnosis of cauda equina, rather than conus medullaris, lesions.

Herniated Lumbar Disk

Disk protrusion involves the L4-5 and L5-S1 interspaces in a majority of cases and the L3-4 space much less frequently. Lesions at the remaining higher or lower levels should revise diagnostic possibilities of uncomplicated herniation. The protruding disk tends to compress the lumbosacral roots slightly above the level of their respective foramina before their lateral deviation toward the exit. A herniated disk at the L4-5 intervertebral space, therefore, compresses the L-5 root,

which emerges under the L-5 vertebra. Similarly, a disk protrusion between the L-5 and S-1 vertebrae damages the S-1 root, exiting the interspace below. As mentioned earlier, cervical disk herniation at the C6-7 space compresses the C-7 root, which exits above the C-7 vertebra. Thus, in both the cervical and lumbar regions, the root most frequently subjected to damage carries the same number as the vertebra below the herniated disk.

Clinical symptoms consist of weakness in the affected myotomes and pain in the appropriate dermatomes, aggravated by leg-raising or other maneuvers that stretch the root. Patients may have pure sensory or pure motor symptoms. In rare instances, fiber hypertrophy exceeds atrophy, resulting in unilateral calf enlargement following an S-1 radiculopathy.[84]

Electromyographic examinations help confirm the diagnosis and identify the damaged root (see Fig. 13–7, bottom left). Denervation of the paraspinal muscles (see Figs. 13–10 and 13–11) implies a lesion located proximal to the origin of the posterior ramus.[19] Such localized involvement, however, provides no specific etiology, because, for example, metastatic disease can affect the paraspinal muscle.[73] Further, the absence of denervation here does not necessarily exclude the possibility of root compression.[69] In addition to being used diagnostically, series of studies can guide management by substantiating progression or improvement.[61] The clinical course of radiculopathy correlates with electrical abnormalities better than computed tomography.[65]

Following laminectomy, spontaneous activity may persist indefinitely, although it usually diminishes substantially by 3 to 6 months.[34] In one study, focal abnormalities found at least 3 cm lateral to the incision and 4 to 5 cm deep indicated a new lesion.[60] Other findings suggestive of an active radiculopathy in postlaminectomy patients include (1) fibrillation potentials and positive sharp waves at a specific level on the symptomatic side only; (2) a mixture of large and small fibrillation and positive sharp waves segmentally on the symptomatic side but only small sparse spontaneous discharges on the asymptomatic side; and (3) the appearance in serial studies of new spontaneous activity at the suspected level on the symptomatic side.

Paraspinal abnormalities differentiate radiculopathy from diseases of the plexus or peripheral nerve. The exact localization of the involved segment is not possible on this basis alone. Determining the precise level of the lesion, therefore, depends on careful exploration of the affected muscles in the lower limbs. Because of anatomic peculiarities, lesions located much higher than the ordinary disk protrusion may compress the L-5 root or the S-1 root within the cauda equina. For example, a tumor of a high lumbar root may produce confusing clinical features and myelographic abnormalities. In assessing radiculopathy, nerve conduction studies are used in excluding a neuropathy. Reduction in the amplitude of compound muscle action potentials also assists in detection of modest nerve damage. H-reflex studies reveal abnormalities with an S-1, but not with an L-5, radiculopathy. The measures of F-wave latency or dermatomal somatosensory evoked potentials reflect delayed conduction, but their clinical value for early detection of radiculopathy remains to be determined[4,38,39,67] (see Chapter 17.6).

Spinal Stenosis

Lumbar stenosis usually results from central narrowing of the spinal canal with or without associated constriction in the nerve root canals. In a review of 37 patients, stenosis most commonly affected the L-4 or the L-5 level, or both.[97] In 36 patients, electromyography revealed fibrillation potentials and poorly recruiting, polyphasic long-duration motor unit potentials in several leg muscles and to a lesser extent in the paraspinal muscles bilaterally.

Root Avulsion

Intradural avulsion does not involve the lumbosacral roots as often as the cervical roots,[43] although this condition is frequently overlooked in patients with pelvic fractures or sacroiliac dislocation.[53] In these instances, tension in the lumbar and sacral plexuses stretch the root intradurally.[7] Electromyography shows evidence of denervation in the appropriate myotomes, including the paraspinal muscles. Myelography delineates the level of involvement.

5 LUMBOSACRAL PLEXUS

The lumbosacral plexus, often considered a single anatomic entity, consists of lumbar and sacral portions with a connection between them. The division helps in delineating clinical problems that tend to affect each portion independently. A lesion involving the lumbar plexus diminishes the knee reflex and causes sensory loss over the L-2, L-3, and L-4 dermatomes. It also weakens not only the hip flexors and knee extensors but also leg adductors. In contrast, isolated femoral neuropathy spares the obturator-innervated muscles. A lesion of the sacral plexus produces a clinical picture similar to that seen with a sciatic nerve lesion, but with additional involvement of the gluteal muscles, and, at times, the anal sphincter. Traumatic injuries result from fractures of the pelvis or inappropriate traction during orthopedic or other operative manipulations.[7,23]

Neoplasms extending from the rectum, prostate, or cervix often invade the lumbosacral plexus. Metastatic, leukemic, or lymphomatous infiltration gives rise to painful and slowly progressive paralysis. In one series dealing with 85 cases of documented pelvic tumors, plexopathy involved the upper portion, derived from the L-1 to L-4 roots, in 31 percent; the lower portion, formed by the L-4 to S-3 roots, in 51 percent; and both in 18 percent.[56] Clinical features were leg pain, weakness, edema, a rectal mass, and hydronephrosis. Electrophysiologic studies revealed evidence of denervation and reinnervation in electromyography, together with conduction abnormalities of the motor fibers 4 months after onset on the average. In another series of 50 patients, radiation plexopathy caused indolent painless leg weakness early, often bilaterally. In contrast, tumor patients typically had painful unilateral weakness. Electromyography revealed partial denervation and chronic reinnervation in both entities. The tumor group had no myokymic discharges, seen in more than half the cases of radiation plexopathy.[107]

Immune or vascular abnormalities probably play an important role in idiopathic lumbar plexopathies, as is the case for their better described and more frequent counterparts, brachial plexopathies. Acute pain in one or both legs usually precedes the onset of weakness and areflexia, followed by atrophy of affected muscles.[95] In 10 cases of idiopathic lumbosacral plexopathy with follow-up up to an average of 6 years, the patients recovered slowly and often incompletely.[42] In diabetics, symptoms of lumbar plexopathy may represent femoral neuropathy or radiculopathy.[20]

Compression plexopathy may develop with hematomas in hemophiliacs, in those with other coagulopathies, or during anticoagulation therapy.[99] Two anatomically distinct syndromes have been described[21,51]: (1) involvement of the lumbar plexus by hematoma within the psoas muscle[40] and (2) selective compression of the femoral nerve.[126] With a plexus lesion, weakness involves the thigh adductors, hip flexors, and quadriceps. Sensory loss affects the entire anterior thigh, including the area supplied by the lateral femoral cutaneous nerve. In contrast, femoral neuropathy selectively weakens the quadriceps and hip flexors, together with sensory deficits limited to the distribution of the anterior femoral cutaneous and saphenous nerves.[40]

Electromyography of the proximal muscles innervated rostrally in relation to the plexus plays a major role in distinguishing a plexopathy from radiculopathy. The commonly tested muscles include, in addition to the paraspinal muscle, the gluteus maximus, medius, and minimus and iliopsoas. Typical findings consist of poor recruitment of motor unit potentials and fibrillation potentials at rest in the myotomes supplied by the anterior rami of multiple spinal nerves.

Distal nerve stimulation shows reduction in amplitude of the compound muscle or nerve action potentials on the affected side, as compared with the normal side.[1,18] Root stimulation may reveal increased latency across the plexus in the appropriate distribution.[82] F waves may or may not have a prolonged latency (see Fig. 17–12). Involvement of the S-1 root diminishes the amplitude of the H reflex and increases its latency.

REFERENCES

1. Adelman, JU, Goldberg, GS, and Puckett, JD: Postpartum bilateral femoral neuropathy. Obstet Gynecol 42:845–850, 1973.
2. Adler, JB, and Patterson, RL, Jr: Erb's palsy. Long-term results of treatment in eighty-eight cases. J Bone Joint Surg 49A:1052–1064, 1967.
3. Albers, JW, Allen, AA, II, Bastron, JA, and Daube, JR: Limb myokymia. Muscle Nerve 4:494–504, 1981.
4. Aminoff, MJ, Goodin, DS, Parry, GJ, Barbarop, NM, Weinstein, PR, and Rosenblum, ML: Electrophysiologic evaluation of lumbosacral radiculopathies: Electromyography, late responses and somatosensory evoked potentials. Neurology 35:1514–1518, 1985.
5. Arts, WFM, Busch, HFM, Van den Brand, HJ, Jennekens, FGI, Frants, RR, and Stefanko, SZ: Hereditary neuralgic amyotrophy: clinical, genetic, electrophysiological and histopathological studies. J Neurol Sci 62:261–279, 1983.
6. Augustin, P, Verdure, L, and Samson, M: Le syndrome du nerf sus-scapulaire a l'etroit. Rev Neurol (Paris) 132:219–222, 1976.
7. Barnett, HG, and Connolly, ES: Lumbosacral nerve root avulsion: Report of a case and review of the literature. J Traum 15:532–535, 1975.
8. Bartleson, JD, Cohen, MD, Harrington, TM, Goldstein, NP, and Ginsberg, WW: Caude equina syndrome secondary to long-standing ankylosing spondylitis. Ann Neurol 14:662–669, 1983.
9. Beghi, E, Kurland, LT, Mulder, DW, and Nicolosi, A: Brachial plexus neuropathy in the population of Rochester, Minnesota, 1970–1981. Ann Neurol 18:320–323, 1985.
10. Behse, F, Buchthal, F, Carlsen, F, and Knappeis, GG: Hereditary neuropathy with liability to pressure palsies. Electrophysiological and histopathological aspects. Brain 95:777–794, 1972.
11. Berger, AR, Busis, NA, Logigian, EL, Wierzbicka, M, and Shahani, BT: Cervical root stimulation in the diagnosis of radiculopathy. Neurology 37:329–332, 1987.
12. Bloch, SL, Jarrett, MP, Swerdlow, M, and Grayzel, AI: Brachial plexus neuropathy as

the initial presentation of systemic lupus erythematosus. Neurology 29:1633–1634, 1979.

13. Bonney, G: Prognosis in traction lesions of the brachial plexus. J Bone Joint Surg 41B:4–35, 1959.

14. Bonney, G, and Gilliatt, RW: Sensory nerve conduction after traction lesion of the brachial plexus. Proc R Soc Med 51:365–367, 1958.

15. Bosch, EP, Chui, HC, Martin, MA, and Cancilla, PA: Brachial plexus involvement in familial pressure-sensitive neuropathy: Electrophysiological and morphological findings. Ann Neurol 8:620–624, 1980.

16. Bradley, WG: Disorders of Peripheral Nerves. Blackwell Scientific Publications, London, 1974.

17. Bradley, WG, Madrid, R, Thrush, DC, and Campbell, MJ: Recurrent brachial plexus neuropathy. Brain 98:381–398, 1975.

18. Buchthal, A: Femoralisparesen als Komplikation gynäkologischer Operationen. Dtsch Med Wochenschr 98:2024–2027, 1973.

19. Bufalini, C, and Pescatori, G: Posterior cervical electromyography in the diagnosis and prognosis of brachial plexus injuries. J Bone Joint Surg 51B:627–631, 1969.

20. Calverley, JR, and Mulder, DW: Femoral neuropathy. Neurology 10:963–967, 1960.

21. Chiu, WS: The syndrome of retroperitoneal hermorrhage and lumbar plexus neuropathy during anticoagulant therapy. South Med J 60:595–599, 1976.

22. Clein, LJ: Suprascapular entrapment neuropathy. J Neurosurg 43:337–342, 1975.

23. Coles, CC, and Miller, KD, Jr: Traumatic avulsion of the lumbar nerve roots. South Med J 71:334–335, 1978.

24. Conrad, B, and Benecke, R: How valid is the distal sensory nerve action potential for differentiating between radicular and non-radicular nerve lesions? (abstr) Acta Neurol Scand (Suppl 73) 60:122, 1979.

25. Cruz Martinez, A, Barrio, M, Perez Conde, MC, and Gutierrez, AM: Electrophysiological aspects of sensory conduction velocity in healthy adults. 1. Conduction velocity from digit to palm, from palm to wrist, and across the elbow, as a function of age. J Neurol Neurosurg Psychiatry 41:1092–1096, 1978.

26. Cruz Martinez, A, Barrio, M, Perez Conde, MC, and Ferrer, MT: Electrophysiological aspects of sensory conduction velocity in healthy adults. 2. Ratio between the amplitude of sensory evoked potentials at the wrist on stimulating different fingers in both hands. J Neurol Neurosurg Psychiatry 41:1097–1101, 1978b.

27. Daube, JR: Rucksack paralysis. JAMA 208:2447–2452, 1969.

28. Daube, JR: Nerve conduction studies in the thoracic outlet syndrome. Neurology 25:347, 1975.

29. Daube, JR: An electromyographer's view of plexopathy. In Neuromuscular Diseases as Seen by the Electromyographer. Second Annual Continuing Education Course, American Association of Electromyography and Electrodiagnosis, October 4, 1979.

30. Davis, DH, Onofrio, BM, and MacCarty, CS: Brachial plexus injuries. Mayo Clin Proc 53:799–807, 1978.

31. De Grandis, D, Fiaschi, A, Michieli, G, and Mezzina, C: Anomalous reinnervation as a sequel to obstetric brachial plexus palsy. J Neurol Sci 43:127–132, 1979.

32. Di Benedetto, M: Thoracic outlet slowing: A critical evaluation of established criteria for the diagnosis of outlet syndrome by nerve conduction studies. Electromyogr Clin Neurophysiol 17:191–204, 1977.

33. Di Benedetto, M, and Markey, K: Electrodiagnostic localization of traumatic upper trunk brachial plexopathy. Arch Phys Med Rehabil 65:15–17, 1984.

34. Donovan, WH, Dwyer, AP, and Bedbrook, GM: Electromyographic activity in paraspinal musculature in patients with idiopathic scoliosis before and after Harrington instrumentation. Arch Phys Med Rehabil 61:413–417, 1980.

35. Dorfman, LJ: F-wave latency in the cervical-rib-and-band syndrome (corres). Muscle Nerve 2:158–159, 1979.

36. Dunn, HG, Daube, JR, and Gomez, MR: Heredofamilial brachial plexus neuropathy (hereditary neuralgic amyotrophy with brachial predilection) in childhood. Dev Med Child Neurol 20:28–46, 1978.

37. Earl, CJ, Fullerton, PM, Wakefield, GS, and Schutta, HS: Hereditary neuropathy, with liability to pressure palsies. Q J Med 33:481–498, 1964.

38. Eisen, A, Schomer, D, and Melmed, C: The application of F-wave measurements in the differentiation of proximal and distal upper limb entrapments. Neurology 27:662–668, 1977.

39. Eisen, A, Schomer, D, and Melmed, C: An electrophysiological method for examining lumbosacral root compression. Can J Neurol Sci 4:117–123, 1977.

40. Emery, S, and Ochoa, J: Lumbar plexus neuropathy resulting from retroperitoneal hemorrhage. Muscle Nerve 1:330–334, 1978.

41. Ertekin, C, Sarica, Y, and Uckardesler, L: Somatosensory cerebral potentials evoked by stimulation of the lumbo-sacral spinal cord in normal subjects and in patients with conus medullaris and cauda equina lesions. Electroencephalogr Clin Neurophysiol 59:57–66, 1984.

42. Evans, BA, Stevens, JC, and Dyck, PJ: Lumbosacral plexus neuropathy. Neurology 31:1327–1330, 1981.

43. Finney, LA, and Wulfman, WA: Traumatic intradural lumbar nerve root avulsion with associated traction injury to the common peroneal nerve. AJR 84:952–957, 1960.

44. Flaggman, PD, and Kelly, JJ, Jr: Brachial plexus neuropathy: An electrophysiologic evaluation. Arch Neurol 37:160–164, 1980.

45. Fowler, CJ, Kirby, RS, Harrison, MJG, Milroy, EJG, and Turner-Warwick, R: Individual motor unit analysis in the diagnosis of disorders of urethral sphincter innervation.

J Neurol Neurosurg Psychiatry 47:637–641, 1984.

46. Gardner, JH, and Maloney, W: Hereditary brachial and cranial neuritis genetically linked with ocular hypotelorism and syndactyly (abstr). Neurology 18:278, 1968.

47. Geiger, LR, Mancall, EL, Penn, AS, and Tucker, SH: Familial neuralgic amyotrophy: Report of three families with review of the literature. Brain 97:87–102, 1974.

48. Gilliatt, RW: Thoracic outlet compression syndrome. Br Med J 1:1274–1275, 1976.

49. Gilliatt, RW, Le Quesne, PM, Logue, V, and Sumner, AJ: Wasting of the hand associated with a cervical rib or band. J Neurol Neurosurg Psychiatry 33:615–624, 1970.

50. Gilliatt, RW, Willison, REG, Dietz, V, and Williams, IR: Peripheral nerve conduction in patients with a cervical rib and band. Ann Neurol 4:124–129, 1978.

51. Goodfellow, J, Fearn, CBD'A, and Matthews, JM: Iliacus haematoma: A common complication of haemophilia. J Bone Joint Surg 49B:748–756, 1967.

52. Graham, JG, Pye, IF, and McQueen, INF: Brachial plexus injury after median sternotomy. J Neurol Neurosurg Psychiatry 44:621–625, 1981.

53. Harris, WR, Rathbun, JB, Wortzman, G, and Humphrey, G: Avulsion of lumbar roots complicating fracture of the pelvis. J Bone Joint Surg 55A:1436–1442, 1973.

54. Horowitz, SL, and Stewart, JD: Lower motor neuron syndrome following radiotherapy. Can J Neurol Sci 10:56–58, 1983.

55. Jacob, JC, Andermann, F, and Robb, JP: Heredofamilial neuritis with brachial predilection. Neurology 11:1025–1033, 1961.

56. Jaeckle, KA, Young, DF, and Foley, KM: The natural history of lumbosacral plexopathy in cancer. Neurology 35:8–15, 1985.

57. Jaeger, R, and Whiteley, WH: Avulsion of the brachial plexus: Report of six cases. JAMA 153:633–635, 1953.

58. Jelasic, F, and Piepgras, U: Functional restitution after cervical avulsion injury with "typical" myelographic findings. Eur Neurol 11:158–163, 1974.

59. Jerrett, SA, Cuzzone, LJ, and Pasternak, BM: Thoracic outlet syndrome: electrophysiologic reappraisal. Arch Neurol 41:960–963, 1984.

60. Johnson, EW, Burkhart, JA, and Earl, WC: Electromyography in postlaminectomy patients. Arch Phys Med Rehabil 53:407–409, 1972.

61. Johnson, EW, and Fletcher, FR: Lumbosacral radiculopathy: Review of 100 consecutive cases. Arch Phys Med Rehabil 61:321–323, 1981.

62. Kayed, K, and Kass, B: Acute multiple brachial neuropathy and Ehlers-Danlos syndrome. Neurology 29:1620–1621, 1979.

63. Kennedy, WR, and Resch, JA: Paralytic brachial neuritis. Lancet 2:459–462, 1966.

64. Kewalramani, LS, and Taylor, RG: Brachial plexus root avulsion: Role of myelography: Review of diagnostic procedures. J Trauma 15:603–608, 1975.

65. Khatri, BO, Baruah, J, and McQuillen, MP: Correlation of electromyography with computed tomography in evaluation of lower back pain. Arch Neurol 41:594–597, 1984.

66. Kiloh, LG, and Nevin, S: Isolated neuritis of the anterior interosseous nerve. Br Med J 1:850–851, 1952.

67. Kimura, J: A comment (corres). Muscle Nerve 1:250–251, 1978.

68. Kline, DG, Hackett, ER, and Happel, LH: Surgery for lesions of the brachial plexus. Arch Neurol 43:170–181, 1986.

69. Knutsson, B: Comparative value of electromyographic, myelographic and clinical-neurological examinations in diagnosis of lumbar root compression syndrome. Acta Orthop Scand Suppl 49:1–135, 1961.

70. Kómár, J: Eine wichtige Ursache des Schulterschmerzes: Incisura-scapulae-Syndrom. Fortschr Neurol Psychiatr 44:644–648, 1976.

71. Kori, SH, Foley, KM, and Posner, JB: Brachial plexus lesions in patients with cancer: 100 cases. Neurology 31:45–50, 1981.

72. Kremer, RM, and Ahlquist, RE, Jr: Thoracic outlet compression syndrome. Am J Surg 130:612–616, 1975.

73. Laban, MM, Meerschaert, JR, Perez, L, and Goodman, PA: Metastatic disease of the paraspinal muscles: Electromyographic and histopathologic correlation in early detection. Arch Phys Med Rehabil 59:34–37, 1978.

74. Lascelles, RG, Mohr, PD, Neary, D, and Bloor, K: The thoracic outlet syndrome. Brain 100:601–612, 1977.

75. Lederman, RJ, Breuer, AC, Hanson, MR, Furlan, AJ, Loop, FD, Cosgrove, DM, Estafanous, FG, and Greenstreet, RL: Peripheral nervous system complications of coronary artery bypass graft surgery. Ann Neurol 12:297–301, 1982.

76. Lederman, RJ, and Wilbourn, AJ: Brachial plexopathy: recurrent cancer or radiation? Neurology 34:1331–1335, 1984.

77. Levin, KH, and Daube, JR: Spinal cord infarction: another cause of "lumbosacral polyradiculopathy." Neurology 34:389–390, 1984.

78. Lishman, WA, and Russell, WR: The brachial neuropathies. Lancet 2:941–946, 1961.

79. Liveson, JA: Thoracic radiculopathy related to collapsed thoracic vertebral bodies. J Neurol Neurosurg Psychiatry 47:404–406, 1984.

80. London, GW: Normal ulnar nerve conduction velocity across the thoracic outlet: Comparison of two measuring techniques. J Neurol Neurosurg Psychiatry 38:756–760, 1975.

81. Lyon, BB, Hansen, BA, and Mygind, T: Peripheral nerve injury as a complication of axillary arteriography. Acta Neurol Scand 51:29–36, 1975.

82. MacLean, I: Nerve root stimulation to evaluate conduction across the lumbosacral plexus. Acta Neurol Scand (Suppl 73) 60:270, 1979.

83. Martin, WA, and Kraft, GH: Shoulder girdle neuritis: A clinical and electrophysiological evaluation. Milit Med 139:21–25, 1974.

84. Mielke, U, Ricker, K, Emser, W, and Boxler, K: Unilateral calf enlargement following S1 radiculopathy. Muscle Nerve 5:434–438, 1982.

85. Newman, M, and Nelson, N: Digital nerve sensory potentials in lesions of cervical roots and brachial plexus. Can J Neurol Sci 10:252–255, 1983.

86. Olarte, M, and Adams, D: Accessory nerve palsy. J Neurol Neurosurg Psychiatry 40:1113–1116, 1977.

87. Parker, KR, Kane, JT, Wiechers, DO, and Johnson, EW: Electromyographic changes reviewed in chronic spinal arachnoiditis. Arch Phys Med Rehabil 60:320–322, 1979.

88. Parsonage, MJ, and Turner, JWA: Neuralgic amyotrophy. The shoulder-girdle syndrome. Lancet 1:973–978, 1948.

89. Patten, J: Neurological Differential Diagnosis. Springer-Verlag, New York, 1977.

90. Pezzimenti, JF, Bruckner, HW, and DeConti, RC: Paralytic brachial neuritis in Hodgkin's disease. Cancer 31:626–629, 1973.

91. Phillips, L, II: Familial long thoracic nerve palsy: A manifestation of brachial plexus neuropathy. Neurology 36:1251–1253, 1986.

92. Reinstein, L, Twardzik, FG, Russo, GL, and Bose, P: Electromyographic abnormalities in redundant nerve root syndrome of the cauda equina. Arch Phys Med Rehabil 65:270–272, 1984.

93. Rennels, GD, and Ochoa, J: Neuralgic amyotrophy manifesting as anterior interosseous nerve palsy. Muscle Nerve 3:160–164, 1980.

94. Sadler, TR, Jr, Rainer, WG, and Twombley, G: Thoracic outlet compression: Application of positional arteriographic and nerve conduction studies. Am J Surg 130:704–706, 1975.

95. Sander, JE, and Sharp, FR: Lumbosacral plexus neuritis. Neurology 31:470–473, 1981.

96. Schott, GD: A chronic and painless form of idiopathic brachial plexus neuropathy. J Neurol Neurosurg Psychiatry 46:555–557, 1983.

97. Seppalainen, AM, Alaranta, H, and Soini, J: Electromyography in diagnosis of lumbar spinal stenosis. Electromogr Clin Neurophysiol 21:55–66, 1981.

98. Smith, BH, Ramakrishna, T, and Schlagenhauff, RE: Familial brachial neuropathy: Two case reports with discussion. Neurology 21:941–945, 1971.

99. Spiegel, PG, and Meltzer, JL: Femoral-nerve neuropathy secondary to anticoagulation: Report of a case. J Bone Joint Surg 56A:425–427, 1974.

100. Spillane, JD: Localized neuritis of the shoulder girdle: A report of 46 cases in the MEF. Lancet 2:532–535, 1943.

101. Spillane, JD: An Atlas of Clinical Neurology, ed 2. Oxford University Press, London, 1975.

102. Stanwood, JE, and Kraft, GH: Diagnosis and management of brachial plexus injuries. Arch Phys Med Rehabil 52:52–60, 1971.

103. Sunderland, S: The intraneural topography of the radial, median, and ulnar nerves. Brain 68:243–299, 1945.

104. Swift, TR, Leshner, RT, and Gross, JA: Arm-diaphragm synkinesis: Electrodiagnostic studies of aberrant regeneration of phrenic motor neurons. Neurology 30:339–344, 1980.

105. Swift, TR, and Nichols, FT: The droopy shoulder syndrome. Neurology 34:212–215, 1984.

106. Taylor, RA: Heredofamilial mononeuritis multiplex with brachial predilection. Brain 83:113–137, 1960.

107. Thomas, JE, Cascino, TL, and Earle, JD: Differential diagnosis between radiation and tumor plexopathy of the pelvis. Neurology 35:1–7, 1985.

108. Thomas, JE, and Colby, MY, Jr: Radiation-induced or metastatic brachial plexopathy? A diagnostic dilemma. JAMA 222:1392–1395, 1972.

109. Trojaborg, W: Electrophysiological findings in pressure palsy of the brachial plexus. J Neurol Neurosurg Psychiatry 40:1160–1167, 1977.

110. Tsairis, P: Brachial plexus neuropathies. In Dyck, PJ, Thomas, PK, and Lambert, EH (eds): Peripheral Neuropathy, Vol 1. WB Saunders, Philadelphia, 1975, pp 659–681.

111. Tsairis, P, Dyck, PJ, and Mulder, DW: Natural history of brachial plexus neuropathy: Report on 99 patients. Arch Neurol 27:109–117, 1972.

112. Turner, JWA, and Parsonage, MJ: Neuralgic amyotrophy (paralytic brachial neuritis) with special reference to prognosis. Lancet 2:209–212, 1957.

113. Urschel, HC, Jr, Razzuk, MA, Wood, RE, Parekh, M, and Paulson, DL: Objective diagnosis (ulnar nerve conduction velocity) and current therapy of the thoracic outlet syndrome. Ann Thorac Surg 12:608–620, 1971.

114. Walsh, NE, Dumitru, D, Kalantri, A, and Roman, AM, Jr: Brachial neuritis involving the bilateral phrenic nerves. Arch Phys Med Rehabil 68:46–48, 1987.

115. Wanamaker, WM: Firearm recoil palsy. Arch Neurol 31:208–209, 1974.

116. Warot, P, Petit, H, Nuyts, JP, and Meignie, S: Nevrites amyotrophaintes brachiales familiales: Etude de deux familles. Rev Neurol (Paris) 128:281–288, 1973.

117. Warren, J, Gutmann, L, Figueroa, AF, Jr, and Bloor, BM: Electromyographic changes of brachial plexus root avulsion. J Neurosurg 31:137–140, 1969.

118. Waxman, SG: The flexion-adduction sign in neuralgic amyotrophy. Neurology 29:1301–1304, 1979.

119. Weikers, NJ, and Mattson, RH: Acute paralytic brachial neuritis: A clinical and electrodiagnostic study. Neurology 19:1153–1158, 1969.

120. Wiederholt, WC: Hereditary brachial neuropathy. Report of two families. Arch Neurol 30:252–254, 1974.

121. Wilbourn, AJ: Slowing across the thoracic outlet with thoracic outlet syndrome: Fact or fiction? Neurology 34:143, 1984.

122. Wilbourn, AJ, and Lederman, RJ: Evidence

for conduction delay in thoracic outlet syndrome is challenged. N Engl J Med 310:1052–1053, 1984.

123. Wolpow, ER: Brachial plexus neuropathy: Association with desensitizing antiallergy injections. JAMA 234:620–621, 1975.

124. Wulff, CH, and Gilliatt, RW: F waves in patients with hand wasting caused by a cervical rib and band. Muscle Nerve 2:452–457, 1979.

125. Wurmser, P, and Kaeser, HE: Zur neuralgischen Amyotrophie. Schweiz Med Wochenschr 93:1393–1396, 1963.

126. Young, MR, and Norris, JW: Femoral neuropathy during anticoagulant therapy. Neurology 26:1173–1175, 1976.

Chapter 22

POLYNEUROPATHIES

1. INTRODUCTION
2. NEUROPATHIES ASSOCIATED WITH GENERAL MEDICAL CONDITIONS
   Diabetic Neuropathy
   Alcoholic Neuropathy
   Uremic Neuropathy
   Neuropathies in Malignant Conditions
   Neuropathies Associated With Paraproteinemia
   Necrotizing Angiopathy
   Sarcoid Neuropathy
   Others

3. INFLAMMATORY OR INFECTIOUS NEUROPATHIES
   Guillain-Barré Syndrome and Its Variants
   Diphtheric Neuropathy
   Leprosy
   Others

4. METABOLIC NEUROPATHIES
   Nutritional Neuropathies
   Toxic Neuropathies

5. INHERITED NEUROPATHIES
   Charcot-Marie-Tooth Disease (HMSN Types I and II)
   Dejerine-Sottas Disease—Hypertrophic Polyneuropathy (HMSN Type III)
   Refsum's Disease—Hereditary Ataxic Neuropathy (HMSN Type IV)
   Spinocerebellar Degeneration With Neuropathy (HMSN Type V)
   Friedreich's Ataxia
   Acute Intermittent Porphyria
   Pressure-Sensitive Hereditary Neuropathy
   Cerebral Lipidosis
   Hereditary Sensory Neuropathy
   Lipoprotein Neuropathies
   Giant Axonal Neuropathy
   Fabry's Disease
   Familial Amyloid Neuropathy
   Olivopontocerebellar Atrophy
   Others

1 INTRODUCTION

The triad of polyneuropathy consists of sensory changes in a glove-and-stocking distribution, distal weakness, and hyporeflexia. Certain types of neuropathy may show widespread sensory symptoms, and others may begin with more prominent proximal weakness. In general, normal muscle stretch reflexes speak against peripheral neuropathy, but not as an absolute rule. Anatomic localization of a lesion depends on clinical and electrodiagnostic evaluation, although patterns of peripheral nerve involvement can rarely indicate a specific diagnosis of a given disorder. In some patients with an unequivocal diagnosis of polyneuropathy, an extensive search may fail to uncover the exact cause.[262] In one study, an intensive evaluation permitted classification in 76 percent of 205 patients with initially undiagnosed neuropathy; 42 percent having the final diagnoses of inherited disorders; 21 percent, inflammatory demyelinating polyradiculoneuropathy; and 13 percent, neuropathies associated with systemic disorders.[124]

A detailed history often reveals general medical conditions such as diabetes, alcoholism, renal disease, malignancies, sarcoidosis, periarteritis nodosa, amyloidosis, and infectious processes such as diphtheria and leprosy. Inflammatory neuropathies include the Guillain-Barré syndrome and chronic inflammatory demyelinating neuropathy. Metabolic neuropathies result from nutritional deficiencies or toxic effects of drugs or chemicals. The family history provides essential information in establishing the type of inherited conditions associated with polyneuropathy. Sometimes a patient's own account may not provide sufficient information, necessitating independent examination of family members.

Nerve conduction studies and electromyography delineate the extent and distribution of the lesion, and differentiate two major pathologic changes in the nerve fibers (see Chapter 4): axonal degeneration and demyelination. Electrical studies are of limited value in distinguishing clinical types of neuropathies or establishing the exact cause in a given case. The specific diagnosis and therapy[166] depends heavily on clinical and histologic assessments. This chapter will deal with essential characteristics of peripheral neuropathies as they relate to electrophysiologic abnormalities. Interested readers should consult comprehensive reviews available elsewhere.[3,16,38,57,125]

2 NEUROPATHIES ASSOCIATED WITH GENERAL MEDICAL CONDITIONS

This category includes some of the most commonly encountered polyneuropathies. Despite their clear association with a general medical condition, the exact cause of neuropathy remains uncertain.

Diabetic Neuropathy

Diabetes often causes a symmetric polyneuropathy that probably has a metabolic basis. In one attractive hypothesis it is postulated that there is an increased amount of sorbitol in diabetic neural tissue.[147] In hyperglycemia, glucose is shunted through the sorbitol pathway, and the accumulation of sorbitol in Schwann cells may cause osmotic damage with segmental demyelination. In one study, the ulnar motor conduction velocity and F latency improved slightly but significantly after treatment with an aldose reductase inhibitor. This finding would support the sorbitol pathway hypothesis.[133] Other factors considered important in the pathogenesis include insulin deficiency and altered myoinositol metabolism.[394]

An alternative and perhaps more attractive theory suggests that small vessel disease leads to infarcts within the nerve,[309] causing asymmetric types of diabetic neuropathy and diabetic cranial mononeuropathies. The spatial distribution of fiber loss also suggests ischemia, similar to that found in experimental embolization of nerve capillaries.[115,187] Ischemic changes in the nerve presumably result from proliferation of the endothelium in blood vessels and abnormalities of the capillaries.

A wide spectrum of neuropathic processes develops in patients with diabetes mellitus. The most commonly used clinical classification[16] consists of (1) distal symmetric primarily sensory neuropathy, (2) autonomic neuropathy, (3) proximal asymmetric painful motor neuropathy, and (4) cranial mononeuropathies. Pathologic classification separates them into two groups: predominantly large-fiber and predominantly small-fiber diseases. In the large-fiber type,[24,79] segmental demyelination and remyelination predominate.[125] These changes, however, may be secondary to diffuse or multifocal axonal loss, which seems to constitute the primary abnormality.[116] This process would distort the normally linear relationship between internodal length and fiber diameter.[78] In the small-fiber type, it is more certain that the primary impact of the disease falls on the axons with secondary demyelination.[44]

A mixture of the large and small fiber types is common. Loss of both myelinated and unmyelinated fibers may occur, with Schwann cell damage and axonal degeneration proceeding independently.[24] Distal axonopathy in experimental diabetes mellitus of the rat first affects the terminal portions of the susceptible nerves.[45] In some patients, abnormalities in the autonomic nervous system closely parallel changes in the peripheral nervous system.[1,132] In these cases, prominent histologic changes include active axonal degeneration, affecting mainly unmyelinated and small myelinated fibers.[236]

The clinical presentation depends on varying combinations of the two basic types. On the whole, persons with adult-onset diabetes have the large-fiber type, with symptoms consisting of distal paresthesias and peripheral weakness. These patients have disassociated loss of vibratory, position, and two-point discrimination senses, with relative sparing of pain and temperature senses. Vulnerability at the common sites of compression may cause multiple pressure palsies.

The small-fiber type characteristically is common in those with insulin-dependent juvenile diabetes. Dysautonomia and pain are prominent features, as sug-gested by the name autonomic, or painful, diabetic neuropathy. The patient often awakes at night with painful dysesthesias. Charcot's joint, perforating ulcers, and other trophic changes of the feet may develop after severe loss of pain sensation. Impotence and postural hypotension result from involvement of the autonomic nerves. Acute painful neuropathy may follow precipitous weight loss, but severe symptoms subside within 10 months. Histologic studies show degeneration of both myelinated and unmyelinated axons.[10]

Mononeuropathies most often affect the femoral nerve and lumbosacral plexus, but also involve the sciatic, common peroneal, median, ulnar, and cranial nerves.[358] Unilateral femoral neuropathy is a common complication in elderly men with poorly controlled diabetes. Thigh pain precedes wasting of the quadriceps and other proximal muscles of the anterior thigh. Unlike the distal symptoms of diffuse polyneuropathy, the proximal weakness tends to improve with adequate control of the diabetes. This condition, though generally referred to as diabetic amyotrophy, probably represents a form of diabetic mononeuropathy, rather than a separate entity.[11,75] The sudden onset of pain may herald involvement of a major proximal nerve trunk, including the lateral cutaneous nerve of the calf.[140] Diabetic thoracic radiculopathy produces a distinct syndrome characterized by radicular involvement, abdominal or chest pain, weight loss, and relatively good prognosis.[199] This condition may mimic myelopathy.[385] Polyradiculoneuropathy and truncal mononeuropathy[130] may accompany advanced distal polyneuropathy.[352]

Electrophysiologic studies have revealed a number of different abnormalities in diabetic neuropathy.[114,243,244,283–285,287] These include slower nerve conduction velocities and reduced amplitude in diabetics with signs of neuropathy than in those without clinical signs.[153,158,185] In juvenile diabetics, patients with the longest duration of disease have the highest incidence of conduction abnormalities.[131] Studies show a close correlation between clinical signs of neuropathy and the degree of slowing in conduction of the peroneal,[157,165,219,273] sural, and medial dorsal cutaneous nerves.[182] Conduction abnormalities develop diffusely along the entire length of the nerve,[232] but more in distal than in proximal segments[83,206] (see Fig. 17–11). Tibial and peroneal nerves show more slowing than median and ulnar nerves. The disease may affect any nerve, including the phrenic nerve.[395] Distribution of conduction abnormalities suggests preferential involvement of the fastest conducting large myelinated fibers.[105]

Administration of gangliosides may facilitate the reappearance of sensory potentials and increase the amplitude of the compound muscle action potentials, presumably by improving the process of reinnervation.[20] In one study, correction of hyperglycemia resulted in a slight increase in conduction velocity after 6 hours.[367] In another series, nerve conduction velocity improved by 2.5 m/s after a year of glucoregulation with continuous subcutaneous insulin infusion.[126] Despite the encouraging reports, however, such therapy has generally yielded negative results. For example, a carefully controlled study revealed little improvement in conduction at the end of 3 days.[337]

Patients with diabetes have abnormal persistence of sensory evoked potentials during induced ischemia. This finding may herald other electrophysiologic abnormalities.[180] Electrophysiologic estimates of motor unit numbers reveal axonal dysfunction that parallels the severity of the demyelinative process.[168] Electromyography may show fibrillation potentials and positive sharp waves in patients with prominent axonal degeneration. Studies of spinal somatosensory evoked potentials have shown impairment of peripheral as well as central afferent transmission.[86,162] Increased interpeak latencies of the brainstem auditory evoked responses also suggest the presence of a central neuropathy in some cases[103] but not in others.[380]

Alcoholic Neuropathy

Alcohol is one of the most common causes of peripheral neuropathy in the United States. It primarily affects those drinking a large quantity for a number of years, and improves once the patients abstain.[178] In addition to a possible toxic effect of the alcohol itself, dietary insufficiency and impaired absorption may play important roles. Indeed, vitamin B_1 or thiamine deficiency alone can cause similar clinical findings. The pathologic changes include reduced density of large and small myelinated fibers and acute axonal degeneration and regeneration.[22,383] Secondary paranodal demyelination may involve the most distal segment.[251]

Clinical symptoms usually appear insidiously over weeks or months, but sometimes more acutely over a period of a few days. The initial sensory complaints consist of distal pain, paresthesias and dysesthesias, first in the legs and later in the arms. Burning sensations in the extremities resemble those seen in the neuropathy of diabetes. Trophic changes such as plantar ulcers develop when patients subject insensitive tissues to unusual amounts of trauma.[325] More advanced cases involve bilateral footdrop. Distal muscular atrophy affects the extensors more than the flexors. Neuropathic, rather than myopathic, changes predominate in chronically weak and atrophic muscles.[220] Sensory symptoms may respond to daily administration of vitamin B_1, but muscular atrophy tends to persist despite therapy.

Electrophysiologic evaluations demonstrate impaired function of small-caliber motor fibers and large cutaneous sensory fibers.[28] Despite traditional emphasis on the role of conduction velocity, early abnormalities consist of decreased amplitude of sensory nerve and compound muscle action potentials. As in other axonal neuropathies, conduction velocity is slowed in proportion to the loss of evoked sensory and motor responses.[18] The disease usually affects sensory nerve conduction more than motor conduction, with the degree of slowing indicating approximate severity of the polyneuropathy.[249]

Initially, nerve conduction studies reveal either normal or only slightly reduced velocities in most patients.[22,62,63,383] Conduction abnormalities may involve not only the distal segment but also the proximal segment of the nerve.[154] Assessments of sural nerve and late responses improve the diagnostic yield.[92] Electromyography reveals fibrillation potentials and other neuropathic changes. Usually abnormalities involve the lower extremities earlier and more prominently than the upper limbs.

Chronic alcoholics may also have abnormal visual evoked responses[68] and brainstem auditory evoked responses.[69]

Uremic Neuropathy

A variety of neuropathies result from complex effect of renal failure on peripheral neurons, myelin and Schwann cells.[326] The responsible toxins are still unidentified but parathyroid hormone, myoinosital and middle molecules are elevated in uremia. Their effects on the peripheral nervous system await further clarification.[29] Uremic neuropathy often develops in patients with severe chronic renal failure or in patients undergoing chronic hemodialysis. The use of neurotoxic drugs such as nitrofurantoin can contribute to the nerve damage. Histologic findings vary, but the most consistent abnormality is axonal degeneration with secondary segmental demyelination and, less frequently, segmental remyelination.[17,113,169,326,359]

Clinical symptoms of neuropathy usually develop abruptly, with a sudden rise in vibratory threshold as one of the early signs.[280] The lower extremities tend to show earlier and more prominent disturbances than the upper extremities. Some patients have "restless legs" as a presenting symptom.[359] After successful treatment with hemodialysis, vibratory perception returns to normal, followed by improvement in other clinical findings. A distal ischemic neuropathy has developed after the placement of bovine arteriovenous shunts for chronic hemodialysis.[30] Proximal muscle weakness may also appear in uremic patients receiving hemodialysis.[222]

Electrophysiologic findings generally, but not exactly, correlate with the clinical signs, levels of serum creatinine, or pathologic changes of the peripheral nerve.[375,391] Mild electric abnormalities sometimes herald clinical manifestations. Conduction velocities decrease with the deterioration of signs and symptoms and increase with improvement after dialysis or kidney transplant.[55,91,280,281,290,381] Patients with severe renal insufficiency often have motor and sensory conduction abnormalities in all limbs with greater deficits in the peroneal than in the median nerve.[279] As a sensitive indicator of neuropathy, facial nerve latency may rival conduction studies of the peroneal, median, and ulnar nerves.[269] Studies of late responses and sural nerve conduction also reveal a high degree of abnormality.[2] In acute renal failure, the muscle action potential may show a marked reversible reduction in amplitude, presumably as the result of conduction block.[50] In chronic renal failure, such diminution in size of the compound potentials usually signals axonal degeneration, evidenced by fibrillation potentials in electromyography.

Most uremic patients show abnormalities in the pattern shift visual evoked potentials and somatosensory evoked potentials.[322]

Neuropathies in Malignant Conditions

Malignant processes affect the peripheral nerve directly or indirectly. Lymphomas and leukemias may invade or infiltrate through hematogenous spread,[213] whereas nonlymphomatous solid tumors may cause external compression. Occasionally a metastasis may involve the dorsal root ganglia.[186] Neuropathies result from distant effects of lymphoma,[332] bronchogenic carcinoma,[176,357] or, less commonly, tumors of the ovary, testes, penis, stomach, or oral cavity.[299] Pathologic features include (1) neuronal degeneration with secondary peripheral or central axonal changes; (2) demyelination reminiscent of acute or chronic idiopathic polyneuritis;[230] (3) microvasculitis with active wallerian degeneration, causing mononeuritis multiplex;[188] and possibly (4) an opportunistic neuropathic infection.[332]

Remote malignancies usually affect the dorsal root ganglia, but also occasionally the anterior horn cells. Patients have clinical findings of sensory or motor deficits, although mixed involvement is most common. Sensory motor neuropathy represents a group of heterogeneous conditions with overlapping clinical and histologic features.[16] Less common abnormalities include a pure motor neuropathy mimicking the myasthenic syndrome and a polyradiculopathy seen in meningeal carcinomatosis. Systemic cancer may initially cause symptoms of mental neuropathy, such as a numb chin.[246]

Subacute sensory neuropathy of oat cell carcinoma may result in severe sensory loss secondary to dorsal root ganglionitis.[181] In one case, morphometric studies at autopsy showed preferential loss of large-diameter sensory nerve cell bodies, marked loss of large myelinated fibers in the dorsal root and sural nerve, and almost total loss of myelinated fibers in the fasciculus gracilis.[292] The same type of progressive sensory neuropathy, if seen without evidence of cancer,[194] is called chronic idiopathic ataxic neuropathy.[89]

In approximately one third of patients with malignancies clinically latent neuropathies develop.[299] Patients with lung cancer have a slightly higher incidence.[56]

Conduction studies reveal only a mild slowing of sensory[257] or motor fibers,[366] or both,[56] with substantial reduction in amplitude of sensory nerve and muscle action potentials. Electromyography typically shows fibrillation potentials and high-amplitude, long-duration motor unit potentials in atrophic muscles.[299] Small, short-duration polyphasic motor unit potentials occasionally seen in wasted proximal muscles probably result from neuropathic abnormalities of the intermuscular axonal twigs.[19]

Neuropathies Associated With Paraproteinemia

A number of studies have demonstrated clear association between monoclonal proteins and peripheral neuropathy.[53,263,364] Most of these cases involve benign monoclonal gammopathy. Other occasionally encountered syndromes include primary systemic amyloidosis and osteosclerotic myeloma and, less frequently, osteolytic multiple myeloma, Waldenstrom's macroglobulinemia, and gamma heavy chain disease. Table 22–1 summarizes main clinical and laboratory features of these entities. In multiple myeloma and macroglobulinemia, neuropathy may develop as a feature of the

Table 22–1 MAIN FEATURES OF MONOCLONAL PROTEIN-PERIPHERAL NEUROPATHY SYNDROMES

Type of PN	Topography	Weakness	Sensory Loss	Autonomic Loss	Course	CSF Protein	MNCV	Pathology
Benign monoclonal gammopathy (IgG, IgA)	Distal, rarely proximal	++	++	+	Chronic progressive	++	Mild slowing	SD + AD
Benign monoclonal gammopathy (IgM)	Distal, symmetric	++	++	0	Chronic progressive	++	Very slow	SD
Amyloidosis, light chain type	Distal, symmetric	+/++	+++	++	Chronic progressive	+	Mild slowing	AD
Osteosclerotic myeloma	Distal, symmetric	+++	+	0	Chronic progressive	+++	Very slow	SD (+AD)
Waldenstrom's macroglobulinemia	Distal, symmetric	++	++	0	Chronic progressive	++	Very slow	SD (occ'l AD)

MNCV, motor nerve conduction velocity; AD, axonal degeneration; SD, segmental demyelination.
From Kelly,[196] with permission.

underlying disorder or as the result of paraproteins.[311,354,373]

Benign monoclonal gammopathy occurs in 10 percent of all patients with idiopathic peripheral neuropathy.[196] Conversely, in 30 to 70 percent of those with benign monoclonal gammopathy chronic sensory motor neuropathy develops.[190,192,294] The clinical features closely mimic those of chronic inflammatory polyradiculoneuropathy with progressive sensorimotor loss.[117] Plasma exchange rapidly lowers the level of monoclonal antibody, with some recovery of motor function.[338] The characteristic laboratory findings include IgM and less commonly IgG or IgA gammopathy and high cerebrospinal fluid protein. Electrophysiologic and morphologic studies show evidence of demyelination, although a minority has axonal loss as a major finding.[344]

Primary systemic amyloidosis or light-chain type of amyloidosis affects multiple organ systems with symptoms similar to those seen in malignancy or collagen vascular disease. Patients with amyloidosis, however, have plasma cell dyscrasias and amyloidogenic immunoglobulins.[142,155,368] Amyloid accumulates in the flexor retinaculum, causing the carpal tunnel syndrome. Diffuse peripheral neuropathy develops as the result of metabolic or ischemic changes or direct infiltration by amyloid.[81,196,216] The clinical features consist of painful sensory and motor neuropathy with prominent autonomic dysfunction, affecting multiple systems. Axonal degeneration predominates in small myelinated and unmyelinated fibers.[361] Thus, it involves pain, temperature, and autonomic fibers but usually spares the larger motor and proprioceptive fibers. This accounts for the typical dissociated sensory loss with predilection for pain and temperature sense and relative sparing of vibratory and position sense, that is, the reverse of the findings seen in large-fiber diabetic neuropathy.

Electrophysiologic abnormalities include slight slowing of the motor nerve conduction velocity, with mild reduction in amplitude of the compound muscle action potential; absence of the sensory nerve action potentials, with distal stimulation of the ulnar, median, or sural nerve; and evidence of superimposed compression of the median nerve at the wrist. Electromyography reveals evidence of denervation diffusely, but more conspicuously, in the distal muscles of the leg.[198] An in-vitro study of sural nerve compound action potentials has shown a selective reduction of C and A delta potentials in familial amyloid neuropathy.[120] These findings support the view that amyloid neuropathy predominantly causes distal axonal damage first in the sensory and then in the motor fibers.[327]

Skeletal osteosclerotic lesions, though seen in less than 3 percent of myeloma patients as a whole, develop in at least 50 percent of those with myeloma neuropathy.[197] This type of myeloma commonly affects younger patients and takes a benign clinical course that resembles chronic inflammatory demyelinating polyradiculoneuropathy.[104] Electrophysiologic and histologic evidence of prominent demyelination suggests an immunologic effect of the monoclonal protein on a myelin antigen as a precipitating cause. Intraneural injection of patient serum into rat sciatic nerve, however, does not produce a demyelinative lesion.[197] Instead, the morphologic features suggest axonal attenuation or distal axonal degeneration with secondary demyelination,[291] similar to the findings seen in uremic neuropathy or hereditary sensorimotor neuropathy type I.[109]

Patients with osteolytic multiple myeloma may have amyloid neuropathy much like the type seen in systemic amyloidosis without multiple myeloma. These cases show, in addition to prominent distal axonal loss and the carpal tunnel syndrome, atypical features such as radiculopathy and mononeuritis multiplex. In this condition a peripheral neuropathy also develops without amyloidosis in 30 to 40 percent of cases, based on electrophysiologic and histologic findings.[66,196] Diverse clinical and electrophysiologic features resemble various subgroups of carcinomatous peripheral neuropathy. The sensorimotor type shows distal axonal degeneration and mild decrease in nerve conduction velocity. Patients with primary sensory involvement character-istically have a loss of proprioception but little deficit in the motor system.[195] Those with primary motor abnormalities have features similar to those of chronic idio-pathic demyelinating polyradiculoneuro-pathy, with prominent slowing of nerve conduction velocities.

Patients with Waldenström's macro-globulinemia may have a primarily de-myelinating sensorimotor neuropathy of the type commonly associated with benign monoclonal gammopathy. Oc-casional cases, however, show axonal degeneration and amyloid infiltration, as in osteolytic multiple myeloma. Electrophysiologic studies typically re-veal predominant segmental demyelina-tion, but some patients also have the evi-dence of axonal changes as a major finding.[382] Other rare disorders with monoclonal proteins and peripheral neu-ropathy include gamma heavy chain disease and systemic malignancies such as lymphoma, leukemias, benign lymph node hyperplasia, and cryoglo-bulinemia.[66]

Necrotizing Angiopathy

In this disorder, probably related to au-toimmune hypersensitivity, the inflam-matory process involves the small and medium-sized arteries in multiple organ systems, including thoracic and abdomi-nal viscera, joints, muscle, and the ner-vous system. Necrosis of the media gives rise to small aneurysms and thrombosis of the vessels, with palpable nodules along the affected arteries.

The clinical symptoms and signs, which may appear either abruptly or in-sidiously, consist of malaise, fever, sweating, tachycardia, and abdominal and joint pain. In approximately one half of the patients, neuronal disturbances such as diffuse polyneuropathy and mononeuritis multiplex develop. Neurop-athy presumably results from ischemia, caused by thrombosis of the nutrient ar-teries, which are heavily infiltrated with inflammatory cells. The disease may abate spontaneously, despite a generally poor prognosis, with survival of only a

few months to a few years after the onset of clinical symptoms. In one series, 10 of 16 patients had features of mononeuritis multiplex, whereas the remaining 6 had a distal symmetric sensory motor polyneu-ropathy.[208] The conduction velocity is slow in proportion to the reduced ampli-tude of the compound muscle and sen-sory potentials in the affected limbs. Electromyography reveals spontaneous activities in atrophic muscles, as ex-pected in acute or subacute axonal neuropathy.[37]

Sarcoid Neuropathy

Patients with sarcoidosis develop distal sensory-motor polyneuropathy as a rare complication.[100,278] Histologic studies reveal granulomas or inflammatory changes in the epineural and perineural spaces, which lead to periangiitis, pan-angiitis, and axonal degeneration.[289]

Electrophysiologic abnormalities in-clude reduced amplitude of the compound sensory and muscle action potentials and mild slowing in conduction studies[67,289] and prominent fibrillation potentials and positive sharp waves in electromyogra-phy. In one case, morphologic studies confirmed the electrodiagnostic impres-sion of an acute axonal and demyelinat-ing neuropathy.[278]

Others

Peripheral neuropathy accompanies some system atrophies such as Shy-Drager syndrome[148] and the syndrome of skin pigmentation, edema, and hepato-splenomegaly.[356,370] Patients with hypo-thyroidism may have sensory and motor conduction abnormalities diffusely[305] or localized to sites of nerve compression.[333] In systemic lupus erythematosus a pre-dominantly motor or sensory demyelinat-ing polyneuropathy may develop as a presenting feature.[253,267,312] In Sjögren's syndrome, subacute sensory neuropathy primarily affects the distribution of the trigeminal nerve.[242]

The neuropathy associated with poly-cythemia vera involves both large and small myelinated fibers with mild slowing of both motor and sensory conduction.[397] Distal axonal degeneration follows ischemia produced by thromboembolic occlusion of a major proximal limb artery.[389] Multiple sclerosis occasionally accompanies hypertrophic demyelinating neuropathy, with typical nerve conduction changes.[316,321] Critically ill patients may have a severe motor and sensory polyneuropathy of unknown cause.[31] Limb compression during unattended coma may also cause multiple peripheral nerve injuries. The unique combination of swollen limbs, pressure blisters, and myoglobinuria may be associated with the severe axonal loss that results from the compartment syndrome.[340]

3 INFLAMMATORY OR INFECTIOUS NEUROPATHIES

Nerve conduction studies often help establish the diagnosis of the most common disorder in this category, the Guillain-Barré syndrome. Diphtheria and leprosy are now seldom seen in the United States. The acquired immunodeficiency syndrome (AIDS) can give rise to acute inflammatory axonal or demyelinating neuropathy, clinically indistinguishable from the Guillain-Barré syndrome.

Guillain-Barré Syndrome and Its Variants

Though of unknown etiology, the Guillain-Barré syndrome closely resemble experimental allergic neuropathy caused by the injection of Freund's adjuvant with extracts of peripheral nerve. Some patients with this syndrome have human immunodeficiency virus (HIV or HTLV III) infection.[84] Others have demyelinating neuropathy associated with hepatitis B virus infection. In most, however, repeated attempts have failed to isolate infectious agents. These findings support an autoimmune pathogenesis, rather than direct invasion of the nerve by infectious agents.[235] An inflammatory de-

myelinative neuritis affects all levels of the peripheral nervous system, occasionally with retrograde degeneration in the motor cells of the spinal cord or brainstem. In mild cases, pathologic changes may consist only of slight edema of the nerves or roots. The segment of maximal involvement varies from one patient to the next.[13] This helps explain the diversity of clinical findings and of conduction abnormalities in different patients. A substantial proportion of the patients with an initial diagnosis of Guillain-Barré syndrome may be found to have a neuropathy with another cause, particularly heavy metal intoxication.[137]

The clinical findings vary, but certain diagnostic criteria have emerged.[14] In about two thirds of the cases, neurologic symptoms follow a mild, transient infectious process of either the respiratory or, less commonly, the gastrointestinal system. Some patients seem to have other precipitating events.[134] The first symptoms of neuropathy usually appear in about 1 to 2 weeks, when the infection is no longer present. Rarely, the disease takes the form of encephalomyeloradiculoneuropathy, with progressive central nervous system disease, bilateral deafness, and severe sensory motor neuropathy.[387] Weakness initially involves the lower extremities but may rapidly progress to the upper extremities and cranial nerves within a few days. Paralysis of proximal muscles and facial diplegia stand in contrast to the distal weakness seen in other forms of neuropathy. Respiratory problems develop in approximately one half of patients.[320] Some patients have an acute, severe, and prolonged initial illness with quadriplegia in 2 to 5 days, requiring mechanical ventilation over 2 months.[319]

Other early signs include diminished or lost muscle stretch reflexes and minimal sensory loss, despite painful distal paresthesias. Careful testing usually reveals deficiency in vibratory sense, two-point discrimination, and pain perception. Some patients have transient elevation of blood pressure and heart rate as a result of sympathetic hyperactivity.[135] The autonomic dysfunction probably results from axonal degeneration of the vagus and splanchnic nerves, as seen in experimental allergic neuritis.[261,372] The spinal fluid typically contains a high protein level and no cells, although some lymphocytes may be present.

The disease follows an acute or subacute course, with progression up to 6 weeks after onset. The symptoms and signs then plateau for a variable period of time before showing gradual improvement. Some patients improve dramatically after corticosteroid therapy[288] or plasma exchange,[15,143,336] but beneficial effects are not universal.[265] The time course of recovery depends on the extent of demyelination and, perhaps more importantly, axonal degeneration. Some cases have severe axonal loss without inflammation or demyelination.[136] These patients may not regain motor function for 1 to 2 years.

Apart from the slow recovery mentioned above, the disease may continue to worsen beyond 6 weeks, with persistent evidence of ongoing demyelination. This variety, referred to as chronic inflammatory polyradiculopathy, or chronic relapsing polyneuropathy, may even follow a progressive course over several years, with severe disability.[110,308] Patients with this syndrome may have immunoglobulin and complement deposits in the nerve.[90] There is an increased risk of relapse in pregnancy.[254] In some patients chronic asymmetric sensorimotor neuropathy develops, with multifocal conduction delay and persistent conduction block.[226] The clinical features in children may mimic a genetically determined disorder.[343]

Prednisone causes a small but statistically significant improvement over no treatment.[123] Plasma exchange provides useful therapy, especially in cases with features of demyelination, rather than axonal degeneration.[306] Successful treatment with plasma exchange suggests pathogenic humoral factors. Systemic passive transfer of immunoglobulin, in fact, causes demyelinative disease in monkeys, with substantial reduction of conduction velocity.[177]

Chronic demyelinating neuropathy may be associated with a relapsing multifocal central nervous system disorder, the clinical features of which resemble multiple sclerosis.[362] Electrophysiologic studies reveal slowing of peripheral conduction velocity as well as increased central conduction time. These findings suggest the occurrence of combined peripheral and central demyelination like that in chronic relapsing experimental allergic encephalomyelitis and neuritis.

A variant of the Guillain-Barré syndrome originally described by Fisher[141] consists of ataxia of gait, absence of the muscle stretch reflex, and ophthalmoplegia. The findings in one series included normal distal motor nerve conduction velocities, F-wave latencies, and blink reflex and abnormal sensory action potentials.[331] Serial studies in such a case, however, showed a time course of conduction changes identical to those found in the Guillain-Barré syndrome.[184] Electromyography usually reveals only slight abnormalities in the limbs and evidence of facial denervation in this variant. An atypical patient with this syndrome had abnormal pupils and normal eye movements.[390]

In advanced stages nerve conduction studies usually show reduction in velocity by more than 30 to 40 percent from the normal mean value and abnormal temporal dispersion of the compound muscle action potential (see Fig. 5–8). In milder forms, studies may reveal less dramatic changes, because initial weakness commonly results from proximal conduction block without distal abnormalities.[46,47] Indeed, 15 to 20 percent of cases have entirely normal nerve conduction studies distally during the first 1 to 2 weeks.[128,205] Thus, the absence of clear conduction abnormalities by no means precludes the diagnosis.[218]

The initial absence and later delay of the F wave with normal distal conduction (see Fig. 17–6 and Table 17–4) suggest an increased vulnerability of the most proximal, possibly radicular portions of the motor fibers, with little change along the main nerve trunk at the onset of illness.[203,205,207,258] As in any neuropathy, the early changes may also selectively involve the common sites of nerve compression[46,47,218] and the most terminal segment, presumably reflecting the longest distance from the cell body.

Despite the clinical picture of predominantly motor involvement, sensory conduction studies usually show distinct abnormalities of the median and ulnar nerves. Interestingly, the disease tends to spare the sural nerve sensory action potential, often regarded as one of the first affected in other neuropathies.[275] Phrenic nerve conduction time may provide a sensitive measure in predicting impending ventilatory failure.[156] Studies of the blink reflex frequently reveal conduction abnormalities, confirming clinical facial palsy (see Figs. 16–10 and 16–12, and Tables 16–2 and 16–4). Somatosensory evoked potential to median nerve stimulation may demonstrate a proximal conduction delay between Erb's point and the cervical cord in patients with normal sensory conduction distal to Erb's point during the early stages.[46,47,384]

Facial or limb myokymic discharges (see Fig. 13–9B) may appear early in the course of the disease and persist.[247] Otherwise, electromyography usually shows only a reduced interference pattern, indicating neurapraxia without axonal degeneration. Some patients with typical clinical features, however, may have a primarily axonal neuropathy and prominent denervation first detectable 2 to 3 weeks after onset.[136,258] In these cases, reduction in amplitude of compound muscle action potentials with distal stimulation implies a poor prognosis despite the relative normality of calculated conduction velocities.[159] Here, functional recovery depends on axonal regeneration, which takes considerably longer than remyelination. In contrast, proximal conduction block without evidence of axonotmesis generally suggests a good prognosis.

Sequential conduction studies show great variability among different patients and even from one nerve to another in the same patient.[204] After treatment, nerve conduction velocities may or may not revert toward normal values.[336,363] In fact, the nerve conduction velocity may become slower while the patient begins to improve, demonstrating again the lack of strong correlation between clinical symptoms and conduction velocities.[175]

Diphtheric Neuropathy

Prophylactic immunization and early use of immune sera and antibiotics in infected cases have drastically lowered the incidence of diphtheric polyneuropathy in the United States; but outbreaks, though rare, still occur. The exotoxin of Corynebacterium diphtheriae becomes fixed to the nerve and produces segmental demyelination after several weeks. Local paralysis of the palatal muscles may immediately follow an infection of the throat. Neuropathy may also develop in adults after contracting cutaneous diphtheria, which still prevails in the tropics.

The clinical signs resemble those seen in the Guillain-Barré syndrome. The symptoms typically develop 2 to 4 weeks after the initial infection. Patients have a high incidence of lower cranial nerve dysfunction, most notably palatal and pharyngolaryngoesophageal weakness. Blurring of vision results from paralysis of accommodation. The involvement of sensory and motor nerves of the extremities causes paresthesias and weakness of the affected limbs. A rapidly descending paralysis may lead to respiratory problems. The primary pathologic change consists of segmental demyelination involving the sensory and motor fibers. Conduction abnormalities usually begin a few weeks after the onset of neurologic symptoms and peak after clinical recovery has already begun.[214]

Leprosy

An acid-fast bacillus, Mycobacterium leprae, transmits this chronic infectious disease by close and prolonged contact. Although rare in the United States, the disease still prevails in Africa, India, and South and Central America. The organism seems to have predilection toward the great auricular, ulnar, radial, peroneal, facial, and trigeminal nerves. Of the two clinical forms, the lepromatous or neural type causes extensive and widespread granulomatous infiltration of the skin, resulting in the characteristic disfiguration. The diffuse sensory neuropathy seen in this variety results from direct invasion of the nerve trunks by the bacillus. The thickened perineurium, by an overgrowth of connective tissue, compresses the myelin sheath and the axons. In the other type, called the tuberculoid form, more focal involvement of the skin causes patches of the depigmented, maculoanesthetic areas. Here, swelling of the nerves does not necessarily imply direct invasion by the organisms. The two types of clinical presentation commonly overlap without clear separation, giving rise to an intermediate or mixed form.

Clinical features suggest mononeuritis multiplex or slowly progressive diffuse polyneuropathy. Common manifestations include facial palsy involving the upper half of the face, wristdrop, footdrop, and clawhand. Neural leprosy may begin with a small erythematous macule, which soon enlarges, forming anesthetic depigmented areas. The loss of pain and temperature sensation result in ulcerated necrosis of the skin. Palpation of the affected nerve reveals characteristic fusiform swelling caused by an infective granulomatous process. The denervated muscle shows histopathologic findings of fascicular atrophy and inflammatory nodules.[335]

Electrophysiologic abnormalities consist of moderately to markedly slowed motor and sensory conduction not only across enlarged segments[240,355] but also along the unpalpable portions.[260] In one study, sensory nerve conduction studies of the radial nerve showed the best correlation with the clinical findings.[334] Electromyography reveals denervation in the atrophic muscles.[9,101,335]

Others

In paralytic rabies, vascular and inflammatory changes predominate in the central nervous system, but peripheral nerves show segmental demyelination, remyelination, and wallerian degeneration with variable axonal loss.[76] Electrophysiologic abnormalities include slowing of motor and, to a lesser extent,

sensory conduction velocity, a reduced number of motor unit potentials, and fibrillation potentials.[353]

The neuropathy associated with the hypereosinophilic syndrome develops at the onset of marked eosinophilia. It affects both the sensory and motor fibers, with multifocal conduction abnormalities and evidence of severe axonal degeneration.[106]

Migrant sensory neuritis of Wartenberg takes a benign relapsing and remitting course. Movement of the limbs induces a stretch, leading to pain and subsequent loss of sensation in the distribution of individual cutaneous nerves. Stimulation of the affected nerves may elicit small or no sensory action potentials.[248]

A severe sensory motor neuropathy may develop in patients with rheumatoid arthritis or Sjögren's syndrome. Electrophysiologic and sural nerve biopsy studies reveal an axonal neuropathy in these cases.[303]

In AIDS, cytomegalovirus infection causes a severe motor polyradiculopathy by selective destruction of the motor neurons of ventral spinal roots and motor cranial nerves.[21,127] In these patients, electromyography shows severe diffuse denervation but only mildly slowed nerve conduction velocities. Multifocal or distal symmetric inflammatory demyelinating neuropathy may herald the onset of AIDS in some homosexual men with lymphadenopathy.[229]

Meningococcal septicemia may cause a mixed sensory motor neuropathy with predominant loss of axons. Mononeuritis multiplex results from disseminated intravascular coagulation and multiple vascular occlusions.[318] Electrophysiologic studies reveal findings consistent with axonal neuropathy.

4 METABOLIC NEUROPATHIES

Metabolic neuropathies consist of two groups, those associated with nutritional disturbances and those resulting from toxic causes. Neuropathies attributable to a specific nutritional deficiency include beriberi, pellagra, and pernicious anemia. Toxic neuropathies develop after the administration of various drugs or the exposure to chemical substances such as lead or arsenic. Many neuropathies associated with general medical conditions also belong to this broad category.

Nutritional Neuropathies

Children who have insufficient intake of protein or calorie suffer from retarded myelination or segmental demyelination.[77] They have abnormalities of motor and sensory nerve conduction related to the severity of the malnutrition. Severe malabsorption from the blind loop syndrome also causes vitamin E deficiency.[42,386] Alcoholic and paraneoplastic neuropathies result, at least in part, from inadequate food and vitamin intake, although some toxins may also interfere with the metabolism of the nerves. In primary biliary cirrhosis, a sensory neuropathy develops from poor nutrition, xanthomatous infiltrates, or immunologic abnormalities.[71]

Diets deficient in vitamins and other nutritional factors play a major role in the polyneuropathy associated with beriberi, pellagra, pernicious anemia, dysentery, and cachexia.[94] Beriberi or thiamine deficiency causes signs and symptoms similar to those seen in alcoholic polyneuropathy.[179] They consist of pain, paresthesias, sensory loss, and weakness, all affecting the distal segments, and the absence of the stretch reflexes. Similar neuropathy may develop during intended weight reduction.[346] Histologic studies reveal conspicuous axonal degeneration and less prominent demyelination. Pellagra is another deficiency disease involving the vitamin B_1 complex. The condition often affects malnourished patients with chronic alcoholism. The clinical features consist of gastrointestinal symptoms, skin eruptions, and disorders of the peripheral and central nervous systems. Neuropathic characteristics include paresthesias, loss of distal sensation, tenderness of the nerve trunks, hyporeflexia, and mild paralysis.

Pernicious anemia results from deficiency of intrinsic factors in the gastrointestinal secretion that mediate absorption of vitamin B_{12}. As indicated by its alternative name, combined system disease, pathologic changes primarily involve the dorsal and lateral funiculi of the spinal cord. The peripheral nerves also show fragmentation of myelin sheaths and degeneration of axons.[212] The presenting clinical symptoms consist of paresthesias, dysesthesias, and loss of vibration and position sense. The patients commonly have spastic or flaccid paralysis of the lower extremities, with increased stretch reflexes during the early stages, and areflexia as the disease progresses. Untreated patients have reduced conduction velocity, in part because of thiamine deficiency seen in the majority of cases.[85] Patients with prominent axonal degeneration have diffuse spontaneous discharges in electromyography but nearly normal motor nerve conduction velocity.[212] Appropriate treatment arrests the progression of the neuropathy, but residual neurologic abnormalities persist.[252]

Toxic Neuropathies

Based on the presumed site of cellular involvement, toxic neuropathies may be divided into three groups: (1) neuropathy affecting the cell body, especially those of dorsal root ganglion; (2) myelinopathy or Schwannopathy with primary segmental demyelination; and (3) distal axonopathy causing dying-back axonal degeneration. Of these, the first two types include rare acute sensory neuropathy following antibiotic treatment[350] and segmental demyelination by perhexiline maleate used for therapy of angina pectoris.[35] Administration of diphtheria toxin[255] or tetanus toxoid[313] and chronic exposure to lead may also cause myelinopathy. Distal axonopathy, described below, is the most common form of toxic neuropathy.

A variety of drugs and industrial chemicals cause distal axonopathy.[393] The drugs known to be neurotoxic include amiodarone,[70,144,183,245,301] chloramphenicol, cisplatin,[317] dapsone,[211] diphenylhydantoin,[129,310] disulfiram[8,271,293] gold,[193] isoniazid, lithium,[49] metronidazol,[39] misonidazole,[264] nitrofurantoin, nitrous oxide,[221,323] perhexiline maleate,[35,324] phenytoin,[341] pyridoxine,[298,392] thalidomide,[146] and vincristine.[40,61] Some drugs show a characteristic pattern of neuropathic involvement. For example, vincristine causes primarily motor neuropathy, whereas pyridoxine abuse leads to a pure sensory central-peripheral distal axonopathy.[298] Cisplatin, used to treat malignant tumors, induces an axonopathy that bears a great resemblance to the sensory neuropathy sometimes associated with such a neoplasm.

Industrial chemicals causing toxic axonal neuropathy include acrylamide,[272,329] carbon disulfide, inorganic mercury,[4] methyl n-butyl ketone,[6,348] organophosphate parathion,[97,233,378] polychlorinated biphenyl,[72] thallium,[95] triorthocresyl phosphate,[377] and vinyl chloride.[302] In toxic axonal neuropathy, fibers of large diameter are initially affected in the distal segments and subsequently show progression proximally toward the cell body. The pathologic process then spreads to axons of small diameter.

The sudden development of clinical symptoms in distal axonopathy reflects the acuteness of intoxication. In contrast, an insidious onset suggests a chronic low-level exposure. Toxins often affect the nerves of the lower extremities initially because the longer axons are more vulnerable. Patients have distal weakness, hypesthesia, or paresthesia in a glove-and-stocking distribution, and reduced ankle stretch reflexes as early signs. The removal of the neurotoxin leads to a gradual recovery, but the axons, once degenerated, regenerate slowly over months to years, with incomplete return of function. The selection of proper electrophysiologic tests depends largely on the nature of the condition under study.[223,224] A few specific toxins, such as perhexiline maleate, cause demyelination, as evidenced by marked slowing of motor nerve conduction velocity.[35] Most other toxins lead to axonal loss, resulting in reduced amplitude of the compound nerve and muscle action potentials. In these cases, substantial de-

generation of large, fast conducting fibers account for a slight increase in distal latency and a decrease in conduction velocity. Electromyography shows fibrillation potentials and positive sharp waves.

Lead and arsenic, two specific agents responsible for distal axonal neuropathies, merit further attention. The general features of lead poisoning include abdominal cramps, encephalopathy, and the occasional appearance of a blue lead line along the gingival border. The laboratory tests reveal the presence of basophilic stippling of erythrocytes and elevated lead levels. Neuropathy occurs primarily in adults occupationally exposed to lead or following accidental ingestation of contaminated food, but may also affect children with known plumbism or pica.[138] Predominant involvement of motor fibers innervating the extensor muscles of the upper extremities produces bilateral radial nerve palsies without sensory loss. The removal of the toxin causes recovery to take place over a period of several months. Lead produces segmental demyelination in some animal species, possibly because extravasated lead in the interstitial fluid injures the Schwann cells directly.[276] This type of pathologic change does not necessarily characterize the neuropathy seen in human cases.[52] A group of workers exposed to lead had temporally dispersed compound muscle action potentials but normal maximal conduction velocity.[64]

Arsenic poisoning usually results from accidental ingestion of rat poison or exposure to industrial sprays.[139] Polyneuropathy develops several weeks after acute poisoning or more slowly with chronic low-level exposure. Pale, transverse bands bearing the eponym of Mee's lines appear parallel to the lunula in all finger and toe nails about 4 to 6 weeks after ingestion. In one study, serial examination revealed maximal sensory and motor loss within 4 weeks of the estimated time of exposure, and only partial improvement 2 years after the onset of illness.[274] Arsenic is found in the urine during acute exposure and in the hair and nails later. These clinical features resemble those of alcoholic neuropathy, with early loss of stretch reflexes, and painful paresthesias and sensory loss in a glove-and-stocking distribution. Flaccid paralysis may develop later, beginning in the lower extremities, but eventually affecting the upper extremities.

Electrophysiologic studies show progressive slowing of motor conduction velocity[274] and evidence of denervation in electromyography. Timely removal of toxin leads to nearly complete recovery of conduction abnormalities.

5 INHERITED NEUROPATHIES

The seven types[109] of hereditary motor and sensory neuropathy (HMSN) comprise the hypertrophic (type I) and neuronal (type II) varieties of Charcot-Marie-Tooth disease, Dejerine-Sottas disease (type III), Refsum's disease (type IV), and neuropathies associated with spinocerebellar degeneration (type V), optic atrophy (type VI), and retinitis pigmentosa (type VII). Other inherited polyneuropathies include Friedreich's ataxia, acute intermittent porphyria, pressure sensitive hereditary neuropathy, cerebral lipidosis, hereditary sensory neuropathy, lipoprotein neuropathies, giant axonal neuropathies, Fabry's disease, and familial amyloid neuropathy.

Patients with familial demyelinative neuropathy characteristically have uniform conduction slowing of all nerves without signs of conduction block. This stands in sharp contrast to the typical findings in an acquired demyelinative neuropathy with multifocal slowing and differential involvement of various nerves and nerve segments.[225]

Charcot-Marie-Tooth Disease (HMSN Types I and II)

This most common form of hereditary neuropathy usually shows an autosomal dominant inheritance but occasionally autosomal recessive[172] or X-linked dominant pattern.[304] Men tend to have a more severe form of the disease than women, who may have formes frustes.[172] This disorder, although long regarded as a single entity, consists of two different varieties, hypertrophic and neuronal[23,51,118,119,172] Genetic linkage studies provide evidence for heterogeneity.[27,161] A form of HMSN has been described in association with myotonic dystrophy. In this family, even the members who had only peripheral neuropathy had genetic markers for the myotonic dystrophy gene on chromosone 19.[347]

A large group of clinically unequivocal cases shows a bimodal distribution of nerve conduction velocities.[172,360] Some kinships, however, include both the neuronal and hypertrophic types,[7] and some investigators emphasize the existence of an intermediate type.[36,41,48,150,239,304]

HMSN TYPE I

The hypertrophic variety of HMSN affects either sex, but men more commonly than women. The histologic studies reveal enlargement of the peripheral nerves, segmental demyelination and remyelination with onion-bulb formation, and axonal atrophy.[376] Despite the pathologic findings generally in favor of a Schwann cell disorder, axonal atrophy and abnormalities in axonal transport of dopamine beta-hydroxylase suggest a primary neuronal disturbance.[108,200] In a kindred displaying a dominant inheritance, marriage between two heterozygotes resulted in two homozygous offspring. The homozygotes had clinical features reminiscent of classic Dejerine-Sottas disease (HMSN type III).

The symptoms begin insidiously during the first two decades, with musculoskeletal deformities such as high arches or clubfoot. Atrophy initially involves the peroneal musculature, then the thigh and the upper extremities, sparing the trunk and girdle musculature. The classic stork-leg configuration develops only rarely in the hypertrophic type. Bilateral footdrop causes characteristic gait difficulty. The patient has paresthesias, dysethesias, and muscle pain associated with foot deformity. The typical findings include palpable nerves, loss of vibratory and position senses, reduced cutaneous sensations, and diminished stretch reflexes first at the ankle and later diffusely. The disease progresses very slowly over many decades, at times showing spontaneous arrest. Muscle atrophy and weakness may incapacitate the patient but not as a rule. Many investigators consider the Roussy-Lévy syndrome with a static tremor of the hands a variant of this type.

A very slow nerve conduction velocity is a hallmark of the HMSN type I.[82,152,173,217,282] The uncommon recessive forms show slower conduction than the dominant form.[173] The motor conduction velocities in affected family members average less than half those of normal individuals, varying from 9 to 41 m/s, with a mean of 25 m/s.[121] Prolonged terminal latencies in the early stages indicate distally prominent slowing.[163] The disease affects both peripheral and central sensory fibers, as evidenced by delay and reduction of sensory potentials as well as somatosensory evoked potentials.[189] Despite slowing, the degree of temporal dispersion is limited, which indicates a homogeneity of the pathologic process. The degree of conduction abnormality varies little not only among members in the same family but also from one nerve to another in the same patient.[189] Such uniformity helps in differentiating this entity from acute inflammatory polyneuropathy. Conduction abnormalities may herald the clinical onset of neuropathy.[374] Maximal slowing of motor nerve conduction velocities evolve over the first 3 to 5 years of life.[103]

Other conduction abnormalities include absent or delayed F waves.[204] Reduction in the F-wave velocity in the proximal segment matches that of the motor nerve conduction in the distal segment (see Figs. 17–7 and 17–10 and Table 17–3).[202] Blink reflex studies also show increased latencies despite relatively normal strength in the facial muscles (see Figs. 16–10A and 16–12 and Tables 16–2 and 16–4).[201] In many patients, a minor degree of visual pathway involvement is detectable by visual evoked potential studies.[59]

HMSN TYPE II

In the neuronal variety of HMSN, patients have neither hypertrophic nerves nor prominent segmental demyelination. Its inheritance is autosomal dominant, and symptoms and signs appear in early adult life or later. In rare instances, the disease appears sporadically or with autosomal recessive or dominant inheritance in early childhood.[297] A third type of Charcot-Marie-Tooth disease designated as the spinal form is possibly a variant of the neuronal type or distal spinal muscular atrophy.

The clinical features resemble those of the HMSN type I, though much less generalized, with less conspicuous sensory disturbances. As the name peroneal muscular atrophy indicates, the patient develops selective muscular wasting of the legs with limited involvement of the upper extremities in the early stages. An almost total loss of muscle bulk below the knee gives rise to a stork-leg appearance. Despite footdrop with severe weakness of the plantar flexors and clubfoot, the patient often walks fairly well, rarely showing total incapacitation. Some affected individuals have tremors of the hands, but much less commonly than those with HMSN type I.

Electrophysiologic studies reveal mild slowing of the nerve conduction velocities consistent with reduction in amplitude of the compound sensory nerve and muscle action potentials.[26,151,173] Electromyography typically shows large motor unit potentials, fasciculation potentials, fibrillation potentials, and positive sharp waves.[119]

Dejerine-Sottas Disease— Hypertrophic Polyneuropathy (HMSN Type III)

Dejerine[98] and Dejerine and Sottas[99] described this most severe, generalized form of sensory-motor neuropathy, inherited as an autosomal recessive trait.[330] The affected nerves show marked thickening, onion-bulb formation, segmental demyelination, and thinning of the myelin surrounding the nerve. The symptoms appear in infancy, with delayed development of motor skills, especially in walking. The clinical features consist of pes cavus, muscle cramps, incoordination, kyphoscoliosis, weakness, sensory loss, and abducens and facial nerve palsies. Adult patients are often wheelchair-bound, with paraparesis and severe truncal ataxia. Compared with HMSN type I, HMSN type III shows a higher incidence of ataxia, areflexia, and hypertrophic nerves, clinically. Pathologic analysis reveals greater loss of myelinated fibers, a larger number of onion bulbs, each with more lamellae, and a higher ratio of mean axon diameter to fiber diameter.[296] Nerve conduction studies reveal marked slowing of the motor and sensory fibers.

In contrast to the generalized form, rare localized hypertrophic neuropathy consists of isolated mononeuropathy with focal nerve enlargement.[345] This entity represents a localized form of Dejerine-Sottas disease, entrapment neuropathy, or intraneural neurofibroma. In some patients, morphologic findings in the localized areas of enlarged nerves consists of primary perineurial cell hyperplasia or perineurinoma.[268] Nerve conduction studies suggest severe motor and sensory axonal loss with no evidence of slowed conduction velocity. Electromyography also indicates focal axonal loss with evidence of severe denervation limited to the territory of the affected nerve.

Refsum's Disease—Hereditary Ataxic Neuropathy (HMSN Type IV)

Refsum's disease, a rare disorder transmitted by a recessive gene, shows characteristic pathologic changes in the olivocerebellar tracts, anterior horn cells, and the peripheral nerves.[328] The typical clinical features are deafness, anosmia, night blindness with retinitis pigmentosa, skin resembling that in ichthyosis, cerebellar signs, and nystagmus. Involvement of the peripheral nerve causes lightning pain in the legs, wasting of muscles, hyporeflexia, hypotonia, and diminished vibration and position sense. Serum phytanic acid is elevated because of a metabolic defect in the oxidation of branched-chain fatty acid, which for as yet unknown reasons leads to hypertrophic

neuropathy. Dietary restriction of phytol results in considerable improvement of symptoms. Some patients with retinitis pigmentosa and ataxia have a syndrome that clinically resembles Refsum's disease without detectable biochemical abnormality. In these cases, electrophysiologic studies reveal mildly delayed, low-amplitude sensory action potentials but no evidence of hypertrophic neuropathy.[371]

Spinocerebellar Degeneration With Neuropathy (HMSN Type V)

HMSN type V superficially resembles HMSN types I and II, with distal wasting and weakness involving the legs more than the arms.[351] Most patients have an extensor plantar response with normal or increased stretch reflexes in the upper limbs and at the knee but often absent ankle jerks. Electrophysiologic abnormalities include lower-than-normal mean motor and sensory nerve conduction velocity and reduced amplitude of sensory nerve action potentials.[174,259] Sural nerve biopsies show a reduction in the number of myelinated fibers and a normal number of unmyelinated fibers.[259]

Friedreich's Ataxia

This autosomal recessive disorder primarily affects the spinocerebellar tracts, corticospinal tracts, and posterior columns of the spinal cord. In advanced cases, the degeneration also affects the dorsal roots and peripheral nerves. Despite the severe loss of large myelinated fibers, well-preserved unmyelinated C fibers conduct normally.[122,295] The only consistent clinical findings within 5 years of presentation consist of limb and truncal ataxia and absent stretch reflexes in the legs.[171] All patients eventually have dysarthria; signs of pyramidal tract dysfunction in the legs; and loss of joint, position, and vibratory senses. Other less frequent clinical features include kyphosis, scoliosis, pes cavus, distal amyotrophy, optic atrophy, nystagmus, and deafness. On the average, the patients loose the ability to walk by the age of 25 years and become chairbound by the age of 44 years.[171]

Electrophysiologic studies show absent or considerably reduced sensory nerve potentials[256] and normal motor conduction studies except for modest slowing in some patients.[295] Somatosensory evoked potentials may reveal abnormal peripheral as well as central conduction.[102,300] Patients complain of little visual impairment, although most have increased latency or reduced amplitude of the visual evoked potentials.[60,231,300]

Acute Intermittent Porphyria

This is a rare hereditary disorder that belongs to the category called inborn errors of metabolism. A partial defect in hepatic heme synthesis results in overproduction of delta-aminolevulinic acid and porphobilinogen.[369] The disease shows a higher incidence in women, autosomal dominant inheritance, and variable degrees of expression. Clinical features include abdominal pain, vomiting, peripheral neuropathy, neurogenic bladder, seizures, and mental status changes. Skin photosensitivity does not occur. Excessive quantities of porphyrin intermediates excreted in the urine impart a deep red color with formation of polypyrrole from porphobilinogen on exposure to light. The patient experiences acute attacks either spontaneously or after inadvertent administration of barbituates, sulfonamides, or certain other drugs.

Acute axonal neuropathy develops that affects motor fibers regularly and sensory fibers in about 50% of patients. Weakness progresses rapidly, involving the axial muscles more than distal muscles. The sensory loss, though relatively mild, may also predominate proximally.[314] Nerve conduction studies show low amplitude compound action potentials with normal conduction velocities. Electromyography reveals prominent fibrillation potentials and positive sharp waves in

the proximal muscles 1 to 2 weeks after onset.[5,34]

Pressure-Sensitive Hereditary Neuropathy

In this familial disorder of autosomal dominant inheritance,[96] slight traction or compression leads to motor and sensory deficits in an otherwise asymptomatic patient.[250] The histopathologic changes include focal, sausagelike thickening of the myelin sheaths and noncompacted "loose" myelin lamellae together with segmental demyelination and remyelination.[25,238,365] The most prominent feature of the disease consists of pressure-induced, reversible motor weakness, although sensory symptoms may also appear.[107] Compression palsy commonly affects the ulnar, radial, and peroneal nerves, with recovery taking place slowly over weeks or months.

The motor and sensory studies show slowed conduction velocities in paretic limbs but also in some clinically unaffected nerves.[96,365] Evaluations of clinically normal nerves reveal electrophysiologic abnormalities in approximately 50 percent of the patients and some asymptomatic relatives.[88] The slowing of conduction is attributable to pathologically thick myelin sheath, although segmental demyelination also plays a role.[32]

Cerebral Lipidosis

Polyneuropathy is a clinical feature of at least two types of cerebral lipidosis: Krabbe's disease and metachromatic leukodystrophy. In both entities, marked slowing of nerve conduction helps establish the clinical diagnosis, although confirmation comes from a nerve or cerebral biopsy.[149,398]

Histologic studies in Krabbe's globoid cell leukodystrophy reveal diffuse loss of myelin throughout the cerebral white matter and peripheral nerves. Prominent perivascular cuffs appear, consisting of greatly enlarged cells with the accumulation of cerebroside. The affected infants are normal at birth, but within the first few months of life severe neurologic disturbances develop. The disease often follows a fulminant course, with rigidity, head retraction, optic atrophy, bulbar paralysis, a decorticate posture, and finally death before the end of the first year. Neuropathy is usually a late manifestation but occasionally appears as one of the presenting features.[87,227]

In metachromatic leukodystrophy, a deficiency of arylsulfatase leads to abnormal breakdown of myelin. Metachromatic staining properties result from a cerebroside sulfate that accumulates in the nervous tissue. The neurologic signs include spasticity, ataxia, dementia, and neuropathy. The disease usually affects infants, but rarely children[80,167] or adults.[33] Electrophysiologic studies reveal the substantially slowed nerve conduction expected in a demyelinative neuropathy.

Hereditary Sensory Neuropathy

Hereditary sensory neuropathy consists of four distinct entities.[109] Type I, an autosomal dominant condition, is characterized by degeneration of the dorsal root ganglions. In one family, sural nerve biopsies showed a marked loss of all myelinated fibers and a comparable loss of unmyelinated fibers.[93] Clinical findings include loss of pain and temperature sensation, areflexia, and development of ulcers in the lower extremities with almost complete sparing of the upper limbs. The disease tends to progress slowly after its onset in the second decade of life. Deafness, diarrhea, and ataxia occasionally develop in affected individuals. Type II has autosomal recessive inheritance with onset in infancy or early childhood. It affects both upper and lower extremities equally, with a higher incidence of chronic ulceration than in type I.[65] Nerve conduction studies show absent sensory action potentials and borderline slow motor nerve conduction velocities. Type III is the same as familial dysautonomia or the Riley-Day syndrome;[315,358] and Type IV, a rare congenital loss of C fibers with complete insensitivity to pain.[234]

Lipoprotein Neuropathies

Two types of lipoprotein disorders accompany neuropathies. Bassen-Kornzweig syndrome is seen mostly in Jewish children. The clinical features include malabsorption, cerebellar signs, retinitis pigmentosa, acanthocytosis, and virtual absence of beta-lipoprotein in the serum, or a-beta-lipoproteinemia. Diminished stretch reflexes and the absence of position and vibratory senses suggest a peripheral neuropathy. Neurologic signs resemble those of Friedreich's ataxia and Refsum's syndrome. In one histologic study, the sural nerve showed a decreased number of large fibers with a diameter greater than 7 μm, regeneration, and paranodal demyelination.[388]

Electromyographic findings include signs of chronic denervation in distal limb muscles, myotonic discharges, large amplitude, long-duration motor unit potentials, and poor recruitment. Nerve conduction studies reveal reduced amplitude of compound sensory nerve action potentials with slight slowing in maximum conduction velocity distally.[388] Muscle action potentials may be normal or slightly reduced, with normal conduction velocities.[237,266] The fiber diameter spectrum of the sural nerve indicates a loss in the 8- to 12-μm diameter range. Other electrophysiologic studies may reveal a prolonged P_{100} latency of visual evoked potentials and evidence of dorsal column dysfunction in somatosensory evoked potentials.[43]

Tangier disease is associated with a low level of high-density lipoprotein and cholesterol in the serum. The tonsils are enlarged and bright orange from the deposition of cholesterol esters. The skin and rectal mucosa display similar changes. Both myelinated and unmyelinated fibers show degeneration.[209]

Dissociated losses of pain and temperature sensation, not unlike those in syringomyelia, suggest selective involvement of the small fibers.[112] Patients may have a relapsing and remitting mononeuropathy with prominent demyelination and remyelination or slowly progressive neuropathy with advanced axonal degeneration.[307] Conduction studies may reveal abnormal velocities in some but not in others.[149]

Giant Axonal Neuropathy

Children with this progressive peripheral neuropathy[12,58] usually have minor central nervous system involvement and intellectual dysfunction.[270] The accumulation of neurofilamentous material leads to ballooning and degeneration of the axons, affecting the motor more than sensory fibers. Patients characteristically have tightly curled, reddish hair, which must be distinguished from the sparse hair in Menke's kinky hair disease.

Electrophysiologic studies suggest the presence of secondary demyelination triggered by axonal enlargement, but available data are insufficient to characterize the condition. Abnormalities demonstrated by evoked potential studies are consistent with clinical and pathologic findings of central nervous system dysfunction.[241]

Fabry's Disease

This is a multisystem disorder transmitted as an X-linked recessive trait. Because of an inborn error involving glycosphingolipid metabolism, ceramide trihexose is accumulated in various tissues. The enzymatic defect of ceramide trihexosidase affects the skin, the blood vessels, the cornea, and the cell bodies of the dorsal ganglia.[191] Both the central and peripheral nervous systems are affected by deposition of lipid in endothelial and perithelial cells of the vessel walls or perikaryon.[342] Axonal degeneration primarily involves small myelinated and unmyelinated fibers.[145,210] The presenting clinical features include severe burning sensations of the hands and feet.

Nerve conduction studies, though ordinarily normal, may show some slowing in affected men and occasionally in female carriers.[339] Electromyography reveals no abnormalities in most cases.

Familial Amyloid Neuropathy

Signs and symptoms of amyloidosis result from deposits of amyloid around blood vessels and connective tissue in multiple organ systems. Clinical features depend on the organs involved, which commonly include the heart, tongue, gastrointestinal tract, skeletal muscles, and kidneys. Amyloid deposits in the flexor retinaculum may cause the carpal tunnel syndrome in about one quarter of the patients. Familial amyloid neuropathies differ clinically from primary or nonfamilial amyloid neuropathies associated with paraproteinemia (see Chapter 22.2). Neurologic symptoms rarely develop in secondary amyloidosis seen in chronic debilitating inflammatory processes. A form of autosomal dominant amyloidosis prevalent in northern Portugal produces progressive neuropathy involving the legs in young adults. Another milder form of autosomal dominant amyloidosis with neuropathy of the upper extremities primarily affects Swiss families, with an onset later in life. Familial amyloid neuropathy has also involved a kinship of German ancestry.[286]

Olivopontocerebellar Atrophy

A predominantly sensory axonal neuropathy, seen in olivopontocerebellar atrophy, affects those patients with glutamate dehydrogenase deficiency but not those with normal enzymatic activity.[73] Such a distinction may serve as an electrophysiologic marker for differentiating the subtypes. Postmortem examination of one patient revealed olivopontocerebellar atrophy, demyelination of the posterior columns, degeneration of anterior horn and dorsal root ganglion cells, and reduction of myelinated fibers in the sural nerve.[74]

Others

Other rare inherited systemic disorders associated with peripheral neuropathy include sialidosis type I, the cherry-red spot-myoclonus syndrome,[349] cerebrotendinous xanthomatosis,[215] a variety of extrapyramidal syndromes,[54] multiple endocrine neoplasia,[111] Cockayne's syndrome,[160] congenital hypomyelination polyneuropathy,[164,170] chorea-acanthocytosis,[228] adrenomyeloneuropathy (see Fig. 17–4B),[379] infantile neuroaxonal dystrophy,[277] and metachromatic leukodystrophy.[396]

REFERENCES

1. Abraham, RR, Abraham, RM, and Wynn, V: Autonomic and electrophysiological studies in patients with signs or symptoms of diabetic neuropathy. Electroencephalogr Clin Neurophysiol 63:223–230, 1986.
2. Ackil, AA, Shahani, BT, Young, RR, and Rubin, NE: Late response and sural conduction studies: Usefulness in patients with chronic renal failure. Arch Neurol 38:482–485, 1981.
3. Aguayo, AJ, and Karpati, G (eds): Current Topics in Nerve and Muscle Research. Excerpta Medica, Amsterdam, 1979.
4. Albers, JW, Cavender, GD, Levine, SP, and Langolf, GD: Asymptomatic sensorimotor polyneuropathy in workers exposed to elemental mercury. Neurology 32:1168–1174, 1982.
5. Albers, JW, Robertson, WC, and Daube, JR: Electrodiagnostic findings in acute porphyric neuropathy. Muscle Nerve 1:292–296, 1978.
6. Allen, N, Mendell, JR, Billmaier, DJ, Fontaine, RE, and O'Neill, J: Toxic polyneuropathy due to methyl n-butyl ketone. Arch Neurol 32:209–218, 1975.
7. Amick, LD, and Lemmi, H: Electromyographic studies in peroneal muscular atrophy: Charcot-Marie-Tooth disease. Arch Neurol 9:273–284, 1963.
8. Ansbacher, LE, Bosch, EP, and Cancilla, PA: Disulfiram neuropathy: a neurofilamentous distal axonopathy. Neurology 32:424–428, 1982.
9. Antia, NH, Pandya, SS, and Dastur, DK: Nerves in the arm in leprosy. I. Clinical, electrodiagnostic and operative aspects. Int J Leprosy 38:12–29, 1970.
10. Archer, AG, Watkins, PJ, Thomas, PK, Sharma, AK, and Payan, J: The natural history of acute painful neuropathy in diabetes mellitus. J Neurol Neurosurg Psychiatry 46:491–499, 1983.
11. Asbury, AK: Proximal diabetic neuropathy (editorial). Ann Neurol 2:179–180, 1977.
12. Asbury, AK: Neuropathies with filamentous abnormalities. In Aguayo, AJ, and Karpati, G (eds): Current Topics in Nerve and Muscle Research. Excerpta Medica, Amsterdam, 1979, pp 243–254.
13. Asbury, AK, Arnason, BG, and Adams, RD: The inflammatory lesion in idiopathic polyneuritis: Its role in pathogenesis. Medicine 48:173–215, 1969.
14. Asbury, AK, Arnason, BGW, Karp, HR, and McFarlin, DE: Criteria for diagnosis of Guillain-Barré syndrome. Ann Neurol 3:565–566, 1978.
15. Asbury, AK, Fisher R, McKhann, GM, Mobley, W, and Server, A: Guillain-Barré syndrome: Is there a role for plasmapheresis? Neurology 30:1112, 1980.
16. Asbury, AK, and Johnson, PC: Pathology of Peripheral Nerve. WB Saunders, Philadelphia, 1978.
17. Asbury, AK, Victor, M, and Adams, RD: Uremic polyneuropathy. Trans Am Neurol Assoc 87:100–103, 1962.
18. Ballantyne, JP, Hansen, S, Weir, A, Whitehead, JRG, and Mullin, PJ: Quantitative electrophysiological study of alcoholic neuropathy. J Neurol Neurosurg Psychiatry 43:427–432, 1980.
19. Barron, SA, and Heffner, RR, Jr: Weakness in malignancy: Evidence for a remote effect of tumor on distal axons. Ann Neurol 4:268–274, 1978.
20. Bassi, S, Grazia Albizatti, M, Calloni, E, and Frattola, L: Electromyographic study of diabetic and alcoholic polyneuropathic patients treated with gangliosides. Muscle Nerve 5:351–356, 1982.
21. Behar, R, Wiley, C, and McCutchan, JA: Cytomegalovirus polyradiculoneuropathy in acquired immune deficiency syndrome. Neurology 37:557–561, 1987.
22. Behse, F, and Buchthal, F: Alcoholic neuropathy: Clinical, electrophysiological, and biopsy findings. Ann Neurol 2:95–110, 1977.
23. Behse, F, and Buchthal, F: Peroneal muscular atrophy (PMA) and related disorders. II. Histological findings in sural nerves. Brain 100:67–85, 1977.
24. Behse, F, Buchthal, F, and Carlsen, F: Nerve biopsy and conduction studies in diabetic neuropathy. J Neurol Neurosurg Psychiatry 40:1072–1082, 1977.
25. Behse, F, Buchthal, F, Carlsen, F, and Knappeis, GG: Conduction and histopathology of the sural nerve in hereditary neuropathy with liability to pressure palsies. In Desmedt, JE (ed): New Developments in Electromyography and Clinical Neurophysiology, Vol 2. Karger, Basel, 1973, pp 286–287.
26. Berciano, J, Combarros, O, Figols, J, Calleja, J, Cabello, A, Silow, I, and Coria, F: Hereditary motor and sensory neuropathy type II clinicopathological study of a family. Brain 109:897–914, 1986.
27. Bird, TD, Ott, J, Giblett, ER, Chance, PF, Sumi, SM, and Kraft, GH: Genetic linkage evidence for heterogeneity in Charcot-Marie-Tooth neuropathy (HMSN Type I). Ann Neurol 14:679–684, 1983.
28. Blackstock, E, Rushworth, G, and Gath, D: Electrophysiological studies in alcoholism. J Neurol Neurosurg Psychiatry 35:326–334, 1972.
29. Bolton, CF: Peripheral neuropathies associated with chronic renal failure. Can J Neurol Sci 7:89–96, 1980.
30. Bolton, CF, Driedger, AA, and Lindsay, RM: Ischaemic neuropathy in uraemic patients

caused by bovine arteriovenous shunt. J Neurol Neurosurg Psychiatry 42:810–814, 1979.

31. Bolton, CF, Gilbert, JJ, Hahn, AF, and Sibbald, WJ: Polyneuropathy in critically ill patients. J Neurol Neurosurg Psychiatry 47:1223–1231, 1984.

32. Bosch, EP, Chui, HC, Martin, MA, and Cancilla, PA: Brachial plexus involvement in familial pressure sensitive neuropathy: Electrophysiologic and morphologic findings. Ann Neurol 8:620–624, 1980.

33. Bosch, EP, and Hart, MN: Late adult-onset metachromatic leukodystrophy: Dementia and polyneuropathy in a 63-year-old man. Arch Neurol 35:475–477, 1978.

34. Bosch, EP, Pierach, CA, Bossenmaier, I, Cardinal, R, and Thorson, M: Effect of hematin in porphyric neuropathy. Neurology 27:1053–1056, 1977.

35. Bouche, P, Bousser, MG, Peytour, MA, and Cathala, HP: Perhexiline maleate and peripheral neuropathy. Neurology 29:739–743, 1979.

36. Bouche, P, Gherardi, R, Cathala, HP, L'Hermitte, F, and Castaigne, P: Peroneal muscular atrophy. Part I. Clinical and electrophysiological study. J Neurol Sci 61:389–399, 1983.

37. Bouche, P, Leger, JM, Travers, MA, Cathala, HP, and Castaigne, P: Peripheral neuropathy in systemic vasculitis: Clinical and electrophysiologic study of 22 patients. Neurology 36:1598–1602, 1986.

38. Bradley, WG: Disorders of Peripheral Nerves. Blackwell Scientific Publishers, Oxford, 1974.

39. Bradley, WG, Karlsson, IJ, and Rassol, CG: Metronidazole neuropathy. Br Med J 2:610–611, 1977.

40. Bradley, WG, Lassman, LP, Pearce, GW, and Walton, JM: The neuromyopathy of vincristine in man: Clinical, electrophysiological and pathological studies. J Neurol Sci 10:107–131, 1970.

41. Bradley, WG, Madrid, R, and Davis, CJF: The peroneal muscular atrophy syndrome: Clinical, genetic, electrophysiological and nerve biopsy studies. Part 3. Clinical, electrophysiological and pathological correlations. J Neurol Sci 32:123–136, 1977.

42. Brin, MF, Fetell, MR, Green, PHA, Kayden, HJ, Hays, AP, Behrens, MM, and Baker, H: Blind loop syndrome, vitamin E malabsorption, and spinocerebellar degeneration. Neurology 35:338–342, 1985.

43. Brin, MF, Pedley, T, Lovelace, R, Emerson, R, Gouras, P, MacKay, C, Kayden, H, Levy, J, and Baker, H: Electrophysiologic features of abetalipoproteinemia: Functional consequences of vitamin E deficiency. Neurology 36:669–673, 1986.

44. Brown, MJ, Martin, JR, and Asbury, AK: Painful diabetic neuropathy: A morphometric study. Arch Neurol 33:164–171, 1976.

45. Brown, MJ, Sumner, AJ, Greene, DA, Diamond, SM, and Asbury, AK: Distal neuropathy in experimental diabetes mellitus. Ann Neurol 8:168–178, 1980.

46. Brown, WF, and Feasby, TE: Conduction block and denervation in Guillain-Barré polyneuropathy. Brain 107:219–239, 1984.

47. Brown, WF, and Feasby, TE: Sensory evoked potentials in Guillain-Barré polyneuropathy. J Neurol Neurosurg Psychiatry 47:288–291, 1984.

48. Brust, J, Lovelace, R, and Devi, S: Classification of Charcot-Marie-Tooth disorder by electrophysiological studies. Fifth International Congress of Electromyography, Abstracts of Communications, Rochester, Minnesota, 1975.

49. Brust, JCM, Hammer, JS, Challenor, Y, Healton, EB, and Lesser, RP: Acute generalized polyneuropathy accompanying lithium poisoning. Ann Neurol 6:360–362, 1979.

50. Buchthal, F: Electrophysiological abnormalities in metabolic myopathies and neuropathies. Acta Neurol Scand (Suppl 43) 46:129–176, 1970.

51. Buchthal, F, and Behse, F: Peroneal muscular atrophy (PMA) and related disorders. I. Clinical manifestations as related to biopsy findings, nerve conduction and electromyography. Brain 100:41–66, 1977.

52. Buchthal, F, and Behse, F: Electrophysiology and nerve biopsy in men exposed to lead. Br J Indust Med 36:135–147, 1979.

53. Busis, NA, Halperin, JJ, Stefansson, K, Kwiatkowski, DJ, Sagar, SM, Schife, SR, and Logigian, EL: Peripheral neuropathy, high serum IGM and paraproteinemia in mother and son. Neurology 35:679–683, 1985.

54. Byrne, E, Thomas, PK, and Zilkha, KJ: Familial extrapyramidal disease with peripheral neuropathy. J Neurol Neurosurg Psychiatry 45:372–374, 1982.

55. Cadilhac, J, Dapres, G, Fabre, JL, and Mion, C: Follow-up study of motor conduction velocity in uraemic patients treated by hemodialysis. In Desmedt, JE (ed): New Developments in Electromyography and Clinical Neurophysiology, Vol 2. Karger, Basel, 1973, pp 372–380.

56. Campbell, MJ, and Paty, DW: Carcinomatous neuromyopathy. 1. Electrophysiological studies: An electrophysiological and immunological study of patients with carcinoma of the lung. J Neurol Neurosurg Psychiatry 37:131–141, 1974.

57. Canal, N, and Pozza, G (eds): Peripheral Neuropathies. Developments in Neurology, Vol 1. Elsevier/North-Holland Biomedical Press, Amsterdam, 1978.

58. Carpenter, S, Karpati, G, Andermann, F, and Gold, R: Giant axonal neuropathy: A clinically and morphologically distinct neurological disease. Arch Neurol 31:312–316, 1974.

59. Carroll, WM, Jones, SJ, and Halliday, AM: Visual evoked potential abnormalities in Charcot-Marie-Tooth disease and comparison with Friedreich's ataxia. J Neurol Sci 61:123–133, 1983.

60. Carroll, WM, Kriss, A, Baraitser, M, Barrett, G, and Halliday, AM: The incidence and nature of visual pathway involvement in Friedreich's ataxia. Brain 103:413–434, 1980.

61. Casey, EB, Jellife, AM, Le Quesne, PM, and Millett, YL: Vincristine neuropathy: Clinical and electrophysiological observations. Brain 96:69–86, 1973.

62. Casey, EB, and Le Quesne, PM: Electrophysiological evidence for a distal lesion in alcoholic neuropathy. J Neurol Neurosurg Psychiatry 35:624–630, 1972.

63. Casey, EB, and Le Quesne, PM: Alcoholic neuropathy. In Desmedt, JE (ed): New Developments in Electromyography and Clinical Neurophysiology, Vol 2. Karger, Basel, 1973, pp 279–285.

64. Catton, MJ, Harrison, MJG, Fullerton, PM, and Kazantzis, G: Subclinical neuropathy in lead workers. Br Med J 2:80–82, 1970.

65. Cavanagh, NPC, Eames, RA, Galvin, RJ, Brett, EM, and Kelly, RE: Hereditary sensory neuropathy with spastic paraplegia. Brain 102:79–94, 1979.

66. Chad, D, Pariser, K, Bradley, WG, Adelman, LS, and Pinn, VW: The pathogenesis of cryobulinemic neuropathy. Neurology 32:725–729, 1982.

67. Challenor, YB, Felton, CP, and Brust, JCM: Peripheral nerve involvement in sarcoidosis: An electrodiagnostic study. J Neurol Neurosurg Psychiatry 47:1219–1222, 1984.

68. Chang, Y, McLeod, J, Tuck, R, Walsh, J, and Feary, P: Visual evoked responses in chronic alcoholics. J Neurol Neurosurg Psychiatry 49:945–950, 1986.

69. Chang, YW, McLeod, JF, Tuck, RR, and Feary, PA: Brain stem auditory evoked responses in chronic alcoholics. J Neurol Neurosurg Psychiatry 48:1107–1112, 1985.

70. Charness, ME, Morady, F, and Scheinman, MM: Frequent neurologic toxicity associated with amiodarone therapy. Neurology 34:669–671, 1984.

71. Charron, L, Peyronnard, JM, and Marchand, L: Sensory neuropathy associated with primary biliary cirrhosis. Arch Neurol 37:84–87, 1980.

72. Chia, LG, and Chu, FL: A clinical and electrophysiological study of patients with polychlorinated biphenyl poisoning. J Neurol Neurosurg Psychiatry 48:894–901, 1985.

73. Chokroverty, S, Duvoisin, RC, Sachdeo, R, Sage, J, Lepore, F and Nicklas, W: Neurophysiologic study of olivopontocerebellar atrophy with or without glutamate dehydrogenase deficiency. Neurology 35:652–659, 1985.

74. Chokroverty, S, Khedekar, R, Derby, B, Sachdeo, R, Yook, C, Lepore, F, Nicklas, W, and Duvoisin, RC: Pathology of olivopontocerebellar atrophy with glutamate dehydrogenase deficiency. Neurology 34:1451–1455, 1984.

75. Chokroverty, S, Reyes, MG, Rubino, FA, and Tonaki, H: The syndrome of diabetic amyotrophy. Ann Neurol 2:181–194, 1977.

76. Chopra, JS, Banerjee, AK, Murthy, JMK, and Pal, SR: Paralytic rabies: A clinico-pathological study. Brain 103:789–802, 1980.

77. Chopra, JS, Dhand, U, Mehta, S, Bakshi, V, Rana, S, and Mehta, J: Effect of protein calorie malnutrition on peripheral nerves: A clinical, electrophysiological and histopathological study. Brain 190:307–323, 1986.

78. Chopra, JS, Hurwitz, LJ, and Montgomery, DAD: The pathogenesis of sural nerve changes in diabetes mellitus. Brain 92:391–418, 1969.

79. Chopra, JS, Sawhney, BB, and Chakravorty, RN: Pathology and time relationship of peripheral nerve changes in experimental diabetes. J Neurol Sci 32:53–67, 1977.

80. Clark, JR, Miller, RG, and Vidgoff, JM: Juvenile-onset metachromatic leukodystrophy: Biochemical and electrophysiologic studies. Neurology 29:346–353, 1979.

81. Cohen, AS, and Rubinow, A: Amyloid neuropathy. In Dyck, PJ, Thomas, PK, Lambert, EH, and Bunge, R (eds): Peripheral Neuropathy, Vol 2. WB Saunders, Philadelphia, 1984, pp 1866–1898.

82. Combarros, O, Calleja, J, Figols, J, Cabello, A, and Berciano, J: Dominantly inherited motor and sensory neuropathy Type I. J Neurol Sci 61:181–191, 1983.

83. Conrad, B, Aschoff, JC, and Fischler, M: Der diagnostische Wert der F-Wellen-Latenz. J Neurol 210:151–159, 1975.

84. Cornblath, DR, McArthur, JC, Kennedy, PGE, Witte, AS, and Griffin, JW: Inflammatory demyelinating peripheral neuropathies associated with human T-cell lymphotropic virus type III infection. Ann Neurol 21:32–40, 1987.

85. Cox-Klazinga, M, and Endtz, LJ: Peripheral nerve involvement in pernicious anaemia. J Neurol Sci 45:367–371, 1980.

86. Cracco, J, Castells, S, and Mark, E: Spinal somatosensory evoked potentials in juvenile diabetes. Ann Neurol 15:55–58, 1984.

87. Cruz Martinez, A, Ferrer, MT, Fueyo, E, and Galdos, L: Peripheral neuropathy detected on electrophysiological study as first manifestation of metachromatic leucodystrophy in infancy. J Neurol Neurosurg Psychiatry 38:169–174, 1975.

88. Cruz Martinez, A, Perez Conde, MC, Ramon, Y, Cajal, S, and Martinez, A: Recurrent familiar polyneuropathy with liability to pressure palsies: Special regard to electrophysiological aspects of twenty-five members from seven families. Elctromyogr Clin Neurophysiol 17:101–124, 1977.

89. Dalakas, M: Chronic idiopathic ataxic neuropathy. Ann Neurol 19:545–554, 1986.

90. Dalakas, MC, and Engel, WK: Immunoglobulin and complement deposits in nerves of patients with chronic relapsing polyneuropathy. Arch Neurol 37:637–640, 1980.

91. D'Amour, ML, Dufresne, LR, Morin, C, and Slaughter, D: Sensory nerve conduction in chronic uremic patients during the first six months of hemodialysis. Can J Neurol Sci 11:269–271, 1984.

92. D'Amour, ML, Shahani, BT, Young, RR, and Bird, KT: The importance of studying sural nerve conduction and late responses in the evaluation of alcoholic subjects. Neurology 29:1600–1604, 1979.

93. Danan, MJ, and Carpenter, S: Hereditary sen-

sory neuropathy: Biopsy study of an autosomal dominant variety. Neurology 35:1226–1229, 1985.

94. Dastur, DK, Manghani, DK, Osuntokun, BO, Sourander, P, and Kondo, K: Neuromuscular and related changes in malnutrition. J Neurol Sci 55:207–230, 1982.

95. Davis, LE, Standefer, JC, Kornfeld, M, Abercrombie, DM, and Butler, C: Acute thallium poisoning: Toxicological and morphological studies of the nervous system. Ann Neurol 10:38–44, 1981.

96. Debruyne, J, Dehaene, I, and Martin, JJ: Hereditary pressure-sensitive neuropathy. J Neurol Sci 47:385–394, 1980.

97. De Jager, AEJ, Van Weerden, TW, Houthoff, HJ, and De Monchy, JGR: Polyneuropathy after massive exposure to parathion. Neurology 31:603–605, 1981.

98. Dejerine, J: Sur une forme particuliere de maladie de Friedreich avec atrophie musculaire et troubles de la senibilite. C R Soc Biol (Memoires) 42:43–53, 1890.

99. Dejerine, J, and Sottas, J: Sur la nevrite interstitielle, hypertrophique et progressive de l'enfance. C R Soc Biol 45:63–96, 1893.

100. Delaney, P: Neurologic manifestations in sarcoidosis. Review of the literature, with a report of 23 cases. Ann Intern Med 87:336–345, 1977.

101. De Sena, PG: Aspectos neurologicos e electromyographicos de hanseniase. Arq Neuropsiquiat 34:1–17, 1976.

102. Desmedt, JE, and Noel, P: Average cerebral evoked potentials in the evaluation of lesions in the sensory nerves and of the central somatosensory pathway. In Desmedt, JE (ed): New Developments in Electromyography and Clinical Neurophysiology, Vol 2. Karger, Basel, 1973, pp 352–371.

103. Donald, MW, Bird, CE, Lawson, JS, Letemendia, FJJ, Monga, TN, Surridge, DHC, Varette-Cerre, P, Williams, DL, Williams, DML, and Wilson, DL: Delayed auditory brainstem responses in diabetes mellitus. J Neurol Neurosurg Psychiatry 44:641–644, 1981.

104. Donofrio, PD, Albers, JW, Greenberg, HS, and Mitchell, BS: Peripheral neuropathy in osteosclerotic myeloma: Clinical and electrodiagnostic improvement with chemotherapy. Muscle Nerve 7:137–141, 1984.

105. Dorfman, LJ, Cummins, KL, Reaven, GM, Ceranski, J, Greenfield, MS, and Doberne, L: Studies of diabetic polyneuropathy using conduction velocity distribution (DCV) analysis. Neurology 33:773–779, 1983.

106. Dorfman, LJ, Ransom, BR, Forno, LS, and Kelts, A: Neuropathy in the hypereosinophilic syndrome. Muscle Nerve 6:291–298, 1983.

107. Dubi, J, Regli, F, Bischoff, A, Schneider, C, and De Crousaz, G: Recurrent familial neuropathy with liability to pressure palsies: Reports of two cases and ultrastructural nerve study. J Neurol 220:43–55, 1979.

108. Dyck, PJ: Inherited neuronal degeneration and atrophy affecting peripheral motor, sensory, and autonomic neurons. In Dyck, PJ,

Thomas, PK, Lambert, EH, and Bunge, R (eds): Peripheral Neuropathy, Vol 2. WB Saunders, Philadelphia, 1984, pp 1600–1641.

109. Dyck, PJ: Neuronal atrophy and degeneration predominantly affecting peripheral sensory and autonomic neurons. In Dyck, PJ, Thomas, PK, Lambert, EH, and Bunge, R (eds): Peripheral Neuropathy, Vol 2. WB Saunders, Philadelphia, 1984, pp 1557–1559.

110. Dyck, PJ, and Arnason, BGW: Chronic inflammatory demyelinating polyradiculoneuropathy. In Dyck, PJ, Thomas, PK, Lambert, EH, and Bunge, R (eds): Peripheral Neuropathy. WB Saunders, Philadelphia, 1984, pp 2101–2114.

111. Dyck, PJ, Carney, JA, Sizemore, GW, Okazaki, H, Brimijoin, WS, and Lambert, EH: Multiple endocrine neoplasia type 2b: phenotype recognition: Neurological features and their pathological basis. Ann Neurol 6:302–314, 1979.

112. Dyck, PJ, Ellefson, RD, Yao, JK, and Herbert, PN: Adult-onset of Tangier disease: 1. Morphometric and pathologic studies suggesting delayed degradation of neutral lipids after fiber degeneration. J Neuropathol Exp Neurol 37:119–137, 1978.

113. Dyck, PJ, Johnson, WJ, Lambert, EH, and O'Brien, PC: Segmental demyelination secondary to axonal degeneration in uremic neuropathy. Mayo Clin Proc 46:400–431, 1971.

114. Dyck, PJ, Karnes, JL, Daube, JR, O'Brien, P, and Service, FJ: Clinical and neuropathological criteria for the diagnosis and staging of diabetic polyneuropathy. Brain 108:861–880, 1985.

115. Dyck, PJ, Karnes, JL, O'Brien, P, Okazaki, H, Lais, A, and Engelstad, J: The spatial distribution of fiber loss in diabetic polyneuropathy suggests ischemia. Ann Neurol 19:440–449, 1986.

116. Dyck, PJ, Lais, A, Karnes, J, O'Brien, P, and Rizza, R: Fiber loss is primary and multifocal in sural nerves in diabetic polyneuropathy. Ann Neurol 19:425–439, 1986.

117. Dyck, PJ, Lais, AC, Ohta, M, Bastron, JA, Okazaki, H, and Groover, RV: Chronic inflammatory polyradiculopathies. Mayo Clin Proc 50:621–637, 1975.

118. Dyck, PJ, and Lambert, EH: Lower motor and primary sensory neuron diseases with peroneal muscular atrophy. I. Neurologic, genetic, and electrophysiologic findings in hereditary polyneuropathies. Arch Neurol 18:603–618, 1968.

119. Dyck, PJ, and Lambert, EH: Lower motor and primary sensory neuron diseases with peroneal muscular atrophy. II. Neurologic, genetic, and electrophysiologic findings in various neuronal degenerations. Arch Neurol 18:619–625, 1968.

120. Dyck, PJ, and Lambert, EH: Dissociated sensation in amyloidosis: Compound action potential, quantitative histologic and teased-fiber, and electron microscopic studies of sural nerve biopsies. Arch Neurol 20:490–507, 1969.

121. Dyck, PJ, Lambert, EH, and Mulder, DW: Charcot-Marie-Tooth disease: Nerve conduction and clinical studies of a large kinship. Neurology 13:1–11, 1963.

122. Dyck, PJ, Lambert, EH, and Nichols, PC: Quantitative measurement of sensation related to compound action potential and number and sizes of myelinated and unmyelinated fibers of sural nerve in health, Friedreich's ataxia, hereditary sensory neuropathy, and tabes dorsalis. In Remond, A (ed): Handbook of Electroencephalography and Clinical Neurophysiology, Vol 9. Elsevier, Amsterdam, 1971, pp 83–118.

123. Dyck, PJ, O'Brien, PC, Oviatt, KF, Dinapoli, RP, Daube, JR, Bartleson, JD, Mokri, B, Swift, T, Low, PA, and Windebank, AJ: Prednisone improves chronic inflammatory demyelinating polyradiculoneuropathy more than no treatment. Ann Neurol 11:136–141, 1982.

124. Dyck, PJ, Oviatt, KF, and Lambert, EH: Intensive evaluation of referred unclassified neuropathies yields improved diagnosis. Ann Neurol 10:222–226, 1981.

125. Dyck, PJ, Thomas, PK, Lambert, EH, and Bunge, R: Peripheral Neuropathy. WB Saunders, Philadelphia, 1984.

126. Ehle, A, and Raskin, P: Increased nerve conduction in diabetics after a year of improved glucoregulation. J Neurol Sci 74:191–197, 1986.

127. Eidelberg, D, Sotrel, A, Vogel, H, Walker, P, Kleefield, J, and Crumpacker, C, III: Progressive polyradiculopathy in acquired immune deficiency syndrome. Neurology 36:912–916, 1986.

128. Eisen, A, and Humphreys, P: The Guillain-Barré syndrome: A clinical and electrodiagnostic study of 25 cases. Arch Neurol 30:438–443, 1974.

129. Eisen, AA, Woods, JF, and Sherwin, AL: Peripheral nerve function in long-term therapy with diphenylhydantoin: A clinical and electrophysiologic correlation. Neurology 24:411–417, 1974.

130. Ellenberg, M: Diabetic truncal mononeuropathy: A new clinical syndrome. Diabetes Care 1:10–13, 1978.

131. Eng, GD, Hung, W, August, GP, and Smokvina, MD: Nerve conduction velocity determinations in juvenile diabetes: Continuing study of 190 patients. Arch Phys Med Rehabil 57:1–5, 1976.

132. Ewing, DJ, Burt, AA, Williams, IR, Campbell, IW, and Clarke, BF: Peripheral motor nerve function in diabetic autonomic neuropathy. J Neurol Neurosurg Psychiatry 39:453–460, 1976.

133. Fagius, J, and Jameson, S: Effects of aldose reductase inhibitor treatment in diabetic polyneuropathy: A clinical and neurophysiologic study. J Neurol Neurosurg Psychiatry 44:991–1001, 1981.

134. Fagius, J, Osterman, PO, Siden, A, and Wiholm, BE: Guillain-Barré syndrome following zimeldine treatment. J Neurol Neurosurg Psychiatry 48:65–69, 1985.

135. Fagius, J, and Wallin, BG: Microneurographic evidence of excessive sympathetic outflow in the Guillain-Barré syndrome. Brain 106:589–600, 1983.

136. Feasby, TE, Gilbert, JJ, Brown, WF, Bolton, CF, Hahn, AF, Koopman, WJ, and Zochodne, DW: An acute axonal form of Guillain-Barré polyneuropathy. Brain 109:1115–1126, 1986.

137. Feit, H, Tindall, RSA, and Glasberg, M: Sources of error in the diagnosis of Guillain-Barré syndrome. Muscle Nerve 5:111–117, 1982.

138. Feldman, RG, Haddow, J, and Chisolm, JJ: Chronic lead intoxication in urban children: Motor nerve conduction velocity studies. In Desmedt, JE (ed): New Developments in Electromyography and Clinical Neurophysiology, Vol 2. Karger, Basel, 1973, pp 313–317.

139. Feldman, RG, Niles, CA, Kelly-Hayes, M, Sax, DS, Dixon, WJ, Thompson, DJ: Peripheral neuropathy in arsenic smelter workers. Neurology 29:939–944, 1979.

140. Finelli, PF, and DiBenedetto, M: Bilateral involvement of the lateral cutaneous nerve of the calf in a diabetic. Ann Neurol 4:480–481, 1978.

141. Fisher, M: An unusual variant of acute idiopathic polyneuritis (syndrome of ophthalmoplegia, ataxia, and areflexia). N Engl J Med 255:57–65, 1956.

142. Fitting, JW, Bischoff, A, Regli, F, and De Crousaz, G: Neuropathy, amyloidosis, and monoclonal gammopathy. J Neurol Neurosurg Psychiatry 42:193–202, 1979.

143. Fowler, H, Vulpe, M, Marks, G, Egolf, C, and Dau, PC: Recovery from chronic progressive polyneuropathy after treatment with plasma exchange and cyclophosphamide. Lancet 2:1193, 1979.

144. Fraser, AG, McQueen, INF, Watt, AH, and Stephens, MR: Peripheral neuropathy during long term high-dose amiodarone therapy. J Neurol Neurosurg Psychiatry 48:576–578, 1985.

145. Fukuhara, N, Suzuki, M, Fujita, N, and Tsubaki, T: Fabry's disease on the mechanism of the peripheral nerve involvement. Acta Neuropathol (Berl) 33:9–21, 1975.

146. Fullerton, PM, and O'Sullivan, DJ: Thalidomide neuropathy: A clinical, electrophysiological, and histological follow-up study. J Neurol Neurosurg Psychiatry 31:543–551, 1968.

147. Gabbay, KH: The sorbitol pathway and the complications of diabetes. N Engl J Med 288:831–836, 1973.

148. Galassi, G, Nemni, R, Baraldi, A, Gibertoni, M, and Colombo, A: Peripheral neuropathy in multiple system atrophy with autonomic failure. Neurology 32:1116–1121, 1982.

149. Gamstorp, I: Involvement of peripheral nerves in disorders causing progressive cerebral symptoms and signs in infancy and childhood. In Desmedt, JE (ed): New Developments in Electromyography and Clinical Neurophysiology, Vol 2. Karger, Basel, 1973, pp 306–312.

150. Gherardi, R, Bouche, P, Escourolle, R, and

Hauw, JJ: Peroneal muscular atrophy. Part 2. Nerve biopsy studies. J Neurol Sci 61:401–416, 1983.

151. Gilliatt, RW: Peripheral nerve conduction in neurological patients (abstr). J Neurol Neurosurg Psychiatry 22:344, 1959.

152. Gilliatt, RW, and Thomas, PK: Extreme slowing of nerve conduction in peroneal muscular atrophy. Ann Phys Med 4:104–106, 1957.

153. Gilliatt, RW, and Willison, RG: Peripheral nerve conduction in diabetic neuropathy. J Neurol Neurosurg Psychiatry 25:11–18, 1962.

154. Ginzburg, M, Lee, M, Ginzburg, J, and Alba, A: The primary role of the Erb's point-axilla segment in median and ulnar motor nerve conduction determinations in alcoholic neuropathy. J Neurol Sci 72:299–306, 1986.

155. Glenner, GG: Amyloid deposits and amyloidosis: The B-fibrilloses. (Part 1 of 2). N Engl J Med 302:1283–1292, 1980.

156. Gourie-Devi, M, and Ganapathy, GR: Phrenic nerve conduction time in Guillain-Barré syndrome. J Neurol Neurosurg Psychiatry 48:245–249, 1985.

157. Graf, RJ, Halter, JB, Halar, E, and Porte, D, Jr: Nerve conduction abnormalities in untreated maturity-onset diabetes: Relation to levels of fasting plasma glucose and glycosylated hemoglobin. Ann Int Med 90:298–303, 1979.

158. Gregersen, G: Diabetic neuropathy: Influence of age, sex, metabolic control, and duration of diabetes on motor conduction velocity. Neurology 17:972–980, 1967.

159. Gruener, G, Bosch, EP, Strauss, RG, Klugman, M, and Kimura, J: Prediction of early beneficial response to plasma exchange in Guillain-Barré syndrome. Arch Neurol 44:295–298, 1987.

160. Grunnet, ML, Zimmerman, AW, and Lewis, RA: Ultrastructure and electrodiagnosis of peripheral neuropathy in Cockayne's syndrome. Neurology 33:1606–1609, 1983.

161. Guiloff, RJ, Thomas, PK, Contreras, M, Armitage, S, Schwarz, G, and Sedgewick, EM: Linkage of autosomal dominant Type 1 hereditary motor and sensory neuropathy to the duffy locus on chromosome 1. J Neurol Neurosurg Psychiatry 45:669–674, 1982.

162. Gupta, PR, and Dorfman, LJ: Spinal somatosensory conduction in diabetes. Neurology 31:841–845, 1981.

163. Gutmann, L, Fakadej, A, and Riggs, JE: Evolution of nerve conduction abnormalities in children with dominant hypertrophic neuropathy of the Charcot-Marie-Tooth type. Muscle Nerve 6:515–519, 1983.

164. Guzzetta, F, Ferriere, G, and Lyon, G: Congenital hypomyelination polyneuropathy: Pathological findings compared with polyneuropathies starting later in life. Brain 105:395–416, 1982.

165. Halar, EM, Graf, RJ, Halter, JB, and Brozovich, FV, and Soine, TL: Diabetic neuropathy: A clinical, laboratory and electrodiagnostic study. Arch Phys Med Rehabil 63:298–303, 1982.

166. Hallett, M, Tandon, D, and Berardelli, A: Treatment of peripheral neuropathies. J Neurol Neurosurg Psychiatry 48:1193–1207, 1985.

167. Haltia, T, Palo J, Haltia, M, and Icen, A: Juvenile metachromatic leukodystrophy: Clinical, biochemical, and neuropathologic studies in nine new cases. Arch Neurol 37:42–46, 1980.

168. Hansen, S, and Ballantyne, JP: Axonal dysfunction in the neuropathy of diabetes mellitus: A quantitative electrophysiological study. J Neurol Neurosurg Psychiatry 40:555–564, 1977.

169. Hansen, S, and Ballantyne, JP: A quantitative electrophysiological study of uraemic neuropathy: Diabetic and renal neuropathies compared. J Neurol Neurosurg Psychiatry 41:128–134, 1978.

170. Harati, Y, and Butler, IJ: Congenital hypomyelinating neuropathy. J Neurol Neurosurg Psychiatry 48:1269–1276, 1985.

171. Harding, AE: Friedreich's ataxia: A clinical and genetic study of 90 families with an analysis of early diagnostic criteria and intrafamilial clustering of clinical features. Brain 104:589–620, 1981.

172. Harding, AE, and Thomas, PK: The clinical features of hereditary motor and sensory neuropathy types I and II. Brain 103:259–280, 1980.

173. Harding, AE, and Thomas, PK: Autosomal recessive forms of hereditary motor and sensory neuropathy. J Neurol Neurosurg Psychiatry 43:669–678, 1980.

174. Harding, AE, and Thomas, PK: Peroneal muscular atrophy with pyramidal features. J Neurol Neurosurg Psychiatry 47:168–172, 1984.

175. Hausmanowa-Petrusewicz, I, Emeryk, B, Rowinska-Marcinska, K, and Jedrzejowksa, H: Nerve conduction in the Guillain-Barré-Strohl syndrome. J Neurol 220:169–184, 1979.

176. Hawley, RJ, Cohen, MH, Saini, N, and Armbrustmacher, VW: The carcinomatous neuromyopathy of oat cell lung cancer. Ann Neurol 7:65–72, 1980.

177. Heininger, K, Liebert, UG, Toyka, KV, Haneveld, FT, Schwendemann, G, Kolb-Bachofen, V, Ross, H, Cleveland, S, Besinger, UA, Gibbels, E, and Wechsler, W: Chronic inflammatory polyneuropathy: Reduction of nerve conduction velocities in monkeys by systemic passive transfer of immunoglobulin G. J Neurol Sci 66:1–14, 1984.

178. Hillbom, M, and Wennberg, A: Prognosis of alcoholic peripheral neuropathy. J Neurol Neurosurg Psychiatry 47:699–703, 1984.

179. Hong, CZ: Electrodiagnostic findings of persisting polyneuropathy due to previous nutritional deficiency in former prisoners of war. Electromyogr Clin Neurophysiol 26:351–363, 1986.

180. Horowitz, SH, and Ginsberg-Fellner, F: Ischemia and sensory nerve conduction in diabetes mellitus. Neurology 29:695–704, 1979.

181. Horwich, MS, Cho, L, Porro, RS, and Posner, JB: Subacute sensory neuropathy: A remote effect of carcinoma. Ann Neurol 2:7–19, 1977.

182. Izzo, KL, Sobel, E, and Demopoulos, JT: Dia-

betic neuropathy: Electrophysiologic abnormalities of distal lower extremity sensory nerves. Arch Phys Med Rehabil 67:7–11, 1986.

183. Jacobs, JM, and Costa-Jussa, FR: The pathology of amiodarone neurotoxicity. II. Peripheral neuropathy in man. Brain 108:753–769, 1985.

184. Jamal, GA, and MacLeod, WN: Electrophysiologic studies in Miller Fisher syndrome. Neurology 34:685–688, 1984.

185. Johnson, EW, and Waylonis, GW: Facial nerve conduction delay in patients with diabetes mellitus. Arch Phys Med Rehabil 45:131–139, 1964.

186. Johnson, PC: Hematogenous metastases of carcinoma to dorsal root ganglia. Acta Neuropathol (Berl) 38:171–172, 1977.

187. Johnson, PC, Doll, S, and Cromey, D: Pathogenesis of diabetic neuropathy. Ann Neurol 19:450–457, 1986.

188. Johnson, PC, Rolak, LA, Hamilton, RH, and Laguna, JF: Paraneoplastic vasculitis of nerve: A remote effect of cancer. Ann Neurol 5:437–444, 1979.

189. Jones, SJ, Carroll, WM, and Halliday, AM: Peripheral and central sensory nerve conduction in Charcot-Marie-Tooth disease and comparison with Friedreich's ataxia. J Neurol Sci 61:135–248, 1983.

190. Julien, J, Vital, C, Vallat, JM, Ferrer, X, and LeBoutet, MJ: Chronic demyelinating neuropathy with IgM-producing lymphocytes in peripheral nerve and delayed appearance of "benign" monoclonal gammopathy. Neurology 34:1387–1389, 1984.

191. Kahn, P: Anderson-Fabry disease: A histopathological study of three cases with observations on the mechanism of production of pain. J Neurol Neurosurg Psychiatry 36:1053–1062, 1973.

192. Kahn, SN, Riches, PG, and Kohn, J: Paraproteinemia in neurological disease: Incidence association and classification of monoclonal immunoglobulins. J Clin Pathol 33:617–621, 1980.

193. Katrak, SM, Pollock, M, O'Brien, CP, Nukada, H, Allpress, S, Calder, C, Palmer, DG, Grennan, DM, McCormack, PL, and Laurent, MR: Clinical and morphological features of gold neuropathy. Brain 103:671–693, 1980.

194. Kaufman, MD, Hopkins, LC, and Hurwitz, BJ: Progressive sensory neuropathy in patients without carcinoma: A disorder with distinctive clinical and electrophysiological findings. Ann Neurol 9:237–242, 1981.

195. Kelly, JJ, Jr: The electrodiagnostic findings in peripheral neuropathies associated with monoclonal gammopathies. Muscle Nerve 6:504–509, 1983.

196. Kelly, JJ, Jr: Peripheral neuropathies associated with monoclonal proteins: A clinical review. Muscle Nerve 8:138–150, 1985.

197. Kelly, JJ, Jr, Kyle, RA, Miles, JM, and Dyck, PJ: Osteosclerotic myeloma and peripheral neuropathy. Neurology 33:202–210, 1983.

198. Kelly, JJ, Jr, Kyle, RA, O'Brien, PC, and Dyck, PJ: The natural history of peripheral neuropathy in primary systemic amyloidosis. Ann Neurol 6:1–7, 1979.

199. Kikta, DG, Breuer, AC, and Wilbourn, AJ: Thoracic root pain in diabetes: The spectrum of clinical and electromyographic findings. Ann Neurol 11:80–85, 1982.

200. Killian, JM, and Kloepfer, HW: Homozygous expression of a dominant gene for Charcot-Marie-Tooth neuropathy. Ann Neurol 5:515–522, 1979.

201. Kimura, J: An evaluation of the facial and trigeminal nerves in polyneuropathy: Electrodiagnostic study in Charcot-Marie-Tooth disease, Guillain-Barre syndrome, and diabetic neuropathy. Neurology 21:745–752, 1971.

202. Kimura, J: F-wave velocity in the central segment of the median and ulnar nerves: A study in normal subjects and in patients with Charcot-Marie-Tooth disease. Neurology 24:539–546, 1974.

203. Kimura, J: Proximal versus distal slowing of motor nerve conduction velocity in the Guillain-Barre syndrome. Ann Neurol 3:344–350, 1978.

204. Kimura, J, Bosch, P, and Lindsay, GM: F-wave conduction velocity in the central segment of the peroneal and tibial nerves. Arch Phys Med Rehabil 56:492–497, 1975.

205. Kimura, J, and Butzer, JF: F-wave conduction velocity in Guillain-Barre syndrome: Assessment of nerve segment between axilla and spinal cord. Arch Neurol 32:524–529, 1975.

206. Kimura, J, Yamada, T, and Stevland, NP: Distal slowing of motor nerve conduction velocity in diabetic polyneuropathy. J Neurol Sci 42:291–302, 1979.

207. King, D, and Ashby, P: Conduction velocity in the proximal segments of a motor nerve in the Guillain-Barre syndrome. J Neurol Neurosurg Psychiatry 39:538–544, 1976.

208. Kissel, JT, Slivka, AP, Warmolts, JR, and Mendell, JR: The clinical spectrum of necrotizing angiopathy of the peripheral nervous system. Ann Neurol 18:251–257, 1985.

209. Kocen, RS, King, RHM, Thomas, PK, and Haas, LF: Nerve biopsy findings in two cases of Tangier disease. Acta Neuropathol (Berl) 26:317–327, 1973.

210. Kocen, RS, and Thomas, PK: Peripheral nerve involvement in Fabry's disease. Arch Neurol 22:81–88, 1970.

211. Koller, WC, Gehlmann, LK, Malkinson, FD, and Davis, FA: Dapsone-induced peripheral neuropathy. Arch Neurol 34:644–646, 1977.

212. Kosik, KS, Mullins, TF, Bradley, WG, Tempelis, LD, and Cretella, AJ: Coma and axonal degeneration in vitamin B_{12} deficiency. Arch Neurol 37:590–592, 1980.

213. Krendel, DA, Albright, RE, and Graham, DG: Infiltrative polyneuropathy due to acute monoblastic leukemia in hematologic remission. Neurology 37:474–477, 1987.

214. Kurdi, A, and Abdul-Kader, M: Clinical and electrophysiological studies of diphtheritic neuritis in Jordan. J Neurol Sci 42:243–250, 1979.

215. Kuritzky, A, Berginer, VM, and Korczyn, AD: Peripheral neuropathy in cerebrotendinous xanthomatosis. Neurology 29:880–881, 1979.

216. Kyle, RA, and Bayrd, ED: Amyloidosis: Review of 236 cases. Medicine (Baltimore) 54:271–299, 1975.

217. Lambert, EH: Electromyography and electric stimulation of peripheral nerves and muscle. In Clinical Examinations in Neurology, ed 4. Departments of Neurology, Physiology, and Biophysics, Mayo Clinic and Mayo Foundation. WB Saunders, Philadelphia, 1956, pp 298–329.

218. Lambert, EH, and Mulder, DW: Nerve conduction in the Guillain-Barré syndrome (abstr). Electroenceph Algor Clin Neurophysiol 17:86, 1964.

219. Lamontagne, A, and Buchthal, F: Electrophysiological studies in diabetic neuropathy. J Neurol Neurosurg Psychiatry 33:442–452, 1970.

220. Langohr, HD, Wietholter, H, and Peiffer, J: Muscle wasting in chronic alcoholics: Comparative histochemical and biochemical studies. J Neurol Neurosurg Psychiatry 46:248–254, 1983.

221. Layzer, RB, Fishman, RA, and Schafer, JA: Neuropathy following abuse of nitrous oxide. Neurology 28:504–506, 1978.

222. Lazaro, RP, and Kirshner, HS: Proximal muscle weakness in uremia: Case reports and review of the literature. Arch Neurol 37:555–558, 1980.

223. Le Quesne, PM: Neurophysiological investigation of subclinical and minimal toxic neuropathies. Muscle Nerve 1:392–395, 1978.

224. Le Quesne, PM: Neuropathy due to drugs. In Dyck, PJ, Thomas, PK, Lambert, EH, and Bunge, R (eds): Peripheral Neuropathy, Vol 2. WB Saunders, Philadelphia, 1984, pp 2162–2179.

225. Lewis, RA, and Sumner, AJ: The electrodiagnostic distinctions between chronic, familial and acquired demyelinative neuropathies. Neurology 32:592–596, 1982.

226. Lewis, RA, Sumner, AJ, Brown, MJ, and Asbury, AK: Multifocal demyelinating neuropathy with persistent conduction block. Neurology 32:958–964, 1982.

227. Lieberman, JS, Oshtory, M, Taylor, RG, and Dreyfus, PM: Perinatal neuropathy as an early manifestation of Krabbe's disease. Arch Neurol 37:446–447, 1980.

228. Limos, LC, Ohnishi, A, Sakai, T, Fujii, N, Goto, I, and Kuroiwa, Y: "Myopathic" changes in chorea-acanthocytosis: Clinical and histopathological studies. J Neurol Sci 55:49–58, 1982.

229. Lipkin, WI, Parry, G, Kiprov, D, and Abrams, D: Inflammatory neuropathy in homosexual men with lymphadenopathy. Neurology 35:1479–1483, 1985.

230. Lisak, RP, Mitchell, M, Zweiman, B, Orrechio, E, and Asbury, AK: Guillain-Barré syndrome and Hodgkin's disease: Three cases with immunological studies. Ann Neurol 1:72–78, 1977.

231. Livingstone, IR, Mastaglia, FL, Edis, R, and Howe, JW: Visual involvement in Friedreich's ataxia and hereditary spastic ataxia: A clinical and visual evoked response study. Arch Neurol 38:75–79, 1981.

232. Lopez-Alburquerque, T, Ortin, A, Arcaya, J, Cacho, J, and De Portugal-Alvarez, J: Proximal sensory conduction in diabetic patients. Electroencephalogr Clin Neurophysiol 66:25–28, 1987.

233. Lotti, M, Becker, CE, and Aminoff, MJ: Organophosphate polyneuropathy: pathogenesis and prevention. Neurology 34:658–662, 1984.

234. Low, PA, Burke, WJ, and McLeod, JG: Congenital sensory neuropathy with selective loss of small myelinated fibers. Ann Neurol 3:179–182, 1978.

235. Low, PA, Schmelzer, J, Dyck, PJ, and Kelly, JJ, Jr: Endoneurial effects of sera from patients with acute inflammatory polyradiculoneuropathy: Electrophysiologic studies on normal and demyelinated rat nerves. Neurology 32:720–724, 1982.

236. Low, PA, Walsh, JC, Huang, CY, and McLeod, JG: The sympathetic nervous system in diabetic neuropathy: A clinical and pathological study. Brain 98:341–356, 1975.

237. Lowry, NJ, Taylor, MJ, Belknapp, W, and Logan, WJ: Electrophysiological studies in five cases of abetalipoproteinemia. Can J Neurol Sci 11:60–63, 1984.

238. Madrid, R, and Bradley, WG: The pathology of neuropathies with focal thickening of the myelin sheath (tomaculous neuropathy): Studies on the formation of the abnormal myelin sheath. J Neurol Sci 25:415–448, 1975.

239. Madrid, R, Bradley, WG, and Davis, CJF: The peroneal muscular atrophy syndrome. Part 2. Observations on pathological changes in sural nerve biopsies. J Neurol Sci 32:91–122, 1977.

240. Magora, A, Sheskin, J, Sagher, F, and Gonen, B: The condition of the peripheral nerve in leprosy under various forms of treatment: Conduction velocity studies in long-term follow-up. Int J Leprosy 39:639–652, 1971.

241. Majnemer, A, Rosenblatt, B, Watters, G, and Andermann, F: Giant axonal neuropathy: Central abnormalities demonstrated by evoked potentials. Ann Neurol 19:394–396, 1986.

242. Malinow, K, Yannakakis, GD, Glusman, SM, Edlow, DW, Griffin, J, Pestronk, A, Powell, DL, Ramsey-Goldman, R, Eidelman, BH, Medsger, TA, Jr, and Alexander, EL: Subacute sensory neuronopathy secondary to dorsal root ganglionitis in primary Sjögren's syndrome. Ann Neurol 20:535–537, 1986.

243. Martinez, AC: Diabetic neuropathy, topography, general electrophysiologic features, effect of ischemia on nerve evoked potential and frequency of the entrapment neuropathy. Electromyogr Clin Neurophysiol 26:283–295, 1986.

244. Martinez, AC: Spinal evoked potentials and single fibre EMG in diabetic neuropathy. Electromyogr Clin Neurophysiol 26:499–511, 1986.

245. Martinez-Arizala, A, Sobol, SM, McCarty, GE, Nichols, BR, and Rakita, L: Amiodarone neuropathy. Neurology 33:643–645, 1983.

246. Massey, EW, Moore, J, and Schold, SC: Mental neuropathy from systemic cancer. Neurology 31:1277–1281, 1981.

247. Mateer, JE, Gutmann, L, and McComas, CF: Myokymia in Guillain-Barré syndrome. Neurology 33:374–376, 1983.

248. Matthews, WB, and Esiri, M: The migrant sensory neuritis of Wartenberg. J Neurol Neurosurg Psychiatry 46:1–4, 1983.

249. Mawdsley, C, and Mayer, RF: Nerve conduction in alcoholic polyneuropathy. Brain 88:335–356, 1965.

250. Mayer, RF: Hereditary neuropathy manifested by recurring nerve palsies. In Vinken, PJ, and Bruyn, GW (eds): Handbook of Clinical Neurology, Vol 21. American Elsevier, New York, 1975, pp 87–105.

251. Mayer, RF, and Denny-Brown, D: Conduction velocity in peripheral nerve during experimental demyelination in the cat. Neurology 14:714–726, 1964.

252. McCombe, PA, and McLeod, JG: The peripheral neuropathy of vitamin B_{12} deficiency. J Neurol Sci 66:117–126, 1984.

253. McCombe, PA, McLeod, JG, Pollard, JD, Guo, YP, and Ingall, TJ: Peripheral sensorimotor and autonomic neuropathy associated with systemic lupus erythematosus. Brain 110:533–549, 1987.

254. McCombe, PA, McManis, PG, Frith, JA, Pollard, JD, and McLeod, JG: Chronic inflammatory demyelinating polyradiculoneuropathy associated with pregnancy. Ann Neurol 21:102–104, 1987.

255. McDonald, WI: Experimental neuropathy: The use of diphtheria toxin. In Desmedt, JE (ed): New Developments in Electromyography and Clinical Neurophysiology, Vol 2. Karger, Basel, 1973, pp 128–144.

256. McLeod, JG: An electrophysiological and pathological study of peripheral nerves in Friedreich's ataxia. J Neurol Sci 12:333–349, 1971.

257. McLeod, JG: Carcinomatous neuropathy. In Dyck, PJ, Thomas, PK, Lambert, EH, and Bunge, R (eds): Peripheral Neuropathy, Vol 2. WB Saunders, Philadelphia, 1984, pp 2180–2191.

258. McLeod, JG: Electrophysiological studies in the Guillain-Barré syndrome. Ann Neurol 9(Suppl):20–27, 1981.

259. McLeod, JG, and Evans,WA: Peripheral neuropathy in spinocerebellar degenerations. Muscle Nerve 4:51–61, 1981.

260. McLeod, JG, Hargrave, JC, Walsh, JC, Booth, GC, Gye, RS, and Barron, A: Nerve conduction studies in leprosy. Int J Leprosy 43:21–31, 1975.

261. McLeod, JG, and Tuck, RR: Disorders of the autonomic nervous system: Part 1. Pathophysiology and Clinical Features. Ann Neurol 21:419–430, 1987.

262. McLeod, JG, Tuck, RR, Pollard, JD, Cameron, J, and Walsh, JC: Chronic polyneuropathy of undetermined cause. J Neurol Neurosurg Psychiatry 47:530–535, 1984.

263. McLeod, JG, Walsh, JC, and Pollard, D: Neuropathies associated with paraproteinemias and dysproteinemias. In Dyck, PJ, Thomas, PK, Lambert, EH, and Bunge, R (eds): Peripheral Neuropathy, Vol 2. WB Saunders, Philadelphia, 1984, pp 1847–1865.

264. Melgaard, B, Hansen, HS, Kamieniecka, Z, Paulson, OB, Pedersen, AG, Tang, X, and Trojaborg, W: Misonidazole neuropathy: A clinical, electrophysiological, and histological study. Ann Neurol 12:10–17, 1982.

265. Mendell, JR, Kissel, JT, Kennedy, MS, Sahenk, Z, Grinvalsky, HT, Pittman, GL, Kyler, RS, Roelofs, RI, Whitaker, JN, and Bertorini, TE: Plasma exchange and prednisone in Guillaine-Barré syndrome: A controlled randomized trial. Neurology 35:1551–1555, 1985.

266. Miller, RG, Davis, CJF, Illingworth, DR, and Bradley, W: The neuropathy of abetalipoproteinemia. Neurology 30:1286–1291, 1980.

267. Millette, TJ, Subramony, SH, Wee, AS, and Harisdangkul, V: Systemic lupus erythematosus presenting with recurrent acute demyelinating polyneuropathy. Eur Neurol 25:397–402, 1986.

268. Mitsumoto, H, Wilbourn, AJ, and Goren, H: Perineurioma as the cause of localized hypertrophic neuropathy. Muscle Nerve 3:403–412, 1980.

269. Mitz, M, Prakash, AS, Melvin, J, and Piering, W: Motor nerve conduction indicators in uremic neuropathy. Arch Phys Med Rehabil 61:45–48, 1980.

270. Mizuno, Y, Otsuka, S, Takano, Y, Suzuki, Y, Hosaka, A, Kaga, M, and Segawa, M: Giant axonal neuropathy: Combined central and peripheral nervous system disease. Arch Neurol 36:107–108, 1979.

271. Mokri, B, Ohnishi, A, and Dyck, PJ: Disulfiram neuropathy. Neurology 31:730–735, 1981.

272. Morgan-Hughes, JA, Sinclair, S, and Durston, JHJ: The pattern of peripheral nerve regeneration induced by crush in rats with severe acrylamide neuropathy. Brain 97:235–250, 1974.

273. Mulder, DW, Lambert, EH, Bastron, JA, and Sprague, RG: The neuropathies associated with diabetes mellitus: A clinical and electromyographic study of 103 unselected diabetic patients. Neurology 11:275–284, 1961.

274. Murphy, MJ, Lyon, LW, and Taylor, JW: Subacute arsenic neuropathy: Clinical and electrophysiological observations. J Neurol Neurosurg Psychiatry 44:896–900, 1981.

275. Murray, NMF, and Wade, DT: The sural sensory action potential in Guillain-Barré syndrome (corres). Muscle Nerve 3:444, 1980.

276. Myers, RR, Powell, HC, Shapiro, HM, Costello, ML, and Lambert, PW: Changes in endoneurial fluid pressure, permeability, and peripheral nerve ultrastructure in experimental lead neuropathy. Ann Neurol 8:392–401, 1980.

277. Nagashima, K, Suzuki, S, Ichikawa, E, Uchida, S, Honma, T, Kuroume, T, Hirato, J,

Ogawa, A, and Ishida, Y: Infantile neuroaxonal dystrophy: Perinatal onset with symptoms of diencephalic syndrome. Neurology 35:735–738, 1985.

278. Nemni, R, Galassi, G, Cohen, M, Hays, AP, Gould, R, Singh, N, Bressman, S, and Gamboa, ET: Symmetric sarcoid polyneuropathy: Analysis of a sural nerve biopsy. Neurology 31:1217–1223, 1981.

279. Nielsen, VK: The peripheral nerve function in chronic renal failure. V. Sensory and motor conduction velocity. Acta Med Scand 194:445–454, 1973.

280. Nielsen, VK: The peripheral nerve function in chronic renal failure. VII. Longitudinal course during terminal renal failure and regular hemodialysis. Acta Med Scand 195:155–162, 1974.

281. Nielsen, VK: The peripheral nerve function in chronic renal failure. IX. Recovery after renal transplantation. Electrophysiological aspects (sensory and motor nerve conduction). Acta Med Scand 195:171–180, 1974.

282. Nielsen, VK, and Pilgaard, S: On the pathogenesis of Charcot-Marie-Tooth disease: A study of the sensory and motor conduction velocity in the median nerve. Acta Orthop Scand 43:4–18 1972.

283. Noel, P: Diabetic neuropathy. In Desmedt, JE (ed): New Developments in Electromyography and Clinical Neurophysiology, Vol 2. Karger, Basel, 1973, pp 318–332.

284. Noel, P: Sensory nerve conduction in the upper limbs at various stages of diabetic neuropathy. J Neurol Neurosurg Psychiatry 36:786–796, 1973.

285. Noel, P, Lauvaux, JP, and Pirart, J: Upper limbs diabetic neuropathy: A clinical and electrophysiological study. Horm Metab Res 3:386–393, 1971.

286. O'Connor, CR, Rubinow, A, Brandwein, S, and Cohen, AS: Familial amyloid polyneuropathy: A new kinship of German ancestry. Neurology 34:1096–1099, 1984.

287. Odusote, K, Ohwovoriole, A, and Roberts, O: Electrophysiologic quantification of distal polyneuropathy in diabetes. Neurology 35:1432–1437, 1985.

288. Oh, SJ: Subacute demyelinating polyneuropathy responding to corticosteroid treatment. Arch Neurol 35:509–516, 1978.

289. Oh, SJ: Sarcoid polyneuropathy: A histologically proved case. Ann Neurol 7:178–181, 1980.

290. Oh, SJ, Clements, RS, Jr, Lee, YW, and Diethelm, AG: Rapid improvement in nerve conduction velocity following renal transplantation. Ann Neurol 4:369–373, 1978.

291. Ohi, T, Nukada, H, Kyle, RA, and Dyck, PJ: Detection of an axonal abnormality in myeloma neuropathy (abstr). Ann Neurol 14:120, 1983.

292. Ohnishi, A, and Ogawa, M: Preferential loss of large lumbar primary sensory neurons in carcinomatous sensory neuropathy. Ann Neurol 20:102–104, 1986.

293. Olney, RK, and Miller, RG: Peripheral neuropathy associated with disulfiram administration. Muscle Nerve 3:172–175, 1980.

294. Osby, LE, Noring, L, Hast, R, Kjellin, DG, Knutsson, E, and Siden, A: Benign monoclonal gammopathy and peripheral neuropathy. Br J Haematol 51:531–539, 1982.

295. Ouvrier, RA, McLeod, JG, and Conchin, TE: Friedreich's ataxia: Early detection and progression of peripheral nerve abnormalities. J Neurol Sci 55:137–145, 1982.

296. Ouvrier, RA, McLeod, JG, and Conchin, TE: The hypertrophic forms of hereditary motor and sensory neuropathy: A study of hypertrophic Charcot-Marie-Tooth disease (HMSN type I) and Dejerine-Sottas disease (HMSN type III) in childhood. Brain 110:121–148, 1987.

297. Ouvrier, RA, McLeod, JG, Morgan, GJ, Wise, GA, and Conchin, TE: Hereditary motor and sensory neuropathy of neuronal type with onset in early childhood. J Neurol Sci 51:181–197, 1981.

298. Parry, GJ, and Bredesen, DE: Sensory neuropathy with low-dose pyridoxine. Neurology 35:1466–1468, 1985.

299. Paul, T, Katiyar, BC, Misra, S, and Pant, GC: Carcinomatous neuromuscular syndromes: A clinical and quantitative electrophysiological study. Brain 101:53–63, 1978.

300. Pedersen, L, and Trojaborg, W: Visual, auditory and somatosensory pathway involvement in hereditary cerebellar ataxia, Friedreich's ataxia and familial spastic paraplegia. Electroencephalogr Clin Neurophysiol 52:283–297, 1981.

301. Pellissier, JF, Pouget, J, Cros, D, De Victor, B, Serratrice, G, and Toga, M: Peripheral neuropathy induced by amiodarone chlorhydrate: A clinicopathological study. J Neurol Sci 63:251–266, 1984.

302. Perticoni, G, Abbritti, G, Cantisani, T, Bondi, L, and Mauro, L: Polyneuropathy in workers with long exposure to vinyl chloride: Electrophysiological study. Electromyogr Clin Neurophysiol 26:41–47, 1986.

303. Peyronnard, JM, Charron, L, Beaudet, F, and Couture, F: Vasculitic neuropathy in rheumatoid disease and Sjögren syndrome. Neurology 32:839–845, 1982.

304. Phillips, LH, II, Kelly, TE, Schnatterly, P, and Parker, D: Hereditary motor-sensory neuropathy (HMSN): Possible X-linked dominant inheritance. Neurology 35:498–502, 1985.

305. Pollard, JD, McLeod, JG, Angel Honnibal, TG, and Verheijden, MA: Hypothyroid polyneuropathy. J Neurol Sci 53:461–471, 1982.

306. Pollard, JD, McLeod, JG, Gatenby, P, and Kronenberg, H: Prediction of response to plasma exchange in chronic relapsing polyneuropathy. J Neurol Sci 58:269–287, 1983.

307. Pollock, M, Nukada, H, Frith, RW, Simcock, JP, and Allpress, S: Peripheral neuropathy in Tangier disease. Brain 106:911–928, 1983.

308. Prineas, JW, and McLeod, JG: Chronic relapsing polyneuritis. J Neurol Sci 27:427–458, 1976.

309. Raff, MC, and Asbury, AK: Ischemic mononeuropathy and mononeuropathy multiplex in

diabetes mellitus. N Engl J Med 279:17–22, 1968.

310. Ramirez, JA, Mendell, JR, Warmolts, JR, and Griggs, RC: Phenytoin neuropathy: Structural changes in the sural nerve. Ann Neurol 19:162–167, 1986.

311. Read, DJ, Vanhegan, RI, and Matthews, WB: Peripheral neuropathy and benign IgG paraproteinaemia. J Neurol Neurosurg Psychiatry 41:215–219, 1978.

312. Rechthand, E, Cornblath, DR, Stein, BJ, and Meyerhoff, JO: Chronic demyelinating polyneuropathy in systemic lupus erythematosus. Neurology 34:1375–1377, 1984.

313. Reinstein, L, Pargament, JM, and Goodman, JS: Peripheral neuropathy after multiple tetanus toxoid injections. Arch Phys Med Rehabil 63:332–334, 1982.

314. Ridley, A: The neuropathy of acute intermittent porphyria. Q J Med 38:307–333, 1969.

315. Riley, CM, Day, RL, Greeley, DMcL, and Langford, WS: Central autonomic dysfunction with defective lacrimation. I. Report of 5 cases. Paediatrics 3:468–478, 1949.

316. Ro, YI, Alexander, CB, and Oh, SJ: Multiple sclerosis and hypertrophic demyelinating peripheral neuropathy. Muscle Nerve 6:312–316, 1983.

317. Roelofs, RI, Hrushesky, W, Rogin, J, and Rosenberg, L: Peripheral sensory neuropathy and cisplatin chemotherapy. Neurology 34:934–938, 1984.

318. Roig, M, Santamaria, J, Fernandez, E, and Colomer, J: Peripheral neuropathy in meningococcal septicemia. Eur Neurol 24:310–313, 1985.

319. Ropper, AH: Severe acute Guillain-Barré syndrome. Neurology 36:429–432, 1986.

320. Ropper, AH, and Kehne, SM: Guillain-Barré syndrome: Management of respiratory failure. Neurology 35:1662–1665, 1985.

321. Rosenberg, NL, and Bourdette, D: Hypertrophic neuropathy and multiple sclerosis. Neurology 33:1361–1364, 1983.

322. Rossini, PM, Treviso, M, Di Stefano, E, and Di Paola, B: Nervous impulse propagation along peripheral and central fibers in patients with chronic renal failure. Electroencephalalogr Clin Neurophysiol 56:293–303, 1983.

323. Sahenk, Z, Mendell, JR, Couri, D, and Nachtman, J: Polyneuropathy from inhalation of N20 cartridges through a whipped cream dispenser. Neurology 28:485–487, 1978.

324. Said, G: Perhexiline neuropathy: A clinicopathological study. Ann Neurol 3:259–266, 1978.

325. Said, G: A clinicopathologic study of acrodystrophic neuropathies. Muscle Nerve 3:491–501, 1980.

326. Said, G, Boudier, L, Selva, J, Zingraff, J, and Drueke, T: Different patterns of uremic polyneuropathy: Clinicopathologic study. Neurology 33:567–574, 1983.

327. Sales Luis, ML: Electroneurophysiological studies in familial amyloid polyneuropathy—Portuguese type. J Neurol Neurosurg Psychiatry 41:847–850, 1978.

328. Salisachs, P: Ataxia and other data reviewed in Charcot-Marie-Tooth and Refsum's disease. J Neurol Neurosurg Psychiatry 45:1085–1091, 1982.

329. Satchell, PM, McLeod, JG, Harper, B, and Goodman, AH: Abnormalities in the vagus nerve in canine acrylamide neuropathy. J Neurol Neurosurg Psychiatry 45:609–619, 1982.

330. Satran, R: Dejerine-Sottas disease revisited. Arch Neurol 37:67–68, 1980.

331. Sauron, B, Bouche, P, Cathala, HP, Chain, F, and Castaigne, P: Miller Fisher syndrome: Clinical and electrophysiologic evidence of peripheral origin in 10 cases. Neurology 34:953–956, 1984.

332. Schold, SC, Cho, ES, Somasundaram, M, and Posner, JB: Subacute motor neuronopathy: A remote effect of lymphoma. Ann Neurol 5:271–287, 1979.

333. Schwartz, MS, Mackworth-Young, CG, and McKernan, RO: The tarsal tunnel syndrome in hypothyroidism. J Neurol Neurosurg Psychiatry 46:440–442, 1983.

334. Sebille, A: Respective importance of different nerve conduction velocities in leprosy. J Neurol Sci 38:89–95, 1978.

335. Sebille, A, and Gray, F: Electromyographic recording and muscle biopsy in lepromatous leprosy. J Neurol Sci 40:3–10, 1979.

336. Server, AC, Lefkowith, J, Braine, H, and McKhann, GM: Treatment of chronic relapsing inflammatory polyradiculoneuropathy by plasma exchange. Ann Neurol 6:258–261, 1979.

337. Service, FJ, Daube, JR, O'Brien, PC, and Dyck, PJ: Effect of artificial pancreas treatment on peripheral nerve function in diabetes. Neurology 31:1375–1380, 1981.

338. Sherman, WH, Olarte, MR, McKiernan, G, Sweeney, K, Latov, N, and Hays, AP: Plasma exchange treatment of peripheral neuropathy associated with plasma cell dyscrasia. J Neurol Neurosurg Psychiatry 47:813–819, 1984.

339. Sheth, KJ, and Swick, HM: Peripheral nerve conduction in Fabry disease. Ann Neurol 7:319–323, 1980.

340. Shields, R, Jr, Root, K, and Wilbourn, A: Compartment syndromes and compression neuropathies in coma. Neurology 36:1370–1374, 1986.

341. Shorvon, SD, and Reynolds, EH: Anticonvulsant peripheral neuropathy: A clinical and electrophysiological study of patients on single drug treatment with phenytoin, carbamazepine or barbiturates. J Neurol Neurosurg Psychiatry 45:620–626, 1982.

342. Sima, AAF, and Robertson, DM: Involvement of peripheral nerve and muscle in Fabry's disease: Histologic, ultrastructural, and morphometric studies. Arch Neurol 35:291–301, 1978.

343. Sladky, J, Brown, M, and Berman, P: Chronic inflammatory demyelinating polyneuropathy of infancy: A corticosteroid-responsive disorder. Ann Neurol 20:76–81, 1986.

344. Smith, IS, Kahn, SN, Lacey, BW, King, RHM, Eames, RA, Whybrew, DJ, and Thomas, PK: Chronic demyelinating neuropathy associated with benign IgM paraproteinaemia. Brain 106:169–195, 1983.

345. Snyder, M, Cancilla, PA, and Batzdorf, U: Hypertrophic neuropathy simulating a neoplasm of the brachial plexus. Surg Neurol 7:131–134, 1977.

346. Sotaniemi, KA: Slimmer's paralysis: Peroneal neuropathy during weight reduction. J Neurol Neurosurg Psychiatry 47:564–566, 1984.

347. Spaans, F, Jennekens, FGI, Mirandolle, JF, Bijlsma, JB, and De Gast, GC: Myotonic dystrophy associated with hereditary motor and sensory neuropathy. Brain 109:1149–1168, 1986.

348. Spencer, PS, Schaumburg, HH, Raleigh, RL, and Terhaar, CJ: Nervous system degeneration produced by the industrial solvent methyl n-butyl ketone. Arch Neurol 32:219–222, 1975.

349. Steinman, L, Tharp, BR, Dorfman, LJ, Forno, LS, Sogg, RL, Kelts, KA, and O'Brien, JS: Peripheral neuropathy in the cherry-red spot-myoclonus syndrome (sialidosis type I). Ann Neurol 7:450–456, 1980.

350. Sterman, AB, Schaumburg, HH, and Asbury, AK: The acute sensory neuronopathy syndrome: A distinct clinical entity. Ann Neurol 7:354–358, 1980.

351. Stewart, RM, Tunnell, G, and Ehle, E: Familial spastic paraplegia, peroneal neuropathy and crural hypopigmentation: A new neurocutaneous syndrome. Neurology 31:754–757, 1981.

352. Subramony, SH, and Wilbourn, AJ: Diabetic proximal neuropathy. J Neurol Sci 53:293–304, 1982.

353. Swamy, HS, Shankar, SK, Chandra, PS, Aroor, SR, Krishna, AS, and Perumal, VGK: Neurological complications due to beta-propiolactone (BPL)-inactivated antirabies vaccination: Clinical, electrophysiological and therapeutic aspects. J Neurol Sci 63:111–128, 1984.

354. Swash, M, Perrin, J, and Schwartz, MS: Significance of immunoglobulin deposition in peripheral nerve in neuropathies associated with paraproteinaemia. J Neurol Neurosurg Psychiatry 42:179–183, 1979.

355. Swift, TR, Hackett, ER, Shipley, DE, and Miner, KM: The peroneal and tibial nerves in lepromatous leprosy: Clinical and electrophysiologic observations. Int J Leprosy 41:25–34, 1973.

356. Tang, LM, Hsi, MS, Ryu, SJ, and Minauchi, Y: Syndrome of polyneuropathy, skin hyperpigmentation, oedema and hepatosplenomegaly. J Neurol Neurosurg Psychiatry 46:1108–1114, 1983.

357. Teravainen, H, and Larsen, A: Some features of the neuromuscular complications of pulmonary carcinoma. Ann Neurol 2:495–502, 1977.

358. Thomas, PK: Peripheral neuropathy. In Matthews, WB (ed): Recent Advances in Clinical Neurology. Churchill Livingstone, Edinburgh, 1975, pp 253–283.

359. Thomas, PK: Screening for peripheral neuropathy in patients treated by chronic hemodialysis. Muscle Nerve 1:396–399, 1978.

360. Thomas, PK, and Calne, DB: Motor nerve conduction velocity in peroneal muscular atrophy: Evidence for genetic heterogeneity. J Neurol Neurosurg Psychiatry 37:68–75, 1974

361. Thomas, PK, and King, RHM: Peripheral nerve changes in amyloid neuropathy. Brain 97:395–406, 1974.

362. Thomas, PK, Walker, RWH, Rudge, P, Morgan-Hughes, JA, King, RHM, Jacobs, JM, Mills, KR, Ormerod, IEC, Murray, NMF, and McDonald, WI: Chronic demyelinating peripheral neuropathy associated with multifocal central nervous system demyelination. Brain 110:53–76, 1987.

363. Toyka, KV, Augspach, R, Paulus, W, Grabensee, B, and Hein, D: Plasma exchange in polyradiculoneuropathy. Ann Neurol 8:205–206, 1980.

364. Tredici, G, Minazzi, M, and Lampugnani, E: Peripheral neuropathy in angioimmunoblastic lymphadenopathy with dysproteinaemia. J Neurol Neurosurg Psychiatry 42:519–523, 1979.

365. Trockel, U, Schroder, JM, Reiners, KH, Toyka, KV, Goerz, G, and Freund, H-J: Multiple exercise-related mononeuropathy with abdominal colic. J Neurol Sci 60:431–442, 1983.

366. Trojaborg, W, Frantzen, E, and Andersen, I: Peripheral neuropathy and myopathy associated with carcinoma of the lung. Brain 92:71–82, 1969.

367. Troni, W, Carta, Q, Cantello, R, Caselle, MT, and Rainero, I: Peripheral nerve function and metabolic control in diabetes mellitus. Ann Neurol 16:178–183, 1984.

368. Trotter, JL, Engel, WK, and Ignaczak, TF: Amyloidosis with plasma cell dyscrasia: An overlooked cause of adult onset sensorimotor neuropathy. Arch Neurol 34:209–214, 1977.

369. Tschudy, DP, Valsamis, M, and Magnussen, CR: Acute intermittent porphyria: Clinical and selected research aspects. Ann Int Med 83:851–864, 1975.

370. Tsukada, N, Koh, CS, Inoue, A, and Yanagisawa, N: Demyelinating neuropathy associated with hepatitis B virus infection: Detection of immune complexes composed of hepatitis B virus surface antigen. J Neurol Sci 77:203–216, 1987.

371. Tuck, RR, and McLeod, JG: Retinitis pigmentosa, ataxia, and peripheral neuropathy. J Neurol Neurosurg Psychiatry 46:206–213, 1983.

372. Tuck, RR, Pollard, JD, and McLeod, JG: Autonomic neuropathy in experimental allergic neuritis: An electrophysiological and histological study. Brain 104:187–208, 1981.

373. Vallat, JM, Desproges-Gotteron, R, Leboutet, MJ, Loubet, A, Gualde, E, and Treves, R: Cryoglobulinemic neuropathy: A pathological study. Ann Neurol 8:179–185, 1980.

374. Vanasse, M, and Dubowitz, V: Dominantly in-

herited peroneal muscular atrophy (hereditary motor and sensory neuropathy type I) in infancy and childhood. Muscle Nerve 4:26–30, 1981.

375. van der Most van Spijk, D, Hoogland, RA, and Dijkstra, S: Conduction velocities compared and related to degrees of renal insufficiency. In Desmedt, JE (ed): New Developments in Electromyography and Clinical Neurophysiology, Vol 2. Karger, Basel, 1973, pp 381–389.

376. Van Weerden, TW, Houthoff, HJ, Sie, O, and Minderhoud, JM: Variability in nerve biopsy findings in a kinship with dominantly inherited Charcot-Marie-Tooth disease. Muscle Nerve 5:185–196, 1982.

377. Vasilescu, C: Motor nerve conduction velocity and electromyogram in triorthocresyl-phosphate poisoning. Rev Roum Neurol 9:345–350, 1972.

378. Vasilescu, C, Alexianu, M, and Dan, A: Delayed neuropathy after organophosphorus insecticide (dipterex) poisoning: A clinical electrophysiological and nerve biopsy study. J Neurol Neurosurg Psychiatry 47:543–548, 1984.

379. Vercruyssen, A, Martin, JJ, and Mercelis, R: Neurophysiological studies in adrenomyeloneuropathy. J Neurol Sci 56:327–336, 1982.

380. Verma, A, Bisht, MS, and Ahuja, GK: Involvement of central nervous system in diabetes mellitus. J Neurol Neurosurg Psychiatry 47:414–416, 1984.

381. Violante, F, Lorenzi, S, and Fusello, M: Uremic neuropathy: Clinical and neurophysiological investigation of dialysis patients using different chemical membranes. Eur Neurol 24:298–404, 1985.

382. Vital, C, Vallat, JM, Deminierre, C, Loubet, A, and Leboutet, MJ: Peripheral nerve damage during multiple myeloma and Waldenstrom's macroglobulinemia. Cancer 50:1491–1497, 1982.

383. Walsh, JC, and McLeod, JG: Alcoholic neuropathy: An electrophysiological and histological study. J Neurol Sci 10:457–469, 1970.

384. Walsh, JC, Yiannikas, C, and McLeod, JG: Abnormalities of proximal conduction in acute idiopathic polyneuritis: Comparison of short latency evoked potentials and F waves. J Neurol Neurosurg Psychiatry 47:197–200, 1984.

385. Waxman, SG, and Sabin, TD: Diabetic truncal polyneuropathy. Arch Neurol 38:46–47, 1981.

386. Weder, B, Meienberg, O, Wildi, E, and Meier, C: Neurologic disorder of vitamin E deficiency in acquired intestinal malabsorption. Neurology 34:1561–1565, 1984.

387. Wendt, JS, and Burks, JS: An unusual case of encephalomyeloradiculoneuropathy in a young woman. Arch Neurol 38:726–727, 1981.

388. Wichman, A, Buchthal, F, Pezeshkpour, GH, and Gregg, RE: Peripheral neuropathy in abetalipoproteinemia. Neurology 35:1279–1289, 1985.

389. Wilbourn, AJ, Furlan, AJ, Hulley, W, and Rauschhaupt, W: Ischemic monomelic neuropathy. Neurology 33:447–451, 1983.

390. Williams, D, Brust, JCM, Abrams, G, Challenor, Y, and Devereaux, M: Laundry-Guillain-Barré syndrome with abnormal pupils and normal eye movements: A case report. Neurology 29:1033–1036, 1979.

391. Williams, IR, Davison, AM, Mawdsley, C, and Robson, JS: Neuropathy in chronic renal failure. In Desmedt, JE (ed): New Developments in Electromyography and Clinical Neurophysiology, Vol 2. Karger, Basel, 1973, pp 390–399.

392. Windebank, AJ, Low, PA, Blexrud, MD, Schmelzer, JD, and Schaumburg, HH: Pyridoxine neuropathy in rats: Specific degeneration of sensory axons. Neurology 35:1617–1622, 1985.

393. Windebank, AJ, McCall, JT, and Dyck, PJ: Metal neuropathy. In Dyck, PJ, Thomas, PK, Lambert, EH, and Bunge, R (eds): Peripheral Neuropathy, Vol 2. WB Saunders, Philadelphia, 1984, pp 2133–2161.

394. Winegrad, AI, and Greene, DA: Diabetic polyneuropathy: The importance of insulin deficiency, hyperglycemia and alterations in myoinositol metabolism in its pathogenesis. N Engl J Med 295:1416–1421, 1976.

395. Wolf, E, Shochina, M, Fidel, Y, and Gonen, B: Phrenic neuropathy in patients with diabetes mellitus. Electromyogr Clin Neurophysiol 23:523–530, 1983.

396. Wulff, CH, and Trojaborg, W: Adult metachromatic leukodystrophy: Neurophysiologic findings. Neurology 35:1776–1778, 1985.

397. Yiannikas, C, McLeod, JG, and Walsh, JC: Peripheral neuropathy associated with polycythemia vera. Neurology 33:139–143, 1983.

398. Yudell, A, Gomez, MR, Lambert, EH, and Dockerty, MB: The neuropathy of sulfatide lipidosis (metachromatic leukodystrophy). Neurology 17:103–111, 1967.

Chapter **23**

MONONEUROPATHIES AND ENTRAPMENT SYNDROMES

1 INTRODUCTION

Despite the unpredictable nature of traumatic injury, certain individual nerves have predilection for isolated damage. These include the long thoracic, suprascapular, musculocutaneous, and axillary nerves in the shoulder girdle and the lateral femoral cutaneous, femoral, and sciatic nerves in the pelvic girdle. More distally, entrapment syndromes develop at the common sites of chronic or recurrent compression for the radial, median, ulnar, common peroneal, and tibial nerves.[2,95,145] Unusual sites of involvement may suggest rare anomalies such as congenital ring constrictions of peripheral nerve.[118]

The diagnosis of a focal nerve lesion depends on elucidating weakness and atrophy of all muscles supplied by the nerve distal to the lesion. Sensory findings, which usually appear earlier, provide less reliable localizing signs than motor deficits, particularly in the upper limbs, where sensory dermatomes overlap considerably. Electromyographic examination delineates the exact distribution of denervated muscles in localizing a focal nerve lesion. In demyelinative neuropathy a reduced recruitment pattern of motor unit potentials signals a conduction block, despite the absence of axonal loss. The pattern of distribution here also helps elucidate the zone of involvement.

Nerve conduction studies may detect evidence of demyelination, which usually precedes axonal degeneration in a compression neuropathy. Stimulation above and below the suspected site of the lesion will document not only the slowing of conduction velocity but also changes in the amplitude and area of the muscle or nerve action potential as indices of functional block. Such a pattern of abnormalities often helps differentiate an entrapment syndrome from a diffuse neuropathy. This distinction, however, may blur in types of polyneuropathy that in early stages mimic localized disease at the common sites of compression.

2 CRANIAL NERVES

Cranial nerves most frequently assessed in an electromyographic laboratory include the facial and accessory nerves. They both travel superficially and allow easy access to electrical stimulation from the surface. They also innervate the muscles readily approachable by needle or disk electrodes for recording.

Facial Nerve

Bell's palsy affects the facial nerve sporadically in isolated incidences. Patients with a rare familial type may suffer from recurrent episodes, which tend to cause increasing residual weakness after each attack.[7] The cause remains unknown, but swelling and hyperemia in the intraosseous portion of the facial nerve suggest a focal pathologic process during the acute stage. Paralysis of the upper and lower portions of the face develops suddenly, often associated with pain behind the ear. Additional features may include loss of taste in the anterior two thirds of the tongue and hyperacusis on the affected side. At least 80 percent of patients improve quickly without specific therapy.[223] A complete recovery follows a demyelinative form, whereas functions return slowly and poorly after degeneration of the facial nerve. Synkinesis nearly always develops with regeneration (see Fig. 16–9).[89] Patients may complain of sensory signs in the trigeminal distribution in an otherwise typical case of Bell's palsy.[194]

The same principles apply to the electromyographic examination of facial and limb muscles. In the face, however, physiologically small motor unit potentials may mimic fibrillation potentials, and signs of denervation appear early in less than 3 weeks after injury, presumably because of the short nerve length. Serial electrodiagnostic studies help delineate the course of the illness (see Fig. 16–3 and Tables 16–3 and 16–4). The amplitude of the direct response elicited by stimulation of the facial nerve provides the best means of determining prognosis after the fourth to fifth day of onset. An amplitude greater than one-half the control value on the normal side indicates a good prognosis, although late degeneration can still occur. If either R_1 or R_2 of the blink reflex is preserved or previously absent reflexes return while the direct response remains normal, or nearly normal, the patient will almost certainly recover completely (see Fig. 16–7). Thus, the presence of R_1 or R_2 offers reasonable assurance that the remaining axons will survive without undergoing further deterioration. Unfortunately, the reflex rarely recovers during the first few days after onset. In a series of 56 patients who recovered without distal degeneration, the reflex reappeared by the latter half of the first week in 57 percent, by the second week in 67 percent, and by the third week in 89 percent.[89] Other signs for good prognosis include incomplete clinical paresis and the presence of voluntary motor unit potentials in electromyography.

In the absence of substantial nerve degeneration, the latency of the direct response remains unaltered throughout the course on the affected side. In these patients, however, the latency of R_1 of the blink reflex, if present, shows mild delay during the first days and further increases during the latter half of the first week, then remains essentially unchanged up to the fourth week. It then shows a notable recovery during the second month, returning to a normal range during the third or fourth month (see Fig. 16–8). These findings suggest that most patients with Bell's palsy in whom little axonal degeneration develops suffer from a focal demyelination of the facial nerve. If the facial nerve undergoes substantial degeneration, the ultimate recovery depends on the completeness of regeneration. This process generally takes a few months to a few years, resulting almost always in aberrant reinnervation.[91]

Peripheral facial paresis secondary to herpes zoster infection carries a less favorable prognosis. Diabetic patients in whom a facial palsy develops also tend to have a more severe paresis and evidence of substantial denervation.[1] After facial nerve degeneration, autogenous grafting may give good functional recovery, although commonly associated with synkinesis and contracture.[214] Patients with Bannwarth's syndrome may develop unilateral or bilateral facial palsy as part of multiple mononeuritis associated with erythema, pain, elevated cerebrospinal fluid protein, and pleocytosis.[225]

The facial nerve is commonly affected during the course of polyneuropathy such as the Guillain-Barré syndrome and hereditary motor and sensory neuropathy type I (see Figs. 16–10A and 16–12).[84] Patients with acquired demyelinative neuropathy usually have prominent facial paresis, whereas those with hereditary neuropathy have minimal signs and symptoms in the face despite marked delay in conduction. Weakness usually occurs as the consequence of conduction block that typically accompanies acute, fulminating demyelination but not a chronic insidious process. In contrast, slow progression in the hereditary process allows development of compensatory mechanisms for motor function.

Acoustic neuroma and multiple sclerosis deserve special mention with regard to involvement of the facial nerve. The tumor, strategically located at the cerebellopontine angle, may compress not only the facial nerve but also the trigeminal nerve and the pons, that is, the efferent, afferent, and central arcs of the blink reflex.[90,92,110] Thus, the electrically elicited blink reflex shows abnormality in most patients (see Tables 16–2 and 16–4). Peripheral facial palsy may herald other symptoms of multiple sclerosis in young adults (see Fig. 16–11B). Electromyographic examination may reveal myokymic discharges, considered characteristic of this condition (see Fig. 13–9A). The blink reflex study may show an absent or delayed R_1, indicating demyelination of the central reflex arc, which includes the intrapontine portion of the facial nerve.[83,85,86]

Weakness of the orbicularis oculi and frontalis usually distinguishes a peripheral from central facial palsy. This differentiation, however, is not always easy, because of variability in the innervation pattern. In equivocal cases, an increased latency of electrically elicited R_1 will confirm that the lesion involves the facial nerve. The latency of R_1 may increase during an acute stage of contralateral hemispheric lesions if elicited by the glabellar tap.[45] In doubtful cases, paired stimuli minimize the supranuclear effect of reduced excitability to make the shortest latency of R_1 a more accurate measure of conduction along the reflex arc itself. The excitability of R_2 may show significant alteration with a hemispheric lesion, showing either an afferent or efferent pattern (see Chapter 16.6 and Fig. 16–17).

Accessory Nerve

Pressure from a tumor or surgical procedures of the posterior triangle can damage the spinal accessory nerve.[58,150,224] In trapezius palsies, after injury of the accessory nerve, the upper vertebral border of the scapula moves away from the spinal vertebrae. With the lower angle of the scapula relatively fixed by muscles supplied by the C-3 and C-4 roots through the cervical plexus, the whole scapula slips downward. The scapula also rotates from the physiologic position, displacing the superior angle outward and the inferior angle inward. This type of winging tends to worsen by abduction of the arm to the horizontal plane, that displaces the superior angle further laterally. The paralysis of the sternocleidomastoid causes weakness in rotating the face toward the opposite shoulder. The muscle is easily palpable for assessing the degree of atrophy. Bilateral involvement of the muscles makes flexion of the neck difficult. In a sequential study of patients with trapezius palsy, nerve conduction changes revealed evidence of spontaneous regeneration after complete axonal degeneration.[153]

3 NERVES OF THE SHOULDER GIRDLE

Certain peripheral nerves derived directly from the brachial plexus have a predilection for isolated injury by compression or stab wounds. The most commonly affected include the long thoracic, suprascapular, dorsal scapular, axillary, and musculocutaneous nerves.

Long Throracic Nerve

The nerve lies superficially in the supraclavicular region, where it is subject to trauma. In addition to stab wounds, the nerve may suffer direct pressure from a heavy shoulder bag or shoulder braces during surgery.[68] Radical mastectomy may also damage the nerve. Its straight course from origin to insertion also makes it vulnerable to stretch.

The serratus anterior, the only muscle innervated by the long thoracic nerve, functions as a stabilizer of the shoulder in abduction of the arm. It holds the scapula flat against the back by keeping its inner margin fixed to the thorax. With paralysis of this muscle, the patient cannot raise the arm up straight. Because of the unopposed action of the rhomboids and levator scapulae, the superior angle of the scapula is displaced medially, whereas the inferior angle swings laterally. Thus, the scapula rotates in a direction the reverse of that in which it rotates with a lesion of the accessory nerve. In addition, the vertebral border of the lower scapula projects backward, away from the thorax. This tendency, called scapular winging, worsens with the outstretched arm thrust forward. In contrast, winging of the scapula caused by trapezius weakness exaggerates with abduction of the arm laterally. Lesions of the long thoracic nerve give rise to isolated electromyographic abnormalities in the serratus anterior muscle. Conduction studies provide valuable information in distinguishing partial from complete degeneration and in revealing regeneration after wallerian degeneration.[154]

Suprascapular Nerve

Injury may result from pressure on the shoulder, stab wounds above the scapula,[52,191] improper use of crutches,[184] or stretching of the nerve, as may be seen in volleyball players during serving.[44] Rupture of the rotator cuff[77] or downward displacement of the upper trunk may also stretch the nerve anchored at the notch, a mechanism in part responsible for Erb's palsy. Injury to this nerve results in atrophy of the supraspinatus with weakness in initiating abduction of the arm and external rotation of the glenohumeral joint. The weakness of the infraspinatus muscle causes difficulty in external rotation of the arm at the shoulder, despite partial compensation by the teres minor and deltoid, innervated by the axillary nerve. Compressive lesions often induce a poorly defined aching pain along the posterior and lateral aspect of the shoulder joint and the adjacent scapula, supplied by the sensory branches.[160,165,167,168]

Stimulation at Erb's point may reveal an increase in suprascapular nerve latency, measured from the involved supraspinatus or infraspinatus muscle.[77] Electromyographic exploration shows evidence of denervation in the supraspinatus or infraspinatus, or both, but in no other muscles supplied by the C-5 and C-6 roots.

Dorsal Scapular Nerve

With entrapment or other injury of this nerve, the scapula tends to wing on wide abduction of the arm.[132] The patient may complain of pain in the C-5 and C-6 root distribution. The diagnosis depends on electromyographic demonstration of abnormalities restricted to the rhomboids, major and minor, and levator scapulae.

Axillary Nerve

The nerve may undergo degeneration as part of brachial plexus neuritis or as the result of selective injury. A partial nerve palsy sustained in association with fracture or dislocation of the head of the humerus usually makes a full recovery.[109] The lesion after blunt trauma to the shoulder carries a less favorable prognosis.[9] Other causes include the pressure of crutches or hyperextension of the shoulder, as might occur in wrestling. A circumscribed area of numbness develops in the lateral aspect of the upper arm over the belly of the deltoid. Atrophy of this muscle produces obvious flattening of the shoulder and limited abduction of the arm after the first 30 degrees subserved by the supraspinatus. In contrast, a C-5 root lesion weakens all 180 degrees with involvement of both muscles. An isolated lesion of the teres minor often escapes clinical detection in the presence of the infraspinatus, which also rotates the arm outward. Electromyographic abnormalities confined to the teres minor and deltoid help establish an axillary nerve palsy.

Musculocutaneous Nerve

Injuries of this nerve result from fractures or dislocations of the humerus,[109] gunshot or stab wounds, compression of the arm, entrapment by the coracobrachialis muscle, heavy exercise,[14] and rare complications of surgery.[34] Sensory examination reveals numbness along the lateral aspect of the forearm. Paralysis of the biceps results in weakness of elbow flexion, although the brachioradialis partially compensates for this deficit. The biceps stretch reflex is absent. Electromyography shows denervation in the biceps brachii, brachialis, and coracobrachialis. Nerve conduction studies of the musculocutaneous nerve may corroborate the diagnosis.[210]

Vigorous arm exercise with elbow extension and arm pronation may give rise to a compression syndrome of the lateral cutaneous nerve, the distal termination of the musculocutaneous nerve. The patient complains of pain or numbness along the lateral aspect of the distal forearm and tenderness to palpation over the nerve. Nerve conduction studies may show a decreased sensory amplitude and a prolonged distal latency.[43]

4 RADIAL NERVE

Focal damage of this nerve usually results from crush or twisting injury to the

wrist or forearm or from repetitive pronation and supination at work.[28] External
trauma at the spiral groove commonly injures the nerve with or without concomitant fracture of the humerus. A local
compression at this level also results
when an individual, often intoxicated,
falls asleep leaning against some hard
surface or with an arm draped over a
bench, as in the so-called Saturday night
palsy. In newborn infants, the umbilical
cord may play a role in the entrapment.[172] The lesion here usually spares
the triceps, involving all the remaining
extensor muscles of the hand, wrist, and
fingers as well as the brachioradialis. The
sensory losses vary but most often affect
the dorsum of the hand and first two
digits. Nerve injury at the axilla from an
incorrectly used crutch results in weakness of all the radial-innervated muscles
and the loss of the triceps stretch reflex.
Fractures of the head of the radius injure
the nerve more distally.

Compression of the recurrent epicon-
dylar branch gives rise to pain at the
elbow, usually with simultaneous entrapment of the deep branch of the radial
nerve. This syndrome, one of the many
entities commonly known as tennis
elbow, results from repeated indirect
trauma by forceful supination as the predisposing factor. Pain and tenderness localized to the lateral aspect of the elbow
resemble the symptoms of lateral epicon-
dylitis, another condition referred to by
some as tennis elbow. In the entrapment
syndrome, however, additional dysfunctions indicate the involvement of the radial nerve. Subluxation of the head of the
radius may produce a radial nerve palsy.
Superficial radial neuropathy may develop after wearing a tight watch
band.[161] Handcuff-related compression
injuries often involve the sensory fibers of
the radial nerve with or without concomitant involvement of the median or ulnar
nerves at the wrist.[32,105,120]

Conduction studies may reveal the absence of both motor and sensory potentials after a fracture of the humerus or
slowing of conduction along the compression site at the spiral groove. The size of
the muscle or antidromic sensory poten-

tial elicited by stimulus distal to the presumed site of the lesion can differentiate
between neurapraxia and axonotmesis,
although these two types of injury coexist
to a varying degree in most cases. Pressure neuropathy of the radial nerve usually resolves in 6 to 8 weeks, but the recovery takes considerably longer if a
substantial number of axons have degenerated. Electromyographic exploration
helps demonstrate the type and location
of injury (see Figs. 13–13 and 13–16).[209]

Posterior Interosseous Syndrome

The posterior interosseous nerve, as
the terminal motor branch of the radial
nerve in the forearm, penetrates the supinator muscle in its entrance to the forearm. The compression syndrome here
may develop spontaneously or after
closed injuries to the elbow[76] or occasionally in rheumatoid arthritis with synovitis.[122] The entrapment usually involves
the nerve at the arcade of Frohse between
the two heads of the supinator.[140] A lesion at this level causes weakness in the
extensors of the wrist and digits with a
notable sparing of the supinator, which
receives innervation proximal to the site
of compression. The patient complains of
pain over the lateral aspect of the elbow
but no sensory loss. The radial nerve
proper supplies the extensor carpi radialis longus and brevis. Normal contraction of these muscles despite the weak
extensor carpi ulnaris results in characteristic radial deviation of the wrist on
attempted dorsiflexion. The differential
diagnosis includes rupture of the extensor tendons, especially if paralysis affects
only the last three digits, with preservation of the first two. In this case, passive
palmar flexion of the wrist induces no
extension of the metacarpophalangeal
joints.

In addition to electromyographic findings in the denervated muscles, conduction studies may reveal mild abnormalities across the entrapment, especially if
tested with the arm supinated against
resistance.[171]

5 MEDIAN NERVE

The median nerve traverses three common sites of compression along its course. At the elbow, entrapment may occur between the two heads of the pronator teres or more distally with selective involvement of the anterior interosseous branch. The carpal tunnel syndrome results from compression at the distal edge of the transverse carpal ligament or, less commonly, within the intermetacarpal tunnel.

Pronator Teres Syndrome

In 83 percent of dissections, the median nerve pierces the two heads of the pronator teres before passing under it.[6] Trauma, fracture, muscle hypertrophy, or an anomalous fibrous band connecting the pronator teres to the tendinous arch of the flexor digitorum sublimis may injure the nerve at this point, giving rise to the pronator teres syndrome.[40,128] The clinical features include pain and tenderness over the pronator teres, weakness of the flexor pollicis and abductor pollicis brevis, and preservation of forearm pronation. Sensory changes over the thenar eminence help differentiate this entity from the carpal tunnel syndrome, because the sensory branch innervating this area passes superficially to the flexor retinaculum. The conduction studies may reveal mild slowing in the proximal forearm in conjunction with a normal distal latency and sensory nerve action potentials at the wrist.[128] Injection of corticosteroids into the pronator teres may relieve the pain to aid in diagnosis, but definitive treatment requires surgical decompression.[95]

A similar but distinct entrapment may develop as the median nerve traverses the ligament of Struthers, a fibrous band attached to an anomalous spur on the anteromedial aspect of the lower humerus.[119] This ligament may compress the median nerve together with the brachial artery above the elbow, proximal to the innervation of the pronator teres. Weakness and electromyographic abnor-malities of this muscle thus serve to differentiate this condition from the pronator teres syndrome, which usually, but not always, spares the muscle.[3,205] Compression of the brachial artery with full extension of the forearm obliterates the radial pulse.

Anterior Interosseous Syndrome

This entity, also known as the syndrome of Kiloh and Nevin,[81] results from selective injury of the anterior interosseous nerve that branches off the median nerve just distal to the pronator passage, unilaterally or bilaterally.[134] Despite the common presenting symptoms of pain in the forearm or elbow,[50,102,139] examination reveals no distinct sensory abnormalities. Asked to make an OK sign (or money sign in Japan) with the first two digits, the patient will form a triangle instead of a circle—the so-called pinch sign. Spontaneous recovery may take place from 6 weeks to 18 months. Neuralgic amyotrophy caused by lesions in the brachial plexus (see Chapter 21.3) may be manifest as an anterior interosseous nerve palsy,[169] presumably because the spinal root fibers are already rearranged into terminal nerve branch groupings at the brachial plexus.[179] Similarly, partial median nerve compression at an antecubital level can also involve the bundles that form the anterior interosseous nerve, also causing identical clinical features.[217]

Ordinary nerve conduction studies of the median nerve reveal no abnormalities.[143] Stimulation of the anterior interosseous nerve at the elbow may demonstrate a delay in latency of the compound muscle action potential recorded from the pronator quadratus.[134] Electromyographic explorations show evidence of selective denervation in the flexor pollicis longus, flexor digitorum profundus I and II, and pronator quadratus.

Carpal Tunnel Syndrome

The carpal tunnel syndrome ranks by far the first in incidence of all the entrapment neuropathies. The median nerve passes, with nine extrinsic digital flexors, through the tunnel bounded by the carpal bones and transverse ligament, which is attached to the scaphoid, trapezoid, and hamate. Anatomically, the carpal tunnel narrows in cross-section at 2.0 to 2.5 cm distal to the entrance, where it is rigidly bound on three sides by bony structures and roofed by a thickened transverse carpal ligament.[170] Pathologic studies show that a striking reduction in myelinated fiber size takes place under the retinaculum at this point.[207]

Interestingly, even normal subjects have the slowest nerve conduction 2 to 4 cm distal to the origin of the ligament.[88] This finding suggests a mild compression of the median nerve at this particular level in some clinically asymptomatic hands. In fact, a histologic study[136] revealed focal abnormalities at this site in 5 of 12 median nerves at routine autopsy despite the absence of any symptoms suggestive of the carpal tunnel syndrome in life.[53] Certain anatomic peculiarities, such as a smaller cross-sectional area of the tunnel, may predispose these individuals to entrapment neuropathy.[11,27] Any expanding lesion in the closed space of the carpal tunnel enhances such a tendency.

The carpal tunnel syndrome affects women more than men, most commonly in the fifth or sixth decades. The symptoms usually involve the dominant hand[166] or that contralateral to amputation,[163] with higher incidence in those requiring considerable use of the hands occupationally.[48] Symptoms may appear during pregnancy and resolve after delivery. In addition to a common isolated entity seen in most patients,[63] the syndrome may rarely be familial.[13] It may accompany a variety of polyneuropathies and systemic illnesses.[8,62,93,213,227]

Patients with familial amyloidosis have a high incidence of the carpal tunnel syndrome.[99] Certain secondary amyloidoses, especially those associated with multiple myeloma, may also give rise to diffuse neuropathy. Among endocrine disorders, acromegaly[79,146] ranks first, one study reporting 35 of 100 patients with evidence of entrapment neuropathy.[146] The carpal

tunnel syndrome occurs in approximately 23 percent of patients with rheumatoid arthritis,[131,215] often as the initial manifestation of tenosynovitis affecting the wrist flexor. Rheumatoid patients may also develop thenar atrophy from disuse, cervical spine disease, or compression of the ulnar nerve at the elbow. Other disorders associated with the carpal tunnel syndrome include eosinophilic fascitis,[74] myxedema,[130,178] lupus erythematosus,[188] hyperparathyroidism,[212] and toxic shock syndrome.[175]

Nonspecific tenosynovitis also gives rise to compression symptoms in this region. Patients often have other evidence of degenerative arthritis, such as trigger fingers, bursitis, tendinitis, and tennis elbow. In addition to internal entrapment, traumatic conditions may result in acute compression of the median nerve at the wrist. These include Colles' fracture,[107] isolated fracture of capitatum[180] or hamate,[115] acute soft tissue swelling following crushing injury of the hand, and acute intraneural hemorrhage.[65] Most of these cases require emergency decompression of the median nerve.

Paresthesias in the hand frequently awaken the patients at night. The pain often extends to the elbow and not uncommonly to the shoulder, mimicking the clinical features of cervical spine disease or high median nerve compression.[23] The symptoms of proximal lesions exacerbates with manipulation of the neck or shoulder girdle and subsides with the arm at rest. In contrast, the pain in the carpal tunnel syndrome is alleviated by moving the hand. Compression can affect the peripheral autonomic fibers, causing defective vasomotor reflex.[4] Thus, Raynaud's phenomenon may develop, especially in patients with systemic diseases such as rheumatoid arthritis.[108]

Sensory changes vary a great deal in early stages. Hypesthesia involves the first three digits and the radial half of the fourth digit or, not uncommonly, only the second or third digit. The patients may indeed complain of a sensory loss outside the median nerve distribution. In one large series, 83 percent of 384 patients had sensory disturbance, consisting mostly of hypesthesia, often confined to the tip of the third digit.[156] Typically, the sensory changes spare the skin of the thenar eminence because the palmar cutaneous branch arises approximately 3 cm proximal to the carpal tunnel. Examination of the fourth digit usually reveals characteristic sensory splitting into median and ulnar halves, a pattern rarely seen in radiculopathies.

Thanks to early detection, the patient now seldom develops major wasting of thenar muscles, once considered a distinctive feature of the syndrome. Comparison between the affected hand and the normal side, however, often reveals slight weakness. To test the abductor pollicis brevis in relative isolation, the patient presses the thumb upward perpendicular to the plane of the palm. For assessment of the opponens, the patient presses the tip of the thumb against the tip of the little finger. The two heads of the flexor pollicis brevis receive mixed median and ulnar innervation with considerable variation.

Passive flexion or hyperextension of the affected hand at the wrist for more than 1 minute may worsen the symptoms.[155] Wrist flexion may also delay motor or sensory conduction across the wrist.[116,182] Percussion of the median nerve at the wrist causes paresthesia of the digits. This finding, however, has no diagnostic value specific to the carpal tunnel syndrome.[200] In fact, electrophysiologic data localize the compressed segment about 2 to 3 cm distal to the traditional percussion site on the volar aspect of the wrist.[88] The phenomenon originally described by Tinel[208] relates to tapping the proximal stump of an injured nerve to elicit paresthesia as an indication for axonal regeneration and not for entrapment neuropathy.[195,221]

Symptoms of the carpal tunnel syndrome worsen during ischemia of the arm.[56] The factors that determine the degree of such susceptibility include severity of pain and paresthesia but not extent of muscle wasting or duration of symptoms.[46] These findings suggest rapidly reversible changes in the nerve fibers associated with ischemic attacks. Sharply focal structural changes seen in compression neuropathy, however, indicate

that mechanical factors must play an important role in the pathogenesis.[47,144]

Differential diagnosis include polyneuropathy with distally prominent symptoms, high median nerve compression at the elbow, a C-6 radiculopathy, and traumatic injury at the wrist, including handcuff neuropathy.[105] The carpal tunnel syndrome may occur in association with degenerative cervical spine diseases. This combination, called the double-crush syndrome,[211] may represent a chance occurrence of two very common entities. Nonetheless, awareness of this possibility underscores the need for adequate electrophysiologic assessments, because the presence of one condition does not preclude the other. The lateral border of the flexor digitorum sublimis muscle may compress the median nerve against the forearm fascia and other flexor tendons. This rare entity causes symptoms similar to the carpal tunnel syndrome, with additional findings of local tenderness and firmness in the forearm.[49]

Simpson's original work[189] on the carpal tunnel syndrome, demonstrating focal slowing at the wrist, paved the way for clinical use of conduction studies of this entity. Since that time, a number of investigators have published extensive studies.[33,54,75,103,121,155,206] The electrophysiologic procedure has become so sensitive that it not only confirms the clinical diagnosis in most patients but also detects an incidental finding in some asymptomatic subjects. One must interpret test results in the context of patients' symptoms and clinical findings to avoid unnecessary or premature surgical intervention.

Conduction abnormalities often selectively involve the wrist-to-palm segment of the median nerve for both sensory[15–17,26,87,127,218] and motor fibers.[88,173] In one series,[88] palmar stimulation elucidated sensory or motor conduction abnormality in all but 13 (8 percent) of 172 clinically affected hands. Without palmar stimulation, an additional 32 (19 percent) hands would have been regarded as normal. In another study, recording of the orthodromic sensory action potential revealed abnormalities in 53 percent of 72 suspected hands using the conventional criteria and 67 percent of the hands with addition of palmar stimulation.[126]

With serial stimulation from midpalm to distal forearm in 1-cm increments, sensory axons normally show a latency change of 0.16 to 0.21 ms/cm (see Fig. 6–3A and B). In about half of the affected nerves, there is an abrupt latency increase across a 1-cm segment, most commonly 2 to 4 cm distal to the origin of the transverse carpal ligament.[88] In these hands, the focal latency change across the affected 1-cm segment averages more than four times that of the adjoining distal or proximal 1-cm segments (see Fig. 6–3C and D). In the remaining hands, conduction delay affected more than one 1-cm segment across the carpal tunnel but was usually maximal at the site described above. Segmental studies of the motor axons in short increments are technically more demanding because the recurrent course of the thenar nerve varies anatomically from one subject to another.[73]

Previous work emphasized a higher sensitivity of sensory conduction testing, compared with studies of the motor axons.[17,121,206] In our series,[88] however, the sensory and motor axons showed a comparable incidence of abnormalities. In addition, we often encountered selective involvement of motor fibers with normal sensory conduction, or conversely, a delay of sensory conduction with a normal motor latency. Palmar stimulation provides a simple means of differentiating compression by the transverse carpal ligament from diseases of the most terminal segment, as might be expected in a distal neuropathy.[20]

With palmar stimulation, it is also possible to record simultaneously nerve action potential from the digit and that from the median nerve trunk at the wrist. This method has the advantage of instantaneously testing the two segments[111] but depends on the comparison between antidromic sensory potentials, on the one hand, and mixed sensory and motor nerve potentials, on the other. In advanced stages the axons may degenerate distal to the entrapment. Retrograde changes may also occur in the forearm as

a result of a severe compression at the wrist.[5,201] The loss of fast conducting fibers may also lead to slowed conduction velocity proximal to the site of the lesion in advanced stages.

A number of other variations may improve the sensitivity of the motor and sensory conduction studies. The difference between the right and left sides, although useful in unilateral lesions, provides limited help in assessing bilateral compression. Comparison of the median sensory latency with radial or ulnar sensory latency has proven useful as long as the length of the tested segment is kept consistent.[19] An interesting approach along the same lines takes advantage of simultaneous activation of two nerves, for example, median and ulnar sensory potentials from the fourth digit or median and radial sensory potentials from the first digit.[71,72] Two measures compare the terminal latency of the distal segment with the conduction time in the proximal segment adjusted to the same distance (see Chapter 5.4) Of the two, the residual latency increases in patients with the carpal tunnel syndrome,[97] whereas the terminal latency index decreases below the normal range.[82,185]

In advanced cases, electromyographic studies show fibrillation potentials and positive sharp waves in the median-innervated intrinsic hand muscles. Spontaneous rhythmic discharges of motor unit potential, seen in some patients, may have a distal site of origin, in the area of compression.[193] Even with complete denervation of the thenar muscles, the first and second lumbricals may maintain part of their innervation presumably because of the deeper location of their motor funiculi.[29] Some series,[21,183] but not others,[17,63] report a high incidence of electrophysiologic evidence for an ulnar nerve lesion at the wrist in patients with the carpal tunnel syndrome.

Digital Nerve Entrapment

The interdigital nerves supply the skin of the index and middle fingers and half of the ring finger as extensions of the median sensory fibers. These small sensory branches may be compressed against the edge of the deep transverse metacarpal ligament. Entrapment appears in association with trauma, tumor, phalangeal fracture, or inflammation of the metacarpophalangeal joint or tendon.[95] The presenting symptoms include pain in one or two fingers, exacerbated by lateral hyperextension of the affected digits, and tenderness and dysesthesia over the palmar surfaces between the metacarpals. Local infiltration of the steroid may relieve the symptoms and assist in diagnosis.[133] Abnormal median sensory potential may result from unsuspected digital nerve lesions.[69]

6 ULNAR NERVE

Tardy Ulnar Palsy and Cubital Tunnel Syndrome

Ulnar nerve injury results from repeated trauma at the elbow or immobilization of the upper limb during surgery. Originally, tardy ulnar palsy implied antecedent traumatic joint deformity or recurrent subluxation. Many clinicians, however, now use the term for entrapment of the ulnar nerve at the elbow, even without history of trauma. Ulnar nerve palsy at the elbow may constitute part of diffuse neuropathy or develop concomitantly with lower cervical spine disease involving the C-8 and T-1 roots or with the thoracic outlet syndrome.[132] Nerve conduction studies and electromyography help localize the site of major disease in these patients.

Ulnar neuropathy at the elbow results from widely varying causes.[61,106] The compressive lesion at this site can affect different fascicles, but most commonly the terminal digital nerves and the fibers to the hand muscles, which are involved much more frequently than those to the forearm muscles.[199] Some reports emphasize the cubital tunnel syndrome as the most common discrete entity.[39,123] In this condition, nerve entrapment accompanies neither joint deformity nor a history of major trauma.[38,151] A number of factors give rise to entrapment of the

nerve under the aponeurosis connecting the two heads of the flexor carpi ulnaris.[42,95,124,196,204] Here, the nerve has the largest diameter,[22] may show palpable swelling in the ulnar groove, and appears hyperemic at surgery.

The appearance of bilateral ulnar neuropathy, in a large number of patients, suggests a congenital predisposition to this syndrome.[123] In some cases of idiopathic ulnar neuropathy, the asymptomatic contralateral nerve may also show some involvement histologically.[135] In one study[136] routine autopsy revealed focal pathologic change at the aponeurosis in 5 of 12 presumably normal nerves.

Frequent hand use in the elbow flexed position narrows the cubital tunnel and exacerbates the symptoms.[42,123] The earliest clinical features include impairment of sensation over the fifth digit and ulnar half of the fourth digit with weakness and wasting of the first dorsal interosseous and other ulnar-innervated intrinsic hand muscles. Surgical treatment consists of transposition,[64] simple decompression,[125,222] or interfascicular neurolysis.[138] The patient may have some functional recovery if operated upon early.[64,101,137,152] Once a moderate degree of motor deficit has developed, symptoms persist after operative intervention in 30 percent, or more, of patients.[39]

Nerve conduction studies in some cases show localized slowing of motor or sensory conduction velocity across the elbow as compared with the more proximal or distal segments.[189] For diagnostic purposes, the difference must exceed 10 m/s.[38,147] The segment distal to the presumed compression may also show mild slowing.[55] More commonly, reduction in amplitude of the compound muscle action potential suggests conduction block at the site of compression. A drop in motor amplitude greater than 25 percent across the elbow localizes the lesion in this segment.[158] Similarly, studies of mixed nerve action potentials recorded at the axilla show a lower amplitude after stimulation of the nerve below the elbow, as compared with that above the elbow.[55] Stimulating the nerve at multiple sites across the cubital tunnel helps establish the precise location of the lesion.[123]

Electromyography further defines the site of involvement by demonstrating the distribution of denervation. Typically, the cubital tunnel syndrome affects the ulnar half of the flexor digitorum profundus, which receives the nerve supply distal to the aponeurosis, sparing the flexor carpi ulnaris, supplied by a proximal branch. The reverse, however, is not always true, because a proximal lesion can selectively damage the bundle of axons destined for the more distal muscles. In fact, ulnar nerve lesions at any level tend to affect the first dorsal interosseous muscle most consistently.

Compression at Guyon's Canal

The ulnar never enters the hand through Guyon's canal at the wrist.[186] Nerve injury at this level, though less common than at the elbow, gives rise to the clinical features similar to those of tardy ulnar palsy. Sensory deficit, if present, characteristically spares the dorsum of the hand, innervated by the dorsal cutaneous branch, which arises proximal to the wrist. Entrapment in Guyon's canal most frequently results from a ganglion and less so from trauma or rheumatoid arthritis. The ulnar-innervated intrinsic hand muscles show weakness and atrophy as well as electromyographic evidence of denervation. In contrast, the flexor carpi ulnaris and flexor digitorum profundus III and IV function normally. Reduced or absent sensory action potentials of the fourth and fifth digits indicate involvement of the superficial sensory branch. The mixed nerve action potential between wrist and elbow remains normal.

Involvement of the Palmar Branch

Further distally, the deep motor branch may sustain external trauma or compression by a ganglion arising from the carpal articulations.[24,35,67,186] Using the heel of the hand against a crutch causes repeated injuries to this branch, as does an attempt to shut or raise a window by striking the bottom edge with the palm. Compression of the ulnar nerve at the palm has also followed prolonged bicycle riding.[37] Damage distal to the origin of the superficial sensory branch gives rise to no sensory abnormality clinically or electrophysiologically. In cyclist's palsy, however, a severe lesion may also affect the superficial branch supplying the skin of the fourth and fifth digits.[141]

This lesion usually spares the motor branches supplying the hypothenar muscles. Thus, conduction studies reveal no abnormalities between the elbow and wrist and a normal distal latency from the wrist to the abductor digiti minimi. The compound action potential recorded from the first dorsal interosseous, however, may show a prolonged latency and reduced amplitude, compared with the unaffected side. Electromyography shows selective abnormalities of the ulnar innervated intrinsic hand muscles except for the abductor digiti minimi. These findings indicate slowing or block of nerve conduction distal to the origin of the hypothenar branch.[12,36]

Digital Nerve Entrapment

The interdigital branch to the fifth digit receives its sensory fibers solely from the ulnar nerve, whereas branches supplying the fourth digit are formed by anastomosis between the median and ulnar nerves. The clinical symptoms and signs resemble those of the median digital nerve entrapments described earlier.

7 NERVES OF THE PELVIC GIRDLE

Although traumatic injury rarely affects the lumbar plexus because of the protection afforded by the pelvic bones, individual nerves derived from the plexus may sustain isolated damage by either chronic compression or acute injury.

Ilioinguinal Nerve

This nerve may be injured accidentally or during surgery. Patients with ilioinguinal neuropathy complain of pain in the groin region, especially when standing.[95] Pressure immediately medial to the anterior-superior iliac spine causes pain radiating into the crural region. Muscle weakness and increased intra-abdominal tension may lead to formation of direct inguinal hernia.

Genitofemoral Nerve

Selective damage of this nerve may result from trauma to the groin or surgical adhesions. Clinical features include pain in the inguinal region, sensory deficits over the femoral triangle, and absence of the cremasteric reflex.

Lateral Femoral Cutaneous Nerve

Entrapment of this purely sensory nerve causes a condition known as meralgia paresthetica. The damage usually occurs at the anterior superior iliac spine where the nerve emerges from the lateral border of the psoas major and sharply angulates over the inguinal ligament.[95] The precipitating factors include the compression of the nerve by tight belts, corsets, or seatbelts, although the symptoms may develop without obvious cause presumably because of the pressure on the nerve as it penetrates the inguinal fascia. Pathologic changes consist of local demyelination and wallerian degeneration, particularly of the large-diameter fibers.[70]

Clinical diagnosis depends on the characteristic distribution of paresthesias, pain, and objective sensory loss over the anterolateral surface of the thigh without motor weakness.[95] Electrophysiologic studies may reveal slowed sensory conduction across the compression site.[18,177,198] Patients with an L-3 or L-4 herniated disk may also have radiating pain along the lateral aspect of the thigh.

They have motor deficits clinically as well as electromyographically. The lack of objective sensory loss also helps differentiate this entity from meralgia paresthetica.

Femoral Nerve

An intrapelvic lesion of the femoral nerve may result from compression by tumors of the vertebrae, psoas abscesses, retroperitoneal lymphadenopathy, or hematoma.[25,80,96,197,226] Direct trauma may occur with fractures of the femur or cardiac catheterization. Diabetes and vascular disease are also common causes of femoral neuropathy. A complete lesion of the femoral nerve results in (1) inability to flex the thigh on the abdomen or to extend the leg at the knee; (2) reduced or absent knee stretch reflex; and (3) variable sensory loss.[10] Electrophysiologic studies show an increase in femoral nerve latency, reduction in the amplitude of the compound muscle action potential, and evidence of denervation in the appropriate muscles.

In mononeuropathy of the femoral nerve associated with diabetes, the syndrome begins with pain in the anterior aspect of the thigh followed by weakness and atrophy of the quadriceps. In most of these cases, however, careful clinical and electromyographic examination often reveals more widespread involvement in the territory of the L-2 through L-4 roots, suggesting polyradiculopathy.

Saphenous Nerve

The nerve exits from Hunter's subsartorial canal, together with the femoral vessels.[95] Thus, obstructive vascular disease may entrap the nerve at this level, causing pain localized to the medial aspect of the knee as the main clinical feature. It often radiates distally to the medial side of the foot[132] and worsens with any exercise such as climbing stairs. Electrophysiologic studies may reveal slowed saphenous nerve conduction tested either orthodromically[202] or antidromically.[216]

Obturator Nerve

The nerve is often selectively damaged during pregnancy or labor by pressure from a gravid uterus. Other causes of nerve injury include pelvic fracture and surgical procedures for obturator hernia. Nerve entrapment in the obturator canal may also result from increased intra-abdominal pressure. Injury to this nerve weakens the adductors and internal and external rotators of the thigh. Typically, the patient complains of pain in the groin radiating along the medial aspect of the thigh, as well as hypesthesia or dysesthesia over the medial aspect of the upper thigh. Electromyographic studies show evidence of denervation in the gracilis and adductor muscles.

Superior and Inferior Gluteal Nerves

These nerves, situated directly behind the hip joint, are subject to damage by fractures of the upper femur or by misdirected intramuscular injection. Anterior-superior tendinous fibers of the piriformis may compress the superior gluteal nerve, causing buttock pain and tenderness to palpation in the area superolateral to the greater sciatic notch.[162] Compromise of the inferior gluteal nerve documented electromyographically may herald clinical signs of recurrent colorectal carcinoma.[100] Damage to the superior gluteal nerve gives rise to weakness and denervation of gluteus medius and minimus, which adduct and rotate the thigh inward. A lesion of the inferior gluteal nerve compromises the gluteus maximus, which extends, adducts, and rotates the thigh externally.

Sciatic Nerve

A number of conditions cause sciatic nerve injury in the pelvis. These include the direct spread of neoplasm from the genitourinary tract or rectum, neurinoma of the sciatic nerve itself, abscess of the pelvic floor, pressure from a gravid uterus, and fractures of the pelvis, hip, or femur. Sciatic endometriosis may cause cyclic sciatic pain and a sensorimotor mononeuropathy.[176] Misdirected intragluteal injection tends to damage the sciatic nerve selectively. At the level of the infrapiriform foramen, this type of injury may affect the inferior gluteal nerve, posterior femoral cutaneous nerve, or pudenal nerve.[142] Penetrating wound, hip surgery, or insertion of a prosthesis may also traumatize the sciatic nerve. Baker's popliteal cyst is formed by an effusion into the semimembranous bursa. This palpable enlargement, especially with the knee extended, compresses the sciatic, peroneal, tibial, or sural nerves in any combination.[133]

Although rare, the piriformis muscle may entrap the nerve as it exits the pelvis through the greater sciatic notch.[95] The piriformis syndrome, unlike more proximal involvement, spares the gluteus medius, gluteus minimus, tensor fasciae latae, and paraspinal muscles clinically and electromyographically. During prolonged squatting, the sciatic nerve is compressed in the segment between the ischial tuberosity and trochanter major or between the adductor magnus and hamstring muscles.[190] Nerve conduction studies and electromyography help delineate the extent and distribution of abnormality.

For reasons not entirely clear, trauma affecting the sciatic nerve as a whole tends to involve the peroneal component much more frequently than the tibial portion.[129,203] Reaction to injuries may depend on funicular size and disposition of the nerves. The peroneal nerve trunk has less connective tissue and fewer but longer nerve bundles than the tibial nerve. The topical distribution may also make the peroneal division, located laterally and posteriorly, more susceptible than the tibial division to an injection in the buttock.

Footdrop may result either from a lesion of the peroneal nerve distally at the neck of the fibula or more proximally at the level of the sciatic nerve. Differentiation between the two possibilities depends on electromyographic exploration of the hamstring muscles and the posterior compartment of the leg. A more proximal lesion is likely with denervation of the short head of the biceps femoris innervated by the peroneal component of the sciatic nerve or of the tibialis posterior supplied by the tibial nerve. Studies of the H reflex, the F wave, or direct needle stimulation of the nerve at the radicular level and sciatic notch[113] may reveal conduction abnormalities in these cases.

8 COMMON PERONEAL NERVE

Following the separation into individual nerves in the lower thigh, the common peroneal nerve becomes superficial to reach the lateral aspect of the knee.[2] Habitual crossing of the leg compresses the nerve against the head of the fibula at this vulnerable point.[187] Injury here most frequently affects the deep branch, and less commonly, the whole nerve. Though rare, a ganglion in the same location can cause selective involvement of the superficial branch.[187,204] Prolonged squatting may compress the peroneal nerve against the biceps tendon, the lateral head of the gastrocnemius, or the head of the fibula.[94,204] Unilateral peroneal nerve paralysis has developed during intended weight reduction.[192]

Injury to the deep branch weakens the toe and foot dorsiflexors, with sensory changes over the web of skin between the first and second toe. Lesions of the superficial branch affect the evertors, with sensory deficits over most of the dorsum of the foot. The preservation of the ankle reflex and ability to invert the foot normally serve to distinguish a peroneal nerve palsy from a sciatic nerve lesion in patients with footdrop. The tibialis posterior receives L-4 and L-5 root innervation via the tibial nerve. Thus, needle study of this muscle, though technically difficult to isolate, helps differentiate between a peroneal palsy and an L-5 radiculopathy.[66]

A change in amplitude or, less frequently, slowed conduction across the fibular head localizes the site of the lesion. For diagnosis of a focal abnormality based on conduction velocity, it must fall more than 10 m/s, compared with the re-

maining distal segment below the knee. A drop in amplitude by more than 20 percent from distal to proximal stimulation also indicates a localized lesion at the compression site.[157] Reversed amplitude discrepancy suggests the presence of the accessory deep peroneal nerve, which is stimulated by the proximal shock of the knee but not by the distal stimulus at the ankle.[104] Recording from the tibialis anterior, in lieu of an atrophic extensor digitorum brevis, improves accuracy of conduction assessment across the knee in some cases.[30,164,219] Distal stimulation elicits a small and delayed mixed nerve potential above the head of the fibula in mild compression and no responses in advanced stages.

The anterior tarsal tunnel syndrome, rare entrapment of the deep peroneal nerve at the ankle, gives rise to pain on the dorsum of the foot, sensory deficits in the small web area between the first and second toes, and atrophy of the extensor digitorum brevis.[117] An incomplete form affects the motor or sensory fibers selectively after their division under the inferior extensor retinaculum.[98] Nerve conduction studies show increased distal motor latency with stimulation of the deep peroneal nerve proximal to the inferior extensor retinaculum.[174]

Electromyography in the anterior tarsal tunnel syndrome reveals evidence of denervation in the extensor digitorum brevis and other appropriate muscles. Spontaneous discharges in the intrinsic foot muscles, however, may simply reflect chronic nerve damage caused by wearing a tight shoe.[41] The presence of fibrillation potentials, as compared with positive sharp waves, provides a more reliable indicator of true disease.[51]

9 TIBIAL NERVE

This nerve, because of its deep location, rarely sustains injury in the posterior compartment of the thigh or leg. Most often, the nerve is compressed by the flexor retinaculum as it passes behind the medial malleolus.[31,57,78,114] This condition, known as the tarsal tunnel syn-

drome, may result from trauma, tenosynovitis, venous stasis of the posterior tibial
vein, or a ganglion arising from the sub-
talar joint.[112] A patient with a more proximal lesion such as tumor of the tibial
nerve may show signs and symptoms of
the tarsal tunnel syndrome because of
venous thrombosis in the calf.[220] The
clinical features consist of painful dysesthesia and sensory deficits in the toes
and sole, and weakness of the intrinsic
foot muscles. Electromyography reveals
evidence of denervation in the intrinsic
foot muscles supplied by the tibial nerve.

In the tarsal tunnel syndrome, nerve
conduction studies show increased motor
latencies along the medial or lateral plantar nerve with stimulation of the tibial
nerve slightly above the medial malleous.
Additional stimulation of the nerve
slightly below the malleolus may document segmental slowing across the compression site. The calculated conduction
velocity, however, ranges widely, reflecting the short distance between the two
stimuli. Alternatively, serial stimulation
in 1-cm increments along the course of
the nerve may reveal an abrupt change in
waveform in the recorded response together with a disproportionate latency increase at the compression site. Near-
nerve sensory conduction of the medial
and lateral plantar nerve elucidates
slowed velocities and abnormal temporal
dispersion in most cases.[148] These findings indicate a focal segmental demyelination as the primary pathologic process.
Conduction studies on the clinically unaffected side serve as a control.

Although rare, the nerve may be compressed more proximally in the popliteal
fossa or more distally within the abductor
hallucis muscle. A lesion distal to the
flexor retinaculum results in deficit of either medial or lateral plantar branches of
the tibial nerve. The patient complains of
pain and sensory changes in the plantar
aspect of the foot but not in the heel. Useful diagnostic techniques include conduction studies of the medial and lateral
plantar nerves and electromyography of
the intrinsic foot muscles.[60,149]

Chronic compression of the terminal
digital branches under the metatarsal
heads, usually in the third and fourth interspace, gives rise to a syndrome of
painful toes, or Morton's neuroma. The
interdigital nerve syndrome also results
from ligamentous mechanical irritation
with hyperextension of the toes in high-
heeled shoes, hallux valgus deformities,
congenital malformation, rheumatoid arthritis, or any form of trauma.[132] Typically, pain is precipitated in the affected
digits by walking, although the patient
also suffers from spontaneous nocturnal
discomfort.

Sural Nerve

Isolated compression and traumatic
neuropathy of the sural nerve, though infrequent, results from a ganglion,[159]
Baker's cyst,[133] use of a combat boot,[181]
or stretch injury.[59] The sural nerve is frequently biopsied for diagnostic purposes
because of its superficial location. The
sensory innervation differs from one subject to another as the nerve receives
various contributions from the tibial and
peroneal nerves. In general, sensory
changes involve the posterolateral
aspects of the lower third of the leg and
the lateral aspects of the dorsum of the
foot. Nerve conduction studies help delineate the lesion.[59]

REFERENCES

1. Adour, KK, Bell, DN, and Wingerd, J: Bell palsy. Dilemma of diabetes mellitus. Arch Oto-laryngol 99:114–117, 1974.
2. Aguayo, AJ: Neuropathy due to compression and entrapment. In Dyck, PJ, Thomas, PK, and Lambert, EH (eds): Peripheral Neuropathy, Vol. 1. WB Saunders, Philadelphia, 1975.
3. Aiken, BM, and Moritz, MJ: Atypical electro-myographic findings in pronator teres syndrome. Arch Phys Med Rehabil 68:173–175, 1987.
4. Aminoff, MJ: Involvement of peripheral vaso-motor fibres in carpal tunnel syndrome. J Neurol Neurosurg Psychiatry 42:649–655, 1979.
5. Anderson, MH, Fullerton, PM, Gilliatt, RW, and Hern, JEC: Changes in the forearm associated with median nerve compression at the wrist in the guinea-pig. J Neurol Neurosurg Psychiatry 33:70–79, 1970.
6. Anson, BJ: An Atlas of Human Anatomy, ed 2. WB Saunders, Philadelphia, 1963.

7. Auerbach, SH, Depiero, TJ, and Mejlszenkier, J: Familial recurrent peripheral facial palsy. Arch Neurol 38:463–464, 1981.

8. Bastian, FO: Amyloidosis and the carpal tunnel syndrome. Am J Clin Pathol 61:711–717, 1974.

9. Berry, H, and Bril, V: Axillary nerve palsy following blunt trauma to the shoulder region: A clinical and electrophysiological review. J Neurol Neurosurg Psychiatry 45:1027–1032, 1982.

10. Biemond, A: Femoral neuropathy. In Vinken, PJ, and Bruyn, GW (eds): Handbook of Clinical Neurology, Vol 8, Diseases of the Nerves. American Elsevier, New York, 1970.

11. Bleecker, ML, Bohlman, M, Moreland, R, and Tipton, A: Carpal tunnel syndrome: Role of carpal canal size. Neurology 35:1599–1604, 1985.

12. Bouche, P, Esnault, S, Broglin, D, Sedel, L, Cathala, HP, and Laplane, D: Isolated compression of the deep motor branch of the ulnar nerve. Electromyogr Clin Neurophysiol 26:415–422, 1986.

13. Braddom, RL: Familial carpal tunnel syndrome in three generations of a black family. Am J Phys Med 64:5:227–234, 1985.

14. Braddom, RL, and Wolfe, C: Musculocutaneous nerve injury after heavy exercise. Arch Phys Med Rehabil 59:209–293, 1978.

15. Brown, WF, and Yates, SK: Percutaneous localization of conduction abnormalities in human entrapment neuropathies. J Can Neurol Sci, Nov:391–400, 1982.

16. Buchthal, F, and Rosenfalck, A: Sensory conduction from digit to palm and from palm to wrist in the carpal tunnel syndrome. J Neurol Neurosurg Psychiatry 34:243–252, 1971.

17. Buchthal, F, Rosenfalck, A, and Trojaborg, W: Electrophysiologic findings in entrapment of the median nerve at the wrist and elbow. J Neurol Neurosurg Psychiatry 37:340–360, 1974.

18. Butler, ET, Johnson, EW, and Kaye, ZA: Normal conduction velocity in the lateral femoral cutaneous nerve. Arch Phys Med Rehabil 55:31–32, 1974.

19. Carroll, GJ: Comparison of median and radial nerve sensory latencies in the electrophysiological diagnosis of carpal tunnel syndrome. Electroencephalogr Clin Neurophysiol 68:101–106, 1987.

20. Casey, EB, and Le Quesne, PM: Digital nerve action potentials in healthy subjects and in carpal tunnel and diabetic patients. J Neurol Neurosurg Psychiatry 35:612–623, 1972.

21. Cassvan, A, Rosenberg, A, and Rivera, L: Ulnar nerve involvement in carpal tunnel syndrome. Arch Phys Med Rehabil 67:290–292, 1986.

22. Chang, KSF, Low, WD, Chan, ST, Chuang, A, and Poon, KT: Enlargement of the ulnar nerve behind the medial epicondyle. Anat Rec 145:149–153, 1963.

23. Cherington, M: Proximal pain in carpal tunnel syndrome. Arch Surg 108:69, 1974.

24. Cowen, NJ: Hypothenar mass and ulnar neuropathy: A case report. Clin Orthop Rel Res 69:203–206, 1970.

25. Cranberg, L: Femoral neuropathy from iliac hematoma: Report of a case. Neurology 29:1071–1072, 1979.

26. Daube, JR: Percutaneous palmar median nerve stimulation for carpal tunnel syndrome. Electroenceph Clin Neurophysiol 43:139–140, 1977.

27. Dekel, S, and Coates, R: Primary carpal stenosis as a cause of "idiopathic" carpal tunnel syndrome. Lancet 2:1024, 1979.

28. Dellon, A, and MacKinnon, S: Radial sensory nerve entrapment. Arch Neurol 43:833–835, 1986.

29. Desjacques, P, Egloff-Baer, S, and Roth, G: Lumbrical muscles and the carpal tunnel syndrome. Electromyogr Clin Neurophysiol 20:443–450, 1980.

30. Devi, S, Lovelace, RE, and Durate, N: Proximal peroneal nerve conduction velocity: Recording from anterior tibial and peroneus brevis muscles. Ann Neurol 2:116–119, 1977.

31. Di Stefano, V, Sack, JT, Whittaker, R, and Nixon, JE: Tarsal-tunnel syndrome: Review of the literature and two case reports. Clin Orthop Rel Res 88:76–79, 1972.

32. Dorfman, LJ, and Jayaram, AR: Handcuff neuropathy. JAMA 239:957, 1978.

33. Duensing, F, Lowitzsch, K, Thorwirth, V, and Vogel, P: Neurophysiologische Befunde beim Karpaltunnelsyndrom: Korrelationen zum klinischen Befund. Z Neurol 206:267–284, 1974.

34. Dundore, DE, and DeLisa, JA: Musculocutaneous nerve palsy: An isolated complication of surgery. Arch Phys Med Rehabil 60:130–133, 1979.

35. Dupont, C, Cloutier, GE, Prevost, Y, and Dion, MA: Ulnar-tunnel syndrome at the wrist: A report of four cases of ulnar-nerve compression at the wrist. J Bone Joint Surg [Am] 47A:757–761, 1965.

36. Ebeling, P, Gilliatt, RW, and Thomas, PK: A clinical and electrical study of ulnar nerve lesions in the hand. J Neurol Neurosurg Psychiatry 23:1–9, 1960.

37. Eckman, PB, Perlstein, G, and Altrocchi, PH: Ulnar neuropathy in bicycle riders. Arch Neurol 32:130–131, 1975.

38. Eisen, A: Early diagnosis of ulnar nerve palsy: An electrophysiologic study. Neurology 24:256–262, 1974.

39. Eisen, A, and Danon, J: The mild cubital tunnel syndrome: Its natural history and indications for surgical intervention. Neurology 24:608–613, 1974.

40. Esposito, GM: Peripheral entrapment neuropathies of upper extremity. NY State J Med 72:717–724, 1972.

41. Falck, B, and Alaranta, H: Fibrillation potentials, positive sharp waves and fasciculation in the intrinsic muscles of the foot in healthy subjects. J Neurol Neurosurg Psychiatry 46:681–683, 1983.

42. Feindel, W, and Stratford, J: The role of the

cubital tunnel in tardy ulnar palsy. Can J Surg 1:287–300, 1958.

43. Felsenthal, G, Mondell, DL, Reischer, MA, and Mack, RH: Forearm pain secondary to compression syndrome of the lateral cutaneous nerve of the forearm. Arch Phys Med Rehabil 65:139–141, 1984.

44. Ferretti, A, Cerullo, G, and Russo, G: Suprascapular neuropathy in volleyball players. J Bone Joint Surg [Am] 69:260–263, 1987.

45. Fisher, MA, Shahani, BT, and Young, RR: Assessing segmental excitability after acute rostral lesions. II. The blink reflex. Neurology 29:45–50, 1979.

46. Fullerton, PM: The effect of ischaemia on nerve conduction in the carpal tunnel syndrome. J Neurol Neurosurg Psychiatry 26:385–397, 1963.

47. Fullerton, PM, and Gilliatt, RW: Median and ulnar neuropathy in the guinea-pig. J Neurol Neurosurg Psychiatry 30:393–402, 1967.

48. Gainer, JV, Jr, and Nugent, GR: Carpal tunnel syndrome. Report of 430 operations. South Med J 70:325–328, 1977.

49. Gardner, RC: Confirmed case and diagnosis of pseudocarpal-tunnel (sublimis) syndrome. N Engl J Med 282:858, 1970.

50. Gardner-Thorpe, C: Anterior interosseous nerve palsy: Spontaneous recovery in two patients. J Neurol Neurosurg Psychiatry 37:1146–1150, 1974.

51. Gatens, PF, and Saeed, MA: Electromyographic findings in the intrinsic muscles of normal feet. Arch Phys Med Rehabil 63:317–318, 1982.

52. Gelmers, HJ, and Buys, DA: Suprascapular entrapment neuropathy. Acta Neurochir 38:121–124, 1977.

53. Gilliatt, RW: Sensory conduction studies in the early recognition of nerve disorders. Muscle Nerve 1:352–359, 1978.

54. Gilliatt, RW, and Sears, TA: Sensory nerve action potentials in patients with peripheral nerve lesions. J Neurol Neurosurg Psychiatry 21:109–118, 1958.

55. Gilliatt, RW, and Thomas, PK: Changes in nerve conduction with ulnar lesions at the elbow. J Neurol Neurosurg Psychiatry 23:312–320, 1960.

56. Gilliatt, RW, and Wilson, TG: Ischaemic sensory loss in patients with peripheral nerve lesions. J Neurol Neurosurg Psychiatry 17:104–114, 1954.

57. Goodgold, J, Kopell, HP, and Spielholz, NI: The tarsal-tunnel syndrome: Objective diagnostic criteria. N Engl J Med 273:742–745, 1965.

58. Gordon, SL, Graham, WP, III, Black, JT, and Miller, SH: Accessory nerve function after surgical procedures in the posterior triangle. Arch Surg 112:264–268, 1977.

59. Gross, JA, Hamilton, WJ, and Swift, TR: Isolated mechanical lesions of the sural nerve. Muscle Nerve 3:248–249, 1980.

60. Guiloff, RJ, and Sherratt, RM: Sensory conduction in medial plantar nerve: Normal values, clinical applications, and a comparison with the sural and upper limb sensory nerve action potentials in peripheral neuropa-

thy. J Neurol Neurosurg Psychiatry 40:1168–1181, 1977.

61. Hagstrom, P: Ulnar nerve compression at the elbow. Results of surgery in 85 cases. Scand J Plast Reconstr Surg 11:59–62, 1977.

62. Halter, SK, DeLisa, JA, Stolov, WC, Scardapane, D, and Sherrard, DJ: Carpal tunnel syndrome in chronic renal dialysis patients. Arch Phys Med Rehabil 6:197–201, 1981.

63. Harrison, MJG: Lack of evidence of generalized sensory neuropathy in patients with carpal tunnel syndrome. J Neurol Neurosurg Psychiatry 41:957–959, 1978.

64. Harrison, MJG, and Nurick, S: Results of anterior transposition of the ulnar nerve for ulnar neuritis. Br Med J 1:27–29, 1970.

65. Hayden, JW: Median neuropathy in the carpal tunnel caused by spontaneous intraneural hemorrhage. J Bone Joint Surg 46A:1242–1244, 1964.

66. Heffernan, LPM: Electromyographic value of the tibialis posterior muscle. Arch Phys Med Rehabil 60:170–174, 1979.

67. Hunt, JR: Occupation neuritis of the deep palmar branch of the ulnar nerve. J Nerve Ment Dis 35:673–689, 1908.

68. Ilfeld, FW, and Holder, HG: Winged scapula: Case occurring in soldier from knapsack. JAMA 120:448–449, 1942.

69. Jablecki, C, and Nazemi, R: Unsuspected digital nerve lesions responsible for abnormal median sensory responses. Arch Phys Med Rehabil 63:135–138, 1982.

70. Jefferson, D, and Eames, RA: Subclinical entrapment of the lateral femoral cutaneous nerve: An autopsy study. Muscle Nerve 2:145–154, 1979.

71. Johnson, EW, Kukla, RD, Wongsam, RE, and Piedmont, A: Sensory latencies to the ring finger: Normal values and relation to carpal tunnel syndrome. Arch Phys Med Rehabil 62:206–208, 1981.

72. Johnson, EW, Sipski, M, and Lammertse, T: Median and radial sensory latencies to digit I: Normal values and usefulness in carpal tunnel syndrome. Arch Phys Med Rehabil 68:140–141, 1987.

73. Johnson, RK, and Shrewsbury, MM: Anatomical course of the thenar branch of the median nerve—usually in a separate tunnel through the transverse carpal ligament. J Bone Joint Surg 52A:269–273, 1970.

74. Jones, HR, Jr, Beetham, WP, Jr, Silverman, ML, and Margles, SW: Eosinophilic fasciitis and the carpal tunnel syndrome. J Neurol Neurosurg Psychiatry 49:324–327, 1986.

75. Kaeser, HE: Diagnostische Probleme beim Karpaltunnelsyndrom. Dtsch Z Nervenheilk 185:453–470, 1963.

76. Kaplan, PE: Posterior interosseous neuropathies: Natural history. Arch Phys Med Rehabil 65:399–400, 1984.

77. Kaplan, PE, and Kernahan, WT: Rotator cuff rupture: Management with suprascapular neuropathy. Arch Phys Med Rehabil 65:273–275, 1984.

78. Keck, C: The tarsal-tunnel syndrome. J Bone Joint Surg 44A:180–182, 1962.

79. Khaleeli, AA, Levy, RD, Edwards, RHT, McPhail, G, Mills, KR, Round, JM, and Betteridge, DJ: The neuromuscular features of acromegaly: A clinical and pathological study. J Neurol Neurosurg Psychiatry 47:1009–1015, 1984.
80. Khella, L: Femoral nerve palsy: Compression by lymph glands in the inguinal region. Arch Phys Med Rehabil 60:325–326, 1979.
81. Kiloh, LG, and Nevin, S: Isolated neuritis of the anterior interosseous nerve. Br Med J 1:850–851, 1952.
82. Kimura, I, and Ayyar, DR: The carpal tunnel syndrome: Electrophysiological aspects of 639 symptomatic extremities. Electromyogr Clin Neurophysiol 25:151–164, 1985.
83. Kimura, J: Alteration of the orbicularis oculi reflex by pontine lesions: Study in multiple sclerosis. Arch Neurol 22:156–161, 1970.
84. Kimura, J: An evaluation of the facial and trigeminal nerves in polyneuropathy: Electrodiagnostic study in Charcot-Marie-Tooth disease, Guillain-Barré syndrome, and diabetic neuropathy. Neurology 21:745–752, 1971.
85. Kimura, J: The blink reflex as a test for brainstem and higher central nervous system function. In Desmedt, JE (ed): New Developments in Electromyography and Clinical Neurophysiology, Vol 3. Karger, Basel, 1973, pp 682–691.
86. Kimura, J: Electrically elicited blink reflex in diagnosis of multiple sclerosis: Review of 260 patients over a seven-year period. Brain 98:413–426, 1975.
87. Kimura, J: A method for determining median nerve conduction velocity across the carpal tunnel. J Neurol Sci 38:1–10, 1978.
88. Kimura, J: The carpal tunnel syndrome. Localization of conduction abnormalities within the distal segment of the median nerve. Brain 102:619–635, 1979.
89. Kimura, J, Giron, LT, Jr, and Young, SM: Electrophysiological study of Bell palsy: Electrically elicited blink reflex in assessment of prognosis. Arch Otolaryngol 102:140–143, 1976.
90. Kimura, J, and Lyon, LW: Alteration of orbicularis oculi reflex by posterior fossa tumors. J Neurosurg 38:10–16, 1973.
91. Kimura, J, Rodnitzky, RL, and Okawara, S: Electrophysiologic analysis of aberrant regeneration after facial nerve paralysis. Neurology 25:989–993, 1975.
92. Kimura, J, Rodnitzky, RL, and Van Allen, MW: Electrodiagnostic study of trigeminal nerve: Orbicularis oculi reflex and masseter reflex in trigeminal neuralgia, paratrigeminal syndrome, and other lesions of the trigeminal nerve. Neurology 20:574–583, 1970.
93. Klofkorn, RW, and Steigerwald, JC: Carpal tunnel syndrome as the initial manifestation of tuberculosis. Am J Med 60:583–586, 1976.
94. Koller, RL, and Blank, NK: Strawberry picker's palsy. Arch Neurol 37:320, 1980.
95. Kopell, HP, and Thompson, WAL: Peripheral Entrapment Neuropathies, ed 2. Robert E. Krieger, Huntington, NY, 1976.
96. Kounis, NG, Macauley, MB, and Ghorbal, MS: Iliacus hematoma syndrome. Can Med Assoc J 112:872–873, 1975.
97. Kraft, GH, and Halvorson, GA: Median nerve residual latency: Normal value and use in diagnosis of carpal tunnel syndrome. Arch Phys Med Rehabil 64:221–226, 1983.
98. Krause, KH, Witt, T, and Ross, A: The anterior tarsal tunnel syndrome. J Neurol 217:67–74, 1977.
99. Kyle, RA, and Bayrd, ED: Amyloidosis: Review of 236 cases. Medicine 54:271–299, 1975.
100. Laban, MM, Meerschaert, JR, and Taylor, RS: Electromyographic evidence of inferior gluteal nerve compromise: An early representation of recurrent colorectal carcinoma. Arch Phys Med Rehabil 63:33–35, 1982.
101. Laha, RK, and Panchal, PD: Surgical treatment of ulnar neuropathy. Surg Neurol 11:393–398, 1979.
102. Lake, PA: Anterior interosseous nerve syndrome. J Neurosurg 41:306–309, 1974.
103. Lambert, EH: Diagnostic value of electrical stimulation of motor nerves. Electroencephalogr Clin Neurophysiol (Suppl) 22:9–16, 1962.
104. Lambert, EH: The accessory deep peroneal nerve: A common variation in innervation of extensor digitorum brevis. Neurology 19:1169–1176, 1969.
105. Levin, RA, and Felsenthal, G: Handcuff neuropathy: Two unusual cases. Arch Phys Med Rehabil 65:41–43, 1984.
106. Levy, DM, and Apfelberg, DB: Results of anterior transposition for ulnar neuropathy at the elbow. Am J Surg 123:304–308, 1972.
107. Lewis, MH: Median nerve decompression after Colles' fracture. J Bone Joint Surg 60B:195–196, 1969.
108. Lindscheid, RL, Peterson, LFA, and Juergens, JL: Carpal tunnel syndrome associated with vasospasm. J Bone Joint Surg 49A:1141–1146, 1967.
109. Liveson, JA: Nerve lesions associated with shoulder dislocation: An electrodiagnostic study of 11 cases. J Neurol Neurosurg Psychiatry 47:742–744, 1984.
110. Lyon, LW, and Van Allen, MW: Alteration of the orbicularis oculi reflex by acoustic neuroma. Arch Otolaryngol 95:100–103, 1972.
111. Maccabee, PJ, Shahani, BT, and Young, RR: Usefulness of double simultaneous recording (DSR) and F response studies in the diagnosis of carpal tunnel syndrome (CTS). Neurology 30:18P, 1980.
112. MacFarlane, IJA, and Du Toit, SN: A ganglion causing tarsal tunnel syndrome. S Afr Med J 48:2568, 1974.
113. MacLean, IC: Spinal nerve and phrenic nerve studies. American Academy of Neurology Special Course 16, 1979.
114. Mann, RA: Tarsal tunnel syndrome. Orthop Clin North Am 5:109–115, 1974.
115. Manske, PR: Fracture of the hook of the hamate presenting as carpal tunnel. Hand 10:191, 1978.
116. Marin, EL, Vernick, S, and Friedmann, LW: Carpal tunnel syndrome: median nerve stress test. Arch Phys Med Rehabil 64:206–208, 1983.

117. Marinacci, AA: Neurological syndromes of the tarsal tunnels. Bull LA Neurol Soc 33:90–100, 1968.

118. Marlow, N, Jarratt, J, and Hosking, G: Congenital ring constrictions with entrapment neuropathies. J Neurol Neurosurg Psychiatry 44:247–249, 1981.

119. Marquis, JW, Bruwer, AJ, and Keith, HM: Supracondyloid process of the humerus. Proc Staff Meet Mayo Clin 32:691–697, 1957.

120. Massey, EW, and Pleet, AB: Handcuffs and cheiralgia paresthetica. Neurology 28:1312–1313, 1978.

121. Melvin, JL, Schuchmann, JA, and Lanese, RR: Diagnostic specificity of motor and sensory nerve conduction variables in the carpal tunnel syndrome. Arch Phys Med Rehabil 54:69–74, 1973.

122. Millender, LH, Nalebuff, EA, and Holdsworth, DE: Posterior interosseous-nerve syndrome secondary to rheumatoid synovitis. J Bone Joint Surg 55A:753–757, 1973.

123. Miller, RG: The cubital tunnel syndrome: Diagnosis and precise localization. Ann Neurol 6:56–59, 1979.

124. Miller, RG, and Camp, PE: Postoperative ulnar neuropathy. JAMA 242:1636–1639, 1979.

125. Miller, RG, and Hummel, EE: The cubital tunnel syndrome: Treatment with simple decompression. Ann Neurol 7:567–569, 1980.

126. Mills, KR: Orthodromic sensory action potentials from palmar stimulation in the diagnosis of carpal tunnel syndrome. J Neurol Neurosurg Psychiatry 48:250–255, 1985.

127. Monga, TN, Shanks, GL, and Poole, BJ: Sensory palmar stimulation in the diagnosis of carpal tunnel syndrome. Arch Phys Med Rehabil 66:598–600, 1985.

128. Morris, HH, and Peters, BH: Pronator syndrome: Clinical and electrophysiological features in seven cases. J Neurol Neurosurg Psychiatry 39:461–464, 1976.

129. Mumenthaler, M, and Schliack, H (eds): Lasionen Peripherer Nerven, ed 3. Georg Thieme Verlag, Stuttgart, 1977.

130. Murray, IPC, and Simpson, JA: Acroparesthesias in myxedema: A clinical and electromyographic study. Lancet 1:1360, 1958.

131. Nakano, KK: The entrapment neuropathies of rheumatoid arthritis. Orthop Clin North Am 6:837–860, 1975.

132. Nakano, KK: The entrapment neuropathies. Muscle Nerve 1:264–279, 1978.

133. Nakano, KK: Entrapment neuropathy from Baker's cyst. JAMA 239:135, 1978.

134. Nakano, KK, Lundergan, C, and Okihiro, MM: Anterior interosseous nerve syndromes: Diagnostic methods and alternative treatments. Arch Neurol 34:477–480, 1977.

135. Neary, D, and Eames, RA: The pathology of ulnar nerve compression in man. Neuropathol Appl Neurobiol 1:69–88, 1975.

136. Neary, D, Ochoa, J, and Gilliatt, RW: Subclinical entrapment neuropathy in man. J Neurol Sci 24:283–298, 1975.

137. Neblett, C, and Ehni, G: Medial epicondylectomy for ulnar palsy. J Neurosurg 32:55–62, 1970.

138. Neilsen, VK, Osgaard, O, and Trojaborg, W: Interfascicular neurolysis in chronic ulnar nerve lesions at the elbow: An electrophysiological study. J Neurol Neurosurg Psychiatry 43:272–280, 1980.

139. Neundorfer, B, and Kroger, M: The anterior interosseous nerve syndrome. J Neurol 213:347–352, 1976.

140. Nielsen, HO: Posterior interosseous nerve paralysis caused by fibrous band compression at the supinator muscle: A report of four cases. Acta Orthop Scand 47:304–307, 1976.

141. Noth, J, Dietz, V, and Mauritz, KH: Cyclist's palsy. Neurological and EMG study in 4 cases with distal ulnar lesions. J Neurol Sci 47:111–116, 1980.

142. Obach, J, Aragones, JM, and Ruano, D: The infrapiriformis foramen syndrome resulting from intragluteal injection. J Neurol Sci 58:135–142, 1983.

143. O'Brien, MD, and Upton, ARM: Anterior interosseous nerve syndrome: A case report with neurophysiological investigation. J Neurol Neurosurg Psychiatry 35:531–536, 1972.

144. Ochoa, J, and Marotte, L: The nature of the nerve lesion caused by chronic entrapment in the guinea-pig. J Neurol Sci 19:491–495, 1973.

145. Ochoa, J, and Neary, D: Localized hypertrophic neuropathy, intraneural tumour, or chronic nerve entrapment? Lancet 1:632–633, 1975.

146. O'Duffy, JD, Randall, RV, and MacCarty, CS: Median neuropathy (carpal-tunnel syndrome) in acromegaly: A sign of endocrine overactivity. Ann Intern Med 78:379–383, 1973.

147. Odusote, K, and Eisen, A: An electrophysiological quantitation of the cubital tunnel syndrome. Can J Neurol Sci 6:403–410, 1979.

148. Oh, SJ, Kim, HS, and Ahmad, BK: The near-nerve sensory nerve conduction in tarsal tunnel syndrome. J Neurol Neurosurg Psychiatry 48:999–1003, 1985

149. Oh, SJ, Sarala, PK, Kuba, T, and Elmore, RS: Tarsal tunnel syndrome: Electrophysiological study. Ann Neurol 5:327–330, 1979.

150. Olarte, M, and Adams, D: Accessory nerve palsy. J Neurol Neurosurg Psychiatry 40:1113–1116, 1977.

151. Payan, J: Electrophysiological localization of ulnar nerve lesions. J Neurol Neurosurg Psychiatry 32:208–220, 1969.

152. Payan, J: Anterior transposition of the ulnar nerve: An electrophysiological study. J Neurol Neurosurg Psychiatry 33:157–165, 1970.

153. Petra, JE, and Trojaborg, W: Conduction studies along the accessory nerve and follow-up of patients with trapezius palsy. J Neurol Neurosurg Psychiatry 47:630–636, 1984.

154. Petra, JE, and Trojaborg, W: Conduction studies of the long thoracic nerve in serratus anterior palsy of different etiology. Neurology 34:1033–1037, 1984.

155. Phalen, GS: The carpal-tunnel syndrome: Seventeen years experience in diagnosis and treatment of six hundred fifty-four hands. J Bone Joint Surg 48A:211–228, 1966.

156. Phalen, GS: Reflections of 21 years experience

with the carpal tunnel syndrome. JAMA 212:1365–1367, 1970.

157. Pickett, JB: Localizing peroneal nerve lesions to the knee by motor conduction studies. Arch Neurol 41:192–195, 1984.

158. Pickett, JB, and Coleman, LL: Localizing ulnar nerve lesions to the elbow by motor conduction studies. Electromyogr Clin Neurophysiol 24:343–360, 1984.

159. Pringle, RM, Protheroe, K, and Mukherjee, SK: Entrapment neuropathy of the sural nerve. J Bone Joint Surg 56B:465–468, 1974.

160. Rask, MR: Suprascapular nerve entrapment: A report of two cases treated with suprascapular notch resection. Clin Orthop 123:73, 1977.

161. Rask, MR: Watchband superficial radial neurapraxia. JAMA 241:2702, 1979.

162. Rask, MR: Superior gluteal nerve entrapment syndrome. Muscle Nerve 3:304–307, 1980.

163. Reddy, MP: Nerve entrapment syndromes in the upper extremity contralateral to amputation. Arch Phys Med Rehabil 65:24–26, 1984.

164. Redford, JB: Nerve conduction in motor fibers to the anterior tibial muscle in peroneal palsy. Arch Phys Med Rehabil 45:500–504, 1964.

165. Reid, AC, and Hazelton, RA: Suprascapular nerve entrapment in the differential diagnosis of shoulder pain. Lancet 2:477, 1979.

166. Reinstein, L: Hand dominance in carpal tunnel syndrome. Arch Phys Med Rehabil 62:202–203, 1981.

167. Rengachary, SS, Burr, D, Lucas, S, Hassanein, KM, Mohn, MP, and Matzke, H: Suprascapular entrapment neuropathy: A clinical, anatomical and comparative study. Part 1: Clinical study. Neurosurgery 5:441–446, 1979.

168. Rengachary, SS, Neff, JP, Singer, PA, and Brackett, CE: Suprascapular entrapment neuropathy: a clinical, anatomical, and comparative study. Part 2: Anatomical study. Neurosurgery 5:447–451, 1979.

169. Rennels, GD, and Ochoa, J: Neuralgic amyotrophy manifesting as anterior interosseous nerve palsy. Muscle Nerve 3:160–164, 1980.

170. Robbins, H: Anatomical study of the median nerve in the carpal tunnel and etiologies of the carpal-tunnel syndrome. J Bone Joint Surg 45A:953–966, 1963.

171. Rosen, I, and Werner, CO: Neurophysiological investigations of posterior interosseous nerve entrapment causing lateral elbow pain. Electroencephalogr Clin Neurophysiol 50:125–133, 1980.

172. Ross, D, Jones, HR, Jr, Fisher, J, and Konkol, RJ: Isolated radial nerve lesion in the newborn. Neurology 33:1354–1356, 1983.

173. Roth, G: Vitesse de conduction motrice du nerf median dans le canal carpien. Ann Med Phys 13:117–132, 1970.

174. Ruprecht, EO: Befunde bei Neuropathien. In Hopf, HC, and Struppler, A (eds): Electromyographie. Georg Thieme Verlag, Stuttgart, 1974, pp 37–65.

175. Sahs, AL, Helms, CM, and Dubois, C: Carpal tunnel syndrome. Complication of toxic shock syndrome. Arch Neurol 40:414–415, 1983.

176. Salazar-Grueso, E, and Roos, R: Sciatic endometriosis: A treatable sensorimotor mononeuropathy. Neurology 36:1360–1363, 1986.

177. Sarala, PK, Nishihara, T, and Oh, SJ: Meralgia paresthetica: Electrophysiologic study. Arch Phys Med Rehabil 60:30–31, 1979.

178. Scarpalezos, S, Lygidakis, C, Papageorgiou, C, Maliara, S, Koukoulommati, AS, and Koutras, DA: Neural and muscular manifestations of hypothyroidism. Arch Neurol 29:140–144, 1973.

179. Schady, W, Ochoa, JL, Torebjork, HE, and Chen, LS: Peripheral projections of fascicles in the human median nerve. Brain 106:745–760, 1983.

180. Schmitt, O, and Temme, CH: Carpal tunnel syndrome bei pseudarthrosebildung nach isolierter fraktur des os capitatum. Arch Orthop Tramat Surg 93:25–28, 1978.

181. Schuchmann, JA: Isolated sural neuropathy: Report two cases. Arch Phys Med Rehabil 61:329–331, 1980.

182. Schwartz, MS, Gordon, JA, and Swash, M: Slowed nerve conduction with wrist flexion in carpal tunnel syndrome. Ann Neurol 8:69–71, 1980.

183. Sedal, L, McLeod, JG, and Walsh, JC: Ulnar nerve lesions associated with the carpal tunnel syndrome. J Neurol Neurosurg Psychiatry 36:118–123, 1973.

184. Shabas, D, and Scheiber, M: Suprascapular neuropathy related to the use of crutches. Am J Phys Med 65:298–300, 1986.

185. Shahani, BT, Young, RR, Potts, F, and Maccabee, P: Terminal latency index (TLI) and late response studies in motor neuron disease (MND), peripheral neuropathies and entrapment syndromes. Acta Neurol Scand (Suppl) 73:60, 1979.

186. Shea, JD, and McClain, EJ: Ulnar-nerve compression syndromes at and below the wrist. J Bone Joint Surg 51A:1095–1103, 1969.

187. Sidey, JD: Weak ankles. A study of common peroneal entrapment neuropathy. Br Med J 3:623–626, 1969.

188. Sidiq, M, Kirsner, AB, and Sheon, RP: Carpal tunnel syndrome: First manifestation of systemic lupus erythematosus. JAMA 222:1416–1417, 1972.

189. Simpson, JA: Electrical signs in the diagnosis of carpal tunnel and related syndromes. J Neurol Neurosurg Psychiatry 19:275–280, 1956.

190. Singh, A, and Jolly, SS: Wasted leg syndrome (a compression neuropathy of lower limbs). J Assoc Physicians India 11:1031–1037, 1963.

191. Solheim, LF, and Roaas, A: Compression of the suprascapular nerve after fracture of the scapular notch. Acta Orthop Scand 49:338–340, 1978.

192. Sotaniemi, KA: Slimmer's paralysis—peroneal neuropathy during weight reduction. J Neurol Neurosurg Psychiatry 47:564–566, 1984.

193. Spaans, F: Spontaneous rhythmic motor unit potentials in the carpal tunnel syndrome. J Neurol Neurosurg Psychiatry 45:19–28, 1982.

194. Spector, RH, and Schwartzman, RJ: Benign

trigeminal and facial neuropathy. Arch Intern Med 135:992–993, 1975.

195. Spinner, M: Injuries to the Major Branches of Peripheral Nerves of the Forearm, ed 2. WB Saunders, Philadelphia, 1978.

196. Staal, A: The entrapment neuropathies. In Vinken, PJ, and Bruyn, GW (eds): Handbook of Clinical Neurology. North-Holland, Amsterdam, 1970, pp 285–325.

197. Stern, MB, and Spiegel, P: Femoral neuropathy as a complication of heparin anticoagulation therapy. Clin Orthop 106:140–142, 1975.

198. Stevens, A, and Rosselle, N: Sensory nerve conduction velocity of N. cutaneous femoris lateralis. Electromyography 10:397–398, 1970.

199. Stewart, JD: The variable clinical manifestations of ulnar neuropathies at the elbow. J Neurol Neurosurg Psychiatry 50:252–258, 1987.

200. Stewart, JD, and Eisen, A: Tinel's sign and the carpal tunnel syndrome. Br Med J 2:1125–1126, 1978.

201. Stöhr, M, Petruch, F, Scheglmann, K, and Schilling, K: Retrograde changes of nerve fibers with the carpal tunnel syndrome: An electroneurographic investigation. J Neurol 218:287–292, 1978.

202. Stöhr, M, Schumm, F, and Ballier, R: Normal sensory conduction in the saphenous nerve in man. Electroencephalogr Clin Neurophysiol 44:172–178, 1978.

203. Sunderland, S: The relative susceptibility to injury of the medial and lateral popliteal divisions of the sciatic nerve. Br J Surg 41:300–302, 1953.

204. Sunderland, S: Nerves and Nerve Injuries, ed 2. Churchill Livingstone, Edinburgh, 1978.

205. Suranyi, L: Median nerve compression by Struthers ligament. J Neurol Neurosurg Psychiatry 46:1047–1049, 1983.

206. Thomas, JE, Lambert, EH, and Cseuz, KA: Electrodiagnostic aspects of the carpal tunnel syndrome. Arch Neurol 16:635–641, 1967.

207. Thomas, PK, and Fullerton, PM: Nerve fibre size in the carpal tunnel syndrome. J Neurol Neurosurg Psychiatry 26:520–527, 1963.

208. Tinel, J: Le Signe du "Fourmillement" dans les Lesions des Nerfs Peripheriques. Press Med 47:388, October, 1915. Translated into English by Dr. Emanuel B. Kaplan, J. Tinel's "Fourmillement" paper, The "Tingling" sign in peripheral nerve lesions. In Spinner, M (ed): Injuries to the Major Branches of Peripheral Nerves of the Forearm, ed 2. WB Saunders, Philadelphia, 1978.

209. Trojaborg, W: Rate of recovery in motor and sensory fibres of the radial nerve: Clinical and electrophysiological aspects. J Neurol Neurosurg Psychiatry 33:625–638, 1970.

210. Trojaborg, W: Motor and sensory conduction in the musculocutaneous nerve. J Neurol Neurosurg Psychiatry 39:890, 1976.

211. Upton, ARM, and McComas, AJ: The double

212. crush in nerve-entrapment syndromes. Lancet 2:359–362, 1973.

212. Valenta, LJ: Hyperparathyroidism due to parathyroid adenoma and carpal tunnel syndrome. Ann Intern Med 82:541–542, 1975.

213. Vallat, JM, and Dunoyer, J: Familial occurrence of entrapment neuropathies. Arch Neurol 36:323, 1979.

214. Valli, G, Barbieri, S, Sergi, P, Arnaboldi, M, and Scarlato, G: Electromyographic aspects of the facial muscles reinnervation after seventh nerve autogenous grafts. Electromyogr Clin Neurophysiol 21:575–584, 1981.

215. Vemireddi, NK, Redford, JB, and Pombejara, CN: Serial nerve conduction studies in carpal tunnel syndrome secondary to rheumatoid arthritis: Preliminary study. Arch Phys Med Rehabil 60:393–396, 1979.

216. Wainapel, SF, Kim, DJ, and Ebel, A: Conduction studies of the saphenous nerve in healthy subjects. Arch Phys Med Rehabil 59:316–319, 1978.

217. Wertsch, JJ, Sanger, JR, and Matloub, HS: Pseudo-anterior interosseous nerve syndrome. Muscle Nerve 8:68–70, 1985.

218. Wiederholt, WC: Median nerve conduction velocity in sensory fibers through carpal tunnel. Arch Phys Med Rehabil 51:328–330, 1970.

219. Wilbourn, AJ: AAEE Case report #12: Common peroneal mononeuropathy at the fibular head. American Association of Electromyography and Electrodiagnosis, Rochester, Minn, 1986.

220. Wiles, CM, Whitehead, S, Ward, AB, and Fletcher, CDM: Not tarsal tunnel syndrome: Malignant "triton" tumour of the tibial nerve. J Neurol Neurosurg Psychiatry 50:479–481, 1987.

221. Wilkins, RH, and Brody, IA: Tinel's sign (abstr). Arch Neurol 24:573, 1971.

222. Wilson, DH, and Krout, R: Surgery of ulnar neuropathy at the elbow: 16 cases treated by decompression without transposition: Technical note. J Neurosurg 38:780–785, 1973.

223. Wolf, SM, Wagner, JH, Davidson, S, and Forsythe, A: Treatment of Bell palsy with prednisone: A prospective, randomized study. Neurology 28:158–161, 1978.

224. Wright, TA: Accessory spinal nerve injury. Clin Orthop 108:15–18, 1975.

225. Wulff, CH, Hansen, K, Strange, P, and Trojaborg, W: Multiple mononeuritis and radiculopathies with erythema, pain, elevated CSF protein and pleocytosis (Bannwarth's syndrome). J Neurol Neurosurg Psychiatry 46:485–490, 1983.

226. Young, MR, and Norris, JW: Femoral neuropathy during anticoagulant therapy. Neurology 26:1173–1175, 1976.

227. Yu, J, Bendler, EM, and Mentari, A: Neurological disorders associated with carpal tunnel syndrome. Electromyogr Clin Neurophysiol 19:27–32, 1979.

Part VII

Disorders of the Neuromuscular Junction, Myopathies, and Abnormal Muscle Activity

Chapter 24

Myasthenia Gravis and Other Disorders of Neuromuscular Transmission

1 INTRODUCTION

Recent work has elucidated the pathophysiology underlying disorders of neuromuscular transmission, correlating morphologic abnormalities with physiologic alterations in the kinetics of acetylcholine (ACh) release (see Figs. 8–1 and 8–2). Contrary to previous views, current evidence clearly implicates the postsynaptic ACh receptor as the site of pathology in myasthenia gravis. In contrast, presynaptic defects of ACh release characterize the transmission abnormalities in the myasthenic syndrome and in botulism. Although such a dichotomy helps simplify the classification of pathologies, the exact physiologic or morphologic bases of these categories remain unknown. Furthermore, additional diseases of the neuromuscular junction may affect the complex process of chemical transmission at different steps. For example, a congenital defect of acetycholinesterase gives rise to a type of myasthenic syndrome.[35]

Physicians must always consider defects of neuromuscular transmission in any patient presenting with unexplained weakness.[134] Diagnostic possibilities include not only primary diseases of the neuromuscular junction, such as myasthenia gravis, myasthenic syndromes, and botulism, but also abnormalities of the nerve terminals, seen in motor neuron disease and certain types of neuropathy. Electrodiagnostic studies help confirm and categorize the abnormalities of neuromuscular transmission.[83]

2 MYASTHENIA GRAVIS

Myasthenia gravis has an incidence of approximately 1 per 20,000 in the United States,[97] primarily affecting young women in the third decade and middle-aged men in the fifth and sixth decades. Women have a slightly higher overall average incidence by a ratio of 3:2. The disease usually occurs sporadically, although about 5 percent of cases are familial. The symptoms and signs tend to appear between the second and fourth decades. Children account for 11 percent of all patients with myasthenia gravis.[97] They may have some other systemic disease or a seizure tendency in association with this disorder.[129]

Etiologic Considerations

Findings in support of an autoimmune hypothesis[101,125–127] include the development of thymoma in 10 percent and thymic hyperplasia in 70 percent of patients with myasthenia gravis.[40] Patients may also have other potentially immunologic diseases such as thyroiditis, hyperthyroidism, hypothyroidism, polymyositis, systemic lupus erythematosus, and rheumatoid arthritis.[98,108,130] Further, as an experiment of nature, 20 percent of infants born of myasthenic mothers have transient myasthenia following transplacental transfer of antibody. In fact, about 80 percent of patients with myasthenia gravis have antireceptor antibodies.[1,6,68]

Animal studies also suggest the presence of a circulating immunoglobulin and altered cellular immunity. Passive transfer of a certain serum fraction from patients causes myasthenic features in mice histologically as well as electrophysiologically.[139] Injection of nicotinic ACh receptor protein from the electric eel into the rabbit or monkey with Freund's adjuvant sensitizes the animal. After a second injection, many animals develop myasthenic features that improve with the administration of an anticholinesterase.[100,138] Immunization of rats with thymus extracts has, however, failed to produce a myasthenia-like condition.[84]

Histometric studies of motor end-plate ultrastructure[37–39] reveal a reduced size of the nerve terminal area and a simplified postsynaptic membrane with poorly developed folds and clefts. In contrast, mean synaptic vesicle diameter and mean synaptic vesicle count per unit nerve terminal area remain unaltered. Microphysiologic findings indicate reduced sensitivity of the postsynaptic membrane to iontophoretic application of ACh. A decreased number of functional ACh receptors is demonstrated by binding of alpha-bungarotoxin.[42,55,140] Myasthenic muscles contain IgG and complement bound to the postsynaptic membranes. Thus, an immunologic abnormality must play a role in the destruction of the membrane architecture. Experimental studies in mouse muscle further indicate that this process requires a heat-sensitive factor.[71] These observations clearly implicate the ACh receptor in the pathogenesis of myasthenia gravis.

Clinical Signs and Symptoms

The main clinical features consist of weakness and excessive fatiguability of striated muscles.[94] Although usually of insidious onset, the disease may become clinically manifest after acute infection or various surgical procedures, including thymectomy.[61] Symptoms initially appear toward the end of the day or after strenuous exercise. These patients usually have weakness confined to restricted groups of muscles.[105] Involvement of the ocular muscles causes diplopia in about half of the patients. Less frequently, bulbar weakness constitutes the presenting symptom. Paralysis of palatal and pharyngeal muscles, seen in about one third of the patients, results in nasal speech and difficulty in swallowing and chewing. Patients rarely complain of generalized weakness of the trunk and extremities as the initial symptom. Paralysis worsens with elevation of body temperature.[10,11,48,49]

Characteristic physical signs include a wide spectrum of ocular disturbances, ranging from nystagmus to complete ophthalmoplegia. Pupillary dysfunction may develop[70] as an exception to the rule that the disease affects only the striated muscles. Ptosis, if present in an early stage, may alternate between the two sides. Disturbance of ocular muscles, not confined to the distribution of a single nerve, varies from one examination to the next. Weakness of the orbicularis oris and other muscles of the lower face produces a characteristic, expressionless myasthenic face. To compensate for the weakness of the neck extensors, patients support the chin with the hand. This maneuver also facilitates chewing and swallowing at the end of a meal despite the weakened muscles of mastication. Speech may deteriorate with fatigue, showing a flaccid dysarthria. Involvement of the respiratory muscles, common in advanced cases, poses a major threat to life. In some patients, generalized or focal muscular atrophy develops.[95] The sensory examination reveals no abnormality.

The clinical courses vary, often showing remissions and exacerbations. Approximately one third of the patients improve spontaneously, some nearly completely, requiring no further medication. Symptoms often fluctuate without apparent cause, but several circumstances tend to exacerbate the symptoms. These include infection, exposure to heat, emotional stress, thyroid disease, and, perhaps most importantly, overmedication. In some patients, respiratory failure or pneumonia develops. Though

unpredictable, the disease commonly worsens during early pregnancy and improves later. In the mildest form of the disease, weakness is limited to the muscles of the eye. This entity, designated as ocular myasthenia, usually has a benign course. If signs outside the eye have not appeared within 1 year, 90 percent of such patients will have no further progression of symptoms.[96,105]

Semiquantitative assessment is useful in clinical evaluation with serial measurements of sustained upward gaze, grip dynamometry, vital capacity, and arm abduction. If the patient exercises the limb with a pneumatic cuff inflated around the upper arm, myasthenic signs worsen in the rest of the body upon release of the cuff.[145] Early investigators erroneously interpreted this phenomenon to indicate the presence of a circulating toxic substance. The spreading weakness probably results from reduction in serum calcium, which binds with the lactate produced during ischemic exercise.[102] Myasthenic muscles have characteristic hypersensitivity to curare,[115] although the finding is common to any disorders with defective neuromuscular transmission, such as motor neuron disease,[82] ocular myopathy,[74,114] and antibiotic toxicity.[107]

In previously untreated cases, intravenous administration of edrophonium (Tensilon) almost uniformly improves the strength of involved muscles. The usual clinical diagnostic procedure consists of injecting a 2-mg test dose initially, followed by an 8-mg booster dose if the patient shows neither improvement nor adverse reaction. The effect of edrophonium begins within 1 minute and ceases in 5 to 10 minutes. For objective assessment, an injection of normal saline in a double-blind fashion serves as a control. Some patients, especially those with ocular myasthenia, have an equivocal or a false-negative result when tested with short-acting edrophonium. In these cases, the administration of longer acting neostigmine (Prostigmin) may improve strength more appreciably.

Differential diagnoses comprise all diseases characterized by weakness of ocular, bulbar, or extremity muscles.

These include muscular dystrophy, motor neuron disease, progressive bulbar palsy, multiple sclerosis, ophthalmoplegia, pseudobulbar palsy, and psychoneurosis. Patients with myasthenia gravis typically complain of excessive fatigability after exercise. In mild cases, symptoms may appear only after exertion, not uncommonly leading to a mistaken diagnosis of hysteria. A hot bath may worsen the symptoms of myasthenia gravis by lowering the margin of safety in neuromuscular transmission.[10,125] Here, distinction from multiple sclerosis may prove difficult, especially if the patient presents with pseudo-internuclear ophthalmoplegia.[72] Routine muscle biopsy has limited diagnostic value. Type II fiber atrophy, though commonly seen in myasthenia gravis, can also result from disuse or corticosteroid treatment.

Prognosis has improved with the therapeutic regimens, which include thymectomy, steroids, and immunosuppressive drugs.[104] Administration of prednisone, however, may cause acute inhibition of neuromuscular function. Patients then have increased decremental response to repetitive nerve stimulation, reduced twitch tension, and lowered force of maximum voluntary contraction.[77] Plasma exchange induces a rapid improvement of neuromuscular transmission over 1 to 4 weeks in some patients.[90]

Electrophysiologic Tests

Electrophysiologic studies play an important role in establishing the diagnosis of myasthenia gravis. The incidence of decremental response to repetitive nerve stimulation varies widely, from 41 percent in one laboratory[117] to 95 percent in another.[99] In general, 65 to 85 percent of patients show a positive result, after a comprehensive survey from multiple recording sites.[92] To reduce false-negative results, one must study proximal muscles, despite the technical difficulty with movement artifacts. Studies of distal muscles provide more consistent results but less sensitivity. To compromise, one may proceed from the easily immobilizable intrinsic hand muscles to the del-

toid, trapezius, and other shoulder girdle muscles and finally to the facial muscles. Warming the muscle increases the yield of the test (see Fig. 9–2). When conventional studies show equivocal results, ischemic conditions or regional administration of curare may lower the margin of safety sufficiently to produce clear abnormality.[14]

A single stimulus elicits a compound muscle action potential of a normal or only slightly reduced amplitude. The muscle action potentials show a decremental tendency to repetitive stimulation at two to three per second and to a lesser extent at higher rates (see Figs. 8–6 and 8–8). The amplitude drops maximally between the first and second responses of a train with fewer changes for the next few peaks and subsequent partial recovery, or repair. According to generally accepted criteria, at least two muscles should show a reproducible reduction of more than 10 percent between the first response and the smallest of the first five of a train. In our experience, any reproducible decrement should raise suspicion, provided the study reveals a clean tracing free of technical problems.

If repetitive stimulation at two to three per second demonstrates a decrement, intravenous administration of edrophonium will usually normalize the response partially or completely. A brief voluntary exercise for 15 to 30 seconds also repairs a bona fide tendency for decrement during subsequent trains, a phenomenon called posttetanic potentiation. In contrast, amplitude diminution within a train exceeds the preexercise value 2 to 4 minutes later during posttetanic exhaustion. Again, an additional 5 seconds of exercise will partially correct the change. Persisting changes may suggest technical factors, rather than defective neuromuscular transmission. Thus, brief voluntary exercise helps differentiate an abnormal response form a movement artifact (see Fig. 9–4).

Electromyography shows varying in amplitude and configurations of recurring motor unit potentials. Although unpredictable, the initial few discharges tend to decrease progressively in size and duration. Fibrillation potentials and positive sharp waves may be present, indicating the loss of innervation in severely affected muscles. Single fiber electromyography (SFEMG) provides one of the most sensitive measures of myasthenia gravis.[131] Increased jitter may be detected in clinically strong muscles showing no decrement in response to repetitive nerve stimulation. In one study, the severity of disease correlated better with the degree of jitter than with antibody titer to ACh receptor.[62] In another study of 43 mild myasthenics with normal repetitive stimulation tests, abnormalities were detected by SFEMG in 79 percent, by anti-ACh receptor antibodies in 71 percent, and by Lancaster red-green tests in 81 percent.[60] The three tests complemented each other in confirming the diagnosis.

Some patients with myasthenia gravis show the electrophysiologic features more typically associated with the myasthenic syndrome. These cases suggest the existence of an intermediate disorder characterized by defective ACh release as well as diminished numbers of ACh receptors.[26] Microelectrode studies have, however, provided no convincing evidence to support such a contention. In general, ACh release is decreased by low rates of stimulation and increased by high rates of stimulation (see Chapter 8.7). These physiologic phenomena become clinically manifest in the presence of defective neuromuscular transmission (see Fig. 8–8). The size of the first compound muscle potential often dictates the pattern of responses to repetitive stimulation. For example, an initially subnormal response has more room to increase during a train of rapid stimulation, even in patients with myasthenia gravis (see Fig. 8–7). In the same patient, some muscles may demonstrate an abnormal pattern typical of myasthenia gravis, whereas others may show changes reminiscent of the myasthenic syndrome.

3 MYASTHENIC SYNDROME

The myasthenic syndrome, or Lambert-Eaton syndrome,[33] affects men twice as commonly as women, with onset usu-

ally after age 40. The syndrome has also been reported in children 4 and 9 years of age.[9,20] Clear association between malignancy and the syndrome holds the key to elucidating the mechanism that results in defective release of ACh. Perhaps biologically active polypeptides, like botulinum toxin or magnesium ions, block ACh release from the motor terminals by interfering with the utilization of calcium.

Etiologic Considerations

Over 50 percent of the affected patients have small-cell carcinoma of the bronchus, the most common tumor seen in conjunction with this syndrome. Careful search reveals a malignant neoplasm of one kind or another in about 75 percent of men and 25 percent of women, although not necessarily at the time of initial neuromuscular symptoms. These include reticulum cell sarcoma,[113] rectal carcinoma,[18] renal carcinoma,[19] basal cell carcinoma of the skin,[136] leukemia,[124] and malignant thymoma.[67] Systemic disorders associated with the syndrome include thyrotoxicosis,[91] Sjögren's syndrome,[15] rheumatoid arthritis,[136] and other autoimmune disorders.[50] The malignancy may escape detection for many months or, occasionally, for many years after the onset of the myasthenic syndrome. With adequate follow-up, however, only 30 percent of patients remain free of cancer.[66]

Histometric studies of motor end-plate ultrastructure[38,39] have revealed overdevelopment and increased area of the postsynaptic membrane (see Fig. 8–2). The nerve terminal retains a normal mean synaptic vesicle diameter and mean synaptic vesicle density. Routine muscle biopsy shows only nonspecific findings with some type II fiber atrophy and mild inflammatory reactions. In microelectrode studies of excised intercostal muscles, the miniature end-plate potentials (MEPPs) are normal in amplitude but low in frequency.[34] The mean quantum content of the end-plate potential (EPP) is abnormally low at first but increases with repetitive nerve impulses.

These findings suggest either an abnormality in the calcium-dependent release of ACh from the motor nerve terminals or a decreased store of available ACh. Ultrastructural studies show a normal synaptic vesicle number per unit nerve terminal, which tends to discount the possibility of defective storage. Thus, weakness in the myasthenic syndrome probably results from presynaptic abnormalities that lead to a reduced number of ACh quanta released per volley of nerve impulses. In an experimental setting, high magnesium or low calcium ion concentrations induce similar block of neuromuscular transmission. The pathologic abnormalities vary in the myasthenic syndrome. In one patient who had a small decremental response and increased jitter and blocking, for example, histologic studies showed alteration in the number and affinity of junctional ACh receptors and prominent tubular aggregations in muscle fibers.[80]

Clinical Signs and Symptoms

In striking contrast to fatigue phenomena in myasthenia gravis, weakness in the myasthenic syndrome peaks after a rest or immediately after awakening in the morning. Strength tends to improve transiently with brief exercise, although it is not sustained during a prolonged effort. Weakness and fatigability primarily affect the lower extremities, particularly the pelvic girdle and thigh muscles. Thus, the patient has difficulty in climbing stairs and, to a lesser degree, arising from a chair. The abnormality also involves the shoulders and the four extremities but characteristically spares the neck, bulbar, and extraocular musculature. This distribution of weakness is the reverse of the typical pattern seen in myasthenia gravis, with conspicuous bulbar symptoms such as ptosis, diplopia, dysphagia, and dysarthria.

The patient often complains of dryness of the mouth, and less frequently, impotence, paresthesias, and dysautonomia. These symptoms suggest that the defect of ACh release, not restricted to skeletal muscle, may affect the autonomic nervous system as well.[116] Peripheral neuropathy and subacute cerebellar degeneration may develop, probably as epiphenomena of a paraneoplastic syndrome.

Neurologic evaluation reveals marked weakness of proximal muscles in the lower limbs that appreciably improves after exercise. With each successive effort, the resistance needed to overcome the patient's strength increases, giving the examiner a sensation similar to drawing up water from a well with a hand pump.[13] Reduced muscle stretch reflexes and other signs of polyneuropathy may be present. The edrophonium (Tensilon) test ordinarily gives negative or equivocal results, but a small dose of *d*-tubocurarine and decamethonium causes a depolarizing block at the neuromuscular junction.

Guanidine partially corrects defective calcium-dependent ACh release and results in dramatic improvement in strength. The neuromuscular defect also improves partially after the administration of calcium, 4-aminopyridine, aminophylline, or caffeine, which increases the cyclic adenosine monophosphate essential in calcium mobilization in cells.[137] Adverse side effects severely limit the use of 4-aminopyridine in the treatment of patients.[85] Plasma exchange and immunosuppressive drugs may also temporarily alleviate the symptoms.[88] Muscle strength may increase with simultaneous electrophysiologic improvement after long-term prednisone therapy.[132]

Electrophysiologic Tests

As the electrical hallmark of the syndrome, nerve stimulation typically elicits a very small compound muscle action potential (see Figs. 8–8 and 9–8) and, in striking contrast, entirely normal sensory responses. Repetitive stimulation at low rates further diminishes muscle action potentials similar to the decrement seen in myasthenia gravis. Stimulation at high rates or brief voluntary contraction for up to 10 seconds gives rise to a substantial increment, usually exceeding 50 to 200 percent of the baseline value (see Figs. 9–6 and 9–7). Paired stimulation with interstimulus intervals of 5 to 10 ms causes the second response to increase, rather than decrease, as expected in normal muscles. Posttetanic facilitation decays within 20 seconds. During posttetanic exhaustion, which peaks in 2 to 4 minutes, the muscle potential falls below the resting level (see Figs. 9–10 and 9–11).

Nerve stimulation may reveal marked abnormalities even in patients with mild clinical symptoms. Clinical remission after therapy usually accompanies a parallel improvement in serial electrophysiologic studies.[54] Interestingly, patients with mild cases of the myasthenic syndrome complain little of motor dysfunction because posttetanic facilitation during voluntary contraction produces nearly normal strength. Rested muscles, however, show an unequivocal defect of neuromuscular transmission. Nearly all muscles show a mild decrement at low rates and prominent increment at high rates of stimulation. This uniformity stands in sharp contrast to the variability of responses in myasthenia gravis, in which electrical abnormalities are confined to clinically symptomatic muscles. In one reported case, electrophysiologic studies revealed a unique combination of marked depression to single nerve stimulation and facilitation at all rates from 1 to 200 per second.[64] This case may represent a separate entity or variation of the myasthenic syndrome.

Needle studies show varying configurations of repetitive motor unit potentials with an incrementing tendency. As expected, increased jitter and blocking in single fiber studies improve with high rates of stimulation and worsen following rest.[121]

4 MYASTHENIA IN INFANCY

Transient Neonatal Myasthenia

Approximately 15 percent of infants born to myasthenic mothers have neonatal myasthenia gravis.[133,142] This condition presumably results from transpla-

cental transfer of anti-ACh receptor antibodies[87,93] or transient synthesis of receptor antibodies.[69] The onset of clinical weakness on the second or third day coincides with the release of antibodies from hemoglobins to which they are combined at birth.[59] A similar clinical syndrome develops in mice after injection of the IgG serum fraction from patients with myasthenia gravis.[139]

Clinical features during the first few days after birth consist of diffuse hypotonia with difficulty in breathing and sucking, although some infants have selective weakness of the diaphragm.[53] The neonates usually respond to anticholinesterase medication. Symptoms generally disappear when the infant's own immune system has developed in a few weeks,[86,146] but they may occasionally persist beyond 2 months of age.[12] Electrophysiologic studies show characteristic abnormalities in distal muscles as late as 30 days after clinical recovery.[31] An elevated antibody titer against ACh receptor returns to the normal range over a 3-month period.[59]

Other Forms of Congenital Myasthenia

In the absence of maternal transfer, infantile myasthenia gravis may result from acquired autoimmune pathogenesis or nonautoimmune hereditary diseases. The term congenital myasthenia gravis, or familial infantile myasthenia, implies the absence of anti-ACh receptor antibodies in the serum.[144] These patients have a family history of similar disease but otherwise clinically indistinguishable from the autoimmune type.[123] Thus, antibody determinations provide a useful aid in differentiating autoimmune and hereditary myasthenia in infancy. This section will review rare varieties of congenital myasthenia gravis with onset at birth or in childhood and a persistent clinical course.

Congenital type accounts for about 1 percent of all cases of myasthenia gravis. Although the disease begins in infancy, it continues into childhood and adulthood, unlike transient neonatal myasthenia. In many cases, the family history reveals affected siblings, although the mother has no disease. This entity encompasses a variety of specific defects at the neuromuscular junction, with no evidence of an immunologic attack against neuromuscular junctions. Detailed physiologic, chemical, and histologic studies have elucidated a number of types with specific presynaptic or postsynaptic abnormalities (Table 24–1). In general, the disease primarily involves extraocular muscles without associated generalized weakness.[96] Initially mild symptoms slowly progress despite therapy.[78] The infants have respiratory depression at birth and episodic weakness and apnea during the first 2 years[24,44,47] but improve with anticholinesterase medication.[112] These and other syndromes of congenital myasthenia may represent separate pathologic, electrophysiologic, and clinical entities.[143] In-vitro intracellular microelectrode studies have revealed a different mechanism of defective neuromuscular transmission in each of the following entities.

The first type, described in a 15-year-old boy with intermittent ptosis, delayed motor development, and generalized weakness, has three main features: acetylcholinesterase deficiency, small nerve terminals, and reduced ACh release.[35] The patient had a negative result in the edrophonium (Tensilon) test, no serum antibodies to muscle ACh receptors, and absent acetylcholinesterase at the end plates. Nerve terminals averaged one-third to one-fourth normal size. In-vitro microelectrode studies revealed a number of unusual features: normal amplitude but low discharge frequency of MEPPs, a marked reduction in number of ACh quanta released per nerve stimulation, and prolonged MEPP and EPP duration. A single shock to the nerve elicited repetitive discharges, whereas a train of stimuli at 2 and 40 per second gave rise to a decremental response. Needle studies showed temporal variability of the motor unit potentials.

In another form involving one sporadic case and four others from two different families, an abnormal ACh receptor presumably caused a prolonged EPP despite normal muscle acetylcholinesterase.[36] The affected infants had ophthalmoparesis and weakness of neck muscles. Easy fatigability and weakness of shoulder girdle and forearm muscles developed later, in adolescence or adulthood. Single stimuli to motor nerves elicited repetitive muscle action potentials in the proximal and distal muscles tested. In view of normal muscle acetylcholinesterase, the prolonged EPP might result from an abnormal ACh receptor with a prolonged open

Table 24–1 CHARACTERISTICS THAT DIFFERENTIATE NEUROMUSCULAR TRANSMISSION DEFECTS IN MYASTHENIC SYNDROMES

Myasthenic Syndrome	AChR Antibodies	Repetitive Muscle AP to Single Nerve Stimulus	MEPP Duration	MEPP Duration Increased by Esterase Inhibition	MEPP Amplitude	Marked Decrement of EPP and MEPP During 10 Hz Stimulation	Quantum Content
MG	+	−	−	+	↓	−	−
LES	−	−	−	+	−	−	↓
Congenital							
A	−	+	↑	−	↓	−	↓
B	−	+	↑	+	↓	−	−
C	−	−	−	+	−	+	−
D	−	−	−	+	↓	−	−
Dog	−	−	−	+	↓	−	−

MG, myasthenia gravis; LES, Lambert-Eaton syndrome; A, myastenic syndrome with end-plate acetylcholinesterase deficiency, small nerve terminals, and reduced acetylcholine release[35]; B, familial, congenital myasthenic syndrome possibly from an abnormal acetylcholine receptor with prolonged open time[36]; C, familial, congenital myasthenic syndrome possibly from deficient synthesis of acetylcholine[52]; D, familial, congenital myasthenic syndrome with a possible abnormality of acetylcholine receptor synthesis or incorporation in the postsynaptic membrane[65]; Dog, congenital myasthenia in dogs.

time or an abnormal transmitter resistant to muscle acetylcholinesterase.[143]

In a separate entity, probably caused by deficient synthesis of ACh, affected infants had intermittent ptosis, feeding difficulties, dyspnea or apnea, and vomiting.[52] Weakness worsened with febrile illness and during exercise but gradually improved with age. Progressive weakness developed during prolonged nerve stimulation at 10 per second. A brief repetitive nerve stimulation produced no decrement of the muscle action potential. In another full-term infant with similar clinical features, electrodiagnostic studies demonstrated defective neuromuscular transmission characterized by borderline low motor evoked amplitudes, profound decremental responses at all stimulus rates, and moderate facilitation ranging from 50 to 74 percent, 15 seconds after 5 seconds of stimulation at 50 per second.[3] Although not proven, these findings suggest abnormality of ACh resynthesis, mobilization, or storage, rather than defective receptors. In fact, prolonged nerve stimulation induces a temporal decline in EPP and MEPP amplitude in normal muscle after blocking ACh synthesis with hemicholinium.[30] Despite abnormally small synaptic vesicles found in some patients with familial infantile myasthenia, vesicle size cannot be reliably correlated with the MEPP amplitude.[79]

In a case of congenital myasthenic syndrome with a possible abnormality of ACh receptor synthesis, clinical features included ptosis, limb weakness, and easy fatigability since birth.[65] He had a similarly affected brother. Intracellular microelectrode studies revealed low amplitude but normal MEPP duration and frequency, a normal number of ACh quanta released by nerve stimulation, normal store of readily releasable quanta in the nerve terminal, and abnormally low ACh receptor content. In the absence of autoimmunity, the abnormality might result from a defect of the ACh receptor molecule or its synthesis in this type of myasthenia as well as in congenital myasthenia in dogs.

In still another type of congenital myasthenia, with paucity of secondary synaptic clefts, clinical features included weak fetal movements during pregnancy, muscle weakness at birth, multiple contractures of the lower limbs, and myasthenic crisis during febrile illness.[128] Neurophysiologic studies demonstrated a 55 percent decremental response to stimulation at three per second and reversal of this abnormality by administration of endrophonium (Tensilon).

5 BOTULISM

Botulinum Toxin

The exotoxin of Clostridium botulinum has a generalized effect on the neuromuscular junction involving both striated and smooth muscles. Of the six immunologic types of Bacillus botulinus, types A, B, and E account for most human cases. Poisoning by this heat-sensitive toxin usually follows the ingestion of raw or inadequately cooked or canned vegetables, meat, or fish. An infected wound may occasionally harbor the toxins.[29,109] Types A and B usually originate in contaminated canned vegetables and type E in fish products. The mortality is higher with types A or E than with type B.[76]

The incidence of botulism increases at high altitudes, probably because of the lower temperature required to boil water.[21] Botulism bares a great resemblance to the myasthenic syndrome, with marked impairment of ACh release from the nerve terminal.[57] In-vitro studies of the MEPP show extremely low rates of discharge but normal or only slightly reduced amplitude. A small quantum content per volley of nerve impulse results in a markedly decreased EPP.

Clinical Signs and Symptoms

Botulism should be suspected if in several members of a family similar symptoms develop after a shared meal. Isolated cases pose a greater diagnostic challenge. The mouse toxin neutralization test and culture of the suspected food confirm the diagnosis. Ingestion of a large amount of toxin may rapidly result in fatal cardiac or respiratory failure. Some cases of the sudden infant death syndrome may result from botulism, now recognized with increasing frequency in this age group.[56,106] In less severe cases, mild symptoms abate, and, as the rule, complete recovery ensues.

Symptoms appear within 1 to 2 days after consumption of contaminated food but in 1 to 2 weeks after wound inoculation because of the time necessary for the elaboration of toxin. Gastrointestinal dysfunctions such as diarrhea, nausea, and vomiting precede the onset of cranial weakness initially characterized by external ophthalmoplegia and ptosis. Patients may also have failure of convergence, fixed and dilated pupils, dysarthria, dysphagia, or difficulty in mastication.[22] The involvement of intestine and bladder causes constipation and urinary retention.

The disease affects the muscles of the extremities and the trunk later. By then, examination reveals a flaccid and areflexic patient with widespread paralysis. Exercise causes fatigue, though not as prominently as in myasthenia gravis. Unlike the weakness seen in the myasthenic syndrome, muscle strength does not improve with repeated efforts. Likewise, treatment with guanidine fails to enhance recovery from botulism.[58]

Electrophysiologic Tests

Nerve conduction studies show normal amplitude and latency of sensory action potentials. A small compound muscle action potential elicited by single shock further declines with repetitive stimulation at a slow rate. Paired stimuli at interstimulus intervals of less than 10 ms characteristically potentiate the second response by summation of the two EPPs (see Fig. 9–3). This finding stands in sharp contrast to the smaller second response normally elicited during the refractory period by paired shocks of such short interstimulus intervals. In botulism, as in the myasthenic syndrome, the refractory period plays a limited role, because only a small number of muscle

fibers discharge in response to the first stimulus.

Muscle response is facilitated by a fast train of stimuli or during posttetanic potentiation[141] but usually not to the same degree as seen in the myasthenic syndrome (see Fig. 9–9). In infantile botulism, repetitive stimulation at 20 to 50 per second provides the most specific single test, showing an incremental response in over 90 percent of patients.[25] Prolongation of posttetanic facilitation, up to 4 minutes in one case, also constitutes a unique feature of botulism.[41]

The presence of fibrillation potentials may indicate functional denervation caused by limited release of ACh.[45] SFEMG has shown increased jitter and blocking and some reduction in fiber density.[119] SFEMG has also revealed that injection of botulinum toxin for blepharospasm causes abnormal neuromuscular transmission in arm muscles, indicating remote spread of toxin from the site of injection.[118]

6 OTHER DISORDERS

Tick Paralysis

Available data suggest that the neurotoxin affects either the nerve terminial or the neuromuscular junction. The condition, reported worldwide, results from infestation by the gravid female tick, Dermacentor andersoni (wood tick) or Dermacentor paridulis (dog tick) in the United States, and Ixodes holocyclus (scrub tick) in Australia.[103] Most cases involve young children, especially girls with long hair, in spring or summer, when ticks are active.[120] The symptoms and signs begin 5 to 7 days after the tick has become embedded. During this latent period, the organism, attached near the hairline, may remain unnoticed.

Illness begins with general symptoms such as irritability and diarrhea. Weakness initially affects the lower limbs and, within a day, spreads to the upper limbs. Paralyses of the bulbar and respiratory musculature, although now rare, pose a major threat until the removal of the embedded tick. Other features include dysarthria, dysphagia, blurred vision, facial weakness, and reduced muscle stretch reflexes. Occasionally patients complain of numbness and tingling of the extremities. Removal of the tick usually leads to rapid improvement. Application of heat or Vaseline causes the tick to withdraw from the skin, allowing its gentle separation in one piece with a forceps.

Electrophysiologic studies in a few confirmed cases have consistently shown reduced amplitude of the compound muscle action potential.[23,51,81,103,135] In one study[135] muscle action potentials changed little on repetitive stimulation of up to 50 per second. Mildly increased distal motor and sensory latency during the paralytic phase returned to normal after clinical recovery. Persistent weakness and the presence of fibrillation potentials in some cases after the removal of the tick suggest a structural lesion of distal motor axons.[32]

The toxin probably prevents depolarization in the terminal axons by altering the ionic conductance that mediates action potentials in the nerve. Like other potent biotoxins such as tetradotoxin and saxotoxin, tick toxin blocks the inward flux of sodium ions at sensory and motor nerve terminals and at internodes. Tick toxin may also interfere with release of ACh at the nerve terminal,[23] but not with its synthesis or storage.[83] Intracellular studies of hamsters paralyzed by tick toxin, however, have shown normal MEPP size and frequency and normal EPP quantal content.[75]

Effect of Drug or Toxin

The administration of some drugs, notably kanamycin and neomycin and all other polypeptide aminoglycoside antibiotics, may cause abnormalities of neuromuscular transmission.[7] At low rates of repetitive nerve stimulation, the muscle action potentials show a decremental response, but facilitation after exercise typically exceeds that seen in myasthenia gravis. In rats, small-amplitude MEPPs and abnormally low mean EPP quantum content suggest combined presynaptic and postsynaptic effects.[28] Another type of abnormality produced experimentally with hemicholinium impairs ACh synthesis.[30] Myasthenia-like weakness may also develop during procainamide therapy.[89]

The use of penicillamine in rheumatoid arthritis may herald the clinical onset of myasthenia.[8,16,17,43,122] The clinical and electrophysiologic characteristics are indistinguishable from idiopathic myasthenia gravis but improve after discontinuing the drug.[4] The degree of jitter shows a positive correlation with the duration of administration but not the dosage of penicillamine.[2] This disorder and idiopathic autoimmune myasthenia gravis probably share the same pathophysiology that underlies the presence of anti-ACh receptor antibody and resultant quantitative reduction in available junctional anti-ACh receptors.[63] These data suggest that penicillamine produces myasthenia gravis by initiating a new autoimmune response, rather than enhancing ongoing autoimmunity.

Exposure to an organophosphate insecticide causes flaccid paralysis of extremities. Electrophysiologic studies demonstrate repetitive compound muscle action potentials in response to a single stimulus of the nerve; a decremental response at higher rates of stimulation, further accentuated following administration of edrophonium (Tensilon); and normal nerve conduction studies during acute stages.[73]

Lower Motor Neuron Disorders

Defects of neuromuscular transmission are also seen in motor neuron disease and peripheral neuropathies.[82] Experimental studies suggest that the immediately available store of ACh is diminished during regeneration. Alternatively, a defect may lie in the propagation of impulses along the terminal portion of the nerve if the axonal refractory period is abnormally prolonged. In these cases, repetitive stimulation at low rates results in a progressive decrement of the muscle action potential. In contrast to the changes seen in myasthenia gravis, however, this decrement is minimal at low rates, becoming progressively more prominent at faster rates of stimulation.[46] Such phenomena as posttetanic potentiation and exhaustion may also occur.

Muscle Diseases

A decrementing response may also result from increasing muscle membrane refractoriness associated with repetitive discharges in myotonia[5] and periodic paralysis.[111] In these disorders, a decrement occurs regardless of the rate of stimulation. Unlike the pattern seen in myasthenia gravis, the decrement is steadily progressive, with no tendency for the amplitude to plateau or recover at the fifth or sixth stimulus. Immediately after exercise, the muscle action potential is much reduced, because many muscle fibers are refractory. The amplitude returns to resting values in 15 to 30 seconds. Thus, exercise first induces reduction in muscle excitability followed by recovery, as opposed to the initial posttetanic potentiation and subsequent exhaustion seen in myasthenia gravis. The decremental response in myotonia may erroneously suggest defective neuromuscular transmission. Improper interpretation of such findings may be responsible for a few reports of patients with both myotonic dystrophy and myasthenia gravis.

In McArdle's syndrome, weakness increases with exertion, and associated muscle contractures are electrically silent (see Fig. 11–3). The compound muscle action potential progressively decreases in amplitude as contractures develop in response to rapid repetitive stimulation.[110]

REFERENCES

1. Aharonov, A, Abramsky, O, Tarrab-Hazdai, R, and Fuchs, S: Humoral antibodies to acetylcholine receptor in patients with myasthenia gravis. Lancet 2:340–342, 1975.
2. Albers, JW, Beals, CA, and Levine, SP: Neuromuscular transmission in rheumatoid arthri-
tis, with and without penicillamine treatment. Neurology 31:1562–1564, 1981.
3. Albers, JW, Faulkner, JA, Dorovini-Zis, K, Barald, KF, Must, RE, and Ball, RD: Abnormal neuromuscular transmission in an infantile myasthenic syndrome. Ann Neurol 16:28–34, 1984.
4. Albers, JW, Hodach, RJ, Kimmel, DW, and Treacy, WL: Penicillamine-associated myasthenia gravis. Neurology 30:1246–1250, 1980.
5. Aminoff, MJ, Layzer, RB, Satya-Murti, S, and Faden, AI: The declining electrical response of muscle to repetitive nerve stimulation in myotonia. Neurology 27:812–816, 1977.
6. Appel, SH, Almon, RR, and Levy, N: Acetycholine receptor antibodies in myasthenia gravis. N Engl J Med 293:760–761, 1975.
7. Argov, Z, and Mastaglia, FL: Disorders of neuromuscular transmission by drugs. N Engl J Med 301:409–413, 1979.
8. Atcheson, SG, and Ward, JR: Ptosis and weakness after start of D-penicillamine therapy. Ann Intern Med 89:939–940, 1978.
9. Bady, B, Chauplannaz, G, and Carrier, H: Congenital Lambert-Eaton myasthenic syndrome. J Neurol Neurosurg Psychiatry 50:476–478, 1987.
10. Borenstein, S, and Desmedt, JE: Temperature and weather correlates of myasthenic fatigue. Lancet 2:63–66, 1974.
11. Borenstein, S, and Desmedt, JE: Local cooling in myasthenia: Improvement of neuromuscular failure. Arch Neurol 32:152–157, 1975.
12. Branch, CE, Jr, Swift, TR, and Dyken, PR: Prolonged neonatal myasthenia gravis: Electrophysiological studies. Ann Neurol 3:416–418, 1978.
13. Brooke, MH: A Clinician's View of Neuromuscular Diseases. Williams & Wilkins, Baltimore, 1977.
14. Brown, JC, Charlton, JE, and White, DJK: A regional technique for the study of sensitivity to curare in human muscle. J Neurol Neurosurg Psychiatry 38:18–26, 1975.
15. Brown, JW, Nelson, JR, and Herrmann, C, Jr: Sjögren's syndrome with myopathic and myasthenic features. Bull LA Neurol Soc 33:9–20, 1968.
16. Bucknall, RC: Myasthenia associated with D-penicillamine therapy in rheumatoid arthritis. Proc R Soc Med 70 (Suppl 3):114–117, 1977.
17. Bucknall, RC, Dixon, AStJ, Glick, EN, Woodland, J, and Zutshi, DW: Myasthenia gravis associated with penicillamine treatment for rheumatoid arthritis. Br Med J 1:600–602, 1975.
18. Canak, S, Jantsch, H, Kucher, R, Lill, L, Pateisky, K, and Steinbereithmer, K: Zur Kenntnis der Neuromyopathia carcinomatosa mit myasthenischem Syndrom. Der Anaesthesist 13:65–69, 1964.
19. Castaigne, P, Cambier, J, Masson, M, Cathala, HP, and Pierrot-Deseilligny, E: Le syndrome pseudo-myasthénique paranéoplasique de Lambert-Eaton. Ann Med Intern (Paris) 120:313–322, 1969.

20. Chelmicka-Schorr, E, Bernstein, LP, Zurbrugg, EB, and Huttenlocher, PR: Eaton-Lambert syndrome in a 9-year-old girl. Arch Neurol 36:572–574, 1979.

21. Cherington, M: Botulism: Clinical and therapeutic observations. Rocky Mt Med J 69:55–58, 1972.

22. Cherington, M: Botulism: Ten-year experience. Arch Neurol 30:432–437, 1974.

23. Cherington, MA, and Snyder, RD: Tick paralysis. Neurophysiologic studies. N Engl J Med 278:95–97, 1968.

24. Conomy, JP, Levinsohn, M, and Fanaroff, A: Familial infantile myasthenia gravis: a cause of sudden death in young children. J Pediatr 87:428–430, 1975.

25. Cornblath, DR, Sladky, JT, and Sumner, AJ: Clinical electrophysiology of infantile botulism. Muscle Nerve 6:448–452, 1983.

26. Dahl, DS, and Sato, S: Unusual myasthenic state in a teen-age boy. Neurology 24:897–901, 1974.

27. Daube, JR: Disorders of neuromuscular transmission: A review. Arch Phys Med Rehabil 64:195–200, 1983.

28. Daube, JR, and Lambert, EH: Post-activation exhaustion in rat muscle. In Desmedt, JE (ed): New Developments in Electromyography and Clinical Neurophysiology, Vol 1. Karger, Basel, 1973, pp 343–349.

29. de Jesus, PV, Jr, Slater, R, Spitz, LK, and Penn, AS: Neuromuscular physiology of wound botulism. Arch Neurol 29:425–431, 1973.

30. Desmedt, JE: The neuromuscular disorder in myasthenia gravis. II. Presynaptic cholinergic metabolism, myasthenia-like syndromes and a hypothesis. In Desmedt, JE (ed): New Developments in Electromyography and Clinical Neurophysiology, Vol 1. Karger, Basel, 1973, pp 305–342.

31. Desmedt, JE, and Borenstein, S: Time course of neonatal myasthenia gravis and unsuspectedly long duration of neuromuscular block in distal muscles (abstr). N Engl J Med 296:633, 1977.

32. Donat, JR, and Donat, JF: Tick paralysis with persistent weakness and electromyographic abnormalities. Arch Neurol 38:59–61, 1981.

33. Eaton, LM, and Lambert, EH: Electromyography and electric stimulation of nerves in diseases of motor units: Observations on myasthenic syndrome associated with malignant tumors. JAMA 163:1117–1124, 1957.

34. Elmqvist, D, and Lambert, EH: Detailed analysis of neuromuscular transmission in a patient with the myasthenic syndrome sometimes associated with bronchogenic carcinoma. Mayo Clin Proc 43:689–713, 1968.

35. Engel, AG, Lambert, EH, and Gomez, MR: A new myasthenic syndrome with end-plate acetylcholinesterase deficiency, small nerve terminals and reduced acetylcholine release. Ann Neurol 1:315–330, 1977.

36. Engel, AG, Lambert, EH, Mulder, DM, Torres, CF, Sahashi, K, Bertorini, TE, and Whitaker, JN: Investigations of 3 cases of a newly recognized familial, congenital myasthenic syndrome. Ann Neurol 6:146–147, 1979.

37. Engel, AG, Lindstrom, JM, Lambert, EH, and Lennon, VA: Ultrastructural localization of the acetylcholine receptor in myasthenia gravis and in its experimental autoimmune model. Neurology 27:307–315, 1977.

38. Engel, AG, and Santa, T: Histometric analysis of the ultrastructure of the neuromuscular junction in myasthenia gravis and in the myasthenic syndrome. Ann NY Acad Sci 183:46–63, 1971.

39. Engel, AG, and Santa, T: Motor endplate fine structure. In Desmedt, JE (ed): New Developments in Electromyography and Clinical Neurophysiology, Vol 1. Karger, Basel, 1973, pp 196–228.

40. Engel, WK, Festoff, BW, Patten, BM, Swerdlow, ML, Newball, HH, and Thompson, MD: Myasthenia gravis. Ann Intern Med 81:225–246, 1974.

41. Fakadej, AV, and Gutmann, L: Prolongation of post-tetanic facilitation in infant botulism. Muscle Nerve 5:727–729, 1982.

42. Fambrough, DM, Drachman, DB, and Satyamurti, S: Neuromuscular junction in myasthenia gravis: Decreased acetylcholine receptors. Science 182:293–295, 1973.

43. Fawcett, PRW, McLachlan, SM, Nicholson, LVB, Argov, Z, and Mastaglia, FL: D-penicillamine-associated myasthenia gravis: Immunological and electrophysiological studies. Muscle Nerve 5:328–334, 1982.

44. Fenichel, GM: Clinical syndromes of myasthenia in infancy and childhood: A review. Arch Neurol 35:97–103, 1978.

45. Fusfeld, RD: Electromyographic abnormalities in a case of botulism. Bull LA Neurol Soc 35:164–168, 1970.

46. Gilliatt, RW: Applied electrophysiology in nerve and muscle disease. Proc R Soc Med 59:989–993, 1966.

47. Greer, M, and Schotland, M: Mysathenia gravis in the newborn. Pediatrics 26:101–108, 1960.

48. Gutmann, L: Heat exacerbation of myasthenia gravis (abstr). Neurology 28:398, 1978.

49. Gutmann, L: Heat-induced myasthenic crisis. Arch Neurol 37:671–672, 1980.

50. Gutmann, L, Crosby, TW, Takamori, M, and Martin, JD: The Eaton-Lambert syndrome and autoimmune disorders. Am J Med 53:354–356, 1972.

51. Haller, JS, and Fabara, JA: Tick paralysis. Case report with emphasis on neurological toxicity. Am J Dis Child 124:915–917, 1972.

52. Hart, ZH, Sahashi, K, Lambert, EH, Engel, AG, and Lindstrom, JM: A congenital, familial myasthenic syndrome caused by a presynaptic defect of transmitter resynthesis or mobilization. Neurology 29:556–557, 1979.

53. Heckmatt, JZ, Placzek, M, Thompson, AH, Dubowitz, V, and Watson, G: An unusual case of neonatal myasthenia. J Child Neurol 2:63–66, 1987.

54. Ingram, DA, Davis, GR, Schwartz, MS, Traub, M, Newland, AC, and Swash, M: Cancer-asso-

ciated myasthenic (Eaton-Lambert) syndrome: Distribution of abnormality and effect of treatment. J Neurol Neurosurg Psychiatry 47:806–812, 1984.

55. Ito, Y, Miledi, R, Vincent, A, and Newsom-Davis, J: Acetylcholine receptors and end-plate electrophysiology in myasthenia gravis. Brain 101:345–368, 1978.

56. Johnson, RO, Clay, SA, and Arnon, SS: Diagnosis and management of infant botulism. Am J Dis Child 133:586–593, 1979.

57. Kao, I, Drachman, DB, and Price, DL: Botulinum toxin: Mechanism of presynaptic blockade. Science 193:1256–1258, 1976.

58. Kaplan, JE, Davis, LE, Narayan, V, Koster, J, and Katzenstein, D: Botulism, type A, and treatment with guanidine. Ann Neurol 6:69–71, 1979.

59. Keesey, J, Lindstrom, J, and Cokely, H: Anti-acetylcholine receptor antibody in neonatal myasthenia gravis. N Engl J Med 296:55, 1977.

60. Kelly, JJ, Jr, Daube, JR, Lennon, VA, Howard, FM, Jr, and Younge, BR: The laboratory diagnosis of mild myasthenia gravis. Ann Neurol 12:238–242, 1982.

61. Kimura, J, and Van Allen, MW: Post-thymomectomy myasthenia gravis: Report of a case of ocular myasthenia gravis after total removal of a thymoma and review of literature. Neurology 17:413–420, 1967.

62. Konishi, T, Nishitani, H, Matsubara, F, and Ohta, M: Myasthenia gravis: Relation between jitter in single fiber EMG and antibody to acetylcholine receptor. Neurology 31:386–392, 1981.

63. Kuncl, RW, Pestronk, A, Drachman, DB, and Rechthand, E: The pathophysiology of penicillamine-induced myasthenia gravis. Ann Neurol 20:740–744, 1986.

64. Lambert, EH: Defects of neuromuscular transmission in syndromes other than myasthenia gravis. Ann NY Acad Sci 135:367–384, 1966.

65. Lambert, EH: Neuromuscular transmission: Normal and pathologic. American Academy of Neurology Special Course #11: Neurophysiology. April 28, 1981.

66. Lambert, EH, and Rooke, ED: Myasthenic state and lung cancer. In Brain, W, and Norris, FH, Jr (eds): The Remote Effects of Cancer on the Nervous System, Vol 1. Grune & Stratton, New York, 1965, pp 67–80.

67. Lauritzen, M, Smith, T, Fischer-Hansen, B, Sparup, J, and Olesen, J: Eaton-Lambert syndrome and malignant thymoma. Neurology 30:634–638, 1980.

68. Lefvert, AK, Bergstrom, K, Matell, G, Osterman, PO, and Pirskanen, R: Determination of acetylcholine receptor antibody in myasthenia gravis: Clinical usefulness and pathogenetic implications. J Neurol Neurosurg Psychiatry 41:394–403, 1978.

69. Lefvert, AK, and Osterman, PO: Newborn infants to myasthenic mothers: A clinical study and an investigation of acetycholine receptor antibodies in 17 children. Neurology 33:133–138, 1983.

70. Lepore, FE, Sanborn, GE, and Slevin, JT: Pupillary dysfunction in myasthenia gravis. Ann Neurol 6:29–33, 1979.

71. Lerrick, AJ, Wray, D, Vincent, A, and Newsom-Davis, J: Electrophysiological effects of myasthenic serum factors studied in mouse muscle. Ann Neurol 13:186–191, 1983.

72. Lyon, LW, and Van Allen, MW: Orbicularis oculi reflex: Studies in internuclear opthalmoplegia and pseudointernuclear opthalmoplegia. Arch Ophthalmol 87:148–154, 1972.

73. Maselli, R, Jacobsen, J, and Spire, J: Edrophonium: An aid in the diagnosis of acute organophosphate poisoning. Ann Neurol 19:508–510, 1986.

74. Mathew, NT, Jacob, JC, and Chandy, J: Familial ocular myopathy with curare sensitivity. Arch Neruol 22:68–74, 1970.

75. McLennan, H, and Oikawa, I: Changes in function of the neuromuscular junction occurring in tick paralysis. Can J Physiol Pharmacol 50:53–58, 1972.

76. Merson, MH, Hughes, JM, Dowell, VER, Taylor, A, Barker, WH, and Gangarosa, EJ: Current trends in botulism in the United States. JAMA 229:1305–1308, 1974.

77. Miller, R, Milner-Brown, H, and Mirka, A: Prednisone-induced worsening of neuromuscular function in myasthenia gravis. Neurology 36:729–732, 1986.

78. Millichap, JG, and Dodge, PR: Diagnosis and treatment of myasthenia gravis in infancy, childhood, and adolescence. Neurology 10:1007–1014, 1960.

79. Mora, M, Lambert, EH, and Engel, AG: Synaptic vesicle abnormality in familial infantile myasthenia. Neurology 37:206–214, 1987.

80. Morgan-Hughes, JA, Lecky, BRF, Landon, DN, and Murray, NMF: Alterations in the number and affinity of junctional acetylcholine receptors in a myopathy with tubular aggregates: A newly recognized receptor defect. Brain 104:279–295, 1981.

81. Morris, DH, III: Tick paralysis: Electrophysiologic measurements. South Med J 70:121–122, 1977.

82. Mulder, DW, Lambert, EH, and Eaton, LM: Myasthenic syndrome in patients with amyotrophic lateral sclerosis. Neurology 9:627–631, 1959.

83. Murnaghan, MF: Site and mechanism of tick paralysis. Science 131:418–419, 1960.

84. Murphy, A, Drachman, DB, Satya-Murti, S, Pestronk, A, and Eggleston, JC: Critical reexamination of the thymus immunization model of myasthenia gravis. Muscle Nerve 3:293–297, 1980.

85. Murray, NMF, and Newson-Davis, J: Treatment with oral 4-aminopyridine in disorders of neuromuscular transmission. Neurology 31:265–271, 1981.

86. Namba, T, Brown, SB, and Grob, D: Neonatal myasthenia gravis: Report of two cases and review of the literature. Pediatrics 45:488–504, 1970.

87. Nastuk, WL, and Strauss, AJL: Further developments in the search for a neuromuscular

blocking agent in the blood of patients with myasthenia gravis. In Viets, HR (ed): Myasthenia Gravis. Charles C Thomas, Springfield, Ill, 1961, pp 229–237.

88. Newsom-Davis, J, and Murray, NMF: Plasma exchange and immunosuppressive drug treatment in the Lambert-Eaton myasthenic syndrome. Neurology 23:480–485, 1984.

89. Niakan, E, Bertorini, TE, Acchiarod, SR, and Werner, MF: Procainamide-induced myasthenia-like weakness in a patient with peripheral neuropathy. Arch Neurol 38:378–379, 1981.

90. Nielsen, VK, Paulson, OB, Rosenkvist, J, Holsoe, E, and Lefvert, AK: Rapid improvement of myasthenia gravis after plasma exchange. Ann Neurol 11:160–169, 1982.

91. Norris, FH, Jr: Neuromuscular transmission in thyroid disease. Ann Intern Med 64:81–86, 1966.

92. Oh, SJ, Eslami, N, Nishihira, T, Sarala, PK, and Kuba, T: Electrophysiological and clinical correlation in myasthenia gravis. Ann Neurol 12:348–354, 1982.

93. Ohta, M, Matsubara, F, Hayashi, K, Nakao, K, and Nishitani, H: Acetylcholine receptor antibodies in infants of mothers with myasthenia gravis. Neurology 31:1019–1022, 1981.

94. Oosterhuis, HJGH: Studies in myasthenia gravis. Part I. A clinical study of 180 patients. J Neurol Sci 1:512–546, 1964.

95. Oosterhuis, H, and Bethlem, J: Neurogenic muscle involvement in myasthenia gravis: A clinical and histopathological study. J Neurol Neurosurg Psychiatry 36:244–254, 1973.

96. Osserman, KE: Myasthenia Gravis. Grune & Stratton, New York, 1958.

97. Osserman, KE, and Genkins, G: Studies in myasthenia gravis: Review of a twenty-year experience in over 1200 patients. Mt Sinai J Med (NY) 38:497–537, 1971.

98. Osserman, KE, Tsairis, P, and Weiner, LB: Myasthenia gravis and thyroid disease: Clinical and immunologic correlation. Mt Sinai J Med (NY) 34:469–483, 1967.

99. Ozdemir, C, and Young, RR: The results to be expected from electrical testing in the diagnosis of myasthenia gravis. Ann NY Acad Sci 274:203–222, 1976.

100. Patrick, J, and Lindstrom, J: Autoimmune response to acetylcholine receptor. Science 180:871–872, 1973.

101. Patten, BM: Myasthenia gravis: Review of diagnosis and management. Muscle Nerve 1:190–205, 1978.

102. Patten, BM, Oliver, KL, and Engel, WK: Effect of lactate infusions on patients with myasthenia gravis. Neurology 24:986–990, 1974.

103. Pearn, J: Neuromuscular paralysis caused by tick envenomation. J Neurol Sci 34:37–42, 1977.

104. Perez, MC, Buoto, WL, Mercado-Dangullan, C, Bagabaldo, ZG, and Renales, LD: Stable remissions in myasthenia gravis. Neurology 31:32–37, 1981.

105. Perlo, VP, Poskanzer, D, Schwab, RS, Viets, HR, Osserman, KE, and Genkins, G: Myasthenia gravis: Evaluation of treatment in 1355 patients. Neurology 16:431–439, 1966.

106. Pickett, J, Berg, B, Chaplin, E, and Brunstetter-Shaffer, MA: Syndrome of botulism in infancy: Clinical and electrophysiologic study. N Engl J Med 295:770–772, 1976.

107. Pittinger, C, and Adamson, R: Antibiotic blockade of neuromuscular function. Ann Rev Pharm 12:169–184, 1972.

108. Puvanendran, K, Cheah, JS, Naganathan, N, Yeo, PD, and Wong, PK: Neuromuscular transmission in thyrotoxicosis. J Neurol Sci 43:47–57, 1979.

109. Rapoport, S, and Watkins, PB: Descending paralysis resulting from occult wound botulism. Ann Neurol 16:359–361, 1984.

110. Ricker, K, and Mertens, HG: Myasthenic reaction in primary muscle fibre disease. Electroencephalogr Clin Neurophysiol 25:413–414, 1968.

111. Ricker, K, Samland, O, and Peter, A: Elektrische und mechanische Muskelreaktion bei Adynamia episodica und Paramyotonia congenita nach Kälteeinwirkung und Kaliumgabe. J Neurol 208:95–108, 1974.

112. Robertson, WC, Jr, Chun, RWM, and Kornguth, SE: Familial infantile myasthenia. Arch Neurol 37:117–119, 1980.

113. Rooke, ED, Lambert, EH, and Thomas, JE: Ein myasthenisches Syndrom mit enger Beziehung zu gewissen malignen intrathorakalen Tumoren. Dtsch Med Wochenschr 86:1660–1664, 1961.

114. Ross, RT: Ocular myopathy sensitive to curare. Brain 86:67–74, 1963.

115. Rowland, LP, Aranow, H, Jr, and Hoefer, PFA: Observations on the curare test in the differential diagnosis of myasthenia gravis. In Viets, HR (ed): Myasthenia Gravis. Charles C Thomas, Springfield, Ill, 1961, pp 411–434.

116. Rubenstein, AE, Horowitz, SH, and Bender, AN: Cholinergic dysautonomia and Eaton-Lambert syndrome. Neurology 29:720–723, 1979.

117. Sanders, DB, Howard, JF, and Johns, TR: Single-fiber electromyography in myasthenia gravis. Neurology 29:68–76, 1979.

118. Sanders, DB, Massey, EW, and Buckley, EG: Botulinum toxin for blepharospasm: Single-fiber EMG studies. Neurology 36:545–547, 1986.

119. Schiller, HH, and Stalberg, E: Human botulism studied with a single-fiber electromyography. Arch Neurol 35:346–349, 1978.

120. Schmitt, N, Bowmer, EJ, and Gregson, JD: Tick paralysis in British Columbia. Can Med Assoc J 100:417–421, 1969.

121. Schwartz, MS, and Stalberg, E: Myasthenic syndrome studied with single fiber electromyography. Arch Neurol 32:815–817, 1975.

122. Seitz, D, Hopf, HC, Janzen, RWC, and Meyer, W: Penicillamine-induced myasthenia in chronic rheumatoid arthritis (German with English abstract). Dtsch Med Wochenschr 101:1153–1158, 1976.

123. Seybold, ME, and Lindstrom, JM: Myasthenia

gravis in infancy. Neurology 31:476–480, 1981.

124. Shapira, Y, Cividalli, G, Szabo, G, Rozin, R, and Russell, A: A myasthenic syndrome in childhood leukemia. Dev Med Child Neurol 16:668–671, 1974.

125. Simpson, JA: Myasthenia gravis: A new hypothesis. Scot Med J 5:419–436, 1960.

126. Simpson, JA: Myasthenia gravis: A personal view of pathogenesis and mechanism. Part 1. Muscle Nerve 1:45–56, 1978.

127. Simpson, JA: Myasthenia gravis: A personal view of pathogenesis mechanism. Part 2. Muscle Nerve 1:151–156, 1978.

128. Smit, LME, Jennekens, FGI, Veldman, H, and Barth, PG: Paucity of secondary synaptic clefts in a case of congenital myasthenia with multiple contractures: Ultrastructural morphology of a developmental disorder. J Neurol Neurosurg Psychiatry 47:1091–1097, 1984.

129. Snead, OC, III, Benton, JW, Dwyer, D, Morley, BJ, Kemp, GE, Bradley, RJ, and Oh, SJ: Juvenile myasthenia gravis. Neurology 30:732–739, 1980.

130. Spoor, TC, Martinez, AJ, Kennerdell, JS, and Mark, LE: Dysthyroid and myasthenic myopathy of the medial rectus: A clinical pathologic report. Neurology 30:939–944, 1980.

131. Stalberg, E: Clinical electrophysiology in myasthenia gravis. J Neurol Neurosurg Psychiatry 43:622–633, 1980.

132. Streib, EW, and Rothner, AD: Eaton-Lambert myasthenic syndrome: Long-term treatment of three patients with prednisone. Ann Neurol 10:448–453, 1981.

133. Strickroot, FL, Schaeffer, RL, and Bergo, HL: Myasthenia gravis occurring in an infant born of a myasthenic mother. JAMA 120:1207–1209, 1942.

134. Swift, TR: Disorder of neuromuscular transmission other than myasthenia gravis. Muscle Nerve 4:334–353, 1981.

135. Swift, TR, and Ignacio, OJ: Tick paralysis: Electrophysiologic studies. Neurology 25:1130–1133, 1975.

136. Takamori, M: Caffeine, calcium, and Eaton-Lambert syndrome. Arch Neurol 27:285–291, 1972.

137. Takamori, M, Ishii, N, and Mori, M: The role of cyclic 3′, 5′-adenosine monophosphate in neuromuscular transmission. Arch Neurol 29:420–424, 1973.

138. Tarrab-Hazdai, R, Aharonov, A, Silman, I, and Fuchs, S: Experimental autoimmune myasthenia induced in monkeys by purified acetylcholine receptor. Nature 256:128–130, 1975.

139. Toyka, KV, Drachman, DB, Griffin, DE, Pestronk, A, Winkelstei, JA, Fischbeck, KH, Jr, and Kao, I: Myasthenia gravis: Study of humoral immune mechanisms by passive transfer to mice. N Engl J Med 296:125–131, 1977.

140. Tsujihata, M, Hazama, R, Ishii, N, Ide, Y, and Takamori, M: Ultrastructural localization of acetylcholine receptor at the motor endplate: Myasthenia gravis and other neuromuscular diseases. Neurology 30:1203–1211, 1980.

141. Valli, G, Barbieri, S, and Scarlato, G: Neurophysiological tests in human botulism. Electromyogr Clin Neurophysiol 23:3–11, 1983.

142. Viets, HR, and Schwab, RS: Protigmin in the diagnosis of myasthenia gravis. N Engl J Med 213:1280–1283, 1935.

143. Vincent, A, Cull-Candy, SG, Newsom-Davis, J, Trautmann, A, Molenaar, PC, and Polack, RL: Congenital myasthenia: End-plate acetylcholine receptors and electrophysiology in five cases. Muscle Nerve 4:306–318, 1981.

144. Vincent, A, and Newsom-Davis, J: Absence of anti-acetylcholine receptor antibodies in congenital myasthenia gravis. Lancet 1:441–442, 1979.

145. Walker, MB: Myasthenia gravis: A case in which fatigue of the forearm muscles could induce paralysis of the extra-ocular muscles. Proc R Soc Med 31:722, 1938.

146. Wise, GA, and McQuillen, MP: Transient neonatal myasthenia: Clinical and electromyographic studies. Arch Neurol 22:556–565, 1970.

Chapter 25

MYOPATHIES

1 INTRODUCTION

Primary diseases of muscle include genetically determined disorders and those of a toxic or inflammatory nature. Entities traditionally referred to as muscular dystrophies have a clearly delineated mode of genetic transmission and a progressive clinical course, whereas congenital myopathies show a less well-defined pattern of inheritance and a benign clinical course.[39] Some myopathies also result from an inborn error of metabolism as part of a hereditary systemic disorder. In addition, a wide variety of inflammatory processes such as dermatomyositis and polymyositis affect the muscle.

Although in patients with myogenic disorders hypotonia develops as one of the essential features, not all floppy infants have a primary muscle disease. In fact, disorders of the motor unit constitute less than 10 percent of the identifiable causes of weakness during infancy.[174] A disease of the central nervous system commonly produces so-called cerebral hypotonia. Other nonmyogenic etiologies include spinal muscular atrophy, poliomyelitis, inflammatory polyneuropathy, myasthenia gravis, and botulism. Myalgia is also the presenting symptom in patients with a wide variety of disorders.

Differential diagnosis depends on the pattern of inheritance, the distribution of muscle weakness, and the time course of progression. Electromyography and analysis of force help delineate the physiologic mechanism of weakness and fatigue.[82] New techniques include the needle biopsy for histologic and histochemical examination of skeletal muscle. Useful screening tools include determination of creatine kinase (CK), determination of the erythrocyte sedimentation rate, muscle biopsy, electromyography, and muscle strength exercise testing.[157] Electromyographic studies contribute not only in differentiating myogenic from neurogenic paresis but also in delineating the distribution of abnormalities and categorizing dystrophies and myopathies. The patterns classically associated with myopathy may occasionally result from neurogenic involvement. This confusing feature develops in late stages following complex changes of denervation and reinnervation. Nerve conduction studies also mimic a neuropathic process of the motor axons with a reduction in amplitude of compound muscle action potentials and preservation of sensory nerve potentials. Neuromuscular transmission studies show no abnormality in primary disorders of muscles.

2 MUSCULAR DYSTROPHY

Muscular dystrophy comprises a group of inherited muscle diseases with a progressive clinical course from birth or after a variable period of apparently normal infancy. Most types result from a primary myogenic lesion in the form of muscle fiber degeneration,[193,194] although a neurogenic process may play a role in some.[151] Currently accepted classification based on the mode of inheritance and distribution of muscle degeneration includes four main types of muscular dystrophy; Duchenne's, Becker's, facioscapulohumeral, and limb-girdle. Other categories include oculopharyngeal dystrophy, hereditary distal myopathies, muscular dystrophy of the Emery-Dreifuss type, and myotonic dystrophy. Differential diagnosis depends on the clinical features, genetic mode of inheritance, electrophysiologic pattern, and histologic characteristics.

Duchenne's Muscular Dystrophy

Duchenne's dystrophy,[80] also known as the pseudohypertrophic variety, shows a sex-linked recessive inheritance. A phenotypically normal female transmits the disease, although carriers may have partial symptoms.[99] Clinical manifestations in these cases are milder, as explained by the Lyon hypothesis. Symptomatic young girls, if not carriers of Duchenne's dystrophy, may have childhood muscular dystrophy of autosomal recessive inheritance.[106] Molecular biologic techniques have led to the identification of the primary biochemical defect based solely on the chromosomal location of the Duchenne muscular dystrophy locus.[111] The discovery of the protein product, called dystrophin, should prove useful in the diagnosis and characterization of this disorder and eventually the development of therapeutic regimens.

A number of investigators have advocated the neurogenic[9,150–152,176] or vascular[104,153] theories with vigor but without universal acceptance. Others described defects of erythrocyte membranes[146,148,156] but without subsequent confirmation.[195,232] Some suggest possible involvement of calcium metabolism in the dystrophic process.[22,147,161]

According to the currently prevalent theory[160] an inherited biochemical deficiency alters the composition and function of the muscle membrane.[192,196] The main pathologic sequence of events in the early stages consists of repeated episodes of muscle fiber necrosis and regeneration.[44] Incomplete regeneration reduces the number of muscle cells, rendering some fibers hypertrophic and others atrophic.[203] Progressive accumulation of collagen finally replaces the muscle cells.

Proximal weakness of the leg begins during early childhood, although histologic evidence indicates that abnormalities already exist at birth. The child normally attains the initial developmental milestones such as raising his head or sitting upright. Early difficulty in standing or walking may give an erroneous impression of clumsiness. He tends to walk on his toes and with the feet externally rotated. Weakness becomes apparent by age 3 or 4 with inability to run or to climb stairs. The disease progresses slowly and may even remit as natural growth temporarily compensates for the weakness. A steady, downhill course ensues with the development of lumbar lordosis and progressive scoliosis. Frequent falls force the child into a wheelchair, and eventually contractures of the joints prevent movement of the extremities.

Neurologic findings depend on the stage of the illness. Muscles harden with rubbery consistency, which leads to reduced or absent stretch reflexes. The quadriceps are particularly vulnerable, but the muscles of the shoulder girdle also show prominent abnormalities. Later, weakness becomes diffuse, sparing only the extraocular muscles. Other features include macroglossia, pseudohypertrophy of the calves, and, in some, mild mental retardation. A markedly elevated serum level of CK during the first year often heralds the clinical onset of illness.[168] The CK values then fall gradually as the disease advances but never return to normal. Carriers can sometimes be detected by elevated CK levels.[179,191] Cardiac involvement results in typical electrocardiographic abnormalities, which consist of a tall right precordial R wave and a deep limb and precordial Q wave.[62,180]

Electromyographic evaluation reveals characteristic features of myopathy. Insertion of the needle elicits normal or prolonged activity initially but very little potential in the advanced stage, when fibrosis has replaced muscle tissues. Fibrillation potentials and positive sharp waves may be seen early (see Fig. 13–7E) but to a much lesser extent, compared with myositis or motor neuron disease.[43] Low-amplitude, short-duration motor unit potentials result from random loss of muscle fibers. When recruited in abundance (see Figs. 13–15, and 13–18B), these potentials give rise to a characteristic noise resembling a shower of fibrillation potentials. In mildly affected muscles the abnormalities, limited in degree and distribution, could escape detection without careful exploration. Electromyography generally offers only limited help as a test for carrier detection.[31,96,105,162,227]

Becker's Muscular Dystrophy

Becker's sex-linked recessive dystrophy, although a distinct entity, bears a close resemblance to Duchenne's dystrophy.[14,15,86,189] Its clinical features include proximal weakness of the upper and lower extremities, pseudohypertrophy, and elevated levels of serum CK. The patient's initial difficulty involves walking, climbing stairs, and rising from the floor. Compared with Duchenne's dystrophy, the Becker type has a later onset and a considerably longer and milder clinical course, with survival into the middle adulthood. Contractures and skeletal deformities eventually develop; but if present, are not as severe as in Duchenne's dystrophy. Early myocardial disease and myalgia may develop.[134]

Electromyography shows nearly symmetric abnormalities in the proximal muscles. Fibrillation potentials and complex repetitive discharges abound in the paraspinal muscles. Small and polyphasic motor unit potentials recruit early. Each of eight families reviewed had mixed features of myopathy and denervation.[30] Muscle biopsy revealed fiber atrophy and hypertrophy, with many split and angulated fibers and clumps of pyknotic nuclei. All but one of the patients could walk until the age of 16, and most lived beyond age 20.

Facioscapulohumeral Dystrophy

This variety, also known as Landouzy-Dejerine dystrophy, is inherited as an autosomal dominant trait, affecting both sexes equally.[128] The disease typically begins toward the end of the first decade, although the symptoms may appear much earlier, within the first 2 years.[102] The weakness initially involves the muscles of the face and the trapezius, pectoralis, triceps, and biceps. Muscles of the lower extremities are affected later than those of the shoulder girdle. The very slowly progressive deficit causes only minor disability and little alteration of a normal life expectancy. In contrast to Duchenne or Becker dystrophy, the level of serum CK tends to remain normal in this entity.

The patient has bilateral footdrop as the presenting sign in a variety known as scapuloperoneal dystrophy.[223] Some authors prefer the term facioscapulohumeral syndrome,[229] with subdivisions of neurogenic, myopathic, and rare myositic entities.[79,167] Initial myositic features may lead to clinical patterns indistinguishable from the myopathic type after

some months to years. Some patients have congenital absence of the pectoralis, biceps, or brachioradialis.

At the beginning, electromyography may show only limited abnormality, which may escape detection even in clinically weak muscles. In well-advanced cases, low-amplitude, short-duration, polyphasic motor unit potentials are recruited early out of proportion to the degree of muscle force. The presence of spontaneous discharges suggests the neuropathic form of this syndrome. Patients with myasthenia gravis or polymyositis may have weakness in the facioscapulohumeral distribution. Complete electrophysiologic testing should include studies of neuromuscular transmission and paraspinal electromyography to exclude these possibilities.

Limb-Girdle Dystrophy

Limb-girdle dystrophy comprises a group of heterogeneous disorders with progressive weakness of the hip and shoulder muscles. It affects men and women equally, with an autosomal recessive pattern of inheritance. Sporadic cases and a kindred with a rare autosomal dominant pattern have also been described.[52] The illness often begins during the second or third decade with involvement of the pelvis. Weakness soon spreads to the shoulder girdle, typically sparing the facial muscles. Symptoms, restricted to these areas for many years, show only mild progression. Rarely, involvement of the diaphragm is the presenting symptom of a limb-girdle syndrome.[235] Some patients have weakness of only one limb without other characteristic features, or only one muscle, as in quadriceps myopathy.[219] The disease process usually runs a more rapid course in the tibialis anterior than in the plantar flexor muscles.[16] Pseudohypertrophy may or may not occur in the calves and deltoid. Despite eventual confinement to a wheelchair, the patient usually has a normal life span. The serum enzymes may be slightly elevated.

The name, limb-girdle syndrome, appropriately denotes the heterogeneity of this entity, with the subdivision into myogenic and neurogenic types based on clinical, histologic, and electrophysiologic findings (see Fig. 12–8C). In addition, a clinical syndrome of progressive proximal limb-girdle distribution may appear as a secondary manifestation in other well-defined conditions. These include chronic polymyositis,[190] myasthenia gravis, and various metabolic and congenital myopathies, such as late-onset acid maltase deficiency and carnitine deficiency. Spinal muscular atrophy also has a similar distribution of weakness, making clinical differentiation difficult.[85,233]

Review of 18 patients with proximal weakness in the limb-girdle distribution established a firm diagnosis only in four cases, even after histologic evaluation: two with spinal muscular atrophy and two others with muscular dystrophy.[53] Motor innervation patterns suggested spinal muscular atrophy in 4 of the 18 and limb-girdle dystrophy in the others. Electromyographic features revealed myopathic changes in 11, denervation in 3, and inconclusive results in 4. In another series of 20 patients, single-fiber electromyography (SFEMG) confirmed the original diagnosis of the myopathic limb-girdle syndrome in 11 and chronic spinal muscular atrophy in 5 and helped differentiate the other 4 cases into myopathic and neuropathic varieties.[210]

Other Dystrophies

Oculopharyngeal dystrophy, a rare form of progressive ophthalmoplegia, affects French-Canadian families in an autosomal dominant fashion.[231] Progressive ptosis and dysphagia develop late in life, with or without extraocular muscle weakness.[34,170] Muscle biopsy specimens show variation in fiber size, occasional internal nuclei, small angulated fibers, and an intermyofibrillar network with a moth-eaten appearance when stained with oxidative enzyme.[79] Differentiation from myasthenia gravis poses a major problem clinically. Patients with oculopharyngeal dystrophy have absent titers for acetylcholine (ACh) receptor antibody

and negative results in the edrophonium (Tensilon) test.

Electromyographic studies usually reveal no spontaneous activity. Brief, low-amplitude, polyphasic motor unit potentials are recruited early in proximal muscles of the upper extremities.[26] A neurogenic pattern with large motor unit potentials may accompany the myopathic features.[204] Conduction studies are normal except for low-amplitude compound muscle action potentials in the weak muscles. Repetitive nerve stimulation show no decrement of muscle response.

Hereditary distal myopathy, first described by Welander,[234] is a rare autosomal dominant disorder with onset in adulthood.[154] Unlike most other forms of dystrophies, it predominantly affects the distal muscles of the upper and lower limbs. Weakness typically begins in the intrinsic hand muscles, although the initial symptoms may involve the small muscles of the foot. As the disease slowly progresses, the dorsiflexors of the wrist and foot become weak, usually with nearly complete sparing of proximal musculature. Widespread weakness and wasting may occur, especially if the disease appears at an earlier age and worsens rapidly. Most patients have slightly elevated levels of serum CK. Muscle biopsies show vacuolar changes.[81,142] Electromyography demonstrates an abundance of low-amplitude, short-duration motor unit potentials during mild voluntary contraction.

Another type of progressive distal myopathy described in Japan has an autosomal recessive inheritance.[159] The disease affects young adults, with the initial impairment in standing on the tiptoes, followed by difficulty in climbing stairs and standing. Muscle atrophy involves distal muscles in the leg and forearm, sparing intrinsic hand muscles. Electromyography reveals abnormalities consistent with myopathy. Muscle biopsies show severe segmental necrosis of myofibers with regeneration.

In a rare type of muscular dystrophy, called Emery-Dreifuss type, weakness develops in the humeroperoneal distribution.[198] Other features include early contractures with marked restriction of neck and elbow flexion and cardiac abnormalities causing atrial fibrillation, a slow ventricular rate, and exertional dyspnea. Most pedigrees show an X-linked inheritance, but rare families have autosomal dominant transmission.[155] Electromyography and muscle histology show mixed patterns of neurogenic and myopathic changes.

3 CONGENITAL MYOPATHY

A number of congenital conditions are characterized by nonprogressive or only slightly progressive muscular weakness. Some of these entities show morphologically distinctive features in muscle biopsy. These include central core disease, nemaline myopathy, myotubular myopathy, congenital fiber type disproportion, and congenital hypotonia with type I fiber predominance. The diagnosis of these rare conditions depends on histologic examination of the muscle, although morphologic changes may represent the fundamental pathology or secondary manifestations.

Central Core Disease

Central core disease is inherited as an autosomal dominant trait. The infants have hypotonia shortly after birth, delayed developmental milestones, and occasionally congenital hip dislocations.[212] As older children, they cannot keep pace with their peers because of proximal weakness, although they have no distinct muscular atrophy. Neither the patient nor the family may recognize the disease unless skeletal deformities develop. These include lordosis, kyphoscoliosis, and abnormalities of the foot.[222] Malignant hyperthermia has been reported in association with central core disease.[64,87,94] In high-risk patients who require surgery for musculoskeletal defects, preoperative evaluation should include in-vitro tests for this devastating phenomenon, described later in this chapter.

Muscle biopsy shows marked type I fiber predominance. The central region of the muscle fiber contains compact myofibrils devoid of oxidative and phosphorylase enzymes because of the virtual absence of mitochondria.[79] These central areas, referred to as cores, show no histochemical reactivity with the oxidative enzyme. They commonly appear in type I and to a lesser extent in type II fibers, but their absence does not preclude the diagnosis.[163] The resemblance of the cores to target fibers, which usually indicate denervation and reinnervation, supports the disputed notion that the disease may be neurogenic in nature.[171] An increased terminal innervation ratio described in this entity also suggests a neurogenic process.[54,120]

Electrophysiologic findings vary, but tend to suggest a mixed myopathic-neuropathic process. Electromyography usually consists of normal insertional activity, no spontaneous discharges at rest, and small motor unit potentials with early recruitment.[165] Other studies have revealed large and polyphasic potentials[120] with increased fiber density.[58] Nerve conduction studies show reduced amplitude of muscle potentials with either normal[120] or mildly slowed conduction velocity.[114]

Nemaline Myopathy

The disease is inherited as an autosomal dominant trait,[131] causing nonprogressive hypotonia that begins at a very early age. Although considered benign in older children and adults, it may be responsible for early death in the neonate or young infant.[172] In addition to diffuse weakness, the affected child shows dysmorphism with reduced muscle bulk and slender musculature. The clinical features include an elongated face, a high-arched palate, high-arched feet, kyphoscoliosis,[135] and a sometimes scapuloperoneal distribution of weakness. Often, there is a slightly elevated level of serum CK.

Patients and carriers both have a predominance of small type I fibers, seen on muscle biopsy specimens.[17,61] Gomori's trichrome stain shows the characteristic rod-shaped bodies, not apparent with routine methods. These contain material identical to the Z bands of muscle fibers, involving either type I or type II fibers, or both.[209] Nemaline myopathy[211] derives its name from the presence of these rod- or threadlike structures (*nemaline* in Greek) lying under the sarcolemma. The rods, however, also appear in many other conditions as nonspecific findings.

Electromyography may show low-amplitude, short-duration motor unit potentials with early recruitment or, conversely, fibrillation potentials and a decreased number of high-amplitude, long-duration motor unit potentials.[172]

Myotubular, or Centronuclear, Myopathy

In myotubular,[216] or centronuclear, myopathy,[19,207] fetal myotubes persist into adult life. This rare condition shows various modes of inheritance.[12] Three subgroups have emerged, based on the severity and mode of presentation, together with the genetic pattern. These are a severe neonatal X-linked recessive type, a less severe infantile-juvenile autosomal recessive type, and a milder autosomal dominant type.[107] Most patients have hypotonia, ptosis, facial weakness, and extraocular palsy at birth. The disease may also affect proximal or distal muscles of the limbs. Some patients die in infancy from cardiorespiratory failure, whereas others survive until adulthood with little progression and only mildly elevated serum CK. Muscle biopsies show type I fiber atrophy and central nuclei, considered characteristic of fetal muscle. The central part of the fiber, devoid of myofibrils and myofibrillar adenosine triphosphate (ATP), stains poorly with the ATPase reaction. Oxidative enzymes may show increased or decreased activity in the central region.

Electromyographic abnormalities include an excessive number of polyphasic, low-amplitude motor unit potentials, fibrillation potentials, positive sharp waves, and complex repetitive discharges.[107] Myotubular myopathy is the only congenital myopathy consistently associated with spontaneous activity in electromyography.[84] Occasional myotonic discharges may lead to an erroneous diagnosis of myotonic dystrophy, especially in a patient with distal weakness and ptosis.[187] Two sisters with otherwise typical centronuclear myopathy had clinical myotonia.[98] The patients usually have normal motor and sensory nerve conduction studies.

Congenital Fiber Type Disproportion

In normal muscles, type II fibers constitute more than 60 percent of the fibers and type I, 30 to 40 percent. A reversed relationship characterizes the histologic findings in some children with congenital hypotonia.[38,60,118,141] The disease may be genetically determined, with an autosomal dominant mode of inheritance. The infants may have dysmorphic features at birth.[218] Additional signs include contractures as the major source of functional limitation, congenital dislocation of the hip joint secondary to intrauterine hypotonia, and other skeletal abnormalities such as deformities of the feet and kyphoscoliosis. The disease progresses for the first several years and then either stabilizes or improves slightly. The patient has short stature and fails to develop expected motor skills despite normal mental capacity.

The muscle biopsy shows, in addition to fiber type disproportion, small type I fibers, hypertrophic type II fibers, and scattered internal nuclei. The presence of occasional rods suggests possible but unconfirmed relationships between this condition and nemaline myopathy.[130]

Electromyography usually demonstrates low-amplitude, short-duration motor unit potentials with early recruitment. Some patients have fibrillation potentials, positive sharp waves, and large motor unit potentials.[218]

Cystoplasmic Body Myopathy

In this rare congenital condition, weakness characteristically involves the face, neck, and proximal limb muscles, as well as respiratory, spinal, and cardiac muscles. The patient may have scoliosis. Cardiorespiratory failure may follow respiratory infection. The patient has an elevated serum CK level and an abnormal electrocardiogram. Muscle biopsy reveals centrally placed nuclei, necrosis, fibrosis, and cytoplasmic bodies. Electrophysiologic studies show normal nerve conduction studies and abnormal electromyography consistent with myopathy.[177]

4 METABOLIC MYOPATHY

A variety of myopathies result from inborn errors of metabolism.[27] These include certain types of glycogen storage disease and disorders of lipid metabolism. Of the 10 glycogen storage diseases identified to date, prominent muscle involvement occurs only in types II (Pompe's disease), III (Cori-Forbes disease), V (McArdle's disease), and VII (Tarui's disease) glycogenosis.[116] Two other metabolic myopathies, mitochondrial diseases and malignant hyperpyrexia or hyperthermia, deserve brief mention.

Acid Maltase Deficiency (Type II Glycogenosis)

In this condition, inherited as an autosomal recessive, acid maltase deficiency leads to accumulation of glycogen in tissue lysosomes.[6,110] In the infantile type, called Pompe's disease, children develop severe hypotonia shortly after birth and die within the first year of cardiac or respiratory failure.[25] Anterior horn cells contain deposits of glycogen particles, as do other affected organs, such as the heart, tongue, and liver. An enlarged tongue and cardiac abnormalities differentiate this condition from Werdnig-Hoffmann disease.

In the more benign childhood and adult types, the symptoms limited to skeletal muscle mimic those of the limb-girdle syndrome or polymyositis. Patients with the onset of symptoms in childhood have proximal limb and trunk muscle weakness with variable progression. They may

die of respiratory failure before the end of the second decade.[90,144] Increased net muscle protein catabolism plays a part in the pathogenesis because the condition improves with a high-protein diet.[214] In the adult variant, the symptoms begin with insidious limb-girdle weakness during the second or third decade and respiratory difficulty some years later, necessitating tracheostomy.[74,90,220,225] Both types show elevated serum enzyme levels. The muscle biopsy reveals a vacuolar myopathy affecting type I more than type II fibers.[48] Glycogen commonly deposits in the central nervous system, particularly in the infantile form. Tissue cultures have reproduced the enzymatic defect.[7]

Electromyography in infantile form shows increased insertional activity, fibrillation potentials, positive sharp waves, and complex repetitive discharges, as expected from anterior horn cell involvement.[88,90,112]

Severely affected muscles typically lack insertional activity. As one of the few exceptions to the rule (see Chapter 13.3), true myotonic discharges may occur in the absence of clinical myotonia. Mild voluntary contraction recruits polyphasic, low-amplitude, short-duration motor unit potentials in abundance. In contrast to the widespread abnormalities seen in the infantile type, changes are restricted to the gluteal, paraspinal, and other proximal muscles in the adult or late-onset childhood type. A majority of these patients have electromyographic findings of myopathy without fibrillation potentials.[225] Studies of motor and sensory nerve conduction and of neuromuscular transmission reveal no abnormalities, except for reduced amplitude of the compound muscle action potentials.

Debrancher Deficiency (Type III Glycogenosis)

In this disease, inherited as an autosomal recessive trait, the absence of the debrancher enzyme prevents breakdown of glycogen beyond the outer straight glucosyl chains. Consequently, glycogen with short-branched outer chains, called phosphorylase-limit-dextrin, accumulates in the liver and striated and cardiac muscles. Despite the generalized enzymatic defect,[228] the skeletal muscles do not necessarily show weakness on clinical examination.[41,169]

The affected children with hypotonia and proximal weakness fail to thrive. Accumulation of glycogen in the liver causes hepatomegaly, episodes of hypoglycemia, and markedly elevated serum CK. Clinical features of myopathy may develop after hepatic symptoms have abated. The patient may improve in adolescence, despite the enzymatic defect. The distal weakness and wasting sometimes resemble those seen in patients with motor neuron disease.[41,73] Muscle biopsy shows subsarcolemmal periodic acid-Schiff (PAS)-positive vacuoles in type II fibers without histochemical signs of denervation.[73] Electromyography may reveal profuse fibrillation potentials, complex repetitive discharges, and small, short-duration motor unit potentials.[41,73]

Muscle Phosphorylase Deficiency (Type V Glycogenosis)

McArdle[149] first described this rare condition inherited in an autosomal recessive pattern, although others have subsequently reported families with an autosomal dominant pattern.[51] It affects men more frequently than women by a ratio of 4:1.[70] Myophosphorylase deficiency blocks the conversion of muscle glycogen to glucose during heavy exercise under ischemic conditions. The patient typically becomes symptomatic in childhood or adolescence; but, rarely, neither muscle weakness nor cramps develop until late adulthood.[1,132] The differential diagnoses of late-onset and childhood myopathies should always include muscle phosphorylase deficiency.[56]

The abnormality, confined to skeletal muscles, initially causes only nonspecific complaints of mild weakness and fatigue. Sometime during adolescence the patients begin to notice exercise intolerance.[197,199] A heavy muscle contraction or repetitive stimulation of the nerve produces painful cramps that may last for several hours. Associated breakdown of muscle leads to myoglobinuria, causing the urine to become wine-colored. Muscle pain and fatigue may improve during continued exercise if the patient slows down and sustains nonstrenuous activity. This second-wind phenomenon presumably results from increased mobilization of serum free fatty acids as an alternative source of energy.[181] Exposure to cold during exercise may also delay the development of contracture.

Neurologic examination between bouts of muscle cramps initially reveals only mild proximal weakness without apparent muscular wasting. The patient may develop permanent limb-girdle weakness later in life. In advanced stages, mild exercise may precipitate painful muscle cramps, severely limiting the patient's activity. Atypical clinical presentations in adult patients include progressive muscle weakness without exercise-induced contracture.[145] In infants, generalized hypotonia may lead to respiratory insufficiency and early death.[72]

The differential diagnoses include muscle phosphofructokinase deficiency characterized by recurrent myoglobinuria and persistent weakness[20] and Brody's disease, caused by a deficiency of calcium ATPase in sarcoplasmic reticulum.[127] The ischemic exercise test can confirm the diagnosis in suspected cases.[166] The test consists of contracting the forearm muscles under ischemic conditions induced by an inflated pneumatic cuff placed around the arm. The inability to convert glycogen to glucose by anaerobic glycolysis promptly precipitates a muscle cramp. The pathogenesis of the contracture clearly centers around the depletion of high-energy phosphates in the absence of glycogen metabolism. This may prevent the energy-dependent reuptake of calcium by the sarcoplasmic reticulum,[101] but no studies have confirmed such an abnormality.[37] Normally, lactate levels in venous blood should rise with the breakdown of glycogen under ischemic conditions. Patients with McArdle's disease show no rise in the lactate level in the blood drawn from the exercised arm.

In contracture (see Chapter 26.12), electromyography of the cramped muscle reveals no electrical activity despite muscle shortening (see Fig. 11–3).[139] In contrast, the ordinary muscle cramp or spasm shows abundant discharges of motor unit potentials. In one patient, the posttetanic mechanical tension of the contracture reached only 17 percent of the peak tetanic tension, and twitches superimposed on the contracture fell by half, as did the amplitude of the action potentials.[33] Between attacks, electromyography may show no abnormalities or may reveal fibrillation potentials and polyphasic motor unit potentials[72] or spontaneous activity and myopathic features as seen in inflammatory muscle disease.[185] In one study, quantitative analysis of the motor unit potential in the biceps showed a mean duration of 7.1 ms, compared with 9.4 ms in the controls, suggesting myopathic changes.[33] Others have proposed a reduction in the number of motor units but without subsequent confirmation.[226] Motor and sensory nerve conduction studies are normal except for decreased electrical activity during muscle contracture precipitated by repetitive shocks.

Muscle Phosphofructokinase Deficiency (Type VII Glycogenosis)

This disorder, first described by Tarui and associates,[221] results from a defect of muscle phosphofructokinase, which precludes the conversion of fructose-6-phosphate to fructose 1-6 diphosphate. The clinical features include painful muscle contracture and myoglobinuria much like those of McArdle's disease.[3,224] An infant with this syndrome may have, in addition to limb weakness, seizures, cortical blindness, and corneal opacifications.[208] Distinction of this entity from McArdle's disease depends on biochemical or histochemical determination of phosphofructokinase activity in the muscle biopsy. Electromyography reveals no abnormalities between attacks.

Disorders of Lipid Metabolism

Whereas glycogen serves as the major source of energy for rapid strenuous effort, circulating lipid in the form of free fatty acids maintains the energy supply at rest and during prolonged low-intensity exercise. Carnitine palmityl transferase catalyzes the reversible binding of carnitine to plasma fatty acids; once bound, carnitine can transport fatty acids across the mitochondrial membrane for oxidation. Disorders of lipid metabolism include carnitine palmityl transferase deficiency, carnitine deficiency,[75] and other rare conditions such as lipid neuromyopathy with normal carnitine.[8]

Carnitine palmityl transferase deficiency is a rare disorder inherited as an autosomal recessive trait. The patients have painful muscle cramps and, on prolonged exercise or fasting, recurrent episodes of myoglobinuria.[10,46,59,68,71,92,137,205] Oxidation of lipid substrates is impaired because long-chain fatty acids, not coupled to carnitine, cannot shuttle across the inner mitochondrial membrane.[138] The first attack of myoglobinuria appears in adolescence, although muscle pain may develop in early childhood. Although muscle remains strong between attacks, exercise during fasting results in painful cramps. Muscle biopsy may show no abnormalities or only slight excess of intrafiber lipid droplets next to the mitochondria in type I fibers.

Electrophysiologic studies, reported in a few patients, have revealed normal electromyography and normal motor and sensory nerve conduction velocities.[10,92] Carnitine deficiency, probably inherited as an autosomal recessive disorder, holds a distinction as the first biochemical defect identified in muscle lipid metabolism.[5,89] Of the two forms of this condition, the restricted type develops lipid storage predominantly or exclusively in the muscle, leading to a lipid storage myopathy, so called before recognition of the specific biochemical defect.[29,91] Reduced muscle carnitine possibly results from a deficit in carnitine uptake in the muscle, despite normal serum carnitine levels in most patients. In the systemic type, insufficient synthesis lowers the carnitine levels in serum, liver, and muscle.

Carnitine deficiency causes a congenital and slowly progressive myopathy of the limb-girdle type and episodic hepatic insufficiency.[28,125] Severe defects may cause bulbar and respiratory involvement, leading to death at an early age.[28,57,103] Some patients show features of both systemic and muscle carnitine deficiency.[47] Lipid utilization takes place in the mitochondria. This link may explain some overlap between lipid storage myopathies and mitochondrial myopathies.[32] The muscle biopsy reveals an excess of lipid droplets, mostly in the type I fibers, which depend on oxidation of long-chain fatty acids to a greater extent than type II fibers.

In electromyography mild voluntary contractions recruit low-amplitude, short-duration, polyphasic motor unit potentials in abundance. Slightly over half of the patients have fibrillation potentials and other forms of spontaneous activity, such as complex repetitive discharges. Neuropathy may develop in some,[143] but motor and sensory nerve conduction studies and tests of neuromuscular transmission usually reveal no abnormalities.

Mitochondrial Disease

Some systemic disorders characterized by structural changes of the mitochondria cause progressive muscle weakness as a part of complex neurologic manifestation.[78,113,129,164,183,201,213] The most common type, called Kearns-Sayres opthalmoplegia, occurs sporadically, with the clinical signs of ptosis and extraocular palsy appearing during childhood or adolescence.[18,126,206] As indicated by its alternative name, oculocraniosomatic neuromuscular disease with ragged red fibers,[173] characteristic features include progressive weakness of the extraocular muscles, cardiac abnormalities, somatic deficits, and ragged red fibers seen in the muscle biopsy, indicating a mitochondrial abnormality. Progressive weakness and fatigue may accompany a wide vari-

ety of neurologic deficits: for example, pigmentary degeneration of the retina, sensorineuronal deafness, cerebellar degeneration, endocrine abnormalities, sensory motor neuropathy, and demyelinating radiculopathy.[78,100,182] Laboratory studies reveal a moderate increase in cerebrospinal fluid protein and a mild elevation of serum CK.

Electromyography may be normal or mildly abnormal, with low-amplitude, short-duration motor unit potentials that recruit early. Clinically asymptomatic members of the family may have subtle changes consistent with subclinical myopathy by conventional or single-fiber recording.[93] In one series, 10 of 20 patients had abnormalities of nerve conduction studies, although only 5 had clinical features of a mild sensory motor neuropathy. In these patients, sural nerve biopsy revealed reduction in density of myelinated fibers and axonal degeneration affecting myelinated and unmyelinated fibers.[236] In another study, brief periods of exercise as well as repetitive nerve stimulation produced a marked decrease in twitch tension with only a very slight change in the amplitude of the compound action potential. Progressive dissociation between the electrical and mechanical responses suggests a failure of contraction, rather than a disorder of the neuromuscular apparatus.[164]

Malignant Hyperpyrexia, or Hyperthermia

In this rare entity,[65] with an autosomal dominant inheritance, affected individuals have unusual susceptibility to anesthetics in general and the administration of halothane and succinylcholine in particular.[36,175] After the induction of general anesthesia, they develop fasciculations and increased muscle tone. An explosive rise in temperature coincides with the development of muscular rigidity and necrosis. The remarkable hyperpyrexia, metabolic in nature, may result from abnormal depolarization of skeletal muscle by halothane.[95] Patients with malignant hyperthermia characteristically show reduced re-uptake of calcium (Ca^+)

by the sarcoplasmic reticulum.[119] If untreated, they die of metabolic acidosis and recurrent convulsions.

Without the knowledge of a positive family history, clinicians rarely suspect malignant hyperthermia. Susceptible individuals have no symptoms unless subjected to anesthesia. Common physical characteristics include proximal hypertrophy and distal atrophy of the thigh muscles and lumbar lordosis. Some patients have mild weakness of the proximal muscles, diminution of the muscle stretch reflexes, and an elevated serum CK level. The abnormal muscle shows hypersensitivity to caffeine, which normally causes muscle contracture by increasing the concentration of calcium in the sarcoplasm. In an in-vitro screening test for suspected cases, concentrations of halothane and caffeine too low to affect normal muscles produce contracture in specimens obtained from the patients.[35] As mentioned before, malignant hyperthermia may develop in association with central core disease.[64,94]

Nutritional or Toxic Myopathies

Pentazocine abuse masquerades as myopathy with proximal weakness and electromyographic findings of low-amplitude, short-duration polyphasic motor unit potentials.[50] Acute myopathy and myoglobinuria with a markedly elevated CK level may develop after gasoline sniffing, presumably as the result of lead toxicity.[133]

5 ENDOCRINE MYOPATHY

Endocrine myopathies develop in hyperthyroidism, hypothyroidism, parathyroid disease, and adrenal or pituitary dysfunction. Cushing's syndrome secondary to systemic administration of corticosteroids or adrenocorticotropic hormone (ACTH) also causes myopathy.

Thyroid Myopathy

Disorders of thyroid function may lead to a variety of neuromuscular problems, although fulminating systemic features may obscure muscular symptoms. Thyrotoxic myopathy probably ranks the first in incidence, with the majority of patients having some proximal weakness and electromyographic features of myopathy.[123] Myopathy affects men more frequently, even though women have a higher incidence of thyrotoxicosis. Typically, weakness involves the muscles of the shoulder girdle more than those of the pelvic girdle. The patient usually has normal or at times even hyperactive muscle stretch reflexes. Spontaneous muscle twitching and generalized myokymia may develop, but not commonly. Quantitative electromyographic studies have shown low-amplitude, short-duration motor unit potentials, even in the absence of clinically evident muscle weakness.[186] Other neuromuscular conditions commonly associated with thyrotoxicosis include exophthalmic opthalmoplegia, myasthenia gravis, and hypokalemic periodic paralysis.

Hypothyroidism causes proximal muscle weakness, painful muscle spasm, and muscle hypertrophy, especially in children. Characteristic features of myxedema include Hoffmann's sign, or delayed relaxation of contracted muscle. The ankle stretch reflex best demonstrates this change in muscle contractibility—a brisk reflex movement of the foot with a slow return to the resting position. A sharp tap to the muscle with a reflex hammer causes a local ridge of muscle to contract. This phenomenon, called myoedema or mounding of hypothyroidism, is electrically silent.[200]

Electromyography may show increased insertional activity with some repetitive discharges, but no evidence of myotonia. Elevations of serum CK levels, commonly a result of deranged creatine metabolism, do not necessarily imply the presence of myopathy.

Parathyroid Disease

The influx of calcium into axon terminals facilitates the release of ACh at the neuromuscular junction, leading to excitation-contraction coupling. Calcium apparently plays an opposite role at the central junction of axons: a reduction in calcium here results in increased conductance for sodium and potassium, causing instability and hyperexcitability of the cell membrane. Thus, in hypoparathyroidism, chronic hypocalcemia gives rise to tetany, the most dramatic neuromuscular complication. Less frequently, neuromuscular symptoms in hypercalcemia may also result from osteolytic metastases, multiple myeloma, or chronic renal disease.

Varying degrees of proximal muscle weakness develop in patients with hyperparathyroidism,[178,215] usually affecting the pelvic girdle more than the shoulder girdle. Brisk stretch reflexes and occasional extensor plantar responses, combined with axial muscle wasting, raise the diagnostic possibility of motor neuron disease.

Electromyographic changes in tetany include the presence of motor unit potentials in doublets and triplets. Low-amplitude, short-duration motor unit potentials recruit early in weak muscles. There are no spontaneous activities. Nerve conduction studies reveal reduced amplitude of the compound muscle action potentials and normal motor and sensory nerve conduction velocities.

Adrenal and Pituitary Disease

Diseases of the adrenal and pituitary glands may give rise to nonspecific muscle weakness, as in Cushing's syndrome, acromegaly, or Addison's disease. Similar weakness also appears after systemic administration of corticosteroids or ACTH. Steroids reduce the intracellular concentration of potassium, but their relationship to myopathy awaits clarification. Dysfunction of the reticulum or mitochondria may also contribute to the pathogenesis. With preferential weakness of the pelvic girdle and thigh muscles, patients have difficulty rising from a chair of climbing stairs. The neuromuscular symptoms usually improve if the underlying abnormality abates or upon discontinuation of the steroids. Laboratory studies show normal serum enzymes but increased urinary creatine excretion.

Muscle biopsy reveals type II fiber atrophy but neither necrosis nor inflammatory changes, as might be expected from the degree of muscle wasting observed clinically.[184]

The compound muscle action potentials are reduced in amplitude, especially in proximal muscles. Endocrine or steroid myopathy with type II fiber atrophy usually reveals no specific abnormality in electromyography, which only assesses the initially recruited type I fibers. Patients with inflammatory myopathy may develop progressive weakness after prolonged steroid therapy. In this situation, a normal insertional activity and the absence of fibrillation potentials suggest steroid myopathy, rather than exacerbation of the disease. In some cases, low-amplitude, short- duration motor unit potentials may recruit early, but such mild abnormalities generally reverse with withdrawal of steroids.

6 MYOSITIS

A variety of inflammatory processes affect the muscle, including the most frequently encountered polymyositis.[24,67] The patients with dermatomyositis have a skin rash in conjunction with the signs and symptoms of muscle involvement. Despite the usually typical characteristics, the diagnosis poses a considerable challenge in some cases because of the protein clinical presentation. Complicated schemes of classifying inflammatory myositis reflect the uncertainty in whether different clinical forms represent separate entities or a spectrum of the same illness. Subtypes based on the patient's age and underlying disorder include (1) primary idiopathic polymyositis, (2) primary idiopathic dermatomyositis, (3) dermatomyositis (or polymyositis) associated with neoplasia, (4) childhood dermatomyositis (or polymyositis) associated with vasculitis, and (5) polymyositis or dermatomyositis associated with collagen vascular disease.[23] For purposes of this discussion, a brief description suffices to highlight certain clinical features considered characteristic of dermatomyositis and polymyositis as a broad and general category.

Dermatomyositis

The diagnosis of dermatomyositis is made on the basis of the combination of a skin rash and muscular weakness. The symptoms begin at any age, but rarely in adolescence or early adulthood.Thus, the incidence histogram shows a bimodal distribution with peaks in childhood and in the fifth and sixth decades. Dermatomyositis in childhood often accompanies the systemic symptoms of collagen vascular diseases, but the patient rarely suffers from malignancy. In particular, myositis seems to appear in association with Raynaud's phenomenon, lupus erythematosus, polyarteritis nodosa, Sjögren's syndrome, or pneumonitis.

The initial symptoms include such nonspecific systemic manifestations as malaise, fever, anorexia, loss of weight, and features of respiratory infection. Despite the traditional emphasis, pain and tenderness of affected muscle, if present, constitute neither a presenting nor a primary symptom in most cases. Some patients have demonstrable tenderness restricted to the muscles of the shoulder. Vague pains and muscle aches have no specific diagnostic value in this context.

The skin lesions that may precede or follow the onset of weakness consist of a purple-colored rash over the cheeks and eyelids. Particularly prominent discoloration over the upper eyelids usually accompanies periorbital edema. An erythematous rash may also appear in exposed body parts such as the neck, upper chest, knees and hands. The affected skin thickens with a reddish hue, especially over the interphalangeal joints. Telangiectasia may develop over the chest and the back of the hands in advanced stages. In extreme cases, the inflammation renders the skin over the entire body atrophic, edematous, and reddish in color.

Polymyositis

Except for the absence of skin lesions, the signs and symptoms of polymyositis closely resemble those of dermatomyositis. Initial systemic manifestations also bear close resemblance in the two varieties. Children with myositis usually have skin rashes. Polymyositis primarily affects adults with possible underlying conditions such as collagen vascular disease or malignancy.[13] Men have a higher incidence of neoplasms that involve bowel, stomach or lung. Women may have a malignancy of the breast or ovary.

Weakness is the usual presenting symptom, ordinarily progressing slowly over a matter of weeks. The disease may take a fulminating course, crippling the patient during the first week of onset. The initial involvement of pelvic girdle muscles causes difficulty in climbing stairs or rising from a chair. Subsequent paresis of the shoulder girdle renders the patient incapable of lifting objects or combing the hair. In most patients, weakness soon spreads to involve the distal limb muscles.

The disease may begin as a focal process that mimics a localized inflammatory reaction[108] or as paralysis and wasting of only one limb.[140] Weakness of the neck musculature shows predilection for the anterior, rather than the posterior, compartment. The disease may cause dysphagia but spares the extraocular and other bulbar muscles. An extremely focal inflammatory process may involve the diaphragm and intercostal muscles.[21]

The patient has normal muscle stretch reflexes until very late in the course of the disease. Atrophy may escape detection in the deep muscles of the pelvic or shoulder girdle, but not in the orbicularis oculi or other superficial muscles. High-dose steroid therapy retards progression in most patients,[67] but the remission may not last long.[188]

The serum level of CK usually serves as a helpful indicator in determining the diagnosis and clinical course of myositis. Approximately 10 percent of patients with proven diagnoses, however, have no elevation even during acute stages. A normal enzymatic level despite active myositis suggests extensive muscle atrophy in long-standing disease.[23] Enzymes may be spilled from defects in the muscle plasma membrane, as postulated in Duchenne's dystrophy.[168] Alternatively, anastomosis of transverse tubules with terminal cisternae may be responsible for the leakage.[49] Other inconsistent laboratory findings include an elevated erythrocyte sedimentation rate and gamma globulin level.

Muscle biopsy findings include necrosis, phagocytosis, atrophy, degeneration and regeneration of both type I and type II fibers, internal nuclei, vacuolization, random variation of fiber size, mononuclear inflammatory infiltrates and endomysial or perimysial fibrosis.[23] A perifascicular distribution of muscle fiber atrophy presumably implies the interruption of blood supply to the peripherally located fibers.[2,11,23] SFEMG and histochemical investigation have revealed changes of the terminal innervation pattern consistent with reinnervation.[109] Denervation could result either from segmental necrosis of muscle fibers separated from the end-plate region[66] or from involvement of the terminal nerve endings.

A triad of electromyographic abnormalities[42] nearly always appears in untreated myositis, especially in the clinically weak muscles. It consists of (1) fibrillation potentials and positive sharp waves (see Fig. 13–7D), (2) complex repetitive discharges, and (3) polyphasic low-amplitude, short-duration motor unit potentials with early recruitment (see Fig. 13–18A). Certain muscles, however, may remain electrically normal, even in patients with moderately advanced disease. For adequate assessment, therefore, examination should include a number of proximal and distal muscles, with emphasis on those exhibiting moderate weakness clinically. Electromyographic and histologic abnormalities often involve the paraspinal muscles predominantly or selectively.[4,158,217]

A retrospective study of 153 patients with polymyositis or dermatomyositis revealed the following electromyographic abnormalities[24]: (1) low-amplitude, short-duration, polyphasic motor unit potentials (90 percent); (2) fibrillation potentials, positive sharp waves, and insertional irritability (74 percent); (3) complex repetitive discharges (38 percent); (4) a completely normal study with otherwise classic disease (10 percent); and (5) electrical abnormalities confined to the paraspinal muscle with widespread muscle weakness (1.6 percent). In another large series of 98 patients,[67] electromyographic findings consisted of (1) fibrillation potentials, positive sharp waves, and polyphasic, low-amplitude, short-duration motor unit potentials with early recruitment (45 percent); (2) the above changes of motor unit potentials but without spontaneous activity (44 percent); and (3) no abnormalities (11 percent). No correlation emerged between the grade of clinical impairment at the onset of illness and the electromyographic findings.

Muscle action potentials may show a decrement or, less frequently, increment upon repetitive stimulation of the nerve.[115,117,230] Such electrophysiologic abnormalities often accompany clinical features of myasthenia. These patients probably have myasthenia gravis with concomitant inflammatory changes of polymyositis and represent an overlap of these two entities. Indeed, the electrophysiologic and histologic features characteristic of polymyositis commonly occur in patients with severe myasthemia gravis.

Spontaneous activity in polymyositis, unlike wallerian degeneration, diminishes or disappears within a few weeks of successful steroid therapy. Because the time course of this change correlates well with clinical improvement, serial electromyographic evaluation can objectively assess patient response to various therapies. It also helps distinguish a recurrence of myositis from the emergence of steroid myopathy. Clinical recovery generally parallels serial improvement of electromyographic findings. Motor unit potentials show progressive increases in amplitude and duration initially some weeks or months after therapy, followed by diminution of the number of polyphasic units in a year or two.

Myositis Caused by Infectious Agents

Bacterial and viral infections of muscle occur less commonly than dermatomyositis and polymyositis.[55,69,122] Parasitic infection, however, prevails in tropical countries. In trichinosis, Trichinella spiralis preferentially invades the extraocular muscles.[63] In cysticercosis, Taenia solium mostly affects the trunk muscles.[202] Inflammation of muscle may be a feature of sarcoidosis,[77,97] sometimes accompanied by a rash typical of dermatomyositis.[121] Myositis may also develop in association with giant-cell arteritis.[40]

Inclusion body myositis is a distinct but infrequently recognized inflammatory disease of skeletal muscle.[45] The pathologic characteristics consist of rimmed vacuoles containing osmophilic membranous whorls and intracytoplasmic or intranuclear inclusions. Some investigators stress the mixed myopathic and neurogenic aspects.[83] In contrast to dermatomyositis, the disease lacks the features of collagen vascular involvement, but some patients have evidence of associated autoimmune disease.[136] The disease frequently affects distal muscles in men. It responds poorly to corticosteroid treatment but progresses slowly, taking a benign clinical course.

Electromyographic abnormalities, as in other myositic conditions, include fibrillation potentials; positive sharp waves; complex repetitive discharges; and low-amplitude, short-duration motor unit potentials with early recruitment. Most patients have changes suggestive of a mixed neurogenic and myopathic pattern with or without myotonic discharges,[124] but some do not.[136]

7 OTHER MYOPATHIES

Carcinomatous Metastasis

Proximal weakness may accompany electromyographic abnormalities consistent with a myopathy in patients with discrete carcinomatous metastatic deposits in the affected muscle.[76]

Distal Myopathy

Weakness in a primary muscle disease rarely shows a definite distal predilection. Exception to this rule occurs in an adult-onset hereditary myopathy described in Sweden and rare sporadic cases of distal myopathy with early adult onset.[154] The differential diagnoses include myotonic dystrophy and inclusion body myositis, both of which characteristically cause atrophy of the distal, rather than proximal, musculature.

REFERENCES

1. Abarbanel, JM, Bashan, N, Potashnik, R, Osimani, A, Moses, SW, and Herishanu, Y: Adult muscle phosphorylase "B" kinase deficiency. Neurology 36:560–562, 1986.
2. Adams, RD: The pathologic substratum of polymyositis. In Pearson, CM, and Mostofi, FK (eds): The Striated Muscle. Williams & Wilkins, Baltimore, 1973, pp 292–300.
3. Agamanolis, DP, Askari, AD, Dimauro, S, Hays, A, Kumar, K, Lipton, M, and Raynor, A: Muscle phosphofructokinase deficiency: Two cases with unusual polysaccharide accumulation and immunologically active enzyme protein. Muscle Nerve 3:456–467, 1980.
4. Albers, JW, Mitz, M, Sulaiman, AR, and Chang, GJ: Spontaneous electrical activity and muscle biopsy abnormalities in polymyositis and dermatomyositis (corres). Muscle Nerve 2:503, 1979.
5. Angelini, C, Govoni, E, Bragaglia, MM, and Vergani, L: Carnitine deficiency: Acute postpartum crisis. Ann Neurol 4:558–561, 1978.
6. Araoz, C, Sun, CN, Shenefelt, R, and White, HJ: Glycogenosis Type II (Pompe's disease): Ultrastructure of peripheral nerves. Neurology 24:739–742, 1974.
7. Askanas, V, Engel, WK, Dimauro, S, Brooks, BR, and Mehler, M: Adult-onset acid maltase deficiency: Morphologic and biochemical abnormalities reproduced in cultured muscle. N Engl J Med 294:573–578, 1976.
8. Askanas, V, Engel, WK, Kwan, HH, Reddy, NB, Husainy, T, Carlo, J, Siddique, T, Schwartzman, RJ, and Hanns, CJ: Autosomal dominant syndrome of lipid neuromyopathy with normal carnitine: Successful treatment with long-chain fatty-acid-free diet. Neurology 35:66–72, 1985.
9. Ballantyne, JR, and Hansen, S: New method for the estimation of the number of motor units in a muscle. 2. Duchenne, limb-girdle and facioscapulohumeral, and myotonic muscular dystrophies. J Neurol Neurosurg Psychiatry 37:1195–1201, 1974.
10. Bank, WJ, Dimauro, S, Bonilla, E, Capuzzi, DM, and Rowland, LP: A disorder of muscle lipid metabolism and myoglobinuria: Absence of carnitine palmityl transferase. N Engl J Med 292:443–449, 1975.
11. Banker, BQ, and Victor, M: Dermatomyositis

(systemic angiopathy) of childhood. Medicine (Baltimore) 45:261–289, 1966.

12. Barth, PG, Van Wijngaarden, GK, and Bethlem, J: X-linked myotubular myopathy with fatal neonatal asphyxia. Neurology 25:531–536, 1975.

13. Barwick, DD, and Walton, JN: Polymyositis. Am J Med 35:646–660, 1963.

14. Becker, PE: Two new families of benign sex-linked recessive muscular dystrophy. Rev Can Biol 21:551–566, 1962.

15. Becker, PE, and Kiener, F: Eine neue X-chromosomale Muskeldystrophie. Arch Psychiatr Zeit Neurol 193:427–448, 1955.

16. Belanger, AY, and McComas, AJ: Neuromuscular function in limb girdle dystrophy. J Neurol Neurosurg Psychiatry 48:1253–1258, 1985.

17. Bender, AN, and Willner, JP: Nemaline (rod) myopathy: The need for histochemical evaluation of affected families. Ann Neurol 4:37–42, 1978.

18. Berenberg, RA, Pellock, JM, Dimauro, S, Schotland, DL, Bonilla, E, Eastwood, A, Hays, A, Vicale, CT, Behrens, M, Chutorian, A, and Rowland, LP: Lumping or splitting? "Ophthalmoplegia-plus" or Kearns-Sayre syndrome? Ann Neurol 1:37–54, 1977.

19. Bergen, BJ, Carry, MP, Wilson, WB, Barden, MT, and Ringel, SF: Centronuclear myopathy: Extraocular- and limb-muscle findings in an adult. Muscle Nerve 3:165–171, 1980.

20. Bermils, C, Tassin, S, Brucher, JM, and Debarsy, TH: Idiopathic recurrent myoglobinuria and persistent weakness. Neurology 33:1613–1615, 1983.

21. Blumbergs, PC, Byrne, E, and Kakulas, BA: Polymyositis presenting with respiratory failure. J Neurol Sci 65:221–229, 1984.

22. Bodensteiner, JB, and Engel, AG: Intracellular calcium accumulation in Duchenne dystrophy and other myopathies: A study of 567,000 muscle fibers in 114 biopsies. Neurology 28:439–446, 1978.

23. Bohan, A, and Peter, JB: Polymyositis and dermatomyositis. (Part 2.) N Engl J Med 292:403–407, 1975.

24. Bohan, A, Peter, JB, Bowman, RL, and Pearson, CM: A computer-assisted analysis of 153 patients with polymyositis and dermatomyositis. Medicine 56:255–286, 1977.

25. Bordiuk, JM, Legato, MJ, Lovelace, RE, and Blumenthal, S: Pompe's disease: Electromyographic, electron microscopic, and cardiovascular aspects. Arch Neurol 23:113–119, 1970.

26. Bosch, EP, Gowans, JDC, and Munsat, T: Inflammatory myopathy in oculopharyngeal dystrophy. Muscle Nerve 2:73–77, 1979.

27. Bosch, EP, and Munsat, TL: Metabolic myopathies. Med Clin North Am 63:759–782, 1979.

28. Boudin, G, Mikol, J, Guillard, A, and Engel, AG: Fatal systemic carnitine deficiency with lipid storage in skeletal muscle, heart, liver and kidney. J Neurol Sci 30:313–325, 1976.

29. Bradley, WG, Hudgson, P, Gardner-Medwin, D, and Walton, JN: Myopathy associated with abnormal lipid metabolism in skeletal muscle. Lancet 1:495–498, 1969.

30. Bradley, WG, Jones, MZ, Mussini, JM, and Fawcett, PRW: Becker type muscular dystrophy. Muscle Nerve 1:111–132, 1978.

31. Bradley, WG, and Kelemen, J: Editorial: Genetic counseling in Duchenne muscular dystrophy. Muscle Nerve 2:325–328, 1979.

32. Bradley, WG, Tomlinson, BE, and Hardy, M: Further studies of mitochondrial and lipid storage myopathies. J Neurol Sci 35:201–210, 1978.

33. Brandt, NJ, Buchthal, F, Ebbesen, F, Kamieniecka, Z, and Krarup, C: Post-tetanic mechanical tension and evoked action potentials in McArdle's disease. J Neurol Neurosurg Psychiatry 40:920–925, 1977.

34. Bray, GM, Kaarsoo, M, and Ross, RT: Ocular myopathy with dysphagia. Neurology 15:678–684, 1965.

35. Britt, BA, Kalow, W, Gordon, A, Humphrey, JG, and Newcastle, NB: Malignant hyperthermia: An investigation of five patients. Can Anaesth Soc J 20:431–467, 1973.

36. Britt, BA, Kwong, FHF, and Endrenyi, L: The clinical and laboratory features of malignant hyperthermia management — A review. In Henschel, EO (ed): Malignant Hyperthermia: Current Concepts. Appleton-Century-Crofts, New York, 1977, pp 9–45.

37. Brody, IA, Gerber, CJ, and Sidbury, JB, Jr: Relaxing factor in McArdle's disease: Calcium uptake by sarcoplasmic reticulum. Neurology 20:555–558, 1970.

38. Brooke, MH: Congenital fiber type disproportion. In Kakulas, BA (ed): Clinical Studies in Myology. Part 2. Excerpta Medica, Amsterdam, 1973, pp 147–159.

39. Brooke, MH, Carroll, JE, and Ringel, SP: Congenital hypotonia revisited. Muscle Nerve 2:84–100, 1979.

40. Brooke, MH, and Kaplan, H: Muscle pathology in rheumatoid arthritis, polymyalgia rheumatica, and polymyositis. Arch Pathol 94:101–118, 1972.

41. Brunberg, JA, McCormick, WF, and Schochet, SS, Jr: Type III glycogenosis: An adult with diffuse weakness and muscle wasting. Arch Neurol 25:171–178, 1971.

42. Buchthal, F, and Pinelli, P: Muscle action potentials in polymyositis. Neurology 3:424–436, 1953.

43. Buchthal, F, and Rosenfalck, P: Electrophysiological aspects of myopathy with particular reference to progressive muscular dystrophy. In Bourne, GH, and Golarz, MN (eds): Muscular Dystrophy in Man and Animals. Hafner, New York, 1963.

44. Carpenter, S, and Karpati, G: Duchenne muscular dystrophy: Plasma membrane loss initiates muscle cell necrosis unless it is repaired. Brain 102:147–161, 1979.

45. Carpenter, S, Karpati, G, Heller, I, and Eisen, A: Inclusion body myositis: a distinct variety of idiopathic inflammatory myopathy. Neurology 28:8–17, 1978.

46. Carroll, JE, Brooke, MH, Devivo, DC, Kaiser,

KK, and Hagberg, JM: Biochemical and physiologic consequences of carnitine palmityl transferase deficiency. Muscle Nerve 1:103–110, 1978.

47. Carroll, JE, Brooke, MH, Devivo, DC, Shmate, JB, Kratz, R, Ringel, SP, and Hageberg, JM: Carnitine "deficiency": Lack of response to carnitine therapy. Neurology 30:618–626, 1980.

48. Chou, SM, Gutmann, L, Martin, JD, and Kettler, HL: Adult-type acid maltase deficiency: Pathologic features (abstr). Neurology 24:394, 1974.

49. Chou, SM, Nonaka, I, and Voice, GF: Anastomoses of transverse tubules with terminal cisternae in polymyositis. Arch Neurol 37:257–266, 1980.

50. Choucair, AK, and Ziter, FA: Pentazocine abuse masquerading as familial myopathy. Neurology 34:524–527, 1984.

51. Chui, LA, and Munsat, TL: Dominant inheritance of McArdle syndrome. Arch Neurol 33:636–641, 1976.

52. Chutkow, J, Heffner, R, Jr, Kramer, A, and Edwards, J: Adult-onset autosomal dominant limb-girdle muscular dystrophy. Ann Neurol 20:240–248, 1986.

53. Coers, C, and Telerman-Toppet, N: Differential diagnosis of limb-girdle muscular dystrophy and spinal muscular atrophy. Neurology 29:957–972, 1979.

54. Coers, C, Telerman-Toppet, N, Gerard, JM, Szliwowski, H, Bethlem, J, and Wijngaarden, GK: Changes in motor innervation and histochemical pattern of muscle fibers in some congenital myopathies. Neurology 26:1046–1053, 1976.

55. Congy, F, Hauw, JJ, Wang, A, and Moulias, R: Influenzal acute myositis in the elderly. Neurology 30:877–878, 1980.

56. Cornelio, F, Bresolin, N, Dimauro, S, Moro, M, and Balestrini, M: Congenital myopathy due to phosphorylase deficiency. Neurology 33:1383–1385, 1983.

57. Cornelio, F, Didonata, S, Peluchetti, D, Bizzi, A, Bertagndio, B, D'Angelo, A, and Wiesmann, U: Fatal cases of lipid storage myopathy with carnitine deficiency. J Neurol Neurosurg Psychiatry 40:170–178, 1977.

58. Cruz Martinez, A, Ferrer, MT, Lopez-Terradas, JM, Pascual-Castroveigo, I, and Mingo, P: Single fibre electromyography in central core disease. J Neurol Neurosurg Psychiatry 42:662–667, 1979.

59. Cumming, WJK, Hardy, M, Hudgson, P, and Walls, J: Carnitine-palmityl-transferase deficiency. J Neurol Sci 30:247–258, 1976.

60. Curless, RG, and Nelson, MB: Congenital fiber type disproportion in identical twins. Ann Neurol 2:455–459, 1977.

61. Dahl, DS, and Klutzow, FW: Congenital rod disease: Further evidence of innervational abnormalities as the basis for the clinicopathologic features. J Neurol Sci 23:371–385, 1974.

62. Danilowicz, D, Rutkowski, M, Myung, D, and Schively, D: Echocardiography in Duchenne muscular dystrophy. Muscle Nerve 3:298–303, 1980.

63. Davis, MJ, Cilo, M, Plaitakis, A, and Yahr, MD: Trichinosis: Severe myopathic involvement with recovery. Neurology 26:37–40, 1976.

64. Denborough, MA, Dennett, X, and Anderson, RMcD: Central-core disease and malignant hyperpyrexia. Br Med J 1:272–273, 1973.

65. Denborough, MA, Forster, JFA, Lovell, RRN, Maplestone, PA, and Villiers, JD: Anaesthetic deaths in a family. Br J Anaesth 34:395–396, 1962.

66. Desmedt, JE, and Borenstein, S: Relationship of spontaneous fibrillation potentials to muscle fibre segmentation in human muscular dystrophy. Nature 258:531–534, 1975.

67. Devere, R, and Bradley, WG: Polymyositis: Its presentation, morbidity and mortality. Brain 98:637–666, 1975.

68. Didonato, S, Cornelio, F, Pacini, L, Peluchetti, D, Rimoldi, M, and Spreafico, S: Muscle carnitine palmityltransferase deficiency: A case with enzyme deficiency in cultured fibroblasts. Ann Neurol 4:465–467, 1978.

69. Dietzman, DE, Schaller, JG, Ray, CG, and Reed, ME: Acute myositis associated with influenza B infection. Pediatrics 57:255–258, 1976.

70. Dimauro, S, Arnold, S, Miranda, A, and Rowland, LP: McArdle disease: The mystery of reappearing phosphorylase activity in muscle culture—A fetal isoenzyme. Ann Neurol 3:60–66, 1978.

71. Dimauro, S, and Dimauro, PMM: Muscle carnitine palmityltransferase deficiency and myoglobinuria. Science 182:929–931, 1973.

72. Dimauro, S, and Hartlage, PL: Fatal infantile form of muscle phosphorylase deficiency. Neurology 28:1124–1129, 1978.

73. Dimauro, S, Hartwig, GB, Hays, A, Eastwood, AB, Franco, R, Olarte, M, Chang, M, Roses, AD, Fetell, M, Schoenfeldt, RS, and Stern, LS: Debrancher deficiency: Neuromuscular disorder in five adults. Ann Neurol 5:422–436, 1979.

74. Dimauro, S, Stern, LZ, Mehler, M, Nagel, RB, and Payne, C: Adult-onset acid maltase deficiency: A postmortem study. Muscle Nerve 1:27–36, 1978.

75. Dimauro, S, Trevisan, C, and Hays, A: Disorders of lipid metabolism in muscle. Muscle Nerve 3:369–388, 1980.

76. Doshi, R, and Fowler, T: Proximal myopathy due to discrete carcinomatous metastases in muscle. J Neurol Neurosurg Psychiatry 46:358–360, 1983.

77. Douglas, AC, MacLeod, JG, and Matthews, JD: Symptomatic sarcoidosis of skeletal muscle. J Neurol Neurosurg Psychiatry 36:1034–1040, 1973.

78. Drachman, DA: Ophthalmoplegia plus: The neurodegenerative disorders associated with progressive external ophthalmoplegia. Arch Neurol 18:654–674, 1968.

79. Dubowitz, V, and Brooke, MH: Muscle Biopsy: A Modern Approach. WB Saunders, Philadelphia, 1973.

80. Duchenne, De B: (Recherches sur) la paralysie musculaire pseudohypertrophique ou paraly-

sie myo-sclerosique. Archives Generales de Medecine, Asselin, Paris, 1868.

81. Edstrom, L: Histochemical and histopathological changes in skeletal muscle in late-onset hereditary distal myopathy (Welander). J Neurol Sci 26:147–157, 1975.

82. Edwards, RHT: New techniques for studying human muscle function, metabolism, and fatigue. Muscle Nerve 7:599–609, 1984

83. Eisen, A, Berry, K, and Gibson, G: Inclusion body myositis (IBM): Myopathy or neuropathy? Neurology 33:1109–1114, 1983.

84. Elder, GB, Dean, D, McComas, AJ, Paes, B, and Desa, D: Infantile centronuclear myopathy. J Neurol Sci 60:79–88, 1983.

85. Emery, AEH, Hausmanowa-Petrusewicz, I, Davie, AM, Holloway, S, and Skinner, R: International collaborative study of the spinal muscular atrophies. Part I. Analysis of clinical and laboratory data. J Neurol Sci 29:83–94, 1976.

86. Emery, AEH, and Skinner, R: Clinical studies in benign (Becker type) X-linked muscular dystrophy. Clin Genet 10:189–201, 1976.

87. Eng, GD, Epstein, BS, Engel, WK, McKay, DW, and McKay, RJ: Malignant hyperthermia and central core disease in a child with congenital dislocating hips. Arch Neurol 35:189–197, 1978.

88. Engel, AG: Acid maltase deficiency in adults: Studies in four cases of syndrome which may mimic muscular dystrophy or other myopathies. Brain 93:599–616, 1970.

89. Engel, AG, and Angelini, C: Carnitine deficiency of human skeletal muscle with associated lipid storage myopathy: A new syndrome. Science 179:899–902, 1973.

90. Engel, AG, Gomez, MR, Seybold, ME, and Lambert, EH: The spectrum and diagnosis of acid maltase deficiency. Neurology 23:95–106, 1973.

91. Engel, AG, and Siekert, RG: Lipid storage myopathy responsive to prednisone. Arch Neurol 27:174–181, 1972.

92. Engel, WK, Vick, NA, Glueck, CJ, and Levy, RI: A skeletal-muscle disorder associated with intermittent symptoms and a possible defect of lipid metabolism. N Engl J Med 282:697–704, 1970.

93. Fawcett, PRW, Mastaglia, FL, and Mechler, F: Electrophysiological findings including single fiber EMG in a family with mitochondrial myopathy. J Neurol Sci 53:397–410, 1982.

94. Frank, JP, Harati, Y, Butler, IJ, Nelson, TE, and Scott, CI: Central core disease and malignant hyperthermia syndrome. Ann Neurol 7:11–17, 1980.

95. Gallant, EM, Godt, RE, and Gronert, GA: Role of plasma membrane defect of skeletal muscle in malignant hyperthermia. Muscle Nerve 2:491–494, 1979.

96. Gardner-Medwin, D, Pennington, RJ, and Walton, JN: The detection of carriers of X-linked muscular dystrophy genes: A review of some methods studied in Newcastle upon Tyne. J Neurol Sci 13:459–474, 1971.

97. Gardner-Thorpe, C: Muscle weakness due to sarcoid myopathy: Six case reports and an evaluation of steroid therapy. Neurology 22:917–928, 1972.

98. Gil-Peralta, A, Rafel, E, Bautista, J, and Alberca, R: Myotonia in centronuclear myopathy. J Neurol Neurosurg Psychiatry 41:1102–1108, 1978.

99. Gomez, MR, Engel, AG, Dewald, G, and Peterson, HA: Failure of inactivation of Duchenne dystrophy X-chromosome in one of female identical twins. Neurology 27:537–541, 1977.

100. Groothuis, DR, Schulman, S, Wollman, R, Frey, J, and Vick, NA: Demylinating radiculopathy in the Kearns-Sayre syndrome: A clinicopathological study. Ann Neurol 8:373–380, 1980.

101. Gruener, R, McArdle, B, Ryman, BE, and Weller, RD: Contracture of phosphorylase deficient muscle. J Neurol Neurosurg Psychiatry 31:268–283, 1968.

102. Hanson, PA, and Rowland, LP: Mobius syndrome and facioscapulohumeral muscular dystrophy. Arch Neurol 24:31–39, 1971.

103. Hart, ZH, Chang, CH, Dimauro, S, Farooki, Q, and Ayyar, R: Muscle carnitine deficiency and fatal cardiomyopathy. Neurology 28:147–151, 1978.

104. Hathaway, PW, Engel, WK, and Zellweger, H: Experimental myopathy after microarterial embolization: Comparison with childhood X-linked pseudohypertrophic muscular dystrophy. Arch Neurol 22:365–378, 1970.

105. Hausmanowa-Petrusewicz, I, Wierzbica, M, Jozwik, A, Szmidt-Salkowska, E, and Borkowska, J: A nearest neighbour decision rule for EMG detection of carriers of Duchenne muscular dystrophy. Electromyogr Clin Neurophysiol 22:445–457, 1982.

106. Hazama, R, Tsujihata, M, Mori, M, and Mori, K: Muscular dystrophy in six young girls. Neurology 29:1486–1491, 1979.

107. Heckmatt, JZ, Sewry, CA, Hodges, D, and Dubowitz, V: Congenital centronuclear (myotubular) myopathy: A clinical, pathological and genetic study in eight children. Brain 108:941–964, 1986.

108. Heffner, RR, Jr, and Barron, SA: Polymyositis beginning as a focal process. Arch Neurol 38:439–442, 1981.

109. Henriksson, KG, and Stalberg, E: The terminal innervation pattern in polymyositis: A histochemical and SFEMG study. Muscle Nerve 1:3–13, 1978.

110. Hers, HG, and De Barsy, T: Type II glycogenosis (acid maltase deficiency). In Hers, HG, and Van Hoof, F (eds): Lysosomes and Storage Diseases. Academic Press, New York, 1973, pp 197–206.

111. Hoffman, EP, Brown, RH, and Kunkel, LM: Dystrophin: The protein product of the Duchenne muscular dystrophy locus. Cell 51:919–928, 1987.

112. Hogan, GR, Gutmann, L, Schmidt, R, and Gilbert, E: Pompe's disease. Neurology 19:894–900, 1969.

113. Holliday, PL, Climie, ARW, Gilroy, J, and Mahmud, MZ: Mitochondrial myopathy and encephalopathy: Three cases--a deficiency of

NADH-CoQ dehydrogenase? Neurology 33:1619–1622, 1983.

114. Hooshmand, H, Martinez, AJ, and Rosenblum, WI: Arthrogryposis multiplex congenita: Simultaneous involvement of peripheral nerve and skeletal muscle. Arch Neurol 24:561–572, 1971.

115. Hopf, HC, and Thorwirth, V: Myasthenie-Myositis-Myopathie. In Hertel, G, Mertono, HG, Ricker, K, and Schimrigk, K: Myasthenia Gravis und Andere Storumgen der Neuromuskularen Synapse. Thieme, Stuttgart, 1977, pp 142–147.

116. Howell, RR: The glycogen storage diseases. In Stanbury, JB, Wyngaarden, JB, and Fredrickson, DS (eds): The Metabolic Basis of Inherited Disease, ed 3. McGraw-Hill, New York, 1972, pp 149–173.

117. Huffmann, G, and Leven, B: Myasthenia und polymyositis. In Hertel, G, Mertono, HG, Ricker, K, and Schimrigk, K: Myasthenia Gravis und Andere Storumgen der Neuromuskularen Synapse. Thieme, Stuttgart, 1977, pp 147–150.

118. Inokuchi, T, Umezaki, H, and Santa, T: A case of type I muscle fibre hypotrophy and internal nuclei. J Neurol Neurosurg Psychiatry 38:475–482, 1975.

119. Isaacs, H, and Heffron, JJA: Morphological and biochemical defects in muscles of human carriers of the malignant hyperthermia syndrome. Br J Anaesth 47:475–481, 1975.

120. Isaacs, H, Heffron, JJA, and Badenhorst, M: Central core disease: a correlated genetic, histochemical, ultramicroscopic, and biochemical study. J Neurol Neurosurg Psychiatry 38:1177–1186, 1975.

121. Itoh, J, Akiguchi, I, Midorikawa, R, and Kameyama, M: Sarcoid myopathy with typical rash of dermatomyositis. Neurology 30:1118–1121, 1980.

122. Jehn, UW, and Fink, MK: Myositis, myoglobinemia, and myoglobinuria associated with enterovirus echo 9 infection. Arch Neurol 37:457–458, 1980.

123. Johnston, DM: Thyrotoxic myopathy. Arch Dis Child 49:968–969, 1974.

124. Julien, J, Vital, CL, Vallat, JM, Lagueny, A, and Sapina, D: Inclusion body myositis: Clinical, biological and ultrastructural study. J Neurol Sci 55:15–24, 1982.

125. Karpati, G, Carpenter, S, Engel, AG, Watters, G, Allen, J, Rothman, S, Klassen, G, and Mamer, OA: The syndrome of systemic carnitine deficiency: Clinical, morphologic, biochemical and pathophysiologic features. Neurology 25:16–24, 1975.

126. Karpati, G, Carpenter, S, Larbrisseau, A, and Lafontaine, R: The Kearns-Shy syndrome: A multisystem disease with mitochondrial abnormality demonstrated in skeletal muscle and skin. J Neurol Sci 19:133–151, 1973.

127. Karpati, G, Charuk, J, Carpenter, S, Jablecki, C, and Holland, P: Myopathy caused by a deficiency of Ca^{2+}-adenosine triphosphatase in sarcoplasmic reticulum (Brody's disease). Ann Neurol 20:38–49, 1986.

128. Kazakov, VM, Bogorodinsky, DK, Znoyko, ZV, and Skorometz, AA: The facio-scapulo-limb (or the facioscapulohumeral) type of muscular dystrophy: Clinical and genetic study of 200 cases. Eur Neurol 11:236–260, 1974.

129. Kearns, TP: External ophthalmoplegia, pigmentary degeneration of the retina, and cardiomyopathy: A newly recognized syndrome. Trans Am Ophthalmol Soc 63:559–625, 1965.

130. Kinoshita, M, Satoyoshi, E, and Kumagai, M: Familial type I fiber atrophy. J Neurol Sci 25:11–17, 1975.

131. Kondo, K, and Yuasa, T: Genetics of congenital nemaline myopathy. Muscle Nerve 3:308–315, 1980.

132. Kost, GJ, and Verity, MA: A new variant of late-onset myophosphorylase deficiency. Muscle Nerve 3:195–201, 1980.

133. Kovanen, J, Somer, H, and Schroeder, P: Acute myopathy associated with gasoline sniffing. Neurology 33:629–631, 1983.

134. Kuhn, E, Fiehn, W, Schroder, JM, Assmus, H, and Wagner, A: Early myocardial disease and cramping myalgia in Becker-type muscular dystrophy: A kindred. Neurology 29:1144–1149, 1979.

135. Kuitunen, P, Rapola, J, Noponen, AL, and Donner, M: Nemaline myopathy: Report of four cases and review of the literature. Acta Paediatr Scand 61:353–361, 1972.

136. Lane, RJM, Fulthorpe, JJ, and Hudgson, P: Inclusion body myositis: A case with associated collagen vascular disease responding to treatment. J Neurol Neurosurg Psychiatry 48:270–273, 1985.

137. Layzer, RB, Havel, RJ, Becker, N, and McIlroy, MB: Muscle carnitine palmityl transferase deficiency: A case with diabetes and ketonuria. Neurology 27:379–380, 1977.

138. Layzer, RB, Havel, RJ, and McIlroy, MB: Partial deficiency of carnitine palmityltransferase: Physiologic and biochemical consequences. Neurology 30:627–633, 1980.

139. Layzer, RB, and Rowland, LP: Cramps. N Engl J Med 285:31–40, 1971.

140. Lederman, RJ, Salanga, VD, Wilbourn, AJ, Hanson, MR, and Dudley, AW, Jr: Focal inflammatory myopathy. Muscle Nerve 7:142–146, 1984.

141. Lenard, HG, and Goebel, HH: Congenital fibre type disproportion. Neuropadiatrie 6:220–231, 1975.

142. Markesbery, WR, Griggs, RC, Leach, RP, and Lapham, LW: Late onset hereditary distal myopathy. Neurology 24:127–134, 1974.

143. Markesbery, WR, McQuillen, MP, Procopis, PG, Harrison, AR, and Engel, AG: Muscle carnitine deficiency: Association with lipid myopathy, vacuolar neuropathy and vacuolated leukocytes. Arch Neurol 31:320–324, 1974.

144. Martin, JJ, De Barsy, T, and Den Tandt, WR: Acid maltase deficiency in non-identical adult twins: A morphological and biochemical study. J Neurol 213:105–118, 1976.

145. Mastaglia, FL, McCollum, JPK, Larson, PF, and Hudgson, P: Steroid myopathy complicat-

ing McArdle's disease. J Neurol Neurosurg Psychiatry 33:111–120, 1970.

146. Matheson, DW, and Howland, JL: Erythrocyte deformation in human muscular dystrophy. Science 184:165–166, 1974.

147. Maunder-Sewry, CA, Gorodetsky, R, Yarom, R, and Dubowitz, V: Element analysis of skeletal muscle in Duchenne muscular dystrophy using x-ray fluorescence spectrometry. Muscle Nerve 3:502–508, 1980.

148. Mawatari, S, Schonberg, M, and Olarte, M: Biochemical abnormalities of erythrocyte membranes in Duchenne dystrophy: Adenosine triphosphatase and adenyl cyclase. Arch Neurol 33:489–493, 1976.

149. McArdle, B: Myopathy due to a defect in muscle glycogen breakdown. Clin Sci 10:13–33, 1951.

150. McComas, AJ, Sica, REP, and Currie, S: An electrophysiological study of Duchenne dystrophy. J Neurol Neurosurg Psychiatry 34:461–468, 1971.

151. McComas, AJ, Sica, REP, Upton, ARM, and Hamilton, MB: Multiple muscle analysis of motor units in muscular dystrophy. Arch Neurol 30:249–251, 1974.

152. McComas, AJ, Sica, REP, Upton, ARM, and Petito, F: Sick motoneurons and muscle disease. Ann NY Acad Sci 228:261–279, 1974.

153. Mendell, JR, Engel, WK, and Derrer, EC: Duchenne muscular dystrophy: Functional ischemia reproduces its characteristic lesions. Science 172:1143–1145, 1971.

154. Miller, RG, Blank, NK, and Layzer, RB: Sporadic distal myopathy with early adult onset. Ann Neurol 5:220–227, 1979.

155. Miller, RG, Layzer, RB, Mellenthin, MA, Golabj, M, Francoz, RA, and Mall, JC: Emery-Dreifuss muscular dystrophy with autosomal dominant transmission. Neurology 35:1230–1233, 1985.

156. Miller, SE, Roses, AD, and Appel, SH: Erythrocytes in human muscular dystrophy (corres). Science 188:1131, 1975.

157. Mills, KR, and Edwards, RHT: Investigative strategies for muscle pain. J Neurol Sci 58:73–88, 1983.

158. Mitz, M, Albers, JW, Sulaiman, AR, and Chang, GJ: Electromyographic and histologic paraspinal abnormalities in polymyositis/dermatomyositis. Arch Phys Med Rehabil 62:118–121, 1981.

159. Miyoshi, K, Kawai, H, Iwasa, M, Kusaka, K, and Nishino, H: Autosomal recessive distal muscular dystrophy as a new type of progressive muscular dystrophy: Seventeen cases in eight families including an autopsy case. Brain 109:31–54, 1986.

160. Mokri, B, and Engel, AG: Duchenne dystrophy: Electron microscopic findings pointing to a basic or early abnormality in the plasma membrane of the muscle fiber. Neurology 25:1111–1120, 1975.

161. Mollman, JE, Cardenas, JC, and Pleasure, DE: Alteration of calcium transport in Duchenne erythrocytes. Neurology 30:1236–1239, 1980.

162. Moosa, A, Brown, BH, and Dubowitz, V: Quantitative electromyography: Carrier detection in Duchenne type muscular dystrophy using a new automatic technique. J Neurol Neurosurg Psychiatry 35:841–844, 1972.

163. Morgan-Hughes, JA, Brett, EM, Lake, BD, and Tome, FMS: Central core disease or not? Observations on a family with a non-progressive myopathy. Brain 96:527–536, 1973.

164. Morgan-Hughes, JA, Darveniza, P, Kahn, SN, Landon, DN, Sherratt, RM, Land, JM, and Clark, JB: A mitochondrial myopathy characterized by a deficiency in reducible cytochrome b. Brain 100:617–640, 1977.

165. Mrozek, K, Strugalska, M, and Fidzianska, A: A sporadic case of central core disease. J Neurol Sci 10:339–348, 1970.

166. Munsat, TL: A standardized forearm ischemic exercise test. Neurology 20:1171–1178, 1970.

167. Munsat, TL, Piper, D, Cancilla, P, and Mednick, J: Inflammatory myopathy with facioscapulohumeral distribution. Neurology 22:335–347, 1972.

168. Munsat, TL, Serum, TL, Baloh, R, Pearson, CM, and Fowler, W: Serum enzyme alterations in neuromuscular disorders. JAMA 226:1536–1543, 1973.

169. Murase, T, Ikeda, H, Muro, T, Nakao, K, and Sugita, H: Myopathy associated with type III glycogenosis. J Neurol Sci 20:287–295, 1973.

170. Murphy, SF, and Drachman, DB: The oculopharyngeal syndrome. JAMA 203:1003–1008, 1968.

171. Neville, HE, and Brooke, MH: Central core fibers: Structured and unstructured. In Kakulas, BA (ed): Basic Research in Myology. Excerpta Medica, Amsterdam, 1973, pp 497–511.

172. Norton, P, Ellison, P, Sulaiman, AR, and Harb, J: Nemaline myopathy in the neonate. Neurology 33:351–356, 1983.

173. Olson, W, Engel, WK, Walsh, GO, and Einangler, R: Oculocraniosomatic neuromuscular disease with "ragged-red" fibers. Arch Neurol 26:193–211, 1972.

174. Paine, RS: The future of the "floppy infant": A follow-up study of 133 patients. Dev Med Child Neurol 5:115–124, 1963.

175. Palmer, EG, Topel, DG, and Christian, LL: Light and electron microscopy of skeletal muscle from malignant hyperthermia susceptible pigs. In Alderete, JA, and Britt, BA (eds): The Second International Symposium on Malignant Hyperthermia. Grune & Stratton, New York, 1978.

176. Panayiotopoulos, CP, Scarpalezos, S, and Papapetropoulos, T: Electrophysiological estimation of motor units in Duchenne muscular dystrophy. J Neurol Sci 23:89–98, 1974.

177. Patel, H, Berry, K, MacLeod, P, and Dunn, HG: Cytoplasmic body myopathy. J Neurol Sci 60:281–292, 1983.

178. Patten, BM, Bilezikian, JP, Mallette, LE, Prince, A, Engel, WK, and Aurbach, GD: Neuromuscular disease in primary hyperparathyroidism. Ann Intern Med 80:182–193, 1974.

179. Percy, ME, Chang, LS, Murphy, EG, Oss, I, Verellen-Dumoulin, C, and Thompson, MW:

Serum creatine kinase and pyruvate kinase in Duchenne muscular dystrophy carrier detection. Muscle Nerve 2:329–339, 1979.

180. Perloff, JK, Roberts, WC, DeLeon, AC, Jr, and O'Doherty, D: The distinctive electrocardiogram of Duchenne's progressive muscular dystrophy. Am J Med 42:179–188, 1967.

181. Pernow, BB, Havel, RJ, and Jennings, DB: The second wind phenomenon in McArdle's syndrome. Acta Med Scand Suppl 472:294–307, 1967.

182. Peyronnard, JM, Charron, L, Bellavance, A, and Marchand, L: Neuropathy and mitochondrial myopathy. Ann Neurol 7:262–268, 1980.

183. Pezeshkpour, G, Krarup, C, Buchthal, F, Dimauro, S, Bresolin, N, and McBurney, J: Peripheral neuropathy in mitochondrial disease. J Neurol Sci 77:285–304, 1987.

184. Pleasure, DE, Walsh, GO, and Engel, WK: Atrophy of skeletal muscle in patients with Cushing's syndrome. Arch Neurol 22:118–125, 1970.

185. Pourmand, R, Sanders, DB, and Corwin, HM: Late-onset McArdle's disease with unusual electromyographic findings. Arch Neurol 40:374–377, 1983.

186. Puvanendran, K, Cheah, JS, Naganathan, N, Yeo, PPB, and Wong, PK: Thyrotoxic myopathy: A clinical and quantitative analytic electromyographic study. J Neurol Sci 42:441–451, 1979.

187. Radu, H, Killyen, I, Ionasescu, V, and Radu, A: Myotubular (centronuclear) (neuro-) myopathy. 1. Clinical, genetical and morphological studies. Eur Neurol 15:285–300, 1977.

188. Riddoch, D, and Morgan-Hughes, JA: Prognosis in adult polymyositis. J Neurol Sci 26:71–80, 1975.

189. Ringel, SP, Carroll, JE, and Schold, SC: The spectrum of mild X-linked recessive muscular dystrophy. Arch Neurol 34:408–416, 1977.

190. Rose, AL, and Walton, JN: Polymyositis: A survey of 89 cases with particular reference to treatment and prognosis. Brain 89:747–768, 1966.

191. Roses, AD, Roses, MJ, Miller, SE, Hull, KL, and Appel, SH: Carrier detection in Duchenne muscular dystrophy. N Engl J Med 294:193–198, 1976.

192. Rothman, SM, and Bischoff, R: Electrophysiology of Duchenne dystrophy myotubes in tissue culture. Ann Neurol 13:176–179, 1983.

193. Rowland, LP: Are the muscular dystrophies neurogenic? Ann NY Acad Sci 228:224–260, 1974.

194. Rowland, LP: Pathogenesis of muscular dystrophies. Arch Neurol 33:315–321, 1976.

195. Rowland, LP (ed): Pathogenesis of human muscular dystrophies. Proceedings of the 5th International Scientific Conference of the Muscular Dystrophy Association, Excerpta Medica, Amsterdam, 1977.

196. Rowland, LP: Biochemistry of muscle membranes in Duchenne muscular dystrophy. Muscle Nerve 3:3–20, 1980.

197. Rowland, LP, Fahn, S, and Schotland, DL:

198. Rowland, LP, Fetell, M, Olarte, M, Hays, A, Singh, N, and Wanat, FE: Emery-Dreifuss muscular dystrophy. Ann Neurol 5:111–117, 1979

199. Sahn, L, and Magee, KR: Phosphorylase deficiency associated with isometric exercise intolerance. Neurology 26:896–898, 1976.

200. Salick, AI, and Pearson, CM: Electrical silence of myoedema. Neurology 17:899–901, 1967.

201. Sasaki, H, Kuzuhara, S, Kanazawa, I, Nakanishi, T, and Ogata, T: Myoclonus, cerebellar disorder, neuropathy, mitochondrial myopathy and ACTH deficiency. Neurology 33:1288–1293, 1983.

202. Sawhney, BB, Chopra, JS, Banerji, AK, and Wahi, PL: Pseudohypertrophic myopathy in cysticercosis. Neurology 26:270–272, 1976.

203. Schmalbruch, H: Regenerated muscle fibers in Duchenne muscular dystrophy: A serial section study. Neurology 34:60–65, 1984.

204. Schmitt, HP, and Krause, KH: An autopsy study of a familiar oculopharyngeal muscular dystrophy (OPMD) with distal spread and neurogenic involvement. Muscle Nerve 4:296–305, 1981.

205. Scholte, HR, Jennekens, FGI, and Bouvy, JJBJ: Carnitine palmityltransferase II deficiency with normal carnitine palmityltransferase I in skeletal muscle and leucocytes. J Neurol Sci 40:39–51, 1979.

206. Schotland, DL, Dimauro, S, Bonilla, E, Scarpa, A, and Lee, CP: Neuromuscular disorder associated with a defect in mitochondrial energy supply. Arch Neurol 33:475–479, 1976.

207. Serratrice, G, Pellissier, JF, Faugere, MC, and Gastaut, JL: Centronuclear myopathy: Possible central nervous system origin. Muscle Nerve 1:62–69, 1978.

208. Servidei, S, Bonilla, E, Diedrich, RG, Kornfeld, M, Oates, JD, Davidson, M, Vora, S, and Dimauro, S: Fatal infantile form of muscle phosphofructokinase deficiency. Neurology 36:1465–1470, 1986.

209. Shafiq, SA, Dubowitz, V, Peterson, HE De C, and Mihorat, AT: Nemaline myopathy: Report of a fatal case, with histochemical and electron microscopic studies. Brain 90:817–828, 1967.

210. Shields, RW, Jr: Single fiber electromyography in the differential diagnosis of myopathic limb girdle syndromes and chronic spinal muscle atrophy. Muscle Nerve 7:265–272, 1984.

211. Shy, GM, Engel, WK, Somers, JE, and Wanko, T: Nemaline myopathy: A new congenital myopathy. Brain 86:793–810, 1963.

212. Shy, GM, and Magee, KR: A new congenital non-progressive myopathy. Brain 79:610–621, 1956.

213. Shy, GM, Silberberg, DH, Appel, SH, Mishkin, MM, and Godfrey, EH: A generalized disorder of nervous system, skeletal muscle and heart resembling Refsum's disease and Hurler's syndrome. Part I. Clinical, pathologic and bio-

McArdle's disease: Hereditary myopathy due to absence of muscle phosphorylase. Arch Neurol 9:325–342, 1963.

chemical characteristics. Am J Med 42:163–168, 1967.

214. Slonim, AE, Coleman, RA, McElligot, MA, Najjar, J, Hirschhorn, K. Labadie, GU, Mrak, R, and Evans, OB: Improvement of muscle function in acid maltase deficiency by high-protein therapy. Neurology 33:34–38, 1983.

215. Smith, R, and Stern, G: Myopathy, osteomalacia and hyperparathyroidism. Brain 90:593–602, 1967.

216. Spiro, AJ, Shy, GM, and Gonatas, NK: Myotubular myopathy. Arch Neurol 14:1–14, 1966.

217. Streib, EW, Wilbourn, AJ, and Mitsumoto, H: Spontaneous electrical muscle fiber activity in polymyositis and dermatomyositis. Muscle Nerve 2:14–18, 1979.

218. Sulaiman, AR, Swick, HM, and Kinder, DS: Congenital fibre type disproportion with unusual clinicopathologic manifestations. J Neurol Neurosurg Psychiatry 46:175–182, 1983.

219. Swash, M, and Heathfield, KWG: Quadriceps myopathy: A variant of the limb-girdle dystrophy syndrome. J Neurol Neurosurg Psychiatry 46:355–357, 1983.

220. Swash, M, Schwartz, MS, and Apps, MCP: Adult onset acid maltase deficiency: Distribution and progression of clinical and pathological abnormality in a family. J Neurol Sci 68:61–74, 1985.

221. Tarui, S, Okuno, G, Ihura, Y, Tanaka, T, Suda, M, and Nishikawa, M: Phosphofructokinase deficiency in skeletal muscle: A new type of glycogenosis. Biochem Biophys Res Commun 19:517–523, 1965.

222. Telerman-Toppet, N, Gerard, JM, and Coers, C: Central core disease: A study of clinically unaffected muscle. J Neurol Sci 19:207–223, 1973.

223. Thomas, PK, Schott, GD, and Morgan-Hughes, JA: Adult onset scapuloperoneal myopathy. J Neurol Neurosurg Psychiatry 38:1008–1015, 1975.

224. Tobin, WE, Huijing, F, Porro, RS, and Slazman, RT: Muscle phosphofructokinase deficiency. Arch Neurol 28:128–130, 1973.

225. Trend, PSJ, Wiles, CM, Spencer, GT, Morgan-Hughes, JA, and Patrick, AD: Acid maltase deficiency in adults: Diagnosis and management in five cases. Brain 108:845–860, 1985.

226. Upton, ARM, McComas, AJ, and Bianchi, FA: Neuropathy in McArdle's syndrome. N Engl J Med 289:750–751, 1973.

227. Valli, G, Scarlato, G, and Contartese, M: Quantitative electromyography in the detection of the carriers in Duchenne type muscular dystrophy. J Neurol 212:139–149, 1976.

228. Van Hoof, F, and Hers, HG: The subgroups of type III glycogenosis. Eur J Biochem 2:265–270, 1967.

229. Van Wijngaarden, GK, and Bethlem, J: The facioscapulohumeral syndrome. In Kakulas, BA (ed): Clinical Studies in Myology. Part 2. Excerpta Medica, Amsterdam, 1973, pp 498–501.

230. Vasilescu, C, Bucur, G, Petrovici, A, and Florescu, A: Myasthenia in patients with dermatomyositis: Clinical, electrophysiological and ultrastructural studies. J Neurol Sci 38:129–144, 1978.

231. Victor, M, Hayes, and Adams, RD: Oculopharyngeal muscular dystrophy: A familial disease of late life characterized by dysphagia and progressive ptosis of the eyelids. N Engl J Med 267:1267–1272, 1962.

232. Wakayama, Y, Hodson, A, Pleasure, D, Bonilla, E, and Schotland, DL: Alteration in erythrocyte membrane structure in Duchenne muscular dystrophy. Ann Neurol 4:253–256, 1978.

233. Walton, JN: Some changing concepts in neuromuscular disease. In Kakulas, BA (ed): Clinical Studies in Myology. Part 2. Excerpta Medica, Amsterdam, 1973, pp 429–438.

234. Welander, L: Myopathia distalis tarda hereditaria. Acta Med Scand (Suppl 265)141:1–124, 1951.

235. Wolf, E, Shocehnia, M, Ferber, I, and Gonen, B: Phrenic nerve and diaphragmatic involvement in progressive muscular dystrophy. Electromyogr Clin Neurophysiol 21:35–53, 1981.

236. Yiannikas, C, McLeod, J, Pollard, J, and Baverstock, J: Peripheral neuropathy associated with mitochondrial myopathy. Ann Neurol 20:249–257, 1986.

Chapter 26

NEUROMUSCULAR DISEASES CHARACTERIZED BY ABNORMAL MUSCLE ACTIVITY

1 INTRODUCTION

Muscles may stiffen pathologically with lesions involving the central nervous system, peripheral nerve trunk, axon terminal, or muscle membrane. Myotonia, or delayed relaxation of voluntarily or reflexively contracted muscle, occurs in a number of myogenic syndromes. These include myotonic dystrophy, myotonia congenita, paramyotonia congenita, and a form of periodic paralysis. Involuntary muscle contraction also results from disorders of the peripheral nerve, as in myokymia, the Schwartz-Jampel syndrome, and neuromyotonia, or continuous muscle fiber discharge. In still other sustained muscle contractions, spontaneous discharges originate centrally, as in the stiff-man syndrome. Other conditions with abnormal muscle activity include common cramp, contracture, tetanus, tetany, and hemifacial spasm.

Several electrophysiologic techniques help characterize involuntary movement and determine the site of abnormal discharges. Nerve blocks will eliminate abnormal muscular activity originating in the central nervous system or the proximal part of the peripheral nerve. In this instance, repetitive nerve stimulation distal to the block fails to induce the abnormal muscle activity. Discharges from the distal or terminal nerve segment cease after the block of neuromuscular transmission. In contrast, curarization does not affect abnormal discharges originating from intrinsic muscle fibers. Some cramp syndromes display a distinctive pattern of abnormalities on electromyography. Others produce a normal interference pattern, although the subject has no voluntary control over the number and frequency of discharging motor units. In contracture, unlike true cramps, the contracted muscle is electrically silent.

2 MYOTONIA

In myotonia, muscle membrane, once activated, tends to fire repetitively, inducing delayed muscle relaxation. Unlike cramp or spontaneous spasm, however, this type of prolonged muscle contraction causes no pain. Myotonic discharges, provoked by voluntary contraction, muscle percussion, or needle insertion, characteristically wax and wane at varying frequencies up to 150 impulses per second.[200] During volitional activity, myotonia may worsen initially but improve after a warm-up period, typically recurring at the beginning of the next voluntary movement after a period of rest. Percussion myotonia follows a brisk tap over the thenar eminence. Cold aggravates both postactivation and percussion myotonia. Myotonic muscles typically have reduced torques during maximal voluntary contraction and decreased mean amplitude of the compound muscle action potentials.[23] Muscle action potentials decline further with repetitive nerve stimulation (see Chapter 9.8 and 13.3) or after isometric exercise.[207]

Myotonic discharge with or without clinical myotonia develops in a number of metabolic muscle diseases such as hyperkalemic periodic paralysis, acid maltase deficiency,[60] hyperthyroidism,[154] familial granulovacuolar lobular myopathy,[100] and malignant hyperpyrexia (see Chapter 25.4). Myotonia and myositis may also constitute part of the symptom complex seen in multicentric reticulohistiocytosis.[6] Administration of the hypocholesterolemic agent diazocholesterol can induce a myopathy with occasional myotonia.[201] In all these entities, however, myotonia plays neither a predominant nor an essential role as in myotonic dystrophy, myotonia congenita, and paramyotonia.

Although the underlying defect in myotonia remains unknown, recent studies implicate the sarcolemmal membrane.[118,177] Potassium ions (K^+) accumulate in the transverse tubular system during activation of the muscle membrane giving rise to a negative afterpotential (see Chapter 2.3, Fig. 2–3). This degree of depolarization, though normally not large enough to generate an action potential, could initiate repetitive discharges in the myotonic muscle.[2] Such membrane instability may result from abnormally low chloride conductance in the myotonia of goats or those induced experimentally with drugs.[15,35,73,118] In humans, however, low chloride permeability is found only in myotonia congenita and not in myotonic dystrophy. As part of multiorgan involvement, erythrocytes may demonstrate biochemical and biophysical abnormalities.[40,143,226] This is, however, not universally confirmed.[74]

Myotonic Dystrophy

Myotonic dystrophy shows an autosomal dominant trait with a gene located on chromosome 19. Clinical variability suggests incomplete penetrance or anticipation with earlier and more severe symptomatology in successive generations. Some claim that these changes may simply reflect earlier recognition or an inaccurate history. Rare chromosomal anenploidies associated with this disorder include the Klinefelter and Down syndromes.[25] In one family, 8 of 13 members with hereditary motor and sensory neuropathy (HMSN) also had signs of myotonic dystrophy. The syndrome could result from an allelic form of the myotonic dystrophy gene or two closely linked genes on chromosome 19.[202] Typically, the illness begins in adolescence or early adult life. Neuromuscular symptoms consist of weakness and myotonia. Patients may have muscle stiffness and cramps, but distal weakness prompts them to seek medical advice. On questioning, they admit to difficulty with grip release, which they describe as more of an inconvenience than a disability. Weakness may begin in the hands and feet but eventually spreads to involve all the muscles, including the flexors of the neck.

Adult patients typically have a hatchet-faced appearance, which results from relatively selective atrophy of the temporalis and masseter. Prominent wasting of the neck muscles, particularly of the sternocleidomastoids, gives rise to a swan-neck. The head, supported by a slender neck, appears unstable. In recumbency, the patient cannot lift the head from a pillow against gravity. Facial weakness produces a blank expression and ptosis. In the absence of this characteristic appearance, milder cases of myotonic dystrophy may escape detection. Usually, however, grip or percussion myotonia determines the diagnosis. Myotonic phenomena become less prominent as the muscle wasting and weakness advance. Myotonia tends to diminish with continued exercise, and indeed the muscle may become almost normal clinically or electrically after repetitive testing.

Additional features include early frontal baldness, cataracts, gynecomastia, and testicular or ovarian atrophy. The disease affects numerous other systems, as evidenced by cardiac abnormalities, bowel symptoms, respiratory infections, mental disturbances, and low intelligence. Unusual susceptibility to certain medications such as barbiturates increases the risks of general anesthesia.[168] Muscle biopsy reveals type I fiber atrophy and long chains of internal nuclei.

In a distinct entity called congenital myotonic dystrophy, neuromuscular and systemic manifestations develop during the neonatal period in offspring of mildly affected mothers.[86,87,213] Some of these hypotonic infants may or may not have evidence of clinical or electrical myotonia until the age of 5 years or later. Weakness produces a triangular mouth in which the upper lip points upward in the middle. Many children have mental retardation, clubfeet, and diaphragmatic elevation.[29,48] Infants frequently die of respiratory infections. Curiously, congenital myotonia rarely shows paternal inheritance, appearing nearly always in children born to myotonic mothers. Some biopsied specimens have revealed severe deficiency in type IIB fibers.[7] This may develop consequent to lasting myotonic activity, rather than genetic factors.[88]

Electromyography shows myotonic discharges (see Fig. 13–6) in all affected adults and approximately one half of relatives at risk for myotonic dystrophy.[162] In 25 patients from 15 different families,[206] electrical myotonia occurred most frequently in the intrinsic hand muscles and orbicularis oculi, less commonly in tibialis anterior and extensor digitorum muscles, and least frequently in proximal and paraspinal muscles. In adults, the test helps to determine whether a patient with mild distal weakness and atrophy has myotonic dystrophy.[205] Patients with the partial syndrome, however, lack clinical or electromyographic evidence of myotonia.[165] During infancy and early childhood, patients may[213] or may not have characteristic electrical or clinical myotonic phenomena.[54] Electromyography may show a myopathic process with low-amplitude, short-duration, polyphasic motor unit potentials.[36]

Neurogenic features as part of the generalized membrane abnormality include (1) mildly slowed motor nerve conduction velocities,[13,41,98,127,157,175,176] (2) striking reduction in the number of functioning motor units,[12,98,131,132] and (3) occasional hypertrophy of peripheral nerves.[28] The severity of the neuropathic changes does not correlate with the degree of muscular atrophy and weakness.[157] Peripheral nerve morphometry has shown no detectable abnormality in the cutaneous branches of the common peroneal nerve.[163]

Myotonia Congenita

Genetic and clinical characteristics distinguish two different varieties of myotonia congenita. The type originally described by Thomsen[217] in four generations of his own family shows an autosomal dominant trait with equal involvement of both sexes. Myotonia appears in infancy or early childhood but remains mild throughout life. Occasional asymptomatic patients with electromyographic evidence of myotonic discharge may represent sporadic cases of Thomsen's disease. The second, more common type described by Becker[17,20-22] appears in an autosomal recessive fashion but affects men more frequently than women. More severe myotonia develops in the recessive type, although the two varieties otherwise share similar clinical features.[107,211,236]

In a third, rare type of myotonia congenita, the patient may have, in addition to myotonia, painful muscle cramps induced by exercise.[19] In this type, painful muscle stiffness may be provoked by fasting or oral potassium administration and relieved by carbohydrate-containing foods. In some patients, acetazolamide alleviates myotonia dramatically.[220] A contracted muscle shows electrical silence, that is, a contracture, probably resulting from some defect of muscle metabolism.[186,204]

Myotonia often predominates in the lower extremities, causing difficulty in ambulation. Movements begin slowly and with difficulty, especially after prolonged rest. Although motor function improves to a normal level with continued exercise, this warm-up phenomenon induces no systemic effect. Thus, repetitive contraction of one set of muscles does not limber up another set of adjacent muscles. Despite the apparent weakness, muscle power returns to normal once myotonia disappears. Children commonly show retardation of motor development. In some patients muscular hypertrophy develops as a result of continuous involuntary exercise. Their herculean appearance stands in striking contrast to the muscular wasting in myotonic dystrophy. This degree of hypertrophy, however, does not appear as commonly as previously publicized. The disease affects no other systems, allowing the patient to have a normal life expectancy.

Diagnosis depends on family history and clinical features, including readily demonstrable percussion myotonia. In equivocal cases, exposure to cold serves as a useful provocative test. Muscle biopsy reveals the absence of type IIB fibers and the presence of internal nuclei, although to a lesser extent than in myotonic dystrophy.[50] A spin-labeled study has shown increased fluidity of the erythrocyte membrane in myotonia congenita, as compared with myotonic dystrophy.[40]

Electromyography plays an important role in establishing the diagnosis of myotonia. Repetitive nerve stimulation may cause progressive decline in successive evoked muscle action potentials as a result of increased muscle fiber refractoriness (see Chapters 9.8 and 24.6). Unlike in myasthenia gravis, the decrementing tendency continues toward the end of a train, a faster rate of stimulation producing a greater change. This phenomenon occurs in any type of myotonic disorder, but particularly in the Becker variety.[5]

Paramyotonia Congenita

Paramyotonia congenita of Eulenberg,[64] transmitted by a single autosomal dominant gene, affects both sexes equally.[18,21,22,170,218] The symptoms begin at birth or in early childhood, showing no improvement with age. Paradoxically, the myotonia intensifies, rather than remits, with exercise.[85] When exposed to cold, the patient may develop stiffness of the tongue, eyelids, face, and limb muscles. Electrical discharges disappear with cooling, despite increasing muscular stiffness.[147,232] Thus, the cold-induced rigidity may not represent true myotonia. The disorder closely resembles hyperkalemic periodic paralysis. Attacks of flaccid weakness accompanied by myotonia resemble the spells of periodic paralysis. In various members of the same family, intermittent paralysis may occur without myotonia, or vice versa. Laboratory findings include elevated or high normal levels of serum potassium.

Electromyography shows evidence of myotonia and, in some fibrillation potentials on cooling.[85] The compound muscle action potential shows a steady decrement in response to repetitive nerve stimulation.[38,39] Cold induces a substantial fall in amplitude of the compound muscle action potential, worsens decremental response, and virtually abolishes myotonic discharges and voluntary recruitment of motor unit potentials.[209] Stimulation of the nerve shows normal conduction between attacks but fails to elicit muscle action potentials during episodes of paralysis.

3 PERIODIC PARALYSIS

Periodic paralysis results from reversible inexcitability of muscle membrane. Traditional classification distinguishes hypokalemic, hyperkalemic, and normokalemic types on the basis of the serum level of potassium during a paralytic attack. All three categories share a number of clinical features, and changes in serum potassium show no direct cause-and-effect relationship to paralytic events. Indeed, episodes of weakness associated with either hypokalemia or hyperkalmia can occur in a given individual.[46]

More recent classification divides the periodic paralyses into primary hereditary and secondary acquired types.[178] Primary hereditary types consist of hypokalemic periodic paralysis and po-

tassium-sensitive hyperkalemic or normokalemic periodic paralysis. The hyperkalemic or normokalemic type, often accompanied by myotonia, bears great resemblance to paramyotonia congenita. The secondary acquired types include thyrotoxic hypokalemic periodic paralysis, acute or chronic potassium depletion and retention, hypokalemia caused by renal tubular acidosis,[24] and chronic hypernatremia.[122]

During an attack of periodic paralysis direct or indirect stimulation fails to excite the muscle membrane.[33,82,90,198] An end-plate potential persists during the paralytic episodes, but action potentials cease to propagate along the muscle fibers.[84] In the hypokalemic types application of calcium induces normal contraction in the muscle fibers stripped of their outer membranes.[61] Thus, inexcitability must result from dysfunction of muscle membrane, rather than the contractile elements. An important finding common to hypokalemic[90,172] and hyperkalemic periodic paralysis[33] is substantial depolarization of the resting membrane potential, presumably reflecting increased sodium conductance with normal potassium and chloride conductance.[90,156] These observations suggest that persistent inactivation of sodium channels leads to muscle fiber inexcitability at least in hypokalemic periodic paralysis. Interestingly, tetrodotoxin, a sodium channel blocker, cannot reverse the depolarization block.

Hypokalemic Periodic Paralysis

This familial disease, inherited as an autosomal dominant, affects men more than women.[62,75,76] Although variable in onset, episodes of paralysis typically begin in the second decade. During an attack, weakness starts in the legs and gradually spreads to involve all the muscles of the body, with the exception of the ocular muscles, diaphragm, and other respiratory muscles. The episodes characteristically occur after rest, especially on waking in the morning. A heavy carbohydrate meal may precipitate the attack. Each paralytic episode, which may immobilize the patient totally, lasts several hours to a day, but a few days may elapse before complete recovery. These attacks vary in frequency and severity but tend to remit after 35 years of age. Eyelid myotonia, originally described in the hyperkalemic type of periodic paralysis, may also appear in the hypokalemic variety.[169]

Administration of potassium chloride relieves the paralysis. Acetazolamide, which usually prevents paralytic attacks, may worsen the episode in some patients, perhaps because of its kaliopenic effect.[219] Between attacks, the patient has neither clinical nor electrophysiologic abnormalities, except for the occasional development of progressive myopathy.[56,62,160] Hypokalemic myopathy may also result from other conditions associated with potassium loss.[174,179,180] The light microscope reveals few structural abnormalities. Electron-microscopic studies, however, show vacuoles arising from local dilatation of the transverse tubules and sarcoplasmic reticulum.[59]

Electrophysiologic studies during severe paralytic episodes show a reduction in the number of voluntarily recruited motor unit potentials and decreased muscle excitability. Thus, electrical stimulation of the nerve elicits no muscle action potentials. Less severe cases show decreased amplitude of the compound muscle action potentials in proportion to the degree of weakness.[82] Repetitive nerve stimulation at a rate of 10 to 25 per second may produce an incremental response in mildly affected muscles[84] but no change in very weak muscles.[43] Analogous to electrical recovery with repetitive stimulation, muscle strength improves temporarily after gentle exercise, followed by severe rebound weakness.

Hyperkalemic Periodic Paralysis

In this autosomal dominant disorder, which affects the two sexes equally, episodes of flaccid weakness accompany elevated serum potassium.[30,75,76,112,130] The disease begins in infancy or early childhood with spells of generalized hypotonia. Sudden weakness develops after a short period of rest following exercise, upon exposure to cold or after the administration of potassium. Further exercise or administration of carbohydrates temporarily delays what eventually becomes a more severe attack. Paralysis usually lasts less than 1 hour. Weakness probably results from muscle release of potassium, rather than from the high serum level. Myotonia commonly involves the muscles of the face, eyes, and tongue. This finding suggests some linkage between hyperkalemic periodic paralysis, or adynamia episodica hereditaria, and paramyotonia congenita.[52,72,112] Both entities may appear in a single family.

Between attacks, electromyography may reveal only increased insertional activity or show myotonic potentials and complex repetitive discharges. During a paralytic episode, muscle irritability and myotonic discharges increase, although electrical or mechanical stimulation fails to excite the muscle. An abundance of low-amplitude, short-duration motor unit potentials and early recruitment suggest progressive myopathy. In the presence of prominent myotonia, repetitive nerve stimulation may cause a decrement of the evoked muscle action potentials,[120] a tendency accentuated by cooling.[171]

The physiologic mechanism underlying episodic paralysis, though unknown,[116] centers on reduced muscle membrane potential at rest,[49] reversible depolarization during the attacks,[30,33,37] and neural hyperexcitability.[194] Sustained immobility reduces the amplitude and area of electrically elicited compound muscle action potential, with maximal effect occurring after 30 minutes. Prior intense muscle exercise may accentuate this to some degree. It appears to be the electrophysiologic correlate of the characteristic symptom of weakness induced by rest after exercise.[210]

Normokalemic Periodic Paralysis

This very rare condition also seems to have some enigmatic relationship to potassium. Only a few reports have appeared since the original account[164] describing attacks of flaccid quadriplegia in infancy with normal serum levels of potassium. The clinical features closely resemble those of hyperkalemic periodic paralysis,[136] of which the normokalemic type may be a variant.

4 NEUROMYOTONIA

Isaacs[94] originally described two patients with progressive painless stiffness and rigidity of the trunk and extremities. Subsequent authors referred to this entity either as the Isaacs syndrome or, more descriptively, as continuous muscle fiber activity,[95,96,119,121,184,223] neuromyotonia,[135] or neurotonia.[228] Still others used the now abandoned term pseudomyotonia[93,199] to distinguish persistent muscle activity of peripheral nerve origin from true myotonia, which represents disorders of the muscle membrane.

The disease usually appears sporadically, although a few reports describe hereditary forms of sustained muscle activity,[8,10] some occurring in association with neuronal type of Charcot-Marie-Tooth disease.[225] Symptoms begin at any age, although rarely in the neonatal period.[26] In myotonia, abnormal muscle activity occurs only after voluntary or induced muscle contraction. In contrast, patients with neuromyotonia suffer from sustained or repetitive spontaneous activity of the muscle fibers. In addition, the affected muscles stiffen and fail to relax completely after voluntary contraction. The patient has reduced or absent muscle stretch reflexes.

In milder forms of the syndrome, the abnormal activity appears restricted in degree and distribution.[16,187] The patient initially notices muscle twitching, especially in the legs. Asynchronous contraction of single or multiple motor units may produce generalized myokymia.[8] In a severe form, continuous and excessive muscle contraction may give rise to abnormal posture and rigid arms with the wrist flexed and the fingers extended. The patient moves slowly and deliberately, as if imitating a slow-motion picture. Stiffness seems to vary from one movement to the next. Excessive sweating occurs, probably as the result of continuous muscle activity. Laryngeal spasm may develop.[97,117] The patient may have an increased level of gamma-aminobutyric acid in the cerebrospinal fluid.[184]

Electromyography reveals characteristic spontaneous discharges firing rhythmically and continuously in all involved muscle groups. Waveforms of varying configuration appear usually, though not always, at high frequencies up to 300 impulses per second, representing either motor unit or single fiber discharges. A marked decrement in successive amplitude results from inability of the motor unit to follow rapidly recurring nerve impulses. This high-frequency, decrementing discharge produces unique musical sounds, "pings," which differ from other spontaneous potentials including myotonic discharge.[109] During voluntary contraction, many motor units fire successively with overlap. Artificially induced ischemia or electrical stimulation of the nerve may abruptly initiate the spontaneous discharge.

The motor activity persists during sleep, during general or spinal anesthesia, or after procaine block of the peripheral nerve.[26,94,96] Local administration of curare abolishes the activity. Other therapeutic agents include diphenylhydantoin and carbamazepine with beneficial effects in most,[8,10,93,94,96,227] but not all, patients.[16] Microelectrode studies of endplate potentials in an intercostal muscle biopsy have demonstrated normal miniature end-plate potentials and no evidence of quantal squander.[109] The peripheral nerve may show conduction abnormalities.[26,223,233] These findings suggest that the high-frequency discharge originates at the distal motor axon. Members of a family may demonstrate hyperexcitability of motor and sensory neurons.[111] Repetitive afterdischarges may follow each motor nerve stimulus.[10]

5 SCHWARTZ-JAMPEL SYNDROME

Continuous muscle fiber activity occurs in osteochondromuscular dystrophy of autosomal recessive inheritance, originally described by Schwartz and Jampel.[193] The characteristic clinical features include short stature, muscular hypertrophy, diffuse bone disease, ocular and facial anomalies, and severe voluntary and percussion myotonia.[1,70,105,159,181] The muscle biopsy may reveal myopathic and neurogenic features.[66] The defect responsible for the continuous muscle contraction presumably lies in the terminal axons, but it may also involve the muscle component of the neuromuscular junction.[193,214]

Electromyographic studies reveal complex repetitive patterns that resemble the continuous discharge seen in neuromyotonia. Unlike myotonia, the repetitive high-frequency discharges are sustained without waxing or waning. They persist after nerve block or even nerve degeneration. Most, but not all, of the spontaneous activity disappears after administration of curare[214] or succinylcholine.[42] Other features reported include increased insertional activity and absence of the silent period after muscle contraction.

6 MYOKYMIA

Myokymia, first introduced to describe the condition of a patient with leg cramps,[192] initially referred to spontaneous muscle contractions of the calves, thighs, chest, and arms. Others have used the term to include delayed muscle relaxation associated with continuous spontaneous motor unit discharges[77] or, more broadly, manifestation of benign neuromuscular irritability.[80]

Different authors have since applied the name to muscle twitches in a variety of conditions, including lead poisoning, thyrotoxicosis, scleroderma, systemic infections, intoxications, and spinal cord lesions.[187,227] Myokymia of the superior oblique muscle may cause microtremor of the globe, causing oscillopsia.[31,212] Generalized myokymia with impaired muscle relaxation may develop in association with the syndromes of continuous muscle fiber activity,[97,187] restless leg syndrome,[92] and peripheral neuropathy.[233] Thus, myokymia occurs in a heterogeneous group of disorders and probably represents a nonspecific neuronal response to injury. In most limb myokymia, discharges arise focally at the site of a chronic peripheral nerve lesion.[3,4] In one study, carbamazepine (Tegretol) led to nearly total symptomatic relief.[158]

According to the current usage, myokymia has a distinctive clinical appearance and association with certain neurologic disorders. In this entity, spontaneous repetitive contraction involves narrow muscle bands for several seconds. Each segment of muscle, 1 to 2 cm in width, slowly contracts along the longitudinal axis. Independent irregular undulations along different strips give rise to the appearance of a cutaneous "race of worms."

Whereas electromyographic abnormalities vary slightly from one patient to another, the prolonged undulating movements of myokymia all seem to result from brief tetanic contractions of repetitively discharging single or multiple motor units.[151] Interestingly, however, myokymic discharges usually occur alone, without concomitant fibrillation potentials or positive sharp waves. They do not typically wax or wane, despite occasional association with myotonia.[77,222] Neither the clinical myokymia nor the electrical counterpart changes substantially with sleep, volitional movement, rest, percussion, electrical stimulation, or needle movement.[93,158] Xylocaine infusion of a peripheral nerve trunk blocks myokymic discharges.

Facial myokymia[110] typically suggests multiple sclerosis[89] (see Fig. 13–9A) or pontine glioma,[45,215] but also appears after Bell's palsy, polyradiculoneuropathy,[51,81,224,230] (see Fig. 13–9B), or cardiopulmonary arrest.[142] Metastatic tumor that interrupts the supranuclear pathways descending upon the facial nucleus may also give rise to myokymia.[195,231]

Two electromyographic patterns characterize facial myokymic discharges.[166] In the continuous type, rhythmic single or paired discharges of one or a few motor units recur with striking regularity at intervals of 100 to 200 ms. In the discontinuous type, bursts of a single motor unit activity at 30 to 40 impulses per second last for 100 to 900 ms and repeat in regular intervals of 100 ms to 10 seconds (see Fig. 13–9). Although the continuous type tends to suggest multiple sclerosis, and the discontinuous type, brainstem glioma, neither pattern provides exceptions to this clinical specificity.

7 HEMIFACIAL SPASM

Hemifacial spasm may develop spontaneously or as a late complication of Bell's palsy or other disorders of the facial nerve, including compression of the brainstem by a contralateral tumor.[150] Unlike focal convulsive twitches of the face, the spasmodic contractions that often follow blinking consist of simultaneous rapid twitching in several facial muscles. Less commonly, one side of the face may show prolonged contraction with irregular, fluctuating movements. Spontaneous movement of this type nearly exclusively involves the facial muscles, although masticatory spasm may develop in association with facial hemiatrophy.[103,216] Idopathic hemifacial spasms typically occur in middle age, affecting women more than men. Involuntary twitching ordinarily begins in the upper or lower eyelid, spreading gradually to involve the remainder of the orbicularis oculi and other facial muscles. In advanced cases, the spasms increase in severity, frequently resulting in sustained contraction of several muscles on the affected side of face. Volitional activation of one muscle results in synchronous involuntary contraction of other muscles.

Spontaneous bursts of facial movement may result from either hyperexcitability of the facial nucleus after axonal injury[67,139,229] or ectopic excitation at the site of injury.[144,145,149,182,235] In the idiopathic type, vascular compression of the facial nerve may play an important role.[57,65,79,99,125,161] Polygraphic studies reveal progressive diminution of spasmotic movements with deepening sleep stages, revealing the lowest values in REM (rapid eye movement) sleep.[140] Central inhibitory processes may account for this partial decline. On the other hand, inhalation anesthesia, which normally abolishes R_1 and R_2 of the blink reflex, fails to suppress the reflex in hemifacial spasm.[139]

The frequency of repetitive motor unit discharges typically vary between 200 to 400 per second, although some patients have a slower irregular pattern in the range of 20 to 40 per second.[89] The diagnosis of hemifacial spasm depends on visual inspection or electromyographic recording of abnormal movements. In clinically equivocal cases, the electrically elicited blink reflex[9,104,144] may document synkinesis by demonstrating the presence of R_1 and R_2 components not only in the orbicularis oculi but also in the orbicularis oris, platysma, or other muscles innervated by the facial nerve (see Chapter 16.4, Fig. 16–9).

A number of investigators have suggested various pathophysiology underlying the hemifacial spasm. Although the published accounts lack complete accord, the evidence of ephaptic transmission (cross-talk) has gained popularity. Focal slowing secondary to demyelination constitutes an important prerequisite for ephapses in experiment with squid axons.[167] Increased latency of R_1 on the affected side of the face[44,144] provides supportive evidence for this mechanism. Some patients with idiopathic hemifacial spasm, however, have a normal R_1 latency on the involved side.[104]

In the presence of ephaptic transmission, stimulation of the individual facial nerve branches may evoke a delayed muscle response.[63] Thus, stimulation of one branch of the facial nerve and recording from muscles innervated by another branch would allow clear separation between ephaptically activated response and the direct response.[91] In one study, following stimulation of zygomatic or marginal mandibular branch of the facial nerve, simultaneous recording from the orbicularis oculi and mental muscles confirmed transmission of impulses between the two branches.[144,146] If such a lateral spread results from ephapses, the onset latency of the delayed response should equal the antidromic and orthodromic conduction to and from the presumed site of the lesion. When the response was recorded from the orbicularis oculi muscle after electrical stimulation of the marginal mandibular nerve, however, its latency exceeded the sum by a few milliseconds. This finding suggests the involvement of the facial nucleus, rather than the motor fibers in the generation of the delayed response.[139]

Stimulation of the supraorbital nerve normally activates only a fraction of the motoneuron pool destined to innervate the orbicularis oculi muscle. Thus, the size of compound muscle action potential evoked by direct stimulation of the facial nerve far exceeds that of the reflexively activated R_1. Increased amplitude of R_1 found in hemifacial spasm suggests lateral spread of the impulse activating more fibers contained in the zygomatic branch. Synkinetic responses of R_1 and R_2 in the mental muscle, not ordinarily involved in blink reflex, further support the theory of lateral spread of impulses to other fibers. The presence of afteractivity and late activity implies autoexcitation of the involved fibers.[145] All these findings, however, fail to differentiate ephaptic transmission along the motor fibers from hyperexcitability of the facial nucleus. Regardless of the type of physiologic mechanism responsible for synkinesis, the beneficial effects of surgical decompression suggest that the primary site of involvement in hemifacial spasm probably resides in the facial nerve, and not in the nucleus.[11,148] Hyperexcitability of the facial motor neurons, however, could develop secondarily as the result of a peripheral lesion.

Synkinesis found in hemifacial spasm and in some patients after Bell's palsy serves to differentiate these entities from other motor disorders, such as essential blepharospasm, facial dystonia, focal seizures, and focal myokymia. In none of these conditions does stimulation of the supraorbital nerve elicit the blink reflex in facial muscles other than orbicularis oculi. Temporal variability of responses, when present, differentiates hemifacial spasm from postparalytic synkinesis in Bell's palsy.[9]

8 TETANUS

The toxin of Clostridium tetani travels from wound to central nervous system via blood or retrograde axonal transport. After an incubation period of 1 to 2 weeks the patient develops either generalized or localized manifestations of neuromuscular irritability. These include spasm of the masticatory muscles (trismus) and facial grimacing (risus sardonicus). These symptoms may worsen within a few days but improve in several weeks, except for possible chronic manifestations of tetanic contraction. The continuous motor unit discharges seen in electromyography resolve during sleep, with administration of general or spinal anesthesia, and after peripheral nerve block.

Tetanus toxin presumably blocks postsynaptic inhibition in the spinal cord and brainstem, thereby increasing the excitability of the alpha motor neurons.[34] The shortened or absent SP probably results from failure of Renshaw inhibition. This characteristic electrodiagnostic feature of tetanus seldom occurs in other disorders with motor unit hyperactivity.[173,208] Although the exact pathophysiology awaits further clarification, the muscle spasms and rigidity almost certainly result from the effect of tetanus toxin on the central nervous system. Some clinical and electrophysiologic findings suggest peripheral nerve involvement in severe tetanus.[196] Facial nerve conduction studies may[78] or may not[221] show abnormalities. Increased jitter and block in single-fiber electromyography suggest a presynaptic defect of neuromuscular transmission in human tetanus.[68]

9 TETANY

The physiologic term tetanus is also used to describe tetany caused by hypocalcemia and alkalosis.[113] Decreased extracellular calcium increases sodium conductance, which leads to membrane depolarization and repetitive nerve firing. Hypomagnesemia and hyperkalemia also induce carpopedal spasm. Tetanic contraction abates with infusion of curare, but not with peripheral nerve block. Thus, the spontaneous discharge seems to occur at some point along the length of the peripheral nerve. Various maneuvers precipitate clinical or electrical neuromuscular irritability. They include a gentle tap over the facial nerve (Chvostek's sign) or the lateral surface of the fibula (peroneal sign) and artificially induced ischemia of the forearm (Trousseau's sign).

Electromyography reveals grouped motor unit potentials firing asynchronously at a rate of 4 to 15 per second, with periods of relative silence in between.

10 STIFF-MAN SYNDROME

A number of authors have described the clinical features of the stiff-man syndrome.[71,83,123,191] It usually occurs sporadically in adult men and women, but a congenital form also exists.[106,185] Muscle stiffness develops insidiously, progressing from tightness to painful, sustained contraction. The disease has some predilection for the pelvic and shoulder girdle muscles. The tightness of the chest muscles may interfere with breathing and swallowing. Co-contraction of agonistic and antagonistic muscles may immobilize the extremities in unnatural positions. Inversion and plantar flexion of the feet reflect the overpowering force of the posterior versus anterior calf muscles. Movement, either active or passive, aggravates the pain.

The excessive muscle contraction resembles physiologic cramps, although it involves many muscle groups simultaneously and continuously. The stiff-man syndrome may resemble hysteria because of facial grimacing, unusual posture, and complaints of muscle cramps that superficially mimic voluntary contraction. Conspicuous absence of other neurologic abnormalities may strengthen this erroneous impression. Close observation, however, reveals the pathologic nature of the powerful spasms that supersede any voluntary contraction. Indeed, fractures of the long bones have resulted. Various conditions described in association with stiff-man-like pictures include nocturnal myoclonus and epilepsy,[126] focal cortical atrophy with increased spinal fluid gammaglobulin,[123] and diffuse stiffness following ingestion of alcohol.[27]

Electromyography shows a sustained interference pattern consisting of normal motor unit potentials in agonistic as well as antagonistic muscles.[124,203] The persistent electrical activity associated with painful muscle cramps probably originates in the central nervous system. The spasm and spontaneous discharges disappear during sleep, with administration of general or spinal anesthesia, after procaine block of the peripheral nerve or after infusion of curare.[83,134] Increased central excitability leads to enhanced exteroceptive reflexes, including cutaneously elicited responses such as the blink reflex.[133]

The exact neurophysiologic mechanism underlying the abnormal discharge remains unknown. Clinical similarities with chronic tetanus suggest a possible relationship between these two entities.[83] Tetanus toxin causes hyperexcitability of motor units by blocking spinal inhibitory postsynaptic potentials. Similarly, the motor neuron pool may become excessively excitable in the absence of the inhibitory spinal mechanisms in the stiff-man syndrome.[155] Unlike those with tetanus, however, patients with the stiff-man syndrome may have a normal silent period.[124,203] Rigidity and the electrical discharges markedly improve with the administration of baclofen[137,234] or diazepam (Valium), which suppresses interneurons at spinal and supraspinal levels. In contrast, clomipramine injection severely aggravate the clinical symptoms.[133]

11 CRAMPS

Cramps represent briefly sustained, painful or painless involuntary contractions lasting seconds to minutes.[113] This definition excludes such sustained movements seen in tremor, chorea, hemiballismus or myoclonus, and isolated muscle twitches associated with fasciculation potentials or complex repetitive discharges. Painful cramps commonly involve the calf muscles and other flexors of the lower limb in healthy subjects. A cramp starts after maintaining a certain posture for a prolonged period of time and improves by rubbing or lengthening the muscle. Numerous predisposing factors include salt depletion, other causes of hyponatremia, hypocalcemia, and vitamin deficiency. Most cases of cramps in otherwise asymptomatic individuals have no detectable underlying cause.

Skeletal muscle cramps, either spontaneous or induced by ischemia or exercise, are also seen in a broad spectrum of illnesses. For example, muscle cramps constitute an early feature of motor neuron disease. They may also accompany sciatica or peripheral neuropathies. Patients with certain inborn errors of metabolism may complain of exertional cramps, but not as an essential feature. The syndrome of progressive muscle spasms, alopecia, and diarrhea[188,189] affects women more frequently than men. Painful intermittent cramps involve the limb muscles initially, then the neck, trunk, and mastication muscles several years later. These painful muscle spasms originate centrally and, except for normal serum calcium levels, resemble tetany. The symptoms begin at about age 10 and slowly progress, leading to malnutrition and possible death.

Cramps also occur in hereditary[101,115] and sporadic cases[92] of the muscular pain-fasciculation syndrome. The familial variety, with an autosomal dominant inheritance, affects both sexes. The symptoms appear during the first or second decade. Exercise-induced painful cramps, although generalized, predominantly affect the hands and feet. Nonfamilial types also affect either sex, with onset of symptoms during the third to seventh decades. Although painful cramps primarily involve the calves, fasciculations develop in the lower extremities diffusely. The tubular aggregates reported in biopsy specimens may have some relationship with muscle cramps.[114]

Electrically, muscle cramps consist of high-frequency, irregular motor unit discharges at rates ranging from 40 to 60 per second, occasionally reaching 200 to 300 per second.[53,152] They involve a large part of the muscle synchronously, as opposed to asynchronous activation of motor units during voluntary muscle contraction. Despite effective inhibition

by nerve block or spinal anesthesia, repetitive nerve stimulation distal to the block still induces cramping. These findings suggest a peripheral origin. In sporadic cases of muscular pain-fasciculation syndrome, nerve conduction studies may show decreased conduction and increased distal latencies. Needle studies may reveal fibrillation potentials and positive sharp waves. In one study, patients with familial cramps had fasciculation potentials, high-amplitude, long-duration, polyphasic motor unit potentials, and low normal nerve conduction velocities.[115] Transcutaneous nerve stimulation may relieve severe muscle cramps, as reported in a patient with muscle hypertrophy and fasciculation potentials whose cramps disappeared during sleep.[138]

12 CONTRACTURE

The term contracture refers to intense mechanical muscle shortening in the absence of muscle action potentials. Thus, electromyography reveals no electrical activity in the contracted muscle.[55] Ischemia induces contracture most commonly in patients with muscle phosphorylase (see Fig. 11–3) or muscle phophofructokinase deficiencies, but also rarely in those with other conditions.[114,129] In these entities, failure to produce adenosine triphosphate (ATP) prohibits reaccumulation of calcium by the sarcoplasmic reticulum, the essential initial step for muscle relaxation (see Chapter 11.2).

Painless exertional contracture may occur in some patients without enzymatic deficiency.[32] In this entity, caused by a deficiency of calcium-ATPase, sarcoplasmic reticulum has a decreased capacity to accumulate calcium.[102]

Electromyography shows normal activity during voluntary muscle contraction. After strong effort, however, the muscle relaxes only slowly, over a period of 10 seconds. During this period the stiff muscle is electrically silent.[108] Normal motor unit potentials reappear if the patient voluntarily contracts the stiff muscle. Needle insertion or voluntary contraction initiates no myotonic discharge. Painful contracture also occurs in a hereditary myopathy associated with electromyographic signs of generalized myotonia.[186,204]

13 MYOCLONUS

Electrophysiologic techniques are useful in characterizing cortical myoclonus. Abnormal sensory motor cortical discharge can cause a wide range of clinical motor phenomena.[153] Brief muscle jerks probably result from cerebral cortical mechanisms that also give rise to enlarged somatosensory and visual evoked potentials. Premotor cortical potentials, time-locked to the spontaneous or action-induced jerking, precede the muscle activity recorded in electromyography. The side of abnormality in the sensory motor cortex may dictate the varied pattern of motor responses, which includes stimulus-sensitive myoclonus, spontaneous myoclonus, and forced motor epilepsy. Jerk-locked somatosensory evoked potential (SEP) studies reveal relatively enhanced cortical excitability for 20 ms just after the myoclonus, followed by suppression throughout the postmyoclonus period.[197] Similar waveforms and scalp topography suggest that the giant SEPs and myoclonus-related cortical spikes may have common physiologic mechanism.[197] Mutual antagonism between physostigmine and anticholinergic agents in myoclonus implies cholinergic hyperactivity in the pathophysiology of myoclonus.[47]

14 OTHER ABNORMAL MOVEMENTS

Tremor

Early essential tremor qualitatively resembles the 8- to 12-Hz component of physiologic tremor, but advanced essential tremor has a frequency of 4 to 8 Hz.[58] Tremor associated with peripheral neuropathy results from minimal weakness and possibly impairment of the stretch reflex, both of which increase central drive and enhance physiologic tremor.[183] Delayed and enhanced long-latency reflexes may cause postural tremor in late cerebellar atrophy.[128]

Mirror Movement

In congenital mirror movements, electromyographic study shows normal temporal characteristics, response latency, duration, and recruitment pattern on the normal and mirror sides. These findings suggest similar motor command for both voluntary and mirror movements.[69]

Restless Legs Syndrome

Patients with the restless leg syndrome may have periodic movements in sleep, although the frequency decreases from wakefulness to stages 1 and 2.[141]

Dystonia

Peripheral entrapment and brachial plexopathy can give rise to distal, action-induced involuntary postures of the hand with focal dystonia. Such causes of secondary dystonia would include the pronator teres syndrome, radial nerve palsy, lower brachial plexus lesions, and median nerve lesions.[190] Mechanical irritation of the brachial plexus can precipitate rhythmic myoclonus in the arm.[14]

REFERENCES

1. Aberfeld, DC, Hinterbuchner, LP, and Schneider, M: Myotonia, dwarfism, diffuse bone disease, and unusual ocular and facial abnormalities: A new syndrome. Brain 88:313–322, 1965.
2. Adrian, RH, and Bryant, SH: On the repetitive discharge in myotonic muscle fibres. J Physiol (Lond) 240:505–515, 1974.
3. Aho, K, and Sainio, K: Late irradiation-induced lesions of the lumbosacral plexus. Neurology 33:953–955, 1983.
4. Albers, JW, Allen, AA, II, Bastron, JA, and Daube, JR: Limb myokymia. Muscle Nerve 4:494–504, 1981.
5. Aminoff, MJ, Layzer RB, Satya-Murti, S, and

Faden, AI: The declining electrical response of muscle to repetitive nerve stimulation in myotonia. Neurology 27:812–816, 1977.

6. Anderson, TE, Carr, AJ, Chapman, RS, Downie, AW, and MacLean, GD: Myositis and myotonia in a case of multicentric reticulohistiocytosis. Br J Dermatol 80:39–45, 1968.

7. Argov, Z, Gardner-Medwin, D, Johnson, MA, and Mastaglia, FL: Congenital myotonic dystrophy: Fiber type abnormalities in two cases. Arch Neurol 37:693–696, 1980.

8. Ashizawa, T, Butler, IJ, Harati, Y, and Roongta, SM: A dominantly inherited syndrome with continuous motor neuron discharges. Ann Neurol 13:285–290, 1983.

9. Auger, RG: Hemifacial spasm: Clinical and electrophysiologic observations. Neurology 29:1261–1272, 1979.

10. Auger, RG, Daube, JR, Gomez, MR, and Lambert, EH: Hereditary form of sustained muscle activity of peripheral nerve origin causing generalized myokymia and muscle stiffness. Ann Neurol 15:13–21, 1984.

11. Auger, RG, Pieprgras, DG, Laws, ER, and Miller, RH: Microvascular decompression of the facial nerve for hemifacial spasm: Clinical and electrophysiologic observations. Neurology 31:346–350, 1981.

12. Ballantyne, JP, and Hansen, S: New method for the estimation of the number of motor units in a muscle. 2. Duchenne, limb-girdle and facioscapulohumeral, and mytonic muscular dystrophies. J Neurol Neurosurg Psychiatry 37:1195–1201, 1974.

13. Ballantyne, JP, and Hansen, S: Neurogenic influence in muscular dystrophies. In Rowland, LP (ed): Pathogenesis of Human Muscular Dystrophies. Proceedings of the Fifth International Scientific Conference of the Muscular Dystrophy Association. Excerpta Medica, Amsterdam, 1977, pp 187–199.

14. Banks, G, Nielsen, VK, Short, MP, and Kowal, CD: Brachial plexus myoclonus. J Neurol Neurosurg Psychiatry 48:582–584, 1985.

15. Barchi, RL: Mytonia: An evaluation of the chloride hypothesis. Arch Neurol 32:175–180, 1975.

16. Barron, SA, and Heffner, RR, Jr: Continuous muscle fiber activity: A case with unusual clinical features. Arch Neurol 36:520–521, 1979.

17. Becker, PE: Zur Genetik der Myotonien. In Kuhn, E (ed): Progressive Muskeldystrophie Myotonie, Myasthenie. Springer-Verlag, Berlin, 1966, pp 247–255.

18. Becker, PE: Fortschritte der Allegemeinen und Klinischen Humangenetik: Paramyotonia Congenita (Eulenburg), Vol III. Georg Thieme, Stuttgart, 1970.

19. Becker, PE: Genetic approaches to the nosology of muscle disease: Myotonias and similar diseases. In Bergsma, D (ed): The Second Conference on the Clinical Delineation of Birth Defects. Part VII. Muscle. Williams & Wilkins, Baltimore, 1971, pp 52–62.

20. Becker, PE: Generalized non-dystrophic myotonia. The dominant (Thomsen) type and the recently identified recessive type. In Desmedt, JE (ed): New Developments in Electromyography and Clinical Neurophysiology. Karger, Basel, 1973, pp 407–412.

21. Becker, PE: Myotonia congenita and syndromes associated with myotonia: Clinical-genetic studies of the nondystrophic myotonias. In Becker, PE, Lenz, W, Vogel, F, and Wendt, GG (eds): Topics in Human Genetics, Vol 3. Georg Thieme, Stuttgart, 1977.

22. Becker, PE: Syndromes associated with myotonia: Clinical-genetic classification. In Rowland, LP (ed): Pathogenesis of Human Muscular Dystrophies. Proceedings of the Fifth International Scientific Conerence of the Muscular Dystrophy Association. Excerpta Medica, Amsterdam, 1977, pp 699–703.

23. Belanger, AY, and McComas, AJ: Contractile properties of muscles in myotonic dystrophy. J Neurol Neurosurg Psychiatry 46:625–631, 1983.

24. Bennett, RH, and Forman, HR: Hypolkalemic periodic paralysis in chronic toluene exposure. Arch Neurol 37:673, 1980.

25. Bird, TD: Myotonic dystrophy associated with down syndrome (trisomy 21). Neurology 31:440–442, 1981.

26. Black, JT, Garcia-Mullin, R, Good, E, and Brown, S: Muscle rigidity in a newborn due to continuous peripheral nerve hyperactivity. Arch Neurol 27:413–425, 1972.

27. Blank, NK, Meerschaert, JR, and Rieder, MJ: Persistent motor neuron discharges of central origin present in the resting state: A case report of alcohol-induced muscle spasms. Neurology 24:277–281, 1974.

28. Borenstein, S, Noel, P, Jacquy, J, and Flament-Durand, J: Myotonic dystrophy with nerve hypertrophy: Report of a case with electrophysiological and ultrastructural study of the sural nerve. J Neurol Sci 34:87–99, 1977.

29. Bossen, EH, Shelburne, JD, and Verkauf, BS: Respiratory muscle involvement in infantile myotonic dystrophy. Arch Pathol 97:250–252, 1974.

30. Bradley, WG: Adynamia episodica hereditaria: Clinical, pathological and electrophysiological studies in an affected family. Brain 92:345–378, 1969.

31. Breen, LA, Butmann, L, and Riggs, JE: Superior oblique myokymia: A misnomer. J Clin Neuroopthalmol 3:131–132, 1983.

32. Brody, IA: Muscle contracture induced by exercise: A syndrome attributable to decreased relaxing factor. N Engl J Med 281:187–192, 1969.

33. Brooks, JE: Hyperkalemic periodic paralysis. Intracellular electromyographic studies. Arch Neurol 20:13–18, 1969.

34. Brooks, VB, Curtis, DR, and Eccles, JC: The action of tetanus toxin on the inhibition of motoneurones. J Physiol (Lond) 135:655–672, 1957.

35. Bryant, SH: The physiological basis of myotonia. In Rowland, LP (ed): Pathogenesis of Human Muscular Dystrophies. Proceedings of the Fifth International Scientific Conference

of the Muscular Dystrophy Association. Excerpta Medica, Amsterdam, 1977, pp 715–728.

36. Buchthal, F: Diagnostic significance of the myopathic EMG. In Rowland, LP (ed): Pathogenesis of Human Muscular Dystrophies. Proceedings of the Fifth International Conference of the Muscular Dystrophy Association. Excerpta Medica, Amsterdam, 1977, pp 205–218.

37. Buchthal, F, Engbaek, L, and Gamstorp, I: Paresis and hyperexcitability in adynamia episodica hereditaria. Neurology 8:347–351, 1958.

38. Burke, D, Skuse, NF, and Lethlean, AK: Contractile properties of the abductor digiti minimi muscle in paramyotonia congenita. J Neurol Neurosurg Psychiatry 37:894–899, 1974.

39. Burke, D, Skuse, NF, and Lethlean, AK: An analysis of myotonia in paramyotonia congenita. J Neurol Neurosurg Psychiatry 37:900–906, 1974.

40. Butterfield, DA, Chestnut, DB, Appel, SH, and Roses, AD: Spin label study of erythrocyte membrane fluidity in mytonic and Duchenne muscular dystrophy and congenital myotonia. Nature 263:159–161, 1976.

41. Caccia, MR, Negri, S, and Parvis, VP: Myotonic dystrophy with neural involvement. J Neurol Sci 16:253–269, 1972.

42. Cadilhac, J, Baldet, D, Greze, JA, and Duday, H: EMG studies of two family cases of the Schwartz and Jampel syndrome (osteo-chondro-muscular dystrophy with myotonia). Electromyogr Clin Neurophysiol 15:5–12, 1975.

43. Campa, JF, and Sanders, DB: Familial hypokalemic periodic paralysis: Local recovery after nerve stimulation. Arch Neurol 31:110–115, 1974.

44. Carceni, T, and Negri, S: Le reflexe de clignement dans l'hemispasme facial. Primitif: Considérations electromyographiques. Electromyogr Clin Neurophysiol 12:85–89, 1972.

45. Cherington, M, Sadler, KM, and Ryan, DW: Facial myokymia. Surg Neurol 11:478–480, 1979.

46. Chesson, AL, Jr, Schochet, SS, Jr, and Peters, BH: Biphasic periodic paralysis. Arch Neurol 36:700–704, 1979.

47. Chokroverty, S, Manocha, MK, and Duvoisin, RC: A physiologic and pharmacologic study in anticholinergic-responsive essential myoclonus. Neurology 37:608–615, 1987.

48. Chudley, AE, and Barmada, MA: Diaphragmatic elevation in neonatal myotonic dystrophy. Am J Dis Child 133:1182–1185, 1979.

49. Creutzfeldt, OD, Abbott, BC, Fowler, WM, and Pearson, CM: Muscle membrane potentials in episodic adynamia. Electroencephalogr Clin Neurophysiol 15:508–519, 1963.

50. Crews, J, Kaiser, KK, and Brooke, MH: Muscle pathology of myotonia congenita. J Neurol Sci 28:449–457, 1976.

51. Daube, JR, Kelly, JJ, Jr, and Martin, RA: Facial myokymia with polyradiculoneuropathy. Neurology 29:662–669, 1979.

52. Delwaide, PJ, and Penders, CA: Paramyotonie familiale et crises parétiques avec hypokaliémie. Rev Neurol 125:287–298, 1971.

53. Denny-Brown, D, and Foley, JM: Myokymia and the benign fasciculation of muscular cramps. Trans Assoc Am Phys 61:88–96, 1948.

54. Dodge, PR, Gamstorp, I, Byers, RK, and Russell, P: Myotonic dystrophy in infancy and childhood. Pediatrics 35:3–19, 1965.

55. Dyken, ML, Smith, DM, and Peake, RL: An electromyographic diagnostic screening test in McArdle's disease and a case report. Neurology 17:45–50, 1967.

56. Dyken, M, Zeman, W, and Rusche, T: Hypokalemic periodic paralysis. Children with permanent myopathic weakness. Neurology 19:691–699, 1969.

57. Eidelman, BH, Nielsen, VK, Moller, M, and Jannetta, PJ: Vascular compression, hemifacial spasm and multiple cranial neuropathy. Neurology 35:712–716, 1985.

58. Elble, RJ: Physiologic and essential tremor. Neurology 36:225–231, 1986.

59. Engel, AG: Evolution and content of vacuoles in primary hypokalemic periodic paralysis. Mayo Clin Proc 45:774–814, 1970.

60. Engel, AG, Gomez, MR, Seybold, ME, and Lambert, EH: The spectrum and diagnosis of acid maltase deficiency. Neurology 23:95–106, 1973.

61. Engel, AG, and Lambert, EH: Calcium activation of electrically inexcitable muscle fibers in primary hypokalemic periodic paralysis. Neurology 19:851–858, 1969.

62. Engel, AG, Lambert, EH, Rosevear, JW, and Tauxe, WM: Clinical and electromyographic studies in a patient with primary hypokalemic periodic paralysis. Am J Med 38:626–640, 1965.

63. Esslen, E: Der spasmus facialis-eine parabioseerscheinung. Dtsch Nervenheilk 176:149–172, 1959.

64. Eulenburg, A: Ueber eine familiäre, durch 6 Generationen verfolgbare Form congenitaler Paramyotonie. Neurologisches Centralblatt 5:265–272, 1886.

65. Fabinyi, GCA, and Adams, CBT: Hemifacial spasm: Treatment by posterior fossa surgery. J Neurol Neurosurg Psychiatry 41:829–833, 1978.

66. Fariello, R, Meloff, K, Murphy, EG, Reilly, BJ, and Armstrong, D: A case of Schwartz-Jampel syndrome with unusual muscle biopsy findings. Ann Neurol 3:93–96, 1978.

67. Ferguson, JH: Hemifacial spasm and the facial nucleus. Ann Neurol 4:97–103, 1978.

68. Fernandez, JM, Ferrandiz, M, Larrea, L, Ramio, R, and Boada, M: Cephalic tetanus studied with single fibre EMG. J Neurol Neurosurg Psychiatry 46:862–866, 1983.

69. Forget, R, Boghen, D, Attig, E, and Lamarre, Y: Electromyographic studies of congenital mirror movements. Neurology 36:1316–1322, 1986.

70. Fowler, WM, Jr, Layzer, RB, Taylor, RG, Eberle, ED, Sims, GE, Munsat, TL, Philppart,

M, and Ilson, BW: The Schwartz-Jampel syndrome: Its clinical, physiological and histological expressions. J Neurol Sci 22:127–146, 1974.

71. Franck, G, Cornette, M, Grisar, T, Moonen, G, and Gerebtzoff, MA: Le syndrome de l'homme raide: Étude clinique, polygraphique et histoenzymologique. Acta Neurol Belg 74:221–240, 1974.

72. French, EB, and Kilpatrick, R: A variety of paramyotonia congenita. J Neurol Neurosurg Psychiatry 20:40–46, 1957.

73. Furman, RE, and Barchi, RL: The pathophysiology of myotonia produced by aromatic carboxylic acids. Ann Neurol 4:357–365, 1978.

74. Gaffney, BJ, Drachman, DB, Lin, DC, and Tennekoon, G: Spin-label studies of erythrocytes in myotonic dystrophy: No increase in membrane fluidity. Neurology 30:272–276, 1980.

75. Gamstorp, I: Adynamia episodica hereditaria. Acta Paediatrica (Suppl 108)45:43–48, 1956.

76. Gamstorp, I: A study of transient muscular weakness. Acta Neurol Scand 38:3–19, 1962.

77. Gamstorp, I, and Wohlfart, G: A syndrome characterized by myokymia, myotonia, muscular wasting and increased perspiration. Acta Psych Neurol Scand 34:181–194, 1959.

78. Garcia-Mullin, R, and Daroff, RB: Electrophysiological investigations of cephalic tetanus. J Neurol Neurosurg Psychiatry 36:296–301, 1973.

79. Gardner, WJ, and Sava, GA: Hemifacial spasm—A reversible pathophysiologic state. J Neurosurg 19:240–247, 1962.

80. Gardner-Medwin, D, and Walton, JN: Myokymia with impaired muscular relaxation. Lancet 1:127–130, 1969.

81. Gemignani, F, Juvarra, G, and Calzetti, S: Facial myokymia in the course of lymphocytic meningoradiculitis: Case report. Neurology 31:1177–1180, 1981.

82. Gordon, AM, Green, JR, and Lagunoff, D: Studies on a patient with hypokalemic familial periodic paralysis. Am J Med 48:185–195, 1970.

83. Gordon, EE, Januszko, DM, and Kaufman, L: A critical survey of stiff-man syndrome. Am J Med 42:582–599, 1967.

84. Grob, D, Johns, RJ, and Liljestrand, A: Potassium movement in patients with familial periodic paralysis: Relationship to the defect in muscle function. Am J Med 23:356–375, 1957.

85. Haass, A, Ricker, K, Rudel, R, Lehmann-Horn, F, Bohlen, R, Dengler, R, and Mertens, HG: Clinical study of paramyotonia congenita with and without myotonia in a warm environment. Muscle Nerve 4:388–395, 1981.

86. Harper, PS: Congenital myotonic dystrophy in Britain. I. Clinical aspects. Arch Dis Child 50:505–513, 1975.

87. Harper, PS: Congenital myotonic dystrophy in Britain. II. Genetic basis. Arch Dis Child 50:514–521, 1975.

88. Heene, R: Evidence of myotonic origin of type 2B muscle fibre deficiency in myotonia and paramyotonia congenita. J Neurol Sci 76:357–359, 1986.

89. Hjorth, RJ, and Willison, RG: The electromyogram in facial myokymia and hemifacial spasm. J Neurol Sci 20:117–126, 1973.

90. Hofmann, WW, and Smith, RA: Hypokalaemic periodic paralysis studied in vitro. Brain 93:445–474, 1970.

91. Hopf, HC, and Lowitzsch, K: Hemifacial spasm: Location of the lesion by electrophysiological means. Muscle Nerve 5:S84–S88, 1982.

92. Hudson, AJ, Brown, WF, and Gilbert, JJ: The muscular pain-fasciculation syndrome. Neurology 28:1105–1109, 1978.

93. Hughes, RC, and Matthews, WB: Pseudo-myotonia and myokymia. J Neurol Neurosurg Psychiatry 32:11–14, 1969.

94. Isaacs, H: A syndrome of continuous muscle-fibre activity. J Neurol Neurosurg Psychiatry 24:319–325, 1961.

95. Isaacs, H: Continuous muscle fibre activity in an Indian male with additional evidence of terminal motor fibre abnormality. J Neurol Neurosurg Psychiatry 30:126–133, 1967.

96. Isaacs, H, and Heffron, JJA: The syndrome of "continuous muscle-fibre activity" cured: Further studies. J Neurol Neurosurg Psychiatry 37:1231–1235, 1974.

97. Jackson, DL, Satya-Murti, S, Davis, L, and Drachman, B: Isaacs syndrome with laryngeal involvement: An unusual presentation of myokymia. Neurology 29:1612–1615, 1979.

98. Jamal, GA, Weir, AI, Hansen, S, and Ballantyne, JP: Myotonic dystrophy: A reassessment by conventional and more recently introduced neurophysiological techniques. Brain 109:1279–1296, 1986.

99. Jannetta, PJ, Abbasy, M, Maroon, JC, Ramos, FM, and Albin, MS: Etiology and definitive microsurgical treatment of hemifacial spasm: Operative techniques and results in 47 patients. J Neurosurg 47:321–328, 1977.

100. Juguilon, A, Chad, D, Bradley, WG, Adelman, L, Kelemen, J, Bosch, P, and Munsat, TL: Familial granulovacuolor lobular myopathy with electrical myotonia. J Neurol Sci 56:133–140, 1982.

101. Jusic, A, Dogan, S, and Stojanovic, V: Hereditary persistent distal cramps. J Neurol Neurosurg Psychiatry 35:379–384, 1972.

102. Karpati, G, Charuk, J, Carpenter, S, Jablecki, C, and Holland, P: Myopathy caused by a deficiency of Ca^{2+}-adenosine triphosphatase in sarcoplasmic reticulum (Brody's disease). Ann Neurol 20:38–49, 1986.

103. Kaufman, MD: Masticatory spasm in facial hemiatrophy. Ann Neurol 7:585–587, 1980.

104. Kimura, J, Rodnitzky, RL, and Okawara, SH: Electrophysiologic analysis of aberrant regeneration after facial nerve paralysis. Neurology 25:989–993, 1975.

105. Kirschner, BS, and Pachman, LM: IgA deficiency and recurrent pneumonia in the Schwartz-Jampel syndrome. J Pediatrics 88:1060–1061, 1976.

106. Klein, R, Haddow, JE, and Deluca, C: Familial

congenital disorder resembling stiff-man syndrome. Am J Dis Child 124:730–731, 1972.

107. Kuhn, E, Fiehn, W, Seiler, D, and Schroeder, JM: The autosomal recessive (Becker) form of myotonia congenita. Muscle Nerve 2:109–117, 1979.

108. Lambert, EH: Neurophysiological techniques useful in the study of neuromuscular disorders. In Adams, RD, Eaton, LM, and Shy, GM (eds): Neuromuscular Disorders, Vol 38. Williams & Wilkins, Baltimore, 1960, pp 247–273.

109. Lambert, EH: Muscle spasms, cramps, and stiffness. American Academy of Neurology Special Course #17, 1978.

110. Lambert, EH, Love, JG, and Mulder, DW: Facial myokymia and brain tumor: Electromyographic studies. American Association of Electromyography and Electrodiagnosis Newsletter 8:8, 1961.

111. Lance, JW, Burke, D, and Pollard, J: Hyperexcitability of motor and sensory neurons in neuromyotonia. Ann Neurol 5:523–532, 1979.

112. Layzer, RB, Lovelace, RE, and Rowland, LP: Hyperkalemic periodic paralysis. Arch Neurol 16:455–472, 1967.

113. Layzer, RB, and Rowland, LP: Cramps. N Engl J Med 285:31–40, 1971.

114. Lazaro, RP, Fenichel, GM, Kilroy, AW, Saito, A, and Fleischer, S: Cramps, muscle pain, and tubular aggregates. Arch Neurol 37:715–717, 1980.

115. Lazaro, RP, Rollinson, RD, and Fenichel, GM: Familial cramps and muscle pain. Arch Neurol 38:22–24, 1981.

116. Lehmann-Horn, F, Rudel, R, Ricker, K, Lorkovic, H, Dengler, R, and Hopf, HC: Two cases of adynamia episodica hereditaria: in vitro investigation of muscle cell membrane and contraction parameters. Muscle Nerve 6:113–121, 1983.

117. Levinson, S, Canalis, RF, and Kaplan, HJ: Laryngeal spasm complicating pseudomyotonia. Arch Otolaryngol 102:185–187, 1976.

118. Lipicky, RJ: Studies in human myotonic dystrophy. In Rowland, LP (ed): Pathogenesis of Human Muscular Dystrophies. Proceedings of the Fifth International Scientific Conference of the Muscular Dystrophy Association. Excerpta Medica, Amsterdam, 1977, pp 729–738.

119. Lublin, FD, Tsairis, P, Streletz, LJ, Chambers, RA, Riker, WF, Van Pozank, A, and Duckett, SW: Myokymia and impaired muscular relaxation with continuous motor unit activity. J Neurol Neurosurg Psychiatry 42:557–562, 1979.

120. Lundberg, PO, Stalberg, E, and Thiele, B: Paralysis periodica paramyotonica: A clinical and neurophysiological study. J Neurol Sci 21:309–321, 1974.

121. Lutschg, J, Jerusalem, F, Ludin, HP, Vassella, F, and Mumenthaler, M: The syndrome of "continuous muscle fiber activity." Arch Neurol 35:198–205, 1978.

122. Maddy, JA, and Winternitz, WW: Hypothalamic syndrome with hypernatremia and muscular paralysis. Am J Med 51:394–402, 1971.

123. Maida, E, Reisner, T, Summer, K, and Sandro-Eggerth, H: Stiff-man syndrome with abnormalities in CSF and computerized tomography findings: Report of a case. Arch Neurol 37:182–183, 1980.

124. Mamoli, B, Heiss, WD, Maida, E, and Podreka, I: Electrophysiological studies on the "stiff-man" syndrome. J Neurol 217:111–121, 1977.

125. Maroon, JC: Hemifacial spasm: A vascular cause. Arch Neurol 35:481–483, 1978.

126. Martinelli, P, Pazzaglia, P, Montagna, P, Coccagna, G, Rizzuto, N, Simonati, S, and Lungaresi, E: Stiff-man sindrome associated with nocturnal myoclonus and epilepsy. J Neurol Neurosurg Psychiatry 41:458–462, 1978.

127. Martinez, AC, Ferrer, MT, and Conde, MCP: Electrophysiological studies in myotonic dystrophy. 1. Potential motor unit parameters and conduction velocity of the motor and sensory peripheral nerve fibres. Electromyogr Clin Neurophysiol 24:523–534, 1984.

128. Mauritz, KH, Schmitt, C, and Dichgans, J: Delayed and enhanced long latency reflexes as the possible cause of postural tremor in late cerebellar atrophy. Brain 104:97–116, 1981.

129. McArdle, B: Myopathy due to a defect in muscle glycogen breakdown. Clin Sci 10:13–33, 1951.

130. McArdle, B: Adynamia episodica hereditaria and its treatment. Brain 85:121–148, 1962.

131. McComas, AJ, Campbell, MJ, and Sica, REP: Electrophysiological study of dystrophia myotonica. J Neurol Neurosurg Psychiatry 34:132–139, 1971.

132. McComas, AJ, Sica, REP, and Toyonaga, K: Incidence, severity, and time-course of motoneurone dysfunction in myotonic dystrophy: Their significance for an understanding of anticipation. J Neurol Neurosurg Psychiatry 41:882–893, 1978.

133. Meinck, HM, Ricker, K, and Conrad, B: The stiff-man syndrome: New pathophysiological aspects from abnormal exteroceptive reflexes and the response to clomipramine, clonidine, and tizanidine. J Neurol Neurosurg Psychiatry 47:280–287, 1984.

134. Mertens, HG, and Ricker, K: The differential diagnosis of the "stiff-man" syndrome. In Walton, JN, Canal, N, and Scarlato, G (eds): Muscle Diseases. Excerpta Medica, Amsterdam, 1970, pp 635–638.

135. Mertens, HG, and Zschocke, S: Neuromyotonie. Klin Wochenschr 43:917–925, 1965.

136. Meyers, KR, Gilden, DH, Rinaldi, CF, and Hansen, JL: Periodic muscle weakness, normokalemia, and tubular aggregates. Neurology 22:269–279, 1972.

137. Miller, F, and Korsvik, H: Baclofen in the treatment of stiff-man syndrome. Ann Neurol 9:511–512, 1981.

138. Mills, KR, Newham, DJ, and Edwards, RHT: Severe muscle cramps relieved by transcutaneous nerve stimulation: A case report. J Neurol Neurosurg Psychiatry 45:539–542, 1982.

139. Moller, AR, and Jannetta, PJ: Hemifacial spasm: Results of electrophysiologic recording during microvascular decompression operations. Neurology 35:969–974, 1985.

140. Montagna, P, Imbriaco, A, Zucconi, M, Liguori, R, Cirignotta, F, and Lugaresi, E: Hemifacial spasm in sleep. Neurology 36:270–273, 1986.

141. Montplaisir, J, Godbout, R, Boghen, D, De Champlain, J, Young, SN, and Lapierre, G: Familial restless legs with periodic movements in sleep: Electrophysiologic, biochemical and pharmacologic study. Neurology 35:130–134, 1985.

142. Morris, HH, and Estes, ML: Bilateral facial myokymia following cardiopulmonary arrest. Arch Neurol 38:393–394, 1981.

143. Nagano, Y, and Roses, AD: Abnormalities of erythrocyte membranes in myotonic muscular dystrophy manifested in lipid vesicles. Neurology 30:989–991, 1980.

144. Nielsen, VK: Pathophysiology of hemifacial spasm. I. Ephaptic transmission and ectopic excitation. Neurology 34:418–426, 1984.

145. Nielsen, VK: Pathophysiology of hemifacial spasm. II. Lateral spread of the supraorbital nerve reflex. Neurology 34:427–431, 1984.

146. Nielsen, VK: AAEE Minimonograph #23: Electrophysiology of the facial nerve in hemifacial spasm: ectopic/ephaptic excitation. Muscle Nerve 8:545–555, 1985.

147. Nielsen, VK, Friis, ML, and Johnsen, TR: Electromyographic distinction between paramyotonia congenita and myotonia congenita: Effect of cold. Neurology 32:827–832, 1982.

148. Nielsen, VK, and Jannetta, PJ: Hemifacial spasm: electrophysiologic effects of facial nerve decompression. Electroencephalogr Clin Neurophysiol 56:S144, 1983.

149. Nielsen, VK, and Jannetta, PJ: Pathophysiology of hemifacial spasm. III. Effects of facial nerve decompression. Neurology 34:891–897, 1984.

150. Nishi, T, Matsukado, Y, Nagahiro, S, Fukushima, M, and Koga, K: Hemifacial spasm due to contralateral acoustic neuroma: Case report. Neurology 37:339–342, 1987.

151. Norris, FH, Jr: Myokymia (corres). Arch Neurol 34:133, 1977.

152. Norris, FH, Jr, Gasteiger, EL, and Chatfield, PO: An electromyographic study of induced and spontaneous muscle cramps. Electroencephalogr Clin Neurophysiol 9:139–147, 1957.

153. Obeso, JA, Rothwell, JC, and Marsden, CD: The spectrum of cortical myoclonus: From focal reflex jerks to spontaneous motor epilepsy. Brain 108:193–224, 1985.

154. Okuno, T, Mori, K, Furomi, K, Takeoka, T, and Kondo, K: Myotonic dystrophy and hyperthyroidism. Neurology 31:91–93, 1981.

155. Olafson, RA, Mulder, DW, and Howard, FM: "Stiff-man" syndrome: A review of the literature, report of three additional cases and discussion of pathophysiology and therapy. Proc Staff Meet Mayo Clin 39:131–144, 1964.

156. Otsuka, M, and Ohtsuki, I: Mechanism of muscular paralysis by insulin with special reference to periodic paralysis. Am J Physiol 219:1178–1182, 1970.

157. Panayiotopoulos, CP, and Scarpalezos, S: Dystrophia myotonica: Peripheral nerve involvement and pathogenic implications. J Neurol Sci 27:1–16, 1976.

158. Parry-Jones, NO, Stephens, JA, Taylor, A, and Yates, DAH: Myokymia, not myotonia. Br Med J 2:300, 1977.

159. Pavone, L, Mollica, F, Grasso, A, Cao, A, and Gullotta, F: Schwartz-Jampel syndrome in two daughters of first cousins. J Neurol Neurosurg Psychiatry 41:161–169, 1978.

160. Pearson, CM: The periodic paralyses: Differential features and pathological observations in permanent myopathic weakness. Brain 87:341–354, 1963.

161. Pierry, A, and Cameron, M: Clonic hemifacial spasm from posterior fossa arteriovenous malformation. J Neurol Neurosurg Psychiatry 42:670–672, 1979.

162. Polgar, JG, Bradley, WG, Upton, ARM, Anderson, J, Howat, JML, Petito, F, Roberts, DF, and Scopa, J: The early detection of dystrophia myotonica. Brain 95:761–776, 1972.

163. Pollock, M, and Dyck, PJ: Peripheral nerve morphometry in myotonic dystrophy. Arch Neurol 33:33–39, 1976.

164. Poskanzer, DC, and Kerr, DNS: A third type of periodic paralysis, with normokalemia and a favorable response to sodium chloride. Am J Med 31:328–342, 1961.

165. Pryse-Phillips, W, Johnson, GJ, and Larsen, B: Incomplete manifestation of myotonic dystrophy in a large kinship in Labrador. Ann Neurol 11:582–591, 1982.

166. Radu, EW, Skorpil, V, and Kaeser, HE: Facial myokymia. Eur Neurol 13:499–512, 1975.

167. Ramon, F, and Moor, JW: Ephaptic transmission in squid giant axons. Am J Physiol 234:C162–169, 1978.

168. Ravin, M, Newmark, Z, and Saviello, G: Myotonia dystrophica—an anesthetic hazard: Two case reports. Anesth Analg 54:216–218, 1975.

169. Resnick, JS, and Engel, WK: Myotonic lid lag in hypokalaemic periodic paralysis. J Neurol Neurosurg Psychiatry 30:47–51, 1967.

170. Ricker, K, and Meinck, HM: Paramyotonia congenita (Eulenburg): Neurophysiologic studies of a case. Z Neurol 203:13–22, 1972.

171. Ricker, K, Samland, O, and Peter, A: Elektrische und mechanische Muskelreaktion bei Adynamia episodica und Paramyotonia congenita nach Kälteeinwirkung und Kaliumgabe. J Neurol 208:95–108, 1974.

172. Riecker, G, and Bolte, HD: Membranpotentiale einzelner Skeletmuskelzellen bein hypokaliämischer periodischer Muskelparalyse. Klin Wochenschr 44:804–807, 1966.

173. Risk, WS, Bosch, EP, Kimura, J, Cancilla, PA, Fischbeck, KH, and Layzor, RG: Chronic tetanus: Clinical report and histochemistry of muscle. Muscle Nerve 4:363–366, 1981.

174. Rivera, VM: Interpretation of serum creatine phosphokinase. JAMA 225:993–994, 1973.

175. Roohi, F, List, T, and Lovelace, RE: Slow motor nerve conduction in myotonic dys-

trophy. Electromyogr Clin Neurophysiol 21:97–105, 1981.

176. Rossi, B, Sartucci, F, Stefanini, A, Pucci, G, and Bianchi, F: Measurement of motor conduction velocity with Hopf's technique in myotonic dystrophy. J Neurol Neurosurg Psychiatry 46:93–95, 1983.

177. Rowland, LP: Pathogenesis of muscular dystrophies. Arch Neurol 33:315–321, 1976.

178. Rowland, LP, and Layzer, RB: Muscular dystrophies, atrophies, and related diseases. In Baker, AB, and Baker, LH (eds): Clinical Neurology, Vol 3. Harper & Row, Hagerstown, 1973, pp 1–100.

179. Rubenstein, AE, and Wainapel, SF: Acute hypokalemic myopathy in alcoholism: A clinical entity. Arch Neurol 34:553–555, 1977.

180. Ruff, RL: Insulin-induced weakness in hypokalemic myopathy. Ann Neurol 6:139–140, 1979.

181. Saadat, M, Mokfi, H, Vakil, H, and Zial, M: Schwartz syndrome: Myotonia with blepharophimosis and limitation of joints. J Pediatr 81:348–350, 1972.

182. Sadjadpour, K: Postfacial palsy phenomena: Faulty nerve regeneration or ephaptic transmission? Brain Res 95:403–406, 1975.

183. Said, G, Bathien, N, and Cesaro, P: Peripheral neuropathies and tremor. Neurology 32:480–485, 1982.

184. Sakai, T, Hosokawa, S, Shibasaki, H, Goto, I, Kuroiwa, Y, Sonoda, H, and Murai, Y: Syndrome of continuous muscle-fiber activity: Increased CSF gaba and effect of dantrolene. Neurology 33:495–498, 1983.

185. Sander, JE, Layzer, RB, and Goldsobel, AB: Congenital stiff-man syndrome. Ann Neurol 8:195–197, 1980.

186. Sanders, DB: Myotonia congenita with painful muscle contractions. Arch Neurol 33:580–582, 1976.

187. Sarova-Pinchas, I, Goldhammer, Y, and Braham, J: Multifocal myokymia. Muscle Nerve 1:253–254, 1978.

188. Satoyoshi, E: Recurrent muscle spasms of central origin. Trans Am Neurol Assoc 92:153–157, 1967.

189. Satoyoshi, E: A syndrome of progressive muscle spasm, alopecia, and diarrhea. Neurology 28:458–471, 1978.

190. Scherokman, B, Husain, F, Cuetter, A, Jabbari, B, and Maniglia, E: Peripheral dystonia. Arch Neurol 43:830–832, 1986.

191. Schmidt, RT, Stahl, SM, and Spehlmann, R: A pharmacologic study of the stiff-man syndrome: Correlation of clinical symptoms with urinary 3-methoxy-4-hydroxy-phenyl glycol excretion. Neurology 25:622–626, 1975.

192. Schultze, FR: Beiträge zur Muskelpathologie. Dtsch Z Nervenheilk 6:65–75, 1895.

193. Schwartz, O, and Jampel, RS: Congenital blepharophimosis associated with a unique generalized myopathy. Arch Ophthalmol 68:52–57, 1962.

194. Segura, RP, and Petajan, JH: Neural hyperexcitability in hyperkalemic periodic paralysis. Muscle Nerve 2:245–249, 1979.

195. Sethi, PK, Smith, BH, and Kalyanaraman, K: Facial myokymia: A clinicopathological study. J Neurol Neurosurg Psychiatry 37:745–749, 1974.

196. Shahani, M, Dastur, FD, Dastoor, DH, Mondkar, VP, Bharucha, EP, Nair, KG, and Shah, JC: Neuropathy in tetanus. J Neurol Sci 43:173–182, 1979.

197. Shibaskai, H, Yamashita, Y, Neshige, R, Tobimatsu, S, and Fukui, R: Pathogenesis of giant somatosensory evoked potentials in progressive myoclonic epilepsy. Brain 108:225–240, 1985.

198. Shy, GM, Wanko, T, Rowley, PT, and Engel, AG: Studies in familial periodic paralysis. Exp Neurol 3:53–121, 1961.

199. Sigwald, J, Raverdy, P, Fardeau, M, Gremy, F, Mace, DE, Lepinay, A, Bouttier, D, and Danic, M: Pseudo-myotonie: Forme particulière d'hypertonie musculaire a predominance distale. Rev Neurol (Paris) 115:1003–1014, 1966.

200. Simpson, JA: Neuromuscular diseases. In Remond, A (ed): Handbook of Electroencephalography and Clinical Neurophysiology, Vol 16B. Elsevier, Amsterdam, 1973.

201. Somers, JE, and Winer, N: Reversible myopathy and myotonia following administration of a hypocholesterolemic agent. Neurology 16:761–765, 1966.

202. Spaans, F, Jennekens, FGI, Mirandolle, JF, Bijlsma, JB, and De Gast, GC: Myotonic dystrophy associated with hereditary motor and sensory neuropathy. Brain 109:1149–1168, 1986.

203. Stöhr, M, and Heckl, R: Das stiff-man syndrom: Klinische, elektromyographische und pharmakologische Befunde bei einem eigenen Fall. Arch Psychiatr Nervenkr 223:171–180, 1977.

204. Stöhr, M, Schlote, W, Bundschu, HD, and Reichenmiller, NE: Myopathia myotonica: Fallbericht über eine neuartige hereditäre metabolische myopathie. J Neurol 210:41–66, 1975.

205. Streib, EW: AAEE Minimonograph #27: Differential diagnosis of myotonic syndromes. Muscle Nerve 10:603–615, 1987.

206. Streib, EW, and Sun, SF: The distribution of electrical myotonia in myotonic muscular dystrophy. Ann Neurol 14:80–82, 1983.

207. Streib, EW, Sun, SF, and Yarkowsky, T: Transient paresis in myotonic syndromes: A simplified electrophysiologic approach. Muscle Nerve 5:719–723, 1982.

208. Struppler, A, Struppler, E, and Adams, RD: Local tetanus in man. Arch Neurol 8:162–178, 1963.

209. Subramony, SH, Malhotra, CP, and Mishra, SK: Distinguishing paramyotonia congenita and myotonia congenita by electromyography. Muscle Nerve 6:374–379, 1983.

210. Subramony, SH, and Wee, AS: Exercise and rest in hyperkalemic periodic paralysis. Neurology 36:173–177, 1986.

211. Sun, SF, and Streib, EW: Autosomal recessive generalized myotonia. Muscle Nerve 6:143–148, 1983.

212. Susac, JO, Smith, JL, and Schatz, NJ: Superior oblique myokymia. Arch Neurol 29:432–433, 1973.

213. Swift, TR, Ignacio, OJ, and Dyken, PR: Neonatal dystrophia myotonica. Am J Dis Child 129:734–737, 1975.

214. Taylor, RG, Layzer, RB, Davis, HS, and Fowler, WM, Jr: Continuous muscle fiber activity in the Schwartz-Jampel syndrome. Electroencephalogr Clin Neurophysiol 33:497–509, 1972.

215. Tenser, RB, and Corbett, JJ: Myokymia and facial contraction in brain stem glioma: An electromyographic study. Arch Neurol 30:425–427, 1974.

216. Thompson, PD, and Carroll, WM: Hemimasticatory spasm—a peripheral paraoxysmal cranial neuropathy? J Neurol Neurosurg Psychiatry 46:274–276, 1983.

217. Thomsen, J: Tonische Krämpe in willkürlich beweglichen Muskeln in Folge von ererbter psychischer Disposition (Ataxia muscularis?). Arch Psychiatr Nervenkr 6:702–718, 1876.

218. Thrush, DC, Morris, CJ, and Salmon, MV: Paramyotonia congenita: A clinical, histochemical and pathological study. Brain 95:537–552, 1972.

219. Torres, CF, Griggs, RC, Moxley, RT, and Bender, AN: Hypokalemic periodic paralysis exacerbated by acetazolamide. Neurology 31:1423–1428, 1981.

220. Trudell, RG, Kaiser, KK, and Griggs, RC: Acetazolamide-responsive myotonia congenita. Neurology 37:488–491, 1987.

221. Vakil, BJ, Singhal, BS, Pandya, SS, and Irani, PF: Cephalic tetanus. Neurology 23:1091–1096, 1973.

222. Valenstein, E, Watson, RT, and Parker, JL: Myokymia, muscle hypertrophy and percussion "myotonia" in chronic recurrent polyneuropathy. Neurology 28:1130–1134, 1978.

223. Valli, G, Barbieri, S, Cappa, S, Pellergrini, G, and Scarlato, G: Syndromes of abnormal muscular activity: Overlap between continuous muscle fibre activity and the stiff man syndrome. J Neurol Neurosurg Psychiatry 46:241–247, 1983.

224. Vanzandycke, M, Martin, JJ, Vande Gaer, L, and Van den Heyning, P: Facial myokymia in the Guillain-Barre syndrome: a clinicopathologic study. Neurology 32:744–748, 1982.

225. Vasilescu, C, Alexianu, M, and Dan, A: Neuronal type of Charcot-Marie-Tooth disease with a syndrome of continuous motor unit activity. J Neurol Sci 63:11–25, 1984.

226. Vickers, JD, McComas, AJ, and Rathbone, MP: Myotonia muscular dystrophy: Abnormal temperature response of membrane phosphorylation in erythrocyte membranes. Neurology 29:791–796, 1979.

227. Wallis, WE, Van Poznak, A, and Plum F: Generalized muscular stiffness, fasciculations, and myokymia of peripheral nerve origin. Arch Neurol 22:430–439, 1970.

228. Warmolts, JR, and Mendell, JR: Neurotonia: Impulse-induced repetitive discharges in motor nerves in peripheral neuropathy. Ann Neurol 7:245–250, 1980.

229. Wartenberg, R: Hemifacial spasm: A clinical and Pathophysiological Study. Oxford University Press, New York, 1952.

230. Wasserstrom, WR, and Starr, A: Facial myokymia in Guillain-Barré syndrome. Arch Neurol 34:576–577, 1977.

231. Waybright, EA, Gutmann, L, and Chou, SM: Facial myokymia. Pathological features. Arch Neurol 36:244–245, 1979.

232. Wegmüller, E, Ludin, HP, and Mumenthaler, M: Paramyotonia congenita: a clinical, electrophysiological and histological study of 12 patients. J Neurol 220:251–257, 1979.

233. Welch, LK, Appenzeller, O, and Bicknell, JM: Peripheral neuropathy with myokymia, sustained muscular contraction and continuous motor unit activity. Neurology 22:161–169, 1972.

234. Whelan, JL: Baclofen in treatment of the "stiff-man" syndrome. Arch Neurol 37:600–601, 1980.

235. Woltman, HW, Williams, HL, and Lambert, EH: An attempt to relieve hemifacial spasm by neurolysis of the facial nerves: A report of two cases of hemifacial spasm with reflections on the nature of the spasm, the contracture and mass movement. Proc Staff Meet Mayo Clin 26:236–240, 1951.

236. Zellweger, H, Pavone, L, Biondi, A, Cumino, V, Gullotta, F, Hart, M, Ionasescu, V, Mollica, F, and Schieken, R: Autosomal recessive generalized myotonia. Muscle Nerve 3:176–180, 1980.

APPENDICES

Appendix 1

HISTORICAL REVIEW

1 INTRODUCTION

Electrophysiology began toward the end of the 18th century with Galvani's discovery of animal electricity and has since progressed steadily during the past two centuries. Electrophysiologic assessments of muscle and nerve are now considered indispensable in the practice of neurology, physiatrics, and other related clinical disciplines. The historical growth of this medical field may be divided arbitrarily into four relatively distinct but overlapping eras. They represent (1) early developments, (2) classical electrodiagnosis, (3) electromyography and nerve stimulation techniques, and (4) recent developments.

During the first period, ending at about the mid-19th century, the existence of bioelectricity was firmly established by Galvani and others. The basic concepts of electricity were also founded during this period by a series of scientific achievements of Volta and his pupils. The progress in these two branches of science complemented each other despite the initial controversy that arose over the existence of animal electricity. A number of studies in the last half of the 19th century established the relationship between the duration of stimulation and current strength in eliciting muscle contractions. This led to the development of classic electrodiagnosis, the study of muscle response to electrical stimulation as a diagnostic test. The method gained popularity during the first half of this century as the recording apparatus was improved from the capillary electrometer to the string galvanometer.

Modern techniques began with the invention of the cathode ray oscilloscope in 1922 and the concentric needle electrode a few years later. Aided by these technical advances, electromyography became a clinically useful tool. The nerve stimulation technique was then introduced, first for studies of neuromuscular transmission and later for assessments of conduction velocity. Since then, there has been wide application of these techniques, which are now considered conventional. More recently, an increasing number of newer electrophysiologic tests emerged for evaluation of anatomic regions not accessible by the traditional methods. These include studies of human reflexes and other late potentials, recordings of somatosensory and motor evoked potentials, and single-fiber electro-myography.

2 EARLY DEVELOPMENTS

Ancient physicians used electrical discharges from the black torpedo fish for the treatment of headaches and arthritis. It was not until the turn of the 17th century that the world electric was first used by William Gilbert[54] in his book *De Magnete*. Static discharges were also well known after the invention of the Leyden jar by Musshenbroek in 1745. In the same year, Kratzenstein first induced muscle contraction by static electricity. The next year he wrote the first paper on the use of electricity in medical therapy.[80] Many similar studies followed toward the end of the 18th century, each describing muscle contraction induced by electrical stimulation.

It was Galvani who laid the foundation for clinical electrophysiology. After a series of experiments on muscle contraction in frog legs, he introduced the idea that electricity was generated by nervous tissue. This observation was first published in 1791 in his now famous article "De viribus electricitatis in motu musculari commentarius," which appeared in the *Proceedings of the Bologna Academy*.[48] His concept of animal electricity was received with considerable skepticism in his time. Controversy arose chiefly from Volta's belief that the two plates of different metals were responsible for the electricity observed in Galvani's experiments.[127] Fowler[46] agreed with Volta that dissimilar metals and the muscle had to be connected to generate frog current.

Later, Galvani was able to produce muscle contraction by draping the free end of the nerve across the muscle without the use of metals. This finding was reproduced by Humboldt[74] in 1797 and Matteucci[102] in 1844. In the meantime, Volta's conviction that animal electricity was in reality the effect of a very weak artificial current induced by application of two different metals led to the development of the Voltaic pile in 1799. He also noted that muscle contracted only at the closing and opening of the circuit. Although Galvani's view on intrinsic electrical current in frog legs was correct, Volta's new invention was so dramatic and convincing that his view of electricity of metallic origin prevailed. This is understandable, because the Voltaic pile produced all the phenomena attributed to animal electricity by Galvani.[128] Indeed, Galvani's experiment was all but forgotten until much later, when Nobili[109] and Matteucci[101] reported electrical activity from muscle in 1830 and 1842, respectively.

In 1822, Magendie,[96] who is credited for distinguishing between motor and sensory nerves, tried to insert a needle into the nerve for electrical stimulation, a practice soon abandoned because of the patient's discomfort! Sarlandiere,[118] in 1825, was the first to introduce electropuncture for direct electrical activation of muscle. One of Volta's pupils, Marianini,[98] found in 1829 that ascending (negative) current elicited muscle contraction more effectively than descending current. Nobili,[109] in 1830, recognized different stages of excitability, based on the degree of muscle contraction after turning on and off the electrical current supplied by a battery. Later, Erb,[42] in 1883, used this concept clinically in the assessment of abnormal excitabilities of disordered muscles.

According to Licht,[92] Ampere introduced the concept of current flow after witnessing Oersted's 1819 demonstration that a battery, through metallic wire extended from the two poles, acted on a magnetic needle at a distance. In 1831, Henry found the augmenting action of a long coil of wire on direct current; and in the same year Faraday described alternating current induced in a coil of wire by another coil that was periodically charged. In 1833, Duchenne de Boulogne found that a muscle could be stimulated electrically from the skin surface with the use of cloth-covered electrodes. He was also the first to use Faradic current for stimulation.[33]

Carlo Matteucci[101,102] of Pisa demonstrated that stimulation of the nerve proximal to the application of a ligature or section failed to elicit muscle contraction. In his 1838 experiment, published a few years later, he placed the sciatic nerve still connected to the leg muscles on the thigh muscles dissected from the other leg.[101] In this preparation, contraction of the thigh muscles induced movements of the other leg, provided that its sciatic nerve was not insulated from bared muscle. Hence, he detected electrical activity of contracting muscle for the first time using a neuromuscular preparation, the only means available in those days. Inspired by the work of Matteucci, DuBois-Reymond[31] registered action potentials generated in the muscle.[105] In 1851, he identified the action potential of voluntarily contracting arm muscles, using jars of liquid as electrodes.[32] This was perhaps the beginning of electromyography.[106]

In 1850, Helmholtz[63] succeeded in measuring the conduction velocity of the nerve impulse in the frog by mechanically recording the muscle twitch. Using the same procedure, a conduction velocity of 61.0 ± 5.1 m/s was found in the human median nerve.[64] He also determined the conduction rate in sensory nerve of man to be 60 m/s by measuring the difference in reaction time. In 1878, Hermann[65,66] stimulated the brachial plexus in the axilla and recorded a response from the surface of the forearm, which he called action potential. Burdon Sanderson[15] was the first to show in 1895 that this wave of excitation preceded the mechanical response.

3 CLASSICAL ELECTRODIAGNOSIS

Duchenne[34] found that electrical simulation activated certain localized areas of muscle more easily than others. Remak[113] discovered that these points represented entry zones of the muscular nerves. In 1857, Ziemssen[135] carefully mapped out the whole skin surface of the body in agonal patients and proved by dissection immediately after death that the motor points were indeed entrances of the nerve into the muscle. Krause,[81] known for the skin corpuscle that now bears his name, suggested that nerve impulses terminated at the motor points. Kuhne[84] coined the name end plates for the nerve endings of striated muscle.

Hammond[104] translated Meyer's comprehensive discussion on electrical stimulation of the muscle into English. He also found that galvanic current activated the paralytic limb from cerebral disease more easily than the normal limb. In contrast, more current was necessary if paralysis was caused by lesions of the spinal (peripheral) nerve. Baierlacher[3] had noted that diseased muscle responded better to continuous galvanic current than interrupted faradic current. Neumann,[108] however, was the first to recognize that it was the duration that determined the effectiveness of current. Erb also noted failure of the paralyzed muscle to contract in response to frequently interrupted stimuli, and called this phenomenon the reaction of degeneration.[42] His quantitative studies revealed a certain relationship between muscle contraction and current strength. Based on this principle, he assessed excitability of the muscle in various disorders and found marked irritability in tetany. In 1882, he introduced a formula of polar contraction in normal subjects and its reversal in some disease states, thus establishing the foundation for classical electrodiagnosis.

DuBois-Reymond believed that change in current, rather than the absolute value of current strength, determined muscle response. This view prevailed until the end of the 19th century despite mounting evidence to the contrary. In 1870, Engelman showed a relationship between current intensity and duration in eliciting muscle contraction. This finding paved the way for determination of the strength-duration curve in laboratory animals.[90] Hoorweg[72] further challenged the concept of DuBois-Reymond by stating that nerve excitation occurred as a function of stimulus time and intensity, a view vigorously supported by Lapicque.[90] Waller and Watteville[130] also suggested a duration-intensity relationship for optimal stimulation in 1883.

Toward the end of the 19th century, a

few investigators recognized abnormal localization of motor points in degenerated muscles.[30,53] Lewis Jones[91] pointed out that the phenomenon of "displaced motor point" simply represented abnormal sensitivity in regions distinct from the motor point. In 1907, Bordet reported that during passage of a sustained current the critical excitatory level changed less rapidly in the denervated muscle than in normal muscle.[114] This observation led to measurements of accommodation and the galvanic-tetanic ratio, electrodiagnostic tests used widely until recent years.

D'Arsonval's[20] use of a reflecting coil improved the galvanometer built by Sturgeon in 1836. Lippmann[95] introduced the capillary electrometer in 1872. In the meantime, Weiss[134] first attempted to produce a rectangular stimulus pulse, with a device called ballistic rheotome. Lapicque[89,90] developed a more accurate apparatus with a circuit breaker operated by gravity in 1907. Using this instrument, he defined rheobase as the minimal continuous current intensity required for muscle excitation and chronaxie as the minimal current duration required at an intensity twice the rheobase.[90] Lewis Jones[91] constructed a battery of condensors (capacitors) for diagnostic purposes. Using this apparatus, Bourguignon[8] was the first to study chronaxie in man. Plotting strength duration curves for the first time in man, Adrian[1] reported a fairly constant time course in healthy muscles. He also noted a predictable shift in the regenerating muscle during different phases of recovery after degeneration. A constant current stimulator designed by Bauwens[5] improved the accuracy in determining the strength-duration curve.

4 ELECTROMYOGRAPHY AND NERVE STIMULATION TECHNIQUES

Bernstein[6] introduced the term action potential, but Schiff[120] was the first to observe oscillation (fasciculation) of denervated muscle after section of the hypoglossal nerve in 1851. This spontaneous movement ceased if the muscle became atrophic or the nerve regenerated. Fibrillation meant a tremor of denervated muscle in experimental animals, according to Rogowicz[116] and Ricker.[115] In the first electromyography after DuBois-Reymond, Piper[111] recorded voluntary activity of muscles using a string galvanometer. He believed that the muscle activity discharges at a constant frequency independent of the force generated. For him this reflected the rhythm of neural impulses, although others considered the rate of firing to be inherent in the muscle.[49,50] Using the capillary electrometer, Buchanan[12] arrived at the opposite conclusion: that the frequency of the electromyogram shifted substantially during different degrees of contraction. She stated that the study of the interference pattern could not elucidate the mechanism of neural innervation. At the turn of the century, Langley and Kato[88] and Langley[87] studied fibrillation in muscular dystrophy.

The study of muscle action potentials progressed rapidly after the development of sensitive recording apparatus. Braun[9] invented the cathode-ray tube. Later, Einthoven[40] designed the string galvanometer with a fiber of quartz. In 1920, Forbes and Thacher[45] were the first to use the electron tube to amplify the action potential and a string galvanometer to record it. Gasser and Erlanger[51] introduced one of the most important advances in technology, the cathode-ray oscilloscope, which eliminated the mechanical limitation of galvanometers.[52] Their book *Electrical Signs of Nervous Activity* laid the foundation of modern clinical electrophysiology.[43]

In 1925, Liddell and Sherrington[93] proposed the concept of the motor unit. Shortly thereafter, Proebster[112] performed the first clinical electromyography in neurogenic weakness, recording spontaneous potentials in brachial plexus injury and long-standing poliomyelitis. Another major advancement came when Adrian and Bronk[2] introduced the concentric needle electrode in 1929. The use of this electrode made it possible for the first time to record from single motor units. Adrian also initiated the use of a loudspeaker so that electromyographers could use not only visual but also acoustic cues. Motor unit potentials were studied by Denny-Brown[25] in the same year and later by Eccles and Sherrington,[38] Clark,[17] and Hoefer and Putnam.[69]

Invention of the differential amplifier by Matthews[103] in 1934 made the recording of small muscle potentials possible, because it minimized electrical interference from other sources. Lindsley[94] noted unusual fluctuation of motor units in a patient with myasthenia gravis. Further work on denervation potentials came from Brown,[11] who tested the effect of acetylcholine on the denervated muscles. Using a bipolar electrode, Denny-Brown and Pennybacker[27] differentiated fibrillation potentials from fasciculation potentials in 1938, a finding later substantiated by Eccles,[37] who used a refined method. In 1941, Denny-Brown and Nevin[26] recorded myotonic discharges. In the same year, Buchthal and Clemmesen[14] confirmed the electromyographic findings of atrophic muscles.

During the two world wars, the large number of battlefield peripheral nerve injuries increased the need for electrical testing. An accelerated growth of electronic devices such as radar and oscilloscopes enhanced this tendency. At the same time, polio epidemics demanded development of procedures to accurately determine the presence and extent of nerve injury and the status of regeneration. Many fundamental contributions to electromyography and nerve conduction studies came from this combination of circumstances.

Using standardized clinical testing, Weddell, Feinstein, and Pattle[132,133] noted the appearance of spontaneous discharges 18 to 20 days after denervation. Watkins, Brazier, and Schwab[131] recorded similar activities in poliomyelitis from the skin surface at various sites. The following year, Hoefer and Guttman[68] recorded paparaspinal denervation using a surface electrode. They reported that such abnormalities, detected longitudinally, sometimes help localize the level of spinal cord lesions. Around

the same time, Jasper and Notman[76] introduced the monopolar electrode, and Jasper, Johnston, and Geddes[75] built a portable apparatus for electromyography. Further clinical applications of the needle examination were reported in poliomyelitis by Huddleston and Golseth,[73] in lower motor neuron by Golseth and Huddleston,[57] and in nerve root compression by Shea, Woods, and Werden.[121] In 1955, Marinacci[99] published the first book of electromyography since Piper, and Buchthal[13] contributed a monograph 2 years later.

Jolly[78] described abnormal fatigability of the orbicularis oculi muscle to intermittent, direct-current stimulation in myasthenic patients. Harvey and Masland[62] were the first to quantitate this clinical observation by stimulating the nerve repetitively and recording the muscle action potentials. This technique was also applied to the study of myasthenic syndromes.[36] It became an important part of our electrodiagnostic armamentation after standardization by Lambert[86] and Desmedt.[29]

Piper[110] and Munnich[107] first recorded the muscle action potential instead of the muscle twitch for determination of motor nerve conduction. Inspired by Sherrington's work[122] on the stretch reflex, Hoffmann[70,71] demonstrated the monosynaptic reflex in man by stimulating the tibial nerve and recording the muscle action potential from the soleus. Based on latency measures of the H reflex, Schäffer[119] calculated a velocity of 60 to 65 m/s for the human sensory nerve. Interest in nerve injury and repair during the war prompted basic scientists to study conduction velocity of regenerating nerves in experimental animals.[7,44,117] Harvey and Kuffler[60] and Harvey, Kuffler, and Tredway[61] studied peripheral neuritis in man, stimulating the nerve and recording muscle action potentials. It was Hodes, Larrabee, and German[67] who first calculated the conduction velocity, stimulating the nerve at different levels in neurologic patients. Around the same time, Kugelberg[82] used nerve stimulation to study the effect of ischemia on nerve excitability. Cobb and Marshall,[18] extending this work, demonstrated slowed impulse propagation in the ischemic nerve.

Eichler[39] was the first to report percutaneous recording of nerve action potentials in response to electrical stimulation of the median and ulnar nerves in 1937. The averaging technique of sensory nerve conduction studies emerged as a by-product when Dawson[21] was attempting to record cortical potentials by stimulating peripheral nerves in patients with myoclonus. He used photographic superimposition[47] of a number of faint traces to improve the resolution of the recorded response. Dawson and Scott[24] needed the same technique to assess the growth of the sensory action potential of the peripheral nerve with increasing stimulus strength to prove the origin of their cortical potential.[55] Dawson[22,23] subsequently resorted to digital nerve stimulation to differentiate sensory potentials from antidromic impulses in motor fibers. Although some felt that latency measures sufficed,[16] calculation of nerve conduction velocity became an integral part of electrodiagnostic assessment in the 1960s.

These initial studies, started independently in the United States and Europe, soon spread to many countries, resulting in the common use of the whole field of electromyography and nerve conduction measurements. Important contributions came from Magladery and McDougal,[97] Wagman and Lesse,[129] Gilliatt and Wilson,[56] Lambert,[85] Simpson,[123] Buchthal,[13] Thomas, Sears, and Gilliatt,[126] Johnson and Olsen,[77] Kato,[79] Thomas and Lambert,[125] and Desmedt,[28] to name only a few. The First International Congress of Electromyography, held at Pavia, Italy, in 1961, signaled the rapidly growing worldwide interest in this then relatively new branch of medicine.

5 RECENT DEVELOPMENTS

Conventional methods of nerve conduction study mainly dealt with diseases affecting the distal portion of the peripheral nerve in the four extremities and seldom contributed to the investigation of the remainder of the nervous system. Several neurophysiologic techniques have emerged as diagnostic tests in evaluating the function of these less accessible anatomic regions. These include studies of human reflexes and other late responses. Of these, the most extensively investigated have been the H reflex of Hoffmann,[70,71] the F wave of Magladery and McDougal,[97] and the blink reflex of Kugelberg.[83]

Somatosensory evoked potentials provided another electrophysiologic means for study of the central nervous system.[19,28,59] The technique of signal averaging initially helped develop the methods for peripheral sensory conduction and much later those for cerebral evoked potential. The wide availability of minicomputers and averagers has since accelerated the clinical application of this technique in the assessment of the central nervous system. As stated above, this development is of historical interest because Dawson[21] originally used photographic superimposition, a forerunner of electrical averaging, in the study of somatosensory cerebral potentials. With the advent of electrical[100] and magnetic coil stimulators[4] capable of noninvasive excitation of the brain or spinal cord, it is now feasible to study the central motor pathways as well.

Introduction of single-fiber electromyography has made it possible to study electrophysiologic characteristics of individual muscle fibers.[41] This stands in contrast to the conventional use of coaxial or monopolar recording needles for assessment of the motor unit, the smallest functional element of muscle contraction. The technique has since been refined and simplified for research application and clinical use.[124] Some other newer techniques, although directly related to electromyography and nerve conduction studies, have not yet found their way into the clinical laboratory. These include the in-vitro technique of sural nerve conduction studies[35] and electroneurography.[58]

The above outline includes most of the major events that have taken place in the history of clinical electrophysiology of muscle and nerve. Inclusion of further details, although tempting because of a number of intriguing anecdotes, falls outside the scope of this book. Interested readers should consult previous publications on this subject by Mottelay,[106] Marinacci,[99] Licht,[92] Gilliatt,[55] and Brazier.[10]

REFERENCES

1. Adrian, ED: The electrical reactions of muscles before and after nerve injury. Brain 39:1–33, 1916.
2. Adrian, ED, and Bronk, DW: The discharge of impulses in motor nerve fibers. Part II. The frequency of discharge in reflex and voluntary contractions. J Physiol (Lond) 67:119–151, 1929.
3. Baierlacher, E: Beiträge zur therapeutischen Verwerthung des galvanischen Stromes. Aerztliches Intelligenz-Blatt 4:37–45, 1859.
4. Barker, AT, Freestone, IL, Jalinous, T, Merton, PA, and Morton, HB: Magnetic stimulation of the human brain (abstr). J Physiology 369:3P, 1985.
5. Bauwens, P: The thermionic control of electric currents in electro-medical work. Part 2. Proc R Soc Med 34:715–724, 1941.
6. Bernstein, J: Untersuchungen über die Natur des elektrotonischen Zustandes und der negativen Schwankung des Nervenstroms. Arch Anat Physiol 596–637, 1866.
7. Berry, CM, Grundfest, H, and Hinsey, JC: The electrical activity of regenerating nerves in the cat. J Neurophysiol 7:103–115, 1944.
8. Bourguignon, G: La Chronaxie Chez l'Homme. Masson, Paris, 1923.
9. Braun, F: Ueber ein Verfahren zur Demonstration und zum Studium des zeitlichen Verlaufes variabler Strome. Annalen der Physik und Chemie 60:552–559, 1897.
10. Brazier, MAB: The emergence of electrophysiology as an aid to neurology. In Aminoff, MJ (ed): Electrodiagnosis in Clinical Neurology. Churchill Livingstone, New York, 1980, pp 1–22.
11. Brown, GL: The actions of acetylcholine on denervated mammalian and frog's muscle. J Physiol (Lond) 89:438–461, 1937.
12. Buchanan, F: The electrical response of muscle to voluntary, reflex, and artificial stimulation. Q J Exp Physiol 1:211–242, 1908.
13. Buchthal, F: An Introduction to Electromyography. Scandinavian University Books, Copenhagen, 1957.
14. Buchthal, F, and Clemmesen, S: On the differentiation of muscle atrophy by electromyography. Acta Psych Neurol 16:143–181, 1941.
15. Burdon Sanderson, J: The electrical response to stimulation of muscle, and its relation to the mechanical response. J Physiol (Lond) 18:117–159, 1895.
16. Christie, BGB, and Coomes, EN: Normal variation of nerve conduction in three peripheral nerves. Ann Phys Med 5:303–309, 1960.
17. Clark, DA: Muscle counts of motor units: A study in innervation ratios. Am J Physiol 96:296–304, 1931.
18. Cobb, W, and Marshall, J: Repetitive discharges from human motor nerves after ischaemia and their absence after cooling. J Neurol Neurosurg Psychiatry 17:183–188, 1954.
19. Cracco, RQ: The initial positive response: Peripheral nerve stimulation in man. Electroencephalogr Clin Neurophysiol 35:379–386, 1973.
20. D'Arsonval, D: Électricité: Galvanomètre apériodique. Acad Sci Compt Rend 94:1347–1350, 1882.
21. Dawson, GD: Cerebral responses to electrical stimulation of peripheral nerve in man. J Neurol Neurosurg Psychiatry 10:137–140, 1947.
22. Dawson, GD: A summation technique for the detection of small evoked potentials. Electroencephalogr Clin Neurophysiol 6:65–84, 1954.
23. Dawson, GD: The relative excitability and conduction velocity of sensory and motor nerve fibres in man. J Physiol (Lond) 131:436–451, 1956.
24. Dawson, GD, and Scott, JW: The recording of nerve action potentials through skin in man. J Neurol Neurosurg Psychiatry 12:259–267, 1949.
25. Denny-Brown, D: On the nature of postural reflexes. Proc R Soc Lond 104b:252–301, 1929.
26. Denny-Brown, D, and Nevin, S: The phenomenon of myotonia. Part 1. Brain 64:1–16, 1941.
27. Denny-Brown, D, and Pennybacker, JB: Fibrillation and fasciculation in voluntary muscle. Brain 61:311–332, 1938.
28. Desmedt, JE: Somatosensory cerebral evoked potentials in man. In Remond, A (ed): Handbook of Electroencephalography and Clinical Neurophysiology, Vol 9. Elsevier, Amsterdam, 1971.
29. Desmedt, JE: The neuromuscular disorder in myasthenia gravis. 1. Electrical and mechanical response to nerve stimulation in hand muscles. In Desmedt, JE (ed): New Developments in Electromyography and Clinical Neurophysiology, Vol 1, Karger, Basel, 1973, pp 241–304.
30. Doumer, E: Note sur un nouveau signe électrique musculaire. Compt Rend de la Societe Biol 9:656–659, 1891.
31. DuBois-Reymond, E: Vorläufiger Abrifs einer Untersuchung über den sogenannten Froschstrom und über die elektromotorischen Fische. Annalen der Physik Und Chemie 58:1–30, Series 2, 1843.
32. DuBois-Reymond, E: On the time required for the transmission of volition and sensation through the nerves. R Inst Great Britain Proc, Vol 4, 1866, pp 575–593.
33. Duchenne, G: De L'électrisation Localisée et de son application a la Physiologie, a la Patho-

logie et a la Therapeutique. JB Bailliere, Paris, 1855. Translated into English by Tibbits, H, Lindsay and Blakiston, Philadelphia, 1871.

34. Duchenne, G: Physiologie des mouvements demonstrée d'laide de l'experimentation électrique et de l'observations cliniques et applicable a l'étude de paralysies et des déformations, 1867. Translated into English by Kaplan, EB. WB Saunders, Philadelphia, 1959.

35. Dyck, PJ, and Lambert, EH: Numbers and diameters of nerve fibers and compound action potential of sural nerve: Controls and hereditary neuromuscular disorders. Trans Am Neurol Assoc 91:214–217, 1966.

36. Eaton, LM, and Lambert, EH: Electromyography and electric stimulation of nerves in diseases of motor unit. Observations on myasthenic syndrome associated with malignant tumors. JAMA 163:1117–1124, 1957.

37. Eccles, JC: Changes in muscle produced by nerve degeneration. J Med Australia 1:573–575, 1941.

38. Eccles, JC, and Sherrington, CS: Numbers and contraction-values of individual motor units examined in some muscles of the limb. Proc R Soc London 106b:326–356, 1930.

39. Eichler, W: Über die Ableitung der Aktionspotentiale vom menschlichen Nerven in situ. Zeit Biol 98:182–214, 1937.

40. Einthoven, W: Ein neues Galvanometer. Drude's Annalen Physik 12:1059–1071, 1903.

41. Ekstedt, J, and Stålberg, E: A method of recording extracellular action potentials of single muscle fibres and measuring their propagation velocity in voluntarily activated human muscle. Bull Am Assoc Electromyogr Electrodiagn 10:16, 1963.

42. Erb, W: Handbuch der Electrotherapie. FCW Vogel, Leipzig, 1882. Translated into English by Putzel, L. William Wood and Company, New York, 1883.

43. Erlanger, J, and Gasser, HS: Electrical Signs of Nervous Activity. University of Pennsylvania Press, Philadelphia, 1937.

44. Erlanger, J, and Schoepfle, GM: A study of nerve degeneration and regeneration. Am J Physiol 147:550–581, 1946.

45. Forbes, A, and Thacher, C: Amplification of action currents with the electron tube in recording with the string galvanometer. Am J Physiol 52:409–471, 1920.

46. Fowler, R: Experiments and observations relative to the influence lately discovered by M Galvani, and commonly called animal electricity. Printed for T Duncan, P Hill, Robertson and Berry, and G Mudie; and J Johnson, St. Paul's Churchyard, London, 1793.

47. Galambos, R, and Davis, H: The response of single auditory-nerve fibers to acoustic stimulation. J Neurophysiol 6:39–57, 1943.

48. Galvani, L: De Viribus Electrocitatis in Motu Musculari Commentarius. Proc Bologna Academy and Institute of Sciences and Arts 7:363–418, 1791. Translated into English by Green, RM. Elizabeth Licht, Cambridge, 1953.

49. Garten, S: Beiträge zur Kenntnis des Erre-

gungsvorganges im Nerven und Muskel des Warmblüters. Z Biol 52:534–567, 1908.

50. Garten, S: Über die zeitliche Folge der Aktionsströme im menschlichen Muskel bei willkürlicher Innervation und bei Erregung des Nerven durch den konstanten Strom. Z. Biol 55:29–35, 1910.

51. Gasser, HS, and Erlanger, J: A study of the action currents of nerve with the cathode ray oscillograph. Am J Physiol 62:496–524, 1922.

52. Gasser, HS, and Erlanger, J: The nature of conduction of an impulse in the relatively refractory period. Am J Physiol 73:613–635, 1925.

53. Ghilarducci, F: Sur une nouvelle forme de la réaction de dégénérescence. (Réaction de Dégénérscence à distance.) Arch D Electricite Medicale et de Physiotherapie du Cancer 4:17–35, 1896.

54. Gilbert, W: De Magnete, Magneti-Cisqve Corporibvs, et Demagno magnete tellure; Phyfiologia noua, plurimis et argumentis, et experimentis demonstrata. London, 1600. Translated into English by Mottelay, PF. John Wiley & Sons, New York, 1893.

55. Gilliatt, RW: History of nerve conduction studies. In Licht, S (ed): Electrodiagnosis and Electromyography, ed 3. Waverly Press, Baltimore, 1971, pp 412–418.

56. Gilliatt, RW, and Wilson, TG: Ischaemic sensory loss in patients with peripheral nerve lesions. J Neurol Neurosurg Psychiatry 17:104–114, 1954.

57. Golseth, JG, and Huddleston, OL: Electromyographic diagnosis of lower motor neuron disease. Arch Phys Med 30:495–499, 1949.

58. Hagbarth, KE, and Vallbo, AB: Single unit recordings from muscle nerves in human subjects. Acta Physiol Scand 76:321–334, 1969.

59. Halliday, AM: Changes in the form of cerebral evoked responses in man associated with various lesions of the nervous system. Electroencephalogr Clin Neurophysiol 25:178–192, 1967.

60. Harvey, AM, and Kuffler, SW: Motor nerve function with lesions of the peripheral nerves: A quantitative study. Arch Neurol Psychiatry 52:317–322, 1944.

61. Harvey, AM, Kuffler, SW, and Tredway, JB: Peripheral neuritis: Clinical and physiological observations on a series of twenty cases of unknown etiology. Bull Johns Hopkins Hosp 77:83–103, 1945.

62. Harvey, AM, and Masland, RL: The electromyogram in myasthenia gravis. Bull Johns Hopkins Hosp 69:1–13, 1941.

63. Helmholtz, H: Vorläufiger Bericht über die Fortpflanzungsgeschwindigkeit der Nervenreizung. Arch Anat Physiol Wiss Med 71–73, 1850.

64. Helmholtz, H, and Baxt, N: Neue Versuche über die Fortpflanzungsgeschwindigkeit der Reizung in den motorischen Nerven der Menschen. Monatsberichte Der Königlich Preussischen, Akademie der Wissenschaften Zu Berlin, pp 184–191, 1870.

65. Hermann, L: Ueber den Actionsstrom der

Muskeln im lebenden Menschen. Pflugers Arch Ges Physiol 16:410–420, 1878.

66. Hermann, L: Untersuchungen über die Actionsströme des Muskels. Pflugers Arch Ges Physiol 16:191–262, 1878.

67. Hodes, R, Larrabee, MG, and German, W: The human electromyogram in response to nerve stimulation and the conduction velocity of motor axons: Studies on normal and on injured peripheral nerves. Arch Neurol Psychiatry 60:340–365, 1948.

68. Hoefer, PFA, and Guttman, SA: Electromyography as a method for determination of level of lesions in the spinal cord. Arch Neurol Psychiatry 51:415–422, 1944.

69. Hoefer, PFA, and Putnam, TJ: Action potentials of muscles in "spastic" conditions. Arch Neurol Psychiatry 43:1–22, 1940.

70. Hoffmann, P: Über die Beziehungen der Sehnenreflexe zur willkürlichen Bewegung und zum Tonus. Z Biol 68:351–370, 1918.

71. Hoffmann, P: Untersuchungen über die Eigenreflexe (Sehnenreflexe) Menschlicher Muskeln. Julius Springer, Berlin, 1922.

72. Hoorweg, JL: Ueber die elektrische Nervenerregung. Arch Ges Physiol 52:87–108, 1892.

73. Huddleston, OL, and Golseth, JG: Electromyographic studies of paralyzed and paretic muscles in anterior poliomyelitis. Arch Phys Med 29:92, 1948.

74. Humboldt, FA: Versuche über die gereizte Muskel-und Nervenfaser nebst Vermutunge über den chemischen Process des Lebens in der Thier-und Pflanzenwelt, Vol 2. Decker, Posen, und Rottmann, Berlin 1797.

75. Jasper, HH, Johnston, RH, and Geddes, LA: The R.C.A.M.C. Electromyograph, Portable Mark II. National Research Council of Canada, Montreal, 1945.

76. Jasper, H, and Notman, R: Electromyography in Peripheral Nerve Injuries. National Research Council of Canada, Montreal, Report #3, 1944.

77. Johnson, EW, and Olsen, KJ: Clinical value of motor nerve conduction velocity determination. JAMA 172:2030–2035, 1960.

78. Jolly, F: Myasthenia gravis pseudoparalytica. Berliner Klinische Wochenschrift 32:33–34, 1985.

79. Kato, M: The conduction velocity of the ulnar nerve and the spinal reflex time measured by means of the H wave in average adults and athletes. Tohoku J Exp Med 73:74–85, 1960.

80. Kratzenstein, C: Physicalische Briefe I. Nutzen der Electricitat in der Arzeneiwissenschaft. Halle, 1746.

81. Krause, W: Die terminalen Körperchen der einfach sensiblen Nerven. Hahn'sche Hofbuchhandlung, Hannover, 1860.

82. Kugelberg, E: Accommodation in human nerves and its significance for the symptoms in circulatory disturbances and tetany. Acta Physiol Scand (Suppl 24)8:7–105, 1944.

83. Kugelberg, E: Facial reflexes. Brain 75:385, 1952.

84. Kühne, W: Über die Peripherischen Endorgane der Motorischen Nerven. W Engelmann, Leipzig, 1862.

85. Lambert, EH: Electromyography and electric stimulation of peripheral nerves and muscles. In Mayo Clinic: Clinical Examinations in Neurology. WB Saunders, Philadelphia, 1956, pp 287–317.

86. Lambert, EH: Neurophysiological techniques useful in the study of neuromuscular disorders. In Adams, RD, Eaton, LM, and Shy, GM (eds): Neuromuscular Disorders. Williams & Wilkins, Baltimore, 1960, pp 247–273.

87. Langley, JN: Observations on denervated muscle. J Physiol (Lond) 50:335–344, 1916.

88. Langley, JN, and Kato, T: The physiological action of physosstigmine and its action on denervated skeletal muscle. J Physiol (Lond) 49:410–431, 1915.

89. Lapicque, L: Première approximation d'une loi nouvelle de l'excitation électrique basée sur une conception physique du phénomène. Compt Rend Societe Biol 62:615–618, 1907.

90. Lapicque, L: Actualitiés Scientifiques et Industrielles #624. Physiologie Générale Due Système Nerveux. Vol 5, La Chronaxie et ses applications physiologiques. Hermann and Company, Paris, 1938.

91. Lewis Jones, H: The use of condenser discharges in electrical testing. Proc R Soc Med 6:49–61, Part 1, 1913.

92. Licht, S: History of electrodiagnosis. In Licht, S (ed): Electrodiagnosis and Electromyography, ed 3. Waverly Press, Baltimore, 1971, pp 1–23.

93. Liddell, EGT, and Sherrington, CS: Recruitment and some other features of reflex inhibition. Proc R Soc B. 97:488–518, 1925.

94. Lindsley, DB: Myographic and electromyographic studies of myasthenia gravis. Brain 58:470–482, 1935.

95. Lippmann, MG: Unités Électriques Absolues. Georges Carré et C Naud, Paris, 1899.

96. Magendie, F: Expériences sur les fonctions des racines des nerfs rachidiens. J Physiol Exp Pathol 2:276–279, 1822.

97. Magladery, JW, and McDougal, DB, Jr: Electrophysiological studies of nerve and reflex activity in normal man. I. Identification of certain reflexes in the electromyogram and the conduction velocity of peripheral nerve fibres. Bull Johns Hopkins Hosp 86:265–290, 1950.

98. Marianini, S: Memoire sur la secousse qu'éprouvent les animaux au moment où ils cessent de servir d'arc de communication entre les pôles d'un électromoteur, et sur quelques autres phénomènes physiologiques produits par l'électricite. Ann Chimie Physique 40:225–256, Series 2, 1829.

99. Marinacci, AA: Clinical Electromyography. San Lucas Press, Los Angeles, 1955.

100. Marsden, CD, Merton, PA, and Morton, HB: Percutaneous stimulation of spinal cord and brain: Pyramidal tract conduction velocities in man. J Physiol 238:6p, 1982.

101. Matteucci, C: Sur un phénomène physiologique produit par les muscles en contraction. Ann Chimie Physique 6:339–343, 1842.

102. Matteucci, C: Traité des Phénomènes Électro-physiologiques des Animaux. Fortin, Masson, Paris, 1844.

103. Matthews, BHC: A special purpose amplifier. J Physiol (Lond) 81:28–29, 1934.

104. Meyer, M: Electricity in its relation to practical medicine. Translated into English from 3rd German Edition by Hammond, WA. D Appleton and Company, New York, 1869.

105. Morgan, CE: Electro-physiology and Therapeutics. William Wood and Company, New York, 1868.

106. Mottelay, PF: Bibliographical History of Electricity and Magnetism. Charles Griffin and Company, London, 1922.

107. Münnich, F: Über die Leitungsgeschwindigkeit im motorischen Nerven bei Warmblütern. Z Biol 66:1–21, 1916.

108. Neumann, E: Über das verschiedene Verhalten gelähmter Muskeln gegen den constanten und inducirten Strom und die Erklärung desselben. Deutsche Klinik 7:65–69, 1864.

109. Nobili, L: Analyse expérimentale et théorique des phénomènes physiologiques produits par l'électricite sur la grenouille; avec un appendice sur la nature du tétanos et de la paralysie, et sur les moyens de traiter ces deux maladies par l'électricité. Ann Chimie Physique 44:60–94, Series 2, 1830.

110. Piper, H: Weitere Mitteilungen über die Geschwindigkeit der Erregungsleitung im markhaltigen menschlichen Nerven. Pflugers Arch Ges Physiol 127:474–480, 1909.

111. Piper, H: Elektrophysiologie menschlicher Muskeln. Julius Springer, Berlin, 1912.

112. Proebster, R: Über Muskelaktionsströme am Gesunden und Kranken Menschen. Zeit fur Orthopadische Chirurgie (Suppl 2)50:1–154, 1928.

113. Remak, R: Galvanotherapie Der Nerven-und Muskelkrankheiten. August Hirschwald, Berlin, 1858.

114. Richardson, AT, and Wynn Parry, CB: The theory and practice of electrodiagnosis. Ann Phys Med 4:3–16, 1957.

115. Ricker, G: Beiträge zur Lehre von der Atrophie und Hyperplasie. Arch Path Anat Physiol 165:263–282, 1901.

116. Rogowicz, N: Ueber pseudomotorische Einwirkung der Ansa Vieussenii auf die Gesichtsmuskeln. Arch Ges Physiol 36:1–12, 1885.

117. Sanders, FK, and Whitteridge, D: Conduction velocity and myelin thickness in regenerating nerve fibres. J Physiol (Lond) 105:152–174, 1946.

118. Sarlandiere, C: Memories sur l'electro-puncture. L'Auteur and M Delaunay, Paris, 1825.

119. Schäffer, H: Eine neue Methode zur Bestimmung der Leitungsgeschwindigkeit im sensiblen Nerven beim Menschen. Dtsch Z Nervenheik 73:234–243, 1922.

120. Schiff, M: Ueber motorische Lahmung der Zunge. Arch Physiol Heilkunde 10:579–593, 1851.

121. Shea, PA, Woods, WW, and Werden, DH: Electromyography in diagnosis of nerve root compression syndrome. Arch Neurol Psychiatry 64:93–104, 1950.

122. Sherrington, CS: On the proprio-ceptive system, especially in its reflex aspect. Brain 29:467–482, 1906.

123. Simpson, JA: Electrical signs in the diagnosis of carpal tunnel and related syndromes. J Neurol Neurosurg Psychiatry 19:275–280, 1956.

124. Stålberg, E, and Trontelj, J: Single Fibre Electromyography. The Miraville Press Limited, Old Woking, Surrey, UK, 1979.

125. Thomas, JE, and Lambert, EH: Ulnar nerve conduction velocity and H-reflex in infants and children. J Applied Physiol 15:1–9, 1960.

126. Thomas, PK, Sears, TA, and Gilliatt, RW: The range of conduction velocity in normal motor nerve fibres to the small muscles of the hand and foot. J Neurol Neurosurg Psychiatry 22:175–181, 1959.

127. Volta, A: Account of some discoveries made by Mr. Galvani, of Bologna; with experiments and observations on them. Phil Trans R Soc Lond 83:10–44, 1793.

128. Volta, A: Collezione dell'opere del cavliere conte Alessandro Volta. Florence, 1816.

129. Wagman, IH, and Lesse, H: Maximum conduction velocities of motor fibres of ulnar nerve in human subjects of various ages and sizes. J Neurophysiol 15:235–244, 1952.

130. Waller, A, and Watteville, A: On the influence of the galvanic current on the excitability of the motor nerves of man. Phil Trans R Soc Lond, Part 3, 173:961–991, 1883.

131. Watkins, AL, Brazier, MAB, and Schwab, RS: Concepts of muscle dysfunction in poliomyelitis based on electromyographic studies. JAMA 123:188–192, 1943.

132. Weddell, G, Feinstein, B, and Pattle, RE: The clinical application of electromyography. Lancet 1:236–239, 1943.

133. Weddell, G, Feinstein, B, and Pattle, RE: The electrical activity of voluntary muscle in man under normal and pathological conditions. Brain 67:178–257, 1944.

134. Weiss, G: Technique d'Électrophysiologie. Gauthier-Villars, Paris, 1892.

135. Ziemssen, H: Die Electrictat in der Medicin. August Hirschwald, Berlin, 1866.

Appendix 2

FUNDAMENTALS OF ELECTRONICS

1 INTRODUCTION

The electromyographer must have a basic knowledge of electronics to understand physiologic signals, instrumentation, and electrical safety. Familiarity with electronics will help in recognizing and correcting recording problems, selecting and operating new equipment, and applying new techniques in the clinical domain. This appendix briefly introduces the essential topics in electronics for application to electromyography. Interested readers should consult a good text on basic electronics for a more detailed discussion.[4,5,6,8,12]

2 ELECTRICAL CONCEPTS AND MEASURES

Charge

The fundamental electrical concept is *charge.* Natural occurrences like lightning or static cling demonstrate the effects of charge. Physics accurately describes and predicts the behavior of "unit test charges" but does not provide any explanation or model for the source of the phenomenon. "Charge" is a name for observed effects in a theory that developed empirically.

The primary concept of charge is that two polarities of matter exist, called positive and negative. Negatively charged electrons revolve around a positively charged nucleus in all atoms. Other subatomic particles show positive or negative charge or a neutral state. No charge is smaller than the charge on one electron, and all measured amounts of charge are exact multiples of this smallest unit; so charge is quantized. The unit for measuring charge, called a *coulomb*, equals the charge on about 6.25×10^{18} electrons. The symbol "Q" commonly represents the amount of charge in equations.

Charged particles exert force on each other, called the *electrostatic force*, depending on the amount of charge and the distance between them. Charges of opposite polarity exert forces of attraction toward each other, analogous to gravita-

tional attraction. Charges of like polarity exert equivalent forces of repulsion. The *electric field* around a charged region describes the force on a unit test charge at any point. Because of these forces, charges moving in relation to one another either absorb or release energy (work).

Voltage

It requires energy to lift a brick over your head. The mass of the brick moves away from the mass of the earth, storing the work of separation, called *potential energy,* in the earth-brick system. Similarly, separating a system of charges requires (positive or negative) energy, stored as the "electric potential." The energy required per unit charge has dimensions of joules per coulomb or *volts.** Voltage refers to the difference in electric potential energy (for a unit test charge) between two points in space.

Like mechanical potential, voltage is a measure of difference relative to some reference. Lifting bricks has immeasurable effects on the huge mass of the earth; so ground level is often the reference (zero) level for calculating potential energy. Similarly, the earth is a huge sink for the dispersion of charge and is frequently the reference (zero) level for measuring voltage. Voltage is also called electromotive force (EMF), which accounts for the symbol "E" in equations for voltage, but "V" is also commonly used.

Conceptualizations and measurements in electronics use voltage much more frequently than charge. Voltages encountered in common electronic circuits range from a few microvolts (10^{-6}) to a few thousand volts. In electrophysiology, measured potentials (voltages) arise from the separation of charged atoms or molecules within the biochemical structures. Active transport of ions across a cell membrane exemplifies the expenditure of energy to separate charges, giving rise to a voltage difference.

---

*Recall that energy = force × distance (1 joule = 1 newton × 1 meter).

Current

Charge can move from one place to another by the motion of charged particles. Charge imbalance also propagates within conducting materials, perhaps like billiard balls in a row translate an impact. The latter mechanism transfers charge much faster than particle motion. *Current,* measured in *amperes* (also called amps), is the rate of charge flow. One amp of current is the flow of one coulomb per second. Currents typically encountered in common electronic applications range from microamps to several amps.

Resistance

Regardless of how charge propagates through a material, its flow results in some conversion of electric energy into heat. One might think of it as charge-carrying particles colliding with other atoms. The terms conductor, semiconductor, and insulator refer to the ease with which current flows through a material. The term *resistance* quantifies this effect. The loss of electric energy to heat manifests as a decreasing electric potential (voltage) in the direction of the current flow. Resistance is the ratio of this voltage difference between two points to the current flow:

$$\text{Resistance} = \frac{\text{Voltage}}{\text{Current}}$$

from which derives the more familiar form of OHM'S LAW:

Voltage = (Resistance) $\times$ (Current)

and also:

$$\text{Current} = \frac{\text{Voltage}}{\text{Resistance}}$$

Using the symbol "R" for resistance, and the symbol "I" for current ("C" being reserved for capacitance), these forms of OHM'S LAW are often expressed as:

$$R = \frac{E}{I} \qquad E = I \times R \qquad I = \frac{E}{R}$$

A good conductor has a relatively low value of resistance, and a good insulator has a relatively high value of resistance, the value judgment depending on the application. The resistance ratio may vary with temperature, voltage, or current, but often is assumed to be constant, for simplicity. The units of resistance are called *ohms*:

$$1 \text{ Ohm} = \frac{1 \text{ Volt}}{1 \text{ Amp}}$$

Resistances typically involved in common electronic circuits range from almost 0 ohms to several megohms (10^6 ohms). The unit kilohms (10^3 ohms) is also frequently used.

Power

Power is the time rate of energy flow. For steady conditions:

$$\text{Power} = \frac{\text{Energy}}{\text{Time}}$$

or

Energy = (Power) $\times$ (Time)

The unit of power, a *watt,* equals 1 joule of energy per second. From the definition of voltage above, energy in a charge flow equals the voltage (difference) times the amount of charge. Because current is the time rate of charge flow, then:

$$\text{Power} = \frac{\text{Energy}}{\text{Time}}$$

$$= \frac{(\text{Voltage}) \times (\text{Charge})}{\text{Time}}$$

$$= (\text{Voltage}) \times (\text{Current})$$

So the units of power also equal volts times amps, often abbreviated *VA.*

1 Watt = 1 Volt $\times$ 1 Amp

For example, if the headlights of an automobile draw 25 amps from the 12-volt battery, the total headlight power equals 300 watts.

Using Ohm's law, the power ("P") in a resistor is also calculated in the following forms:

$$P = E \times I = E \times \frac{E}{R} = \frac{E^2}{R}$$

$$P = E \times I = IR \times I = I^2R$$

3 ELECTRIC CIRCUITS AND CIRCUIT LAWS

Circuits and Schematics

Car headlights are an example of an electric *circuit,* an interconnection of components such that currents flow in one or more closed loops. Electrical systems take the form of circuits so that charge does not accumulate at any one point. Appendix Figure 2–1 shows the headlight circuit schematically.

A schematic diagram of an electrical circuit shows symbols for the various components and shows how they are interconnected. Most circuits of interest contain at least one source of energy, at least one component to dissipate energy (a *load*), some means of controlling the flow of energy, and conductors that allow energy flow between components (connect the components together). Schematics model real circuits by a number of simplifying approximations.

The solid lines represent ideal (zero-resistance) conductors which interconnect the components. Ideal sources of energy are the constant-voltage source and the constant-current source. A battery is a fair approximation to an ideal voltage source. A fixed resistance models the load that the headlights represent.

Resistors in Parallel

Suppose more headlights were connected across the battery in the circuit of Appendix Figure 2–1 above. The sche-

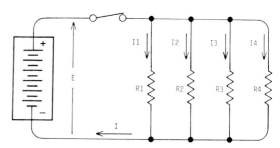

Appendix Figure 2–2. Resistances in parallel. A schematic of the headlight circuit with more lights.

matic of the circuit could be drawn as in Appendix Figure 2–2. It would seem reasonable that the total current from the battery would equal the sum of the individual load currents. Indeed, at any circuit *node,* a point where two or more conductors connect, charge does not accumulate. This leads to one of Kirchhoff's laws for electric circuits:

"The sum of all currents into a node equals the sum of all currents leaving a node."

Resistances connected end to end, as in Appendix Figure 2–2, are in *parallel.* Each resistance has the same voltage across it. So the current in each resistor can be calculated by Ohm's law, giving the total current from the battery as:

$$I = I_1 + I_2 + I_3 + I_4$$

$$I = \frac{E}{R_1} + \frac{E}{R_2} + \frac{E}{R_3} + \frac{E}{R_4}$$

rearranging:

$$\frac{I}{E} = \frac{1}{R_1} + \frac{1}{R_2} + \frac{1}{R_3} + \frac{1}{R_4}$$

or:

$$\frac{E}{I} = \frac{1}{\left(\dfrac{1}{R_1}\right) + \left(\dfrac{1}{R_2}\right) + \left(\dfrac{1}{R_3}\right) + \left(\dfrac{1}{R_4}\right)}$$

From Ohm's law, the above expression for E/I equals the effective resistance of the whole circuit in Appendix Figure 2–2. In general, the equivalent resistance of "n" resistors connected in parallel equals:

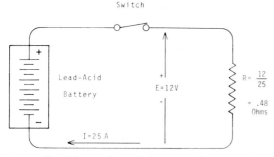

Appendix Figure 2–1. Schematic diagram of a headlight circuit. The switch controls the current by opening and closing the conducting path. The zigzag line is a symbol for resistance.

$$R_{equiv} = \cfrac{1}{\left(\cfrac{1}{R_1}\right) + \left(\cfrac{1}{R_2}\right) + \ldots + \left(\cfrac{1}{R_n}\right)}$$

which, for only two resistors, simplifies to:

$$R_{equiv} = \cfrac{1}{\left(\cfrac{1}{R_1}\right) + \left(\cfrac{1}{R_2}\right)} = \frac{R_1 \times R_2}{R_1 + R_2}$$

Note that the equivalent resistance of two or more resistors in parallel is always less than any one of the individual resistors. If there are more paths along which current can flow, there is less equivalent resistance. The total power in the circuit, the sum of the power in each individual resistance, equals the power calculated for the equivalent resistance.

Resistors in Series

The circuit of Appendix Figure 2–3 shows several resistors connected to a battery in *series*. Series connection of two components means they have a node in common that does not connect anywhere else. By Kirchhoff's current law, above, the same current must flow in all components connected in series. Another of Kirchhoff's circuit laws states:

"Around any closed loop, the algebraic sum of the voltage differences between nodes equals zero."

This is analogous to the principle of physics that the potential energy of an object

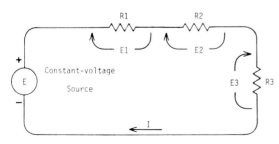

Appendix Figure 2–3. Resistors in series. The direction of the current follows the positive-current convention, as it does in Appendix Figures 2–1 and 2–2.

depends only on its height, and not on the path it followed to get there. Similarly, with some node as the reference point (zero voltage), the voltage at any node does not depend on the circuit path followed for computing it.

To apply Kirchhoff's voltage law, one must establish a convention for the polarity of voltages in relation to the current. First, one assumes a positive direction for the loop current. Engineers often use the "positive-current" convention, that current entering a resistor makes that end of the resistor positive. Many electronics texts will use the "negative-current" convention, that current entering a resistor makes that end negative. The polarity of the convention is irrelevant as long as it is consistent. Following either convention and using Kirchhoff's voltage law, above, results in a correct magnitude for the current, with a negative value if the assumed direction was wrong. Using this current value in the same convention will yield correct polarities for all component voltages.

To apply Kirchhoff's voltage law to the series circuit of Appendix Figure 2–3, one follows the direction of assumed current around the loop and adds the voltages algebraically. A voltage source has a fixed voltage across it regardless of the current magnitude or direction through it. From Appendix Figure 2–3 this process yields:

$$E_1 \quad + \quad E_2 \quad + \quad E_3 \quad - E = 0$$
$$(I \times R_1) + (I \times R_2) + (I \times R_3) - E = 0$$

Rearranging:

$$E = (I \times R_1) + (I \times R_2) + (I \times R_3)$$

or

$$E = I \times (R_1 + R_2 + R_3)$$

Therefore, from Ohm's law:

$$R_{equiv} = R_1 + R_2 + R_3$$

or in general for "n" resistors in series:

$$R_{equiv} = R_1 + R_2 + \ldots + R_n$$

Again, the total power in the circuit, the sum of the power in each resistor equals the power in the equivalent resistance.

Voltage Dividers

In the series circuit, the total voltage across the resistors equals the applied voltage from the battery. With the same current in all resistors, the voltage across each is proportional to its resistance. The applied voltage is "divided up" proportionately to the respective resistances. Taking the negative battery node in Appendix Figure 2–3 as the zero reference point, the voltage across R_3 is given by:

$$V_{R_3} = E \times \frac{(R_3)}{(R_1 + R_2 + R_3)} = k\ E$$

A fraction of the voltage applied to the series circuit of resistors appears across R_3. This frequently used *voltage divider* arrangement provides a voltage output that is always a fixed fraction of the voltage input.

4 CAPACITANCE

When a nonconducting region of space separates two conducting regions, charge cannot flow through the nonconducting medium. Within the conducting regions, charge can flow freely and distributes so there are no voltage gradients. If the charge (net charge imbalance) in one conducting region differs from the charge in the other, a voltage gradient or *electric field* exists across the insulating medium. For a steady charge difference, a fixed voltage difference is established between the conducting regions.

The physical properties of the nonconducting material and the geometry of the regions determine the amount of voltage for a given charge. The constant charge-to-voltage ratio is called the *capacitance*. Different insulating materials, *dielectrics* like air, glass, or plastics, affect the capacitance, compared with the intrinsic ratio of a vacuum. The electric field polarizes atoms or molecules of a dielectric material. Their alignment with the field reduces the voltage for a given charge, increasing the capacitance ratio. Some materials yield several thousand times the capacitance of a vacuum.

A *capacitor,* a two-terminal circuit ele-

ment, provides a certain amount of capacitance between its terminals. While many geometries of construction are used, the capacitor is often conceptualized as two parallel, rectangular plates of metal separated by an insulator. As in Appendix Figure 2–4, the schematic symbol for a capacitor is two separated, parallel plates. The unit of capacitance, a *farad,* equals 1 coulomb per volt. A 1-farad parallel-plate capacitor with 1 mm air dielectric would have plates about 10.5 km square. More common units of capacitance are the microfarad ($\mu F = 10^6$ F), nanofarad (nF = 10^{-9} F), and picofarad (pF = 10^{-12} F).

Connecting a capacitor across a voltage source causes a momentary surge of current while one plate acquires a positive charge and the other, a negative charge. When the voltage across the capacitor equals the voltage source, no current flows. When the voltage source is disconnected, the charges remain on the plates and the voltage remains across them. If connected to a resistor, the charged capacitor can supply some current until its charges dissipate (equilibrate).

Connected across a constant-current source, the charge pumped into a capacitor equals the current times time. So charge and voltage increase linearly with time. A current of 1 amp charges a 1-farad capacitor linearly to 1 volt in 1 second, a total charge of 1 coulomb. Current is the rate of charge flow; so current in a capacitor is proportional to the rate of

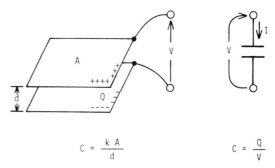

Appendix Figure 2–4. The parallel-plate capacitor and the schematic symbol for capacitance. Capacitance is proportional to the area of the plates and inversely proportional to the distance between them. The constant depends on the insulating material between the plates.

voltage change across it. Or the voltage across a capacitor is proportional to the integral of the current through it. This is another way to define a capacitor.

The property of having this voltage/current relationship, or the ability to store charge, is useful in many electronic circuits. A capacitance tends to oppose rapid changes of voltage across it, because that requires large currents. A certain amount of capacitance exists between any two insulated conductors, for example, between power lines on a pole and the earth. This "stray" capacitance must frequently be considered in electronic circuits. In the electrophysiology of excitable membranes, the capacitance of the membrane plays a considerable part. The very thin membrane, separating regions of fluid with a charge differential, forms a relatively large capacitance between the interior and the exterior of the cell: on the order of 1 μF/cm². This cell-membrane capacitance plays a major role in the time course of cell depolarization and repolarization.

RC Time-Constant Circuit

Consider a charged capacitor suddenly connected in parallel with a resistor (see App. Fig. 2–5). At any instant, the current equals the voltage divided by the resistance. From the definition of capacitance, the rate of voltage decline equals the rate of charge decline divided by the capacitance. So at any instant, the rate of

voltage decline equals the current divided by the capacitance. The charge on the capacitor will dissipate through the resistor until the voltage and current both go to zero. The rate of discharge will be greater initially and will also go to zero. Expressed mathematically:

$$i(t) = \frac{v(t)}{R}$$
$$\frac{dv}{dt} = -\frac{i(t)}{C}$$

leads to:
$$\frac{dv}{dt} = -\frac{v(t)}{C}$$

The solution of this *differential equation* for the voltage decline during discharge is an exponential function of time (shown in App. Fig. 2–6):

$v(t) = v(0)\, e^{-\frac{t}{RC}}$ assuming the resistor is connected at t = 0

$i(t) = \dfrac{v(t)}{R}$

$i(t) = i(0)\, e^{-\frac{t}{RC}}$

where the number "e" ($\sim$2.7183 . . .) is a special constant such that:

$$\frac{d[e^t]}{dt} = e^t$$

The factor RC, resistance times capacitance, has units of seconds and is called the *time constant* of the circuit, or of the exponential equation. The time constant equals the time it would take the voltage or current to reach zero if the discharge maintained its initial rate. Instead, the rate declines, and the discharge theoretically takes an infinite time to reach zero,

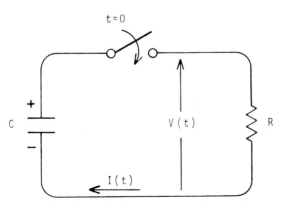

Appendix Figure 2–5. Schematic of the RC discharge circuit. The switch closes at time t = 0.

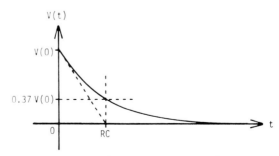

Appendix Figure 2–6. RC discharge voltage curve. After one time constant, the voltage is about 37 percent of its initial value.

although, practically, it approaches zero in about five time-constants. At the end of any interval of one time-constant, the voltage is about 37 percent of its value at the beginning. Therefore, after five time-constants, the voltage will be less than 1 percent of its initial value.

Capacitors in Parallel

Consider two capacitors connected in parallel, as in Appendix Figure 2–7. The voltage across both capacitors must be the same. The total charge in the combination is the sum of the charges on each capacitor. So the equivalent capacitance of two capacitors in parallel, the total charge divided by the voltage, equals the sum of the individual capacitances.

Capacitors in Series

Consider two capacitors connected in series, as in Appendix Figure 2–8. Any current in one capacitor must pass through the other, so the charges on each capacitor must be the same. This charge, Q, is the integral of current over all time up to the present, and so is also the charge in the equivalent capacitance. Dividing this charge by the total voltage across the combination, yields the equivalent capacitance:

$$C_{equiv} = \frac{C_1 \times C_2}{C_1 + C_2}$$

The charge on a capacitor represents some stored energy, equal to the work expended to move the charge there. In the ideal (lossless) capacitor, this amount of energy is available for release to the rest

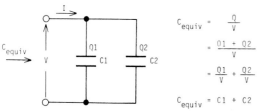

Appendix Figure 2–7. Capacitors in parallel. The equivalent capacitance is the sum of the individual capacitances.

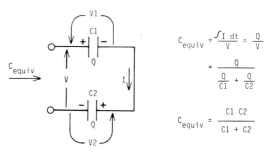

Appendix Figure 2–8. Capacitances in series. The equivalent capacitance is less than the smallest, as is the case with resistances in parallel.

of the electric circuit. It can be shown that the energy stored in a capacitor with capacitance "C," voltage "V," and charge "Q" is:

$$\frac{1}{2} CV^2 = \frac{1}{2} QV = \frac{1}{2} \frac{Q^2}{C}$$

5 INDUCTANCE

Magnetic Fields and Magnetism

A moving charge has an associated magnetic field. "Magnetic field" has no better theoretical explanation than "charge." Like charge, it has axiomatic descriptions in terms of observed forces and electrical interactions. Historically, the "laws" of magnetics arose empirically to form a consistent theory of the phenomena. A magnetic field exerts force on certain *ferromagnetic* metals and compounds. Why this occurs in certain materials involves the way the atoms, which have spinning charges, align themselves with a magnetic field. Some fundamental mechanism, called *magnetism*, couples force between charged particles in motion.

So a current flow has a magnetic field. Also, when a moving charge encounters a magnetic field from another source, it experiences a force. Certain geometries of current allow mathematically tractable magnetic field solutions—for example, current flowing in a line, as in a wire, or current flowing around a cylinder, as in a coil of wire. A magnetic field distribution specifies, for any point in space, the force on a "unit magnetic dipole."

The magnetic field intensity at a point is directly proportional to the current. A steady current has a time-invariant magnetic field. However, establishing this field stores energy in some mechanism, because energy is absorbed if the current is increased or released if the current is decreasing. Energy is "stored in the magnetic field," or the magnetic field "collapses." Current and magnetic field energy have a relationship quantitatively analogous to velocity and kinetic energy. Taking a mass from rest to some velocity absorbs energy, but no energy is required to maintain that velocity. An opposing force that reduces the velocity transfers energy into that force mechanism. The property of an electric circuit equivalent to the mass in this analogy is called *inductance.*

Magnetic Inductance

The name inductance comes from "induce." A time-varying magnetic field will induce current in a closed conducting loop. A varying current in one coil induces a voltage across the open ends of another coil in the same field. This is called *mutual inductance.* A changing magnetic field will also induce a current flow within any conducting material in the region of the field. Magnetic stimulation in electromyography relies on this principal to induce excitation current within the body fluid.

The increasing magnetic field of a coil with increasing current induces a voltage across the same coil, with a polarity that reflects energy absorption from the rest of the circuit. The decreasing magnetic field of a decreasing current induces a voltage across the coil, with a polarity that reflects energy return to the rest of the circuit. This phenomenon is called *self-inductance.*

An inductor is a two-terminal circuit element providing a certain amount of inductance between its terminals. Generally made from coiling some wire around a form or core, the two ends of the wire

coil become the two terminals. A coiled wire is the schematic symbol for inductance, as in Appendix Figure 2–9. The common symbol for amount of inductance in equations is "L" (derivation unknown). Coiling a wire increases the inductance to a useful level, although any conductor carrying current has some inductance. An ideal inductor has zero resistance between the terminals, and the inductance value is independent of current. In practical inductors, the wire has some resistance, and the core material has different magnetic properties at different field strengths. Inductors are less frequently seen in most electronic circuits than capacitors.

With zero resistance in an ideal inductor, the only voltage across its terminals is that induced by a changing magnetic field. If we assume negligible magnetic fields from any other circuits, the inductor voltage is directly proportional to its rate of current change. Inductance, the ratio of voltage over the rate of current change, has units of volts per amp per second or volt-seconds per amp, called *henries.* One henry of inductance has a 1-volt differential when its current has a gradient of 1 amp per second. This is a large unit in many applications (except power transformers), and the units of millihenry and microhenry are commonly used. A coil of 50 turns of wire on a nonmagnetic core 2 cm long and 1 cm^2 in area has an inductance of about 15 microhenries.

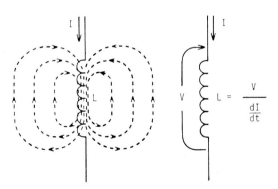

$$L = \frac{V}{\frac{dI}{dt}}$$

Appendix Figure 2–9. An inductance and its schematic symbol. Because of energy storage in its magnetic field, the voltage across an inductance is proportional to the rate of change of its current.

A coil in a vacuum has a certain intrinsic inductance for a given geometry. The same coil wound around various materials may have more or less inductance than in a vacuum, depending on how the atoms interact with a magnetic field and how well the material conducts induced currents. Analogous to the dielectric constant in capacitors, the property of *magnetic permeability* changes the amount of energy stored for a given current. Nonconducting ferromagnetic materials have high permeability. Some materials have relative permeabilities of several thousand. Inductors wound on high-permeability cores have their magnetic fields almost entirely concentrated within the core, a useful property. However, the magnetic permeability of materials varies greatly with magnetic field strength; the core tends to "saturate" and loose permeability as the field strength increases. This makes the inductance vary with current and makes circuits using such an inductor nonlinear.

RL Time-Constant Circuit

As a circuit element, the ideal inductor has a voltage proportional to the derivative of its current, or a current proportional to the integral of its voltage. This voltage/current relationship is another way of defining or conceptualizing an inductor. Inductance in a circuit tends to oppose rapid changes in current, because that requires large voltages. Consider a series circuit of a resistor and an inductor (App. Fig. 2–10) suddenly connected to

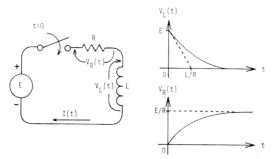

Appendix Figure 2–10. RL time-constant circuit. The current in the inductor rises exponentially to its final value.

voltage source (at $t = 0$). The sum of the resistor and inductor voltages equals the source, a constant. The inductor voltage equals the source voltage minus the current times the resistance. The inductor voltage also equals the derivative of the current times the inductance. Analogous to the capacitor discharge (above), a differential equation describes the resulting current, an exponential rise to the final value. This is expressed mathematically:

$$v_L(t) + v_R(t) = E$$
$$v_R(t) = i(t) \times R$$
$$v_L(t) = \frac{d[i(t)]}{dt} \times L$$

which leads to:

$$i(t) + \left(\frac{L}{R}\right)\frac{di}{dt} = \frac{E}{R}$$

whose solution is:

$$i(t) = \frac{E}{R}[1 - e^{-\frac{t}{\tau}}]$$

$$v_L(t) = E\, e^{-t}$$

where $T = L/R$ is the time-constant of the circuit.

At first, the current is zero, and the full voltage appears across the inductor. The current in an inductance cannot change instantaneously. Then the current rises exponentially to its final value of E/R, while the inductor voltage goes to zero.

Inductors in Series and Parallel

Consider circuits that have two inductors in parallel or in series, under the condition that the two fields do not significantly overlap (not coupled). Two inductors in series have the same current, the same derivative of current, and the same polarity of voltage in relation to the current. Therefore, the voltage across the series combination is the sum of the voltages across each, and the equivalent inductance equals the sum of the individual inductances (the same relationship as resistors in series).

By analysis similar to that used for resistors in parallel, with rate of current change instead of current, the equivalent inductance of two inductors in parallel is given by:

$$L_{equiv} = \frac{L_1 \times L_2}{L_1 + L_2}$$

which is the same relationship as for resistors in parallel.

The magnetic field around an inductor carrying current represents some stored energy, equal to the work required to establish the field. In an ideal (lossless) inductor, this same amount of energy is available for release to the rest of the electric circuit. It can be shown that the energy stored in inductance "L," with current "I," equals:

$$\frac{1}{2} LI^2$$

Transformers

Two coils sufficiently close together that their magnetic fields occupy significant common space have mutual inductance between the separate coil circuits. A changing current in one coil induces a voltage in the other. The *transformer*, a common electronic circuit component, utilizes this effect. When both coils are wound on a highly permeable "core," the energy coupling between the two becomes very efficient. Power transfers from one coil to the other with little loss. Transformers proportionately increase or decrease voltages or currents, and they couple energy from one circuit to another without an electrically conducting path.

An "ideal" transformer, a four-terminal circuit element, multiplies the voltage across two of the terminals by a constant, the *turns-ratio*, to the other two terminals. Since power remains the same, the current is divided by the same constant. Two of the terminals are one coil, often called the "primary" winding, and the other two terminals are the "secondary" coil or winding, with infinite resistance between the windings (App. Fig. 2–11).

Practical transformers have limitations of power loss, maximum power capability, and frequency of fluctuations. Real windings have some resistance in the wire. Some core materials lose their ef-

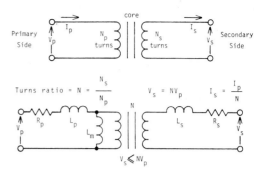

Appendix Figure 2–11. Transformer symbols. The ideal transformer and a simple linear model of a real transformer.

fective permeability at high frequencies of field fluctuation. At very low frequencies, losses become greater than the energy transfer, and transformers become impractical. A steady current in some coil does not induce any voltage in the other.

6 AC CIRCUITS

The term "AC," for *alternating current*, has two meanings in electronics. The literal meaning refers to voltages or currents which reverse in polarity at regular intervals, especially sinusoidal waveforms. The output of a rotating generator or alternator has a sinusoidal shape. A coil rotating in a fixed magnetic field generates voltage proportional to the sine of the angle between the coil plane and the field. This kind of AC, as shown in Appendix Figure 2–12, is characterized by a frequency, an amplitude, and a "phase." The phase specifies the time shift of the waveform, in degrees of angle, relative to a reference sine wave of the same frequency.

Another common meaning for "AC" in electronics is that portion of a fluctuating voltage or current with zero average value over a long time, as opposed to the "DC" (for *direct current*) component, the long-term average value. AC fluctuations could be complex, random, or nonperiodic. For example, the potential between a pair of skin electrodes has a non-zero average value attributable to metal/electrolyte interfaces. Subtracting this average value

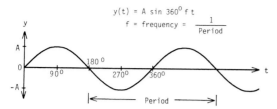

Appendix Figure 2–12. The sine function.

leaves the AC component, a varying potential including biopotentials, noise, and interference.[3]

AC Circuit Laws

DC circuit theory, the circuit laws and calculations considered above, extends to circuits excited by AC sources. A sinusoidal source causes sinusoidal voltages and currents of the same frequency throughout any linear circuit. One can represent such values in the circuit by amplitude and phase information only. The common measure of AC amplitude, the "RMS value," stands for *root-mean-square*, the square root of the time average of the waveform squared. The RMS amplitude of a voltage or current equals the constant (DC) magnitude that has the same power, the same heating effect in a resistor. Referring to an ordinary outlet as "110 volts" means that the AC potential has an RMS value of 110 volts. This sinusoidal voltage typically has a frequency of 60 cycles per second (called *hertz*), with peak voltages of about +155 volts and −155 volts during the cycle.

Impedance and Reactance

In purely resistive AC circuits, the phase of all voltages and currents remains the same. One can solve for the AC values exactly as with DC circuits, by using RMS amplitudes. For example, in the headlight circuit of Appendix Figure 2–1, if the voltage source was 12 volts AC (RMS), then the current would be 25 amps AC (RMS), and the average power would still be 300 watts.

If the circuit contains capacitors or in-

ductors, however, the analysis gets more complex. AC voltages or currents from sinusoidal sources have the same frequency but have different phases throughout the circuit. Thus, RMS amplitudes alone do not specify the AC values, and RMS values of different phases do not add or subtract directly.

If a sinusoidal current passes through a capacitor, the AC voltage across the capacitor "lags" the current in phase by 90 degrees. When the current is crossing zero, reversing polarity, the voltage is at a peak, reversing slope. When the current is at a peak, the voltage is crossing zero, the point of maximum slope. One could also say the current "leads" the voltage by 90 degrees, as shown in Appendix Figure 2–13.

For an inductor the roles of voltage and current are reversed from above. The voltage leads the current, or the current lags the voltage, by 90 degrees, as shown in Appendix Figure 2–14.

For any component or combination of components in an AC circuit, the ratio of voltage to current is called the *impedance*, analogous to DC resistance in Ohm's law. Whereas resistance is a constant, impedance is a two-dimensional quantity, requiring the specification of magnitude and phase angle, both of which may vary with frequency. The impedance of a pure capacitor or inductor is called a *reactance*. An arbitrary impedance (any phase angle) can be divided into resistive and reactive components.

The magnitude of inductive reactance increases with increasing frequency while the phase remains +90 degrees. A more rapid current variation through an inductor (at constant amplitude) induces

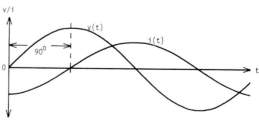

Appendix Figure 2–13. AC voltage and current in a capacitor. The voltage lags the current by 90 degrees.

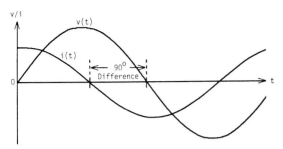

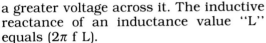

Appendix Figure 2–14. AC voltage and current in an inductor. The voltage leads the current by 90 degrees.

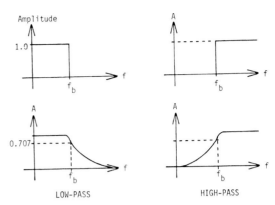

Appendix Figure 2–15. Ideal and practical filter response curves. **A.** Low-pass. **B.** High-pass.

a greater voltage across it. The inductive reactance of an inductance value "L" equals $(2\pi f L)$.

The magnitude of capacitive reactance decreases with increasing frequency while the phase remains −90 degrees. A more rapid voltage variation across a capacitor (at constant amplitude) requires greater current flow. The capacitive reactance of a capacitance value "C" equals $1/(2\pi f C)$.

AC Power

An ideal reactance does not dissipate any energy. Energy may be stored or released, but none is lost. The instantaneous power in a capacitor or inductor, the instantaneous voltage times current, can be positive or negative, but the average power equals zero. Distributed resistance accounts for the power loss in real reactances.

7 FILTERS

In electronics, a filter usually means a circuit that passes some bands of frequency while attenuating others. The effects of filters are often displayed in the "frequency domain" by graphing the output magnitude or the attenuation ratio versus frequency for constant-amplitude sine wave inputs. Examples of electronic filters are bass and treble tone controls or graphic equalizers in stereo music systems. The most common types of filters

are low-pass (high-cut), high-pass (low-cut), band-pass (low- and high-cut), and notch (center-cut) filters. Appendix Figure 2–15 shows real and ideal response curves for high-pass and low-pass filters.

Complex signals composed of a spectrum of frequencies, such as a voice signal or a compound action potential, can often only be described as a graph of component magnitudes versus frequency. Multiplying such a graph times the attenuation curve of a filter point by point in frequency yields the frequency spectrum of the output if such a signal passed through the filter, as shown in Appendix Figure 2–16.

The simple RC network of Appendix Figure 2–17A forms a low-pass filter. Appendix Figure 2–17B shows its attenuation curve. At very low frequencies the capacitor has high impedance and causes

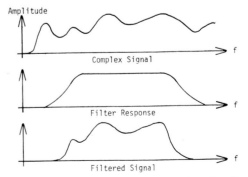

Appendix Figure 2–16. Frequency-domain effects of band-pass filtering.

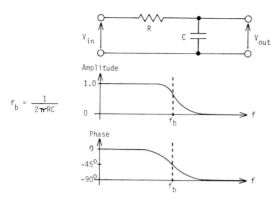

$$f_b = \frac{1}{2\pi RC}$$

Appendix Figure 2–17. RC low-pass filter-network. **A.** Schematic. **B.** Attenuation curve. **C.** Phase curve.

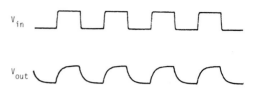

Appendix Figure 2–18. Time-domain effects of low-pass filtering. Note the slowing of abrupt transitions in the square-wave (calibrating) signal, creating a delay.

negligible attenuation. At very high frequencies the capacitor impedance approaches zero, as does the output magnitude. The transition from pass-band to stop-band occurs gradually, with no sudden discontinuities in real filters. The frequency where the attenuation ratio equals 0.707 (−3 dB) is called the "break" or "corner" frequency, where output power equals one-half input power in loads of equal resistance. This is also the frequency where the magnitude of the capacitive reactance equals the resistance, leading to the break frequency equation in Appendix Figure 2–17B. This corner frequency is generally taken as the cutoff point, making the pass band, or *bandwidth,* of the low-pass filter from DC (0 Hz) to the break frequency.

In order to specify a filter response curve completely, one must also specify the phase of the output relative to the sine wave input at each frequency. Appendix Figure 2–17C shows the phase response of the RC low-pass filter. Note that significant phase shift occurs at frequencies where the amplitude attenuation is still relatively insignificant. A negative (lagging) phase indicates a delay in the sine wave response and, indeed, in the time response to a transient signal. Low-pass filters increase the latency of fast peaks and limit the speed of transition at the output, or the rise and fall times of a "square-wave" input. Appendix Figure 2–18 shows the effect of low-pass filtering on a calibrating signal.

The RC network of Appendix Figure 2–19A forms a simple high-pass filter, with the attenuation curve shown in Appendix Figure 2–19B. At very high frequencies the capacitor has low impedance and causes negligible attenuation. At very low frequencies the capacitor impedance becomes very large, and the output amplitude approaches zero. The break frequency of this high-pass filter has the same value as the RC low-pass filter (above).

The phase response of this high-pass filter, Appendix Figure 2–19C, has a phase lead of 45 degrees at the corner frequency, an effective negative delay for steady-state sine wave inputs. This apparent anticipation is indeed seen as reduced latency for transient signals with high-pass filtering, not that the circuit could create a negative delay, but because the attenuation of slowly varying components causes the response to peak earlier at reduced amplitude. High-pass filters suppress a slowly varying baseline shift and cause a droop in the response to

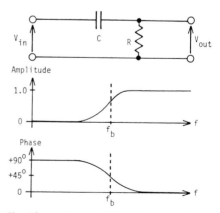

Appendix Figure 2–19. RC high-pass filter network. **A.** Schematic. **B.** Attenuation curve. **C.** Phase curve.

square-wave signals, such as the calibration signal in Appendix Figure 2–20.

These simple RC high- and low-pass filters, called *first order*, have an attenuation slope in the stop band proportional to frequency; attenuation doubles at each octave of frequency. Higher order filters can have more abrupt descent into the stop band, but also have greater phase shift in the passband and sharper phase transitions near the corner frequency. Higher-order or multistage filters can have complex, biphasic responses to sharp transitions or spikes, which may mimic or mask physiologic responses.

Band-pass filters are low-pass and high-pass filters combined, with overlapping pass bands in the center. With the two corner frequencies far apart, a wide passband, frequencies in the middle have little attenuation or phase distortion. As the two corners become close together, making a narrow passband, phase shifts become significant and complex in the passband, causing distortion. Sharp LC band-pass filters are used at radio frequencies for tuning. Amplifiers for EMG and other electrophysiology use wide band-pass filters, with adjustable low- and high-frequency cutoffs, to eliminate baseline shifts, undesirable components, and excessive noise.[7,9,11]

Notch filters pass all frequencies except a small band. The common notch filter encountered in electrophysiology is the "60-Hz filter," generally optional to reduce power line interference. Appendix Figure 2–21 shows a typical 60-Hz notch filter amplitude and phase response. While good filters have extremely narrow amplitude notches, the phase distortion can be significant over a much broader band. Use of notch filters in electromyography should be limited to cases where no recording would be obtained otherwise, and measurements qualified in that light.[10]

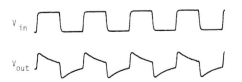

Appendix Figure 2–20. Effect of high-pass filter on square-wave (calibrating) signal.

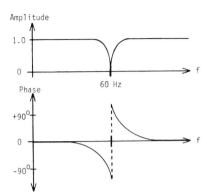

Appendix Figure 2–21. Amplitude and phase curves for a 60-Hz notch filter.

8 SOLID-STATE DEVICES

Active and Passive Circuit Elements

Passive devices, resistors, capacitors, inductors, and transformers, have a constant proportionality between the voltage and current at their terminals, at least within a range of linearity, and they add no power to a circuit. An active device has voltage/current relationships that can vary in response to some circuit parameter, and they can add power to the circuit.

Diodes

An ideal *diode*, a two-terminal nonlinear device, has zero resistance ("short circuit") for current flowing in one direction and infinite resistance ("open circuit") for current flowing in the other direction. Therefore, one terminal is distinguished from the other. Real diodes have some resistance to current in the "forward" direction, typically nonlinear with current; and they have some leakage current and breakdown voltage in the "reverse" direction. The name "diode" carries over from the days of vacuum tubes, when a tube with two electrodes implemented this element. Today, most diodes are implemented in the solid state, in crystals of semiconducting material like silicon, germanium, or others, doped in different regions with various elements to alter their conduction properties.

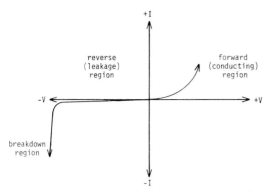

Appendix Figure 2–22. V/I curve of a semiconductor diode.

Graphs of current versus voltage, V/I curves, visually describe the characteristics of nonlinear devices. Appendix Figure 2–22 shows the V/I curve of a semiconductor diode. In the forward direction, the current is essentially an exponential function of voltage. In the reverse direction, a small leakage current flows unless the reverse voltage becomes sufficient to cause breakdown of the diode.

Diodes find frequent use in electronic circuits to restrict current flow to predominantly one direction. Applied to an AC source, this creates a unidirectional supply which can be filtered and regulated to become a DC source. This conversion of AC power into DC power is called *rectification.* Diodes can switch currents between different circuit paths; and they can implement simple logic functions. Special diodes also find use as light emitters (LEDs), light detectors, voltage regulators, temperature sensors, and voltage-variable capacitors.

Transistors

The name "transistor" was a contraction of "transfer resistor," referring to a model whereby current in one loop modulated the resistance of another loop. This effect enables the transistor to amplify the control-loop current.

Transistors are made in crystals of pure semiconducting elements, predominantly silicon, by diffusing impurity elements into different regions of the crystalline structure. In "bipolar" transistors, a small input current facilitates current flow in the output circuit, and thus the input current variations can be multiplied several hundred times in the output circuit. "Field-effect" transistors (FETs) employ a different mechanism. The input voltage creates an electric field, which modulates the transistor conductivity in the output circuit, allowing large output currents to be controlled with very little input current (or power).

Transistors replaced vacuum tube amplifers because of their smaller size and much greater power efficiency. These attributes made electronic applications, such as computers, practical, reliable, and inexpensive where vacuum tube circuits would be very impractical.

Integrated Circuits

Integrated circuits contain many transistors, diodes, resistors, and small capacitors in a single silicon crystal "chip" with interconnections to form a complex circuit block. Using processes with very small geometries, hundreds of thousands of such components can be integrated on chips several millimeters square. Functions available as integrated circuits include logic blocks, amplifiers, microprocessors, memory blocks, speech synthesizers, and filters.

Circuit integration has many advantages. Complex functions occupy a small space, with few external interconnections. Less stray capacitance allows lower power levels and higher speeds. This results in greater reliability at a lower cost and easier maintenance by replacement. A host of standard integrated circuits solve many design problems with a building-block approach. Current technology is making it more practical to implement new applications with custom integrated circuits.

9　DIGITAL ELECTRONICS

Digital and Analog Circuits

An electrical circuit used for *analog* purposes means that a voltage or current is proportional to some measurement that varies in a continuous (smooth) fashion. Transducers provide analog electrical signals from various physical phenomena such as pressure, oxygen concentration, light, temperature, muscle force, and so forth. Biopotentials are analog electrical fluctuations proportional to electrochemical activities.

An electrical circuit ascribed to *digital* purposes has a discrete number of "states" represented by voltages or currents falling within a designated range of values. For example, a wire from a switch to monitor the position of a microwave oven door could have a potential in the range of 0 to 2 volts with the door closed, and in the range of 3 to 5 volts with the door open. The circuit design should make the "door state" signal well within the specified ranges over all reasonable conditions of variability, such as temperature, supply voltage, and manufacturing tolerances. The range of 2 to 3 volts would be an indeterminate band indicating abnormal operation or failure.

From this one can see that a "digital" voltage represents much less information than an "analog" voltage, but the digital voltage conveys its information with much greater reliability and accuracy. The most commonly used digital circuits have just two states, variously named on/off, true/false, high/low, or active/inactive. A digital system could assign three or more states to an electrical quantity, but that would reduce reliability and increase complexity. Instead, to convey more information, more digital circuits are used simultaneously. The major advantage of a digital system is its immunity to electrical noise, interference, and component tolerances. The major disadvantage of digital circuits is that they limit variables to a discrete number of values.

By nature, many applications lend themselves well to digital representations. Integer arithmetic involves numbers as a series of digits; each digit has a discrete number of values. Many operations of machines or processes occur as a number of states. A common furnace thermostat is a good example of a digital circuit, because the furnace fire is either on or off to regulate temperature, not proportionally controlled. Digital circuits can perform the mathematical "logic" in-

volved in many control procedures: "IF the door is open, AND the start button is pressed, THEN inhibit the microwave power generation."

Mathematical Logic

Boolean algebra, the mathematics of variables having only two states, is often called *logic*, when considering the states as "true" or "false." Using voltages to represent these states, digital circuits can perform Boolean operations on variables. A *combinational* logic system has variables derived only from operations on the current states of other variables. *Sequential* logic involves variables depending also on the past states of variables. Introducing the concept of past states requires the system to have memory and a sense of time passage, a clock.

The basic operations of combinational logic, *AND, OR,* and *NOT,* together form more complex operations. The *AND* operator on two variables says:

If A is true and B is true, only then (A *AND* B) is true. The *OR* operator on two variables says:

If A is true or B is true (or both), only then (A *OR* B) is true. The *NOT* operation inverts one variable:

If A is true, then (*NOT* A) is false;
If A is false, then (*NOT* A) is true.
A combination of these gives the *EXCLUSIVE-OR* operation:

If A is true or B is true, but not both, only then (A *EXCLUSIVE-OR* B) is true.

(A *EXCLUSIVE-OR* B) = (A *OR* B) *AND* [(*NOT* A) or (*NOT* B)]

Large systems of combinational and sequential circuits can implement very complex logic functions, such as a digital watch or a computer.

Binary Number System

Our decimal number system uses one of 10 characters (0 to 9) in a digit place and as many digit places as necessary to represent a number. Equally valid are other number systems with more or less characters in the digit set. The binary number system has only two characters, 0 and 1, and thus requires many more digits to represent a number than the decimal system. Each digit place of a binary representation is called a *bit*, from the contraction of "binary digit."

Computer systems do counting and arithmetic in the binary system because of the reliability of on/off digital circuits. Modern systems make this use of binary completely transparent to the user. Data can be input and output in decimal, freeing the typical user from any need to understand binary representations, or any other number systems. It may be useful to understand the powers of two, as these quantities occasionally surface. For instance, 8 bits can represent 256 combinations; 10 bits, 1024 ("1K"); 16 bits, 65,536; and so forth.

Converting Between Analog and Digital Representations

Converting an analog voltage into a digital representation requires an assemblage of analog and digital circuits called an A-to-D converter, which generates a binary number proportional to the instantaneous value of the analog input voltage. The digital representation includes only a finite number of discrete values according to the number of bits implemented. Dividing the maximum analog input range by the number of digital output combinations gives the 1-bit resolution of the converter, the *digitizing error* of the process. For example, a furnace thermostat makes an A-to-D conversion of the room temperature (minus the set point) into a 1-bit (on/off) control signal. Biopotential averaging equipment often makes 8-bit or 10-bit conversions of amplified electrode signals. This digital value represents the amplitude of the biopotential at one instant in time. Repeating the conversions at sufficiently rapid rates allows the waveform over a limited interval to be approximated by an array of digital values. Digital circuits can then store and manipulate the waveform as a set of numbers.[1]

The A-to-D conversion process requires some amount of time, setting the minimum time between samples, and the maximum frequency resolution, of the analog waveform. The sampling speed determines the memory requirements to store an analog signal as a set of sample values, or the maximum interval one can store in a given amount of memory.[2]

D-to-A conversion, converting a digital representation into a proportional analog voltage, results in only a discrete number of steps in the "analog" output, of course. Uses for D-to-A conversion include driving a plotter, displaying a digitized signal on an cathode-ray tube (CRT) screen, and setting the intensity of a stimulus by means of software.

REFERENCES

1. Cooper, R, Osselton, JW, and Shaw, JC: EEG Technology, ed 3. Butterworth & Co., Boston, 1980.
2. Gans, BM: Signal Extraction and Analysis. A Primer for Clinical Electromyographers. Mini-monograph #12, American Association of Electromyography and Electrodiagnosis, 1979.
3. Geddes, LA: Electrodes and the Measurement of Bioelectric Events. John WIley & Sons, New York, 1972.
4. Grob, B: Basic Electronics. McGraw-Hill, New York, 1977.
5. Heath Company: Heath Continuing Education Series in Electronics. Heath Company, Benton Harbor, Michigan.
6. Herrick, CN and Deem, BR: Introduction to Electronics. Goodyear Publishing Company Inc., Pacific Palisades, California, 1973.
7. McGill, KC, et al.: On the Nature and Elimination of Stimulus Artifact In Nerve Signals Evoked and Recorded Using Surface Electrodes. IEEE Trans Biomed Eng BME-29:129, 1982.
8. Mottershead, A: Introduction to Electricity and Electronics. John Wiley & Sons, New York, 1982.
9. Reiner, S and Bogoft, JB: Instrumentation. In Johnson, EW (ED): Practical Electromyography. Williams & Wilkins, Baltimore, 1980.
10. Stolov, W: Instrumentation and Measurement in Electrodiagnosis. Minimonograph #16, American Association of Electromyography and Electrodiagnosis, Rochester, Minnesota, 1981.
11. Walker, DD and Kimura, J: A Fast-Recovery Electrode Amplifier for Electrophysiology. Electroenceph Clin Neurophysiol 45:789, 1978.
12. Yanof, HM: Biomedical Electronics, ed 2. FA Davis, Philadelphia, 1972.

Appendix 3

ELECTRICAL SAFETY

1 INTRODUCTION

All personnel involved in recording bioelectric potentials should be knowledgeable about electrical safety. Electrical safety recommendations are concerned with detecting or preventing dangerous situations. Violating safety standards or neglecting inspections could invalidate insurance coverage or accreditations. If an accident occurs, individuals or institutions could face charges of negligence or malpractice. In addition, some of the measures intended to ensure safety also reduce artifacts and interference in the recording.

The standards and recommendations for electrical safety have changed frequently over the last 10 years. Electromyographers and staff should understand not only the most current regulation, but also the theory of electrical safety.

2 THE ELECTRICAL HAZARD SITUATION

In the United States, the common electrical power is distributed from transformers as 120 VAC (volts alternating current) at 60 Hz frequency, as shown in Appendix Figure 3–1. The center wire of this transformer supply, called the "neutral" or "cold line," connects to the earth ("ground"). The other wires from the transformer, called the "hot" lines, have 120 VAC with respect to the neutral, and thus to the earth (240 VAC is available between the two hot phases for high-power equipment). Touching any hot line while in contact with some conductive path to the earth would cause a shock.

In healthy people, the sensation of shock from a steady application of 60 Hz AC occurs from about 1 mA of current and above. So from a 120 VAC source, a conductive path of 120 kΩ impedance or less could cause a shock. The impedance

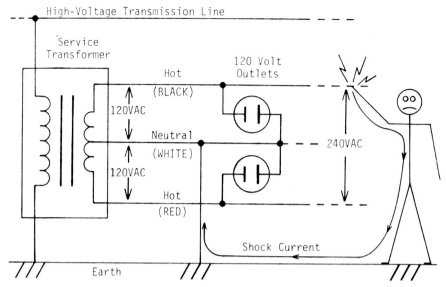

Appendix Figure 3–1. 120 VAC power distribution circuit. The hot lines will supply current through any path to ground.

from one hand to the other in grasping wires is on the order of 50 kΩ, attributable almost entirely to the dry surface layer of the skin. For currents on this order, the sensation of shock causes a jerk-back reflex. At somewhat higher currents, the shock itself may stimulate nerves or muscles. A still higher current may tetanize muscles so that the person cannot release the shock source. If enough current flows across the body, it may induce cardiac fibrillation.

In debilitated patients or those with the surface layer of skin penetrated by a conductor, such as a needle, much less current causes a serious shock. In these "electrosensitive patients," as little as 50 μA can cause cardiac fibrillation because of a direct path via a cardiac catheter, for example.[4] This has also been termed "microshock."

Any line-powered equipment, and especially devices that have a metallic case, could be a safety hazard. Capacitance between the case and wiring, fluids spilled in the machine, or failed insulation could provide an accessible conductive path to the hot line. "Leakage current" will flow through such paths to any earth ground. The hazard is greater, therefore, in areas where earth grounds abound, such as bathrooms, kitchens, basements, out-

doors, or other wet areas. Most metal plumbing pipes are good earth grounds.

Appendix Figure 3–2 shows equipment leakage paths to its chassis. With good insulation, the resistive path should conduct very little. The capacitive path is always present to some degree and accounts for most of the "normal" leakage current. A hazard occurs if these leakage paths become sufficient to conduct an unsafe level of current.

To reduce the hazard of leakage current, modern wiring systems incorporate a separate earth ground wire in the outlets and power cords, sometimes called the "third-wire ground," or "safety ground." These outlets and plugs have three pins: neutral, hot, and earth ground. The earth ground wire connects to the chassis of the equipment and any other exposed metal, conducting the leakage current to earth.[5] The chassis remains at ground potential, and no current flows to a grounded person, as in Appendix Figure 3–3.

The abundance of equipment and fluids in the hospital environment demands a three-wire grounded electrical system for safety. The patient may also have abnormal susceptibility to shock because of health conditions or invasive attachments. Any equipment that is elec-

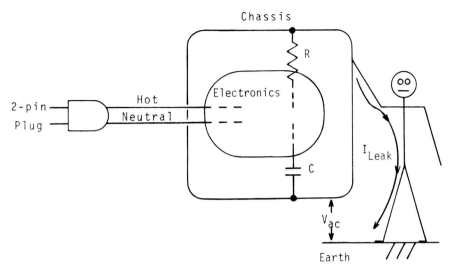

Appendix Figure 3–2. Equipment leakage paths between the hot line and the chassis. A ground person touching this chassis would conduct the leakage current to earth.

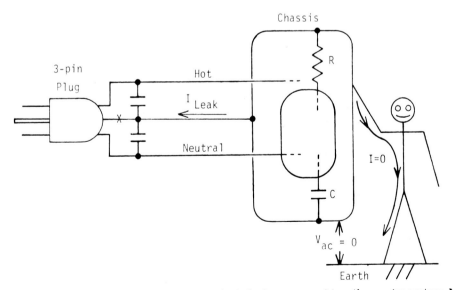

Appendix Figure 3–3. The safety ground wire conducts leakage current in a three-wire system. Notice that the capacitance between the hot and ground wires in the power cord adds some leakage current. If the ground connection is broken at the plug ("X"), then this capacitance becomes an additional leakage path to the chassis.

trically connected to the body presents a much greater risk because the patient cannot quickly break the conductive path by reflex.

3 THE PRIMARY SAFETY PROBLEM — LEAKAGE CURRENT AND LOSS OF GROUND

The third-wire "safety" ground basically solves the problem of hazard from leakage current, provided that no components of the system fail. If this connection opens somehow, then leakage current could flow through a patient or operator to some other ground. If, in addition, the leakage current of the equipment is above safe limits, then a hazardous situation exists. Safety standards and recommendations attempt to prevent or detect this possibility by testing the integrity of the ground system and the leakage current levels in the absence of a ground. Appendix Tables 3–1 and 3–2 list some kinds of faults that can lead to electrical hazards.

4 ADDITIONAL SAFETY CONCERNS

Several pieces of equipment in the same room with a patient increase the likelihood of electrical hazard, especially if more than one is connected to the patient. This situation occurs commonly with multiple monitors in operating rooms or intensive care units. Such a situation may call for portable electromyog-

Table 3–1 COMMON KINDS OF FAULTS THAT COULD RESULT IN LOSS OF GROUND

Broken ground pin on equipment power cord
Broken ground wire in power cord
Poor ground connection inside equipment
Poor earth ground connection to outlet
Weak contact tension between outlet and plug
Corroded, bent, or broken pins on power cord or outlet
Ground system defeated with two-pin adaptors or extension cords
Use of equipment in old or faulty wiring systems

Table 3–2 COMMON KINDS OF FAULTS THAT COULD RESULT IN EXCESSIVE LEAKAGE CURRENT

Failed insulation in equipment or cord
Fluids spilled in or on equipment
Use of extension cords on equipment
Improperly wired outlets—reversed polarity or reversed neutral/ground
Electrical faults in equipment circuits
Unapproved equipment

raphy, or patients may come to the laboratory with ancillary equipment attached.

If any one machine has loss of ground, then a patient touching it or connected to it would have their whole body at some AC potential above ground, due to its leakage current. The proximity of other machines increases the likelihood that the patient could also touch a ground, becoming a path for that leakage current. If they are already connected to another machine, then these connections may conduct to ground at that AC potential. In fact, some equipment may ground the patient directly.

Another hazardous situation could arise from multiple equipment if the earth grounds at their various outlets have some AC potential difference between them. As little as 50 mV difference between the grounds could cause a hazardous current to flow through a patient from one ground connection to the other. Voltage between different grounds could result from fault currents flowing in the ground, improper wiring, or magnetic induction from other wiring. The wiring in patient areas should use a concept known as the "equipotential ground bus." In this system, each of the receptacles in one room has a separate ground connection to a common point. That point ties to earth ground by a wire that does not connect anywhere else.[7]

The above hazards are greatly reduced if all the patient connections from all the instruments are "isolated." An isolated connection will not conduct more than 20 μA even if its potential is 120 VAC to ground. Patient leads are isolated by using nonconductive coupling methods or current-limiting devices. Any isolation can fail if subjected to voltages above its rating, and any isolated circuit has some small leakage current to ground.

Battery-powered devices have no connection to earth and no hot power wires to leak. However, battery devices are not necessarily completely safe. Under fault conditions they can supply enough current to endanger the patient, for example, when fluid is spilled in a battery-powered instrument. Hence, patient connections should still have current-limiting devices, especially with electrosensitive patients.[9]

5 SAFETY REGULATION DOCUMENTS

The following are some of the documents that regulate the manufacture and maintenance of equipment as well as the wiring of hospitals, homes, and private offices:

National Fire Protection Association (NFPA), 1981
Underwriters Laboratories (UL), Inc., 1980
Joint Commission on Accreditation of Hospitals (JCAH), 1982
Veterans Administration (VA), 1978[12]

The section entitled *Electricity in Health Care Facilities* of the National Electrical Code of the NFPA (1981) specifies standards for the wiring of examination or care areas. It requires that all patient areas have three-pin grounded outlets, where the earth grounds are connected with a separate third wire. The use of metal conduits, raceways, or junction boxes to supply the earth ground connections is *not* adequate. Some older wiring systems may not comply with this specification even though they have three-pin outlets.[8]

UL's *Standard for Medical and Dental Equipment*, *UL544* contains specifications for the performance of equipment. These include a variety of electrical and mechanical safety standards, as well as labeling and documentation requirements. To be listed as UL544, equipment must be submitted to UL for testing. The use of equipment without a UL rating

may invalidate accreditations or insurance coverage. Equipment that is UL rated will bear appropriate stickers or insignia, usually on a rear panel where electrical ratings are listed.[11]

The JCAH's *Functional Safety and Sanitation* (1982) and Veterans Administration *Circular 10-77-111* (1982) specifies the requirements for safety inspections. These are summarized in *Hospital Electrical Standards Symposium*, of the American Society of Hospital Engineering (1981). These documents require records of periodic safety inspections.[6]

6 PROTOCOL FOR LABORATORY SAFETY

Safe laboratory protocol involves understanding, prevention, inspection, and record keeping. The laboratory director has the ultimate responsibility for the establishment and execution of safety protocols. Personnel should have some formal training in electrical safety theory and practices and should receive annual reviews. Ideally, staff should understand the basis of electrical safety, so they can react to unfamiliar situations.

Routine practices of prevention can avoid or detect hazardous situations. All electrophysiologic examinations should follow such practices as part of a written protocol. This is especially important in portable recordings. Appendix Table 3–3 lists some common prevention measures.

Periodic electrical inspections of equipment and wiring are a required part of safety protocol. Standards and guidelines set by the JCAH, the NFPA, or the National Electrical Code may apply. If such inspection services are not available, laboratory personnel may have to perform these tests themselves. A good safety test meter may cost from $500 to $1000 and requires some training in its use.

Outlets and wiring in patient areas should be checked at least once a year. Checking every outlet for absence of ground connection or reversal of hot and neutral tests the wiring. The longer slot of the outlet should be neutral, and the shorter slot hot. The ground pin of each

Table 3–3 PREVENTING HAZARDS

Remove any ungrounded devices (two-wire power cords) and nonessential battery-powered devices from patient areas: TV, radio, clock, lamp, tape player.

Keep liquids away from equipment. Spills on or in instruments can increase leakage current, corrode ground connections, or cause equipment failures. Electrode creams contain conductive electrolytes that can destroy electronics and corrode metals.

Inspect all plugs for tightness in outlets. All the plugs and outlets should be HOSPITAL GRADE, identified with a green dot. They have better retention, contact, and wear properties.

Always pull plugs straight out of outlets when unplugging, not to the side or wiggling. Of course, never pull plugs out by the power cord.

Unplug equipment before moving it. Jerking the power cord may break the wire or insulation and may damage the pins in the plug or outlet. Report any such accidents immediately, for proper testing and repair of the equipment *and the outlet*.

Check daily for wear or damage to power cords and plugs.

Never use extension cords on equipment, even three-wire extension cords. The added length of power cord increases the capacitive leakage current between the hot wire and ground. The extra set of contacts increases the chances of the ground connection failing.

In familiar and unfamiliar settings, verify that all equipment near a patient connects only to outlets in the same room.

Never use two-pin outlets or two-pin adaptors.

Never turn the main power to equipment on or off while it is connected to a patient. During these transitions the electronics may not function normally.

Locate the ground electrode on the same side of the body as the recording and stimulating electrodes, unless recording requirements absolutely dictate otherwise. This prevents leakage/fault currents from flowing across the body, where they might affect the heart. When multiple instruments are directly connected to a patient, all the grounds should be on the same side of the body. This is especially important if any of the ground leads is not isolated.

---

outlet should be at neutral potential and have a resistance to a common ground point of not more than 0.1 Ω in sensitive patient areas, or 0.2 Ω elsewhere. The voltage on outlet grounds, relative to a common ground point, should not exceed 20 mV RMS in sensitive patient areas, 50 mV elsewhere.[10]

Measuring the force required to extract a pin from each outlet contact tests the contact tension. This should be greater than 8 ounces. *Hospital grade* outlets and plugs have greater initial retention and longer wear. Even these require periodic replacement.

Equipment must also be periodically tested for ground integrity and leakage current. The resistance between the instrument chassis and the ground pin on the power cord should not exceed 0.1 Ω while pulling and bending the cord in all directions for detection of intermittent or weak connections.

Instruments require testing for leakage current to the chassis and each of the patient leads, using a standard impedance to simulate the body in a leakage circuit. The standard impedance equals about 1000 Ω at 60 Hz.[1] Leakage current to chassis, the RMS value in microamperes at 60 Hz, is measured with the ground to the instrument open and the standard impedance connected between the chassis and ground, under the conditions of equipment turned on and turned off, and with the hot/neutral supply normal and reversed. The worst-case leakage should not exceed 100 μA if the patient ground lead has an isolator, or 50 μA if the patient ground lead connects directly to chassis. For electrosensitive patients, the limit is 20 μA.[2]

The leakage current of patient leads, including patient ground, is measured with the standard impedance between the lead connection and ground, under the conditions of equipment on and off, normal and reverse line, and with the instrument ground connected and open. Worst case lead leakage should not exceed 20 μA for electrosensitive patients, or otherwise, 50 μA.

Isolated inputs are also tested for leakage with their potential at 120 VAC, by connecting the standard impedance between the hot line and the lead input connection. (Do not try this test with nonisolated inputs!) Under all of the above conditions, the worst-case leakage of isolated inputs should not exceed 20 μA. A good safety meter has provision for readily making all these types of measurements.

The JCAH *Accreditation Manual for Hospitals* requires that protocols and procedures be established for these inspections and that records of the periodic tests be kept. Inspected equipment should bear a dated safety sticker.

7 SPECIAL SAFETY DEVICES AND CIRCUITS

Isolated Power Systems

In isolated power systems, the transformer supplying the 120 VAC is *not* connected to earth. Then the power lines are no longer "hot" and "neutral," but "float" with respect to earth; that is, neither line has more than some small leakage conductance to earth. A grounded person could touch either one of the power lines directly and only conduct the leakage current of the system.

Isolated power systems are commonly found in operating rooms. They usually include some monitoring circuits that sound an alarm if leakage limits are exceeded. Isolation-monitoring circuits may cause interference on the power lines, which can cause artifacts in recording equipment. Leakages of large isolated power systems are typically on the order of 1 mA, which is excessive for patient protection. Patient-connected equipment must still have safe leakage limits when used on isolated power.[3]

Isolated power systems are also found in recording equipment that has a number of different line-powered devices, such as computer-based equipment with printers, terminals, monitors, and so forth. The total leakage current of all these devices at the common power cord would exceed safe limits. An internal isolation transformer reduces the total power-cord leakage of the equipment to that of the transformer.

Ground Fault Interrupters

The ground fault interrupter (GFI), a device in the power wiring, senses the amount of line current flowing to earth and shuts off the power if this current to earth exceeds a trip level, usually about 4 mA. Ground fault interrupters are becoming common, and may even be required, for new wiring installations in bathrooms, kitchens, garages, and outdoors. For hospitals, however, their trip level is much too high to be adequate for patient protection in all cases. Also, it may be undesirable to have power interrupted if critical care or life support equipment is in use.

Redundant Grounding

For additional safety against loss of ground, some equipment uses a redundant ground wire, independent of the power cord. Intensive care units and operating rooms typically have redundant-ground panels for this connection. With a redundant ground connected, the instrument remains grounded, even if the power cord ground fails. This is recommended on equipment for routine portable use.

REFERENCES

1. AAMI: Safe Current Limits for Electromedical Apparatus, Association for the Advancement of Medical Instrumentation (AAMI). Arlington, Virginia, 1978.

2. AAMI: Interim Rationale Statement for the American National Standard, Safe Current Limits for Electromedical Apparatus. Association for the Advancement of Medical Instrumentation (AAMI), Arlington, Virginia, 1980.

3. American Society for Hospital Engineering: Hospital Electrical Standards Compendium. American Society for Hospital Engineering, Chicago, Illinois, 1981.

4. Dalziel, CF: Electric Shock Hazards. IEEE Spectrum 9 (2):41, 1972.

5. Hatch, DJ, and Raber, MB: Grounding and Safety. IEEE Trans Biomed Eng BME-22:62, 1975.

6. Joint Commission on Accreditation of Hospitals: Functional Safety and Sanitation. In: Accreditation Manual for Hospitals. Joint Commission on Accreditation of Hospitals, Chicago, Illinois, 1982.

7. McPartland, JF, McPartland, JM, McPartland, GI: McGraw-Hill's National Electrical Code Handbook, ed 17. McGraw-Hill, New York, 1981.

8. National Fire Protection Association: Article 517. Health Care Facilities. In National Electrical Code, NFPA 70-1981. National Fire Protection Association, Boston, Massachusetts, 1981.

9. Seaba, P: Electrical Safety. Am J EEG Technol 20:1, 1980.

10. Strong, P: Grounding-safety. In: Biophysical Measurements, (Tektronix #062-1247-00) Tektronix, Inc., Beaverton, Oregon, 1973.

11. Underwriters Laboratories, Inc.: Standard for Medical and Dental Equipment, UL544, ed 2. Underwriters Laboratories, Inc., Northbrook, Illinois, 1980.

12. Veterans Administration: Veterans Administration Documents on Electrical Safety and Service Manuals. J Clin Eng 3:64, 1978

ACKNOWLEDGMENT

Much of the material in this Appendix reflects the work of Mr. Peter J. Seaba, MSEE, who coauthored this section in the first edition.

Appendix 4

AAEE GLOSSARY OF TERMS IN CLINICAL ELECTROMYOGRAPHY*

FOREWORD

One of the objectives of the American Association of Electromyography and Electrodiagnosis (AAEE) is the publication of information to increase and to extend the knowledge of electromyographers. In 1974, the Board of Directors of the AAEE established a Nomenclature Committee with the task of compiling and defining a list of terms used in electromyography. The resultant Glossary was published by the AAEE in 1980. The Glossary was widely accepted and helped to standardize the terms used in clinical reports and in scientific publications.

Subsequent advances in electromyography necessitated a review and revision of the 1980 Glossary. In 1983, a new Nomenclature Committee was created by the AAEE. Every term in the 1980 Glossary was reviewed; some old terms were redefined, a few were deleted, and some new terms were added. Also new to this 1987 Glossary are illustrations of se-

---

*Compiled by the Nomenclature Committee of the American Association of Electromyography and Electrodiagnosis. Second Edition © AAEE 1987, published in *Muscle & Nerve*, Volume 10, Number 8S/Supplement, October 1987. Reproduced and modified in part by permission.

lected waveforms and lists of terms grouped by subject.

AAEE Nomenclature Committee

Charles K. Jablecki, M.D., Chairman
Charles F. Bolton, M.D.
Walter G. Bradley, D.M.
William F. Brown, M.D.
Fritz Buchthal, M.D.
Roger Q. Cracco, M.D.
Ernest W. Johnson, M.D.
George H. Kraft, M.D.
Edward H. Lambert, M.D., Ph.D.
Hans O. Lüders, M.D., Ph.D.
Dong M. Ma, M.D.
John A. Simpson, M.D.
Erik V. Stålberg, M.D., Ph.D.

INTRODUCTION

In all areas of science, terms should be precisely defined and standardized. Terms should be used consistently so that one scientist in a field can speak or write to another without ambiguity. The need for definitions exists in electromyography because there are numerous clinical investigators conducting studies. By agreeing upon terminology, investigators can understand and verify the findings of others. It is suggested that the terms in this glossary be used by authors of papers for publication in electromyography and by clinical electromyographers for patient reports.

The first edition of this Glossary was prepared and published by the American Association of Electromyography and Electrodiagnosis (AAEE) in 1980. In 1983, the International Federation of Societies for Electroencephalography and Clinical Neurophysiology (IFSECN) published an adaptation of that glossary for its members. This second edition of the AAEE Glossary was compiled after an extensive review of the first AAEE Glossary, of the changes made by the IFSCEN, of new terms suggested by AAEE members, and of the recent literature. The following definitions are the result of considerable deliberation. In some cases, the committee members compromised and

retained terms which have been in use for such a long time that it was agreed that they should remain as they are, even though they are not ideal.

This glossary is presented in four sections. In Section I, all terms are listed in alphabetical order and are defined. The alphabetical presentation permits electromyographers to use the glossary efficiently to prepare and to review reports. An asterisk adjacent to a term indicates that an illustration of that waveform is contained in Section II. In Section III, terms are grouped by subject without definition to permit the systematic review of related terms.

SECTION I:
ALPHABETICAL LIST OF TERMS WITH DEFINITIONS

*A wave A compound action potential evoked consistently from a muscle by submaximal electric stimuli to the nerve and frequently abolished by supramaximal stimuli. The amplitude of the A wave is similar to that of the F wave, but the latency is more constant. The A wave usually occurs before the F wave, but may occur afterwards. The A wave is due to normal or pathologic axonal branching. Compare the *F wave*.

absolute refractory period See *refractory period*.

accommodation True accommodation in neuronal physiology is a rise in the threshold transmembrane depolarization required to initiate a spike when depolarization is slow or a subthreshold depolarization is maintained. In the older literature, accommodation described the observation that the final intensity of current applied in a slowly rising fashion to stimulate a nerve was greater than the intensity of a pulse of current required to stimulate the same nerve. The latter may largely be an artifact of the nerve sheath and bears little relation to true accommodation as measured intracellularly.

accommodation curve See *strength-duration curve*.

---

*Illustration in Section II.

action current The electric currents associated with an *action potential*.

action potential (AP) The brief regenerative electric potential that propagates along a single axon or muscle fiber membrane. The action potential is an all-or-none phenomenon; whenever the stimulus is at or above threshold, the action potential generated has a constant size and configuration. See also *compound action potential, motor unit action potential.*

active electrode Synonymous with *exploring electrode*. See *recording electrode*.

adaptation A decline in the frequency of the spike discharge as typically recorded from sensory axons in response to a maintained stimulus.

AEPs See *auditory evoked potentials*.

afterdischarge The continuation of an impulse train in a neuron, axon or muscle fiber following the termination of an applied stimulus. The number of extra impulses and their periodicity in the train may vary depending on the circumstances.

afterpotential The membrane potential between the end of the spike and the time when the membrane potential is restored to its resting value. The membrane during this period may be depolarized or hyperpolarized.

amplitude With reference to an *action potential*, the maximum voltage difference between two points, usually baseline to peak or peak to peak. By convention, the amplitude of the *compound muscle action potential* is measured from the baseline to the most negative peak. In contrast, the amplitude of a *compound sensory nerve action potential, motor unit potential, fibrillation potential, positive sharp wave, fasciculation potential*, and most other *action potentials* is measured from the most positive peak to the most negative peak

anodal block A local block of nerve conduction caused by *hyperpolarization* of the nerve cell membrane by an electric stimulus. See *stimulating electrode*.

anode The positive terminal of a source of electric current.

antidromic Propagation of an impulse in the direction opposite to physiologic conduction; e.g., conduction along motor nerve fibers away from the muscle and conduction along sensory fibers away from the spinal cord. Contrast with *orthodromic*.

AP See *action potential*.

artifact (also artefact) A voltage change generated by a biologic or nonbiologic source other than the ones of interest. The *stimulus artifact* is the potential recorded at the time the stimulus is applied and includes the *electric* or *shock artifact*, which represents cutaneous spread of stimulating current to the recording electrode. The stimulus and shock artifacts usually precede the activity of interest. A *movement artifact* refers to a change in the recorded activity caused by movement of the recording electrodes.

auditory evoked potentials (AEPs). Electric waveforms of biologic origin elicited in response to sound stimuli. AEPs are classified by their latency as short-latency brainstem AEPs (BAEPs) with a latency of up to 10 ms, middle-latency AEPs with a latency of 10–50 ms, and long-latency AEPs with a latency of over 50 ms. See *brainstem auditory evoked potentials*.

axon reflex Use of term discouraged as it is incorrect. No reflex is considered to be involved. See preferred term, *A wave*.

axon response See preferred term, *A wave*.

axon wave See *A wave*.

axonotmesis Nerve injury characterized by disruption of the axon and myelin sheath, but with preservation of the supporting connective tissue, resulting in axonal degeneration distal to the injury site.

backfiring Discharge of an antidromically activated motor neuron.

BAEPs See *brainstem auditory evoked potentials*.

BAERs Abbreviation for *brainstem auditory evoked responses*. See preferred term, *brainstem auditory evoked potentials*.

baseline The potential recorded from a biologic system while the system is at rest.

benign fasciculation Use of term discouraged to describe a firing pattern of fasciculation potentials. The term has been used to describe a clinical syndrome and/or the presence of fasciculations in nonprogressive neuromuscular disorders. See *fasciculation potential.*

BERs Abbreviation for *brainstem auditory evoked responses.* See preferred term, *brainstem auditory evoked potentials.*

bifilar needle recording electrode *Recording electrode* that measures variations in voltage between the bare tips of two insulated wires cemented side by side in a steel cannula. The bare tips of the electrodes are flush with the level of the cannula. The latter may be grounded.

biphasic action potential An *action potential* with two phases.

biphasic end-plate activity See *end-plate activity* (*biphasic*).

bipolar needle recording electrode See preferred term, *needle bifilar recording electrode*

bipolar stimulating electrode See *stimulating electrode.*

bizarre high-frequency discharge See preferred term, *complex repetitive discharge.*

bizarre repetitive discharge See preferred term, *complex repetitive discharge.*

bizarre repetitive potential See preferred term, *complex repetitive discharge.*

blink reflex See *blink responses.*

blink response Strictly defined, one of the *blink responses.* See *blink responses.*

*blink responses *Compound muscle action potentials* evoked from orbicularis oculi muscles as a result of brief electric or mechanical stimuli to the cutaneous area innervated by the supraorbital (or less commonly, the infraorbital) branch of the trigeminal nerve. Typically, there is an early compound muscle action potential (*R1 wave*) ipsilateral to the stimulation site with a latency of about 10 ms and a bilateral late compound muscle action potential

(*R2 wave*) with a latency of approximately 30 ms. Generally, only the *R2 wave* is associated with a visible twitch of the orbicularis oculi. The configuration, amplitude, duration, and latency of the two components, along with the sites of recording and the sites of stimulation, should be specified. *R1* and *R2 waves* are probably oligosynaptic and polysynaptic brainstem reflexes, respectively, together called the *blink reflex*, with the afferent arc provided by the sensory branches of the trigeminal nerve and the efferent arc provided by the facial nerve motor fibers.

*brainstem auditory evoked potentials (BAEPs) Electric waveforms of biologic origin elicited in response to sound stimuli. The normal BAEP consists of a sequence of up to seven waves, named I to VII, which occur during the first 10 ms after the onset of the stimulus and have positive polarity at the vertex of the head.

brainstem auditory evoked responses (BAERs, BERs) See preferred term, *brainstem auditory evoked potentials.*

BSAPs Abbreviation for brief, small, abundant potentials. Use of term is discouraged. It is used to describe a recruitment pattern of brief-duration, small-amplitude, overly abundant motor unit action potentials. Quantitative measurements of motor unit potential duration, amplitude, numbers of phases, and recruitment frequency are to be preferred to qualitative descriptions such as this. See *motor unit action potential.*

BSAPPs Abbreviation for brief, small abundant, polyphasic potentials. Use of term is discouraged. It is used to describe a recruitment pattern of brief-duration, small-amplitude, overly abundant, polyphasic motor unit action potentials. Quantitative measurements of motor unit potential duration, amplitude, numbers of phases, and recruitment frequency are to be preferred to qualitative descriptions such as this. See *motor unit action potential.*

cathode The negative terminal of a source of electric current.

central electromyography (central EMG) Use of electromyographic record-

---

*Illustration in Section II.

ing techniques to study reflexes and the control of movement by the spinal cord and brain.

chronaxie (also chronaxy) See *strength-duration curve.*

clinical electromyography Synonymous with *electroneuromyography.* Used to refer to all electrodiagnostic studies of human peripheral nerves and muscle. See also *electromyography* and *nerve conduction studies.*

coaxial needle electrode See synonym, *concentric needle electrode.*

collision When used with reference to nerve conduction studies, the interaction of two action potentials propagated toward each other from opposite directions on the same nerve fiber so that the refractory periods of the two potentials prevent propagation past each other.

complex action potential See preferred term, *serrated action potential.*

complex motor unit action potential A *motor unit action potential* that is polyphasic or serrated. See preferred terms, *polyphasic action potential* or *serrated action potential.*

*complex repetitive discharge Polyphasic or serrated action potentials that may begin spontaneously or after a needle movement. They have a uniform frequency, shape, and amplitude, with abrupt onset, cessation, or change in configuration. Amplitude ranges from 100 μV to 1 mV and frequency of discharge from 5 to 100 Hz. This term is preferred to *bizarre high frequency discharge, bizarre repetitive discharge, bizarre repetitive potential, near constant frequency trains, pseudomyotonic discharge* and *synchronized fibrillation.*

compound action potential See *compound mixed nerve action potential, compound motor nerve action potential, compound nerve action potential, compound sensory nerve action potential,* and *compound muscle action potential.*

compound mixed nerve action potential (compound mixed NAP) A compound nerve action potential is considered to have been evoked from afferent and efferent fibers if the recording electrodes detect activity on a mixed nerve with the electric stimulus applied to a segment of the nerve which contains both afferent and efferent fibers. The amplitude, latency, duration, and phases should be noted.

compound motor nerve action potential (compound motor NAP) A compound nerve action potential is considered to have been evoked from efferent fibers to a muscle if the recording electrodes detect activity only in a motor nerve or a motor branch of a mixed nerve, or if the electric stimulus is applied only to such a nerve or a ventral root. The amplitude, latency, duration, and phases should be noted. See *compound nerve action potential.*

compound muscle action potential (CMAP) The summation of nearly synchronous muscle fiber action potentials recorded from a muscle commonly produced by stimulation of the nerve supplying the muscle either directly or indirectly. Baseline-to-peak amplitude, duration, and latency of the negative phase should be noted, along with details of the method of stimulation and recording. Use of specific named potentials is recommended, e.g., *M wave, F wave, H wave, T wave, A wave* and *R1 wave* or *R2 wave* (*blink responses*).

compound nerve action potential (compound NAP) The summation of nearly synchronous nerve fiber action potentials recorded from a nerve trunk, commonly produced by stimulation of the nerve directly or indirectly. Details of the method of stimulation and recording should be specified, together with the fiber type (sensory, motor, or mixed).

*compound sensory nerve action potential (compound SNAP) A compound nerve action potential is considered to have been evoked from afferent fibers if the recording electrodes detect activity only in a sensory nerve or in a sensory branch of a mixed nerve, or if the electric stimulus is applied to a sensory nerve or a dorsal nerve root, or an adequate stimulus is applied syn-

---

*Illustration in Section II.

chronously to sensory receptors. The amplitude, latency, duration, and configuration should be noted. Generally, the amplitude is measured as the maximum peak-to-peak voltage, the latency as either the *latency* to the initial deflection or the *peak latency* to the negative peak, and the duration as the interval from the first deflection of the waveform from the baseline to its final return to the baseline. The compound sensory nerve action potential has been referred to as the *sensory response* or *sensory potential.*

concentric needle electrode *Recording electrode* that measures an electric potential difference between the bare tip of an insulated wire, usually stainless steel, silver or platinum, and the bare shaft of a steel cannula through which it is inserted. The bare tip of the central wire (exploring electrode) is flush with the level of the cannula (reference electrode).

conditioning stimulus See *paired stimuli.*

conduction block Failure of an action potential to be conducted past a particular point in the nervous system whereas conduction is possible below the point of the block. Conduction block is documented by demonstration of a reduction in the area of an evoked potential greater than that normally seen with electric stimulation at two different points on a nerve trunk; anatomic variations of nerve pathways and technical factors related to nerve stimulation must be excluded as the cause of the reduction in area.

conduction distance See *conduction velocity.*

conduction time See *conduction velocity.*

conduction velocity (CV) Speed of propagation of an *action potential* along a nerve or muscle fiber. The nerve fibers studied (motor, sensory, autonomic, or mixed) should be specified. For a nerve trunk, the maximum conduction velocity is calculated from the *latency* of the evoked potential (muscle or nerve) at maximal or supra-

maximal intensity of stimulation at two different points. The distance between the two points (*conduction distance*) is divided by the difference between the corresponding latencies (*conduction time*). The calculated velocity represents the conduction velocity of the fastest fibers and is expressed as meters per second (m/s). As commonly used, the term *conduction velocity* refers to the maximum conduction velocity. By specialized techniques, the conduction velocity of other fibers can be determined as well and should be specified, e.g., minimum conduction velocity.

contraction A voluntary or involuntary reversible muscle shortening that may or may not be accompanied by *action potentials* from muscle. This term is to be contrasted with the term *contracture*, which refers to a condition of fixed muscle shortening.

contraction fasciculation Rhythmic, visible twitching of a muscle with weak voluntary or postural contraction. The phenomenon occurs in neuromuscular disorders in which the motor unit territory is enlarged and the tissue covering the muscle is thin.

contracture The term is used to refer to immobility of a joint due to fixed muscle shortening. Contrast *contraction*. The term has also been used to refer to an electrically silent, involuntary state of maintained muscle contraction, as seen in phosphorylase deficiency, for which the preferred term is *muscle cramp.*

coupled discharge See preferred term, *satellite potential*

cps (also c/s) See *cycles per second.*

*****cramp discharge** Involuntary repetitive firing of *motor unit action potentials* at a high frequency (up to 150 Hz) in a large area of muscles, usually associated with painful muscle contraction. Both the discharge frequency and the number of *motor unit action potentials* firing increase gradually during development and both subside gradually with cessation. See *muscle cramp.*

c/s (also cps) See *cycles per second.*

CV See *conduction velocity.*

cycles per second Unit of frequency. (cps or c/s). See also *hertz* (Hz).

---

*Illustration in Section II.

decremental response See preferred term, *decrementing response.*

*****decrementing response** A reproducible decline in the amplitude and/or area of the *M wave* of successive responses to *repetitive nerve stimulation.* The rate of stimulation and the total number of stimuli should be specified. Decrementing responses with disorders of neuromuscular transmission are most reliably seen with slow rates (2-5 Hz) of nerve stimulation. A decrementing response with repetitive nerve stimulation commonly occurs in disorders of neuromuscular transmission, but can also be seen in some neuropathies, myopathies, and motor neuron disease. An artifact resembling a decrementing response can result from movement of the stimulating or recording electrodes during repetitive nerve stimulation. Contrast with *incrementing response.*

delay As originally used in clinical electromyography, delay referred to the time between the beginning of the horizontal sweep of the oscilloscope and the onset of an applied stimulus. The term is also used to refer to an information storage device (delay line) used to display events occurring before a trigger signal.

denervation potential This term has been used to describe a *fibrillation potential.* The use of this term is discouraged because fibrillation potentials may occur in settings where transient muscle membrane instability occurs in the absence of denervation, e.g., hyperkalemia periodic paralysis. See preferred term, *fibrillation potential.*

depolarization See *polarization.*

depolarization block Failure of an excitable cell to respond to a stimulus because of depolarization of the cell membrane.

discharge Refers to the firing of one or more excitable elements (neurons, axons or muscle fibers) and as conventionally applied refers to the all-or-none potentials only. Synonymous with *action potential.*

discharge frequency The rate of repetition of potentials. When potentials occur in groups, the rate of recurrence of the group and the rate of repetition of the individual components in the groups should be specified. See also *firing rate.*

discrete activity See *interference pattern.*

distal latency See *motor latency* and *sensory latency.*

double discharge Two action potentials (*motor unit action potential, fibrillation potential*) of the same form and nearly the same amplitude, occurring consistently in the same relationship to one another at intervals of 2 to 20 ms. Contrast with *paired discharge.*

doublet Synonymous with *double discharge.*

duration The time during which something exists or acts. (1) The total duration of individual potential *waveforms* is defined as the interval from the beginning of the first deflection from the baseline to its final return to the baseline, unless otherwise specified. If only part of the waveform duration is measured, the points of measurement should be specified. For example, the duration of the *M wave* may refer to the interval from the deflection of the first negative phase from the baseline to its return to the baseline. (2) The duration of a single electric stimulus refers to the interval of the applied current or voltage. (3) The duration of recurring stimuli or action potentials refers to the interval from the beginning to the end of the series.

earth electrode Synonymous with *ground electrode.*

EDX See *electrodiagnosis.*

electric artifact See *artifact.*

electric inactivity Absence of identifiable electric activity in a structure or organ under investigation. See preferred term, *electric silence.*

electric silence The absence of measurable electric activity due to biologic or nonbiologic sources. The sensitivity and signal-to-noise level of the recording system should be specified.

electrode A conducting device used to

---

*Illustration in Section II.

record an electric potential (*recording electrode*) or to apply an electric current (*stimulating electrode*). In addition to the *ground electrode* used in clinical recordings, two electrodes are always required either to record an electric potential or to apply an electric current. Depending on the relative size and location of the electrodes, however, the stimulating or recording condition may be referred to as *monopolar* or *unipolar*. See *ground electrode, recording electrode,* and *stimulating electrode*. Also see specific needle electrode configurations: *monopolar, unipolar, concentric, bifilar recording, bipolar stimulating, multilead, single fiber,* and *macro-EMG needle electrodes.*

electrodiagnosis (EDX) The recording and analysis of responses of nerves and muscles to electric stimulation and the identification of patterns of insertion, spontaneous, involuntary and voluntary action potentials in muscle and nerve tissue. See also *electromyography, electroneurography, electroneuromyography,* and *evoked potential studies.*

electrodiagnostic medicine A specific area of medical practice in which a physician uses information from the clinical history, observations from the physical examination, and the techniques of *electrodiagnosis* to diagnose and treat neuromuscular disorders. See *electrodiagnosis.*

electromyelography The recording and study of electric activity from the spinal cord and/or from the cauda equina.

electromyogram The record obtained by *electromyography.*

electromyograph Equipment used to activate, record, process and display nerve and muscle action potentials for the purpose of evaluating nerve and muscle function.

electromyography (EMG) Strictly defined, the recording and study of insertion, spontaneous, and voluntary electric activity of muscle. It is commonly used to refer to nerve conduction studies as well. See also *clinical electromyography* and *electroneuromyography.*

electroneurography (ENG) The recording and study of the action potentials of peripheral nerves. Synonymous with *nerve conduction studies.*

electroneuromyography (ENMG) The combined studies of *electromyography* and *electroneurography.* Synonymous with *clinical electromyography.*

EMG See *electromyography.*

***end-plate activity** Spontaneous electric activity recorded with a needle electrode close to muscle end-plates. May be either of two forms:

1. *Monophasic:* Low-amplitude (10–20 μV), short-duration (0.5-1 ms), monophasic (negative) potentials that occur in a dense, steady pattern and are restricted to a localized area of the muscle. Because of the multitude of different potentials occurring, the exact frequency, although appearing to be high, cannot be defined. These nonpropagated potentials are probably miniature end-plate potentials recorded extracellularly. This form of end-plate activity has been referred to as *end-plate noise* or *sea shell sound (sea shell noise* or *roar).*

2. *Biphasic:* Moderate-amplitude (100–300 μV), short-duration (2–4 ms), biphasic (negative-positive) spike potentials that occur irregularly in short bursts with a high frequency (50–100 Hz), restricted to a localized area within the muscle. These propagated potentials are generated by muscle fibers excited by activity in nerve terminals. These potentials have been referred to as *biphasic spike potentials, end-plate spikes,* and, incorrectly, *nerve potentials.*

end-plate noise See *end-plate activity (monophasic).*

end-plate potential (EPP) The graded nonpropagated membrane potential induced in the postsynaptic membrane of the muscle fiber by the action of acetylcholine released in response to an action potential in the presynaptic axon terminal.

---

*Illustration in Section II.

end-plate spike See *end-plate activity* (*biphasic*).

end-plate zone The region in a muscle where the neuromuscular junctions of the skeletal muscle fibers are concentrated.

ENG See *electroneurography*.

ENMG See *electroneuromyography*.

EPP See *end-plate potential*.

EPSP See *excitatory postsynaptic potential*.

evoked compound muscle action potential See *compound muscle action potential*.

evoked potential Electric waveform elicited by and temporally related to a stimulus, most commonly an electric stimulus delivered to a sensory receptor or nerve, or applied directly to a discrete area of the brain, spinal cord, or muscle. See *auditory evoked potential, brainstem auditory evoked potential, spinal evoked potential, somatosensory evoked potential, visual evoked potential, compound muscle action potential,* and *compound sensory nerve action potential.*

evoked potential studies Recording and analysis of electric waveforms of biologic origin elicited in response to electric or physiologic stimuli. Generally used to refer to studies of waveforms generated in the peripheral and central nervous system, whereas *nerve conduction studies* refers to studies of waveforms generated in the peripheral nervous system. There are two systems for naming complex waveforms in which multiple components can be distinguished. In the first system, the different components are labeled PI or NI for the initial positive and negative potentials, respectively, and PII, NII, PIII, NIII, etc., for subsequent positive and negative potentials. In the second system, the components are specified by polarity and average peak latency in normal subjects to the nearest millisecond. The first nomenclature principle has been used in an abbreviated form to identify the seven positive components (I–VII) of the normal *brainstem*

auditory evoked potential. The second nomenclature principle has been used to identify the positive and negative components of *visual evoked potentials* (N75, P100) and *somatosensory evoked potentials* (P9, P11, P13, P14, N20, P23). Regardless of the nomenclature system, it is possible under standardized conditions to establish normal ranges of amplitude, duration, and latency of the individual components of these *evoked potentials.* The difficulty with the second system is that the latencies of components of evoked potentials depend upon the length of the pathways in the neural tissues. Thus the components of an SEP recorded in a child have different average latencies from the same components of an SEP recorded in an adult. Despite this problem, there is no better system available for naming these components at this time. See *auditory evoked potentials, brainstem auditory evoked potentials, visual evoked potentials, somatosensory evoked potentials.*

evoked response Tautology. Use of term discouraged. See preferred term, *evoked potential.*

excitability Capacity to be activated by or react to a stimulus.

excitatory postsynaptic potential (EPSP) A local, graded depolarization of a neuron in response to activation by a nerve terminal of a synapse. Contrast with *inhibitory postsynaptic potential.*

exploring electrode Synonymous with *active electrode.* See *recording electrode.*

F reflex See preferred term, *F wave.*

F response Synonymous with *F wave.* See preferred term, *F wave.*

*F wave A *compound action potential* evoked intermittently from a muscle by a supramaximal electric stimulus to the nerve. Compared with the maximal amplitude *M wave* of the same muscle, the F wave has a smaller amplitude (1–5% of the *M wave*), variable configuration and a longer, more variable latency. The F wave can be found in many muscles of the upper and lower

---

*Illustration in Section II.

extremities, and the latency is longer with more distal sites of stimulation. The F wave is due to antidromic activation of motor neurons. It was named by Magladery and McDougal in 1950. Compare the *H wave* and the *A wave*.

*facilitation Improvement of neuromuscular transmission which results in the activation of previously inactive muscle fibers. Facilitation may be identified in several ways:

1. *Incrementing response:* A reproducible increase in the amplitude associated with an increase in the area of successive electric responses (*M waves*) during *repetitive nerve stimulation.*

2. *Postactivation* or *posttetanic facilitation:* Nerve stimulation studies performed within a few seconds after a brief period (2–15 s) of nerve stimulation producing *tetanus* or after a strong voluntary contraction may show changes in the configuration of the *M wave(s)* compared to the results of identical studies of the rested neuromuscular junction as follows:

   a. *Repair of the decrement:* A diminution of the decrementing response seen with slow rates (2–5 Hz) of *repetitive nerve stimulation.*

   b. *Increment after exercise:* An increase in the amplitude associated with an increase in the area of the M wave elicited by a single supramaximal stimulus.

Facilitation should be distinguished from pseudofacilitation. *Pseudofacilitation* occurs in normal subjects with *repetitive nerve stimulation* at high (20–50 Hz) rates or after strong volitional contraction, and probably reflects a reduction in the temporal dispersion of the summation of a constant number of muscle fiber action potentials. *Pseudofacilitation* produces a response characterized by an increase in the amplitude of the successive M waves with a corresponding decrease in the duration of the M wave resulting in no change in the area of the negative phase of the successive M waves.

far-field potential Electric activity of biologic origin generated at a considerable distance from the recording electrodes. Use of the terms *near-field potential* and *far-field potential* is discouraged because all potentials in clinical neurophysiology are recorded at some distance from the generator and there is no consistent distinction between the two terms.

fasciculation The random, spontaneous twitching of a group of muscle fibers or a motor unit. This twitch may produce movement of the overlying skin (limb), mucous membrane (tongue), or digits. The electric activity associated with the spontaneous contraction is called the *fasciculation potential*. See also *myokymia*. Historically the term *fibrillation* has been used to describe fine twitching of muscle fibers visible through the skin or mucous membrane, but this usage is no longer acceptable.

*fasciculation potential The electric potential often associated with a visible *fasciculation* which has the configuration of a *motor unit action potential* but which occurs spontaneously. Most commonly these potentials occur sporadically and are termed "single fasciculation potentials." Occasionally, the potentials occur as a grouped discharge and are termed a "brief repetitive discharge." The occurrence of repetitive firing of adjacent fasciculation potentials, when numerous, may produce an undulating movement of muscle (see *myokymia*). Use of the terms *benign fasciculation* and *malignant fasciculation* is discouraged. Instead, the configuration of the potentials, peak-to-peak amplitude, duration, number of phases, and stability of configuration, in addition to frequency of occurrence, should be specified.

fatigue Generally, a state of depressed responsiveness resulting from protracted activity and requiring an appreciable recovery time. Muscle fatigue is a reduction in the force of contraction of muscle fibers and follows repeated vol-

---

*Illustration in Section II.

untary contraction or direct electric stimulation of the muscle.

fiber density (1) Anatomically, fiber density is a measure of the number of muscle or nerve fibers per unit area. (2) In *single fiber electromyography*, the fiber density is the mean number of *muscle fiber action potentials* fulfilling amplitude and rise time criteria belonging to one motor unit within the recording area of the *single fiber needle electrode* encountered during a systematic search in the weakly, voluntarily contracted muscle. See also *single fiber electromyography, single fiber needle electrode.*

fibrillation The spontaneous contractions of individual muscle fibers which are not visible through the skin. This term has been used loosely in electromyography for the preferred term, *fibrillation potential.*

fibrillation potential The electric activity associated with a spontaneously contracting (fibrillating) muscle fiber. It is the action potential of a single muscle fiber. The action potentials may occur spontaneously or after movement of the needle electrode. The potentials usually fire at a constant rate, although a small proportion fire irregularly. Classically, the potentials are biphasic spikes of short duration (usually less than 5 ms) with an initial positive phase and a peak-to-peak amplitude of less than 1 mV. When recorded with concentric or monopolar needle electrodes, the firing rate has a wide range (1–50 Hz) and often decreases just before cessation of an individual discharge. A high-pitched regular sound is associated with the discharge of fibrillation potentials and has been described in the old literature as "rain on a tin roof." In addition to this classic form of fibrillation potentials, *positive sharp waves* may also be recorded from fibrillating muscle fibers when the potential arises from an area immediately adjacent to the needle electrode.

firing pattern Qualitative and quantitative descriptions of the sequence of discharge of potential waveforms recorded from muscle or nerve.

firing rate Frequency of repetition of a potential. The relationship of the frequency to the occurrence of other potentials and the force of muscle contraction may be described. See also *discharge frequency.*

frequency Number of complete cycles of a repetitive waveform in one second. Measured in *hertz* (Hz) or *cycles per second* (cps or c/s).

frequency analysis Determination of the range of frequencies composing a potential waveform, with a measurement of the absolute or relative amplitude of each component frequency.

full interference pattern See *interference pattern.*

functional refractory period See *refractory period.*

G1, G2 Synonymous with *Grid 1, Grid 2*, and newer terms, *Input Terminal 1, Input Terminal 2.* See *recording electrode.*

"giant" motor unit action potential Use of term discouraged. It refers to a *motor unit action potential* with a peak-to-peak amplitude and duration much greater than the range recorded in corresponding muscles in normal subjects of similar age. Quantitative measurements of amplitude and duration are preferable.

Grid 1 Synonymous with *G1*, *Input Terminal 1*, or *active* or *exploring electrode.* See *recording electrode.*

Grid 2 Synonymous with *G2*, *Input Terminal 2*, or *reference electrode.* See *recording electrode.*

ground electrode An electrode connected to the patient and to a large conducting body (such as the earth) used as a common return for an electric circuit and as an arbitrary zero potential reference point.

grouped discharge The term has been used historically to describe three phenomena: (1) irregular, voluntary grouping of *motor unit action potentials* as seen in a tremulous muscular contraction, (2) involuntary grouping of *motor unit action potentials* as seen in *myokymia*, (3) general term to describe repeated firing of *motor unit action potentials.* See preferred term, *repetitive discharge.*

H reflex Abbreviation for Hoffmann reflex. See *H wave*.

H response See preferred term *H wave*.

*__H wave__ A compound muscle action potential having a consistent latency evoked regularly, when present, from a muscle by an electric stimulus to the nerve. It is regularly found only in a limited group of physiologic extensors, particularly the calf muscles. The H wave is most easily obtained with the cathode positioned proximal to the anode. Compared with the maximum amplitude *M wave* of the same muscle, the H wave has a smaller amplitude, a longer latency, and a lower optimal stimulus intensity. The latency is longer with more distal sites of stimulation. A stimulus intensity sufficient to elicit a maximal amplitude M wave reduces or abolishes the H wave. The H wave is thought to be due to a spinal reflex, the Hoffmann reflex, with electric stimulation of afferent fibers in the mixed nerve to the muscle and activation of motor neurons to the muscle through a monosynaptic connection in the spinal cord. The reflex and wave are named in honor of Hoffmann's description (1918). Compare the *F wave*.

habituation Decrease in size of a reflex motor response to an afferent stimulus when the latter is repeated, especially at regular and recurring short intervals.

hertz (Hz) Unit of frequency equal to *cycles per second*.

Hoffmann reflex See *H wave*.

hyperpolarization See *polarization*.

Hz See *hertz*.

increased insertion activity See *insertion activity*.

*__increment after exercise__ See *facilitation*.

incremental response See preferred term, *incrementing response*.

*__incrementing response__ A reproducible increase in amplitude and/or area of successive responses (M wave) to *repetitive nerve stimulation*. The rate of stimulation and the number of stimuli should be specified. An incrementing

response is commonly seen in two situations. First, in normal subjects the configuration of the M wave may change with repetitive nerve stimulation so that the amplitude progressively increases as the duration deceases, but the area of the M wave remains the same. This phenomenon is termed *pseudofacilitation*. Second, in disorders of neuromuscular transmission, the configuration of the M wave may change with repetitive nerve stimulation so that the amplitude progressively increases as the duration remains the same or increases, and the area of the M wave increases. This phenomenon is termed *facilitation*. Contrast with *decrementing response*.

indifferent electrode Synonymous with *reference electrode*. Use of term discouraged. See *recording electrode*.

inhibitory postsynaptic potential (IPSP) A local graded hyperpolarization of a neuron in response to activation at a synapse by a nerve terminal. Contrast with *excitatory postsynaptic potential*.

injury potential The potential difference between a normal region of the surface of a nerve or muscle and a region that has been injured; also called a demarcation potential. The injury potential approximates the potential across the membrane because the injured surface is almost at the potential of the inside of the cell.

Input Terminal 1 The input terminal of the differential amplifier at which negativity, relative to the other input terminal, produces an upward deflection on the graphic display. Synonymous with *active* or *exploring electrode* (or older term, *Grid 1*). See *recording electrode*.

Input Terminal 2 The input terminal of the differential amplifier at which negativity, relative to the other input terminal, produces a downward deflection on the graphic display. Synonymous with *reference electrode* (or older term, *Grid 2*). See *recording electrode*.

*__insertion activity__ Electric activity caused by insertion or movement of a needle electrode. The amount of the activity may be described as normal,

---

*Illustration in Section II.

reduced, increased (prolonged), with a description of the waveform and repetitive rate.

interdischarge interval Time between consecutive discharges of the same potential. Measurements should be made between the corresponding points on each waveform.

interference Unwanted electric activity arising outside the system being studied.

*interference pattern Electric activity recorded from a muscle with a needle electrode during maximal voluntary effort. A *full interference pattern implies that no individual motor unit action potentials* can be clearly identified. A *reduced interference pattern* (*intermediate pattern*) is one in which some of the individual MUAPs may be identified while other individual MUAPs cannot be identified because of overlap. The term *discrete activity* is used to describe the electric activity recorded when each of several different MUAPs can be identified. The term *single unit pattern* is used to describe a single MUAP, firing at a rapid rate (should be specified) during maximum voluntary effort. The force of contraction associated with the interference pattern should be specified. See also *recruitment pattern.*

intermediate interference pattern See *interference pattern.*

International 10-20 System A system of electrode placement on the scalp in which electrodes are placed either 10% or 20% of the total distance between the nasion and inion in the sagittal plane, and between right and left preauricular points in the coronal plane.

interpeak interval Difference between the peak latencies of two components of a waveform.

interpotential interval Time between two different potentials. Measurement should be made between the corresponding parts on each waveform.

involuntary activity *Motor unit potentials* that are not under voluntary con-

trol. The condition under which they occur should be described, e.g., spontaneous or reflex potentials and, if elicited by a stimulus, the nature of the stimulus. Contrast with *spontaneous activity.*

IPSP See *inhibitory postsynaptic potential.*

irregular potential See preferred term, *serrated action potential.*

iterative discharge See preferred term, *repetitive discharge.*

*jitter Synonymous with single fiber electromyographic jitter. Jitter is the variability with consecutive discharges of the *interpotential interval* between two muscle fiber action potentials belonging to the same motor unit. It is usually expressed quantitatively as the mean value of the difference between the interpotential intervals of successive discharges (the mean consecutive difference, MCD). Under certain conditions, jitter is expressed as the mean value of the difference between interpotential intervals arranged in the order of decreasing interdischarge intervals (the mean sorted difference, MSD).

Jolly test A technique described by Jolly (1895), who applied an electric current to excite a motor nerve while recording the force of muscle contraction. Harvey and Masland (1941) refined the technique by recording the M wave evoked by repetitive, supramaximal nerve stimulation to detect a defect of neuromuscular transmission. Use of the term is discouraged. See preferred term, *repetitive nerve stimulation.*

late component (of a motor unit action potential) See preferred term, *satellite potential.*

late response A general term used to describe an evoked potential having a longer latency than the *M wave.* See *A wave, F wave, H wave, T wave.*

latency Interval between the onset of a stimulus and the onset of a response. Thus the term *onset latency* is a tautology and should not be used. The *peak latency* is the interval between the onset of a stimulus and a specified peak of the evoked potential.

latency of activation The time required

---

*Illustration in Section II.

for an electric stimulus to depolarize a nerve fiber (or bundle of fibers as in a nerve trunk) beyond threshold and to initiate a regenerative action potential in the fiber(s). This time is usually on the order of 0.1 ms or less. An equivalent term now rarely used in the literature is the "utilization time."

latent period See synonym, *latency.*

linked potential See preferred term, *satellite potential.*

long-latency SEP That portion of a *somatosensory evoked potential* normally occurring at a time greater than 100 ms after stimulation of a nerve in the upper extremity at the wrist, or the lower extremity at the knee or ankle.

M response See synonym, *M wave.*

*M wave A *compound action potential* evoked from a muscle by a single electric stimulus to its motor nerve. By convention, the M wave elicited by supramaximal stimulation is used for motor nerve conduction studies. Ideally, the recording electrodes should be placed so that the initial deflection of the evoked potential is negative. The *latency,* commonly called the *motor latency,* is the latency (ms) to the onset of the first phase (positive or negative) of the M wave. The amplitude (MV) is the baseline-to-peak amplitude of the first negative phase, unless otherwise specified. The *duration* (ms) refers to the duration of the first negative phase, unless otherwise specified. Normally, the configuration of the M wave (usually biphasic) is quite stable with repeated stimuli at slow rates (1–5 Hz). See *repetitive nerve stimulation.*

macro motor unit action potential (macro MUAP) The average electric activity of that part of an anatomic motor unit that is within the recording range of a *macro-EMG electrode.* The potential is characterized by its consistent appearance when the small recording surface of the macro-EMG electrode is positioned to record action potentials from one muscle fiber. The following parameters can be specified quantitatively: (1) maximal peak-to-peak ampli-

tude, (2) area contained under the waveform, (3) number of phases.

macro MUAP See *macro motor unit action potential.*

*macroelectromyography (macro-EMG) General term referring to the technique and conditions that approximate recording of all *muscle fiber action potentials* arising from the same motor unit.

macro-EMG See *macroelectromyography.*

macro-EMG needle electrode A modified *single fiber electromyography* electrode insulated to within 15 mm from the tip and with a small recording surface (25 μm in diameter) 7.5 mm from the tip.

malignant fasciculation Use of term discouraged to describe a firing pattern of fasciculation potentials. Historically, the term was used to describe large, polyphasic fasciculation potentials firing at a slow rate. This pattern has been seen in progressive motor neuron disease, but the relationship is not exclusive. See *fasciculation potential.*

maximal stimulus See *stimulus.*

maximum conduction velocity See *conduction velocity.*

MCD Abbreviation for mean consecutive difference. See *jitter.*

mean consecutive difference (MCD) See *jitter.*

membrane instability Tendency of a cell membrane to depolarize spontaneously, with mechanical irritation, or after voluntary activation.

MEPP Miniature end-plate potential.

microneurography The technique of recording peripheral nerve action potentials in man by means of intraneural electrodes.

midlatency SEP That portion of the waveforms of a *somatosensory evoked potential* normally occurring within 25–100 ms after stimulation of a nerve in the upper extremity at the wrist, within 40–100 ms after stimulation of a nerve in the lower extremity at the knee, and within 50–100 ms after stimulation of a nerve in the lower extremity at the ankle.

miniature end-plate potential (MEPP) The postsynaptic muscle fiber potentials produced through the spontaneous release of individual quanta of acetylcholine from the presynaptic axon terminals. As recorded with conventional concentric needle electrodes inserted in the end-plate zone, such potentials are characteristically monophasic, negative, of relatively short duration (less than 5 ms) and generally less than 20 μV in amplitude.

MNCV Abbreviation for *motor nerve conduction velocity*. See *conduction velocity*.

monophasic action potential See *action potential* with one phase.

monophasic end-plate activity *See* end-plate activity (*monophasic*).

monopolar needle recording electrode A solid wire, usually stainless steel, usually coated, except at its tip, with an insulating material. Variations in voltage between the tip of the needle (active or exploring electrode) positioned in a muscle and a conductive plate on the skin surface or a bare needle in subcutaneous tissue (reference electrode) are measured. By convention, this recording condition is referred to as a monopolar needle electrode recording. It should be emphasized, however, that potential differences are always recorded between two electrodes.

motor latency Interval between the onset of a stimulus and the onset of the resultant *compound muscle action potential* (*M wave*). The term may be qualified, as *proximal motor latency* or *distal motor latency*, depending on the relative position of the stimulus.

motor nerve conduction velocity (MNCV) See *conduction velocity*.

motor point The point over a muscle where a contraction of a muscle may be elicited by a minimal-intensity, short-duration electric stimulus. The motor point corresponds anatomically to the location of the terminal portion of the motor nerve fibers (end-plate zone).

motor response (1) The compound muscle action potential (*M wave*) recorded over a muscle with stimulation of the nerve to the muscle, (2) the muscle twitch or contraction elicited by stimulation of the nerve to a muscle, (3) the muscle twitch elicited by the muscle stretch reflex.

motor unit The anatomic unit of an anterior horn cell, its axon, the neuromuscular junctions, and all of the muscle fibers innervated by the axon.

*motor unit action potential (MUAP) Action potential reflecting the electric activity of a single anatomic motor unit. It is the compound action potential of those muscle fibers within the recording range of an electrode. With voluntary muscle contraction, the action potential is characterized by its consistent appearance with, and relationship to, the force of contraction. The following parameters should be specified, quantitatively if possible, after the recording electrode is placed so as to minimize the *rise time* (which by convention should be less than 0.5 ms):

1. Configuration
   a. *Amplitude*, peak-to-peak (μV or mV).
   b. *Duration*, total (ms).
   c. Number of *phases* (*monophasic, biphasic, triphasic, tetraphasic, polyphasic*).
   d. Sign of each *phase* (negative, positive).
   e. Number of *turns*.
   f. Variation of shape, if any, with consecutive discharges.
   g. Presence of *satellite* (*linked*) *potentials*, if any.
2. *Recruitment* characteristics
   a. Threshold of activation (first recruited, low threshold, high threshold).
   b. *Onset frequency* (Hz).
   c. *Recruitment frequency* (Hz) or *recruitment interval* (ms) of individual potentials.

Descriptive terms implying diagnostic significance are not recommended, e.g., *myopathic, neuropathic, regeneration, nascent, giant, BSAP,* and *BSAPP*. See *polyphasic action potential, serrated action potential*.

motor unit fraction See *scanning EMG*.

---

*Illustration in Section II.

motor unit potential (MUP) See synonym, *motor unit action potential.*

motor unit territory The area in a muscle over which the muscle fibers belonging to an individual motor unit are distributed.

movement artifact See *artifact.*

MSD Abbreviation for mean sorted difference. See *jitter.*

MUAP See *motor unit action potential.*

multielectrode See *multilead electrode.*

multilead electrode Three or more insulated wires inserted through a common metal cannula with their bared tips at an aperture in the cannula and flush with the outer circumference of the cannula. The arrangement of the bare tips relative to the axis of the cannula and the distance between each tip should be specified.

multiple discharge Four or more *motor unit action potentials* of the same form and nearly the same amplitude occurring consistently in the same relationship to one another and generated by this same axon or muscle fiber. See *double* and *triple discharge.*

multiplet See *multiple discharge.*

MUP Abbreviation for *motor unit potential.* See preferred term, *motor unit action potential.*

muscle action potential Term commonly used to refer to a *compound muscle action potential.*

muscle cramp Most commonly, an involuntary, painful muscle *contraction* associated with electric activity (See *cramp discharge*). Muscle cramps may be accompanied by other types of *repetitive discharges*, and in some metabolic myopathies (McArdle's disease) the painful, contracted muscles may show *electric silence.*

muscle fiber action potential Action potential recorded from a single muscle fiber.

muscle fiber conduction velocity The speed of propagation of a single *muscle fiber action potential*, usually expressed as meters per second. The muscle fiber conduction velocity is usu-

ally less than most nerve conduction velocities, varies with the rate of discharge of the muscle fiber, and requires special techniques for measurement.

muscle stretch reflex Activation of a muscle which follows stretch of the muscle, e.g., by percussion of a muscle tendon.

myoedema Focal muscle contraction produced by muscle percussion and not associated with propagated electric activity; may be seen in hypothyroidism (myxedema) and chronic malnutrition.

myokymia Continuous quivering or undulating movement of surface and overlying skin and mucous membrane associated with spontaneous, repetitive discharge of *motor unit potentials.* See *myokymic discharge*, *fasciculation*, and *fasciculation potential.*

***myokymic discharge** Motor unit action potentials* that fire repetitively and may be associated with clinical myokymia. Two firing patterns have been described. Commonly, the discharge is a brief, repetitive firing of single units for a short period (up to a few seconds) at a uniform rate (2–60 Hz) followed by a short period (up to a few seconds) of silence, with repetition of the same sequence for a particular potential. Less commonly, the potential recurs continuously at a fairly uniform firing rate (1–5 Hz). Myokymic discharges are a subclass of *grouped discharges* and *repetitive discharges.*

myopathic motor unit potential Use of term discouraged. It has been used to refer to low-amplitude, short-duration, polyphasic *motor unit action potentials.* The term incorrectly implies specific diagnostic significance of a motor unit potential configuration. See *motor unit action potential.*

myopathic recruitment Use of term discouraged. It has been used to describe an increase in the number of and firing rate of *motor unit action potentials* compared with normal for the strength of muscle contraction.

myotonia The clinical observation of delayed relaxation of muscle after voluntary contraction or percussion. The delayed relaxation may be electrically silent, or accompanied by propagated

---

*Illustration in Section II.

electric activity, such as *myotonic discharge, complex repetitive discharge, or neuromyotonic discharge.*

*myotonic discharge Repetitive discharge at rates of 20–80 Hz are of two types: (1) biphasic (positive-negative) spike potentials less than 5 ms in duration resembling *fibrillation potentials.* (2) positive waves of 5–20 ms in duration resembling *positive sharp waves.* Both potential forms are recorded after needle insertion, after voluntary muscle contraction or after muscle percussion, and are due to independent, repetitive discharges of single muscle fibers. The amplitude and frequency of the potentials must both wax and wane to be identified as myotonic discharges. This change produces a characteristic musical sound in the audio display of the electromyograph due to the corresponding change in pitch, which has been likened to the sound of a "dive bomber." Contrast with *waning discharge.*

myotonic potential See preferred term, *myotonic discharge.*

NAP Abbreviation for *nerve action potential.* See *compound nerve action potential.*

nascent motor unit potential From the Latin nascens, to be born. Use of term is discouraged as it incorrectly implies diagnostic significance of a motor unit potential configuration. Term has been used to refer to very low-amplitude, long-duration, highly polyphasic motor unit potentials observed during early states of reinnervation of muscle. See *motor unit action potential.*

NCS See *nerve conduction studies.*

NCV Abbreviation for *nerve conduction velocity.* See *conduction velocity.*

near constant frequency trains See preferred term, *complex repetitive discharge.*

near-field potential Electric activity of biologic origin generated near the recording electrodes. Use of the terms *near-field potential* and *far-field potential* is discouraged because all potentials in clinical neurophysiology are recorded at some distance from the generator and there is no consistent distinction between the two terms.

needle electrode An electrode for recording or stimulating, shaped like a needle. See specific electrodes: *bifilar (bipolar) needle recording electrode, concentric needle electrode, macro-EMG needle electrode, monopolar needle electrode, multilead electrode, single fiber needle electrode* and *stimulating electrode.*

nerve action potential (NAP) Strictly defined, refers to an action potential recorded from a single nerve fiber. The term is commonly used to refer to the compound nerve action potential. See *compound nerve action potential.*

nerve conduction studies (NCS) Synonymous with *electroneurography.* Recording and analysis of electric *waveforms* of biologic origin elicited in response to electric or physiologic *stimuli.* Generally *nerve conduction studies* refer to studies of waveforms generated in the peripheral nervous system, whereas *evoked potential studies* refer to studies of waveforms generated in both the peripheral and central nervous system. The waveforms recorded in *nerve conduction studies* are *compound sensory nerve action potentials* and *compound muscle action potentials.* The *compound sensory nerve action potentials* are generally referred to as *sensory nerve action potentials.* The *compound muscle action potentials* are generally referred to by letters which have historical origins: *M wave, F wave, H wave, T wave, A wave, R1 wave,* and *R2 wave.* It is possible under standardized conditions to establish normal ranges of amplitude, duration, and latencies of these *evoked potentials* and to calculate the maximum conduction velocity of sensory and motor nerves.

nerve conduction velocity (NCV) Loosely used to refer to the maximum nerve conduction velocity. See *conduction velocity.*

nerve fiber action potential Action potential recorded from a single nerve fiber.

nerve potential Equivalent to *nerve ac-*

---

*Illustration in Section II.

tion potential. Also commonly, but inaccurately, used to refer to the biphasic form of *end-plate activity.* The latter use is incorrect because muscle fibers, not nerve fibers, are the source of these potentials.

nerve trunk action potential See preferred term, *compound nerve action potential*

neurapraxia Failure of nerve conduction, usually reversible, due to metabolic or microstructural abnormalities without disruption of the axon. See preferred electrodiagnostic term, *conduction block.*

neuromyotonia Clinical syndrome of continuous muscle fiber activity manifested as continuous muscle rippling and stiffness. The accompanying electric activity may be intermittent or continuous. Terms used to describe related clinical syndromes are continuous muscle fiber activity, Isaac syndrome, Isaac-Merton syndrome, quantal squander syndrome, generalized myokymia, pseudomyotonia, normocalcemic tetany and neurotonia.

*****neuromyotonic discharge** Bursts of *motor unit action potentials* which originate in the motor axons firing at high rates (150–300 Hz) for a few seconds, and which often start and stop abruptly. The amplitude of the response typically wanes. Discharges may occur spontaneously or be initiated by needle movement, voluntary effort and ischemia or percussion of a nerve. These discharges should be distinguished from *myotonic discharges* and *complex repetitive discharges.*

neuropathic motor unit potential Use of term discouraged. It was used to refer to abnormally high-amplitude, long-duration, polyphasic *motor unit action potentials.* The term incorrectly implies a specific diagnostic significance of a motor unit potential configuration. See *motor unit action potential.*

neuropathic recruitment Use of term discouraged. It has been used to describe a recruitment pattern with a decreased number of *motor unit action*

potentials firing at a rapid rate. See preferred terms, *reduced interference pattern, discrete activity, single unit pattern.*

neurotmesis Partial or complete severance of a nerve, with disruption of the axons, their myelin sheaths and the supporting connective tissue, resulting in degeneration of the axons distal to the injury site.

noise Strictly defined, potentials produced by electrodes, cables, amplifier or storage media and unrelated to the potentials of biologic origin. The term has been used loosely to refer to one form of *end-plate activity.*

onset frequency The lowest stable frequency of firing for a single *motor unit action potential* that can be voluntarily maintained by a subject.

onset latency Tautology. See *latency.*

order of activation The sequence of appearance of different *motor unit action potentials* with increasing strength of voluntary contraction. See *recruitment.*

orthodromic Propagation of an impulse in the direction the same as physiologic conduction; e.g., conduction along motor nerve fibers towards the muscle and conduction along sensory nerve fibers towards the spinal cord. Contrast with *antidromic.*

paired discharge Two action potentials occurring consistently in the same relationship with each other. Contrast with *double discharge.*

paired response Use of term discouraged. See preferred term, *paired discharge.*

paired stimuli Two consecutive stimuli. The time interval between the two stimuli and the intensity of each stimulus should be specified. The first stimulus is called the *conditioning stimulus* and the second stimulus is the *test stimulus.* The *conditioning stimulus* may modify the tissue excitability, which can then be evaluated by the response to the test stimulus.

parasite potential See preferred term, *satellite potential.*

peak latency Interval between the onset of a stimulus and a specified peak of the evoked potential.

---

*Illustration in Section II.

phase That portion of a *wave* between the departure from, and the return to, the *baseline*.

polarization As used in neurophysiology, the presence of an electric potential difference across an excitable cell membrane. The potential across the membrane of a cell when it is not excited by an input or spontaneously active is termed the *resting potential*; it is at a stationary nonequilibrium state with regard to the electric potential difference across the membrane. *Depolarization* describes a reduction in the magnitude of the polarization toward the zero potential while *hyperpolarization* refers to an increase in the magnitude of the polarization relative to the resting potential. *Repolarization* describes an increase in polarization from the depolarized state toward, but not above, the normal resting potential.

polyphasic action potential An *action potential* having five or more phases. See *phase*. Contrast with *serrated action potential*.

*****positive sharp wave** A biphasic, positive-negative *action potential* initiated by needle movement and recurring in a uniform, regular pattern at a rate of 1–50 Hz; the discharge frequency may decrease slightly just before cessation of discharge. The initial positive deflection is rapid (<1 ms), its duration is usually less than 5 ms, and the amplitude is up to 1 mV. The negative phase is of low amplitude, with a duration of 10–100 ms. A sequence of positive sharp waves is commonly referred to as a *train of positive sharp waves*. Positive sharp waves can be recorded from the damaged area of fibrillating muscle fibers. Its configuration may result from the position of the needle electrode which is felt to be adjacent to the depolarized segment of a muscle fiber injured by the electrode. Note that the positive sharp waveform is not specific for muscle fiber damage. *Motor unit action potentials* and potentials in *myotonic discharges* may have the configuration of positive sharp waves.

positive wave Loosely defined, the term refers to a positive sharp wave. See *positive sharp wave*.

*****postactivation depression** A descriptive term indicating a reduction in the amplitude associated with a reduction in the area of the M wave(s) in response to a single *stimulus* or *train of stimuli* which occurs a few minutes after a brief (30–60 s), strong voluntary contraction or a period of *repetitive nerve stimulation* that produces *tetanus*. *Postactivation exhaustion* refers to the cellular mechanisms responsible for the observed phenomenon of *postactivation depression*.

postactivation exhaustion A reduction in the safety factor (margin) of neuromuscular transmission after sustained activity of the neuromuscular junction. The changes in the configuration of the M wave due to *postactivation exhaustion* are referred to as *postactivation depression*.

postactivation facilitation See *facilitation*.

postactivation potentiation Refers to the increase in the force of contraction (mechanical response) after *tetanus* or strong voluntary contraction. Contrast *postactivation facilitation*.

posttetanic facilitation See *facilitation*.

posttetanic potentiation The incrementing mechanical response of muscle during and after *repetitive nerve stimulation* without a change in the amplitude of the action potential. In spinal cord physiology, the term has been used to describe enhancement of excitability or reflex outflow of the central nervous system following a long period of high-frequency stimulation. This phenomenon has been described in the mammalian spinal cord, where it lasts minutes or even hours.

potential A physical variable created by differences in charges, measurable in volts, that exists between two points. Most biologically produced potentials arise from the difference in charge between two sides of a cell membrane. See *polarization*.

potentiation Physiologically, the enhancement of a response. Some au-

---

*Illustration in Section II.

thors use the term *potentiation* to describe the incrementing mechanical response of muscle elicited by *repetitive nerve stimulation*, i.e., *posttetanic potentiation*, and the term *facilitation* to describe the incrementing electric response elicited by *repetitive nerve stimulation*, i.e, *postactivation facilitation*.

prolonged insertion activity See *insertion activity.*

propagation velocity of a muscle fiber The speed of transmission of a muscle fiber action potential.

proximal latency See *motor latency* and *sensory latency.*

*****pseudofacilitation** See *facilitation.*

pseudomyotonic discharge Use of term discouraged. It has been used to refer to different phenomena, including (1) *complex repetitive discharges*, and (2) *repetitive discharges* that do not wax or wane in both frequency and amplitude, and end abruptly. These latter discharges may be seen in disorders such as polymyositis in addition to disorders with *myotonic discharges*. See preferred term, *waning discharge.*

pseudopolyphasic action potential Use of term discouraged. See preferred term, *serrated action potential*

R1, R2 waves See *blink responses.*

recording electrode Device used to record electric potential difference. All electric recordings require two *electrodes*. The recording electrode close to the source of the activity to be recorded is called the *active* or *exploring electrode*, and the other recording electrode is called the *reference electrode*. Active electrode is synonymous with *Input Terminal I* (or older terms *Grid 1*, and *G1*) and the reference electrode with *Input Terminal 2* (or older terms *Grid 2*, and *G2*).

In some recordings, it is not certain which electrode is closer to the source of the biologic activity, i.e., recording with a *bifilar (bipolar) needle electrode*. In this situation, it is convenient to refer to one electrode as Input Electrode 1 and the other electrode as Input Electrode 2.

---

*Illustration in Section II.

By present convention, a potential difference that is negative at the active electrode (Input Terminal 1) relative to the reference electrode (Input Terminal 2) causes an upward deflection on the oscilloscope screen. The term "monopolar recording" is not recommended, because all recording requires two electrodes; however, it is commonly used to describe the use of an intramuscular needle exploring electrode in combination with a surface disk or subcutaneous needle reference electrode. A similar combination of needle electrodes has been used to record nerve activity and also has been referred to as "monopolar recording."

recruitment The successive activation of the same and additional motor units with increasing strength of voluntary muscle contraction. See *motor unit action potential.*

recruitment frequency Firing rate of a *motor unit action potential (MUAP)* when a different MUAP first appears with gradually increasing strength of voluntary muscle contraction. This parameter is essential to assessment of *recruitment pattern.*

recruitment interval The *interdischarge interval* between two consecutive discharges of a *motor unit action potential (MUAP)* when a different MUAP first appears with gradually increasing strength of voluntary muscle contraction. The reciprocal of the recruitment interval is the *recruitment frequency.*

*****recruitment pattern** A qualitative and/or quantitative description of the sequence of appearance of *motor unit action potentials* with increasing strength of voluntary muscle contraction. The *recruitment frequency* and *recruitment interval* are two quantitative measures commonly used. See *interference pattern* for qualitative terms commonly used.

reduced insertion activity See *insertion activity.*

reduced interference pattern See *interference pattern.*

reference electrode See *recording electrode.*

reflex A stereotyped *motor response* elicited by a sensory *stimulus.*

refractory period The *absolute refractory period* is the period following an *action potential* during which no stimulus, however strong, evokes a further response. The *relative refractory period* is the period following an *action potential* during which a stimulus must be abnormally large to evoke a second response. The *functional refractory period* is the period following an *action potential* during which a second *action potential* cannot yet excite the given region.

regeneration motor unit potential Use of term discouraged. See *motor unit action potential.*

relative refractory period See *refractory period.*

***repair of the decrement** See *facilitation.*

repetitive discharge General term for the recurrence of an *action potential* with the same or nearly the same form. The term may refer to recurring potentials recorded in muscle at rest, during voluntary contraction, or in response to single nerve stimulus. See *double discharge, triple discharge, multiple discharge, myokymic discharge, myotonic discharge, complex repetitive discharge.*

***repetitive nerve stimulation** The technique of repeated supramaximal stimulations of a nerve while recording M waves from muscles innervated by the nerve. The number of stimuli and the frequency of stimulation should be specified. Activation procedures performed prior to the test should be specified, e.g., sustained voluntary contraction or contraction induced by nerve stimulation. If the test was performed after an activation procedure, the time elapsed after the activation procedure was completed should also be specified. The technique is commonly used to assess the integrity of neuromuscular transmission. For a description of specific patterns of responses, see the terms *incrementing response, decrementing response, facilitation* and *postactivation depression*

repolarization See *polarization.*

---

*Illustration in Section II.

residual latency Refers to the calculated time difference between the measured distal latency of a motor nerve and the expected distal latency, calculated by dividing the distance between the stimulus cathode and the active recording electrode by the maximum conduction velocity measured in a more proximal segment of a nerve. The residual latency is due in part to neuromuscular transmission time and to slowing of conduction in terminal axons due to decreasing diameter and the presence of unmyelinated segments.

response Used to describe an activity elicited by a *stimulus.*

resting membrane potential Voltage across the membrane of an excitable cell at rest. See *polarization.*

rheobase See *strength-duration curve.*

rise time The interval from the onset of a change of a potential to its peak. The method of measurement should be specified.

***satellite potential** A small action potential separated from the main MUAP by an isoelectric interval and firing in a time-locked relationship to the main *action potential.* These potentials usually follow, but may *precede,* the main action potential. Also called *late component, parasite potential, linked potential,* and *coupled discharge* (less preferred terms).

scanning EMG A technique by which an electromyographic electrode is advanced in defined steps through muscle while a separate SFEMG electrode is used to trigger both the oscilloscope-sweep and the advancement device. This recording technique provides temporal and spatial information about the motor unit. Distinct maxima in the recorded activity are considered to be generated by muscle fibers innervated by a common branch of the axon. These groups of fibers form a *motor unit fraction.*

sea shell sound (sea shell roar or noise) Use of term discouraged. See *end-plate activity, monophasic.*

sensory delay See preferred terms, *sensory latency* and *sensory peak latency.*

sensory latency Interval between the onset of a stimulus and the onset of the *compound sensory nerve action potential.* This term has been loosely used to refer to the *sensory peak latency.* The term may be qualified as *proximal sensory latency* or *distal sensory latency*, depending on the relative position of the stimulus.

sensory nerve action potential (SNAP) See *compound sensory nerve action potential.*

sensory nerve conduction velocity See *conduction velocity.*

sensory peak latency Interval between the onset of a *stimulus* and the peak of the negative phase of the *compound sensory nerve action potential.* Note that the term *latency* refers to the interval between the onset of a stimulus and the onset of a response.

sensory potential Used to refer to the compound sensory nerve action potential. See *compound sensory nerve action potential.*

sensory response Used to refer to a sensory evoked potential, e.g., *compound sensory nerve action potential*

SEP See *somatosensory evoked potential.*

serrated action potential An action potential waveform with several changes in direction (*turns*) which do not cross the baseline. This term is preferred to the terms *complex action potential* and *pseudopolyphasic action potential.* See also *turn* and *polyphasic action potential.*

SFEMG See *single fiber electromyography.*

shock artifact See *artifact.*

***short-latency somatosensory evoked potential** (SSEP) That portion of the waveforms of a *somatosensory evoked potential* normally occurring within 25 ms after stimulation of the median nerve in the upper extremity at the wrist, 40 ms after stimulation of the common peroneal nerve in the lower extremity at the knee, and 50 ms after stimulation of the posterior tibial nerve in the lower extremity at the ankle.

1. *Median nerve SSEPs:* Normal short-latency response components to median nerve stimulation are designated P$\overline{9}$, P$\overline{11}$, P$\overline{13}$, P$\overline{14}$, N$\overline{20}$, and P$\overline{23}$ in records taken between scalp and noncephalic reference electrodes, and N$\overline{9}$, N$\overline{11}$, N$\overline{13}$, and N$\overline{14}$ in cervical spine-scalp derivation. It should be emphasized that potentials having opposite polarity but similar latency in spine-scalp and scalp-noncephalic reference derivations do not necessarily have identical generator sources.

2. *Common peroneal nerve SSEPs:* Normal short-latency response components, to common peroneal stimulation are designated P$\overline{27}$ and N$\overline{35}$ in records taken between scalp and noncephalic reference electrodes, and L3 and T12 from a cervical spine-scalp derivation.

3. *Posterior tibial nerve SSEPs:* Normal short-latency response components to posterior tibial nerve stimulation are designated as the P$\overline{F}$ potential in the popliteal fossa, P$\overline{37}$ and N$\overline{45}$ waves in records taken between scalp and noncephalic reference electrode, and L3 and T12 potentials from a cervical spine-scalp derivation.

silent period A pause in the electric activity of a muscle such as that seen after rapid unloading of a muscle.

***single fiber electromyography** (SFEMG) General term referring to the technique and conditions that permit recording of a single *muscle fiber action potential.* See *single fiber needle electrode* and *jitter.*

single fiber EMG See *single fiber electromyography.*

single fiber needle electrode A needle *electrode* with a small recording surface (usually 25 μm in diameter) permitting the recording of single muscle fiber action potentials between the active recording surface and the cannula. See *single fiber electromyography.*

single unit pattern See *interference pattern.*

SNAP Abbreviation for *sensory nerve action potential.* See *compound sensory nerve action potential.*

somatosensory evoked potentials (SEPs) Electric waveforms of biologic

---

*Illustration in Section II.

origin elicited by electric stimulation or physiologic activation of peripheral sensory fibers, for example, the median nerve, common peroneal nerve, or posterior tibial nerve. The normal SEP is a complex waveform with several components which are specified by polarity and average peak latency. The polarity and latency of individual components depend upon (1) subject variables, such as age, sex, (2) stimulus characteristics, such as intensity, rate of stimulation, and (3) recording parameters, such as amplifier time constants, electrode placement, electrode combinations. See *short-latency SEPs.*

spike (1) In cellular neurophysiology, a short-lived (usually in the range of 1–3 ms), all-or-none change in membrane potential that arises when a graded response passes a threshold. (2) The electric record of a nerve impulse or similar event in muscle or elsewhere. (3) In clinical EEG recordings, a wave with duration less than 80 ms (usually 15–80 ms).

spinal evoked potential Electric waveforms of biologic origin recorded over the sacral, lumbar, thoracic or cervical spine in response to electric stimulation or physiologic activation of peripheral sensory fibers. See preferred term, *somatosensory evoked potential.*

spontaneous activity Electric activity recorded from muscle or nerve at rest after insertion activity has subsided and when there is no voluntary contraction or external stimulus. Compare with *involuntary activity.*

SSEP See *short-latency somatosensory evoked potential.*

staircase phenomenon The progressive increase in the force of a muscle contraction observed in response to continued low rates of direct or indirect muscle stimulation.

stigmatic electrode Of historic interest. Used by Sherrington for *active* or *exploring electrode.*

stimulating electrode Device used to apply electric current. All electric stimulation requires two electrodes; the negative terminal is termed the *cathode* and the positive terminal, the *anode.* By convention, the stimulating electrodes are called "*bipolar*" if they are encased or attached together. Stimulating electrodes are called "*monopolar*" if they are not encased or attached together. Electric stimulation for *nerve conduction studies* generally requires application of the cathode to produce depolarization of the nerve trunk fibers. If the anode is inadvertently placed between the cathode and the recording electrodes, a focal block of nerve conduction (*anodal block*) may occur and cause a technically unsatisfactory study.

stimulus Any external agent, state, or change that is capable of influencing the activity of a cell, tissue, or organism. In clinical *nerve conduction studies*, an electric stimulus is generally applied to a nerve or muscle. The electric stimulus may be described in absolute terms or with respect to the evoked potential of the nerve or muscle. In absolute terms, the electric stimulus is defined by a duration (ms), a waveform (square, exponential, linear, etc.) and a strength or intensity measured in voltage (V) or current (mA). With respect to the evoked potential, the stimulus may be graded as subthreshold, threshold, submaximal, maximal, or supramaximal. A *threshold stimulus* is that stimulus just sufficient to produce a detectable response. Stimuli less than the threshold stimulus are termed *subthreshold.* The *maximal stimulus* is the stimulus intensity after which a further increase in the stimulus intensity causes no increase in the amplitude of the evoked potential. Stimuli of intensity below this level but above threshold are *submaximal.* Stimuli of intensity greater than the maximal stimulus are termed *supramaximal.* Ordinarily, supramaximal stimuli are used for nerve conduction studies. By convention, an electric stimulus of approximately 20% greater voltage/current than required for the maximal stimulus may be used for supramaximal stimulation. The frequency, number, and duration of a series of stimuli should be specified.

stimulus artifact See *artifact.*

strength-duration curve Graphic presentation of the relationship between the intensity (Y axis) and various durations (X axis) of the threshold electric stimulus for a muscle with the stimulating cathode positioned over the *motor point*. The *rheobase* is the intensity of an electric current of infinite duration necessary to produce a minimal visible twitch of a muscle when applied to the motor point. In clinical practice, a duration of 300 ms is used to determine the rheobase. The *chronaxie* is the time required for an electric current twice the *rheobase* to elicit the first visible muscle twitch.

submaximal stimulus. See *stimulus.*

subthreshold stimulus See *stimulus.*

supramaximal stimulus See *stimulus.*

surface electrode Conducting device for stimulating or recording placed on a skin surface. The material (metal, fabric), configuration (disk, ring), size, and separation should be specified. See *electrode* (*ground, recording, stimulating*).

synchronized fibrillation See preferred term, *complex repetitive discharge.*

*T wave A compound action potential evoked from a muscle by rapid stretch of its tendon, as part of the muscle stretch reflex.

temporal dispersion Relative desynchronization of components of a compound action potential due to different rates of conduction of each synchronously evoked component from the stimulation point to the recording electrode.

terminal latency Synonymous with preferred term, *distal latency.* See *motor latency* and *sensory latency.*

test stimulus See *paired stimuli.*

tetanic contraction The contraction produced in a muscle through repetitive maximal direct or indirect stimulation at a sufficiently high frequency to produce a smooth summation of successive maximum twitches. The term may also be applied to maximum voluntary contractions in which the firing frequencies of most or all of the compo-

nent motor units are sufficiently high that successive twitches of individual motor units fuse smoothly. Their tensions all combine to produce a steady, smooth maximum contraction of the whole muscle.

tetanus The continuous contraction of muscle caused by repetitive stimulation or discharge of nerve or muscle. Contrast *tetany.*

tetany A clinical syndrome manifested by muscle twitching, cramps, and carpal and pedal spasm. These clinical signs are manifestations of peripheral and central nervous system nerve irritability from several causes. In these conditions, *repetitive discharges* (*double discharge, triple discharge, multiple discharge*) occur frequently with voluntary activation of *motor unit action potentials* or may appear as *spontaneous activity* and are enhanced by systemic alkalosis or local ischemia.

tetraphasic action potential *Action potential* with four phases.

threshold The level at which a clear and abrupt transition occurs from one state to another. The term is generally used to refer to the voltage level at which an *action potential* is initiated in a single axon or a group of axons. It is also operationally defined as the intensity that produced a response in about 50% of equivalent trials.

threshold stimulus See *stimulus.*

train of positive sharp waves See *positive sharp wave.*

train of stimuli A group of stimuli. The duration of the group or the number of stimuli and the frequency of the stimuli should be specified.

triphasic action potential *Action potential* with three phases.

triple discharge Three *motor unit action potentials* of the same form and nearly the same amplitude, occurring consistently in the same relationship to one another and generated by this same axon or muscle fiber. The interval between the second and the third action potential often exceeds that between the first two, and both are usually in the range of 2–20 ms.

triplet See *triple discharge.*

turn Point of change in direction in the

---

*Illustration in Section II.

waveform and the magnitude of the voltage change following the turning point. It is not necessary that the voltage change passes through the baseline. The minimal excursion required to constitute a change should be specified.

unipolar needle electrode See synonym, *monopolar needle recording electrode.*

utilization time See preferred term, *latency of activation.*

VEPs See *visual evoked potentials.*

VERs Abbreviation for *visual evoked responses.* See *visual evoked potentials.*

*****visual evoked potentials** (VEPs) Electric waveforms of biologic origin are recorded over the cerebrum and elicited by light stimuli. VEPs are classified by stimulus rate as transient or steady state VEPs, and can be further divided by presentation mode. The normal transient VEP to checkerboard pattern reversal or shift has a major positive occipital peak at about 100 ms (P100), often preceded by a negative peak (N75). The precise range of normal values for the latency and amplitude of P100 depends on several factors: (1) subject variables, such as age, sex, and visual acuity, (2) stimulus characteristics, such as type of stimulator, full-field or half-field stimulation, check size, contrast and luminescence, and (3) recording parameters, such as placement and combination of recording electrodes.

visual evoked responses (VERs) See *visual evoked potentials.*

volitional activity See *voluntary activity.*

voltage Potential difference between two recording sites.

volume conduction Spread of current from a potential source through a conducting medium, such as the body tissues.

voluntary activity In electromyography, the electric activity recorded from a muscle with consciously controlled muscle contraction. The effort made to contract the muscle should be specified

relative to that of a corresponding normal muscle, e.g., minimal, moderate, or maximal. If the recording remains isoelectric during the attempted contraction of the muscle and artifacts have been excluded, it can be concluded that there is no voluntary activity.

waning discharge General term referring to a *repetitive discharge* that gradually decreases in frequency or amplitude before cessation. Contrast with *myotonic discharge.*

wave An undulating line constituting a graphic representation of a change, e.g., a changing electric potential difference. See *A wave, F wave, H wave,* and *M wave.*

waveform The shape of a *wave.* The term is often used synonymously with wave.

SECTION II:
ILLUSTRATIONS OF SELECTED WAVEFORMS

4–1. Compound sensory nerve action potentials

4–2. Short-latency SEPs of the median nerve

4–3. Short-latency SEPs of the common peroneal nerve

4–4. Short-latency SEPs of the posterior tibial nerve

4–5. Visual evoked potential

4–6. Brainstem auditory evoked potential

4–7. M wave

4–8. F wave

4–9. H wave

4–10. A wave

4–11. T wave

4–12. Blink responses

4–13. Repetitive nerve stimulation: normal response

4–14. Repetitive nerve stimulation: decrementing response

4–15. Repetitive nerve stimulation: incrementing response

4–16. Repetitive nerve stimulation: facilitation, increment after exercise, repair of the decrement, postactivation depression

---

*Illustration in Section II.

Each illustration is accompanied by a complete explanation which is the same as that given in the glossary. The definitions have been repeated fully with the illustrations so that readers do not need to refer back and forth between the illustrations and definitions.

The illustrations have been modified and adapted from material submitted by members of the AAEE. The illustrations of the short-latency somatosensory evoked potentials were reproduced from the *Journal of Clinical Neurophysiology* (1978; 1:41–53) with permission of the journal editor and the authors.

COMPOUND SENSORY NERVE ACTION POTENTIALS

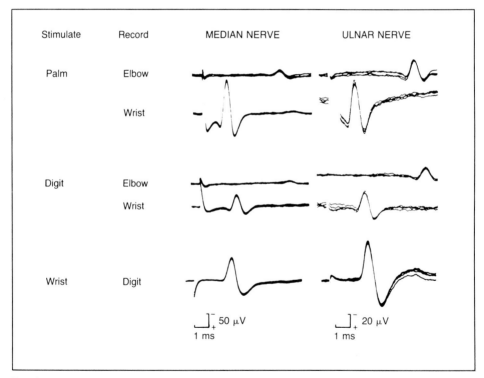

Appendix Figure 4–1. Compound sensory nerve action potentials recorded with surface electrodes in a normal subject. A compound nerve action potential is considered to have been evoked from afferent fibers if the recording electrodes detect activity only in a sensory nerve or in a sensory branch of a mixed nerve, or if the electric stimulus is applied to a sensory nerve or a dorsal nerve root, or an adequate stimulus is applied synchronously to sensory receptors. The amplitude, latency, duration, and configuration should be noted. Generally, the amplitude is measured as the maximum peak-to-peak voltage, the latency as either the *latency* to the initial deflection or the *peak latency* to the negative peak, and the duration as the interval from the first deflection of the waveform from the baseline to its final return to the baseline. The compound sensory nerve action potential has been referred to as the *sensory response* or *sensory potential*.

SHORT-LATENCY SOMATOSENSORY EVOKED POTENTIALS

MEDIAN NERVE

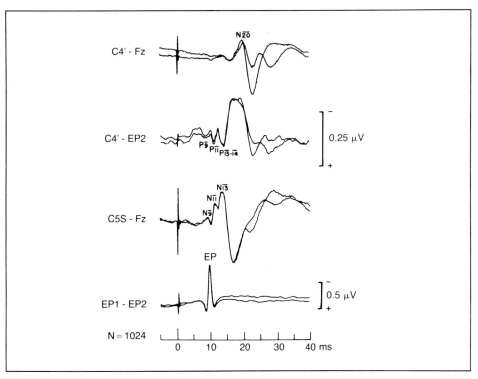

Appendix Figure 4–2. *Short-latency somatosensory evoked potentials* elicited by electric stimulation of the median nerve at the wrist (MN-SSEPs) occur within 25 ms of the stimulus in normal subjects. Normal short-latency response components to median nerve stimulation are designated P9, P11, P13, P14, N20, and P23 in records taken between scalp and noncephalic reference electrodes, and N9, N11, N13, and N14 in cervical spine-scalp derivation. It should be emphasized that potentials having opposite polarity but similar latency in spine-scalp and scalp-noncephalic reference derivations do not necessarily have identical generator sources. The C4' designation indicates that the recording scalp electrode was placed 2 cm posterior to the International 10-20 C4 electrode location.

SHORT-LATENCY SOMATOSENSORY EVOKED POTENTIALS

COMMON PERONEAL NERVE

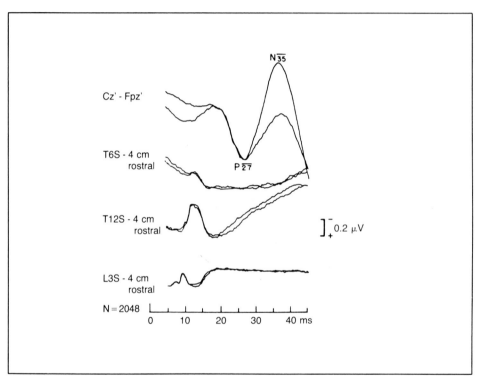

Appendix Figure 4–3. *Short-latency somatosensory evoked potentials* elicited by stimulation of the common peroneal nerve at the knee (CPN-SSEPs) occur within 40 ms of the stimulus in normal subjects. It is suggested that individual response components be designated as follows: (1) Spine components: L3 and T12 spine potentials. (2) Scalp components: P27 and N35. The Cz' and Fpz' designations indicate that the recording scalp electrode was placed 2 cm posterior to the International 10-20 Cz and Fpz electrode locations.

SHORT-LATENCY SOMATOSENSORY EVOKED POTENTIALS

POSTERIOR TIBIAL NERVE

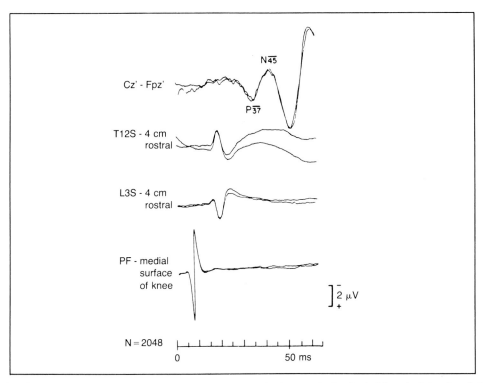

Appendix Figure 4–4. *Short-latency somatosensory evoked potentials* elicited by electric stimulation of the posterior tibial nerve (PTN-SSEPs) at the ankle occur within 50 ms of the stimulus in normal subjects. It is suggested that individual response components be designated as follows: (1) Nerve trunk (tibial nerve) component in the popliteal fossa: PF potential. (2) Spine components: L3 and T12 potentials. (3) Scalp components: P37 and N45 waves. The Cz' and Fpz' designations indicate that the recording scalp electrode was placed 2 cm posterior to the International 10-20 Cz and Fpz electrode locations.

Appendix Figure 4–5. Visual evoked potential (VEP). Normal occipital VEP to checkerboard pattern reversal stimulation recorded between occipital (01) and vertex (Cz) electrodes showing N75, P100 and N175 peaks. Visual evoked potentials are electric waveforms of biologic origin recorded over the cerebrum and elicited by light stimuli. VEPs are classified by stimulus rate as transient or steady state VEPs, and can be further divided by presentation mode. The normal transient VEP to checkerboard pattern reversal or shift has a major positive occipital peak at about 100 ms (P100), often preceded by a negative peak (N75). The precise range of normal values for the latency and amplitude of P100 depends on several factors: (1) subject variables, such as age, sex, and visual acuity, (2) stimulus characteristics, such as type of stimulator, full-field or half-field stimulation, check size, contrast and luminescence, and (3) recording parameters, such as placement and combination of recording electrodes.

VISUAL EVOKED POTENTIAL

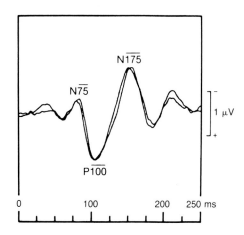

**BRAINSTEM AUDITORY
EVOKED POTENTIAL**

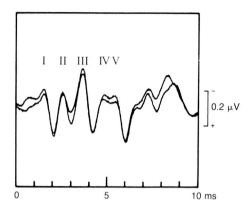

Appendix Figure 4–6. Brainstem auditory evoked potential (BAEP). Normal BAEP to stimulation of the left ear, recorded between left ear (A2) and vertex (Cz) electrodes. Brainstem auditory evoked potentials are electric waveforms of biologic origin elicited in response to sound stimuli. The normal BAEP consists of a sequence of up to seven waves, named I to VII, which occur during the first 10 ms after the onset of the stimulus and have positive polarity at the vertex of the head. In this recording, negativity in Input Terminal 1 or positivity in Input Terminal 2 causes an upward deflection.

M WAVE

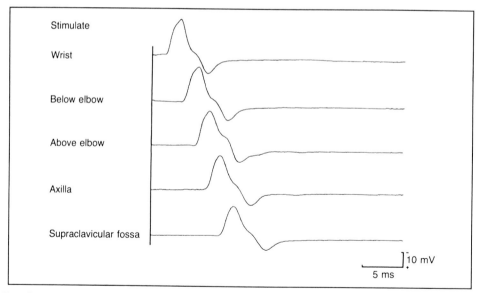

Appendix Figure 4–7. M waves recorded with surface electrodes over the abductor digiti quinti muscle elicited by electric stimulation of the ulnar nerve at several levels. The M wave is a *compound action potential* evoked from a muscle by a single electric stimulus to its motor nerve. By convention, the M wave elicited by supramaximal stimulation is used for motor nerve conduction studies. Ideally, the recording electrodes should be placed so that the initial deflection of the evoked potential is negative. The *latency*, commonly called the *motor latency*, is the latency (ms) to the onset of the first phase (positive or negative) of the M wave. The amplitude (mV) is the baseline-to-peak amplitude of the first negative phase, unless otherwise specified. The *duration* (ms) refers to the duration of the first negative phase, unless otherwise specified. Normally, the configuration of the M wave (usually biphasic) is quite stable with repeated stimuli at slow rates (1–5 Hz). See *repetitive nerve stimulation*.

F WAVE

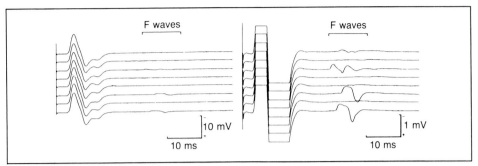

Appendix Figure 4–8. F waves recorded with surface electrodes over the abductor digiti quinti muscle elicited by electric stimulation of the ulnar nerve at the wrist with two different gain settings. The F wave is a compound action potential evoked intermittently from a muscle by a supramaximal electric stimulus to the nerve. Compared with the maximal amplitude *M wave* of the same muscle, the F wave has a smaller amplitude (1–5% of the M wave), variable configuration and a longer, more variable latency. The F wave can be found in many muscles of the upper and lower extremities, and the latency is longer with more distal sites of stimulation. The F wave is due to antidromic activation of motor neurons. It was named by Magladery and McDougal in 1950. Compare the *H wave* and the *A wave*.

H WAVE

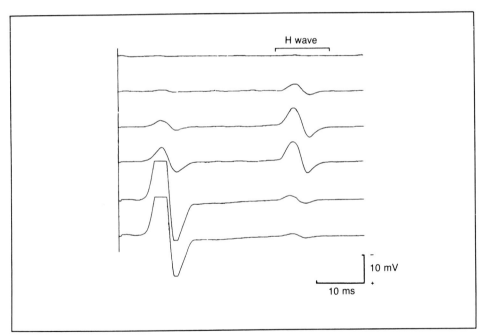

Appendix Figure 4–9. H waves recorded with surface electrodes over the soleus muscle elicited by electric stimulation of the posterior tibial nerve at the knee. The stimulus intensity was gradually increased (top tracing to bottom tracing). The H wave is a compound muscle action potential having a consistent latency evoked regularly, when present, from a muscle by an electric stimulus to the nerve. It is regularly found only in a limited group of physiologic extensors, particularly the calf muscles. The *H wave* is most easily obtained with the cathode positioned proximal to the anode. Compared with the maximum amplitude *M wave* of the same muscle, the H wave has a smaller amplitude, a longer latency, and a lower optimal stimulus intensity. The latency is longer with more distal sites of stimulation. A stimulus intensity sufficient to elicit a maximal amplitude M wave reduces or abolishes the H wave. The H wave is thought to be due to a spinal reflex, the Hoffmann reflex, with electric stimulation of afferent fibers in the mixed nerve to the muscle and activation of motor neurons to the muscle through a monosynaptic connection in the spinal cord. The reflex and wave are named in honor of Hoffmann's description in 1918. Compare the *F wave*.

A WAVE

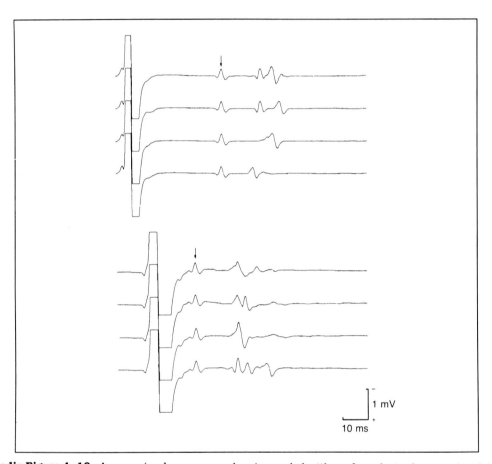

Appendix Figure 4–10. A waves (under arrow markers) recorded with surface electrodes over the abductor hallucis brevis elicited by electric stimulation of the posterior tibial nerve at the level of the ankle (top four traces) and at the level of the knee (bottom four traces). The A wave is a compound action potential evoked consistently from a muscle by submaximal electric stimuli to the nerve and frequently abolished by supra-maximal stimuli. The amplitude of the A wave is similar to that of the F wave, but the latency is more constant. The A wave usually occurs before the F wave, but may occur afterwards. The A wave is due to normal or pathologic axonal branching.Compare the *F wave.*

T WAVE

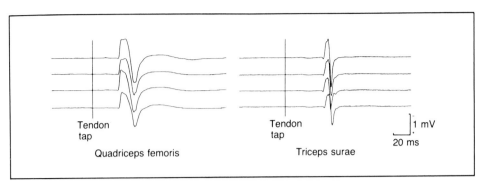

Appendix Figure 4–11. The T wave is a compound action potential evoked from a muscle by rapid stretch of its tendon, as part of the muscle stretch reflex. The T waves were recorded with surface electrodes over the quadriceps femoris (left tracings) and triceps surae (right tracings) and elicited by stretching the muscles by tapping the corresponding tendon.

BLINK RESPONSES

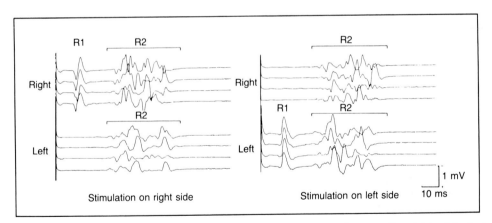

Appendix Figure 4–12. Blink responses recorded with surface electrodes over the right orbicularis oculi (upper tracings) and left orbicularis oculi (lower tracings) elicited by electric stimulation of the supraorbital nerve on the right (left tracings) and on the left (right tracings). The blink responses are *compound muscle action potentials* evoked from orbicularis oculi muscles as a result of brief electric or mechanical stimuli to the cutaneous area innervated by the supraorbital (or less commonly, the infraorbital) branch of the trigeminal nerve. Typically, there is an early compound muscle action potential (*R1 wave*) ipsilateral to the stimulation site with a latency of about 10 ms and a bilateral late compound muscle action potential (*R2 wave*) with a latency of approximately 30 ms. Generally, only the *R2 wave* is associated with a visible twitch of the orbicularis oculi. The configuration, amplitude, duration, and latency of the two components, along with the sites of recording and the sites of stimulation, should be specified. *R1* and *R2* waves are probably oligosynaptic and polysynaptic brainstem reflexes, respectively, together called the *blink reflex*, with the afferent arc provided by the sensory branches of the trigeminal nerve and the efferent arc provided by the facial nerve motor fibers.

REPETITIVE NERVE STIMULATION

NORMAL RESPONSE

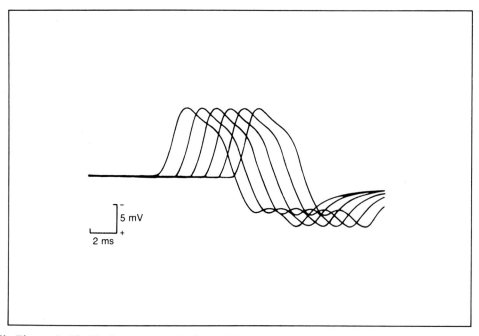

Appendix Figure 4–13. Study in a normal subject. The successive M waves are displayed to the right. The M waves were recorded with surface electrodes over the hypothenar eminence (abductor digiti quinti) during ulnar nerve stimulation at a rate of 3 Hz. Note the configuration of the successive M waves is unchanged. *Repetitive nerve stimulation* is a technique of repeated supramaximal stimulations of a nerve while recording M waves from muscles innervated by the nerve. The number of stimuli and the frequency of stimulation should be specified. Activation procedures performed prior to the test should be specified, e.g., sustained voluntary contraction or contraction induced by nerve stimulation. If the test was performed after an activation procedure, the time elapsed after the activation procedure was completed should also be specified. The technique is commonly used to assess the integrity of neuromuscular transmission. For a description of specific patterns of responses, see the terms *incrementing response, decrementing response, facilitation* and *postactivation depression.*

REPETITIVE NERVE STIMULATION

DECREMENTING RESPONSE

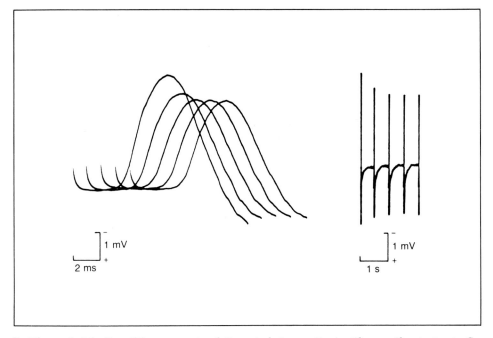

Appendix Figure 4–14. Repetitive nerve simulation study in a patient with myasthenia gravis. Successive M waves were recorded with surface electrodes over the rested cheek (nasalis) muscle during repetitive facial nerve stimulation at a rate of 2 Hz, with a display to permit measurement of the amplitude and duration of the negative phase (left) or peak-to-peak amplitude (right). A *decrementing response* is a reproducible decline in the amplitude and/or area of the *M wave* of successive responses to *repetitive nerve stimulation*. The rate of stimulation and the total number of stimuli should be specified. Decrementing responses with disorders of neuromuscular transmission are most reliably seen with slow rates (2–5 Hz) of nerve stimulation. A decrementing response with repetitive nerve stimulation commonly occurs in disorders of neuromuscular transmission, but can also be seen in some neuropathies, myopathies, and motor neuron disease. An artifact resembling a decrementing response can result from movement of the stimulating or recording electrodes during repetitive nerve stimulation. Contrast with *incrementing response.*

REPETITIVE NERVE STIMULATION

INCREMENTING RESPONSE

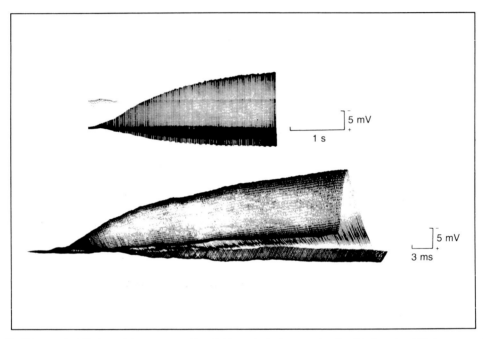

Appendix Figure 4–15. Repetitive nerve stimulation study in a patient with Lambert-Eaton myasthenic syndrome (LEMS). An incrementing response was recorded with surface electrodes over the hypothenar eminence (abductor digiti quinti) during repetitive ulnar nerve stimulation at a rate of 50 Hz with a display to permit measurement of the peak-to-peak amplitude (top) or amplitude and duration of the negative phase (bottom). An *incrementing response* is a reproducible increase in amplitude and/or area of successive responses (M wave) to *repetitive nerve stimulation*. The rate of stimulation and the number of stimuli should be specified. An incrementing response is commonly seen in two situations. First, in normal subjects the configuration of the M wave may change with repetitive nerve stimulation so that the amplitude progressively increases as the duration decreases, but the area of the M wave remains the same. This phenomenon is termed *pseudofacilitation*. Second, in disorders of neuromuscular transmission, the configuration of the M wave may change with repetitive nerve stimulation so that the amplitude progressively increases as the duration remains the same or increases, and the area of the M wave increases. This phenomenon is termed *facilitation*. Contrast with *decrementing response*.

REPETITIVE NERVE STIMULATION

NORMAL (N), MYASTHENIA GRAVIS (MG),
LAMBERT-EATON MYASTHENIC SYNDROME (LEMS)

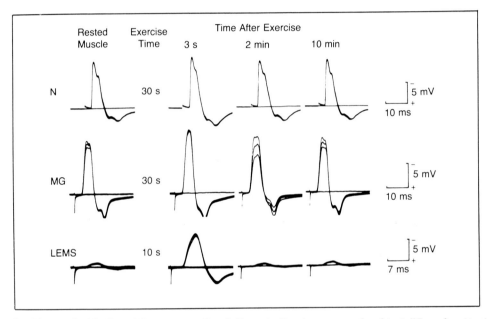

Appendix Figure 4–16. Repetitive nerve stimulation studies in a normal subject (N) and patients with myasthenia gravis (MG) and Lambert-Eaton myasthenic syndrome (LEMS). Three successive M waves were elicited by repetitive nerve stimulation at a rate of 2 Hz. The three responses were superimposed. This method of display emphasizes a change in the configuration of successive responses, but does not permit identification of the order of the responses. In each superimposed display of three responses where the configuration did change, the highest amplitude response was the first response, and the lowest amplitude response was the third response. After testing the rested muscle, the muscle was forcefully contracted for 10 to 30 seconds (exercise time). The repetitive nerve stimulation was carried out again 3 seconds, 2 minutes, and 10 minutes after the exercise ended. The results illustrate *facilitation* and *postactivation depression*.

REPETITIVE NERVE STIMULATION

PSEUDOFACILITATION

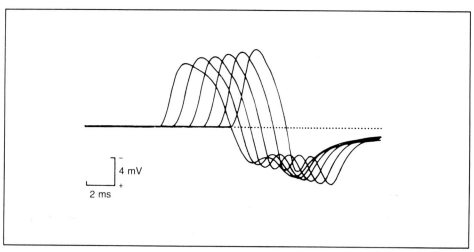

Appendix Figure 4–17. Repetitive nerve stimulation study in a normal subject. The successive M waves were recorded with surface electrodes over the hypothenar eminence (abductor digiti quinti) during ulnar nerve stimulation at a rate of 30 Hz. *Pseudofacilitation* may occur in normal subjects with *repetitive nerve stimulation* at high (20–50 Hz) rates or after strong volitional contraction, and probably reflects a reduction in the temporal dispersion of the summation of a constant number of muscle fiber action potentials due to increases in the propagation velocity of action potentials of muscle cells with repeated activation. *Pseudofacilitation* should be distinguished from *facilitation*. The recording shows an *incrementing response* characterized by an increase in the amplitude of the successive M waves with a corresponding decrease in the duration of the M wave resulting in no change in the area of the negative phase of the successive M waves.

INSERTION ACTIVITY

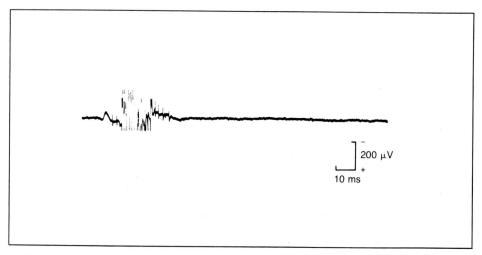

Appendix Figure 4–18. Insertion activity in a normal subject. *Insertion activity* is the electric activity caused by insertion or movement of a needle electrode. The amount of the activity may be described as normal, reduced, increased (prolonged), with a description of the waveform and repetitive rate.

END-PLATE ACTIVITY

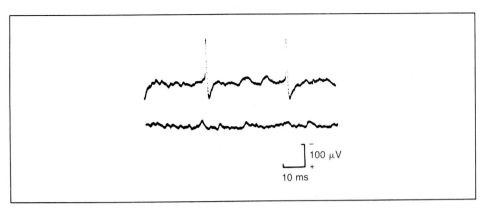

Appendix Figure 4–19. Spontaneous electric activity recorded with a needle electrode close to muscle end-plates. May be either of two forms: 1. *Monophasic* (upper and lower traces): Low-amplitude (10–20 μV), short-duration (0.5–1 ms), monophasic (negative) potentials that occur in a dense, steady pattern and are restricted to a localized area of the muscle. Because of the multitude of different potentials occurring, the exact frequency, although appearing to be high, cannot be defined. These nonpropagated potentials are probably *miniature end-plate potentials* recorded extracellularly. This form of end-plate activity has been referred to as *end-plate noise* or *sea shell sound* (*sea shell noise* or *roar*). 2. *Biphasic* (upper trace): Moderate-amplitude (100–300 μV), short-duration (2–4 ms), biphasic (negative-positive) spike potentials that occur irregularly in short bursts with a high frequency (50–100 Hz), restricted to a localized area within the muscle. These propagated potentials are generated by muscle fibers excited by activity in nerve terminals. These potentials have been referred to as *biphasic spike potentials*, *end-plate spikes*, and, incorrectly, *nerve potentials.*

FIBRILLATION POTENTIAL

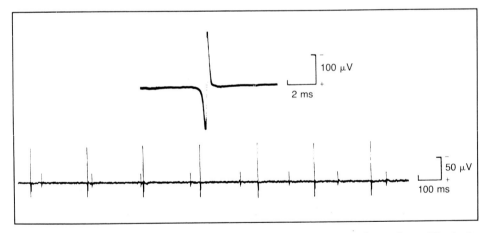

Appendix Figure 4–20. The top trace shows a single *fibrillation potential* waveform. The bottom trace shows the pattern of discharge of two other *fibrillation potentials* which differ with respect to amplitude and discharge frequency. A *fibrillation potential* is the electric activity associated with a spontaneously contracting (fibrillating) muscle fiber. It is the action potential of a single muscle fiber. The action potentials may occur spontaneously or after movement of the needle electrode. The potentials usually fire at a constant rate, although a small proportion fire irregularly. Classically, the potentials are biphasic spikes of short duration (usually less than 5 ms) with an initial positive phase and a peak-to-peak amplitude of less that 1 mV. When recorded with concentric or monopolar needle electrode, the firing rate has a wide range (1–50 Hz) and often decreases just before cessation of an individual discharge. A high-pitched regular sound is associated with the discharge of fibrillation potentials and has been described in the old literature as "rain on a tin roof." In addition to this classic form of fibrillation potentials, *positive sharp waves* may also be recorded from fibrillating muscle fibers when the potential arises from an area immediately adjacent to the needle electrode.

POSITIVE SHARP WAVE

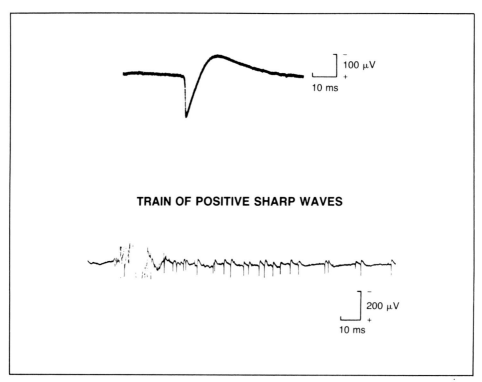

Appendix Figure 4–21. The top trace shows a single *positive sharp wave*. The bottom trace shows the pattern of initial discharge of a number of different *positive sharp waves* after movement of the recording needle electrode in denervated muscle. A *positive sharp wave* is a biphasic, positive-negative *action potential* initiated by needle movement and recurring in a uniform, regular pattern at a rate of 1 to 50 Hz; the discharge frequency may decrease slightly just before cessation of discharge. The initial positive deflection is rapid (<1 ms), its duration is usually less than 5 ms, and the amplitude is up to 1 mV. The negative phase is of low amplitude, with a duration of 10 to 100 ms. A sequence of positive sharp waves is commonly referred to as a *train of positive sharp waves*. Positive sharp waves can be recorded from the damaged area of fibrillating muscle fibers. Its configuration may result from the position of the needle electrode which is felt to be adjacent to the depolarized segment of a muscle fiber injured by the electrode. Note that the positive sharp waveform is not specific for muscle fiber damage. *Motor unit action potentials* and potentials in *myotonic discharges* may have the configuration of positive sharp waves.

MYOTONIC DISCHARGE

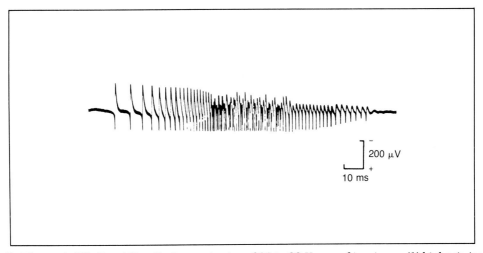

Appendix Figure 4–22. Repetitive discharge at rates of 20 to 80 Hz are of two types: (1) biphasic (positive-negative) spike potentials less than 5 ms in duration resembling *fibrillation potentials*, (2) positive waves of 5 to 20 ms in duration resembling *positive sharp waves*. Both potential forms are recorded after needle insertion, after voluntary muscle contraction or after muscle percussion, and are due to independent, repetitive discharges of single muscle fibers. The amplitude and frequency of the potentials must both wax and wane to be identified as myotonic discharges. This change produces a characteristic musical sound in the audio display of the electromyograph due to the corresponding change in pitch, which has been likened to the sound of a "dive bomber." Contrast with *waning discharge*.

COMPLEX REPETITIVE DISCHARGE

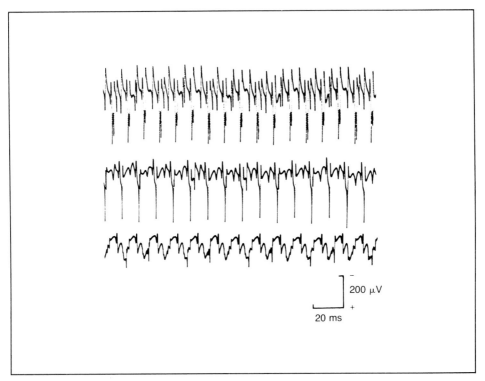

Appendix Figure 4–23. A *complex repetitive discharge* is a polyphasic or serrated action potential that may begin spontaneously or after a needle movement. They have a uniform frequency, shape, and amplitude, with abrupt onset, cessation, or change in configuration. Amplitude ranges from 100 μV to 1 mV and frequency of discharge from 5 to 100 Hz. This term is preferred to *bizarre high frequency discharge, bizarre repetitive discharge, bizarre repetitive potential, near constant frequency trains, pseudomyotonic discharge* and *synchronized fibrillation.*

FASCICULATION POTENTIAL

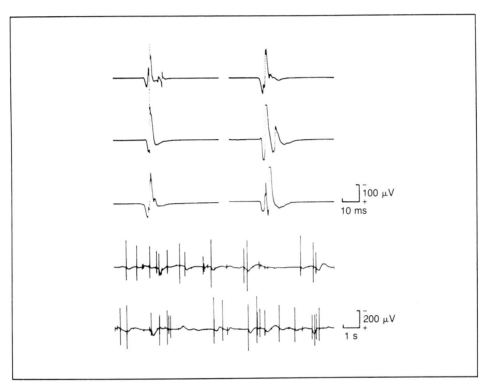

Appendix Figure 4–24. Six different *fasciculation potentials* are displayed in the top traces, with a time scale to permit characterization of the individual waveforms. The bottom two traces display *fasciculation potentials* with a time scale to demonstrate the random discharge pattern. A *fasciculation potential* is the electric potential often associated with a visible *fasciculation* which has the configuration of a *motor unit action potential* but which occurs spontaneously. Most commonly these potentials occur sporadically and are termed "single fasciculation potentials." Ocassionally, the potentials occur as a grouped discharge and are termed a "brief repetitive discharge." The occurrence of repetitive firing of adjacent fasciculation potentials, when numerous, may produce an undulating movement of muscle (see *myokymia*). Use of the terms *benign fasciculation* and *malignant fasciculation* is discouraged. Instead, the configuration of the potentials, peak-to-peak amplitude, duration, number of phases, and stability of configuration, in addition to frequency of occurrence, should be specified.

MYOKYMIC DISCHARGE

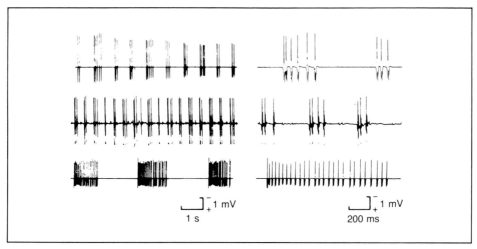

Appendix Figure 4–25. Tracings of three different *myokymic discharges* displayed with a time scale (left) to illustrate the firing pattern and with a different time scale (right) to illustrate that the individual potentials have the configuration of a *motor unit action potential.* A *myokymic discharge* is a group of *motor unit action potentials* that fire repetitively and may be associated with clinical myokymia. Two firing patterns have been described. Commonly, the discharge is a brief, repetitive firing of single units for a short period (up to a few seconds) at a uniform rate (2–60 Hz) followed by a short period (up to a few seconds) of silence, with repetition of the same sequence for a particular potential. Less commonly, the potential recurs continuously at a fairly uniform firing rate (1–5 Hz). Myokymic discharges are a subclass of *grouped discharges* and *repetitive discharges.*

NEUROMYOTONIC DISCHARGE

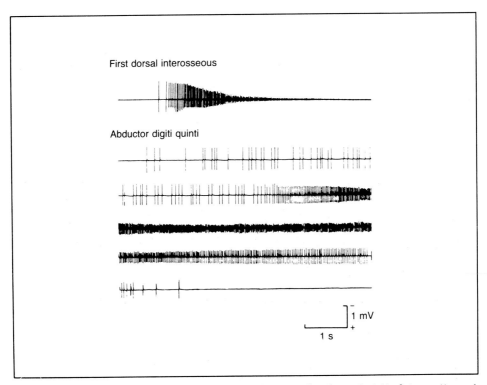

First dorsal interosseous

Abductor digiti quinti

1 mV

1 s

Appendix Figure 4–26. The time scale was chosen to illustrate the characteristic firing pattern. A *neuro-myotonic discharge* is a burst of *motor unit action potentials* which originate in the motor axons firing at high rates (150–300 Hz) for a few seconds, and often start and stop abruptly. The amplitude of the response typically wanes. Discharges may occur spontaneously or be initiated by needle movement, voluntary effort and ischemia or percussion of a nerve. These discharges should be distinguished from *myotonic discharges* and *complex repetitive discharges*.

CRAMP DISCHARGE

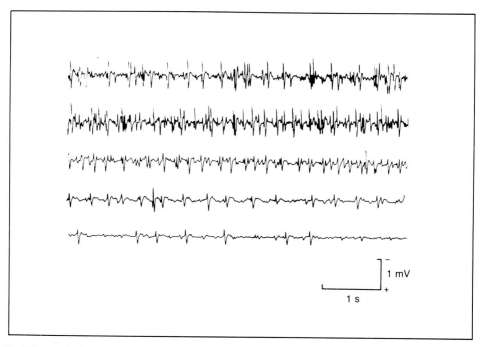

Appendix Figure 4–27. A *cramp discharge* arises from the involuntary repetitive firing of *motor unit action potentials* at a high frequency (up to 150 Hz) in a large area of muscle, usually associated with painful muscle contraction. Both the discharge frequency and the number of *motor unit action potentials* firing increase gradually during development and both subside gradually with cessation. See *muscle cramp.*

MOTOR UNIT ACTION POTENTIALS

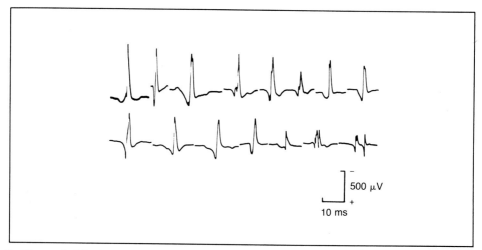

Appendix Figure 4–28. A *motor unit action potential* (MUAP) is the action potential reflecting the electric activity of a single anatomic motor unit. It is the compound action potential of those muscle fibers within the recording range of an electrode. With voluntary muscle contraction, the action potential is characterized by its consistent appearance with, and relationship to, the force of contraction. The following parameters should be specified, quantitatively if possible, after the recording electrode is placed so as to minimize the *rise time* (which by convention should be less than 0.5 ms).

SATELLITE POTENTIAL

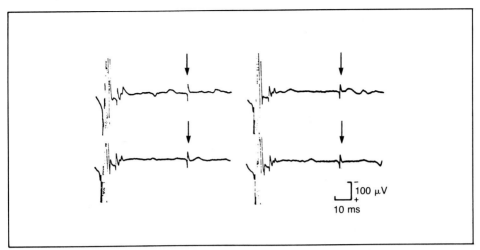

Appendix Figure 4–29. Four tracings of the same *motor unit action potential* indicated by the arrow. A *satellite potential* is a small action potential separated from the main MUAP by an isoelectric interval and firing in a time-locked relationship to the main action potential. These potentials usually follow, but may proceed, the main action potential. Also called *late component, parasite potential, linked potential,* and *coupled discharge (less preferred terms).*

RECRUITMENT PATTERN

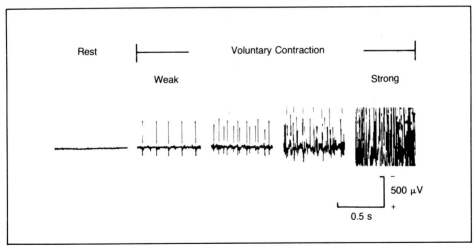

Appendix Figure 4–30. Recruitment pattern and interference pattern. *Recruitment* refers to the successive activation of the same and new motor units with increasing strength of voluntary muscle contraction. The *recruitment pattern* is a qualitative and/or quantitative description of the sequence of appearance of *motor unit action potentials* with increasing strength of voluntary muscle contraction. The *recruitment frequency* and *recruitment interval* are two quantitative measures commonly used. The *interference pattern* is the electric activity recorded from a muscle with a needle electrode during maximal voluntary effort. A *full interference pattern* implies that no individual *motor unit action potential* can be clearly identified (see tracing on far right). A *reduced interference pattern* (*intermediate pattern*) is one in which some of the individual MUAPs may be identified while other individual MUAPs cannot be identified because of overlap. The term *discrete activity* is used to describe the electric activity recorded when each of several different MUAPs can be identified. The term *single unit pattern* is used to describe a single MUAP, firing at rapid rate (should be specified) during maximum voluntary effort. The force of contraction associated with the interference pattern should be specified.

SINGLE FIBER ELECTROMYOGRAPHY

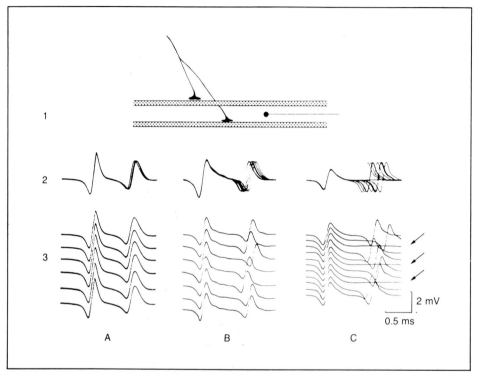

Appendix Figure 4–31. Single fiber electromyography—jitter. Schematic representation of the location of the recording surface of single fiber needle electrode recording from two muscle fibers innervated by the same motor neuron (row 1). Consecutive discharges of a potential pair are shown in a superimposed display (row 2) and in a raster display (row 3). The potential pairs were recorded from the extensor digitorum communis of a patient with myasthenia gravis and show normal *jitter* (column A), increased *jitter* (column B), and increased *jitter* and impulse blocking (column C, arrows). *Jitter* is synonymous with "single fiber electromyographic jitter." Jitter is the variability with consecutive discharges of the *interpotential interval* between two muscle fiber action potentials belonging to the same motor unit. It is usually expressed quantitatively as the mean value of the difference between the interpotential intervals of successive discharges (the mean consecutive difference, abbr. MCD). Under certain conditions, jitter is expressed as the mean value of the difference between interpotential intervals arranged in the order of decreasing interdischarge intervals (the mean sorted difference, abbr. MSD).

MACROELECTROMYOGRAPHY

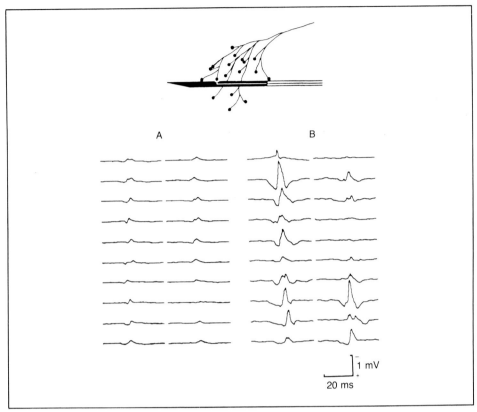

Appendix Figure 4–32. Macroelectromyography (macro-EMG). Schematic representation of the location of the recording surface of the macroelectromyography electrode recording from all the muscle fibers innervated by the same motor neuron (upper diagram). Muscle fiber action potentials recorded by the technique of macroelectromyography (lower traces) from a healthy subject (column A) and from a patient with amyotrophic lateral sclerosis (column B). *Macroelectromyography* is a general term referring to the technique and conditions that approximate recording of all *muscle fiber action potentials* arising from the same motor unit.

SECTION III:
TERMS GROUPED BY SUBJECT WITHOUT DEFINITION

The AAEE Nomenclature Committee felt that electromyography terms should be presented in two ways as follows: the conventional alphabetical list (Section I) and a list of the same terms grouped by subject (Section III).

This listing of the terms of electromyography by subject should be particularly useful for students and physicians who are new to the discipline. It may also help more experienced electromyographers to understand the logic behind the choices of terms which were made by the Committee.

In several instances, one term has been chosen as the preferred expression to describe a phenomenon for which several terms appear in the literature. The glossary is inclusive and, in the following list of terms grouped by subject, the preferred terms are listed first in the small groupings of like terms.

Basic Neurophysiology Terminology

The definition of these terms is based upon their usage in neurophysiology literature.

Action current
Action potential
Muscle fiber action potential
Nerve fiber action potential

Refractory period
Absolute refractory period
Relative refractory period
Functional refractory period

Voltage
Potential
Resting membrane potential
Threshold
Membrane instability
Polarization
Depolarization
Depolarization block
Hyperpolarization
Repolarization
Afterpotential
Injury potential

Baseline
Noise
Interference

Wave
Waveform
Spike

Near-field potential
Far-field potential

Discharge
Afterdischarge
Adaptation

Frequency
Cycles per second
Hertz
Frequency analysis

Anode
Cathode

Excitatory postsynaptic potential
Inhibitory postsynaptic potential

End-plate potential
Miniature end-plate potential
End-plate zone

Accommodation
Accommodation curve

Excitability
Reflex
Muscle stretch reflex
Habituation
Fatigue
Silent period
Backfiring

Volume conduction
Tetanic contraction
Staircase phenomenon

Latency of activation
Utilization time

Motor unit
Motor unit territory

General Terminology

The Board of Directors of the AAEE selected the term "Electrodiagnostic Medicine" to describe the area of medical practice in which a physician uses information from the clinical history, observations from the physical examination, and the techniques of nerve conduction studies and electromyography to diagnose and treat neuromuscular disorders.

Electrodiagnosis
Electrodiagnostic medicine

Nerve conduction studies
Evoked potential studies

Electromyography
Electromyograph
Electromyogram
Electroneurography
Microneurography

Electroneuromyography

Clinical electromyography

Central electromyography

International 10–20 system

Equipment Terminology

Some of the terminology related to equipment dates back to the early descriptions of amplifiers in which one input was referred to as "Grid 1" or "G1" and the other input was called "Grid 2" or "G2." In studies of activities generated by the central nervous system in response to peripheral nerve stimulation (e.g., somatosensory evoked potentials), this convention is preserved by the terms "Input Terminal 1" and "Input Terminal 2" because the exact site of the origin of the recorded activity is not known. In nerve conduction studies and electromyography, the electrodes which lead to the input terminals of the amplifier can be referred to as "Input Terminals 1 and 2," but more commonly they are referred to as the "active electrode" and the "reference electrode," respectively, because the source of the electric activity is better understood.

Electrode
Surface electrode
Needle electrode
Bifilar needle recording electrode
Coaxial needle electrode
Concentric needle recording electrode
Monopolar needle electrode
Unipolar needle electrode
Multilead electrode
Multielectrode

Stimulating electrode
Anodal block

Recording electrode
Active electrode
Exploring electrode
Stigmatic electrode
Reference electrode
Indifferent electrode
Input Terminal 1
Input Terminal 2
Grid 1, Grid 2
G1, G2

Ground electrode
Earth electrode

Single fiber needle electrode

Macro-EMG electrode

Stimulus Terminology

In performing nerve conduction studies, it is important to identify the direction of propagation of the stimulus (antidromic or orthodromic), the intensity of the stimulus relative to the response (subthreshold, submaximal, or supramaximal), and the number of stimuli. The terms related to strength-duration curves

are included here solely for historic purposes because these tests are now rarely used.

Antidromic
Orthodromic

Stimulus
Threshold stimulus
Maximal stimulus
Subthreshold stimulus
Submaximal stimulus
Supramaximal stimulus

Paired stimuli
Conditioning stimulus
Test stimulus

Strength-duration curve
Chronaxie
Rheobase

Artifact
Stimulus artifact
Electric artifact
Shock artifact
Movement artifact

Response Terminology

The terms in this section refer to the electric activity recorded from peripheral nerve and muscle and from the central nervous system in response to physiologic, mechanical or electric stimuli. Historically, the terms chosen to describe these responses often implied physiologic mechanisms which, in some cases, subsequent investigations have disproved. In other cases, the term chosen has also been used to describe more than one phenomenon. To solve those problems, the Nomenclature Committee recommends that some waveforms be referred to by terms (letters) that are specific and unbiased. For example, the term "M wave" specifically refers to the compound muscle action potential recorded over a muscle directly in response to electric nerve stimulation. This term is preferred to the term "motor response" which may mean either an M wave or the contractile movement of the muscle. For similar reasons, the terms "F wave" and "H wave" were chosen to refer to the late responses elicited indirectly from a muscle by electric stimulation of the nerve. The terms "A wave" and "T wave" are introduced to replace the terms "axon reflex" and "tendon reflex."

The terminology to describe short-latency somatosensory evoked potentials is based on the recommendation in the American EEG Society's Clinical Evoked Potentials Guideline (*J Clin Neurophysiol* 1:41–53, 1984).

Evoked potential

Motor point

Motor response

Compound muscle action potential
Evoked compound muscle action
  potential
Muscle action potential

*M wave
M response

Late response

*F wave
F response

*H wave
H response
H reflex
Hoffmann reflex

*A wave
Axon reflex

*T wave

*R1 wave
*R2 wave
Blink responses
Blink reflex

Compound nerve action potential
Nerve action potential
Nerve trunk action potential

Compound mixed nerve action potential
Compound motor nerve action potential

*Compound sensory nerve action
  potential
Sensory response
Sensory potential
Sensory nerve action potential
Compound action potential

Amplitude
Conduction block

Duration
Temporal dispersion

Latency
Distal latency
Proximal latency
Latent period
Peak latency

Latency of activation

Motor latency
Terminal latency

Residual latency

Sensory latency

Sensory peak latency
Sensory delay

Conduction velocity
Nerve conduction velocity
Motor nerve conduction velocity
Sensory nerve conduction velocity
Conduction time
Conduction distance
Maximum conduction velocity

Muscle fiber conduction velocity

Brainstem auditory evoked potential

Brainstem auditory evoked response

Spinal evoked potential

*Visual evoked potential
Visual evoked response

*Somatosenaory evoked potential (SEP)
Short-latency SEP (SSEP)
    *Median nerve SSEP
    *Common peroneal nerve SSEP
    *Posterior tibial nerve SSEP
Midlatency SEP
Long-latency SEP
Interpeak interval

Repetitive Nerve Stimulation Terminology

Repetitive nerve stimulation has gained widespread acceptance as a valid and reproducible clinical technique to assess the integrity of neuromuscular transmission. However, abnormal results of repetitive nerve stimulation studies may also be seen in primary disorders of nerve and muscle, as well as in primary disorders of neuromuscular transmission. Therefore, it is important to be certain that the results of the studies are described completely so that the basis of the conclusion can be reviewed. Descriptive terms such as decrementing re-

---

*Illustration in Section II.

sponse, incrementing response, repair of the decrement, and increment after exercise should be used to describe the results. Quantitative values indicating the magnitude of the change, as well as the method of calculation, should be included in the report.

*Repetitive nerve stimulation
  Jolly test
  Train of stimuli

*Decrementing response
  Decremental response

*Repair of the decrement

*Postactivation depression
  Postactivation exhaustion

*Incrementing response
  Incremental response

*Increment after exercise

*Facilitation
  Postactivation facilitation
  Posttetanic facilitation

  Potentiation
  Postactivation potentiation
  Posttetanic potentiation

*Pseudofacilitation

Needle Examination Terminology

This section is the largest section of the glossary. It includes the range of activities which are observed in muscle with a needle electrode. The activities can be subdivided into insertion activity, spontaneous activity, involuntary activity and voluntary activity. In several cases, different terms have been used in the literature to describe the same phenomena. The committee has made an effort to select the one term which is preferred for each phenomenon. For example, the term "complex repetitive discharge" was chosen to characterize the electric discharge which has two or more different components (complex) and which repeats regularly (repetitive). Other terms which have been used to describe the same activity

are "bizarre high-frequency discharge," "bizarre repetitive discharge," and "bizarre repetitive potential." These latter terms were not chosen since the word "bizarre" is a relative one and it has a negative connotation. The term "pseudomyotonic discharge" has also been used to describe complex repetitive discharges but is to be avoided because there are other electric phenomena which resemble myotonia, for example, waning discharges. Two more terms which have been used to describe complex repetitive discharges are "near constant frequency train" and "synchronized fibrillation."

Occasionally a term which describes a clinical phenomenon is used incorrectly to describe an electric phenomenon. In order to make clear the distinction between them, both terms have been included in this glossary. Examples of these pairs would be

  fasciculation-fasciculation potential
  myokymia-myokymic discharge
  neuromyotonia-neuromyotonic
      discharge
  muscle cramp-cramp discharge
  myotonia-myotonic discharge.

It is important for physicians to use each term in these sets correctly and specifically. For example, it would be incorrect to describe "myotonic discharge" as "myotonia" or vice versa. Not all delayed muscle relaxation (myotonia) is accompanied by "myotonic discharges," and not all "myotonic discharges" are accompanied by visible, delayed muscle relaxation.

The term "motor unit action potential" is preferred to the term "motor unit potential" to describe the synchronized muscle fiber action potentials belonging to one motor unit. This recommendation is in keeping with the origins of the term in the basic neurophysiology laboratory.

Attention is called to the terms "recruitment frequency" and "recruitment interval" which provide more quantitative descriptions of recruitment than the older terms "single unit pattern," "discrete activity," "reduced interference pattern," and "full interference pattern." Many electromyographers now assess the number of motor unit action potentials available in the muscle from the recruit-

---

*Illustration in Section II.

ment frequency or recruitment intervals, and report the results directly as a normal number of motor unit action potentials, or as a mild, moderate, moderately severe, or severe decrease in the number of motor unit action potentials.

*Insertion activity
 Reduced insertion activity
 Increased insertion activity
 Prolonged insertion activity

 Electric silence
 Electric inactivity

 Spontaneous activity
 Involuntary activity

*End-plate activity
 End-plate noise
 End-plate spike
 Nerve potential
 Sea shell sound (sea shell roar or noise)

 Fibrillation
*Fibrillation potential
 Denervation potential

*Positive sharp wave
 Positive wave
 Trains of positive sharp waves

 Motor unit
*Motor unit action potential
 Motor unit potential
 MUAP
 MUP

 Amplitude
 Duration
 Rise Time
 Phase
 Monophasic action potential
 Biphasic action potential
 Triphasic action potential
 Tetraphasic action potential
 Polyphasic action potential
 Serrated action potential
 Turn
 Irregular potential
 Complex motor unit action potential

*Satellite potential
 Late component (of a motor unit action
   potential)
 Coupled discharge
 Linked potential
 Parasite potential

Neuropathic motor unit potential
"Giant" motor unit action potential

Myopathic motor unit potential
BSAP
BSAPP

Nascent motor unit potential

Recruitment
*Recruitment pattern
 Recruitment frequency
 Recruitment interval
 Firing rate
 Firing pattern
 Discharge frequency
 Order of activation
 Onset frequency
*Interference pattern
 Full interference pattern
 Reduced interference pattern
 Intermediate interference pattern

*Complex repetitive discharge
 Bizarre high-frequency discharge
 Bizarre repetitive discharge
 Bizarre repetitive potential
 Pseudomyotonic discharge
 Synchronized fibrillation
 Near constant frequency trains

 Fasciculation
*Fasciculation potential
 Benign fasciculation
 Malignant fasciculation
 Contraction fasciculation

 Repetitive discharge
 Grouped discharge
 Iterative discharge
 Double discharge
 Doublet
 Triple discharge
 Triplet
 Multiple discharge
 Multiplet

 Tetanus
 Tetany

 Myokymia
*Myokymic discharge

 Muscle cramp
*Cramp discharge

 Neuromyotonia
*Neuromyotonic discharge

 Myotonia
*Myotonic discharge
 Myotonic potential
 Pseudomyotonic discharge

---

*Illustration in Section II.

Waning discharge

Voluntary activity
Volitional activity

Contraction
Contracture
Myoedema

Discrete activity
Single unit pattern

Neuropathic recruitment
Myopathic recruitment

SFEMG and Macro-EMG Terminology

Recent modifications of recording electrodes have led to the development of single fiber electromyography, macroelectromyography and scanning electromyography. Because these techniques are used in clinical neurophysiology laboratories, terminology related to them is included in this glossary.

*Single fiber electromyography
Single fiber EMG
SFEMG

*Jitter
MCD
MSD

Fiber density

Interpotential interval
Interdischarge interval

Propagation velocity of a muscle fiber

*Macroelectromyography
Macro-EMG

Macro motor unit action potential
Macro MUAP
Macro-EMG needle electrode

Scanning EMG
Motor unit fraction

---

*Illustration in Section II.

INDEX

Figures and tables are indicated in parentheses by an "F" or a "T," respectively, followed by the figure or table number and the page number. Page numbers preceded by a "G" refer to terms in the *AAEE Glossary of Terms in Clinical Electromyography* (Appendix 4).